D1462788

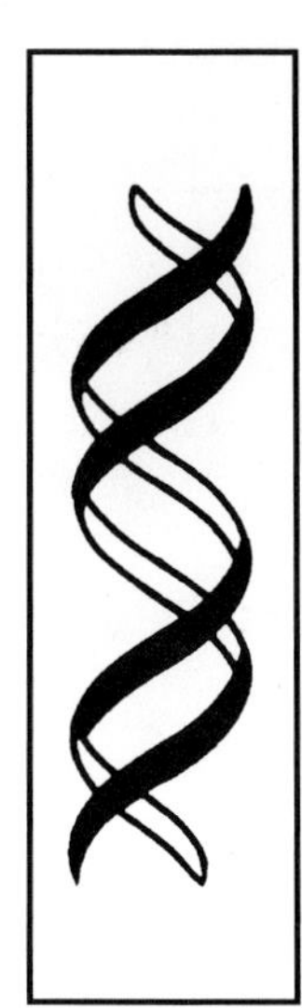

Genetic Dissection of Complex Traits

Second Edition 2008

Advances in Genetics, Volume 60

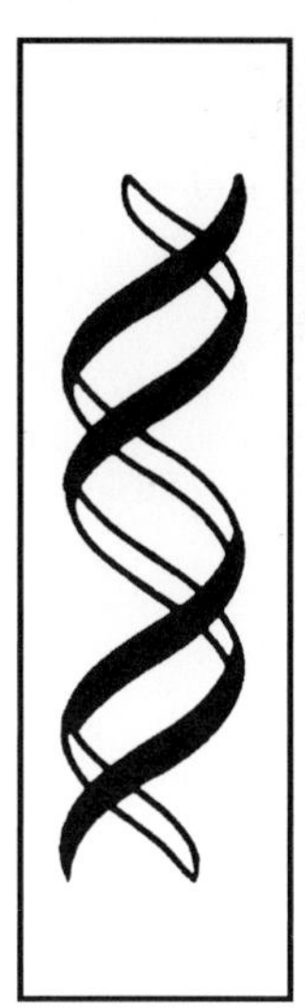

Genetic Dissection of Complex Traits

Second Edition 2008

Edited by

D. C. Rao
Washington University School of Medicine
Division of Biostatistics
St. Louis, Missouri

C. Charles Gu
Washington University School of Medicine
Division of Biostatistics
St. Louis, Missouri

ELSEVIER

AMSTERDAM • BOSTON • HEIDELBERG • LONDON
NEW YORK • OXFORD • PARIS • SAN DIEGO
SAN FRANCISCO • SINGAPORE • SYDNEY • TOKYO
Academic Press is an imprint of Elsevier

Academic Press is an imprint of Elsevier
84 Theobald's Road, London WC1X 8RR, UK
Radarweg 29, PO Box 211, 1000 AE Amsterdam, The Netherlands
Linacre House, Jordan Hill, Oxford OX2 8DP, UK
30 Corporate Drive, Suite 400, Burlington, MA 01803, USA
525 B Street, Suite 1900, San Diego, CA 92101-4495, USA

First edition 2000 isbn 978-0-12-017642-7; vol. 42 in this series
Second edition 2008

ISBN: 978-0-12-373883-7
ISSN: 0065-2660

For information on all Academic Press publications
visit our website at books.elsevier.com

Printed and bound in USA
08 09 10 11 12 10 9 8 7 6 5 4 3 2 1

Contents

Part II: Linkage and Association Analysis 153

CONTRIBUTORS

Numbers in parentheses indicate the pages on which the authors' contributions begin.

Laura Almasy (175) Department of Genetics, Southwest Foundation for Biomedical Research, San Antonio, Texas 78245

Themistocles L. Assimes (437) Department of Medicine, Division of Cardiovascular Medicine, Stanford University, Stanford, California 94305–5406

Lisa F. Barcellos (253) Division of Epidemiology, School of Public Health, University of California, Berkeley, California 94720

Laura M. Beskow (505) Institute for Genome Sciences and Policy, Center for Genome Ethics, Law, and Policy, Duke University, Durham, North Carolina 27708

John Blangero (175) Department of Genetics, Southwest Foundation for Biomedical Research, San Antonio, Texas 78245

Ingrid B. Borecki (51) Division of Statistical Genomics, Washington University School of Medicine, St. Louis, Missouri 63108

Ulrich Broeckel (107) Department of Pediatrics, Medicine and Physiology, Children's Hospital of Wisconsin, Human and Molecular Genetics Center, Medical College of Wisconsin, Milwaukee, Wisconsin 53226

C. Charles Gu (407) Division of Biostatistics and Department of Genetics, Washington University School of Medicine, St. Louis, Missouri 63110

Jonathan Corbett (155) Department of Genetics, Washington University School of Medicine, St. Louis, Missouri 63110

David G. Cox (465) Program in Molecular and Genetic Epidemiology, Harvard School of Public Health, Boston, Massachusetts 02115

Charles R. Farber (571) Department of Medicine, Division of Cardiology, BH-307 CHS, University of California, Los Angeles, California 90095

Susanne B. Haga (505) Institute for Genome Sciences and Policy, Center for Genome Ethics, Law, and Policy, Duke University, Durham, North Carolina 27708

John P. A. Ioannidis (311) Department of Medicine, Tufts University School of Medicine, Boston, Massachusetts 02111; Department of Hygiene and Epidemiology, University of Ioannina School of Medicine, Ioannina, Greece 45110

Howard J. Jacob (655) Human and Molecular Genetics Center, Medical College of Wisconsin, Milwaukee, Wisconsin 53226

Peter Kraft (465) Program in Molecular and Genetic Epidemiology, Harvard School of Public Health, Boston, Massachusetts 02115, and Department of Biostatistics, Harvard School of Public Health, Boston, Massachusetts 02115

Anne E. Kwitek (655) Human and Molecular Genetics Center, Medical College of Wisconsin, Milwaukee, Wisconsin 53226

Nan M. Laird (219) Department of Biostatistics, Harvard School of Public Health, Boston, Massachusetts 02115

Jean-Marc Lalouel (701) Department of Human Genetics, The University of Utah School of Medicine, Salt Lake City, Utah

Christoph Lange (219) Department of Biostatistics, Harvard School of Public Health, Boston, Massachusetts 02115

Jozef Lazar (655) Human and Molecular Genetics Center, Medical College of Wisconsin, Milwaukee, Wisconsin 53226

Na Li (637) Division of Biostatistics, School of Public Health, University of Minnesota, Minneapolis, Minnesota 55455

Nianjun Liu (335) Section on Statistical Genetics, Department of Biostatistics, University of Alabama at Birmingham, Birmingham, Alabama 35294

Paul Lott (701) Department of Human Genetics, The University of Utah School of Medicine, Salt Lake City, Utah

Aldons J. Lusis (571) Department of Medicine, Division of Cardiology, BH-307 CHS, University of California, Los Angeles, California 90095; Department of Microbiology, Immunology and Molecular Genetics, University of California, Los Angeles, California 90095; Department of Human Genetics, University of California, Los Angeles, California 90095

Nathalie Malo (195) Department of Psychiatry, University of California at San Diego, La Jolla, California 92093; The Center for Human Genetics and Genomics, University of California at San Diego, La Jolla, California 92093; Scripps Genomic Medicine and Department of Molecular and Experimental Medicine, The Scripps Research Institute, La Jolla, California 92037

Karen Maresso (107) Department of Pediatrics, Medicine and Physiology Children's Hospital of Wisconsin, Human and Molecular Genetics Center, Medical College of Wisconsin, Milwaukee, Wisconsin 53226

Michael B. Miller (141) Division of Epidemiology and Community Health, School of Public Health, and Institute of Human Genetics, University of Minnesota, Minneapolis, Minnesota 55454

Carol Moreno (655) Human and Molecular Genetics Center, Medical College of Wisconsin, Milwaukee, Wisconsin 53226

Newton E. Morton (727) Human Genetics Division, University of Southampton, Southampton SO16 6YD, United Kingdom

Balasubramanian Narasimhan (437) Departments of Health Research and Policy and Statistics, Stanford University, Stanford, California 94305–5405

Adam B. Olshen (437) Department of Epidemiology and Biostatistics, Memorial Sloan-Kettering Cancer Center, New York, New York 10021

Richard A. Olshen (437) Departments of Health Research and Policy, Statistics, and Electrical Engineering, Stanford University, Stanford, California 94305–5405

Michael A. Province (51) Division of Statistical Genomics, Washington University School of Medicine, St. Louis, Missouri 63108

D. C. Rao (3, 141, 293, 407) Division of Biostatistics and Departments of Genetics, Psychiatry, and Mathematics, Washington University in St. Louis, School of Medicine, St. Louis, Missouri 63110; Department of Mathematics, Washington University in St. Louis, Missouri 63130

John P. Rice (155) Department of Psychiatry and Department of Genetics, Washington University School of Medicine, St. Louis, Missouri 63110

Treva K. Rice (35, 293) Division of Biostatistics and Department of Psychiatry, Washington University School of Medicine, St. Louis, Missouri 63110

Neil Risch (547) Institute for Human Genetics, University of California, San Francisco, California 94143

Andreas Rohrwasser (701) Department of Human Genetics, The University of Utah School of Medicine, Salt Lake City, Utah

Nancy L. Saccone (155) Department of Genetics, Washington University School of Medicine, St. Louis, Missouri 63110

Georgia Salanti (311) Department of Hygiene and Epidemiology, University of Ioannina School of Medicine, Ioannina 45110, Greece

Eric E. Schadt (603) Rosetta Inpharmatics, LLC, a Wholly Owned Subsidiary of Merck and Co., Inc., Seattle, Washington 98109

Nicholas J. Schork (195, 293) Department of Psychiatry, University of California at San Diego, La Jolla, California 92093; Department of Family and Preventive Medicine, University of California at San Diego, La Jolla, California 92093; The Center for Human Genetics and Genomics, University of California at San Diego, La Jolla, California 92093; The Stein Institute for Research on Aging, University of California at San Diego, La Jolla, California 92093; Scripps Genomic Medicine and Department of Molecular and Experimental Medicine, The Scripps Research Institute, La Jolla, California 92037; Scripps Genomic Medicine and Scripps Research Institute, La Jolla, California 92069

Karen Schwander (141) Division of Biostatistics, Washington University School of Medicine, St. Louis, Missouri 63141

Hua Tang (547) Department of Genetics, Stanford University, Stanford, California 94305

Glenys Thomson (253) Department of Integrative Biology, University of California, Berkeley, California 94720

Hemant K. Tiwari (75) Department of Biostatistics, Section on Statistical Genetics, University of Alabama at Birmingham, Birmingham, Alabama 35294; Section on Statistical Genetics, Department of Biostatistics, University of Alabama at Birmingham, 1665 University Boulevard, Birmingham, Alabama 35294

Thomas A. Trikalinos (311) Department of Medicine, Tufts University School of Medicine, Boston, Massachusetts 02111

Ana M. Valdes (253) Twin Research Unit, King's College School of Medicine, London, United Kingdom

Robert B. Weiss (701) Department of Human Genetics, The University of Utah School of Medicine, Salt Lake City, Utah

Jennifer Wessel (195) Department of Psychiatry, University of California at San Diego, La Jolla, California 92093; Department of Family and Preventive Medicine, University of California at San Diego, La Jolla, California 92093; The Center for Human Genetics and Genomics, University of California at San Diego, La Jolla, California 92093; Scripps Genomic Medicine and Department of Molecular and Experimental Medicine, The Scripps Research Institute, La Jolla, California 92037

Mary K. Wojczynski (75) Department of Biostatistics, Section on Statistical Genetics, University of Alabama at Birmingham, Birmingham, Alabama 35294

Kai Yu (407) Division of Cancer Epidemiology and Genetics, National Cancer Institute, Bethesda, Maryland 20892

Bin Zhang (603) Rosetta Inpharmatics, LLC, a Wholly Owned Subsidiary of Merck and Co., Inc., Seattle, Washington 98109

Kui Zhang (335) Section on Statistical Genetics, Department of Biostatistics, University of Alabama at Birmingham, Birmingham, Alabama 35294

Hongyu Zhao (335) Department of Epidemiology and Public Health and Department of Genetics, Yale University School of Medicine, New Haven, Connecticut 06520

Jun Zhu (603) Rosetta Inpharmatics, LLC, a Wholly Owned Subsidiary of Merck and Co., Inc., Seattle, Washington 98109

Xiaofeng Zhu (547) Department of Epidemiology and Biostatistics, Case Western Reserve University, Cleveland, Ohio 44106

Elias Zintzaras (311) Department of Biomathematics, University of Thessaly School of Medicine, Larissa, 41222 Greece

Preface to First Edition

This volume is based on a symposium held in honor of one of the founding fathers of genetic epidemiology, Newton E. Morton, on the occasion of his 70th birthday. We hope that it constitutes a fitting tribute to the man who continues to make pioneering contributions to the field.

The primary goal of this volume is to ask how best to even partially achieve the genetic dissection of complex traits, which do not have simple, single-gene causes. Toward that goal, this volume documents state-of-the-art methods and strategies and provides guidelines for undertaking the genetic dissection of complex traits. Is genetic dissection of complex traits achievable? The answer seems to be a resounding "yes," now more than ever before. This is an exciting time to be a genetic epidemiologist, with unprecedented new opportunities unfolding that we could only dream of a few short years ago. At the same time, we must recognize the limitations of some of the current approaches. It is generally recognized that genes and environments affect most human biological processes in complex and often interacting ways. Investigators the world over may be divided by differences in language and methodologies, but they are united in the conviction that genetic dissection of complex traits, though a formidable challenge, is possible.

Recent failures have actually strengthened our resolve to succeed. The recent years have witnessed a flurry of activity, in the development of novel methods as well as promising new strategies. Therefore, the time is ripe to undertake a realistic evaluation of contemporary methods, to critique their real utility, and to project promising new directions. Toward this end, the current volume, written by leading experts in genetic epidemiology, has three objectives: first, to provide a comprehensive and well-balanced review intended to quickly bring scientists and students up to speed with the state of the art in an important and rapidly growing field; second, to place contemporary methodologies in their proper perspective by including critical evaluations of their real value; and finally, to project promising new directions for the future. To assist investigators in asking precise questions about complex phenomena, the chapter authors endeavor to convey balanced opinions about the various methods.

The genetic dissection of complex traits is a challenge to the practitioners who seek to uncover the genetic architecture underlying complex phenotypes, as well as to those who advocate specific approaches and methodologies. Now that even well-designed and thoughtful investigations are beginning to produce ambiguous results, we are coming to realize that unanticipated challenges underlie what were considered hitherto to be very promising approaches. Thus, we need new strategies, as well as a dose of humility, as we apply them thoughtfully in ways that anticipate failures and frustrations.

This book contains 32 chapters, divided into 9 sections, and an appendix It fills an important void by including a comprehensive account of contemporary methods and a detailed overview of current methodological trends. Section I that summarizes briefly Newton Morton's impact on science is supplemented by a complete list of his research contributions in the appendix. In Section II , we provide an overview of the methods for genetic dissection of complex traits and a succinct summary of the fundamental concepts of heritability, linkage, and association. In Section III , we cover phenotypic and genotypic issues, such as quality and refinement, and discuss ways of handling multivariate phenotypes. In Section IV , we discuss the most powerful model-based methodology for linkage analysis, including a critical discussion of its strengths and weaknesses. In Section V , we present contemporary and promising model-free methods including variance components methods and transmission disequilibrium test (TDT). Section VI, presents a comprehensive discussion of emerging new methodologies with considerable potential, including meta-analysis, classification methods, neural networks, and genome partitioning methods. In Section VII , we discuss optimum strategies for mapping complex trait loci, including gene–gene and gene–environment interactions, and special studies of population isolates. In Section VIII , we deal with the thorny issues of multiple comparisons and significance levels. Finally, in Section IX , we offer some thoughts for the millennium.

This book can be used as a handbook for a wide audience of pre- and postdoctoral scientists, methodologists who seek an overview of some of the latest thinking in this area, and, most importantly, the scores of investigators who seek to evaluate the etiological basis of complex traits. It can be used as a reference book for upper-level undergraduate students, as well as a textbook for graduate students majoring in quantitative aspects of human genetics or in genetic epidemiology. It also provides an excellent account for statisticians interested in methodological opportunities. We believe that it represents an important resource that will have some longevity as we march forward in the new millennium.

As the saying goes, "Wisdom comes from experience, and experience comes from making mistakes." Surely we have all made our share of mistakes, hence are experienced, and may now hope that wisdom is just waiting for us to claim it! We have tried to fill this volume with the wisdom the authors have accumulated from their combined experience. Perhaps the most important sign of wisdom is the humility with which we recognize the limitations of some of our current approaches. Predicting the future is a perilous exercise, especially when it comes to complex traits. It is nearly impossible for us to know today which methods will have been most useful in dissecting the genetic architecture of complex traits when we look back 10–20 years from now. We can only judge what appear to be the most promising approaches from our current perspective, and that is what we have tried to document here. No doubt there will be many surprises ahead. Genetic dissection of complex traits is the greatest challenge in genetics at the start of the new millennium, and we hope that the wisdom conveyed herein will shed some light. What a time it is to be a genetic epidemiologist!

D. C. Rao and M. A. Province

Preface to Second Edition

A lot has happened since the publication of the first edition 7 years back. Genetic dissection of complex traits seemed daunting at that time. While it still is not easy at the current time, the premise has changed quite dramatically. Genome-wide linkage analysis with a few hundred markers was a standard tool back then, whereas genome-wide association (GWA) scans with a million markers are the norm now. Linkage disequilibrium among markers was almost irrelevant then, but it constitutes now a major opportunity as well as challenge for multimarker analysis. Single-gene studies were the standard at the time, while pathway-based approaches and even networks of pathways appear within our reach soon.

The mega advances in genomic science and technology are responsible for this sea of change. Completion of the Human Genome Project and the subsequent Hapmap Project has revolutionized the field by categorizing several million common SNPs in humans, which made it possible to undertake GWA scans. The avalanche of large-scale databases descending upon us will surely pose challenges in terms of computational resources, analysis of the data, interpretation of the results, and how best to integrate various pieces of the genetic puzzle so that one may uncover the complete genetic architecture of common complex traits. As we begin this long but exciting journey into the future, it becomes even more important for us to follow efficient and state-of-the-art methods and best practices including the often neglected Ethical, Legal, and Social Issues (ELSI). All in all, these are exciting times, with multiple GWA scans completed and many more under way.

Despite these changes, then and now, the primary goal remains the same, that of genetic dissection of complex traits, which do not have simple, single-gene causes. Toward that goal, this edition documents state-of-the-art methods that represent a paradigm shift since the first edition. There is an amazing degree of optimism and consensus that genetic dissection of complex traits, though still a formidable challenge, appears to be within our reach. The current edition, written by leading experts in genetic epidemiology and related fields, provides a comprehensive and well-balanced review of standard and emerging methods and approaches intended to quickly bring scientists and students up to speed with the state of the art in an important and rapidly growing field.

This volume contains 25 chapters, divided into 5 sections. Section I provides an overview of the methods for genetic dissection of complex traits and a succinct summary of the fundamental concepts of heritability, linkage and association, definition of phenotype, genotyping platforms, and a brief discussion of genotyping errors. Section II covers linkage and association analysis methods in detail, including DNA sequence-based association methods and family-based association methods. Section III discusses special topics, including meta-analysis, haplotype analysis, optimum study designs for GWA studies, and ELSI issues. Section IV presents promising new topics including admixture mapping, gene expression analysis, systems biology, and comparative genomics. Finally, Section V discusses outstanding challenges underlying the process of genetic dissection for complex traits, from genetics to mechanism of disease liability, and an overview of the outlook in the post-HapMap era.

This volume can be used as a standard reference by a wide audience of pre- and postdoctoral scientists, methodologists who seek an overview of some of the latest developments in this area, and, most importantly, the scores of investigators who seek to evaluate the genetic basis of complex traits. It can be used as a reference book for upper-level undergraduate students, as well as a textbook for graduate students majoring in quantitative aspects of human genetics or in genetic epidemiology. It also provides an excellent account for statisticians interested in methodological opportunities. We believe that it represents an important resource that will have some longevity as we march forward into the age of GWAs and personalized medicine.

As Rao and Province concluded in the first edition, predicting the future is a perilous exercise, especially when it comes to complex traits. While it is difficult to predict even today which methods will have been most useful in dissecting the genetic architecture of complex traits when we look back 5–10 years from now, we have reason to hope that this volume includes them. Genetic dissection of complex traits continues to be the greatest challenge in genetics. However, the future looks more promising and exciting than ever before.

D. C. Rao and C. Charles Gu

Acknowledgments

The following individuals generously contributed their time for prompt reviewing of the individual chapters, and often at very short notice. Their services and support are greatly appreciated: Jim Cheverud, Georg Ehret, Charles Gu, Jim Hixson, B. M. Knoppers, Aldi Kraja, Jean-Marc Lalouel, Zhaohai Li, Newton Morton, Kari North, Jim Pankow, Haydeh Payami, John Rice, Treva Rice, Daniel Schaid, Eric Schadt, Nik Schork, Yan Sun, Yun Ju Sung, Hemant Tiwari, Joe Terwilliger, Hongyu Zhao, and Xiaofeng Zhu.

Finally, our sincere thanks go to Matthew Brown for the extraordinary editorial assistance he provided in all areas of the project, always ensuring that the project was making progress in a timely manner. Last but not least, it was a pleasure working with Luna Han, Sally Cheney, and Pat Gonzalez at the Academic Press. Their diligence is responsible for the attractive layout and timely publication of this Second Edition.

The Editors

Part I

OVERVIEW AND FUNDAMENTALS

1

An Overview of the Genetic Dissection of Complex Traits

D. C. Rao

Division of Biostatistics and Departments of Genetics, Psychiatry, and Mathematics, Washington University in St. Louis, School of Medicine, St. Louis, Missouri 63110

Advances in Genetics, Vol. 60
0065-2660/08 $35.00
DOI: 10.1016/S0065-2660(07)00401-4

ABSTRACT

Thanks to the recent revolutionary genomic advances such as the International HapMap consortium, resolution of the genetic architecture of common complex traits is beginning to look hopeful. While demonstrating the feasibility of genome-wide association (GWA) studies, the pathbreaking Wellcome Trust Case Control Consortium (WTCCC) study also serves to underscore the critical importance of very large sample sizes and draws attention to potential problems, which need to be addressed as part of the study design. Even the large WTCCC study had vastly inadequate power for several of the associations reported (and confirmed) and, therefore, most of the regions harboring relevant associations may not be identified anytime soon. This chapter provides an overview of some of the key developments in the methodological approaches to genetic dissection of common complex traits. Constrained Bayesian networks are suggested as especially useful for analysis of pathway-based SNPs. Likewise, composite likelihood is suggested as a promising method for modeling complex systems. It discusses the key steps in a study design, with an emphasis on GWA studies. Potential limitations highlighted by the WTCCC GWA study are discussed, including problems associated with massive genotype imputation, analysis of pooled national samples, shared controls, and the critical role of interactions. GWA studies clearly need massive sample sizes that are only possible through genuine collaborations. After all, for common complex traits, the question is not whether we can find some pieces of the puzzle, but how large and what kind of a sample we need to (nearly) solve the genetic puzzle.

I. INTRODUCTION

Genetic dissection has been successfully pursued in plants and model organisms although the job was quite challenging even in experimental organisms owing to technological limitations of the day (e.g., see Hummel *et al.*, 1966; McClintock, 1984). Against this background, evaluating the genetic basis of human traits seemed daunting at first. With technological, methodological, genomic, and computational advances, evaluating the genetic basis of simple Mendelian traits and Mendelian-like traits has been largely successful despite notable challenges (e.g., see Poirier *et al.*, 1993; Scheuner *et al.*, 2004; Zerres *et al.*, 1994; Zhang *et al.*, 1994). Many of these achievements have relied on the method of linkage analysis, which was greatly enhanced by the numerous pioneering contributions of Newton E. Morton over the decades. In particular, the *lo*garithm of *od*ds (LOD) score method (Morton, 1955) has been singularly recognized as a pivotal contribution, and constitutes the basis of most linkage studies. It certainly represents a major milestone in the genetic dissection of human traits, benefiting from the computational enhancements by Elston and Stewart (1971) which made LIPED, a widely used model-based linkage analysis package, possible (Ott, 1974). The LINKAGE package took it a step further (Lathrop *et al.*, 1984), with numerous subsequent computational enhancements. These methods, called (strongly) model-based because they require specification of the trait segregation model, have been widely used over the years for investigating the genetic basis of Mendelian traits. Much of the success owed partly to the fact that relatively simple segregation patterns were observable for Mendelian-like traits. Rice *et al.* (2008a, this volume) review model-based linkage analysis methods, and Rice (2008, this volume) and Borecki and Province (2008, this volume) discuss fundamentals of the underlying methods.

Beginning in the 1980s, investigators increasingly realized that complex traits pose unique challenges that require more sophisticated methods. With the advent of molecular revolution, new methods were sought for investigating the genetic basis of complex traits. The realization that complex traits derive from multiple genetic and nongenetic determinants, including interactions among them, led the way to the so-called "model-free" methods for linkage analysis ("model-free" simply implies that the method does not require explicit specification of a trait model). These methods were spearheaded by Haseman and Elston (1972) way before genome-wide scans were even conceived. Almasy and Blangero (2008, this volume) review the current status of model-free linkage analysis. These developments and molecular advances led to a series of genome-wide linkage scans with several hundred microsatellite markers in the 1990s and early years of this century, with the promise of discovering the genetic architecture of complex traits. For a while, it seemed even fashionable to undertake genome-wide linkage scans even though the end did not seem inevitable. Perhaps a degree of uncritical optimism led the scientific world to believe that with more genome-wide linkage scans somehow the puzzle will be solved. Often, even large-scale investigations ended up with disappointing results. Despite some successes (e.g., see Barroso *et al.*, 1999; Clee *et al.*, 2006; Easton *et al.*, 1995; Hanis *et al.*, 1996; Van Eerdewegh *et al.*, 2002), it is fair to conclude that genome-wide linkage scans may not be ideal for common complex traits. They certainly did not deliver the promise. It is also fair to note that most of the linkage scans in the 1990s used suboptimal study designs based on very small family structures such as sib pairs, whereas the few successful studies were based on quantitative traits in large pedigrees. Finally, linkage analysis may not be an ideal strategy for the initial phase of gene discovery for common complex traits. However, we believe that integration of linkage information into other forums of gene discovery can be helpful.

There is convergence of knowledge that complex traits derive from multiple etiologic factors with small effect sizes. Therefore, methodologies meant for detecting genes with large effects (major genes) are unlikely to be successful with complex traits, as the experience of recent years has shown. The genetic component of many complex traits is oligogenic (a few genes, each with a moderate effect) or even polygenic (many genes, each with a small effect). Even though the individual effect of a gene may be small, collectively the multiple etiologic factors and interactions with other genes and/or environments can make a substantial contribution to the final manifestation of the trait. Failure to recognize and accommodate such interactions may often mask the effects of the very genes we seek. Therefore, so as to unmask the gene effects and aid in the discovery of disease/trait genes, we must pay attention to all relevant aspects of gene discovery, including the available technologies, study design, optimal methods of analysis, and interpretation of the results. Although large sample sizes are essential, the brute force of very large sample sizes alone may not achieve the desired goal.

The mega advances in genomic science offer much hope for the future. Completion of the Human Genome Project (Lander *et al.*, 2001; McPherson *et al.*, 2001) has led to the era of the single nucleotide polymorphisms (SNPs) that paved the way to the conceptual realization that the human genome is organized into haplotype blocks in some fashion. The subsequent HapMap Project and the International HapMap consortium (HapMap 2001, 2005) have revolutionized the field with over 3.7 million SNPs as of early 2007. Investigations as to how these SNPs are organized into linkage disequilibrium (LD) blocks led to the original concept of tagging SNPs (even though not all algorithms for selection of tagSNPs are based exclusively on LD structure). This in turn made it possible to undertake genome-wide association (GWA) scans, thus making it scalable at the level of the genome. Even though GWA scans have been pursued with microsatellites (e.g., Gu *et al.*, 2007a; Pekkarinen *et al.*, 1998), its feasibility as a systematic approach was enhanced as a consequence of the HapMap Project. GWA scans started with a modest 10,000 SNPs, soon to be followed by 100,000, half a million, and now one million tagSNPs. Out of 3 billion nucleotides in the human genome, it is estimated that there are about 7 million SNPs with a minor allele frequency $\geq$0.5%. Thus, the density and coverage are bound to increase over time in major ethnic populations. These are exciting times, with several

GWA scans completed and several more under way. One would hope that the GWA scans of the day are not a repetition of the experience with genome-wide linkage scans of the last decade in so far as their utility is concerned. The pathbreaking WTCCC study offers much hope in this regard (The Wellcome Trust Case Control Consortium, 2007).

Finally, the avalanche of mega databases descending upon us will pose challenges in terms of computer and computational resources, analysis of the data, interpretation of the results, and how best to integrate various pieces of the genetic puzzle so that one may uncover the complete genetic architecture of common complex traits. As we begin this long but exciting journey into the future, it becomes even more important for us to follow efficient and state of the art methods and best practices including the often neglected ethical, legal, and social issues (ELSI) (see Haga and Beskow, 2008, this volume). In this volume, leading experts in the field discuss and summarize forward looking methods and practices including designing GWA studies (see Kraft and Cox, 2008, this volume), integrating and synthesizing results from multiple studies through novel meta-analysis methods (see Trikalinos *et al.*, 2008, this volume), admixture mapping (Zhu *et al.*, 2008b, this volume), integrating information from gene expression studies (Farber and Lusis, 2008, this volume), integration of systems biology (see Zhu *et al.*, 2008a, this volume), and a new look at how comparative genomics can enhance human disease gene discovery. This volume is concluded by an expert discussion of outstanding challenges in the study of mechanism of disease by Lalouel and colleagues (Rohrwasser *et al.*, 2008, this volume) and in the study of complex inheritance in the post-HapMap era (Morton, 2008, this volume).

II. CHALLENGES ARISING FROM COMPLEX TRAITS

Genetic dissection presupposes that there indeed exist some trait genes. Investigators undertaking an evaluation of the genetic architecture of a disease/trait usually ensure, or should ensure, that there is significant familial aggregation, with additional evidence showing that at least part of the familial resemblance is genetic. With the advent of the molecular revolution, it may be tempting to bypass the first step and jumping straight into genome-wide scans, with large numbers of genetic markers, whose goal is often twofold: first, to identify and locate genomic regions that may harbor genes influencing the trait variability and second, to thus demonstrate the very existence of trait genes with detectable effect sizes. We believe that a fundamental investigation into the causes of variation is still very useful in interpreting results pertaining to phenotype–genotype correlation analysis.

Some investigators believe that incorporating the full complexity of causation into our analytical models is necessary and important if we are to succeed in finding the genes and understanding their effects. Unfortunately, such complexity often renders the models intractable or indeterminate. Often, the lack of data on appropriate family structures and/or on the relevant interacting determinants make it impossible to even entertain full-blown models. Therefore, despite the awareness that identification of important interactions involving multiple genetic and nongenetic determinants is necessary for the detection of the very genes we are seeking, the complex reality is often approximated by much simpler models that are tractable. Using such simple models, one can investigate the degree of familial resemblance for the phenotype (e.g., see Rice, 2008, this volume). It is important to note that, when data are limited to certain family structures like intact nuclear families, it may not be possible to determine the relative roles of genetic and environmental factors.

Complex traits pose several challenges. First, they involve multiple genes with (supposedly) modest effect sizes of individual genes even under the best of circumstances. Detection of such small effect sizes of individual genes is very hard as experience suggests, and may not be detectable even with substantially large sample sizes. Second, it is readily acknowledged that complex traits involve gene–gene and gene–environment interactions, which are largely ignored in data analyses owing to the complications they pose. Unless the marginal gene effects capture interaction effects either by design or otherwise, ignoring interactions can hurt gene discovery efforts. Third, at least some physiological and metabolic traits like obesity involve imprinting effects (e.g., Dong *et al.*, 2005; Gorlova *et al.*, 2003; Guo *et al.*, 2006), which are largely ignored. Modeling imprinting effects in such cases should improve gene discovery. In fact, Sung and Rao (2007) have shown that misspecification of imprinting effects substantially decreases power of linkage analysis. Fourth, at a practical level, we may not be capturing the full effect of an individual gene because we are not measuring its effect in an optimal way. For example, if a gene effect is maximized in a certain optimal age window of an individual (when it exerts its maximal effect), we will only be capturing its average effect (over all ages) by ignoring the age dimension (temporal trend in gene effect). If we are able to incorporate age into the analysis method, we expect much increased power for the same sample size (Shi and Rao, 2007). Fifth, differential frequency of the trait-causing genetic variants and/or interactions with varying environments in different populations makes it difficult to detect uniform effects in all populations. While replication in multiple populations is essential for pruning false discoveries, we must recognize that sometimes we may be unwittingly throwing the baby with the bath water. Against such odds, how does one go about discovering the genetic architecture underlying complex traits? Numerous experts in the field offer promising approaches in this edited volume (Rao and Gu, 2008).

III. STUDY DESIGN

Study design is perhaps the single most important issue in the planning of any genetic study. Some would argue that choice of analysis methods is less important than a carefully designed study. Features such as the feasibility of the study and necessary sample size and statistical power depend critically on the study design. It is important that all the available information about the disease/trait be used fully when decisions are made about the sampling schemes, sampling units, and analytical methods. More information should lead to better designs. Several chapters in this volume discuss some aspects of study design, while designing GWA scans is discussed at length by Kraft and Cox (2008, this volume). The major steps involved in a study design are discussed below.

A. Phenotype

Although definition of the phenotype may at first seem to be a trivial issue, serious thought should be given to whether the current definition of the phenotype, however expertly done originally, is still the right one to use in gene finding studies. After all, our goal is one of finding the trait genes, not routinely following a traditional approach. Different definitions of the phenotype do lead to different results. Certain definitions tend to dwarf the signal (when they are far removed from the gene product, see Fig. 1.1) while others might at least have the potential to enhance the signal (when they are closer to the gene product, see Fig. 1.1). The fundamental idea behind a linkage-based approach to gene finding is one of evaluating the correlation between the degree of phenotypic similarity and genotypic similarity among relatives. Needless to say, this relationship will be weakened if either type of similarity (phenotypic or genotypic similarity) is underestimated. In particular, to avoid underestimating the phenotypic similarity, or at least to minimize the extent of underestimation, we must require

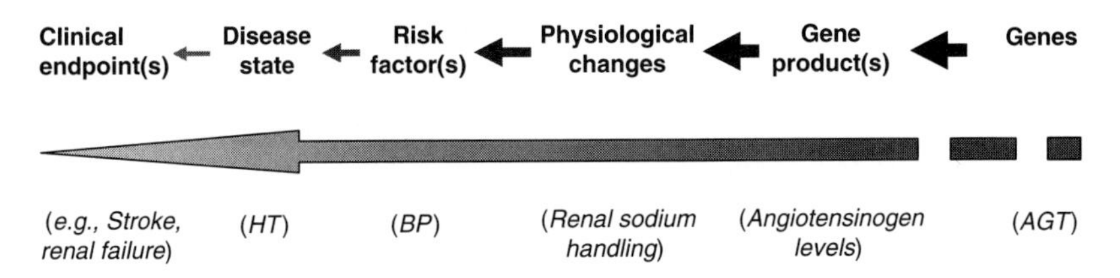

Figure 1.1. A representation of how far removed a given disease (phenotype) and the resulting clinical endpoint may be from the underlying gene(s), thus illustrating the challenges involved in gene discovery. The example shows the steps involved in the process from the angiotensinogen gene (AGT) to hypertension (HT) to stroke or renal failure. BP denotes blood pressure.

that the phenotype be reasonably highly reproducible (e.g., see Rao, 1998). As Wojczynski and Tiwari (2008, this volume) argue, definition of the phenotype is crucial in designing any genetic study, especially an association study. They review different types of phenotypes such as discrete or continuous and discuss the impact of issues like diagnostic error (misclassification) in case-control studies and measurement error in continuous traits. They show that the power of a study depends heavily on the phenotype measured and that misclassification or measurement error can dramatically reduce the power. They also suggest some potential remedies.

B. Genotype

Genetic studies pay due attention to important issues such as which type of markers to use, the coverage and density of markers, and the type of samples to base genotyping on (such as case-control or family units). With the advent of large-scale genotyping using automated robotics, data management, quality control, and informatics issues need special attention. Integrating, maneuvering, and leveraging extensively large databases requires special skills and expertise in bioinformatics. Maresso and Broeckel (2008, this volume) review and evaluate various platforms for mass throughput genotyping with SNPs, including GWA scans. With the advent of the Human Genome Project and the International HapMap Project, mass-throughput genotyping platforms are now available at affordable prices. These platforms have made GWA scans a reality and hopefully bring us closer than ever to elucidating the genetic architecture of complex disease, while genotyping error rates have been impressively controlled in some platforms, which is not the case with all genotyping studies. Genotyping errors need special attention because they can have a significant effect on the power of a study (e.g., see Rao, 1998). Miller *et al.* (2008, this volume) discuss the impact of genotyping errors on genetic analysis. They focus on the problems of misspecification of familial relationships and misclassification of genotypes, allele shifting in multibatch processing, sample mix ups, and how those kinds of errors can be detected and corrected using readily available methods and software.

C. Linkage versus association

A linkage study analyzes the cosegregation of a disease trait (locus) and a marker in families, while an association study investigates the co-occurrence (coexistence) of the disease trait and the marker in individuals. However, the premise of a genetic association is based on the hope that association induced by LD will lead us to the genetic variant, and "spurious" association will be excluded by other means of study (such as using family-based controls and replication studies).

Whether one wishes to focus on association or linkage (and association), it would be desirable to follow similar study design principles. As we discussed earlier, linkage analysis seems to be losing favor, although we believe that it can be helpful in interpreting association results from massive GWA scans as well as in directing dense follow-up studies. A decade after Risch and Merikangas (1996) advocated that, GWA scans are finally becoming a reality.

D. Sampling type

Important sampling considerations relate to the sampling unit (e.g., unrelated cases and controls, sib pairs, or families) and the sampling method (random or ascertained sample), which depends on the type of study being conducted (linkage vs association). Selection of a population is an important issue (e.g., see Terwilliger and Lee, 2007), although sometimes investigators have little choice in this regard. For example, for linkage studies of complex traits, sib pairs of one type or another are commonly used in conjunction with model-free methods of analysis. Sampling larger sibships yields more power per sampled subject than sampling-independent sib pairs (Todorov *et al.*, 1997). Extremely discordant sib pairs (Eaves, 1994; Risch and Zhang, 1995) or extremely discordant and extremely concordant sib pairs (Gu *et al.*, 1996) can reduce the required sample size without hurting the power. In general, larger families have more power and any selective sampling strategy could be used to enhance power, except that some are more efficient than others in achieving the desired goal. Feasibility, cost, and reproducibility of the phenotype(s) should all be taken into account when choosing a sampling unit. Unlike for linkage studies that always require families, association studies are carried out using multiple sampling approaches. Case-control and family-based approaches constitute the most popular designs, and each has its merits and limitations. When population stratification is not present, some consider the case-control design as the most powerful and cost effective study. However, on the face of substratification to varying degrees in the study population, some investigators still favor studies of unrelated individuals with appropriate genomic control, while others prefer family-based studies. A number of chapters in this volume deal with this issue directly or indirectly (e.g., see Rice *et al.*, 2008a; Almasy and Blangero, 2008; Schork *et al.*, 2008; Laird and Lange, 2008; and Zhu *et al.*, 2008b, this volume).

E. Sample size, significance level, and power

In general, the power of a study $(1-\beta)$ is a function of the sample size (N), the significance level (α), and the effect size of the gene (or the locus-specific heritability, h_g^2) (assuming that all other things are identical, such as the sampling unit). For a given effect size, appropriate choice of any two of the other three variables (α, β, and N) determines the third variable.

$$\text{Power} = (1 - \beta) \approx f(N, \alpha, h_g^2)$$

For a given sample size (e.g., after a family study has been carried out, the N is fixed), the power increases (and the rate of false negatives, β, decreases) with the significance level. Unfortunately, we cannot increase the power simply by increasing the significance level because false positives also increase with the significance level. This is especially true for complex traits whose effect sizes are modest, thus rendering the issue of sample size and power even more important. Choice of a significance level also depends on the type of study and the number of hypothesis tests conducted. There is a huge difference between a linkage study based on hundreds of microsatellite markers or a few thousand SNPs and an association study based on several hundred thousand SNPs. It is clear that both types of errors need to be minimized as much as possible at the stage of designing a study. Therefore, the required sample size should be calculated so as to yield power as high as possible (at least 80%) when using an appropriate significance level. Because no single center may be able to recruit large samples, well-designed and coordinated multicenter studies are particularly attractive for this purpose. Several chapters discuss sample size and power calculations in different contexts (e.g., see Almasy and Blangero, 2008; Kraft and Cox, 2008; Laird and Lange, 2008; Rice *et al.*, 2008a; Schork *et al.*, 2008; Zhu *et al.*, 2008b, this volume).

IV. ANALYTICAL METHODS

Study design and analysis methods are tied to each other, and the available methods are extensive for carrying out linkage and association analysis. Although linkage studies are losing favor, it is not clear to what extent this owes to the fact that suboptimal study designs (such as sib pairs and small families) were used in most of the genome-wide linkage scans. Also, novel methods continue to be developed for optimal extraction of linkage information. While some simple association methods have been scaled to the level of the GWA scans, this continues to be work in progress. As more and more GWA data are released, we anticipate that the low hanging fruit will be harvested quickly using available methods, while novel methods will continue to be developed for getting the most from the massive datasets. A collection of standard and evolving methods is discussed by several leading experts in the field, which are briefly outlined here.

A. Familial aggregation and genetic effects

Familial resemblance may be estimated in terms of correlations among family members. Generalized heritability quantifies the strength of the familial resemblance and represents the percentage of variance in a phenotype that is explained by all additive familial effects including additive genetic effects and those of the familial environment. On the other hand, genetic heritability quantifies the percentage of phenotypic variance due to additive genetic effects. The type of family study (e.g., nuclear families or large pedigrees) determines which of these sources are resolvable. For example, genetic and familial environmental effects cannot be resolved in intact nuclear families consisting of parents and offspring because they share both genes and environments. However, extended pedigrees, twin and adoption study designs, and migrant studies allow separation of the heritable effects. See Rice (2008, this volume) for details.

B. Linkage, association, and admixture mapping

1. Model-based linkage analysis

Morton (1955) revolutionized the field and unified the early methods for linkage analysis with his seminal article in 1955. The method, based on (strong) assumptions about the trait inheritance model (and hence the name, "model-based"), is broadly concerned with issues of power and the posterior probability of linkage, ensuring that a reported linkage has a high probability of being a true linkage. In addition, the method is sequential so that LOD score tables could be combined from published reports to pool data for analysis (e.g., as done by Keats *et al.*, 1979; Rao *et al.*, 1978, 1979a). This approach has been remarkably successful for 50 years in identifying disease genes for Mendelian disorders. After discussing these issues, Rice *et al.* (2008a, this volume) consider complex disorders where the maximum LOD score statistic shares some of the advantages of the traditional LOD score method, but is limited by unknown power. Also, optimal pooling of the maximum LOD scores across studies and/or families requires access to the primary data.

2. Model-free linkage analysis

Model-free linkage methods, which do not require assumptions about the way the phenotype is inherited, are based on identifying genomic regions in which patterns of allele sharing among family members correspond to patterns of phenotype correlation among those family members. There is a large literature of excellent model-free methods, including variance components methods (see, e.g., Abecasis *et al.*, 2002; Almasy and Blangero, 1998; Amos, 1994; Fulker and Cardon, 1994; Goldgar, 1990; Heath, 1997; Kruglyak and Lander, 1995). Two general classes of model-free linkage methods are discussed by Almasy and Blangero (2008, this volume), relative pair methods designed primarily for analysis of discrete traits and variance component methods designed primarily for analysis of quantitative traits. These methods have been used to identify several genes influencing complex traits and remain viable approaches for the future.

3. Potential inadequacies with linkage analysis methods

Failure of linkage analysis in gene finding studies to date may have a lot to do with potential inadequacies in the methods of analysis. For example, at least some complex traits like obesity are known to involve imprinting effects (e.g., see Dong *et al.*, 2005; Gorlova *et al.*, 2003; Guo *et al.*, 2006), which are ignored in most of the published linkage analyses. Modeling imprinting effects, when relevant, can improve gene discovery. To study the impact of imprinting on linkage analysis, Sung and Rao (2007) simulated pedigree data sets where imprinting effects contributed variously to the quantitative-trait locus (QTL) variance (0%, 25%, 50%, or 75%). To study the effects of misspecification of imprinting, they analyzed each data set four times, each time assuming a different level of the imprinting effect. The model with the correct level of imprinting always provided the highest LOD scores. The incorrect model with no imprinting effects provided the lowest LOD scores. They reported that, when the incorrect level of imprinting was assumed, 60% of the time the max LOD scores were less than −2, which would generally lead to exclusion of linkage. These results show that accounting for imprinting can substantially improve linkage detection. Conversely, misspecification of imprinting effects can substantially decrease the power of linkage analysis, including erroneous exclusion of genomic regions.

Likewise, there is ample biological evidence suggesting that the effects of some trait causing genes may vary with the age of the individual when the phenotype was measured (e.g., see Cheverud *et al.*, 1983). In continuation of the previous work on temporal trends in unmeasured genotypes (Province and Rao, 1985; Rao *et al.*, 1975), Shi and Rao (2007) have recently extended the variance components of linkage analysis method by incorporating temporal trends in QTL effects as well as in residual polygenic effects. By parsimoniously modeling the trends using Gaussian functions, they showed that ignoring temporal trends in QTL effects, when present, can substantially reduce the power of linkage analysis. In several realistic situations, the power gain was severalfold higher when temporal trends were modeled in the QTL. In light of these experiences with imprinting and temporal trends in particular, it seems that linkage analysis methods have not been optimized for addressing some of the real complexities of complex traits. Therefore, some of the negative experience with linkage analysis may in part be due to such inadequacies in the methodology.

4. Association analysis

As Laird and Lange (2008, this volume) suggest, traditional epidemiologic study designs such as case-control or cohort designs can be used with genetic association studies. However, family-based designs are preferred in the presence of population substructure. Although family-based designs need special methods of analysis, there are distinct advantages such as being able to test for linkage and association. They also offer some protection against the multiple comparisons problem. Laird and Lange describe some basic study designs, as well as general approaches to analysis for qualitative, quantitative, and complex traits, including the available software. It is important to note that, while family-based designs may offer some protection against increased type I error in the presence of population stratification, the bigger problem with population stratification may be increased heterogeneity which in turn inflates type II errors (missing real associations).

Schork *et al.* (2008, this volume) discuss association analysis based on DNA sequence. Anticipating that cost-effective and high-throughput DNA sequencing will become available in the future, they consider an alternative to genotyping-based approaches. They consider the use of an analysis technique, termed Multivariate Distance Matrix Regression (or MDMR) analysis, to evaluate associations between DNA sequence information and quantitative traits. The potential of the method is evaluated through computer simulation, showing that the procedure has promise for future association studies.

5. Search for additional genes

Thomson and Barcellos (2008, this volume) discuss methods for searching for additional disease loci in a genomic region by taking into account the known loci. Once one or more genes and SNPs are identified, how does one detect additional disease genes in the region? The power of different methods for this purpose will vary depending on the specific genetic and environmental features of the disease under consideration. The existing programs are not adequate to handle the magnitude and complexity of the analyses needed. Further, even with modern computers, one cannot study every possible combination of genetic markers and their haplotypes across the genome, or even within a genomic region. Thomson and Barcellos recommend a multiple strategy for evaluating whether additional genes are involved in a given genomic region where some gene(s) and SNPs have already been identified.

6. Admixture mapping

Admixture mapping, or mapping by admixture linkage disequilibrium (MALD), is a mapping strategy that has gained considerable popularity in recent years for mapping disease genes. As discussed by Zhu *et al.* (2008b, this volume), the method exploits the long-range LD generated by admixture between genetically distinct ancestral populations. Admixture mapping requires much fewer numbers of markers as compared to case-control association studies, and is more robust to allelic heterogeneity. It can also be more powerful, and can achieve higher mapping resolution than the traditional linkage studies provided that the underlying trait variants occur at sufficiently different frequencies in the ancestral populations. Zhu *et al.* describe the recent methodological advances and software developments, review applications to real data, and comment on the future of this mapping method.

C. Dense SNPs and haplotype analysis

1. Interactions

Assimes *et al.* (2008, this volume) discuss association analysis involving the main effects of genes, their epistatic interactions, as well as interactions with environmental covariates. Prediction of complex disease is especially difficult, though it would be relatively easier if the phase information is available. The ultimate goal of their approaches is to predict the phenotype within a specific window of time. They discuss five modern approaches to two-class association/classification. The classification trees (as in CART®, Breiman *et al.*, 1984) play an important role in discovering interactions among predictors, and accordingly, two of the five approaches discussed by the authors are based on CART. Some would argue that when interactions influence a complex phenotype significantly, nonrandom ascertainment of extreme phenotype should capture the interaction effects as part of the marginal effects of genes (i.e., overestimating the effect sizes of the genes). In that case, modeling interaction effects may not be as critical for gene discovery itself as it would be for an accurate understanding of the phenotype–genotype relationships.

2. HapMap and LD structures

Gu *et al.* (2008, this volume) discuss characterization of LD structures and the utility of HapMap in genetic association Studies. As the authors discuss, the observed distribution of and variation in LD with respect to the evolutionary history and disease transmission in a population is the driving force behind the current wave of GWA studies. Extensive literature covers topics from haplotype analysis (e.g., see Liu *et al.*, 2008, this volume) that utilizes local LD structures to genome-wide organization of LD blocks that led to the development of the International HapMap Project and panels of "tagSNPs" used by current GWA studies. Gu *et al.* (2008, this volume) examine the scenarios where each of the major types of analysis methods may be applicable and where the current popular genotyping platforms for GWA might come short. They discuss current association analysis methods by emphasizing their reliance on the local LD structures or on the global organization of the LD structures, and highlight the importance of and the need to consider individual marker information content in large-scale association mapping.

3. Haplotype analysis

As Liu *et al.* (2008, this volume) argue, association methods based on LD offer a promising approach for detecting genetic variations that are responsible for complex human diseases. Although methods based on individual SNPs may lead to significant findings, it is generally accepted that methods based on haplotypes comprising multiple SNPs on the same inherited chromosome may provide additional power and also provide insights into the factors influencing the dependency among genetic markers. Such insights will likely provide information essential for understanding human evolution and also for identifying *cis*-interactions between two or more causal variants. Because obtaining haplotype information directly from experiments can be cost prohibitive in most studies, especially in large-scale studies, haplotype analysis presents many unique challenges. Liu *et al.* focus on two main issues: haplotype inference and haplotype association analysis. They begin with a detailed review of methods for haplotype inference using unrelated individuals as well as related individuals from pedigrees, followed by a number of statistical methods that employ haplotype information in association analysis. They also discuss the advantages and limitations of different haplotype methods.

4. Haplotype similarity and marker ambiguity

Recognizing that the boundary of common haplotype blocks in HapMap constructions involves a certain degree of ambiguity, Gu *et al.* (2007b) presented a method to address the issue at the level of individual SNP markers. They introduced a measure called the *marker ambiguity score* (MAS), and showed that the MAS method can be used to assess the level of boundary ambiguity caused by varying ethnic background, sample sizes for HapMap construction, and disease aggregation. They found a striking difference in the overall patterns of block boundary distributions in blacks and whites, and subtle changes in block structures that agree with the evolutionary history of the two populations. They suggested that a sample size of 200 or more subjects is desirable for "stable" HapMap construction. Based on real data, they also demonstrated that there are subtle differences in block boundaries in disease populations versus normal controls. This approach based on the MAS concept has the potential to quantify the information content of individual markers, which may have important implications for the design of efficient GWA studies.

D. Composite likelihoods

Genetic modeling of high dimensional data poses an enormous challenge to standard likelihood methods. For example, although multivariate linkage and association analyses would be more powerful for detecting genetic effects (especially in the presence of pleiotropic effects) than univariate analyses, the exact maximum likelihood methods pose a serious challenge with large numbers of correlated phenotypes. As an approximation to the standard likelihood, "composite likelihood" method was developed that makes valid statistical inference feasible even in large systems (Lindsay, 1988). The composite likelihood is given by the product of several component likelihoods of much reduced complexity. For example, the joint likelihood of an N-variate phenotype may be approximated by the composite likelihood obtained as the product of all pairwise bivariate likelihoods. As an approximation, composite likelihood offers a successful strategy for certain applications. Li (2008, this volume) describes the method of composite likelihood, its advantages and potential challenges. He also reviews a few applications of the method in genetic studies, specifically in estimating population genetic parameters such as recombination rate and in multilocus LD mapping of disease genes with some discussion about future research directions (also see Morton, 2008, this volume).

E. Multiple testing

As Rice *et al.* (2008b, this volume) discuss, the availability of high-throughput genotyping technologies and massive amounts of marker data for testing genetic associations is a dual-edged sword. On the one hand, there is the possibility that the causative gene (or one in strong LD with) will be found among those tested for association. But, on the other hand, testing large numbers of markers for association gives rise to the multiple testing problem. The traditional solution is to use the Bonferroni correction or something similar. This works well with small numbers of tests. However, association studies are beginning to explode, with hundreds of thousands of tests as in GWA scans with 500,000 to 1 million SNPs, and at such a large scale, a Bonferroni type correction is not an attractive solution because of LD between close by SNPs. In fact, if a Bonferroni correction is used, it will likely result in a large false negative rate. How best to balance between false positives and false negatives in large-scale association studies remains a challenge. See also Kraft and Cox (2008, this volume).

V. SPECIAL TOPICS

A. Pathway-based association studies and Bayesian networks

Association analysis is often performed using single phenotypes and single markers or haplotype analysis of multiple SNPs in adjoining (short) regions or candidate genes. Although the importance of gene–gene and gene–environment interactions has long been recognized, most analyses are carried out using one gene at a time. This is certainly a valid approach; however, the power of detecting significant genetic associations may be considerably enhanced by simultaneous consideration of markers (SNPs) at multiple genes and/or multiple phenotypes. Unfortunately, arbitrarily large numbers of genes and markers would pose real problems in terms of modeling all possible interactions in a routine manner. Biology comes to our rescue, whereby a simultaneous consideration may be first limited to those multiple genes that are involved in a biological pathway. Such an analysis will be a lot more complex as compared to analysis of single genes in isolation of each other; however, the level of complexity would be vastly simplified as compared to the joint analysis of arbitrarily large numbers of genes and markers.

Exploring interrelationships among several variables benefits from the intellectual foundations created several decades earlier in terms of structural equation models or what is more commonly known as path analysis (Li, 1975; Wright, 1931). For example, a path analysis model (and software called PATH) was developed some three decades earlier for analyzing interrelationships among sets of variables (Rao *et al.*, 1977). Following the first application to an analysis of academic performance in American schools (Rao *et al.*, 1977), the methodology was extended to incorporate multiple layers of (potentially) "causal" relationships using alternative statistical methods along with software called WRIGHT (Rao *et al.*, 1979b). These early developments provide a foundation for more general methodological approaches called Bayesian networks (BNs).

BNs can be used to investigate the relationships among a given set of markers and other measured variables (e.g., multiple correlated phenotypes, environmental variables, intermediate phenotypes). BN is a graphical modeling of the joint multivariate probability distribution of all variables (Heckerman, 1996). Rodin *et al.* (2005, 2008) have considered applications in genetics, where large numbers of markers, phenotypes, and covariates were analyzed simultaneously. In a BN, connections or edges between variables (nodes) indicate associations (dependencies). The thickness of an edge is roughly proportional to the strength of the association, which can be evaluated using either posterior probability ratio test or bootstrap. Edge direction (arrow) indicates that the model with a given direction has higher likelihood ratio than the reverse direction. However, this by itself does not necessarily imply causation. Edges connecting SNPs are indicative of LD. Edges between SNPs and phenotypic nodes may suggest causal associations. If an SNP and an environmental variable appear in proximity of a phenotype, possible gene–environment interactions are indicated. These can be further scrutinized by computing a local conditional probability table for the phenotype node (see Heckerman, 1996). BN findings have been shown to be fairly robust, making them attractive for the analysis of a large number of SNPs simultaneously (Rodin *et al.*, 2008).

We believe that BN is especially attractive for analysis of pathway-based SNP associations where certain constraints would be logical. For example, Fig. 1.2 depicts an example of part of a pathway where marker (SNP) G_1 contributes to X_1, G_2 contributes to X_2, and X_2 is also partly influenced by X_1. However, in terms of the time sequence, G_2 does not effect X_1 and nor may G_1 directly influence X_2. The underlying pathway suggests that, in a BN analysis of the four variables, we must prohibit any direct connections (edges) between G_1 and X_2 and also between G_2 and X_1. Likewise, the arrows must be constrained to be in the directions shown. Such a constrained BN (c-BN) will involve only a smaller subset of the parameter space and should enhance the biological relevance of an otherwise pure statistical analysis of data. Zhu *et al.* (2008a, this volume) discuss a systems biology approach to drug discovery that uses BN analysis.

B. Gene expression and systems biology

As Farber and Lusis (2008, this volume) argue, identifying gene expression changes that are either causing a given disease or reacting to the disease promises to significantly enhance our understanding of common disorders. When applied to samples representing disease and normal states, microarray-based expression profiling can identify differentially expressed genes that may play a role in the disease or predict progression or severity of the disease. Farber and Lusis suggest

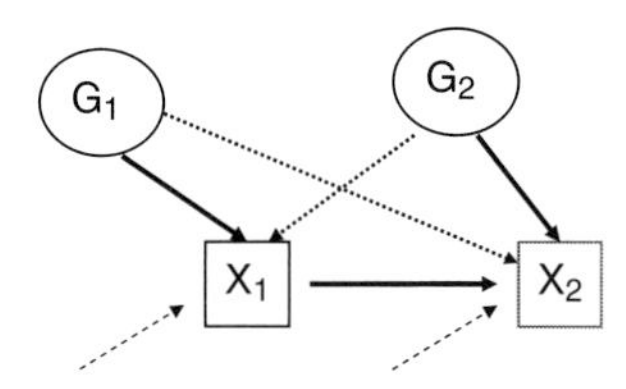

Figure 1.2. Constrained Bayesian networks and pathway-based association analysis. The figure illustrates part of a pathway where four variables are measures (G_1, G_2, X_1, and X_2). Phenotypic measurements (Xs) reflect time-sequence dependence (where X1 influences X2 but not the other way around). Likewise, G_1 influences X_1 (but not X_2 directly) and G_2 influences X_2 (but not X_1). Whereas unconstrained Bayesian network analysis will consider all possible edges shown by solid and dotted lines (in which the arrows could be reversed), constrained BN analysis should only consider the solid edges and also restrict the direction of influence as shown.

that the integration of genetics and gene expression is promising for uncovering common genetic variations that control a particular disease. In animal models this approach has already been used successfully to advance our knowledge about novel disease mechanisms. Farber and Lusis provide an overview of DNA microarray technologies and discuss ways in which microarray expression data can be combined with more traditional experimental approaches to dissect the genetic basis of disease.

Taking a systems biology approach, Zhu *et al.* (2008a, this volume) treat a given disorder such as obesity as a disorder of the system, potentially involving several pathways operating in many different tissues, and ultimately not only giving rise to different disease subtypes, but to different comorbidities as well. The integration of large-scale data such as gene expression, genotypic data, clinical data, and other biologically relevant information will be critical if we ever hope to understand more fully how genetic and environmental perturbations to a given system lead to complex traits. If common forms of disorders like obesity, diabetes, and heart disease represent states of a network, focusing on single gene perturbations are unlikely to reveal the complete picture and hence the most effective ways to treat or prevent these diseases. The integration of a diversity of data in this setting is key because no single dimension may be able to provide the complete answer. For example, the identification of DNA polymorphisms associated with diseases like obesity and diabetes only represents the beginning in a long series of steps needed to elucidate disease pathways and to establish the specific role individual genes may play in the process. While these types of genetic discoveries provide a peek into the underlying pathways, they are usually devoid of context, and therefore elucidating the functional role of such genes may take much longer.

The integration of diverse sets of molecular data now being generated in various populations should move us closer to the systems level necessary to fully understand the complexity of common complex diseases like obesity and diabetes. As Zhu *et al.* suggest, further study and experimentation are needed to demonstrate more convincingly to understand the state of a given molecular network, interactions among molecular networks, and how the states of such networks change in response to different genetic and environmental contexts.

C. Comparative genomics

Originally, comparative genomics was geared toward defining synteny of genes among species. As the human genome project accelerated, there was an increase in the number of tools and means to make comparisons, culminating in having the genomic sequence for a large number of organisms spanning the evolutionary tree. With this level of resolution and a long history of comparative biology and comparative genetics, Moreno *et al.* (2008, this volume) suggest that it is now possible to use comparative genomics to build or select better animal models and to facilitate gene discovery. Comparative genomics takes advantage of the functional genetic information from other organisms so as to apply it to the study of human physiology and disease. It facilitates identification of genes and regulatory regions for acquiring knowledge about gene functions. The current state of comparative genomics and the available tools are discussed by Moreno *et al.* in the context of developing animal model systems that reflect the clinical picture.

VI. GWA STUDIES

A. Recent GWA studies

A series of GWA studies has appeared during last year and many more are under way. The Wellcome Trust Case Control Consortium (2007; abbreviated WTCCC) study is arguably a groundbreaking contribution in many respects. The study genotyped 500,000 SNPs using the Affymetrix platform in 2000 cases for each of seven common diseases and 3000 shared controls. Using the genotypes at the 500,000 SNPs, the LD structure observed in the HapMap, and a Hidden Markov Model, they have imputed genotypes at 2.2 million SNPs not measured in the study. The imputation method has been validated by genotyping 10,000 SNPs from the Illumina platform in 1500 of the 3000 shared controls, reporting an impressive 98.4% concordance. This study demonstrates the feasibility and reality of GWA studies, reporting 25 strong hits, of which 24 have been replicated. In addition to confirming the involvement of 12 genes for which associations have been reported previously, the study also reported 13 novel associations of which 12 have been replicated by the investigators. Notable among the findings is that it identified, for the first time, a genetic link between type 1 diabetes and Crohn's disease and that one or more hits have been identified for six of the seven diseases considered (hypertension being the exception).

Despite the sheer size and the voluminous information contributed by the WTCCC study, it raises some important issues and concerns. First, while imputation has long been used in statistical literature for deriving some pieces of missing information, imputing far greater number of SNPs (2.2 million) based on a much smaller set of SNPs observed (0.5 million) is a very different situation. Although use of the 2.7 million SNPs in the HapMap for this purpose provides added comfort, the later is only based on a very small sample. Likewise, validating the imputation methodology using 10,000 SNPs in 1500 controls is very impressive. However, while the validation experiment conforms to the standard statistical practice of imputation (namely, using a larger set of SNPs to predict genotypes at a much smaller subset of SNPs), imputing 2.2 million SNPs based on 0.5 million SNPs constitutes an extrapolation by several orders of magnitude.

It would be more comparable if, say, 100,000 of the measured SNPs are used to impute the remaining 400,000 SNPs to check the validity of imputing at the scale used in the WTCCC study. Observation of a high concordance in such an experiment would add much comfort. When imputed SNPs show significant associations, it is necessary to genotype the imputed SNPs and verify the result.

Second, while pooling data and samples from different national samples vastly increases the sample size, which is often necessary for achieving adequate power, using the pooled samples in aggregate may actually hurt the power owing to potential heterogeneity in the pooled database. Stratifying the 2000 samples of cases into a few relatively more homogeneous groups and pooling the results from separate stratified analyses may boost the power. Failure to do this may have led to increased false negatives (but not increased false positives). Disease heterogeneity might partly explain why strong associations were not detected for, say, hypertension. In addition, using aggregate pooled samples also raises potential issues of population substructure. Although excellent care was exercised in controlling for major sources of substructure, subtle levels could increase false positives. Replication of 24 of the 25 strong hits argues against such possibility although it is not entirely ruled out.

Third, while the concept of shared controls is cost effective and very attractive in terms of not having to identify multiple series of controls, it involves potential limitations. It is much more challenging to ensure that all controls are free of all seven diseases and, even if they are free of disease at the time of observation (which appears to be true in at least half of the controls), they may not have gone through the risk period yet. Thus, it is possible that some of the so-called controls may also carry the disease-prone allele, which will reduce power.

Finally, a systematic analysis incorporating gene–gene and gene–environment interactions will likely identify additional susceptibility genes. In any case, the current WTCCC study has succeeded in harvesting at least the low hanging fruit and has enormous potential for uncovering many more associated markers. The inherent potential will be realizable once the outstanding issues are resolved.

B. Designing GWA studies

Remarkable advances in high-throughput genotyping and the availability of large-scale genotyping platforms (see Maresso and Broeckel, 2008, this volume) together with a flood of data on human genetic variation from the Human Genome and HapMap Projects have made GWA studies attractive and feasible. However, researchers designing such studies face a number of challenges, including how to avoid subtle systematic biases, how to achieve sufficient statistical power to distinguish modest association signals from chance associations, how best to handle population substructure issues, and how to select the best genotyping platform, which at the current time has significant budgetary implications, among others. In many situations, it remains prohibitively expensive to genotype all the desired samples using a genome-wide genotyping array, so multistage designs can be an attractive cost-saving measure. Kraft and Cox (2008, this volume) review some of the basic design principles for genetic association studies, discuss the properties of fixed genome-wide and custom genotyping arrays as they relate to study design, present a theoretical framework and practical tools for power calculations and for multistage designs, and discuss the limitations of multistage designs. Despite outstanding challenges, the paradigm WTCCC study based on 17,000 samples demonstrates both the feasibility and some of the inherent problems. Finally, as shown by Altshuler and Daly (2007), even in the large WTCCC study based on 2000 cases and 3000 (shared) controls, the power to obtain a genome-wide $P < 10^{-8}$ was <1% for several of the associations reported (and confirmed). Based on this, we should only anticipate that different GWAs will report partially overlapping sets of associations and that most of the regions harboring relevant associations may not be identified anytime soon.

C. On transferability of genome-wide tagSNPs

A general issue facing LD-based association studies is that the detection of an association may be compromised when tagSNPs are chosen from data in one population sample and used in another sample. As de Bakker *et al.* (2006) argue, it is important to know how well tags picked from the HapMap DNA samples capture the variation in other population samples. Recent literature has provided detailed analyses of the issue of transferability of tagSNPs, which are usually selected based on HapMap data, and sometimes on their performance in other populations (de Bakker *et al.*, 2006; Service *et al.*, 2007; Willer *et al.*, 2006). While these studies generally concluded in support of transferability of the tagSNPs, the authors acknowledged the limitations of existing studies, and pointed out that "the field is still awaiting the results of studies testing the performance in various populations of tagSNPs chosen from all of the SNPs available from HapMap" (Service *et al.*, 2007). The issue of transferability is an important one because many ongoing and upcoming GWA studies rely critically on the validity and feasibility of using a universal core set of tagSNPs. Gu *et al.* (2007c) revisit the topic and highlight the important issues that need our immediate attention. As they conclude based on a review of the literature, transferability is not an easy issue that requires further work. So long as investigators using a core set of genome-wide tagSNPs understand some of the limitations and remain cognizant to potential implications, its continued use offers the best approach yet because alternatives are prohibitively more expensive at the current time.

D. Follow-up studies

When conducting a GWA study, ideally one would prefer to massively genotype as large a sample as possible such as in the WTCCC study. However, one is often limited by budgetary considerations and must limit the sample size subject to satisfactory power considerations. In that regard, as Kraft and Cox (2008, this volume) suggest, a two-stage design is desirable at this time. A moderately large sample is genotyped first in stage 1, with a subsequent follow-up study to scrutinize a subset of the most promising SNPs in stage 2. In such a design, there arise multiple choices for selecting the subset of SNPs for further evaluation in stage 2. While selecting the subset of SNPs based entirely on single SNP association results (*p* value) offers one approach, we advocate using a mixed approach whereby a large fraction of the subset is selected based on the single SNP association *p* values whereas the remaining SNPs are selected based on additional criteria such as the available linkage evidence in certain genomic regions and clustering of the single SNP results although they may not meet a conservative significance threshold. Even if one's own study does not have linkage information, we suggest integrating the relevant linkage evidence available in the literature when interpreting the single SNP association analyses and selecting a subset of SNPs. Figure 1.3 illustrates the selection of a subset of 5000 SNPs from a stage 1 GWA scan using 500,000 SNPs. Such selection may be supplemented with additional SNPs of relevance (as soon in the last step in Fig. 1.3), such as flanking markers for each of the 5000 when available. One may also add additional SNPs highly suggested from external sources such as those reported in other studies for the same or a highly correlated phenotype, nonsynonymous SNPs, etc. Also, because the tagSNPs have been preselected based on HapMap data, which may or may not represent the full variability available in the study sample, resequencing of the special population under consideration (such as a hypertensive sample) may yield novel SNPs, which may be relevant for the current study. Thus, one may compile a set of, say, 20,000 SNPs for the follow-up study, which includes 5000 of the original SNPs. Since the marker landscape is constantly evolving, any selection of SNPs must take into consideration the state of the art available at the time of conducting a study.

VII. EFFICIENT STRATEGIES FOR ENHANCING GENE DISCOVERY

A. Lumping and splitting

A common approach to enhance the power of any study is to utilize larger sample sizes such as in the WTCCC study. Fortunately, the concept of multicenter genetic and family studies has rapidly evolved as a means of generating large samples

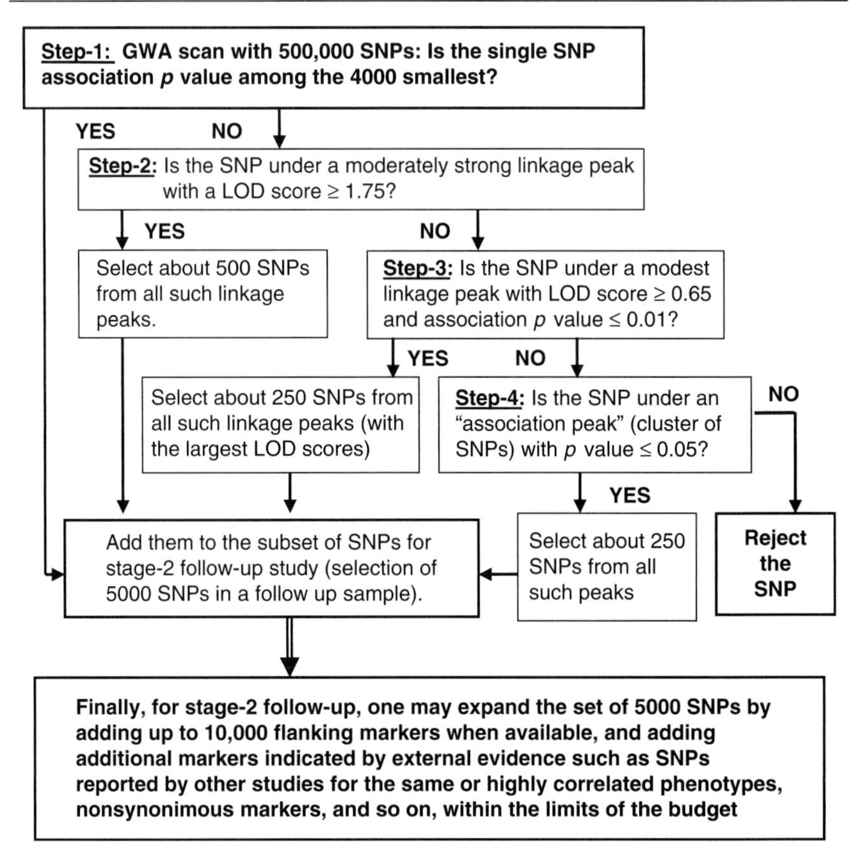

Figure 1.3. Genome-Wide Association (GWA) Study using 500,000 SNP chips: Flowchart shows how single SNP association analysis and linkage analysis (from within the study or from the literature) are used to select a subset of 5000 SNPs for stage 2 follow-up study.

of family data collected using standardized protocols (e.g., Higgins *et al.*, 1996). Even in preplanned collaborations of this type where common protocols are used and data collection is standardized, one must remain cognizant that the frequency and distribution of risk factors—both genetic and environmental—may be well different among the participating study centers. Clearly, pooling data from different studies that are conducted independently without any standardization across the studies encompasses even greater challenges, as there may be considerable differences in the sampling strategy, in the phenotypic measurement, or in the ancillary information available for subclassification of a phenotype or the pooled data. When it is not possible to pool the data directly, it may be useful to pool the results from different studies. Some of these issues have been considered in the development of meta-analytic methods for pooling results from multiple studies.

For complex traits where we expect etiologic heterogeneity, one may argue that pooling data across studies may make it even more heterogeneous. Ideally, one wishes to maximize the signal to noise ratio by analyzing the largest possible subsamples sharing the same predominant etiologic factor(s). Strategies that enable investigators to subdivide the pooled data into relatively more homogeneous subgroups are extremely desirable. One approach is to stratify the aggregate sample into multiple subgroups. By analyzing each subgroup separately, the results may be pooled using appropriate meta-analysis methods. Such a lumping (pooling data from multiple studies) and splitting (stratified subgroups) strategy can help.

Although lumping and splitting is very attractive from a practical point of view, it is not the only desirable strategy for enhancing gene finding. Very large pedigrees of genetic isolates such as the Icelandic study are very promising. It is not clear as to which of the two approaches is more feasible and/or more cost effective, or which has greater potential for finding genes.

B. Meta-analysis

The term "meta-analysis" is used for a wide variety of statistical procedures developed for pooling and summarizing results from multiple studies (see Olkin, 1995 for a review). For early applications of meta-analysis techniques to genetic studies, see Li and Rao (1996), Rice (1998), and Gu *et al.* (1998). For meta-analysis of linkage results, Gu *et al.* (1998) proposed using the proportion of alleles shared IBD at a marker locus by a sib pair (with specified trait outcomes) as the common effect, and presented methods for pooling results from model-free sib pair analyses. A random effects model was used to characterize the among-study variability, and a weighted estimate of the overall effect and the variance components were given using the weighted least-squares method. A heterogeneity test was also proposed to assess variability among studies. Applications of meta-analysis in linkage and association studies have evolved rapidly and the methodology will likely see even greater applications as we move increasingly to GWA studies. The current state of the art is reviewed by Trikalinos *et al.* (2008, this volume). As suggested by the authors, the methodology is applicable to a variety of study designs in genetics, from family-based linkage studies and population-based association studies to genome-wide scans and GWA studies. By combining linkage or association evidence from many studies, statistical power may be increased and more precise estimates obtained. The methodology also provides a framework for assessment of between-study heterogeneity across the various studies. As the methods are used most often in a retrospective manner, meta-analysis is subject to a variety of selection biases that may undermine its validity. Trikalinos *et al.* discuss methods for differentiating genuine between-study heterogeneity from systematic errors and biases.

C. Multivariate phenotypes

Another characteristic feature of some complex traits is that the effects of individual loci some times manifest in multiple correlated traits, and this additional information can be exploited using appropriate multivariate methods of analysis. Almasy *et al.* (1997) have used multivariate linkage analysis to separate out direct causal effects from indirect effects through pleiotropy. Todorov *et al.* (1998) proposed a treatment of multiple traits using structural relationships among multiple phenotypes to differentiate direct causal effects from secondary influence of genes. Other multivariate methods such as principal component analysis and factor analysis are often used to reduce the dimensionality of data (e.g., Bartholomew, 1987). With respect to genetic studies, the dimensionality reducing methods can be used to construct a few "summary phenotypes" (explaining most of the variance) from a large number of correlated traits, which can be used in turn for genetic analysis. Alternatively, full multivariate methods for simultaneous analysis of multiple traits may be used with greater power (Ghosh and Majumder, 2001). As we undertake more and more GWA studies for genetic dissection of certain primary phenotypes, we can also use secondary phenotypes highly correlated with the primary phenotype(s) for interpreting and corroborating the GWA findings. For example, if an individual SNP shows strong significance with a primary phenotype but does not show even suggestive evidence in highly correlated phenotypes, the result would be less convincing. On the other hand, the result may be more appealing even if the primary result is not very strong so long as there is a good pattern of evidence coming from correlated phenotypes. Used this way, analysis of multivariate phenotypes may avoid the complications of joint analysis of large numbers of correlated phenotypes even while using the information in scrutinizing the primary results.

VIII. DISCUSSION

The study of human disease has fully entered the genome era. Major accomplishments from the Human Genome Project and the HapMap Project coupled with major technological advances make it possible to dream big in terms of evaluating the genetic underpinnings of common complex diseases and disease related traits. With unprecedented new opportunities unfolding, the future promises to be both exciting and challenging for the genetic epidemiologist.

Large-scale genome-wide linkage scans of common complex traits yielded disappointing results for the most part, making one wonder whether linkage is a right approach for this purpose. There are two reasons why we believe that linkage analysis can play a useful role even with common complex traits.

First, linkage evidence can be useful in interpreting results from GWA studies as discussed in this chapter as well as for establishing cosegregation in the end. Second, despite the fact that linkage lacks appeal for common complex traits with very small individual gene effects, we have argued in this chapter that there are some methodological shortcomings in terms of extracting the most information from genome-wide linkage scans. In particular, we have shown that misspecification of temporal (age) trends in QTL effects and imprinting effects can substantially decrease the power of linkage analysis. There is existing biological evidence suggesting that such effects are real for certain traits (e.g., see Cheverud *et al.*, 1983). In light of this, it may be premature to abandon linkage altogether.

Unquestionably, genetic dissection of complex traits has entered the GWA era, with several exciting findings to date and many more to come. As demonstrated by the pathbreaking Wellcome Trust Case Control Consortium study (2007) and as argued compellingly by Altshuler and Daly (2007), GWA studies require very large sample sizes that are achievable only through prospective and retrospective collaborations. While it is often not desirable to analyze data pooled from many disparate studies in aggregate, we believe that the "lumping and splitting" strategy, which requires collaborative efforts, holds promise. There is tremendous opportunity for meaningful collaborations, and this may be the limiting factor in terms of whether or not we succeed. Only when investigators interact actively without bars can the added benefits of constructive synergism be pitted against the complex challenges, such as the WTCCC study (2007). After all, for most complex traits, the question is not whether there are genes but only when and how they might be found.

With massive databases becoming available in the public domain, it opens up extraordinary opportunities for data-based and biology-based modeling of complex phenomena. In the pursuit of these opportunities, we must be resolute in terms of abiding by the best ELSI practices. While unguided explorations may be analogous to looking for a needle in a haystack, we should remember that we have the ultimate ally in terms of the underlying biological pathways and systems biology. Integrating biology with the databases should serve like a magnet for explicating the needle. It is more than a small comfort knowing that there are many needles in the haystack.

Acknowledgments

This work was partly supported by a grant from the National Institute of General Medical Sciences (GM 28719) and a grant from the National Heart, Lung, and Blood Institute (HL 54473) of the National Institutes of Health. The author is grateful to several colleagues for help and stimulating discussions, notably, C. Charles Gu, Steve Hunt, Shamika Ketkar, Aldi Kraja, Jean-Marc Lalouel, Gang Shi, Joseph Terwilliger, and Jia Zhao.

References

Abecasis, G. R., Cherny, S. S., Cookson, W. O., and Cardon, L. R. (2002). Merlin-rapid analysis of dense genetic maps using sparse gene flow trees. *Nat. Genet.* **30,** 97–101.

Almasy, L., and Blangero, J. (1998). Multipoint quantitative trait linkage analysis in general pedigrees. *Am. J. Hum. Genet.* **62,** 1198–1211.

Almasy, L., and Blangero, J. (2008). Contemporary model-free methods for linkage analysis. *In* "Genetic Dissection of Complex Traits" (D. C. Rao and C. C. Gu, eds.), 2nd Edition, pp. 175–193. Academic Press, San Diego.

Almasy, L., Dyer, T. D., and Blangero, J. (1997). Bivariate quantitative trait linkage analysis: Pleiotropy versus co-incident linkages. *Genet. Epidemiol.* **14,** 953–958.

Altshuler, D., and Daly, M. (2007). Guilt beyond a reasonable doubt. *Nat. Genet.* **39,** 813–815.

Amos, C. I. (1994). Robust variance-components approach for assessing genetic linkage in pedigrees. *Am. J. Hum. Genet.* **54,** 535–543.

Assimes, T. L., Olshen, A. B., Narasimhan, B., and Olshen, R. A. (2008). Associations among multiple markers and complex disease: Models, algorithms, and applications. *In* "Genetic Dissection of Complex Traits" (D. C. Rao and C. C. Gu, eds.), 2nd Edition, pp. 437–464. Academic Press, San Diego.

Barroso, I., Gurnell, M., Crowley, V. E., Agostini, M., Schwabe, J. W., Soos, M. A., Maslen, G., Williams, T. D., Lewis, H., Schafer, A. J., *et al.* (1999). Dominant negative mutations in human PPARgamma associated with severe insulin resistance, diabetes mellitus and hypertension. *Nature* **402,** 880–883.

Bartholomew, D. J. (1987). "Latent Variable Models and Factor Analysis." Oxford University Press, Oxford, UK.

Borecki, I. B., and Province, M. A. (2008). Linkage and association: Basic concepts. *In* "Genetic Dissection of Complex Traits" (D. C. Rao and C. C. Gu, eds.), 2nd Edition, pp. 51–74. Academic Press, San Diego.

Breiman, L., Friedman, J. H., Olshen, R. A., and Stone, C. H. (1984). "Classification and Regression Trees." Wadsworth International Group Inc., Belmont, CA.

Cheverud, J. M., Rutledge, J. J., and Atchley, W. R. (1983). Quantitative genetics of development: Genetic correlations among age-specific trait values and the evolution of ontogeny. *Evolution* **37,** 895–905.

Clee, S. M., Yandell, B. S., Schueler, K. M., Rabaglia, M. E., Richards, O. C., Raines, S. M., Cabara, E. A., Klass, D. M., Mui, E. T.-K., Stapleton, D. S., *et al.* (2006). Positional cloning of Sorcs1, a type 2 diabetes quantitative trait locus. *Nat. Genet.* **38,** 688–693.

de Bakker, P. I., Burtt, N. P., Graham, R. R., Guiducci, C., Yelensky, R., Drake, J. A., Bersaglieri, T., Penney, K. L., Butler, J., Young, S., *et al.* (2006). Transferability of tag SNPs in genetic association studies in multiple populations. *Nat. Genet.* **38,** 1298–1303.

Dong, C. H., Li, W. D., Geller, F., Lei, L., Li, D., Gorlova, O. Y., Hebebrand, J., Amos, C. I., Nicholls, R. D., and Price, R. A. (2005). Possible genomic imprinting of three human obesity-related genetic loci. *Am. J. Hum. Genet.* **76,** 427–437.

Easton, D. F., Ford, D., and Bishop, D. T. (1995). Breast and ovarian cancer incidence in BRCA1-mutation carriers. Breast Cancer Linkage Consortium. *Am. J. Hum. Genet.* **56,** 265–271.

Eaves, L. J. (1994). Effect of genetic architecture on the power of human linkage studies to resolve the contribution of quantitative trait loci. *Heredity* **72,** 175–192.

Elston, R. C., and Stewart, J. (1971). A general model for genetic analysis of pedigree data. *Hum. Hered.* **21,** 523–542.

Farber, C. R., and Lusis, A. J. (2008). Integrating global gene expression analysis and genetics. *In* "Genetic Dissection of Complex Traits" (D. C. Rao and C. C. Gu, eds.), 2nd Edition, pp. 571–601. Academic Press, San Diego.

Fulker, D. W., and Cardon, L. R. (1994). A sib-pair approach to interval mapping of quantitative trait loci. *Am. J. Hum. Genet.* **54,** 1092–1103.

Ghosh, S., and Majumder, P. P. (2001). Deciphering the genetic architecture of a multivariate phenotype. *Adv. Genet.* **42,** 323–347.

Goldgar, D. (1990). Multipoint analysis of human quantitative genetic variation. *Am. J. Hum. Genet.* **47,** 957–967.

Gorlova, O. Y., Amos, C. I., Wang, N. W., Shete, S., Turner, S., and Boerwinkle, E. (2003). Genetic linkage and imprinting effects on body mass index in children and young adults. *Eur. J. Hum. Genet.* **11,** 425–432.

Gu, C., Todorov, A. A., and Rao, D. C. (1996). Combining extremely concordant sibpairs with extremely discordant sibpairs provides a cost-effective way to linkage analysis of QTL. *Genet. Epidemiol.* **13,** 513–533.

Gu, C., Province, M. A., Todorov, A. A., and Rao, D. C. (1998). Meta-analysis methodology for combining non-parametric sibpair linkage results: Genetic homogeneity and identical markers. *Genet. Epidemiol.* **15,** 609–626.

Gu, C. C., Hunt, S. C., Kardia, S., Turner, S. T., Chakravarti, A., Schork, N., Olshen, R., Curb, D., Jaquish, C., Boerwinkle, E., *et al.* (2007a). An investigation of genome-wide associations of hypertension with microsatellite markers in the Family Blood Pressure Program (FBPP). *Human Genetic.* **121,** 577–590.

Gu, C. C., Yu, K., and Boerwinkle, E. (2007b). Measuring marker information content by the ambiguity of block boundaries observed in dense SNP data. *Ann. Hum. Genet.* **71,** 127–140.

Gu, C. C., Yu, K., Ketkar, S., Templeton, A. R., and Rao, D. C. (2007c). On transferability of genome-wide tagSNPs. *Genet. Epidemiol.* (2007 Sept 25 Epub ahead of print).

Gu, C. C., Yu, K., and Rao, D. C. (2008). Characterization of LD structures and the utility of HapMap in genetic association studies. *In* "Genetic Dissection of Complex Traits" (D. C. Rao and C. C. Gu, eds.), 2nd Edition, pp. 407–435. Academic Press, San Diego.

Guo, Y. F., Shen, H., Liu, Y. J., Wang, W., Xiong, D. H., Xiao, P., Liu, Y. Z., Zhao, L. J., Recker, R. R., and Deng, H. W. (2006). Assessment of genetic linkage and parent-of-origin effects on obesity. *J. Clin. Endocrinol. Metab.* **91,** 4001–4005.

Haga, S. B., and Beskow, L. M. (2008). Ethical, legal, and social implications of biobanks for genetics research. *In* "Genetic Dissection of Complex Traits" (D. C. Rao and C. C. Gu, eds.), 2nd Edition, pp. 505–544. Academic Press, San Diego.

Hanis, C. L., Boerwinkle, E., Chakraborty, R., Ellsworth, D. L., Concannon, P., Stirling, B., Morrison, V. A., Wapelhorst, B., Spielman, R. S., Gogolin-Ewens, K. J., *et al.* (1996). A genome-wide search for human non-insulin-dependent (type 2) diabetes genes reveals a major susceptibility locus on chromosome 2. *Nat. Genet.* **13**(2), 161–166.

Haseman, J. K., and Elston, R. C. (1972). The investigation of linkage between a quantitative trait and a marker locus. *Behav. Genet.* **2,** 3–19.

Heath, S. C. (1997). (2005) Loki Version 2. 4. 5. http://www.stat.washington.edu/thompson/Genepi/Loki.shtml

Heckerman, D. A. (1996). Tutorial on learning with Bayesian networks. Microsoft Research Technical Report.

Higgins, M., Province, M. A., Heiss, G., Eckfeldt, J., Ellison, R. C., Folsom, A. R., Rao, D. C., Sprafka, J. M., and Williams, R. (1996). The NHLBI Family Heart Study: Objectives and design. *Genet. Epidemiol.* **143,** 1219–1228.

Hummel, K. P., Dickie, M. M., and Coleman, D. L. (1966). Diabetes, a new mutation in the mouse. *Science* **153,** 1127–1128.

Keats, B. J. B., Morton, N. E., Rao, D. C., and Williams, W. R. (1979). "A Source Book for Linkage in Man." The Johns Hopkins University Press, Baltimore, USA.

Kraft, P., and Cox, D. G. (2008). Optimum study designs for genome-wide association studies. *In* "Genetic Dissection of Complex Traits" (D. C. Rao and C. C. Gu, eds.), 2nd Edition, pp. 465–504. Academic Press, San Diego.

Kruglyak, L., and Lander, E. (1995). Complete multipoint sib pair analysis of qualitative and quantitative traits. *Am. J. Hum. Genet.* **57,** 439–454.

Laird, N. M., and Lange, C. (2008). Family-based methods for linkage and association analysis. *In* "Genetic Dissection of Complex Traits" (D. C. Rao and C. C. Gu, eds.), 2nd Edition, pp. 219–252. Academic Press, San Diego.

Lander, E. S., Linton, L. M., Birren, B., Nusbaum, C., Zody, M. C., Baldwin, J., Devon, K., Dewar, K., Doyle, M., FitzHugh, W., *et al.* (2001). International Human Genome Sequencing Consortium. Initial sequencing and analysis of the human genome. *Nature* **409,** 860–921.

Lathrop, G. M., Lalouel, J. M., Julier, C., and Ott, J. (1984). Strategies for multilocus linkage analysis in humans. *Proc. Natl. Acad. Sci. USA* **81,** 3443–3446.

Li, C. C. (1975). "Path Analysis—A primer." Boxwood Press, Pacific Grove, CA.

Li, N. (2008). The promise of composite likelihood methods for addressing computationally intensive challenges. *In* "Genetic Dissection of Complex Traits" (D. C. Rao and C. C. Gu, eds.), 2nd Edition, pp. 637–654. Academic Press, San Diego.

Li, Z., and Rao, D. C. (1996). A random effect model for meta-analysis of multiple quantitative sibpair linkage studies. *Genet. Epidemiol.* **13,** 377–383.

Lindsay, B. (1988). Composite likelihood methods. *Contemporary Mathematics.* **80,** 221–239.

Liu, N., Zhang, K., and Zhao, H. (2008). Haplotype association analysis. *In* "Genetic Dissection of Complex Traits" (D. C. Rao and C. C. Gu, eds.), 2nd Edition, pp. 335–405. Academic Press, San Diego.

Maresso, K., and Broeckel, U. (2008). Genotyping platforms for mass throughput genotyping with SNPs, including human genome-wide scans. *In* "Genetic Dissection of Complex Traits" (D. C. Rao and C. C. Gu, eds.), 2nd Edition, pp. 107–139. Academic Press, San Diego.

McClintock, B. (1984). The significance of responses of the genome to challenge. *Science* **226,** 792–801.

McPherson, J. D., Marra, M., Hillier, L., Waterston, R. H., Chinwalla, A., Wallis, J., Sekhon, M., Wylie, K., Mardis, E. R., Wilson, R. K., *et al.* for the International Human Genome Mapping Consortium (2001). A physical map of the human genome. *Nature* **409,** 934–941.

Miller, M. B., Schwander, K., and Rao, D. C. (2008). Genotyping errors and their impact on genetic analysis. *In* "Genetic Dissection of Complex Traits" (D. C. Rao and C. C. Gu, eds.), 2nd Edition, pp. 141–152. Academic Press, San Diego.

Moreno, C., Lazar, J., Jacob, H. J., and Kwitek, A. E. (2008). Comparative genomics for detecting human disease genes. *In* "Genetic Dissection of Complex Traits" (D. C. Rao and C. C. Gu, eds.), 2nd Edition, pp. 655–697. Academic Press, San Diego.

Morton, N. E. (1955). Sequential tests for the detection of linkage. *Am. J. Hum. Genet.* **7,** 277–318.

Morton, N. E. (2008). Into the post-HapMap era. *In* "Genetic Dissection of Complex Traits" (D. C. Rao and C. C. Gu, eds.), 2nd Edition, pp. 727–742. Academic Press, San Diego.

Olkin, I. (1995). Statistical and theoretical consideration in meta-analysis. *J. Clin. Epidemiol.* **48,** 133–146.

Ott, J. (1974). Estimation of the recombination fraction in human pedigrees: Efficient computation of the likelihood for human linkage studies. *Am. J. Hum. Genet.* **26,** 588–597.

Pekkarinen, P., Hovotta, I., Hakola, P., Jarvi, O., Kestila, M., Adolfsson, R., Holmgren, G., Nylander, P. O., Tranebjaerg, L., *et al.* (1998). Assignment of the locus for PLO-SL, a frontal-lobe dementia with bone cysts, to 19q13. *Am. J. Hum. Genet.* **62,** 362–372.

Poirier, J., Davignon, J., Bouthillier, D., Kogan, S., Bertrand, P., and Gauthier, S. (1993). Apolipoprotein E polymorphism and Alzheimer's disease. *Lancet* **342,** 697–699.

Province, M. A., and Rao, D. C. (1985). A new model for the resolution of cultural and biological inheritance in the presence of temporal trends: Application to systolic blood pressure. *Genet. Epidemiol.* **2,** 363–374.

Rao, D. C. (1998). CAT scans, PET scans, and genomic scans. *Genet. Epidemiol.* **15,** 1–18.

Rao, D. C., and Gu, C. C. (2008). *In* "Genetic Dissection of Complex Traits" (D. C. Rao and C. C. Gu, eds.), 2nd Edition, Elsevier, Inc, New York, New York.

Rao, D. C., MacLean, C., Morton, N. E., and Yee, S. (1975). Analysis of family resemblance. V. Height and weight in Northeastern Brazil. *Am. J. Hum. Genet.* **27,** 509–520.

Rao, D. C., Morton, N. E., Elston, R. C., and Yee, S. (1977). Causal analysis of academic performance. *Behav Genet.* **7,** 147–159.

Rao, D. C., Keats, B. J. B., Morton, N. E., Yee, S., and Lew, R. (1978). Variability in human linkage data. *Am. J. Hum. Genet.* **30,** 516–529.

Rao, D. C., Keats, B. J. B., Lalouel, J. M., Morton, N. E., and Yee, S. (1979a). A maximum likelihood map of chromosome l. *Am. J. Hum. Genet.* **31,** 680–696.

Rao, D. C., Morton, N. E., and Yee, S. (1979b). Causal Analysis with three generations of latent causes. "Population Genetics laboratory Technical Report." University of Hawaii, Honolulu.

Rice, J. P. (1998). The role of meta-analysis in linkage studies of complex traits. *Am. J. Med. Genet.* **74,** 112–114.

Rice, T. K. (2008). Familial resemblance and heritability. *In* "Genetic Dissection of Complex Traits" (D. C. Rao and C. C. Gu, eds.), 2nd Edition, pp. 35–49. Academic Press, San Diego.

Rice, J. P., Saccone, N. L., and Corbett, J. (2008a). Model-based methods for linkage analysis. *In* "Genetic Dissection of Complex Traits" (D. C. Rao and C. C. Gu, eds.), 2nd Edition, pp. 155–173. Academic Press, San Diego.

Rice, T. K., Schork, N. J., and Rao, D. C. (2008b). Methods for handling multiple testing. *In* "Genetic Dissection of Complex Traits" (D. C. Rao and C. C. Gu, eds.), 2nd Edition, pp. 293–308. Academic Press, San Diego.

Risch, N., and Zhang, H. (1995). Extreme discordant sibpairs for mapping quantitative trait loci in humans. *Science* **268,** 1584–1589.

Risch, N., and Merikangas, K. (1996). The future of genetic studies of complex human diseases. *Science* **273,** 1516–1517.

Rodin, A., Brown, A., Clark, A. G., Sing, C. F., and Boerwinkle, E. (2005). Mining genetic epidemiology data with Bayesian networks: Application to ApoE gene variants and plasma lipid levels. *J. Comput. Biol.* **12,** 1–11.

Rodin, A., Litvinenko, A., and Boerwinkle, E. (2008). "Exploring genetic epidemiology, data with Bayesian netwones, Hnad Book of Statistics," Vol 30. Elsevier 2008 (in press).

Rohrwasser, A., Lott, P., Weiss, R. B., and Lalouel, J. M. (2008). From genetics to mechanism of disease. *In* "Genetic Dissection of Complex Traits" (D. C. Rao and C. C. Gu, eds.), 2nd Edition, pp. 701–726. Academic Press, San Diego.

Scheuner, M. T., Yoon, P. W., and Khoury, M. J. (2004). Contribution of Mendelian disorders to common chronic disease: opportunities for recognition, intervention, and prevention. *Am. J. Med. Genet. C Semin. Med. Genet.* **125,** 50–65.

Schork, N. J., Wessel, J., and Malo, N. (2008). DNA sequence-based phenotypic association analysis. *In* "Genetic Dissection of Complex Traits" (D. C. Rao and C. C. Gu, eds.), 2nd Edition, pp. 195–217. Academic Press, San Diego.

Service, S., Sabatti, C., and Freimer, N. (2007). International Collaborative Group on Isolated Populations, Tag SNPs chosen from HapMap perform well in several population isolates. *Genet. Epidemiol.* **31,** 189–194.

Shi, G., and Rao, D. C. (2007). Ignoring temporal trends in genetic effects substantially reduces power of quantitative trait linkage analysis. *Genet. Epidemiol.* (2007 Aug 15 Epub ahead of print).

Sung, Y. J., and Rao, D. C. (2007). Ignoring imprinting effects can severely jeopardize detection of linkage. *Presented at the American Society of Human Genetics Meeting, San Diego, California.*

Terwilliger, J., and Lee, J. (2007). Natural experiments in human gene mapping: The intersection of anthropological genetics and genetic epidemiology. *In* "Anthropological Genetics: Theory, Methods, and Applications" (M. H. Crawford, ed.), Chapter 3, pp. 38–76. Cambridge University Press, London.

The International HapMap Consortium (2005). A haplotype map of the human genome. *Nature* **437,** 1299–1320.

The Wellcome Trust Case Control Consortium Study (2007). Genome-wide association study of 14,000 cases of seven common diseases and 3,000 shared controls. *Nature* **447,** 661–678.

Thomson, G., and Barcellos, L. F. (2008). Searching for additional disease loci in a genomic region. *In* "Genetic Dissection of Complex Traits" (D. C. Rao and C. C. Gu, eds.), 2nd Edition, pp. 253–291. Academic Press, San Diego.

Todorov, A. A., Province, M. A., Borecki, I. B., and Rao, D. C. (1997). Trade-off between sibship size and sampling scheme for detecting quantitative trait loci. *Hum. Hered.* **47,** 1–5.

Todorov, A. A., Vogler, G. P., Gu, C., Province, M. A., Li, Z., Heath, A. C., and Rao, D. C. (1998). Testing causal hypotheses in multivariate linkage analysis of quantitative traits: general formulation and application to sibpair data. *Genet. Epidemiol.* **15,** 263–278.

Trikalinos, T. A., Salanti, G., Zintzaras, E., and Ioannidis, J. P. (2008). Meta-analysis methods. *In* "Genetic Dissection of Complex Traits" (D. C. Rao and C. C. Gu, eds.), 2nd Edition, pp. 311–334. Academic Press, San Diego.

Van Eerdewegh, P., Little, R. D., Dupuis, J., Del Mastro, R. G., Falls, K., Simon, J., Torrey, D., Pandit, S., McKenny, J., Braunschweiger, K., *et al.* (2002). Association of the ADAM33 gene with asthma and bronchial hyper-responsiveness. *Nature* **418**(6896), 426–430.

Willer, C. J., Scott, L. J., Bonnycastle, L. L., Jackson, A. U., Chines, P., Pruim, R., Bark, C. W., Tsai, Y. Y., Pugh, E. W., Doheny, K. F., *et al.* (2006). Tag SNP selection for Finnish individuals based on the CEPH Utah HapMap database. *Genet. Epidemiol.* **30,** 180–190.

Wojczynski, M. K., and Tiwari, H. K. (2008). Definition of phenotype. *In* "Genetic Dissection of Complex Traits" (D. C. Rao and C. C. Gu, eds.), 2nd Edition, pp. 75–105. Academic Press, San Diego.

Wright, S. (1931). Statistical methods in biology. *J. Am. Stat. Assoc. Suppl.* **26,** 155–163.

Zerres, K., Mucher, G., Bachner, L., Deschennes, G., Eggermann, T., Kaarianinen, H., Knapp, M., Lennert, T., Misselwitz, J., von Muhlendahl, K. E., *et al.* (1994). Mapping of the gene for autosomal recessive polycystic kidney disease (ARPKD) to chromosome 6p21-cen. *Nat. Genet.* **7,** 429–432.

Zhang, Y., Proenca, R., Maffei, M., Barone, M., Leopold, L., and Friedman, J. M. (1994). Positional cloning of the mouse obese gene and its human homologue. *Nature* **372,** 425–432.

Zhu, J., Zhang, B., and Schadt, E. E. (2008a). A systems biology approach to drug discovery. *In* "Genetic Dissection of Complex Traits" (D. C. Rao and C. C. Gu, eds.), 2nd Edition, pp. 603–635. Academic Press, San Diego.

Zhu, X., Tang, H., and Risch, N. (2008b). Admixture mapping and the role of population structure for localizing disease genes. *In* "Genetic Dissection of Complex Traits" (D. C. Rao and C. C. Gu, eds.), 2nd Edition, pp. 547–569. Academic Press, San Diego.

2

Familial Resemblance and Heritability

Treva K. Rice
Division of Biostatistics and Department of Psychiatry, Washington University School of Medicine, St. Louis, Missouri 63110

ABSTRACT

Familial resemblance arises when family members are more similar than unrelated pairs of individuals, and may be estimated in terms of correlations or covariances among family members. *Multifactorial heritability* (or *generalized heritability*) quantifies the strength of the familial resemblance and represents the percentage of

0065-2660/08 $35.00

DOI: 10.1016/S0065-2660(07)00402-6

variance that is due to all additive familial effects including additive genetic and those of the familial environment. However, the traditional concept of heritability, which may be more appropriately called the *genetic heritability*, represents only the percentage of phenotypic variance due to additive genetic effects. Resolving the various sources of familial resemblance entails other issues. For example, there may be major gene effects that are largely or entirely nonadditive, temporal or developmental trends, and gene–gene (epistasis) and gene–environment interactions. The design of a family study determines which of these sources are resolvable. For example, in nuclear families consisting of parents and offspring, the genetic and familial environmental effects are not resolvable because these relatives share both genes and environments. However, extended pedigree and twin and adoption designs allow separation of the heritable effects and, possibly, more complex etiologies, including interactions. Various factors affect the estimation and interpretability of heritabilities, for example, assumptions regarding linearity and additivity, assortative mating, and the underlying distribution of the data. Nonnormality of the data can lead to errors in hypothesis testing, although it yields reasonably unbiased estimates. Fortunately, these and other complications can be directly modeled in many of the sophisticated software packages available today in genetic epidemiology.

I. INTRODUCTION

One of the earliest statistical concepts, correlations, was conceived and advanced by Karl Pearson and Francis Galton almost simultaneously with the idea of the quantifying genetic resemblance in close relatives (see time line in Table 2.1). These two ideas fed off and complemented one another even before the term *heritability* was coined, as Fisher recounted in his seminal 1925 work:

> One of the earliest and most striking successes of the method of correlation was in the biometrical study of inheritance. At a time when nothing was known of the mechanism of inheritance, or of the structure of the germinal material, it was possible by this method to demonstrate the existence of inheritance, and to "measure its intensity;" and this in an organism in which experimental breeding could not be practiced, namely, Man.
>
> (Fisher, 1925, p. 175)

Although the fundamental concepts of *familial resemblance* and *heritability* reviewed in this chapter were created for an earlier era of genetics when it was not possible to directly measure genes in humans, these ideas remain at the cornerstone of genetic epidemiology even today. Indeed, some of the most recent and powerful

Table 2.1. Selected Highlights Showing Mathematical Development of Heritability Models

Year	Development
1866	*Gregor Mendel:* Experiments demonstrate that hereditary units in pea plants remain distinct from parent to child, laws of segregation and independence
1869	*Galton's "Hereditary Genius":* Francis Galton's work introducing *heritability* of eminence; number of eminent relatives of eminent men dropped off when going from first degree to second degree to third degree; proposed adoption studies to separate heredity from environment effects
1875	*Galton's "The History of Twins":* Francis Galton demonstrates usefulness in inferring relative influence of nature (heredity) and nurture (environment), using twins reared apart and together and concluding evidence favored nature rather than nurture
1888	*Correlation and Regression Line:* Karl Pearson (Galton's student) formalized the methods
1894	*Discontinuous Variations:* William Bateson; *Mathematical Theory of Evolution:* Karl Pearson
1900	*Rediscovery of Mendel's Work:* Hugo DeVries, Carl Correns, and Erich von Tschermak independently rediscover Mendel's laws of inheritance
1902	*1st Description of a Mendelian Inherited Disease:* Archibald Garrod reports alkaptonuria is inherited according to Mendelian (recessive) rules
1909	*1st Use of "Gene," "Genotype," and "Phenotype":* Wilhelm Johannsen uses "gene" to describe Mendelian unit of heredity, with genotype and phenotype differentiating between genetic trait vs outward appearance
1918	*Fisher:* Additive loci, RA Fisher publishes work "On the *correlation between relatives* on the supposition of Mendelian Inheritance," begins modern synthesis of genetics and evolutionary biology. Concluded that "many small, equal and additive loci" result in Gaussian distribution for a phenotype, that is, continuous variable can also be consistent with Mendelian principles. First to use diffusion equations to calculate distribution of gene frequencies among populations. *Laid foundation for biometrical genetics and modern quantitative genetics*, introduced analysis of variance (*variance components analysis*)
1918–1921	*Wright:* Sewell Wright develops and applies the method of *path analysis* to study systems of inbreeding and mating, although not widely used because of relative lack of statistical formalization
1928	*Burks classic IQ analysis:* First post-Wright path analytic study of parent–child resemblance for IQ using adoption design
1932	*Haldane:* Publishes a major series of papers on the mathematical theory of natural selection
1960	*Falconer:* Falconer's formula for computing heritability using monozygotic and dizygotic twins
1960s	*Genetic Epidemiology:* Merger of population/statistical/mathematical genetics and classical/molecular epidemiology. Pioneers include Newton Morton, Douglas Falconer, and Robert C. Elston
1965–1967	*Multifactorial/polygenic:* Falconer introduces idea of normally distributed, quantitative trait as "liability" for genetically determined disorder; Falconer's multifactorial liability (multifactorial threshold) model
1974	*Formalization of Path Analysis in Genetic Epidemiology:* Morton (1974) and Rao *et al.* (1974) present formalized equations and hypothesis tests for models of familial resemblance using path analysis with environmental index
1978	*TAU and BETA models of family resemblance presented:* Rice *et al.*, 1978

developments in linkage equilibrium and disequilibrium analyses rely heavily on the estimation and partitioning of heritable components due to different measured sources in the form of variance components models, such as a linked genetic marker and a measured candidate gene. Even approaches that do not rely upon the estimation of heritability per se often find it useful to "translate" effect sizes onto the heritability scale because it is such a convenient concept. Here we review the methods for estimating the magnitudes of unmeasured and measured effects and for testing hypotheses. We also review some of the relevant study designs and the corresponding statistical methods, factors affecting the estimation, and extensions developed to model more complex etiologic factors. This provides a foundation for the extensions that incorporate linkage and association developed in later chapters.

II. FAMILIAL RESEMBLANCE AND HERITABILITY

A. Family resemblance

Familial resemblance arises when relatives who share genes and/or environmental factors exhibit greater phenotypic similarity than do unrelated individuals. The extent of the familial resemblance can be measured by familial correlations (e.g., sibling, parent–offspring, and spouse). In general, biological relatives such as siblings share both genes and familial environments in common. Thus, familial resemblance can be a function of shared genes, shared environments, or both. In contrast, under the assumption that there are no inbreeding or assortative mating effects, spouse pairs have no genes in common but they do share common environments. Therefore, the magnitude of the spouse correlation provides an indication of the importance of the familial environment.

The accuracy and reliability with which a trait is measured can influence these correlations. If the measurement is not reliable or accurate (i.e., high measurement error), then familial resemblance may be underestimated. On the other hand, certain types of measurement practices can lead to a false assertion of familial resemblance. For example, interrater differences or day-to-day variability in measurement can inflate familial resemblance if entire families are measured by the same rater and/or on the same day.

B. Heritability

The traditional concept of *heritability*, which in retrospect may be called the *genetic heritability*, was developed in the context of quantitative genetics to index the relative contribution of genetic factors to trait variability. In fact, estimation of heritability arose under two methodologically different schools of thought. One was based on correlation and regression methods and was developed by

Wright (1921) and later formalized by Li (1955). This method used path analysis to estimate heritability (Burks, 1928; Morton, 1974; Rao *et al.*, 1974; Rice *et al.*, 1978). The second was based on analysis of variance methods by estimating variance components, as originally developed by Fisher (1925). However, under equivalent assumptions, each of these methods yields comparable results. (See Table 2.1 for a time line for many of these methodological developments.)

Under both schools, heritability was conceived in reference to a polygenic model, that is, one in which a large number of genes, each with a small, linear, and additive effect, influence the phenotypic variability. Under a pure polygenic model, the phenotype (P) is a function of genetic (G) and environment (E) effects (i.e., $P = G + E$), usually expressed in terms of variance components ($V_P = V_G + V_E$). The *broad-sense genetic heritability* is defined as the proportion of the total phenotype variance that is due to genetic effects ($h_B^2 = V_G/V_P$). This genetic variance can be further decomposed into additive effects and dominance deviations ($V_G = V_A + V_D$), which gives rise to the *narrow-sense genetic heritability*, defined as the proportion of total phenotypic variance due strictly to additive genetic effects ($h_N^2 = V_A/V_P$). Likewise, the environmental component can be decomposed into common familial (C) and random nonfamilial (R) environmental effects (i.e., $V_P = V_A + V_D + V_C + V_R$). Thus, analogous to genetic heritability, the *cultural heritability* or familial environmental variance component may be defined as $c^2 = V_C/V_P$, which is assumed to be due to a large number of linear and additive familial environmental effects.

We note that the heritabilities are not absolute measurements, but rather reflect the amount of genotypic variation as compared to environmental variation; consequently, the estimates are specific to a given population. For example, heritability is generally larger when the genetic background is more diverse thereby increasing the genetic variance (e.g., using outbred populations). Heritability can also be increased by minimizing environmental effects that decreases the environmental variance (e.g., measuring blood pressure under standardized fitness and/or sodium/nutritional conditions). This means that heritabilities for the same trait may vary quite widely across populations where the relative importance of these components is very different.

C. Risk-based estimates of heritability

Another statistical method of indexing the heritability is based on probabilities or relative risk ratios. These methods are most often (but not exclusively) applied to qualitative data, for example, what is the risk of being affected by some disease such as diabetes or obesity. In this section, we first delineate the statistical concept of a relative risk ratio, and then expanded it to address the genetic concept of the risk to relatives (λ_R), which is a valid substitute for heritability if one assumes that the phenotypic variability is replaced by some unobserved liability threshold.

The classical relative risk ratio is expressed in terms of the risk of an event occurring relative to exposure. More generally, the relative risk (RR) is a ratio of the probability of an event occurring in an exposed group relative to the probability in a nonexposed or control group.

$$RR = \frac{p_{\text{exposed}}}{p_{\text{control}}}$$

The null hypothesis is that the relative risk ratio is 1, reflecting no difference in risk between the exposed and control groups. A relative risk that is greater than or less than 1 means the event is more or less likely to occur in the exposed than in the control group. In order to judge the significance of the deviation from the expected null value of 1, the log of the RR is used to better approximate a normal distribution, permitting the construction of a symmetrical confidence interval around the estimate. These confidence intervals vary depending on whether signal or noise is greater and on the sample size.

In genetics we are more interested in the *risk to relatives* of an affected individual than in the *relative risk* ratio per se. Both of these concepts are commonly referred to as relative risks, but with distinctly different nuances. In genetic epidemiology, we expect a disease to be more prevalent in siblings of affected individuals as compared to the general population (or to siblings of unaffected individuals) if the trait is influenced by genes. Alternatively stated, we are interested in the risk to relatives of an affected individual or proband rather than the traditional relative risk as explained above. The risk to relatives of a proband is commonly denoted as lambda-R (λ_R) and is defined as the risk ratio for a relative of type R of an affected individual compared with the population prevalence of the disease (Risch, 1990a). Here, type R relatives most commonly refer to siblings, but also can refer to other first-degree (e.g., offspring or parents) or second-degree relatives (e.g., grandparents), and so on. In this case, one assumes an unobserved liability where individuals are affected when their liability crosses a particular threshold. The threshold is defined as the point beyond which the area in the tail equals the population prevalence. Under this formulation, replacing the phenotypic variability with that of the unobserved liability will render the definition of heritability valid. If λ_R is significantly greater than 1 then we can infer that familial factors such as genes explain greater risk and that the trait is therefore heritable. We can also infer a correspondence between the magnitudes of the heritability and λ, that is, the greater the λ_R, the larger the heritability.

These λ_Rs for different classes of relatives can be used not only for inferring heritability but also for evaluating the power of linkage analysis (see Chapters by Rice *et al.*, and Almasy and John Blangero, this volume). These inferences are based on the underlying assumption that values of λ_R should change by predictable quantities for given degrees of relationships. That is, the risk for first-degree relatives should be higher than that for second-degree relatives and so on. For example, Risch (1990a) showed that for either single locus or additive multilocus traits, the value of $(\lambda_R - 1)$ should decrease by a factor of two with each degree of relationship. However, if the value of $(\lambda_R - 1)$ decreases more rapidly with degree of relationship, then the gene action is more likely due to multiple loci with epistasis. The risk ratios, in conjunction with the recombination fraction theta (θ), are also used to make inferences regarding the power to detect linkage for disease-susceptibility loci using relative pair methods both for qualitative (Risch, 1990b,c) and for quantitative traits (Gu and Rao, 1997a,b). For example, when λ is large and θ is small, distant relatives offer more power, but sibling pairs (first-degree relatives) are preferred when λ is small.

III. STUDY DESIGNS AND MULTIFACTORIAL MODELS

The specificity of a given model in terms of what parameters can be estimated depends on the study design or the types of relatives included (e.g., twin, nuclear families, pedigrees, adoptions). Under some designs, the familial components (G and C) are not resolvable and are estimated as a single heritable component that represents all additive effects that are transmitted from parents to offspring. The assumptions underlying these study designs are also critical and are reviewed here.

A. Nuclear families

Nuclear families consisting of parents and their biological offspring are commonly used for investigating familial aggregation. It is often assumed that there is no inbreeding or assortative mating, that the genetic variation is additive with no dominance or epistasis (gene–gene interaction), and that there is no genotype–environment interaction or correlation, although each of these effects can be modeled and estimated provided that informative data are available. In the absence of additional information, nuclear family data cannot resolve familial resemblance into genetic versus cultural heritabilities because family members share both genes and familial environments. Thus, because heritability estimates from nuclear families are confounded with familial environmental effects, they measure the maximal effect of genes. The simplest estimator of the *maximal heritability* (or *generalized heritability*) under this model is given by twice the average correlation for first-degree relatives (parent–offspring and sibling) because they share half (on the average) of their genes in common.

Spouse resemblance, if not accounted for, can lead to bias in both genetic and cultural heritabilities (McGue *et al.*, 1989). However, in the absence of additional information such as from twins (Eaves *et al.*, 1989) or multiple measures on spouses (Heath and Eaves, 1985), it is difficult to distinguish whether the marital resemblance is due to phenotypic assortment or to correlated antecedents. In any case, heritability (or generalized heritability or multifactorial heritability) may be estimated by using the heuristic equation that automatically adjusts the estimate for the spouse correlation (Rice *et al.*, 1997)

$$\frac{(r_{\text{sibling}} + r_{\text{parent-offspring}})(1 + r_{\text{spouse}})}{1 + r_{\text{spouse}} + 2r_{\text{spouse}}r_{\text{parent}} - \text{offspring}}$$

where $r_{\text{parent–offspring}}$ is the average parent–offspring correlation, r_{sibling} is the average sibling correlation, and r_{spouse} is the spouse correlation.

1. Estimation of familial correlations and hypothesis testing

Various software packages are readily available for estimating the familial correlations using maximum likelihood procedures. Maximum likelihood methods are based on the assumption that the phenotypes of all members within a family jointly follow a multivariate normal distribution (Hopper and Mathews, 1982). Hypotheses can be tested by using the likelihood ratio test, which is the difference in log-likelihoods between a general model and a reduced model multiplied by negative two. The likelihood ratio test is asymptotically distributed as a χ^2, with the degrees of freedom being given by the difference in the number of parameters estimated in the two models.

In the simple nuclear family design, there are four types of individuals (fathers = f, mothers = m, sons = s, and daughters = d) leading to eight correlations: four parent–offspring (fs, fd, ms, md), three sibling (ss, dd, sd), and one spouse (fm). The most relevant hypothesis to test is whether the correlations are significantly different from zero. This hypothesis is tested by comparing the log-likelihood of a model in which the correlations are fixed to zero with that of the general model in which all eight correlations are estimated. Similarly, hypotheses regarding the influence of sex may be tested by equating certain correlations (e.g., fs = fd = ms = md, ss = dd = sd) and comparing the result to the case in which the correlations are not equated.

B. Extended pedigrees

Genetic heritability can be more accurately estimated in extended pedigrees with their richness of relative pairs of different degrees. This data structure permits estimation of the genetic heritability without contamination by shared environmental influences for at least two reasons. First, extended family members are unlikely to share environmental influences to any great extent; second, with a reasonable variety of relationships of differing degrees, there is a precise structuring of the expected covariances under a polygenic model. While analysis of extended pedigrees does involve increased computational demands, the advent of faster computers alleviates this constraint.

C. Twins

Twins provide one of the simplest designs for resolving genetic and cultural inheritance (Eaves, 1977). Monozygotic (MZ) twin pairs have 100% of their genes in common, while dizygotic (DZ) pairs are only as similar as full siblings and have 50% of their genes in common. In addition to the assumptions outlined for nuclear families, the common environment is assumed to be the same for both types of twins. The simplest estimates of genetic and cultural heritabilities with twin data use Falconer's formula: $h^2 = 2(r_{MZ} - r_{DZ})$ and $c^2 = 2r_{DZ} - r_{MZ}$, where r_{MZ} and r_{DZ} are the twin correlations (Falconer, 1960). If the common environmental effect is greater for MZ than for DZ twins, as some would argue, h^2 would be overestimated and c^2 would be underestimated.

D. Adoptions

A powerful design for assessing the proportion of variance due to genetic and cultural sources is the full adoption study. It is assumed that the resemblance between an adopted child and the biological parents is due only to genetic effects, while that between the adopted child and the adoptive parent is only familial environmental in origin. The full adoption design is rarely used, however, because it is difficult to obtain both biological parents of the adoptee. A noteworthy exception that is a complete adoption design is the Colorado Adoption Project (Plomin *et al.*, 1990). Important assumptions in adoption studies are that the adoption occurs immediately after birth (which justifies that the resemblance between adopted child and biological parents is entirely due to genetic effects), that the adoptive families are representative, and that there is no selective placement.

E. Modeling extensions

Model extensions have been developed for complex etiologies and to take advantage of additional phenotype information. For example, changes in familial resemblance may arise through temporal and/or developmental trends in the genetic and cultural heritabilities (Province and Rao, 1985a). A variety of mechanisms can cause temporal variations. For example, there may be more than one set of genes acting independently over time, or genes having different effects at various points in development, or a variable lag time between gene action and observed product, or even environmental triggers of gene action. Models that take these complex etiologies into account, often using repeated measurements (e.g., Boomsma and Molenaar, 1987), have been developed for several data types including family (e.g., Hopper and Mathews, 1982; Province and Rao, 1985b), twin (e.g., Eaves *et al.*, 1986), and adoption (e.g., Phillips and Fulker, 1989).

The familial models also have been extended to address multiple correlated phenotypes (e.g., Blangero and Konigsberg, 1991; Lange and Boehnke, 1983; Vogler *et al.*, 1987). The covariation among traits can result from common genetic effects (pleiotropy), from common (familial) or nonfamilial environmental effects, or even from the direct phenotypic influence of one trait on another. These hypotheses may be explored with a simple analysis of a single phenotype by contrasting the effects of different covariate adjustments. A change in the magnitude of the familial resemblance of trait X before and after adjusting for trait Y can provide indirect evidence of pleiotropy.

Another important concern when modeling complex inheritance arises on account of gene–environment interactions (Ottman, 1990). The phenotype corresponding to a particular genotype may depend, in part, on exposure to certain environmental factors. There are numerous examples of gene–environment interaction models (e.g., Blangero *et al.*, 1990; Cavalli-Sforza and Feldman, 1978; Martin and Eaves, 1977). In all cases, the data and family structure types dictate the complexity of the model needed.

A newer innovation in estimating polygenic heritability separates the effects of nature (genetics and heredity) from nurture (culture and environment) using marker data to index the actual degree of genetic relatedness among relative pairs, for example siblings (Visscher *et al.*, 2006; Xu, 2006). In this case, the markers are not used in individual tests (as in linkage or association analysis) but rather are used in a genome-wide (or global) context. While the expectation is that on the average siblings share half of their genes in common, some will actually share more and others will share less. Also recall that sibling pairs share not only genes but also common family environments that can inflate heritability estimates. The method of Visscher *et al.* (2006) uses the proportions of identical-by-descent (IBD) genes that are actually shared between sibpairs to index the shared genetic component (free of any inflation due to shared environments). The actual genome-wide pair-wise IBDs varied in the example of Visscher *et al.* (2006) from 0.374 to 0.617 (mean of 0.498 and a standard deviation of 0.036). The association between the phenotypic similarity and the amount of genome-wide IBD sharing between sibs can be used to calculate unbiased genetic covariances and heritability estimates. This method is referred as "assumption-free" because it is entirely a within-family approach, that is requires no assumptions about the source of the covariances as in twin, adoption, and nuclear family designs. However, as the number of loci in the genome-wide scan increases, the variance (and hence power) decreases. If the maximum possible variance at a single locus is 1/8, then the variance of the average IBDs across k loci is $1/(8k)$. So, as k increases the variance and the power decreases, and consequently the number of sibling pairs needed must increase. For example, at the genome-wide level, 2500 sibling pairs are needed to estimate a heritability of 80% with 90% power assuming 85 independent loci (Visscher *et al.* 2006). Consequently, the method is typically applied at the chromosome level (which increases power because the variance is larger as compared to genome wide), using maximum likelihood methods to jointly estimate the 22 variance components. This procedure is valid if it is assumed that all loci in the genome-wide scan contribute equally to the polygenic variance. A more general multipoint method that increases power (Goldgar, 1990) divides the genome into many regions (over 350 for a 10 cM density per region) and jointly estimates the IBD variances in the multiple regions. Provided there is sufficient data, the major advantage of this method is that a simple sibling design can be used to estimate genetic variances and divide them into additive and nonadditive sources, with some assurance that the estimates are unbiased with regard to environmental sources.

F. Factors affecting heritability estimation

Violations of model assumptions can potentially affect any estimation procedure. For example, linearity and additivity are fundamental assumptions of most polygenic or multifactorial models, and complex traits are likely to be affected by factors that act nonadditively. Assumptions regarding assortative mating have been discussed. Additional assumptions about the underlying distributional properties of the data can be critical. For example, Rao *et al.* (1987) have shown that even relatively large departures from multivariate normality can yield reasonably unbiased parameter estimates, while errors in hypothesis testing can occur even with moderate departures from normality. In many situations—for example, when nonnormality is due to group differences like male versus female—such deviations from normality may be controlled with suitable data transformations or by specifically modeling the group effect. While data transformations will not guarantee multivariate normality, they will tend to minimize the impact.

Various factors can lead to nonnormally distributed data, and if the source is recognized, these factors can be specifically modeled. For example, the presence of a major gene leads to nonnormality. Either admixture analysis or segregation analysis is an appropriate model for investigating this possibility (e.g., Blangero and Konigsberg, 1991; Lalouel and Morton, 1981). Nonnormality also may be induced via gene–environment interaction (Pooni and Jinks, 1976). In fact, nonnormality because of group differences (e.g., sex) can be modeled by considering group as an "environmental" condition that interacts with the genotype (Konigsberg *et al.*, 1991). Finally, a variety of methods have been developed to correct for nonnormality due to the effects of ascertainment, whether the selection conditions are well defined or whether they are ambiguous (e.g., Boehnke and Lange, 1984; Hanis and Chakraborty, 1984; Rao *et al.*, 1988).

IV. DISCUSSION

In 1986, Morton (1986) wrote: "Genetic epidemiology is now assimilating the rapid advances in molecular biology that promise a complete linkage map and molecular definition of disease loci.... When they are complete our morning will have passed into the afternoon of molecular epidemiology."

Our morning of genetic epidemiology has revealed that complex traits generally involve additive and nonadditive interactions among several genes and environmental factors, none of which may entail large effects. Are the methods discussed here likely to be useful in the new millennium? We believe that they will be useful for at least two reasons. First, a preliminary understanding of the magnitude of the heritability of a complex trait is necessary and useful before a massive and expensive analysis involving large-scale genotyping is undertaken. Often, estimates of heritability or even more specific model parameters are needed for power computations. Second, a preliminary understanding of the pattern of familial correlations and heritability provides a basis for undertaking more appropriate modeling of the myriad of etiological effects. By assessing familial resemblance and heritability both before and after accounting for the effects of measured genotypes, we can seek evidence for the involvement of any additional familial effects. Eventually, the molecular variants responsible for the disease or the complex risk factor will be found, which should lead us into our afternoon of molecular epidemiology.

Acknowledgments

This work was supported in part by a grant from the National Institute of General Medical Sciences (GM 28719) and a grant from the National Heart, Lung, and Blood Institute (HL54473), National Institutes of Health.

References

Boehnke, M., and Lange, K. (1984). Ascertainment and goodness of fit of variance component models for pedigree data. *Prog. Clin. Biol. Res.* **147,** 173–192.

Boomsma, D. I., and Molenaar, P. C. (1987). The genetic analysis of repeated measures. I. Simplex models. *Behav. Genet.* **17,** 111–123.

Blangero, J., and Konigsberg, L. W. (1991). Multivariate segregation analysis using the mixed model. *Genet. Epidemiol.* **8,** 299–316.

Blangero, J., MacCluer, J. W., Kammerer, C. M., Mott, G. E., Dyer, T. D., and McGill, H. C., Jr. (1990). Genetic analysis of apolipoprotein A-I in two dietary environments. *Am. J. Hum. Genet.* **47,** 414–428.

Burks, B. S. (1928). The relative influence of nature and nurture upon mental development: A comparative study of foster parent-foster child resemblance and true parent-true child resemblance. *In* "27th Yearbook of the National Society for the Study of Education. Part 1," pp. 219–316. Bloomington Public School Publishing Co, Bloomington.

Cavalli-Sforza, L. L., and Feldman, M. W. (1978). The evolution of continuous variation. III. Joint transmission of genotype, phenotype and environment. *Genetics* **90,** 391–425.

Eaves, L. J. (1977). Inferring the causes of human variation. *J. R. Statist. Soc. B* **140,** 324–355.

Eaves, L. J., Long, J., and Heath, A. C. (1986). A theory of developmental change in quantitative phenotypes applied to cognitive development. *Behav. Genet.* **16,** 143–162.

Eaves, L. J., Fulker, D. W., and Heath, A. C. (1989). The effects of social homogamy and cultural inheritance on the covariances of twins and their parents: A LISREL model. *Behav. Genet.* **19,** 113–122.

Falconer, D. S. (1960). "Introduction to Quantitative Genetics." Ronald Press, New York. [4th Edition (1996) Addison, Wesley, Longman, Harlow, Essex, UK.].

Fisher, R. A. (1925). "Statistical Methods for Research Worker." Oliver and Boyd, Edinburgh. (http://psychology.yorku.ca/Fisher/Methods/).

Goldgar, D. E. (1990). Multipoint analysis of human quantitative genetic variation. *Am. J. Hum. Genet.* **47,** 957–967.

Gu, C., and Rao, D. C. (1997a). A linkage strategy for detection of human quantitative-trait loci. I. Generalized relative risk ratios and power of sib pairs with extreme trait values. *Am. J. Hum. Genet.* **61,** 200–210.

Gu, C., and Rao, D. C. (1997b). A linkage strategy for detection of human quantitative-trait loci. II. Optimization of study designs based on extreme isb pairs and generalized relative risk ratios. *Am. J. Hum. Genet.* **61,** 211–222.

Hanis, C. L., and Chakraborty, R. (1984). Nonrandom sampling in human genetics: Familial correlations. *I. M. A. J. Math. Appl. Med. Biol.* **1,** 193–213.

Heath, A. C., and Eaves, L. J. (1985). Resolving the effects of phenotype and social background on mate selection. *Behav. Genet.* **15,** 15–30.

Hopper, J. L., and Mathews, J. D. (1982). Extensions to multivariate normal models for pedigree analysis. *Ann. Hum. Genet.* **46,** 373–383.

Konigsberg, L. W., Blangero, J., Kammerer, C. M., and Mott, G. E. (1991). Mixed model segregation analysis of LDL-c concentration with genotype-covariate interaction. *Genet. Epidemiol.* **8,** 69–80.

Lalouel, J. M., and Morton, N. E. (1981). Complex segregation analysis with pointers. *Hum. Hered.* **31,** 312–321.

Lange, K., and Boehnke, M. (1983). Extensions to pedigree analysis. IV. Covariance components models for multivariate traits. *Am. J. Med. Genet.* **14,** 513–524.

Li, C. C. (1955). "Population Genetics." University of Chicago Press, Chicago.

Martin, N. G., and Eaves, L. J. (1977). The genetical analysis of covariance structure. *Heredity* **38,** 79–95.

McGue, M., Wette, R., and Rao, D. C. (1989). Path analysis under generalized marital resemblance: Evaluation of the assumptions underlying the mixed homogamy model by the Monte Carlo method. *Genet. Epidemiol.* **6,** 373–388.

Morton, N. E. (1974). Analysis of family resemblance. I. Introduction. *Am. J. Hum. Genet.* **26,** 318–330.

Morton, N. E. (1986). Foundations of genetic epidemiology. *J. Genet.* **65,** 205–212.

Ottman, R. (1990). An epidemiologic approach to gene-environment interaction. *Genet. Epidemiol.* **7,** 177–185.

Phillips, K., and Fulker, D. W. (1989). Quantitative genetic analysis of longitudinal trends in adoption designs with application to IQ in the Colorado Adoption Project. *Behav. Genet.* **19,** 621–658.

Plomin, R., DeFries, J. C., and McClearn, G. E. (1990). "Behavior Genetics: A Primer." 2nd Edition, Freeman, New York.

Pooni, H., and Jinks, J. L. (1976). The efficiency and optimal size of triple test cross designs for detecting epistatic variation. *Heredity* **36,** 215–227.

Province, M. A., and Rao, D. C. (1985a). A new model for the resolution of cultural and biological inheritance in the presence of temporal trends: Application to systolic blood pressure. *Genet. Epidemiol.* **2,** 363–374.

Province, M. A., and Rao, D. C. (1985b). Path analysis of family resemblance with temporal trends: Applications to height, weight and Quetelet index in northeastern Brazil. *Am. J. Hum. Genet.* **37,** 178–192.

Rao, D. C., Morton, N. E., and Yee, S. (1974). Analysis of family resemblance. II. Linear model for family correlation. *Am. J. Hum. Genet.* **26,** 331–359.

Rao, D. C., Vogler, G. P., Borecki, I. B., Province, M. A., and Russell, J. M. (1987). Robustness of path analysis of family resemblance against deviations from multivariate normality. *Hum. Hered.* **37,** 107–112.

Rao, D. C., Wette, R., and Ewens, W. J. (1988). Multifactorial analysis of family data ascertained through truncation: A comparative evaluation of two methods of statistical inference. *Am. J. Hum. Genet.* **42,** 506–515.

Rice, J., Cloninger, C. R., and Reich, T. (1978). Multifactorial inheritance with cultural transmission and assortative mating. I. Description and basic properties of the unitary models. *Am. J. Hum. Genet.* **30,** 618–643.

Rice, T., Despres, J. P., Daw, E. W., Gagnon, J., Borecki, I. B., Perusse, L., Leon, A. S., Skinner, J. S., Wilmore, J. H., Rao, D. C., *et al.* (1997). Familial resemblance for abdominal visceral fat: The HERITAGE family study. *Int. J. Obes.* **21,** 1024–1031.

Risch, N. (1990a). Linkage strategies for genetically complex traits. I. Multilocus models. *Am. J. Hum. Genet.* **46,** 222–228.

Risch, N. (1990b). Linkage strategies for genetically complex traits. II. The power of affected relative pairs. *Am. J. Hum. Genet.* **46,** 229–241.

Risch, N. (1990c). Linkage strategies for genetically complex traits. III. The effect of marker polymorphism on analysis of affected relative pairs. *Am. J. Hum. Genet.* **46,** 242–253.

Visscher, P. M., Medland, S. E., Ferreira, M. A. R., Morley, K. I., Zhu, G., Cornes, B. K., Montgomery, G. W., and Martin, N. G. (2006). Assumption-free estimation of heritability from genome-wide identity-by-descent sharing between full siblings. *PLoS Genet.* **2,** 0316–0325.
Vogler, G. P., Rao, D. C., Laskarzewski, P. M., Glueck, C. J., and Russell, J. M. (1987). Multivariate analysis of lipoprotein cholesterol fractions. *Am. J. Epidemiol.* **125,** 706–719.
Wright, S. (1921). Correlation and causation. *J. Agric. Res.* **20,** 557–585.
Xu, S. (2006). Separating nurture from nature in estimating heritability. *Heredity* **97,** 256–257.

Online sources (Further reading)

http://www.genome.gov/Pages/Education/GeneticTimeline.pdf
http://www.accessexcellence.org/AE/AEPC/WWC/1994/geneticstln.html
http://library.thinkquest.org/C0111983/timeline.html
http://en.wikipedia.org/wiki/History_of_genetics
http://www.dorak.info/epi/genetepi.html

3

Linkage and Association: Basic Concepts

Ingrid B. Borecki and Michael A. Province
Division of Statistical Genomics, Washington University School of Medicine, St. Louis, Missouri 63108

ABSTRACT

Recent progress in characterizing human genetic variation and its organization in several ethnically diverse groups coupled with advances in computational algorithms and computing power have yielded unprecedented opportunities to interrogate the human genome. It is now possible and commonplace to conduct genomic scans using large panels of markers to identify locations that likely harbor disease genes and to investigate interactions with other genetic or environmental factors. The primary analytic approaches for gene mapping, discovery, and statistical characterization of the genotype–phenotype relationship are linkage and

Advances in Genetics, Vol. 60
0065-2660/08 $35.00
DOI: 10.1016/S0065-2660(07)00403-8

association studies. Here, we review basic concepts, the biological and population genetic basis of linkage and linkage disequilibrium, and the common approaches used to assess these relationships with the goal of identifying loci influencing complex phenotypes of biomedical importance that could provide a rational basis for advancing genomic medicine. Further, we comment on the state of the science, recent successes, and the possible shortcomings of these approaches. Finally, we highlight some of the major challenges that arise from these investigative approaches and those that are inherent in the nature of complex traits. Many of these topics will be elaborated in the chapters that follow.

I. INTRODUCTION

Understanding the genetic and etiologic basis of human traits, diseases, and their risk factors remains the holy grail of modern human genetics. Complex traits—such as coronary heart disease, diabetes, certain cancers, atopic, and inflammatory diseases—are of substantial public health importance because of their relatively high prevalence in the population and the associated morbidities and mortalities. The goals of genetic medicine in the twenty-first century include gaining insights into the genes and pathways influencing these important diseases, thus furthering our basic science knowledge, developing diagnostic and prognostic tools, and suggesting either targets for drug development or other intervention strategies. By making progress in these domains, we advance toward "personalized medicine" that would take into account the individual's genotypes as well as his/her exposures and behavioral patterns to recommend the optimal strategies for health maintenance or disease treatment.

Encouraged by the success in mapping genes for Mendelian disorders such as Duchenne muscular dystrophy, cystic fibrosis, and Huntington's disease, investigators have been turning their interest and efforts to mapping genes influencing complex traits, so called because of the anticipated complexity of their underlying etiology. Rather than being determined by (relatively rare) highly penetrant monogenic mechanisms, complex traits (i.e., diseases and their risk factors) are rather determined by several to many genes, environmental exposures, and interactions among these factors. Efforts to investigate complex traits initially adopted similar strategies to those employed in previous successful efforts to map Mendelian disorders, namely, seeking to identify one trait locus at a time. There are two methodological approaches in common use toward this goal: linkage and association studies.

In linkage studies, we seek to identify trait loci that cosegregate with a specific genomic region, tagged by polymorphic markers, within families. By contrast, in association studies, we seek a correlation between a specific genetic variation and trait variation in a sample of individuals, therefore implicating a causal role for the variant. The approaches are illustrated in Fig. 3.1. In linkage, the trait locus we seek is not measured, while markers and phenotypes are. Linkage analysis is based on the relationship between markers and phenotypes, which implicates the presence of a trait locus in proximity to the marker locus. The location of the trait locus relative to the marker is expressed by the genetic distance that is a function of the recombination fraction. Note that because the trait locus is not measured, its location and the penetrance are confounded, unless analysis is restricted to penetrant affected individuals only. By contrast, in association studies, the locus variation itself has a direct effect on trait variation (situation A). In this situation, association studies are more powerful than linkage because the causal risk factor is measured. However, modern exploratory strategies including genome-wide (GW) scanning make it less likely that causal variants are measured directly, and more likely that neighboring polymorphisms in linkage disequilibrium (LD) are found to be associated (situation B). In this case, the power to detect a trait locus is, again, dependent on the strength of the LD, similar to the recombination fraction in linkage analysis.

GW linkage studies have been in use over the last two decades and have produced some useful results (e.g., Blacker *et al.*, 2003; Hanis *et al.*, 1996; Johnson *et al.*, 2005; Rioux *et al.*, 2001). Attention in now turning to GW association studies (GWAS) and enthusiasm is bolstered by some early success (e.g., Frayling *et al.*, 2007; Klein *et al.*, 2005). While it is likely that neither of these approaches alone is sufficient to achieve the goals of personalized medicine, they remain powerful and complementary tools to crack the genetic underpinnings of complex traits.

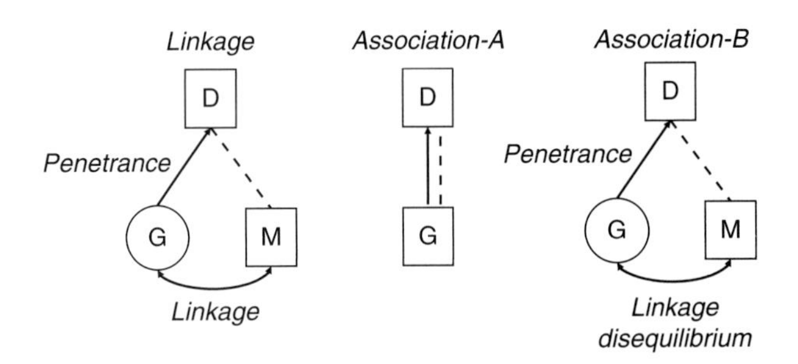

Figure 3.1. Conceptual differences between linkage and association studies.

II. HISTORICAL PERSPECTIVE

One hundred and forty-one years ago a Bohemian monk announced the results of his breeding experiments with the common garden pea (*Pisum sativum*). Manuscript Gregor Mendel (1866) was published the following year and remained virtually unnoticed and unappreciated until it was "rediscovered" 34 years later by three botanists whose own experiments led them to the same conclusions that Mendel had reached more than a generation earlier.

Mendel's experiments revealed two principles that have been elevated to the status of "laws." Mendel's first principle dealt with the segregation of alleles at a single locus, and his second postulate—the law of independent random assortment—dealt with the joint behavior of alleles at two loci. Linkage is, in fact, a violation of Mendel's second law of independent assortment—when two loci demonstrate reduced recombination and cosegregate within families, it is owing to their physical proximity along a stretch of DNA and they are said to be "linked." This was demonstrated by Morgan (1911) who had the good fortune of studying sex-linked traits (Morgan called them sex-limited) in *Drosophila melanogaster*. Morgan was the one who proposed that genes are linked as a consequence of lying close together on a chromosome. He also recognized that the formation of chiasmata during synapsis, a phenomenon described 2 years earlier by Janssens (1909), provided an explanation for the occurrence of recombinants. Sturtevant realized that if Morgan's explanation was correct, recombination could be exploited to construct a linear gene map that would reflect a gene's position, relative to other genes, along a chromosome. Within 2 years Sturtevant (1913) had produced the first such map for *Drosophila*. With the publication of Sturtevant's seminal paper, all of the essentials were in place for what, today, we label "linkage analysis."

For the next four decades, linkage studies in species such as *D. melanogaster* made slow but steady progress. By comparison, linkage studies in our species languished primarily because of the lack of easily assayable markers and the inability (and inappropriateness) of conducting experimental crosses. Indeed, by the time Sturtevant had produced the first linkage map for the common fruit fly, only one useful human marker was known—the ABO blood groups (Landsteiner, 1900)—and the details of its inheritance would not be fully understood for another 11 years when Bernstein (1924) eloquently demonstrated the system's three-allele structure.

Even while the structure of DNA and the molecular processes of recombination were not yet described, the statistical methods for the analysis of linkage were under development. By the 1970s, appropriate statistical methodology for linkage analysis in human including sib pair methods as well as parametric linkage analysis were in place. Increasing utility of computers enabled the creation of the first widely used linkage program LIPED (Ott, 1974). However, a usable and informative map of the human genome was still lacking.

Tremendous progress since 1980 in the discovery of various types of polymorphisms in the human genome and in high-throughput (and affordable) methods of whole genome typing has ushered in the era of GW surveys in search of genes underlying important human traits and diseases. The spectacular accomplishments in human genetic studies of the last quarter century can be attributed to advances in molecular genetics (that allow the rapid and reproducible genotyping of hundreds of thousands of markers), the development of efficient computational algorithms, and the unprecedented progress in computer science rather than to the development of new theoretical or statistical methods which for the most part antedate the discovery of DNA's structure.

III. FUNDAMENTALS AND METHODS

A. Linkage analysis

Linkage refers to the physical proximity of loci along a chromosome. Two loci are *linked* because of their physical connection by a stretch of DNA; they are sufficiently close together such that their alleles tend to cosegregate within families—a departure from Mendel's second law of independent assortment. Such cosegregating haplotypes are broken up by the process of recombination; the probability of a recombination between two loci becomes small with decreasing distance between them; and, conversely, recombination occurs more frequently between loci that are farther apart. Thus, recombination is a function of the distance between the two loci, although it is not a simple linear relationship. The parameter θ is defined as the frequency of an odd number of recombinations between two loci; gametes resulting from an even number of recombinations cannot be recognized as recombinants because they look like the parental types and are therefore scored as nonrecombinants. For two loci far apart on the same chromosome or on two different chromosomes, $\theta = 0.5$ and independent assortment obtains. That is, recombinant and nonrecombinant gametes occur with equal frequency. In linkage analysis, we seek to identify loci for which $\theta < 0.5$.

The expected frequency distribution of gametes under the alternative hypothesis of linkage and the null hypothesis of independent assortment from a completely informative parent is shown in Fig. 3.2. As a class, recombinants occur with frequency θ, while nonrecombinants occur with frequency $(1 - \theta)$, the probability of *not* having a recombination. The simplest estimator of θ is the sample proportion of gametes that are recombinants. However, because the number of offspring in human families is relatively small, it is necessary to find a way to combine information on the distribution of recombinants across families and, also, to consider how to statistically test for the presence of linkage.

Doubly heterozygous parent

A1 B1

A2 B2

Gametes	Type	Probability	
		Linkage	No linkage
A1 B1	Parental	1/2 (1− θ)	1/4
A2 B2	Parental	1/2 (1− θ)	1/4
A1 B2	Recombinant	1/2 θ	1/4
A2 B1	Recombinant	1/2 θ	1/4

Figure 3.2. The distribution of gametes from a fully informative parent genotype under a linkage hypothesis and under the null.

Morton (1955) proposed a method that has turned out to be one of his earliest and most enduring contributions to human genetics. He introduced the concept of the *lo*garithm of an o*d*ds ratio (LOD) score. The odds ratio is the probability of observing the specific genotypes in the family given linkage at a particular recombination fraction versus the same probability computed conditional on independent assortment. Thus, high values of the odds ratio would favor the linkage hypothesis, while values close to 1 favor independent assortment. The log (to the base 10) of the odds ratio is taken for convenience—it has more desirable scale properties and provides an easy way to combine information across families (or studies) by simply summing the LOD scores computed from each family (or study). While Morton proposed this statistic for use in a sequential test, it has been more typically applied to fixed samples. Nonetheless, the original "landmarks" for significance suggested by Morton still have meaning to many investigators for the interpretation of their results. A LOD score of 3 was taken as statistically significant evidence for linkage at the test recombination fraction; this means that the linkage hypothesis is 10^3 times more likely than the hypothesis that the two loci are not linked. Conversion of the LOD score to a likelihood ratio test simply converts the likelihood ratio from log to the base 10 to the natural log scale (LOD $\times$ [2ln 10]). The resulting likelihood ratio test is asymptotically distributed as χ^2 with 1 degree of freedom; thus, a LOD of 3 is associated with $p \approx 0.0001$. According to Morton (1955), a LOD score of 1.5 or greater is considered to be "suggestive" ($p < 0.004$) and, at the opposite end of the spectrum, a LOD score of −2 or less is considered evidence *against* linkage (odds of 1:100 or less in favor of the linkage hypothesis).

This proved to be an extremely powerful method for linkage analysis whose utility was galvanized when Elston and Stewart (1971) provided the algorithm by which the likelihood of more complex pedigrees could be computed and when Ott (1974) provided a user-friendly computer program to quickly compute LOD scores in arbitrary pedigrees. In addition, iterative methods were implemented to obtain a maximum likelihood estimate of the recombination fraction. The key to the power of model-based or "parametric" LOD score method is the ability to distinguish recombinants from nonrecombinants; thus, this method has been ideal for developing the map of the human genome where, for molecular markers, the genotype and phenotype are equivalent. LOD score analysis also has been used to great advantage for mapping single locus Mendelian disorders with accommodation for such complicating factors such as dominance, reduced penetrance, phenocopies, and variable age-at-onset. Even genetic heterogeneity could be resolved as demonstrated by Morton (1956) in the analysis of linkage between Rh and elliptocytosis. However, these complications take a toll on the power of a particular study because of the uncertainty in inferring the underlying trait genotype from the phenotype. Further, recombination is confounded by reduced penetrance, and misspecification of the mode of inheritance and the genetic model parameters can lead to further reductions in power. This LOD score method is discussed in greater detail in subsequent chapters.

Complex phenotypes present novel challenges because, by definition, there is no simple mode of inheritance. Moreover, some traits of interest such as cardiovascular disease, prostate cancer, or Alzheimer's disease have a late age-at-onset and it is difficult or impossible to obtain multigenerational pedigrees without a prospective approach. These problems stimulated renewed interest in sib pair methods (Penrose, 1935) that rely on patterns of allele-sharing to infer linkage, although in contemporary applications, this approach can be applied as well to more distant relative pairs and intact pedigrees (Almasy and Blangero, 1998; Curtis and Sham, 1994; Kruglyak *et al.*, 1996; Weeks and Lange, 1988). The expected pattern of allele-sharing at a locus in relative pairs of a particular class is given by the *coefficient of consanguinity*, which is defined as the probability that an allele drawn at random, one from each of the two individuals, will be identical by descent (ibd).

For sib pairs, the distribution of alleles shared ibd is shown in Table 3.1. In one quarter of the pairs, sibs share none of their alleles at a particular locus ibd; in half the pairs, they share one allele ibd; and in the remaining quarter of the pairs, they share two alleles ibd, which leads to the general expectation that the probability that an allele is ibd in sibs is one half. Note that what is relevant for linkage analysis is the inheritance (or coinheritance) of alleles at adjacent loci; therefore, it is of critical importance to determine whether the alleles are ibd (i.e., copies of the same parental allele) or only identical by state (i.e., only appearing the same, but not necessarily copies of the same parental allele). For example, let us consider a pair of sibs' genotypes *AB* and *AC*. The *A* allele is identical by state, but without information on the parental mating type, it is uncertain whether the *A* is ibd. If the parents are *AD* × *BC*, then the *A* is ibd, coming from the paternal A. In contrast, if the parental mating type is *AB* × AC, then the *A* allele is identical by state only, each sibling inheriting a different

Table 3.1. The Expected Distribution of Alleles Shared IBD in Sib Pairs from Parents with the Completely Informative Mating Type of $AB \times CD$

Sibship genotype		IBD		
		0	1	2
AC	AC			X
AC	AD		X	
AC	BC		X	
AC	BD	X		
AD	AC		X	
AD	AD			X
AD	BC	X		
AD	BD		X	
BC	AC		X	
BC	AD	X		
BC	BC			X
BC	BD		X	
BD	AC	X		
BD	AD		X	
BD	BC		X	
BD	BD			X
Total		4 1/4	8 1/2	4 1/4

copy of the A allele. In the absence of information on parental mating type, the probability that the allele in question is shared ibd can be computed by considering all possible mating types; rarer alleles that occur in each member of a sib pair are more likely ibd than common alleles. To reiterate, only ibd information is useful for linkage studies; however, identity by state information becomes equivalent to ibd as the marker becomes increasingly polymorphic and completely informative. It should be noted that identity by state information forms the basis of tests of association, as the key here is that allele identity (state) is important in predicting phenotype. Thus, because unrelated individuals share no genes ibd (when there is no inbreeding), such samples provide *no* information for linkage.

The linkage test is based on the relationship between phenotypic similarity and π, the proportion of genes shared ibd for sib pairs (or other pairs of relatives). For disease traits, sibling pairs that are both *affected* are ascertained for study, thereby obviating the problem of incomplete penetrance. The idea is that sib pairs will tend to share a greater proportion of marker alleles ibd if the marker is linked to the disease locus. That is, the null hypothesis of no linkage is $\pi = 0.5$ while the alternative hypothesis is $\pi > 0.5$. Many statistical tests have been developed to evaluate this hypothesis (e.g., Blackwelder and Elston, 1985; Kruglyak *et al.*, 1996; Risch, 1990; Suarez and Van Eerdewegh, 1984; Suarez *et al.*, 1978). For quantitative traits, one can look at phenotypic similarity directly without ascertaining the sibships—simply, under the linkage hypothesis, sibs that share a greater proportion of alleles ibd will have more similar phenotypes whereas, under the null, the similarity in sibs' phenotypes will have no relationship with the degree of allele-sharing. It should be noted, however, that ascertaining sib pairs can considerably improve the power for detection of quantitative trait loci (Risch and Zhang, 1995). While Haseman and Elston (1972) used the squared phenotypic difference in sibs, current more favorable methods look at the phenotypic covariance or correlation as a function of π (Almasy and Blangero, 1998; Amos, 1994; Province *et al.*, 2003), using a variance components approach. These methods are referred to as "model-free" or "robust" because there are no assumptions regarding the underlying genetic model, although, certainly, there are the usual assumptions about the distribution of the test statistic. Such model-free methods are elaborated in subsequent chapters.

B. Association analysis

Association at the population level between a marker of genomic variation and the phenotype of interest can arise under either of two circumstances: (1) if the functional variant is measured directly or (2) if the marker variant is in LD with the actual functional variant. In the former case, genotypes can be characterized in a number of ways—for example, electrophoretically, immunologically, or directly at the DNA sequence level—and such assessment gives rise to a "measured genotype," a term coined by Boerwinkle *et al.* (1986). In this case, one can directly test the effect of the measured genotype on the phenotypic outcome. However, if screening markers are rather used, then the test of association relies on LD, which is a population concept.

Let us consider two linked loci, A and B. The pairwise combinations of one allele from locus A with one allele from locus B constitute the gametic haplotypes. Let us assume that a new mutation at locus A, A_i, occurs on a chromosome that has the allele B_j. The gametes produced in the next generation are of two kinds, recombinants and nonrecombinants; if there is no recombination between the loci, then the original haplotype (A_iB_j) is preserved and propagated intact. Recombination breaks up this haplotype such that, eventually, the mutation A_i would occur on haplotypes involving the other alleles at the B locus in proportion to their relative frequency—this is *linkage equilibrium*. However, initially, the haplotype A_iB_j occurs at a disproportionately high rate and produces a situation known as LD. In a randomly mating population, the approach to equilibrium is a function of the recombination fraction (θ); the departure from the equilibrium value is reduced by a fraction equal to θ in each generation. For unlinked loci, the haplotype frequency goes halfway to the equilibrium value each generation. Thus, the tighter the linkage, the longer the disequilibrium (association) will persist in the population. It is useful to talk about the time required to go halfway to the final equilibrium value, similar to the concept of a half-life for radioactive decay. For unlinked loci, the median time is 1 generation, when $\theta = 0.10$ it is about 7 generations, when $\theta = 0.01$ it is 69 generations, and when $\theta = 0.001$ it is about 693 generations. With these examples, it is clear that two loci must be very close in order to have appreciable disequilibrium that may be useful for association studies, and that the prospects for detecting LD is less for ancient mutations than for more recent mutations (Kruglyak, 1999). *De novo* mutations are not the only mechanism by which LD can be generated. Other population phenomena and evolutionary forces such as drift, migration, admixture of populations with distinct evolutionary histories, and rapid population expansion also play an important role in shaping the patterns of LD.

Recent studies of the patterns of LD among single nucleotide polymorphisms (SNPs) in the human genome have revealed some remarkably reproducible results within major population groups; this observation forms the basic premise for the human Haplotype Map (HapMap) project (http://www.hapmap.org/)—an effort to characterize the patterns of LD and their effects on the organization of local regions of the genome in blocks of intercorrelated SNPs. Studies have been carried out in four populations of Asian (Chinese and Japanese), African, and European ancestry. Differences have been observed among these groups in marker allele frequencies as well as in LD patterns; for example, haplotype blocks in individuals of African descent tend to be smaller and greater in number than those in European/Caucasian populations, consistent with the relative evolutionary ages of these progenitor populations. However, the patterns of LD within these groups have been remarkably stable over different samples of individuals, supporting their utility in surveying the genome. Because of the allelic correlations across markers within haplotype blocks, fewer markers have to be typed in order to characterize the common variation in the genome, the so-called tag SNPs. These advances in our understanding of the structure of the human genome have made it possible to construct panels of the SNPs to survey the entire genome, as well as to provide a resource for probing the variation in specific genomic regions, as in evaluations of candidate genes.

There are several standard methods to test for association. In principle, the effect of measured genotypes can be assessed for any given phenotypic outcome, either qualitative (e.g., disease presence or absence) or quantitative. The traditional epidemiological case-control study was among the first approaches utilized to determine whether a particular genetic variant is associated with increased risk of disease. Early on, Woolf (1955) proposed a relative risk statistic that could be used to assess genotype-dependent risk. However, a persistent concern regarding these studies is the adequacy of the matching of cases and controls. In particular, population stratification (admixture of progenitor populations) can produce false positive associations when there are differences in the frequency of the trait and the frequency of marker alleles between the progenitor populations.

Because of this effect, particularly coupled with unprecedented multiple comparisons issues that accompany whole genome scan studies, Falk and Rubenstein (1987) suggested a method for assessing relative risk that uses family-based controls, obviating this source of potential error. Basically, the method uses the parental alleles or haplotypes *not* transmitted to affected offspring as the control sample. Thus, in the fully informative mating shown in Fig. 3.3, the "case" genotype is *AC* and the "control" genotype is *BD*. Many different test statistics have been proposed, including the haplotype relative risk (Falk and Rubenstein, 1987; Terwilliger and Ott, 1992), affected family-based controls (Schaid and Sommer, 1994), and the transmission disequilibrium test (TDT) introduced by Spielman *et al.* (1993). In particular, the TDT gained wide popularity in the mid and late 1990s; this method also focuses on specific alleles transmitted to affected offspring, in the presence of linkage. This test requires genotype information on trios of individuals, namely, affected children and both biological parents, and at least one parent must be heterozygous to be informative, such that the identification of the transmitted and nontransmitted alleles can be ascertained. The proposed test statistic is a McNemar's χ^2 test that tests the Mendelian null hypothesis that the putative disease-associated allele is transmitted 50% of the time from heterozygous parents against the alternative hypothesis that the disease-associated allele will be transmitted more often to affected offspring. Ewens and Spielman (1995) demonstrated that, indeed, the TDT is not affected by population stratification or admixture.

Many investigators have extended the TDT to other relevant situations including multiallelic marker loci (Bickeboeller and Clerget-Darpoux, 1995; Rice *et al.*, 1995; Sham and Curtis, 1995) and missing parents (Sun *et al.*, 1999). In situations where parental data may be absent or impossible to obtain

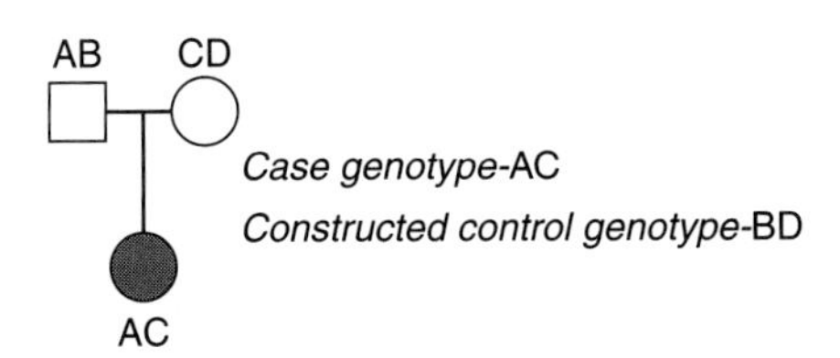

Figure 3.3. Construction of family-based controls for association studies.

(such as for late-onset disorders), the use of discordant siblings (Boehnke and Langefeld, 1998; Hovarth and Laird, 1998; Spielman and Ewens, 1998) can provide information to impute the missing parental genotypes.

Development of association tests for quantitative traits has proceeded on a parallel track. Analysis of variance classifying subjects by their measured genotype can reveal effects of a test locus on quantitative variation. A lucid demonstration of this approach was presented by Boerwinkle *et al.* (1987) who examined the effect of apo E genotypes—composed of isoelectric variants ε_2, ε_3, and ε_4—on lipids and lipoproteins. In this statistical approach, it is straightforward to simultaneously control for the effects of covariates, thereby improving the power to detect associations. These types of analyses are usually aimed at samples of unrelated individuals and treat the measured genotype simply as a fixed main effect using standard statistical methods. However, there are extensions to these variance components models that use the information from family data to simultaneously model residual familial resemblance (Almasy and Blangero, 1998; Boerwinkle *et al.*, 1986; Province *et al.*, 2003). Alternatively, mixed model approaches estimating a cluster correlation among family members, generalized estimating equations, or bootstrap resampling methods all have been used to accommodate the correlations among family members while using simple linear models for analysis (e.g., analysis of variance, regression, or logistic regression).

While marker maps of the human genome have moved now to large panels of diallelic markers (SNPs), it occasionally remains of interest to evaluate multiallelic polymorphisms (e.g., HLA), which poses some additional analytic issues. The larger number of putative risk alleles to test leads to an increase in the degrees of freedom in a single test seeking whether any one of the alleles is overtransmitted to affected offspring. This leads to a consequent loss of power and statistical problems with sparse cells. One way to reduce the number of classes is to sequentially test each allele, collapsing all other alleles into a single class, reducing the locus to a diallelic situation. However, this leads to series of 1 df tests. Remedies to this situation have been proposed by Almasy and Blangero (1998) who proposed modeling the marker genotype as a random variable rather than a fixed effect. Terwilliger (1995) also has proposed a likelihood-based approach for tests of disequilibrium that accommodates multiallelic markers, where the test becomes more conservative as the number of marker alleles increases, while maintaining high power.

The concept of family-based tests of association also has been attractive to investigators studying quantitative traits. Regression and analysis of variance-based methods have been proposed (Abecasis *et al.*, 2000; Lange *et al.*, 2004; Van Steen and Lange, 2005). In general, these methods allow for analysis of families and pedigrees, tests under dominant, additive, and recessive models or without making any assumptions on the mode of inheritance, as well as tests of haplotype effects. While these methods indeed protect against false positives due to population stratification, they rely on heterozygosity and informative transmissions, making them somewhat less powerful than simple linear models accommodating familial correlations.

In recent years, the scientific pendulum has swung back in favor of the unrelated case-control study. With this design it is practical to recruit, phenotype and genotype large numbers of subjects, and is much easier to execute and analyze than family-based designs. It is also well known from epidemiological research that the case-control design is one of the most efficient and cost effective ways of finding signals, although it will tend to exaggerate effect sizes over that found in the population at large for two reasons. First, it contrasts the tails of distributions eliminating the more moderate phenotype subjects by design. Second, by taking the extremely significant SNPs from a genome scan, the phenomenon of the "winner's curse" comes into play (see the end of this section) that will also tends to inflate the estimates of effect-size (e.g., Garner, 2007; Zöllner and Pritchard, 2007). In recent years, the "fatal flaw" of case-control studies regarding false positives and population stratification has been addressed with two related methods: genomic control and structured association (see Garner, 2007). Although these methods make slightly different statistical assumptions, both use information from SNPs, ideally those showing larger differences in allele frequencies between progenitor populations, to correct for potential stratification. The key idea is that most SNPs typed are under the null for any given trait, so that any overdispersion of statistics or hidden (latent) substructure can be estimated and used as a correction factor. This means that the same data that generate the GW association in cases and controls can also be used to provide a correction for population stratification.

In summary, there has been a vigorous development of association methods to detect complex trait loci with the goal of exploiting LD (when it exists) to detect and map loci with relatively small effect. The promise of this approach for identifying common trait variants in LD with haplotype blocks or directly measured is great.

Definition of "the Winner's Curse": The term comes from auction bidding, and denotes the phenomenon that the winner of an item tends to be the one who overbids, and thus is "cursed" with overpaying beyond its true value. In GW scanning, a similar phenomenon occurs in that if there are multiple GW scans "bidding" in a region, the one that "wins" by finding the highest significance in that region tends to be the study in which the randomness of the data overestimates the effect. Thus, the original discovery LOD score is usually overestimating the true effect, raising expectations for replication unrealistically high. Put another way, the power to replicate a finding is thus much lower than that for the initial discovery, which means that much larger sample sizes are needed for replication than expected.

IV. SEVEN STAGES OF RELATIONSHIP WITH GW STUDIES

The enthusiasm for GW linkage and association studies for gene discovery can be traced through the seven stages of a romantic relationship, summarized in Table 3.2. These are Attraction, Infatuation, Boredom, Disillusionment, Cynicism, Rediscovery, and finally, Acceptance. Because GW linkage became more practical to carry out earlier in history, it has progressed further down this pathway and is therefore at a more mature stage today than GW association, which is still in the early stages.

For GW linkage, the initial "attraction" stage might be said to have begun in 1955 when Newton Morton in one fell swoop developed the statistical framework and proposed a statistic for quantifying support for a linkage hypothesis —the LOD score—which was practical to calculate in families. It also had the attractive property that it was additive across families and studies, allowing the accumulation of evidence across all the world's studies. At the same time he offered his famous standards of LOD > 3 as significant evidence for linkage and LOD < -2 as significant evidence against linkage, which won him the very first Allan Award in Human Genetics. This likelihood- (and sequential-) based approach was (and still is) both elegant and powerful. But the real "infatuation" stage with the GW linkage method did not arrive until 1980 (Botstein *et al.*, 1980), with the suggestion that restriction fragment length polymorphisms were

Table 3.2. Seven Stages of a Romantic Relationship: Linkage vs. Association Genome-Wide (GW) Scans

7 Stages of a romantic relationship	GW linkage	GW association
Attraction	1955 Newton Morton LOD Score	1996 Risch and Merikangas declare GWAS feasible
Infatuation	1980 White *et al.*, propose a human linkage map based on RFLPs	2002 HapMap Begins
Boredom	1984 Mammalian Genotyping Service (Marshfield) Begins	*? 2005 Affymetrix & Illumina 500K chips?*
Disillusionment	1995 Lander & Kruglyak set "impossible dream" GWLS	
Cynicism	1996 Risch & Merikangas declare linkage dead	
Rediscovery	2007 Continued use of linkage results to prioritize fine mapping and association studies	
Acceptance	?	?

sufficiently ubiquitously distributed throughout the genome to provide a usable map of the human genome. This gave geneticists for the first time the means to practically tag the genome with highly polymorphic markers, uniformly distributed, to scan the genome for linkage. Now there were probes covering the genome rather than at a few dozen locations corresponding to the blood group antigenic markers and red blood cell enzyme polymorphisms. This technology quickly evolved into a high throughput enterprise, and by 1984, the Mammalian Genotyping Service was established at Marshfield, WI that obtained a contract from the National Heart Lung and Blood Institute to do all GW linkage scan genotyping for its major family studies. This provided a fast, high-quality product that was affordable for many large genetic epidemiological studies of thousands of subjects and hundreds of pedigrees at a time. Ironically, as we look back in hindsight, the creation of MGS can be argued as marking the beginning of the "boredom" stage for GW linkage, as soon it was standard operating procedure to propose doing a GW linkage scan for any and all traits, without needing to put much scientific thought into the process. It was the perfect scientific plan for a statistical mind: a true global fishing expedition masquerading as hypothesis testing. "Disillusionment" began to take root when the task of gene discovery proved more difficult than imagined initially. As a harbinger of the challenges to be encountered in the analysis of complex traits, the initial exciting report of a schizophrenia susceptibility locus on chromosome 5q11–13 (Sherrington *et al.*, 1988) failed to replicate in numerous studies and was ultimately deemed a type I error (McGuffin *et al.*, 1990). Soon followed the famous 1995 *Nature Genetics* paper of Lander and Kruglyak in which they established the GW significance criterion for linkage of LOD > 3.64, which proved to be an "impossible dream" for studies of most complex traits to achieve. Those that did achieve that level of significance were still challenged by failure to consistently replicate their findings, which may be due to several factors, including "the winners curse" genetic heterogeneity, and/or the intricacy of complex trait pathways. There seemed to be a little low hanging fruit for GW linkage while at the same time Lander and Kruglyak's more stringent criteria made it nearly impossible to harvest any fruit from the higher branches, which soon lead to the "cynicism" stage. This is marked by the *Science* paper of Risch and Merikangas (1996), in which they pointed out the vastly lower power for linkage to resolve gene location compared to direct association and advocated a switch in strategy.

The Risch and Merikangas (1996) *Science* publication can also be said to mark the beginning of the "attraction" stage for GWAS, as it suggested that such simpler studies of unrelated cases and control subjects (rather than the more difficult family based designs) might be a much more direct, powerful design for GW scanning. The discovery of LD blocks, regions of the genome in tight correlation with one another even in distantly related randomly selected subjects, allowed for the possibility that a much smaller number of SNPs could be typed to tag the common variation of genome. Thus, the "infatuation" stage for GW association can be said to have begun in 2002 with the beginning of the HapMap project that quickly provided the local genomic topology of LD, needed to make GW association scanning possible. GW SNP chip technology soon put the GWAS into the grasp of many existing case-control and epidemiological studies including large numbers of subjects, including several large consortia of studies such as the Genetic Association Information Network and the Genes and Environment Initiative. Much low hanging fruit is already being harvested (e.g., Bierut *et al.*, 2007; Gudmundsson *et al.*, 2007; Yeager *et al.*, 2007). It is only speculation at this point as to whether we will look back 10 years from now and mark the announcement of the commercial Affymetrix and Illumina GW SNP chips as the initiation of the "boredom" stage for GW association scans or whether we will still be productively engaged in the infatuation stage, enthralled to be finding many novel complex trait genes.

It should be noted that in the present fervor for GWAS, the multiplicity of statistical tests pertaining to the large number of markers typed makes interpretation of the results challenging. Appropriate critical values for identifying "significant" (and hopefully real) signals are disputable, thus placing much more reliance on replication, with all the concomitant challenges for complex traits (i.e., the "winner's curse," heterogeneity, sampling, and population differences). In this context, some investigators have "rediscovered" the utility of linkage analyses that provide information regarding genomic regions, albeit broad, that likely contain trait loci. With linkage regions serving as prior hypotheses, the magnitude of the multiple comparisons problem can be mitigated and attention turned to regions of the genome likely harboring trait loci. Thus, perhaps there is already some rediscovery of the charms of linkage analysis.

V. CONTEMPORARY APPROACHES

Understanding the genetic underpinnings of human diseases and risk factors is the key to making personalized medicine a reality. Factors influencing the feasibility of finding loci affecting complex traits by linkage and/or association studies include (1) the ability to measure the trait reproducibly on a large number of subjects, (2) demonstration that the trait is significantly heritable and is likely influenced by genetic factors, and (3) an appropriate study design and analysis methodology that will yield sufficient power to detect loci with plausible effect sizes.

There are two general strategies for identifying complex trait loci. For the last 30 years or so, the candidate gene approach has been employed extensively because it is motivated by and takes advantage of what is known about the trait biology. For example, reasonable candidate genes for obesity might be appetite-regulating factors, or for diabetes, members of the insulin signaling system. Historically, this strategy was employed with focus on one candidate gene at a time because of the relatively expensive investment of research resources that were required for genotyping until recently. In the last few years, largely spurred by the dramatic reduction in mass genotyping costs as a bonus value-added consequence of the HapMap project, the modern successor to the candidate gene approach is the candidate pathway strategy (e.g., Hazra *et al.*, 2007; Pharoah *et al.*, 2007; Schafmayer *et al.*, 2007). This approach recognizes that genes work together in complex ways and that to understand their interplay, it is necessary to measure all of them simultaneously in a systems biology framework. Boutique pathway chips can now be designed and utilized in thousands of subjects. Scientifically, the candidate gene/pathway strategy can be characterized as a hypothesis-testing approach because of the biological foundation.

In contemporary usage, candidate genes are evaluated by association studies of a group of SNPs placed throughout the gene. The HapMap provides the resource for choosing SNPs to type: typically, any exonic SNPs will be evaluated, in addition to tag SNPs in the 5′ promoter region, within splice junctions, and in the 3′ untranslated regions, because these regions are more likely to harbor functional mutations. However, for this very reason, these sequences tend to be conserved, and most polymorphisms within genes are in the introns. Interestingly, several recent studies have reported associations with intronic SNPs suggesting that there may be important regulatory elements near these locations (cf Cleynen *et al.*, 2007). Tag SNPs marking common haplotypes throughout the gene can be chosen to reduce the genotyping burden. Then, the marginal effects of each SNP as well as haplotype effects can be assessed either in samples of unrelated individuals or in family data, as described above.

A genome scan is a completely different kind of experiment. In this case, anonymous polymorphisms that are uniformly distributed throughout the genome are in turn tested for the presence of a trait locus at each of hundreds of locations (in the case of linkage) or hundreds of thousands to millions of locations (in the case of association). In this sense, a genomic scan is a hypothesis-generating approach. The importance of genome scan experiments is that they represent a means to detect previously unsuspected trait loci, potentially pushing the boundaries of our knowledge in revolutionary ways.

Genome scans can be carried out using either a linkage or an association approach and the map requirements differ. Whereas the typical marker map for linkage studies included approximately 300–400 highly polymorphic microsatellite markers with an average density of 5–15 cM (or 6,000–10,000 uniformly distributed, independent SNPs with roughly equivalent information), Kruglyak (1999) estimated that approximately 500,000 SNPs are required for whole-genome association studies because a useful level of LD is unlikely to extend over distances greater than roughly 3 kb in the general population. GW SNP chips are now widely available providing coverage of this density. At the writing of this chapter, the Affymetrix 500 K chip consists of half a million SNPs of convenience randomly spaced across the genome, while the 650 K Illumina has selected haplotype-tag SNPs based upon the HapMap, to provide more efficient coverage of the block structure of the genome. Both companies are expected to release their own versions of a million SNP chip later this year. Some of the early GW association scans have yielded some promising "low hanging fruit" [e.g., macular degeneration (Klein *et al.*, 2005) and obesity (Frayling *et al.*, 2007)]. Currently, many such studies are under way to identify risk loci influencing multiple diseases including cardiovascular disease (including all its risk factor domains of lipid metabolism, hypertension, glucose metabolism, obesity, and inflammation), prostate and breast cancer among others, inflammatory diseases (such as inflammatory bowel disease, psoriasis, and asthma), autoimmune diseases (such as rheumatoid arthritis), susceptibility to infectious diseases (such as cervical cancer, malaria), and various psychiatric disorders and addictions.

There are some potential weaknesses of this approach. First, functional SNPs not in LD with any typed markers have precisely zero power to be detected, if it is not itself typed. This possibility has been demonstrated by studies of the factor 7 structural gene in relations to serum levels of factor VII (Soria *et al.*, 2005). Complete resequencing of the gene in all subjects in the study revealed a single SNP significantly associated with trait variability that was not in LD with any other SNPs in the gene and, thus, its effect would not have been detected had it not been typed. Furthermore, selection of SNPs from the HapMap places emphasis on SNPs with a minimum allele frequency of 5% that tag common haplotypes, exposing an underlying assumption that common diseases will be determined by common genetic variants (Peng and Kimmel, 2007; Smith and Lusis, 2002). While this may be true in some cases, another viable hypothesis is that collections of less common variants may underlie some diseases. It is also not clear whether the haplotype block structure as determined in by the HapMap project will live up to its promised potential to identify trait variants (Terwilliger and Hiekkalinna, 2006). Nonetheless, data on the type of variation that exists in the human genome is rapidly emerging as more GWAS are being completed. Significant patterns of LD exist in the genome and are being exploited, yet the patterns may be more complex and subtle than they appear at first examination. Certainly, the properties and utility of whole-genome LD mapping is one of the most exciting areas of research right now. We will shortly know for many diseases what is possible to find. What will be more difficult to assess is what these studies might be missing.

VI. CHALLENGES AND ISSUES

A. Multiple comparisons and type I and type II errors

One of the obvious problems in interpreting the results of a genome scan is sorting out the true positive signals from the false ones. Fundamentally, this problem arises because a genome scan is not consistent with a traditional hypothesis testing framework upon which statistical inference is based. One simple way to control the rate of false positives is to use a Bonferroni correction; however, it is difficult to estimate the number of independent tests because the markers for GW scans are correlated due to either linkage/disequilibrium or both, and test statistics can be interpolated even between markers as if a continuous map were available. Under the assumption of a continuous map, Lander and Kruglyak (1995) advocated fairly stringent critical values for tests of linkage, which would be orders of magnitude smaller for GWAS. However, adopting such high critical values comes at a penalty of decreased power. The power to detect loci of small to modest effect—which is what is anticipated for complex trait loci—is critical for the success of this endeavor. Alternative approaches have suggested such use of the false discovery rate (Benjamini and Hockberg, 1995)—the percent of statistical tests deemed significant which are false positives—as well as more promising and sophisticated methods such as those employing sequential analysis that allows for control of both type I and type II errors simultaneously (Province, 2000). Nonetheless, these issues have brought practitioners in the field back to the fundamental of replication, despite the challenges of the "winners curse." Because of the variability in allele frequencies and LD patterns among populations and studies, whether replication should be aimed at the particular SNP, the haplotype block, or the gene itself is under debate. Particularly given that many of these traits will have genetic heterogeneity, the design of studies, the delineation of the phenotype, the role of covariate, and interacting factors all will undoubtedly play a role in the ability to successfully "replicate" a putative association.

B. Genetic heterogeneity

When Morton (1962) first argued that the "... detection of genetic heterogeneity is the principal purpose of a linkage test in man," the only documented autosomal example known at the time was the heterogeneity between the Rh blood groups and elliptocytosis (Morton, 1956). To many, this heterogeneity seemed more of a curiosity than an expected finding since elliptocytosis appeared to be a simple and clinically homogeneous phenotype. Since then more powerful heterogeneity tests have been developed (Risch, 1988; Smith, 1962) and many more examples of "locus heterogeneity" are known. Alleles at 11 unlinked loci, for instance, are known to cause autosomal dominant retinitis pigmentosia (RP).

Alleles at another seven unlinked loci, in homozygotes or compound heterozygotes, can cause recessive RP. In addition, four X-linked RP loci (one dominant, three recessive) have been identified (Heckenlively and Daiger, 1996). Other well-known examples with documented locus heterogeneity include maturity-onset diabetes of the young (Froguel and Velho, 1999), prolongation of the QT interval (Viskin, 1999), early-onset Alzheimer's disease (Selkoe, 1996), breast cancer (Hall *et al.*, 1990; Wooster *et al.*, 1994), epilepsy (De Marco *et al.*, 2007), rheumatoid arthritis (Criswell *et al.*, 2007), among many others. These data support the idea that genetic heterogeneity may be one of the prominent features of complex phenotypes. Morton's suggestion that careful linkage analysis may be a powerful tool to dissect locus heterogeneity was extremely insightful in its time, and may well serve in contemporary studies. However, the task for complex traits poses additional challenges in that the underlying loci may be neither necessary nor sufficient to cause disease, may have modest marginal effects on quantitative variation, or may be involved in complex patterns of interaction with other loci or environmental factors. Nonetheless, strategies like those used in the examples cited above may prove similarly useful in the analysis of complex traits.

In the chapters that follow, various authors will address the challenges and opportunities in mapping and characterizing complex trait loci. While the opportunities have never been greater—with the imminent availability of human genome sequence, technical advances that make high-throughput genotyping a reality, improved computational power coupled with novel and powerful statistical methods—the challenges also remain great, as is the promise of improved health and quality of life. It is an exciting time in human genetics.

Acknowledgments

This work was supported in part by (GWAS) HL087700 (IBB, MAP). A special thanks to Brian Suarez for the interesting early historical perspective.

References

Abecasis, G. R., Cardon, L. R., and Cookson, W. O. (2000). A general test of association for quantitative traits in nuclear families. *Am. J. Hum. Genet.* **66**(1), 279–292.

Almasy, L., and Blangero, J. (1998). Multipoint quantitative-trait linkage analysis in general pedigrees. *Am. J. Hum. Genet.* **62,** 1198–1211.

Amos, C. I. (1994). Robust variance-components approach for assessing genetic linkage in pedigrees. *Am. J. Hum. Genet.* **54,** 535–543.

Benjamini, Y., and Hochberg, Y. (1995). Controlling the false discovery rate: A practical and powerful approach to multiple testing. *J. R. Statist. Soc. B.* **57,** 289–300.

Bernstein, F. (1924). Ergebnisse cinerbiostatischen zusammenfassenden Betrachtung uber die erblichen Blutstrukturen des Menschen. *Klin. Wschr.* **3,** 1495–1497.

Bickeboeller, H., and Clerget-Darpoux, F. (1995). Statistical properties of the allelic and genotypic transmission/disequilibrium test for multiallelic markers. *Genet. Epidemiol.* **12,** 865–870.

Bierut, L. J., Madden, P. A., Breslau, N., Johnson, E. O., Hatsukami, D., Pomerleau, O. F., Swan, G. E., Rutter, J., Bertelsen, S., Fox, L., Fugman, D., Goate, A. M., *et al.* (2007). Novel genes identified in a high-density genome wide association study for nicotine dependence. *Hum. Mol. Genet.* **16**(1), 24–35.

Blacker, D., Bertram, L., Saunders, A. J., Moscarillo, T. J., Albert, M. S., Wiener, H., Perry, R. T., Collins, J. S., Harrell, L. E., Go, R. C., *et al.* (2003). Results of a high-resolution genome screen of 437 Alzheimer's disease families. *Hum. Mol. Genet.* **12**(1), 23–32.

Blackwelder, W. C., and Elston, R. C. (1985). A comparison of sib-pair linkage tests for disease susceptibility loci. *Genet. Epidemiol.* **2,** 85–97.

Boehnke, M., and Langefeld, C. D. (1998). Genetic association mapping based on discordant sib pairs: The discordant-alleles test. *Am. Hum. Genet.* **62**(4), 950–961.

Boerwinkle, E., Chakraborty, R., and Sing, C. F. (1986). The use of measured genotype information in the analysis of quantitative phenotypes in man. I. Models and analytical methods. *Ann. Hum. Genet.* **50,** 181–194.

Boerwinkle, E., Visvikis, S., Welsh, D., Steinmetz, J., Hanash, S. M., and Sing, C. F. (1987). The use of measured genotype information in the analysis of quantitative phenotypes in man. II. The role of the apolipoprotein E polymorphism in determining levels, variability, and covariability of cholesterol, betalipoprotein, and triglycerides in a sample of unrelated individuals. *Am. J. Med. Genet.* **27,** 567–582.

Botstein, D., White, R. L., Skolnick, M., and Davis, R. W. (1980). Construction of a genetic linkage map in man using restriction fragment length polymorphisms. *Am. J. Hum. Genet.* **32**(3), 314–331.

Cleynen, I., Brants, J. R., Peeters, K., Deckers, R., Debiec-Rychter, M., Sciot, R., Van de Ven, W. J., and Petit, M. M. (2007). HMGA2 regulates transcription of the Imp2 gene via an intronic regulatory element in cooperation with nuclear factor-{kappa}B. *Mol. Cancer Res.* **5**(4), 363–372.

Criswell, L. A., Chen, W. V., Jawaheer, D., Lum, R. F., Wener, M. H., Gu, X., Gregersen, P. K., and Amos, C. I. (2007). Dissecting the heterogeneity of rheumatoid arthritis through linkage analysis of quantitative traits. *Arthritis Rheum.* **56**(1), 58–68.

Curtis, D., and Sham, P. C. (1994). Using risk calculation to implement an extended relative pair analysis. *Ann. Hum. Genet.* **58,** 151–162.

De Marco, E. V., Gambardella, A., Annesi, F., Labate, A., Carrideo, S., Forabosco, P., Civitelli, D., Candiano, I. C., Tarantino, P., Annesi, G., and Quattrone, A. (2007). Further evidence of genetic heterogeneity in families with autosomal dominant nocturnal frontal lobe epilepsy. *Epilepsy Res.* **74**(1), 70–73.

Elston, R. C., and Stewart, J. (1971). A general model for the genetic analysis of pedigree data. *Hum. Hered.* **21,** 523–542.

Ewens, W., and Spielman, R. S. (1995). The transmission/disequilibrium test: History, subdivision and admixture. *Am. J. Hum. Genet.* **57,** 455–464.

Falk, C. T., and Rubenstein, P. (1987). Haplotype relative risks: An easy reliable way to construct a proper control sample for risk calculations. *Ann. Hum. Genet.* **51,** 227–233.

Frayling, T. M., Timpson, N. J., Weedon, M. N., Zeggini, E., Freathy, R. M., Lindgren, C. M., Perry, J. R., Elliott, K. S., Lango, H., Rayner, N. W., Shields, B., Harries, L. W., *et al.* (2007). A common variant in the *FTO* gene is associated with body mass index and predisposes to childhood and adult obesity. *Science* **316**(5826), 889–894.

Froguel, P., and Velho, G. (1999). Molecular genetics of maturity onset diabetes of the young. *Trends Endocrinol. Metab.* **10,** 121–125.

Garner, C. (2007). Upward bias in odds ratio estimates from genome-wide association studies. *Genet. Epidemiol.* **31**(4), 288–295.

Gudmundsson, J., Sulem, P., Manolescu, A., Amundadottir, L. T., Gudbjartsson, D., Helgason, A., Rafnar, T., Bergthorsson, J. T., Agnarsson, B. A., Baker, A., Sigurdsson, A., Benediktsdottir, K. R., *et al.* (2007). Genome-wide association study identifies a second prostate cancer susceptibility variant at 8q24. *Nat. Genet.* **39**(5), 631–637.

Hall, J. M., Lee, M. K., Newman, B., Morrow, J. E., Anderson, L. A., Huey, B., and King, M. C. (1990). Linkage of early-onset familial breast cancer to chromosome 17q21. *Science* **250,** 1684–1689.

Hanis, C. L., Boerwinkle, E., Chakraborty, R., Ellsworth, D. L., Concannon, P., Stirling, B., Morrison, V. A., Wapelhorst, B., Spielman, R. S., Gogolin-Ewens, K. J., *et al.* (1996). Related articles, links a genome-wide search for human non-insulin-dependent (type 2) diabetes genes reveals a major susceptibility locus on chromosome 2. *Nat. Genet.* **13**(2), 161–166.

Haseman, J. K., and Elston, R. C. (1972). The investigation of linkage between a quantitative trait and a marker locus. *Behav. Genet.* **2,** 3–19.

Hazra, A., Wu, K., Kraft, P., Fuchs, C. S., Giovannucci, E. L., and Hunter, D. J. (2007). Twenty-four non-synonymous polymorphisms in the one-carbon metabolic pathway and risk of colorectal adenoma in the Nurses' Health Study. *Carcinogenesis* **28**(7), 1510–1519.

Heckenlively, J. R., and Daigler, S. P. (1996). Hereditary retinal and choroidal degenerations. *In* "Emory and Rimoin's Principles and Practice of Medical Genetics" (D. L. Rimoin, J. M. Connor, and R. E. Pyeritz, eds.). 3rd Edition, Churchill-Livingstone, Edinburgh.

Hovarth, S., and Laird, N. M. (1998). A discordant-sibship test for disequilibrium and linkage: No need for parental data. *Am. J. Hum. Genet.* **63,** 1886–1897.

Janssens, F. A. (1909). La theorie de la chiasmatypie. *La Cellule.* **25,** 389–411.

Johnson, L., Luke, A., Adeyemo, A., Deng, H. W., Mitchell, B. D., Comuzzie, A. G., Cole, S. A., Blangero, J., Perola, M., and Teare, M. D. (2005). Related articles, links meta-analysis of five genome-wide linkage studies for body mass index reveals significant evidence for linkage to chromosome 8p. *Int. J. Obes. (Lond).* **29**(4), 413–419.

Klein, R. J., Zeiss, C., Chew, E. Y., Tsai, J. Y., Sackler, R. S., Haynes, C., Henning, A. K., SanGiovanni, J. P., Mane, S. M., Mayne, S. T., *et al.* (2005). Complement factor H polymorphism in age-related macular degeneration. *Science* **308**(5720), 385–389.

Kruglyak, L. (1999). Prospects for whole-genome linkage disequilibrium mapping of common disease genes. *Nat. Genet.* **22,** 130–144.

Kruglyak, L., Daly, M. J., Reeve-Daly, M. P., and Lander, E. S. (1996). Parametric and nonparametric linkage analysis: A unified multipoint approach. *Am. J. Hum. Genet.* **58,** 1347–1363.

Lander, E., and Kruglyak, L. (1995). Genetic dissection of complex traits: Guidelines for interpreting and reporting linkage results. *Nat. Genet.* **11,** 241–247.

Landsteiner, K. (1900). Zur Kenntnis der antifermentativen, lytischen und agglutinierenden Wirkungen des Blutserums und der Lymphe. *Zbl. Bakt.* **27,** 357–362.

Lange, C., van Steen, K., Andrew, T., Lyon, H., DeMeo, D. L., Raby, B., Murphy, A., Silverman, E. K., MacGregor, A., Weiss, S. T., *et al.* (2004). Related articles, links a family-based association test for repeatedly measured quantitative traits adjusting for unknown environmental and/or polygenic effects. *Stat. Appl. Genet. Mol. Biol.* 3 Article17. Epub 2004 Aug 12.

McGuffin, P., Sargeant, M., Hetti, G., Tidmarsh, S., Whatley, S., and Marchbanks, R. M. (1990). Exclusion of a schizophrenia susceptibility gene from the chromosome 5q11-q13 region: New data and a reanalysis of previous reports. *Am. J. Hum. Genet.* **47**(3), 524–535.

Mendel, G. (1866). Versuche uber Pflanzer-hybriden. *Verh. Naturf. Verein Brunn* **4,** 3–47.

Morgan, T. H. (1911). An attempt to analyze the constitution of the chromosomes on the basis of sex-limited inheritance in Drosophila. *J. Exp. Zool.* **11,** 365–412.

Morton, N. E. (1955). Sequential tests for the detection of linkage. *Am. J. Hum. Genet.* **7,** 277–318.

Morton, N. E. (1956). The detection and estimation of linkage between the genes for elliptocytosis and the Rh blood type. *Am. J. Hum. Genet.* **8,** 80–96.

Morton, N. E. (1962). Segregation and linkage. *In* "Methodology in Human Genetics" (W. J. Burdette, ed.), pp. 17–52. Holden Day, San Francisco.

Ott, J. (1974). Estimation of the recombination fraction in human pedigrees: Efficient computation of the likelihood for human linkage studies. *Am. J. Hum. Genet.* **26,** 588–597.

Peng, B., and Kimmel, M. (2007). Simulations provide support for the common disease-common variant hypothesis. *Genetics* **175**(2), 763–776. Epub 2006 Dec 6.

Penrose, L. S. (1935). The detection of autosomal linkage in data which consist of pairs of brothers and sisters of unspecified parentage. *Ann. Eugen. (Lond.)* **6,** 133–138.

Pharoah, P. D., Tyrer, J., Dunning, A. M., Easton, D. F., and Ponder, B. A. SEARCH Investigators (2007). Association between common variation in 120 candidate genes and breast cancer risk. *PLoS Genet.* **3**(3), e42. [Epub ahead of print] PMID: 17367212.

Province, M. A. (2000). A single, sequential, genome-wide test to simultaneously identify all promising areas in a linkage scan. *Genet. Epidemiol.* **19,** 301–322.

Province, M. A., Rice, T., Borecki, I. B., Gu, C., and Rao, D. C. (2003). A multivariate and multilocus variance components approach using structural relationships to assess quantitaive trait linkage via SEGPATH. *Genet. Epidemiol.* **24**(2), 128–138.

Rice, J. P., Neuman, R. J., Hoshaw, S. L., Daw, E. W., and Gu, C. (1995). TDT with covariates and genomic screens with mod scores: Their behavior on simulated data. *Genet. Epidemiol.* **12,** 659–664.

Risch, N. (1988). A new statistical test for linkage heterogeneity. *Am. J. Hum. Genet.* **42,** 353–364.

Risch, N. (1990). Linkage strategies for genetically complex traits. I. Multilocus models. *Am. J. Hum. Genet.* **46,** 222–228.

Risch, N., and Merikangas, K. (1996). The future of genetic studies of complex human diseases. *Science* **273**(5281), 1516–1517.

Risch, N., and Zhang, H. (1995). Extreme discordant sib pairs for mapping quantitative trait loci in humans. *Science* **268,** 1584–1589.

Rioux, J. D., Daly, M. J., Silverberg, M. S., Lindblad, K., Steinhart, H., Cohen, Z., Delmonte, T., Kocher, K., Miller, K., Guschwan, S., *et al.* (2001). Related articles, links genetic variation in the 5q31 cytokine gene cluster confers susceptibility to Crohn disease. *Nat. Genet.* **29**(2), 223–228.

Schafmayer, C., Buch, S., Egberts, J. H., Franke, A., Brosch, M., El Sharawy, A., Conring, M., Koschnick, M., Schwiedernoch, S., Katalinic, A., Kremer, B., Fölsch, U. R., *et al.* (2007). Genetic investigation of DNA-repair pathway genes PMS2, MLH1, MSH2, MSH6, MUTYH, OGG1 and MTH1 in sporadic colon cancer. *Int. J. Cancer* **121**(3), 555–558.

Schaid, D. J., and Sommer, S. S. (1994). Comparison of statistics for candidate gene studies using cases and parents. *Am. J. Hum. Genet.* **55,** 402–409.

Selkoe, D. J. (1996). Amyloid B-protein and the genetics of Alzheimer's disease. *J. Biol. Chem.* **271,** 18295–18298.

Sham, P. C., and Curtis, D. (1995). An extended transmission/disequilibrium test (TDT) for multi-allele marker loci. *Ann. Hum. Genet.* **59,** 323–336.

Sherrington, R., Brynjolfsson, J., Petursson, H., Potter, M., Dudleston, K., Barraclough, B., Wasmuth, J., Dobbs, M., and Gurling, H. (1988). Localization of a susceptibility locus for schizophrenia on chromosome 5. *Nature* **336**(6195), 164–167.

Smith, C. A. B. (1962). Homogeneity test for linkage data. *Proc Sec. Int. Congr. Hum. Genet.* **1,** 212–213.

Smith, D. J., and Lusis, A. J. (2002). The allelic structure of common disease. *Hum. Mol. Genet.* **11**(20), 2455–2461.

Soria, J. M., Almasy, L., Souto, J. C., Sabater-Lleal, M., Fontcuberta, J., and Blangero, J. (2005). The F7 gene and clotting factor VII levels: Dissection of a human quantitative trait locus. *Hum. Biol.* **77**(5), 561–575.

Spielman, R. S., and Ewens, W. J. (1998). A sibship test for linkage in the presence of association: The sib transmission/disequilibrium test. *Am. J. Hum. Genet.* **62,** 450–458.

Spielman, R. S., McGinnis, R. E., and Ewens, W. J. (1993). Transmission test for linkage disequilibrium: The insulin gene region and insulin-dependent diabetes mellitus (IDDM). *Am. J. Hum. Genet.* **52,** 506–516.

Sturtevant, A. H. (1913). The linear arrangement of six sex-linked factors in Drosophila, as shown by their mode of association. *J. Exp. Zool.* **14,** 43–59.

Suarez, B. K., and Van Eerdewegh, P. (1984). A comparison of three affected-sib-pair scoring methods to detect HLA-linked disease susceptibility genes. *Am. J. Med. Genet.* **18,** 135–146.

Suarez, B. K., Rice, J., and Reich, T. (1978). The generalized sib pair IBD distributions: Its use in the detection of linkage. *Ann. Hum. Genet. Lond.* **42,** 87–94.

Sun, F., Flanders, W. D., Yang, Q., and Khoury, M. J. (1999). Transmission disequilibrium test (TDT) when only one parent is available: The 1-TDT. *Am. J. Epidemiol.* **150,** 97–104.

Terwilliger, J. D. (1995). A powerful likelihood method for the analysis of linkage disequilibrium between trait loci and one or more polymorphic marker loci. *Am. J. Hum. Genet.* **56,** 777–787.

Terwilliger, J. D., and Hiekkalinna, T. (2006). An utter refutation of the "Fundamental Theorem of the HapMap." *Eur. J. Hum. Genet.* **14**(4), 426–437.

Terwilliger, J. D., and Ott, J. (1992). A haplotype-based "Haplotype Relative Risk" approach to detecting alleleic associations. *Hum. Hered.* **42,** 337–346.

Van Steen, K., and Lange, C. (2005). PBAT: A comprehensive software package for genome-wide association analysis of complex family-based studies. *Hum. Genomics* **2**(1), 67–69.

Viskin, S. (1999). Long QT syndromes and torsade de points. *Lancet* **354,** 1626–1633.

Weeks, D. E., and Lange, K. (1988). The affected-pedigree-member method of linkage analysis. *Am. J. Hum. Genet.* **42,** 315–326.

Woolf, B. (1955). On estimating the relation between blood group and disease. *Ann. Hum. Genet.* **19,** 251–253.

Wooster, R., Neuhausen, S. L., Mangion, J., Quirk, Y., Ford, D., Collins, W., Nguyen, K., Seal, S., Tran, T., Averill, D., *et al.* (1994). Localization of a breast cancer susceptibility gene, BRCA2, to chromosome 13q 12–13. *Science* **265,** 2088–2090.

Yeager, M., Orr, N., Hayes, R. B., Jacobs, K. B., Kraft, P., Wacholder, S., Minichiello, M. J., Fearnhead, P., Yu, K., Chatterjee, N., Wang, Z., Welch, R., *et al.* (2007). Genome-wide association study of prostate cancer identifies a second risk locus at 8q24. *Nat. Genet.* **39**(5), 645–649.

Zöllner, S., and Pritchard, J. K. (2007). Overcoming the Winner's Curse: Estimating penetrance parameters from case-control data. *Am. J. Hum. Genet.* **80,** 605–615.

4

Definition of Phenotype

Mary K. Wojczynski* and Hemant K. Tiwari*,†
*Department of Biostatistics, Section on Statistical Genetics, University of Alabama at Birmingham, Birmingham, Alabama 35294
†Section on Statistical Genetics, Department of Biostatistics, University of Alabama at Birmingham, 1665 University Boulevard, Birmingham, Alabama 35294

Advances in Genetics, Vol. 60

0065-2660/08 $35.00
DOI: 10.1016/S0065-2660(07)00404-X

ABSTRACT

Definition of the phenotype is crucial in designing any genetic study, especially an association study, intended to detect the disease predisposing genes. In this chapter, we review the different types of phenotypes such as discrete or continuous and discuss the issues impacting on the phenotype definition related to study design, specifically, the impact of diagnostic error (misclassification) in case-control studies and measurement error in continuous traits. We show that the power of a study depends heavily on the phenotype measured and that misclassification or measurement error can dramatically reduce the power. We also suggest some possible responses to these challenges.

I. INTRODUCTION

A phenotype is the observable expression of an individual's genotype. Thus, while genotypes affect proteins, cells, and biological pathways, phenotypes are most readily observed as appearances, signs, and symptoms of disease (Schulze and McMahon, 2004). Further, while an individual's genotype is fairly stable over a lifetime, an individual's phenotype is dynamic, influenced by both the environment and the underlying genotype, including interactions between them (Schulze and McMahon, 2004; Winawer, 2006). Thus, a phenotype may be difficult to measure because the disease may be vaguely defined or poorly measured, may encompass several different underlying disorders each with its own genetic and/or environmental influence, or may have unacknowledged environmental influences.

Defining the phenotype is imperative in any research, including genetic research. The phenotype definition will include information from epidemiology, biology, molecular biology, and computational methods in order to group signs and symptoms of a disorder together to analyze possible genetic influences on the phenotype (Winawer, 2006). Thus, the definition, measurement, and validity of phenotyping need to be standardized in genetic studies to increase the quality of research and the reproducibility of linkage and association studies (Schulze and McMahon, 2004). A good phenotype definition aids with the goal of genetic epidemiological research, which is to map and identify genes that modify risk for common diseases (Schulze and McMahon, 2004). To accomplish this goal, an accurate phenotype definition (i.e., one that has high specificity) is necessary to minimize the number of genes that may affect risk of disease, and thus to increase the odds of finding these genes (Winawer, 2006). Additionally, because the phenotype is the expression of the underlying biology, a precise (highly specific) definition is necessary in order to further understand the pathophysiological mechanisms involved with disease processes (Bel, 2004). Thought should therefore be put into the underlying biology of a phenotype when attempting to define it for a study (Winawer, 2006). In genetic research, definition of a phenotype usually consists of developing a standard checklist of examination items, questions, and laboratory data that must be collected from every study participant (Schulze and McMahon, 2004) in a standardized manner.

Complex disease phenotype definitions are problematic due to issues of heterogeneity, gene–environment interactions, and gene–gene interactions (Gottesman and Gould, 2003). Heterogeneity of complex disease phenotypes stems from the fact that complex diseases have many possible manifestations, different biological paths leading to disease diagnosis, differences in severity of symptoms, symptomologies that wax and wane over time, and differences in disease penetrance (Bickeboller *et al.*, 2005; Verkerk *et al.*, 2006; Winawer, 2006). Allowing all these expressions of the disease to be used as a definition of the same disease leads to heterogeneity and may affect the results of a genetic study. The heterogeneity may lead to false positive associations, associations that cannot be replicated when a more refined phenotype is used, or null association findings due to a diluting effect of the precise phenotype (Bickeboller *et al.*, 2005).

Further, these diseases are called complex because they do not follow Mendelian inheritance and usually involve multiple genes, each with a relatively small influence on the phenotype. Such diseases may be clinically multifactorial, with each facet having its own genetic influences. Two common examples are metabolic syndrome (abdominal obesity, high serum triglycerides, low HDL cholesterol, high blood pressure, high fasting blood glucose) and coronary disease (myocardial infarction, atherosclerosis, and sudden cardiac death). Additionally, a complex disease may have multiple causes, some of which may have a genetic influence while others do not. An example is atherosclerosis which is caused by a combination of lipid accumulation, inflammation, endothelial disruption, and thrombosis (Ellsworth and Manolio, 1999). Taken together, these issues highlight the importance of clearly defining a disease phenotype. In order for researchers to understand the genetic contribution to any complex disorder where the relationship between clinical phenotype/expression and genetic mechanism is not well understood or established, it is imperative to have a precise phenotype definition that encompasses a strict case definition that has high specificity (Winawer, 2006).

In this chapter, we wish to highlight further the issues involved in defining a phenotype for study and how different phenotype definitions may affect aspects of the study design and analysis. We will also highlight some methods for trying to refine multivariate phenotypes into more homogeneous phenotypes.

II. PHENOTYPE CHOICES/OPTIONS

A. Discrete versus continuous traits

Before we illustrate the issues that arise in defining phenotypes, it will be helpful to clarify a key conceptual distinction between discrete and continuous traits. In some genetic studies, traits of particular interest and most clinical relevance such as cancer, hypertension, or arthritis tend to be discrete (Newton-Cheh and Hirschhorn, 2005)—that is, they are either present or absent. Sometimes these traits are investigated as the main phenotype of interest in the study, but more likely they can be (1) contributors to disease risk (e.g., hypertension, a predictor of heart disease, myocardial infarction, and other cardiovascular diseases), (2) attributes of the phenotype that lead to a narrower phenotype definition or clinical entity (e.g., measures of insulin secretion and insulin sensitivity for diabetes), or (3) members of a set of diagnostic criteria (e.g., fasting glucose for diabetes, clinical criteria for depression, obesity, diabetes, and hypertension for metabolic syndrome) (Newton-Cheh and Hirschhorn, 2005).

A quantitative trait, by contrast, is a phenotype that is continuous, thereby having a range of possible values. Often, these quantitative traits are used as such, but a quantitative trait can also be categorized or dichotomized by using a predefined threshold value. Most often these values are clinically meaningful; sometimes, however, especially for genetic studies, only the individuals in the extremes are used in order to try to maximize power and obtain a definitive distinction between diseased and nondiseased.

B. Study base differences

Phenotype definition is important in association studies due to lack of replication of initial findings. For example, Horikawa *et al.* (2000) published the first positive association study between SNP-43 of the Calpain 10 gene and type 2 diabetes. This study was performed in a Mexican–American population and included 233 cases of type 2 diabetes and 101 non–diabetic controls. After this initial report of a positive association, other studies in Table 4.1 followed, each with different populations and different numbers of cases and controls, but most using type 2 diabetes as the phenotype of interest. Examining this table demonstrates the mix of positive and negative results that are obtained for many association studies, which is a common problem in analyses of complex traits. Further, these studies demonstrate the obvious heterogeneity at the population level, based on the different ethnicities studied. Even when comparing results among a single ethnic group (e.g., the Japanese, Finnish, and Mexican–American populations), there are both positive and negative studies. So how are these discrepancies explained? One aspect not included in Table 4.1 is how type 2 diabetes was defined for each study

Table 4.1. Studies Examining the Association Between SNP-43 of the Calpain 10 Gene and Type 2 Diabetes

References	Type of analysis	Phenotype	Ethnicity	Result	Comments
Tsuchiya *et al.* (2006)	Meta-analysis	Type 2 diabetes	Europeans	Positive	Heterogeneity attenuated effect; heterogeneity due to use of populations previously used for linkage; when these families removed from analysis, heterogeneity was removed; results shown are for case-control association
Song *et al.* (2004)	Meta-analysis	Type 2 diabetes	European, Asian, Americans (Mexican–Americans, Pima Indians, African–Americans, Samoans)	Positive	Result for recessive model only Weak association with significant heterogeneity
Weedon *et al.* (2003)	Meta-analysis	Type 2 diabetes	British, Chinese, Japanese, Finnish, Couth Indian, Mexican–American	Positive	
Horikawa *et al.* (2000)	Association	Type 2 diabetes	Mexican–American and Finnish	Positive	FIRST reported association
Evans *et al.* (2001)	Association	Type 2 diabetes	British (United Kingdom)	Mixed	Used three different UK populations; one positive, two null results
Rasmussen *et al.* (2002)	Association	Type 2 diabetes	Scandinavian	Null	
Elbein *et al.* (2002)	Family-based association	Type 2 diabetes	United States (northern European ancestry)	Null	
Fingerlin *et al.* (2002)	Association	Type 2 diabetes	Finnish	Null	781 cases and 408 controls

(*Continues*)

Table 4.1. (*Continued*)

References	Type of analysis	Phenotype	Ethnicity	Result	Comments
Malecki *et al.* (2002)	Association	Type 2 diabetes	Polish	Positive	229 cases and 148 controls
Orho-Melander *et al.* (2002)	Association	Type 2 diabetes	Finnish	Positive	395 cases and 298 controls
Iwasaki *et al.* (2005)	Association	Type 2 diabetes	Japanese	Null	927 patients and 929 controls
Baier *et al.* (2000)	Association	Insulin resistance	Pima Indians	Positive	Association with type 2 diabetes as well
Cassell *et al.* (2002)	Association	Type 2 diabetes	South Indian (Chennai, India)	Positive	Families; haplotype and association of haplotype components 85 cases and 323 controls
Daimon *et al.* (2002)	Association	Type 2 diabetes	Japanese	Null	81 cases and 81 controls
Garant *et al.* (2002)	Association	Type 2 diabetes	African–Americans	Positive	Prospective: 166 incident cases and 993 controls Cross-sectional: 269 cases and 1159 controls
Hegele *et al.* (2001)	Association	Type 2 diabetes	Oji-Cree (Northern Ontario)	Null	121 cases and 468 controls
Horikawa *et al.* (2003)	Association	Type 2 diabetes	Japanese	Null	177 cases and 172 controls
Lynn *et al.* (2002)	Association	Plasma glucose	British	Positive	285 nondiabetics undergoing oral glucose tolerance test
Sun *et al.* (2002)	Association	Type 2 diabetes	Han Chinese	Null	173 cases and 152 controls
Tsai *et al.* (2001)	Association	Type 2 diabetes	Polynesian Samoans	Null	172 cases and 96 controls
Wang *et al.* (2002b)	Association	Type 2 diabetes	Chinese	Null	

(i.e., phenotyping method) and also how the recruitment of subjects (i.e., sampling scheme) into these studies took place. These two areas can lead to substantial differences in the study base (the population giving rise to cases and controls) and have an effect on the results on the association studies (Page *et al.*, 2003).

Differences in the study base can be better demonstrated in Table 4.2, examining the association between the α-adducin Gly460Trp polymorphism and hypertension. Again, the results of these studies are mixed between demonstrating and not demonstrating an association. These studies were performed in many different ethnic groups, but even within a single ethnic group, there are both positive and negative studies. Additionally, most of these studies have fairly small sample sizes, but the main difference this table demonstrates is how the definition of the study base, which is based on phenotype definition, may affect study results. As this table demonstrates, study base populations for these studies ranged from hypertension drug treatment trials, hypertensive patients, salt-sensitive hypertensives, salt-resistant hypertensives, general population cohorts, hospital-based studies, studies among the elderly, and studies using tails of the blood pressure distribution. These are very heterogeneous populations, all having positive and negative impacts on a study; however, from an analytical point of view, it is difficult to compare the results of these studies because the study bases are heterogeneous, and the heterogeneity may influence the study results. Thus, as not all studies are performed using the same study base, how do we come to a consensus as to the impact of the α-adducin Gly460Trp polymorphism on hypertension? Further, these studies also examined different hypertension outcomes to demonstrate associations with hypertension, and some used different blood pressure thresholds to define hypertension. Therefore, combining these studies to obtain an overall picture of the association between the α-adducin Gly460Trp polymorphism and hypertension is ill-advised because the studies are too heterogeneous, making it difficult to arrive at valid results and conclusions.

C. Clinical heterogeneity

Most diseases of interest also display clinical heterogeneity. This means that there is variation in such clinical characteristics as asymptomatic versus symptomatic individuals, diagnostic classification schemes, age at onset, severity, family history, diagnosis being almost any combination of symptoms, atypical features, and features common to more than one phenotype (Winawer, 2006). Additionally, clinical manifestations of a disorder often change over the life of an individual. A prime example of this is seen in the study of mental disorders for which there is no biological test. Because psychiatric illnesses are defined categorically rather than quantitatively (i.e., outside the normal range of variation) (Bearden and Freimer, 2006), the diagnoses for most disorders are based on the classification schemas in the Diagnostic and Statistical Manual, 4th edition

Table 4.2. Association Studies of Hypertension and α-Adducin Gly460Trp Polymorphism

References	Population	Sample size	Ethnicity	Results
Castejon *et al.* (2003)	Salt-sensitive and salt-resistant hypertensives	126	Venezuelan	No association with blood pressure
Psaty *et al.* (2002)	HMO-treated hypertensive patients; cases had history of myocardial infarction or stroke	323 cases; 715 controls	United States	Genotype not associated with mean levels of blood pressure in either cases or controls
Beeks *et al.* (2004)	Hypertension patients randomized to high- or low-salt diet	117	Dutch	Renal hemodynamic outcomes Effective renal plasma flow and glomerular filtration rate lower in those with variant allele on low-sodium diet but no effect among those on high-sodium diet
Schelleman *et al.* (2006)	General practitioners Doetinchem Cohort Study	625 hypertensives (490 untreated and 135 treated)	Dutch	No significant influence of variant allele on blood pressure for any drug or without drug
Schelleman *et al.* (2004)	Rotterdam study (population-based prospective cohort)	3025 hypertensives	The Netherlands	Variant allele had no effect on blood pressure levels among subjects using diuretics, β-blockers, calcium antagonists, or angiotensin-converting enzyme inhibitors
Staessen *et al.* (2001)	Population-based cohort	678 normotensives followed for 9.1 years; 1461 caucasians	Belgium	Variant allele carriers experience larger blood pressure changes in response to sodium loading or diuretic therapy
Wang *et al.* (2002a)	Population-based cohort	1848	White Caucasians	Prevalence of hypertension not associated with genotype or allele frequency at α-adducin

Shioji *et al.* (2004)	Population-based cohort	867 men; 1013 women	Japanese	Homozygous variant associated with hypertension in women ($p = 0.0420$)
Sugimoto *et al.* (2002)	Rural population-based cohort	1490 Ohasama Town	Japanese	Polymorphism is associated with low renin hypertension; association only demonstrated when subdivide entire cohort
Province *et al.* (2000)	Population-based; 60 years old and older	822 white; 572 black	United States	Variant allele significantly increased odds of hypertension in whites, which remained after adjustment; variant allele had protective effect against hypertension among blacks
Shin *et al.* (2004)	Population-based	903	Korean	Polymorphism not associated with hypertension
Sciarrone *et al.* (2003)	Never treated hypertensives	87		Variant carrier had better response to hydrochlorothiazide treatment
Yazdanpanah *et al.* (2006)	Caucasians with type II diabetes, 55 years old and older	599	Rotterdam Study; The Netherlands	Prevalence of hypertension was 2.57 times higher among homozygous variant compared to homozygous wild type Homozygous variant carriers were 1.83 times higher risk of mortality
Clark *et al.* (2000)	White, less than 64 years old with essential hypertension; age- and gender-matched normotensive controls	128	Glasgow, Scottish	No evidence suggesting association between hypertension and polymorphism Population used individuals at blood pressure extremes

(*Continues*)

Table 4.2. *(Continued)*

References	Population	Sample size	Ethnicity	Results
Ranade *et al.* (2000)	Family-based linkage and association; systolic blood pressure $\geq$ 160, diastolic blood pressure $\geq$ 95 and two medications or systolic blood pressure $\geq$ 140, diastolic blood pressure $\geq$ 90 and one medication	411 families; 1061 individuals	Chinese and Japanese ancestry (Grandparents) but now living in United States (Hawaii and San Francisco)	Chinese ancestry: systolic blood pressure demonstrated association with blood pressure when using recessive model, but no association with diastolic blood pressure Japanese ancestry: no association with any model Combined: no association with α-adducin
Bray *et al.* (2000)	Population at large, relatively young, not selected for salt-sensitivity; family-based linkage and association	281 nuclear families; 774 siblings (5–37 years old) 380 parents (26–57 years old)	Rochester, Minnesota	Polymorphism not associated with any measure of blood pressure level in either parents or siblings Method: maximum likelihood to partition variance into genetic and environmental and other influences
Schork *et al.* (2000)	Families; urban African–American and Rural White	295 African–Americans	United States	No evidence significant effect of polymorphism on blood pressure variation in either sample
		279 White		No association α-adducin genotypes and antihypertensive use
Larson *et al.* (2000)	Population-based prospective cohort	904 African–Americans	United States (Jackson, MS; atherosclerosis risk in communities study)	No association between polymorphism and hypertension status

Mulatero *et al.* (2006)	Retrospective cross sectional; acromegaly patients	100	Northern Italy	Polymorphism had no effect on risk of hypertension
Alam *et al.* (2000)	Elderly (61–89); isolated systolic hypertension: systolic blood pressure $\geq$ 160, diastolic blood pressure $\geq$ 90 Normal: systolic blood pressure $<$ 140, diastolic blood pressure $<$ 90	Isolated systolic hypertension: 87 Normal: 124	Australian caucasians	After adjustment for confounders, no statistically significant association between genotype and systolic blood pressure, diastolic blood pressure, pulse pressure, or diagnosis of isolated systolic hypertension
He *et al.* (2001)	Case-control; hypertensive: systolic blood pressure $\geq$ 150, diastolic blood pressure $\geq$ 95	138 cases 121 controls	Chinese	No significant association between polymorphism and blood pressure variation
Melander *et al.* (2000)	Hypertensive: blood pressure of 160/95	374 primary hypertensive	Scandinavian (Sweden and Finland)	Lower frequency of variant allele in hypertensive patients than in normotensives
	Hypertensive patients and normotensives	419 normotensives		Argues against pathogenic role of this allele in primary hypertension Suggests that this allele may be in linkage disequilibrium with another variant that increases susceptibility to hypertensive
Mulatero *et al.* (2002)	Primary aldosteronism patients	167	Italian	α-Adducin genotype strong independent predictor of systolic blood pressure and diastolic blood pressure

(*Continues*)

Table 4.2. (*Continued*)

References	Population	Sample size	Ethnicity	Results
Turner *et al.* (2003)	4 week treatment trial for essential hypertension; ages 30–59.9	291 African–Americans	Unites States (Atlanta and Minnesota)	Blood pressure response to hydrochlorothiazide
		294 White		α-Adducin polymorphism made no statistically significant contribution to prediction of blood pressure response
		All nonhispanic		Genotype and allele frequency differed significantly between African–Americans and Whites
Ju *et al.* (2003)	Population-based (blood pressure 140/90)	748	Chinese	Positive association between α-adducin polymorphism and essential hypertension in Chinese
Ciechanowicz *et al.* (2001)	Thin hypertensives, salt-sensitive and salt-resistant	44 salt-sensitive	Polish	Blood pressure response to 1 week subacute salt loading
		24 salt-resistant		No significant difference in genotype and allele frequency of α-adducing between salt-sensitive and salt-resistant subjects
				No significant difference in change of blood pressure in regard to genotype
Cusi *et al.* (1997)	8 week diuretic treatment	143	Caucasian	Variant allele associated with greater blood pressure response
Matayoshi *et al.* (2004)	2 month diuretic treatment	76	Asian	No association of polymorphism and blood pressure

(DSM-IV), the major classification system for psychiatric diagnoses. These definitions of disease are based on symptom clusters, and usually if individuals display more than x symptoms in the cluster, they are then considered diseased. The main problem with this system is that symptom clusters do not necessarily describe homogeneous disorders; instead, the same diagnosis may be applied to cases that actually result from slightly different biological pathways with different pathophysiological processes involved, including both genetic and environmental contributors (Hasler *et al.*, 2006). This diagnostic consistency therefore may mask underlying heterogeneity at the biological and genetic levels (Schulze *et al.*, 2006). While this approach helps to clinically manage patients, more specific criteria may be necessary for research. As a result, there may be a different phenotype assignment for clinically affected individuals based on the "research diagnosis" that employs different methods for accurately and specifically assigning a phenotype for each particular research project (Schulze and McMahon, 2004). Ultimately, this is meant to try to refine each phenotype of a research study in an attempt to decrease the clinical variability among cases while maintaining high heritability of the disease (Schulze *et al.*, 2006).

Further, some disorders are multifactorial and do not have a biochemical marker, anatomic variant, or test for disease. In these cases, the phenotype may be defined using subjective assessments, through the use of diagnostic criteria, or by satisfying a predetermined number of symptoms. Either using the diagnostic criteria or establishing some diagnostic threshold (i.e., a "diseased" individual has x of a possible y underlying symptoms) can lead to imprecision and heterogeneity in the phenotype (Bearden and Freimer, 2006). As a way to correct this heterogeneity, one might focus solely on those satisfying certain criteria on the list, thereby minimizing heterogeneity and obtaining a more consistent disease definition. However, this approach again may affect sample size and thus power, so it is important also to consider the hypothesis of interest in this choice. In other words, the exploratory hypothesis can be used to help focus and limit the definition of the phenotype.

D. Endophenotype

The concept of an endophenotype has its roots in the field of psychiatry, and is defined as an internal phenotype that is heritable, is measurable, and lies as an intermediate between the disease and the gene (Gottesman and Gould, 2003; Gottesman and Shields, 1973). An endophenotype is a special case of a phenotype where the endophenotype is thought to have a closer relationship to the biological processes involved in the culmination of the disease process. The assumption and rationale for using an endophenotype is the thought that the number of genes affecting endophenotypic variation is smaller than the number of genes affecting the full disease phenotypic variation, and that the gene affecting endophenotypes has a larger effect size, on average. Thus, an endophenotype represents a more elementary phenomenon (Bearden and Freimer, 2006; Hasler *et al.*, 2006). Endophenotypes have also been called "intermediate phenotypes," "biological markers," "subclinical traits," and "vulnerability markers," all of which may not necessarily reflect a genetic predisposition to disease but may rather reflect associated findings (Bearden and Freimer, 2006; Gottesman and Gould, 2003). A distinction can be drawn between the term biological marker and endophenotype, in that biological markers signify differences that may not have an underlying genetic etiology whereas endophenotype should be used only when heritability has been established (Gottesman and Gould, 2003).

Ideally, an endophenotype would attempt to have a monogenic etiology, thus only affected by one underlying gene; however, the reality is that most endophenotypes are not refined and still are affected by many genes (Gottesman and Gould, 2003). Further, the fact that an endophenotype is strongly correlated with a disease phenotype does not ensure that the endophenotype is strongly correlated at the genetic level (Schulze and McMahon, 2004). The hypothesized relationship of an endophenotype to the biological process thought to underlie the disease of interest must be able to be tested (Bearden and Freimer, 2006). Thus, simply because one can quantitatively measure an endophenotype does not ensure that the endophenotype is an actual improvement over using the phenotype with which it is hypothesized to be associated (Bearden and Freimer, 2006).

In genetic research, the use of endophenotypes does offer a fruitful alternative or complement to studies of complex disease phenotypes (Bearden and Freimer, 2006) if the endophenotype displays certain criteria. These criteria are that the endophenotype is associated with disease in the population, the endophenotype is moderately to highly heritable, the presence of the endophenotype does not depend on symptoms of disease, the endophenotype is known to cosegregate with disease within a family, the prevalence of the endophenotype is higher in nonaffected family members than in the general population, and the endophenotype can be objectively and reliably measured (Bearden and Freimer, 2006; Gottesman and Gould, 2003). Further, for this approach to be an improvement over current methods, coherent sets of multiple traits must be defined and measured in many individuals in families (Bearden and Freimer, 2006).

E. Composite phenotype

Often we are required to select the individuals from large data sets with a multivariate phenotype such that the subsample is best representative of the sample and preserves the correlation structure among the multivariate phenotype. For example, an investigator has obesity-related multivariate phenotypic data consisting of body mass index (BMI), skinfolds (subscapular, suprailiac, thigh, triceps), waist circumference, percent body fat on large pedigrees for genetic study; however, he/she does not have resources to genotype all individuals. The question is often asked which phenotype or combination of phenotypes should be used in determining the subsample. In most situations, investigators use a primary phenotype of interest to create a subsample by using an extreme sampling design (i.e., selecting only individuals from both extremes of the distribution using a fixed threshold). This sampling design does provide optimal power for the primary phenotype, but not necessarily for other correlated phenotypes, which may also be of interest. Here, we propose a potential selection strategy using all correlated phenotypes and incorporating the correlation structure in the pedigrees. For example, in seeking genes for a trait, Y_i, our aim is to derive a composite score (phenotype), Z, for an individual, when multiple measurements are available on that individual and/or their relatives so that the new composite phenotype will be a better index on which to select subsets of people for study and/or to use as a phenotype in analysis. Here we define "better" as a situation in which selecting on and/or analyzing Z as opposed to Y_i leads to greater power to detect genes influencing Y_i than does selecting on and/or analyzing Y_i itself. There are a number of ways one can generate a composite score. For example, one can obtain a univariate value from multivariate data by calculating a breeding value as it is commonly used by the plant research community, by defining a propensity score in missing data situations, or by employing data reduction technique(s) used in multivariate analyses such as principal components analysis, factor analysis, and so on. Here, we provide a short description of a potentially useful measure to estimate an individual value based on multivariate data following Franklin and Allison (1992).

Consider k correlated phenotypes Y_i measured on individuals in a sample. Then the minimum variance estimator of combined phenotypes Y_i for the jth individual is given by

$$\hat{Y}_j = \frac{\sum_{i=1}^{k} W_i \hat{Y}_{ij}}{\sum_{i=1}^{k} W_i}$$

where k = the number of observed variables for individual j, Y_{ij} = the standardized ith variable for the jth individual, W_j = the weight for the ith variable.

Without loss of generality, we can assume Y_1 is the primary trait. Let ρ_{1i} be the genetic correlation between ith secondary variable and primary variable Y_1 (e.g., BMI can be considered as a primary variable and percent body fat or skinfold measurements' can be considered as secondary variables). Then following Franklin and Allison (1992), we can define $W_1 = 1/\sqrt{1 - h_1^2}$ and the weight for the correlated secondary nth variable is defined as $W_n = 1/\sqrt{1 - \rho_{1n}^2}$. Note that the weights are a function of the heritability of the primary trait and/or genetic correlations between primary and secondary traits will allow us to capture the genetic variance of each trait in the composite score. We can further refine the combined score by using a variance shrinkage method, and the new updated combined score is given by $Y_j = \hat{Y}_j\sqrt{1 - SE^2_{\hat{Y}_j}}$. Note that this score can easily accommodate missing data and also preserves the correlation between primary and secondary traits.

Another possible way to select the subsample is to use the path-analytic method, which is similar to the method described above and follows the path diagram given in Fig. 4.1.

Let $Y_1, Y_2, \ldots, Y_k$ be correlated traits where Y_1 is a primary trait and Y_i is a secondary traits for $i = 2, \ldots, k$, as above. Assume G_i is a major locus governing the trait locus Y_i, $i = 1, \ldots, k$. Let h_1^2 be the heritability of primary trait, h_i^2 the heritability of secondary trait for $i = 2, \ldots, k$, r_{1i} the pearson correlation between primary trait, Y_1, and secondary trait, Y_i, for $i = 2, \ldots, k$, and ρ_{1i} the genetic correlation between primary trait, X_1, and secondary trait, X_j, for $j = 2, \ldots, k$.

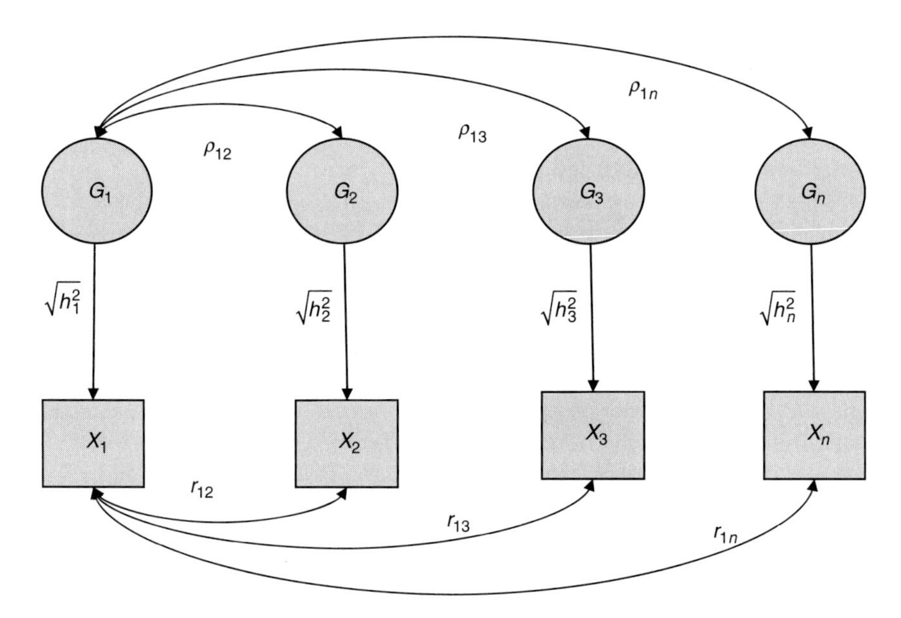

Figure 4.1. Path diagram describing the relationships among correlated traits and genetic loci.

We can write a composite score based on the trait values of all traits as a weighted linear combination of the trait values and is given as:

$$\hat{Y}_j = \frac{\sum_{i=1}^{n} W_i Y_{ij}}{\sum_{i=1}^{n} W_i},$$

Where $W_1 = 1/\sqrt{1 - h_1^2}$ and $W_i = 1/\sqrt{1 - h_i^2 \rho_{1i}^2}$.

One of the advantages of using these methods is that they indirectly incorporate missing data on any variable. Also, they preserve the correlation among the variables.

Another important issue often encountered by investigators is whether they should recruit as many affected cases for case-control study as they can even though the affected cases recruited in this way may constitute a heterogeneous group or whether they should instead insist on a stricter case definition to maintain homogeneity. Chen and Lee (2005) give a simple rule to enable one to decide when one can combine two similar phenotypes to increase the sample size for the study to achieve higher power. Assume there are two types of affected cases. The allele frequencies of the first and second type are p_1 and p_2, respectively. Also, assume we have a control group with an allele frequency of p_0. The number of individuals are n_1 and n_2 corresponding to first and second type of affected cases and n_0 is the number of controls. Chen and Lee (2005) proposed a decision rule where

$$\text{if } \frac{p_2 - p_0}{p_1 - p_0} > \frac{n_1 + n_2}{n_2} \times \frac{\sqrt{\{1/(n_1 + n_2)\} + \{1/n_0\}}}{\sqrt{(1/n_1) + (1/n_0)}} - \frac{n_1}{n_2}$$

then both types of affected cases should be recruited; otherwise, only the first type should be recruited. The decision rule provides a higher statistical power.

However, this analysis assumes that the same gene is affecting both phenotypes. The values of the allele frequencies could be obtained from a literature survey or could be estimated in the pilot study. Chen and Lee (2005) give two examples to describe the decision rule. Suppose the primary goal of the study is to locate gene(s) for cardiovascular diseases. The investigator has the choice to recruit patients with myocardial infarction and/or patients with stroke as a case group. Note that both types of patients can be considered as having cardiovascular diseases. Should the researcher recruit both types of patients to increase the sample size? Or, for the sake of homogeneity, should the researcher recruit only the patients with either myocardial infarction or stroke? Suppose the investigator has 300 patients with myocardial infarction ($n_1 = 300$) and 300 patients with stroke ($n_2 = 300$), and 600 controls. Using a literature survey he/she finds the allele frequencies are $p_1 = 0.158$, $p_2 = 0.149$, and $p_0 = 0.095$ (Helgadottir *et al.*, 2004). In this situation, the above decision rule indicates that the investigator should combine both groups of patients. As a second example, suppose the goal of the study is to find gene(s) for prostate cancer, and it is known that the genetic contributions are larger in earlier-onset prostate cancer cases than in later-onset cases. Suppose the investigator has at his/her disposal, 20 early-onset prostate cancer cases, 80 later-onset cases, and 100 controls. The literature survey reveals $p_1 = 0.430$, $p_2 = 0.090$, and $p_0 = 0.017$ (Carter *et al.*, 1992; Gronberg *et al.*, 1997; Schaid *et al.*, 1998). In this situation, the investigator should not combine both types of patients because the value of left-hand side of the decision rule (0.177) is less than the right-hand side (0.472) and combining the data will result in loss of power.

F. Narrow versus broad phenotype definition

A discrete trait may also be used to narrow a phenotype. However, caution must be used in order to minimize loss of power due to small sample sizes, inappropriate choices of cutoffs used to obtain a discrete phenotype from the continuous phenotypes, and ineffective or inaccurate refinements of broader phenotypes based on a priori hypotheses (Schulze and McMahon, 2004; Winawer, 2006).

III. STUDY DESIGN ISSUES

A. Measurement error

The use of variable methods to assess and measure phenotypes contributes nontrivial sources of variability to any genetic study (Bearden and Freimer, 2006). Thus in order to confirm a phenotype and reduce the potential for misclassification, it is important to use standard phenotype definitions as well as secondary data sources (such as medical records) or consensus committee agreement for a valid diagnosis (Ellsworth and Manolio, 1999). A valid diagnosis/phenotype is more likely to correspond to an underlying biological entity than an invalid one (Schulze and McMahon, 2004). Validity is the concept of knowing what is being measured and how accurately it is being measured (Schulze and McMahon, 2004). Further, most phenotypes can be defined by more than one instrument, each instrument using slightly different measures for the same phenotype that adds variability in phenotype definition and makes cross-study comparisons difficult. Therefore, standardized, structured instruments for collecting clinical data used to define phenotypes should be employed. Additionally, the reliability (Schulze and McMahon, 2004) and validity of these instruments should be assessed.

The nontrivial nature of using valid, reliable, standardized instruments is demonstrated in Table 4.3 which examines the association of the serotonin transporter variable number of tandem repeats (*SLC6A4* VNTR) with major depressive disorder. In this table, the instruments used to classify the study population into cases and controls are defined, and while most use standardized instruments (SADS-L, SCAN interview, and DSM-IV criteria), there are still problems with these populations and studies. First, most did not screen the controls for psychiatric disorders; second, those using in- or outpatients used clinical diagnoses, not necessarily DSM-IV criteria; and third, the major depression diagnosis used for cases ranges from major depressive episode to recurrent major depression to lifetime major depressive disorder. As this table shows, all this variability between the studies leaves one with a murky picture of the true association, if any, between the serotonin transporter polymorphism and major depressive disorder, even when some studies are using standardized instruments. Clearly there would be significant advantages if investigators used known valid, standardized, reliable instruments across all studies of a disorder, though there would of course be serious logistical difficulties in enforcing such a standard.

Recently, Wong *et al.* (2003) investigated the effect of measurement error on power of the gene–environment interaction study for continuous traits. They used the classical measurement error model to describe the uncertainty in measurements in the outcome and the exposure. They found that the magnitude of the impact of measurement precision on the power to detect gene–environment interaction on continuous traits would suggest that smaller studies with better measurements may be preferable to very large studies with less precise measurement. The greater impact on power came not from the genotyping errors but from the precision with which the exposure and outcome were estimated. In their Table 4.3, they showed that the studies with poor measurements of exposure and outcome variables (i.e., ρ_{Ty} = correlation between observed and true value of the outcome = 0.3; ρ_{Ty} = correlation between observed and true value of the exposure = 0.4), a sample size of 926,208 individuals will be required to detect a gene–environment interaction that was detectable in 1836 individuals with reasonably accurate measurement ($\rho_{Ty} = 0.9$ and $\rho_{Ty} = 0.8$), assuming the genotyping error is 2.5%.

B. Misclassification

In order to reduce the number of false-positive results or loss of power, it is necessary to minimize misclassification of the phenotype by using a narrow, precise phenotype definition (Bickeboller *et al.*, 2005). Misclassification due to poorly defined phenotypes can make it impossible to detect a true association or to identify spurious associations stemming from biased classification of the phenotype or confounding of the phenotype (Ellsworth and Manolio, 1999).

Table 4.3. Major Depressive Disorder and Serotonin Transporter (SLC6A4 VNTR) Association Studies

References	Population	Sample size	Ethnicity	Results
Ogilvie *et al.* (1996)	Major depression in- and out-patients	39 cases	English and Scottish	Significant genotype distribution difference between MDE patients and controls
	Cases: DSM IV Controls: SADS-L (if screened)	123 unscreened controls 71 screened controls		Patients had higher frequency of being homozygous or heterozygous for nine repeat allele
Rees *et al.* (1997)	Major depression in- and out-patients SCAN interview (DSM IV) Controls not screened	80 lifetime cases 121 controls	United Kingdom	Genotype and allele frequencies not different; no association between polymorphism and major depression
Gutierrez *et al.* (1998)	Out-patients	74 cases (personal interview) 84 controls	Spanish	No association between polymorphism and major depression
Furlong *et al.* (1998)	In- and out-patients SADS-L	125 cases 174 controls	United Kingdom	No association between polymorphism and major depression
Syagailo *et al.* (2001)	DSM-IV	74 recurrent MD cases 229 controls	German	No association between polymorphism and major depression
Schulze *et al.* (2000)	In-patients	146 cases (42 MDE, 104 recurrent) 101 controls	German	Significant genotype frequency difference among women with recurrent MDD

Further, heterogeneity of both the phenotype and its underlying component causes can also lead to misclassification of the phenotype, which needs to be addressed (Ellsworth and Manolio, 1999). Rice *et al.* (2000, 2001) examined the impact of misclassification (diagnostic) error on the heritability and risk ratio of a dichotomous trait. They have shown that there is a minor impact of reduced sensitivity (probability of correctly diagnosing a true case) on the observed heritability and risk ratio when the specificity (probability of correctly diagnosing a true control) is held fixed and sensitivity is reduced. In contrast, when sensitivity is fixed and specificity is reduced, this could have a dramatic effect in reducing the power compared to the situation of no diagnostic error. Edwards *et al.* (2005) further investigated the effect of phenotypic errors for case-control genetic association studies and provided the power and sample size calculations in the presence of phenotype errors. The phenotype misclassification errors can substantially decrease the power to detect an association between the trait and marker locus (Edwards *et al.*, 2005; Zheng and Tian, 2005). The freely available program, PAWE-PH, can calculate power or sample size in the presence of phenotype misclassification errors in order to investigate an association between trait and marker loci for case-control studies (Edwards *et al.*, 2005; Gordon *et al.*, 2002). Table 4.4a, which gives power calculations with and without phenotypic errors for various conditions, demonstrates different aspects of sample size and power assuming equal number of cases and controls. We assumed that the difference between minor allele frequency for cases and controls to be fixed at 0.1. In particular, we assumed 0.05 corresponds to allele frequency in cases and 0.15 corresponds to allele frequency in controls. We also assumed Hardy Weinberg Equilibrium (HWE) holds in both cases and controls. We used two different disease prevalence figures of 0.005 and 0.01 and two different sample sizes of 200 and 800. Errors were introduced by assuming the conditional probability of observing a control individual if the given individual is truly affected ($\theta = 0$, 0.01, and 0.05), and the conditional probability of observing a case individual if the given individual is truly unaffected ($\phi = 0$, 0.01, and 0.05). In Table 4.4a, when $\phi = 0$, the power is independent of the parameter θ and disease prevalence as also observed by Edwards *et al.* (2005). The loss of power is considerable as the value of ϕ increases for given θ and K. We also investigated the sample size required with phenotypic errors to achieve the power of 99.11% (corresponding to 200 cases and 200 controls without any phenotypic errors) with the same parameters as in Table 4.4a. Table 4.4b provides the required sample size to keep the same power achieved without any phenotype errors corresponding to 200 cases and 200 controls. Note that we need 31,474 cases and 31,474 controls to maintain the same power of 99.11%. However, the required increase in the sample size is not as severe for higher disease prevalence.

Table 4.4a. Power Calculation With and Without Phenotypic Error Assuming the Difference in True Minor Allele Frequency is 0.1 with Given Parameters and Fixed Significance Level of 0.05

Prevalence of the disease	*p* (observed control | true affected)	*p* (observed case | true unaffected)	# Cases = 200 # Controls = 200	# Cases = 800 # Controls = 800
0.005	0	0	99.11	100
		0.01	22.27	71.05
		0.05	6.06	9.41
	0.01	0	99.11	100
		0.01	21.99	70.39
		0.05	6.04	9.32
	0.05	0	99.12	100
		0.01	20.94	67.65
		0.05	5.95	8.97
0.01	0	0	99.11	100
		0.01	47.39	97.84
		0.05	8.77	78.30
	0.01	0	99.11	100
		0.01	46.94	97.71
		0.05	8.70	21.38
	0.05	0	99.11	100
		0.01	45.10	97.14
		0.05	8.42	20.08

Table 4.4b. Sample Size Needed to Achieve the Same Power Achieved by Without any Phenotypic Errors for Sample Size of 200 Cases and 200 Controls Assuming Difference in True Minor Allele Frequency is 0.1 With Given Parameters and Fixed Significance Level of 0.05

Prevalence of the disease	*p* (observed control | true affected)	*p* (observed case | true unaffected)	Sample size needed
0.005	0	0.01	2215
		0.05	31,474
	0.01	0.01	2247
		0.05	32,097
	0.05	0.01	2384
		0.05	34,797
0.01	0	0.01	938
		0.05	9134
	0.01	0.01	948
		0.05	9303
	0.05	0.01	993
		0.05	10,032

C. Loss due to dichotomization/categorization

Continuous phenotypes may also be split into different categories. An example of this is BMI that can be used either as the continuous phenotype or categorized into known categories based on the National Heart, Lung, and Blood Institute guidelines, for example, as high BMI versus low BMI. The validity and power of this type of extreme sampling approach was investigated by Alf and Abraham (1975). Both phenotype definitions are acceptable; however, there may be a loss of power when dichotomizing a continuous trait. When this choice presents itself to a researcher, the decision as to whether to use the continuous or dichotomized phenotype should be made based on the hypothesis or question that the researcher wants to answer. However, Alf and Abrahams (1975) investigated the use of extreme groups in assessing relationships between outcome and predictor values. Here, we compared three sampling and analysis strategies using simulations.

Strategy 1: Selecting subsample on the basis of extreme phenotypic values and correlating with genotypic values using regression analysis.
Strategy 2: Selecting a subsample on the basis of extreme phenotypic values and comparing allele frequency between extreme groups.
Strategy 3. Randomly selecting a subsample of the same size as the extreme sample in strategies 1 and 2.

We performed simulations to check the validity of the tests and to compare the power of the tests. We simulated 1000 individuals with R^2 of 0.0 corresponding to the null situation and locus-specific R^2 of 0.005 and 0.01 for nonnull situation. Note that the simulation procedure corresponds to the simplest situation where only locus-specific effects are modeled using the correlation between the trait and marker loci and the remaining variance is explained by the random environment effect. Five hundred individuals were selected with extreme phenotype belonging to the first and third quartiles for strategies 1 and 2. For strategy 3, we selected 500 individuals randomly from 1000 individuals. We assumed allele frequency of 0.2 for all simulations. Table 4.5 gives the type 1 error rates and the power corresponding to $h^2 = 0.005, 0.01$ at significance level of 0.05 and 0.01. Note that both strategies are valid as indicated by the type 1 error rates. However, the regression analysis using the continuous variable in the extreme sample gives more power compared with testing the allele frequency difference in two extreme groups.

In summary, categorical/threshold/dichotomous phenotypes such as hypertension (yes/no) or elevated triglycerides (yes/no) are useful; however, the use of these measures compared to their continuous counterparts, such as blood pressure and triglyceride level, reduces power and precision. Continuous phenotypes are able to improve the power of the study to detect an association

Table 4.5. Comparisons of Type I Error Rates and Powers for Extreme Group Analysis

	Type I error rate		Power for $R^2 = 0.005$		Power for $R^2 = 0.01$	
Type of data and analysis	$\alpha = 0.05$	$\alpha = 0.01$	$\alpha = 0.05$	$\alpha = 0.01$	$\alpha = 0.05$	$\alpha = 0.01$
Regression analysis on extreme groups	0.046	0.012	0.667	0.427	0.920	0.777
Test for difference in allele frequencies between extreme groups	0.053	0.010	0.642	0.411	0.909	0.757
Random 500 individuals of 1000	0.048	0.011	0.401	0.208	0.704	0.457

and add to the precision of the study. Further, there are methods that have now been developed and designed to deal efficiently with continuous traits (Ellsworth and Manolio, 1999). While there is a loss of power in dichotomizing a continuous phenotype, there may also be an advantage to this strategy. The advantage is demonstrated when only a subset of the continuous distribution is of interest, or when a subset of the continuous distribution better reflects the underlying pathology. When one of these situations occurs, the reasons and specific definition of the subset of the continuous phenotype should be explicitly stated a priori to any analysis or data collections in order to minimize bias. Furthermore, if this is addressed during the planning stages of the study, it would allow for additional subject recruitment, allowing for an appropriately powered design.

D. Effect of narrowing phenotype

Phenotype heterogeneity in genetic studies can lead to both false-positive and false-negative studies because the disorder under study is not pure. Thus, there are approaches aimed at minimizing phenotype heterogeneity. To minimize misclassification, it is best to use precise disease definitions if they are available. This may also help to facilitate comparisons across studies, as use of the same precise definition limits interstudy heterogeneity (Ellsworth and Manolio, 1999). However, precise phenotype definitions may not be available, and also may not always eliminate disease phenotype heterogeneity, but they may limit generalizability of study results and impede subject recruitment. Thus, there is a balancing act that must be undertaken when deciding on a phenotype definition to use for a genetic research study.

The issue of phenotype definition being a balancing act also can affect the sample size of a study. If, in an attempt to reduce the effects of genetic heterogeneity, a narrow phenotype definition is used, this will more than likely lead to a smaller sample size for the study; however, if a slightly broader phenotype definition is used, the effect of a smaller sample size may not be an issue (Winawer, 2006). Further, phenotypic heterogeneity may not necessarily imply genetic heterogeneity. But if genetic heterogeneity does underlie the phenotypic heterogeneity, then the use of a narrow phenotype definition can increase the power to detect linkage; however, if allele sharing does not differ between different phenotype definitions, power may be lost because a narrow phenotype definition effectively limits sample size (Winawer, 2006).

E. Multiple comparisons

Finally, even a well thought-out study using valid methods of phenotyping and reducing a multivariate phenotype to a few key variables can still suffer from chance findings due to multiple testing (Schulze and McMahon, 2004). For example, if there are 10 known polymorphisms for the gene of interest, which is associated with 10 traits in the study (i.e., 100 contrasts), 5 of these would be expected to be significant at $p < 0.05$ by chance alone. Thus, if more than five significant results are discovered, there is not a way to determine which result is a biological reality and which is a statistical artifact (Talmud and Humphries, 2001). One way to reduce multiple testing is to combine all the multivariate measures into one composite phenotype, thus using all possible phenotypic information in one measurement and having a low correlation between the newly defined phenotypes, if the phenotypes were defined using a principal components analysis (Bearden and Freimer, 2006). Another option is to set up a priori hypotheses of interest that are to be tested (based on known biological pathways and relationships or where the coded protein is centrally involved or the polymorphism causes a nonsynonymous amino acid change) and distinguish these analyses from secondary hypothesis generating analyses (Talmud and Humphries, 2001). Additionally, confirmation of any positive findings in a second independent study is the best way to reduce the possibility of spurious associations (Talmud and Humphries, 2001). Rice and Rao (2008) summarize standard methods of correcting for multiple comparisons (see Chapter by Rice *et al.*, this volume).

IV. DISCUSSION

As this chapter demonstrates, the choice of a phenotype is of critical importance. More clearly defined criteria for selection and definition are needed to better navigate the multitude of possible analytical paths from multivariate measures to univariate phenotypes (Bickeboller *et al.*, 2005). Further, simply having a homogeneous phenotype may oversimplify the issue because multiple genes and mutations may contribute to multiple measures in the same basic pathway (Schulze and McMahon, 2004).

To further advance our knowledge of disease processes and pathophysiological mechanisms, precise phenotype definition is important and necessary (Bel, 2004). However, defining the phenotype is a balancing act. If the phenotype is too narrowly defined, this will restrict sample sizes and may inadvertently emphasize atypical forms of illness attributable to nongenetic factors; conversely, if the phenotype is too broadly defined, the phenotype may encompass heterogeneous conditions and complicate gene discovery and mapping (Schulze *et al.*, 2006). Further, a phenotype definition should be in proximity to a gene product, should use an objective and reproducible measurement, should encompass a minimum number of underlying causes, and should demonstrate the strongest evidence of heritability (Ellsworth and Manolio, 1999).

Acknowledgments

H.K.T. is supported in part by NIH grants: R21LM008791, R01DK52431, P20RR01643, and R01HL055673. M.K.W. is supported by *UAB Statistical Genetics Post-Doctoral Training Program* T32HL072757. We would like to thank Amit Patki for performing simulations for power analysis in Section III.C. In addition, we would like to thank Drs. D. C. Rao (partly supported by GM 28719), Donna K. Arnett (partly supported by R01HL055673), and David B. Allison (partly supported by R01DK52431) for their useful suggestion and discussions.

References

Alam, S., Liyou, N., Davis, D., Tresillian, M., and Johnson, A. G. (2000). The 460Trp polymorphism of the human alpha-adducin gene is not associated with isolated systolic hypertension in elderly Australian Caucasians. *J. Hum. Hypertens* **14**(3), 199–203.

Alf, E. F., Jr., and Abrahams, N. M. (1975). The use of extreme groups in assessing relationships. *Psychometrika* **40**(4), 563–572.

Baier, L. J., Permana, P. A., Yang, X., Pratley, R. E., Hanson, R. L., Shen, G. Q., Mott, D., Knowler, W. C., Cox, N. J., Horikawa, Y., Oda, N., Bell, G. I., *et al.* (2000). A calpain-10 gene polymorphism is associated with reduced muscle mRNA levels and insulin resistance. *J. Clin. Invest.* **106,** R69–R73.

Bearden, C. E., and Freimer, N. B. (2006). Endophenotypes for psychiatric disorders: Ready for primetime? *Trends. Genet.* **22**(6), 306–313.

Beeks, E., van der Klauw, M. M., Kroon, A. A., Spiering, W., Fuss-Lejeune, M. J., and de Leeuw, P. W. (2004). Alpha-adducin Gly460Trp polymorphism and renal hemodynamics in essential hypertension. *Hypertension* **44**(4), 419–423.

Bel, E. H. (2004). Clinical phenotypes of asthma. *Curr. Opin. Pulm. Med.* **10**(1), 44–50.

Bickeboller, H., Bailey, J. N., Papanicolaou, G. J., Rosenberger, A., and Viel, K. R. (2005). Dissection of heterogeneous phenotypes for quantitative trait mapping. *Genet. Epidemiol.* **29**(Suppl 1), 41–47.

Bray, M. S., Li, L., Turner, S. T., Kardia, S. L., and Boerwinkle, E. (2000). Association and linkage analysis of the alpha-adducin gene and blood pressure. *Am. J. Hypertens* **13**(6 Pt 1), 699–703.

Carter, B. S., Beaty, T. H., Steinberg, G. D., Childs, B., and Walsh, P. C. (1992). Mendelian inheritance of familial prostate cancer. *Proc. Natl. Acad. Sci. USA* **89**(8), 3367–3371.

Cassell, P. G., Jackson, A. E., North, B. V., Evans, J. C., Syndercombe, C., Phillips, C., Ramachandran, A., Snehalatha, C., Gelding, S. V., Vijayaravaghan, S., Curtis, D., and Hitman, G. A. (2002). Haplotype combinations of calpain 10 gene polymorphisms associate with increased risk of impaired glucose tolerance and type 2 diabetes in South Indians. *Diabetes* **51**, 1622–1628.

Castejon, A. M., Alfieri, A. B., Hoffmann, I. S., Rathinavelu, A., and Cubeddu, L. X. (2003). Alpha-adducin polymorphism, salt sensitivity, nitric oxide excretion, and cardiovascular risk factors in normotensive Hispanics. *Am. J. Hypertens* **16**(12), 1018–1024.

Chen, C. F., and Lee, W. C. (2005). Case recruitment in genetic association studies: Larger sample size or greater homogeneity? *Int. J. Epidemiol.* **34**(3), 711.

Ciechanowicz, A., Widecka, K., Drozd, R., Adler, G., Cyrylowski, L., and Czekalski, S. (2001). Lack of association between Gly460Trp polymorphism of alpha-adducin gene and salt sensitivity of blood pressure in Polish hypertensives. *Kidney Blood Press. Res.* **24**(3), 201–206.

Clark, C. J., Davies, E., Anderson, N. H., Farmer, R., Friel, E. C., Fraser, R., and Connell, J. M. (2000). Alpha-adducin and angiotensin I-converting enzyme polymorphisms in essential hypertension. *Hypertension* **36**(6), 990–994.

Cusi, D., Barlassina, C., Azzani, T., Casari, G., Citterio, L., Devoto, M., Glorioso, N., Lanzani, C., Manunta, P., Righetti, M., Rivera, R., Stella, P., *et al.* (1997). Polymorphisms of alpha-adducin and salt sensitivity in patients with essential hypertension. *Lancet* **349**(9062), 1353–1357.

Daimon, M., Oizumi, T., Saitoh, T., Kameda, W., Yamaguchi, H., Ohnuma, H., Igarashi, M., Manaka, H., and Kato, T. (2002). Calpain 10 gene polymorphisms are related, not to type 2 diabetes, but to increased serum cholesterol in Japanese. *Diabetes Res. Clin. Pract.* **56**, 147–152.

Edwards, B. J., Haynes, C., Levenstien, M. A., Finch, S. J., and Gordon, D. (2005). Power and sample size calculations in the presence of phenotype errors for case/control genetic association studies. *BMC Genet.* **6**(1), 18.

Elbein, S. C., Chu, W., Ren, Q., Hemphill, C., Schay, J., Cox, N. J., Hanis, C. L., and Hasstedt, S. J. (2002). Role of calpain-10 gene variants in familial type 2 diabetes in Caucasians. *J. Clin. Endocrinol. Metab.* **87**, 650–654.

Ellsworth, D. L., and Manolio, T. A. (1999). The emerging importance of genetics in epidemiologic research II. Issues in study design and gene mapping. *Ann. Epidemiol.* **9**(2), 75–90.

Evans, J. C., Frayling, T. M., Cassell, P. G., Saker, P. J., Hitman, G. A., Walker, M., Levy, J. C., O'Rahilly, S., Rao, P. V., Bennett, A. J., Jones, E. C., Menzel, S., *et al.* (2001). Studies of association between the gene for calpain-10 and type 2 diabetes mellitus in the United Kingdom. *Am. J. Hum. Genet.* **69**, 544–552.

Fingerlin, T. E., Erdos, M. R., Watanabe, R. M., Wiles, K. R., Stringham, H. M., Mohlke, K. L., Silander, K., Valle, T. T., Buchanan, T. A., Tuomilehto, J., Bergman, R. N., Boehnke, M., *et al.* (2002). Variation in three single nucleotide polymorphisms in the calpain-10 gene not associated with type 2 diabetes in a large Finnish cohort. *Diabetes* **51**, 1644–1648.

Franklin, R. D., and Allison, D. B. (1992). The Rev: An IBM BASIC program for Bayesian test interpretation. *Behav. Res. Methods Instrum. Comput.* **24**, 491–492.

Furlong, R. A., Ho, L., Walsh, C., Rubinsztein, J. S., Jain, S., Paykel, E. S., Easton, D. F., and Rubinsztein, D. C. (1998). Analysis and meta-analysis of two serotonin transporter gene polymorphisms in bipolar and unipolar affective disorders. *Am. J. Med. Genet.* **81**(1), 58–63.

Garant, M. J., Kao, W. H., Brancati, F., Coresh, J., Rami, T. M., Hanis, C. L., Boerwinkle, E., and Shuldiner, A. R. (2002). SNP43 of CAPN10 and the risk of type 2 diabetes in African–Americans: The atherosclerosis risk in Communities Study. *Diabetes* **51**, 231–237.

Gordon, D., Finch, S. J., Nothnagel, M., and Ott, J. (2002). Power and sample size calculations for case-control genetic association tests when errors present: Application to single nucleotide polymorphisms. *Hum. Hered.* **54,** 22–33.

Gottesman, I. I., and Gould, T. D. (2003). The endophenotype concept in psychiatry: Etymology and strategic intentions. *Am. J. Psychiatry* **160**(4), 636–645.

Gottesman, I. I., and Shields, J. (1973). Genetic theorizing and schizophrenia. *Br. J. Psychiatry* **122**(566), 15–30.

Gronberg, H., Damber, L., Damber, J. E., and Iselius, L. (1997). Segregation analysis of prostate cancer in Sweden: Support for dominant inheritance. *Am. J. Epidemiol.* **146**(7), 552–557.

Gutierrez, B., Pintor, L., Gasto, C., Rosa, A., Bertranpetit, J., Vieta, E., and Fananas, L. (1998). Variability in the serotonin transporter gene and increased risk for major depression with melancholia. *Hum. Genet.* **103**(3), 319–322.

Hasler, G., Drevets, W. C., Gould, T. D., Gottesman, I. I., and Manji, H. K. (2006). Toward constructing an endophenotype strategy for bipolar disorders. *Biol. Psychiatry* **60**(2), 93–105.

He, X., Zhu, D. L., Chu, S. L., Jin, L., Xiong, M. M., Wang, G. L., Zhang, W. Z., Zhou, H. F., Mao, S. Y., Zhan, Y. M., Zhuang, Q. N., Liu, X. M., *et al.* (2001). Alpha-Adducin gene and essential hypertension in China. *Clin. Exp. Hypertens* **23**(7), 579–589.

Hegele, R. A., Harris, S. B., Zinman, B., Hanley, A. J., and Cao, H. (2001). Absence of association of type 2 diabetes with CAPN10 and PC-1 polymorphisms in Oji-Cree. *Diabetes Care* **24,** 1498–1499.

Helgadottir, A., Manolescu, A., Thorleifsson, G., Gretarsdottir, S., Jonsdottir, H., Thorsteinsdottir, U., Samani, N. J., Gudmundsson, G., Grant, S. F., and Thorgeirsson, G. (2004). The gene encoding 5-lipoxygenase activating protein confers risk of myocardial infarction and stroke. *Nat. Genet. Mar.* **36**(3), 233–239.

Horikawa, Y., Oda, N., Cox, N. J., Li, X., Orho-Melander, M., Hara, M., Hinokio, Y., Lindner, T. H., Mashima, H., Schwarz, P. E., Bosque-Plata, L., Horikawa, Y., *et al.* (2000). Genetic variation in the gene encoding calpain-10 is associated with type 2 diabetes mellitus. *Nat. Genet.* **26**(2), 163–175.

Horikawa, Y., Oda, N., Yu, L., Imamura, S., Fujiwara, K., Makino, M., Seino, Y., Itoh, M., and Takeda, J. (2003). Genetic variations in calpain-10 gene are not a major factor in the occurrence of type 2 diabetes in Japanese. *J. Clin. Endocrinol. Metab.* **88,** 244–247.

Iwasaki, N., Horikawa, Y., Tsuchiya, T., Kitamura, Y., Nakamura, T., Tanizawa, Y., Oka, Y., Hara, K., Kadowaki, T., Awata, T., Honda, M., Yamashita, K., *et al.* (2005). Genetic variants in the calpain-10 gene and the development of type 2 diabetes in the Japanese population. *J. Hum. Genet.* **50,** 92–98.

Ju, Z., Zhang, H., Sun, K., Song, Y., Lu, H., Hui, R., and Huang, X. (2003). Alpha-adducin gene polymorphism is associated with essential hypertension in Chinese: A case-control and family-based study. *J. Hypertens* **21**(10), 1861–1868.

Larson, N., Hutchinson, R., and Boerwinkle, E. (2000). Lack of association of 3 functional gene variants with hypertension in African Americans. *Hypertension* **35**(6), 1297–1300.

Lynn, S., Evans, J. C., White, C., Frayling, T. M., Hattersley, A. T., Turnbull, D. M., Horikawa, Y., Cox, N. J., Bell, G. I., and Walker, M. (2002). Variation in the calpain-10 gene affects blood glucose levels in the British population. *Diabetes* **51**(1), 247–250.

Malecki, M. T., Moczulski, D. K., Klupa, T., Wanic, K., Cyganek, K., Frey, J., and Sieradzki, J. (2002). Homozygous combination of calpain 10 gene haplotypes is associated with type 2 diabetes mellitus in a Polish population. *Eur. J. Endocrinol.* **146,** 695–699.

Matayoshi, T., Kamide, K., Takiuchi, S., Yoshii, M., Miwa, Y., Takami, Y., Tanaka, C., Banno, M., Horio, T., Nakamura, S., Nakahama, H., Yoshihara, F., *et al.* (2004). The thiazide-sensitive Na (+)-Cl(–) cotransporter gene, C1784T, and adrenergic receptor-beta3 gene, T727C, may be gene polymorphisms susceptible to the antihypertensive effect of thiazide diuretics. *Hypertens Res.* **27** (11), 821–833.

Melander, O., Bengtsson, K., Orho-Melander, M., Lindblad, U., Forsblom, C., Rastam, L., Groop, L., and Hulthen, U. L. (2000). Role of the Gly460Trp polymorphism of the alpha-adducin gene in primary hypertension in Scandinavians. *J. Hum. Hypertens* **14**(1), 43–46.

Mulatero, P., Williams, T. A., Milan, A., Paglieri, C., Rabbia, F., Fallo, F., and Veglio, F. (2002). Blood pressure in patients with primary aldosteronism is influenced by bradykinin B(2) receptor and alpha-adducin gene polymorphisms. *J. Clin. Endocrinol. Metab.* **87**(7), 3337–3343.

Mulatero, P., Veglio, F., Maffei, P., Bondanelli, M., Bovio, S., Daffara, F., Leotta, G., Angeli, A., Calvo, C., Martini, C., *et al.* (2006). CYP11B2–344T/C gene polymorphism and blood pressure in patients with acromegaly. *J. Clin. Endocrinol. Metab.* **91**(12), 5008–5012.

Newton-Cheh, C., and Hirschhorn, J. N. (2005). Genetic association studies of complex traits: Design and analysis issues. *Mutat. Res.* **573**(1–2), 54–69.

Ogilvie, A. D., Battersby, S., Bubb, V. J., Fink, G., Harmar, A. J., Goodwim, G. M., and Smith, C. A. (1996). Polymorphism in serotonin transporter gene associated with susceptibility to major depression. *Lancet* **347**(9003), 731–733.

Orho-Melander, M., Klannemark, M., Svensson, M. K., Ridderstrale, M., Lindgren, C. M., and Groop, L. (2002). Variants in the calpain-10 gene predispose to insulin resistance and elevated free fatty acid levels. *Diabetes* **51**, 2658–2664.

Page, G. P., George, V., Go, R. C., Page, P. Z., and Allison, D. B. (2003). "Are we there yet?": Deciding when one has demonstrated specific genetic causation in complex diseases and quantitative traits. *Am. J. Hum. Genet.* **73**(4), 711–719.

Province, M. A., Arnett, D. K., Hunt, S. C., Leiendecker-Foster, C., Eckfeldt, J. H., Oberman, A., Ellison, R. C., Heiss, G., Mockrin, S. C., and Williams, R. R. (2000). Association between the alpha-adducin gene and hypertension in the HyperGEN Study. *Am. J. Hypertens* **13**(6 Pt 1), 710–718.

Psaty, B. M., Smith, N. L., Heckbert, S. R., Vos, H. L., Lemaitre, R. N., Reiner, A. P., Siscovick, D. S., Bis, J., Lumley, T., Longstreth, W. T., Jr., and Rosendaal, F. R. (2002). Diuretic therapy, the alpha-adducin gene variant, and the risk of myocardial infarction or stroke in persons with treated hypertension. *JAMA* **287**(13), 1680–1689.

Ranade, K., Hsuing, A. C., Wu, K. D., Chang, M. S., Chen, Y. T., Hebert, J., Chen, Y. I., Olshen, R., Curb, D., Dzau, V., Botstein, D., Cox, D., *et al.* (2000). Lack of evidence for an association between alpha-adducin and blood pressure regulation in Asian populations. *Am. J. Hypertens* **13** (6 Pt 1), 704–709.

Rasmussen, S. K., Urhammer, S. A., Berglund, L., Jensen, J. N., Hansen, L., Echwald, S. M., Borch-Johnsen, K., Horikawa, Y., Mashima, H., Lithell, H., Cox, N. J., Hansen, T., *et al.* (2002). Variants within the calpain-10 gene on chromosome 2q37 (NIDDM1) and relationships to type 2 diabetes, insulin resistance, and impaired acute insulin secretion among Scandinavian Caucasians. *Diabetes* **51**, 3561–3567.

Rees, M., Norton, N., Jones, I., McCandless, F., Scourfield, J., Holmans, P., Moorhead, S., Feldman, E., Sadler, S., Cole, T., Redman, K., Earmer, A., *et al.* (1997). Association studies of bipolar disorder at the human serotonin transporter gene (hSERT; 5HTT). *Mol. Psychiatry* **2**(5), 398–402.

Rice, J. P., Saccone, N. L., and Suarez, B. K. (2000). The design of studies for investigating linkage and association. *In* "Analysis of Multifactorial Diseases" (T. Bishop and P. Sham, eds.), pp. 37–62. Academic Press, Oxford.

Rice, J. P., Saccone, N. L., and Rasmussen, E. (2001). Definition of the phenotype. *Adv. Genet.* **42**, 69–76.

Rice, T. K., and Rao, D. C. (2008). Methods for handling multiple testing. *In* "Genetic Dissection of Complex Traits" (D. C. Rao and C. C. Gu, eds.), 2nd Edition. 295–310. Elsevier.

Schaid, D. J., McDonnell, S. K., Blute, M. L., and Thibodeau, S. N. (1998). Evidence for autosomal dominant inheritance of prostate cancer. *Am. J. Hum. Genet.* **62**(6), 1425–1438.

Schelleman, H., Stricker, B. H., de Boer, A., Kroon, A. A., Verschuren, M. W., Van Duijn, C. M., Psaty, B. M., and Klungel, O. H. (2004). Drug-gene interactions between genetic polymorphisms and antihypertensive therapy. *Drugs* **64**(16), 1801–1816.

Schelleman, H., Stricker, B. H., Verschuren, W. M., de Boer, A., Kroon, A. A., de Leeuw, P. W., Kromhout, D., and Klungel, O. H. (2006). Interactions between five candidate genes and antihypertensive drug therapy on blood pressure. *Pharmacogenomics. J.* **6**(1), 22–26.

Schork, N. J., Chakravarti, A., Thiel, B., Fornage, M., Jacob, H. J., Cai, R., Rotimi, C. N., Cooper, R. S., and Weder, A. B. (2000). Lack of association between a biallelic polymorphism in the adducin gene and blood pressure in whites and African Americans. *Am. J. Hypertens* **13**(6 Pt 1), 693–698.

Schulze, T. G., and McMahon, F. J. (2004). Defining the phenotype in human genetic studies: Forward genetics and reverse phenotyping. *Hum. Hered.* **58**(3–4), 131–138.

Schulze, T. G., Hedeker, D., Zandi, P., Rietschel, M., and McMahon, F. J. (2006). What is familial about familial bipolar disorder? Resemblance among relatives across a broad spectrum of phenotypic characteristics. *Arch. Gen. Psychiatry* **63**(12), 1368–1376.

Schulze, T. G., Muller, D. J., Krauss, H., Scherk, H., Ohlraun, S., Syagailo, Y. V., Windemuth, C., Neidt, H., Grassle, M., Papassotiropoulos, A., Heun, R., Nothen, M. M., *et al.* (2000). Association between a functional polymorphism in the monoamine oxidase a gene promoter and major depressive disorder. *Am. J. Med. Genet.* **96**(6), 801–803.

Sciarrone, M. T., Stella, P., Barlassina, C., Manunta, P., Lanzani, C., Bianchi, G., and Cusi, D. (2003). ACE and alpha-adducin polymorphism as markers of individual response to diuretic therapy. *Hypertension* **41**(3), 398–403.

Shin, M. H., Chung, E. K., Kim, H. N., Park, K. S., Nam, H. S., Kweon, S. S., and Choi, J. S. (2004). Alpha-adducin Gly460Trp polymorphism and essential hypertension in Korea. *J. Korean Med. Sci.* **19**(6), 812–814.

Shioji, K., Kokubo, Y., Mannami, T., Inamoto, N., Morisaki, H., Mino, Y., Tagoi, N., Yasui, N., and Iwaii, N. (2004). Association between hypertension and the alpha-adducin, beta1-adrenoreceptor, and G-protein beta3 subunit genes in the Japanese population; the Suita study. *Hypertens Res.* **27** (1), 31–37.

Song, Y., Niu, T., Manson, J. E., Kwiatkowski, D. J., and Liu, S. (2004). Are variants in the CAPN10 gene related to risk of type 2 diabetes? A quantitative assessment of population and family-based association studies. *Am. J. Hum. Genet.* **74**(2), 208–222.

Staessen, J. A., Wang, J. G., Brand, E., Barlassina, C., Birkenhager, W. H., Herrmann, S. M., Fagard, R., Tizzoni, L., and Bianchi, G. (2001). Effects of three candidate genes on prevalence and incidence of hypertension in a Caucasian population. *J. Hypertens* **19**(8), 1349–1358.

Sugimoto, K., Hozawa, A., Katsuya, T., Matsubara, M., Ohkubo, T., Tsuji, I., Motone, M., Higaki, J., Hisamachi, S., Imai, Y., and Ogihara, T. (2002). Alpha-adducin Gly460Trp polymorphism is associated with low renin hypertension in younger subjects in the Ohasama study. *J. Hypertens* **20** (9), 1779–1784.

Sun, H. X., Zhang, K. X., Du, W. N., Shi, J. X., Jiang, Z. W., Sun, H., Zuo, J., Huang, W., Chen, Z., Shen, Y., Yao, Z. J., Qiang, B. Q., *et al.* (2002). Single nucleotide polymorphisms in CAPN10 gene of Chinese people and its correlation with type 2 diabetes mellitus in Han people of northern China. *Biomed. Environ. Sci.* **15,** 75–82.

Syagailo, Y. V., Stober, G., Grassle, M., Reimer, E., Knapp, M., Jungkunz, G., Okladnova, O., Meyer, J., and Lesch, K. P. (2001). Association analysis of the functional monoamine oxidase A gene promoter polymorphism in psychiatric disorders. *Am. J. Med. Genet.* **105**(2), 168–171.

Talmud, P. J., and Humphries, S. E. (2001). Genetic polymorphisms, lipoproteins and coronary artery disease risk. *Curr. Opin. Lipidol.* **12**(4), 405–409.

Tsai, H. J., Sun, G., Weeks, D. E., Kaushal, R., Wolujewicz, M., McGarvey, S. T., Tufa, J., Viali, S., and Deka, R. (2001). Type 2 diabetes and three calpain-10 gene polymorphisms in Samoans: No evidence of association. *Am. J. Hum. Genet.* **69,** 1236–1244.

Tsuchiya, T., Schwarz, P. E., Bosque-Plata, L. D., Geoffrey, Hayes M., Dina, C., Froguel, P., Wayne Towers, G., Fischer, S., Temelkova-Kurktschiev, T., Rietzsch, H., Graessler, J., Vcelak, J., *et al.* (2006). Association of the calpain-10 gene with type 2 diabetes in Europeans: Results of pooled and meta-analyses. *Mol. Genet. Metab.* **89**(1–2), 174–184.

Turner, S. T., Chapman, A. B., Schwartz, G. L., and Boerwinkle, E. (2003). Effects of endothelial nitric oxide synthase, alpha-adducin, and other candidate gene polymorphisms on blood pressure response to hydrochlorothiazide. *Am. J. Hypertens* **16**(10), 834–839.

Verkerk, A. J., Cath, D. C., van der Linde, H. C., Both, J., Heutink, P., Breedveld, G., Aulchenko, Y. S., and Oostra, B. A. (2006). Genetic and clinical analysis of a large Dutch Gilles de la Tourette family. *Mol. Psychiatry* **11**(10), 954–964.

Wang, J. G., Staessen, J. A., Barlassina, C., Fagard, R., Kuznetsova, T., Struijker-Boudier, H. A., Zagato, L., Citterio, L., Messaggio, E., and Bianchi, G. (2002a). Association between hypertension and variation in the alpha- and beta-adducin genes in a white population. *Kidney Int.* **62**(6), 2152–2159.

Wang, Y., Xiang, K., Zheng, T., Jia, W., Shen, K., and Li, J. (2002b). [The UCSNP44 variation of calpain 10 gene on NIDDM1 locus and its impact on plasma glucose levels in type 2 diabetic patients]. *Zhonghua Yi Xue Za Zhi* **82,** 613–616.

Weedon, M. N., Schwarz, P. E., Horikawa, Y., Iwasaki, N., Illig, T., Holle, R., Rathmann, W., Selisko, T., Schulze, J., Owen, K. R., Evans, J., Bosque-Plata, L., *et al.* (2003). Meta-analysis and a large association study confirm a role for calpain-10 variation in type 2 diabetes susceptibility. *Am. J. Hum. Genet.* **73**(5), 1208–1212.

Winawer, M. R. (2006). Phenotype definition in epilepsy. *Epilepsy Behav.* **8**(3), 462–476.

Wong, M. Y., Day, N. E., Luan, J. A., Chan, K. P., and Wareham, N. J. (2003). The detection of gene-environment interaction for continuous traits: Should we deal with measurement error by bigger studies or better measurement? *Int. J. Epidemiol.* **32**(1), 51–57.

Yazdanpanah, M., Sayed-Tabatabaei, F. A., Hofman, A., Aulchenko, Y. S., Oostra, B. A., Stricker, B. H., Pols, H. A., Lamberts, S. W., Witteman, J. C., Janssen, J. A., and van Duijn, C. M. (2006). The alpha-adducin gene is associated with macrovascular complications and mortality in patients with type 2 diabetes. *Diabetes* **55**(10), 2922–2927.

Zheng, G., and Tian, X. (2005). The impact of diagnostic error on testing genetic association in case-control studies. *Stat. Med.* **24,** 869–882.

5

Genotyping Platforms for Mass-Throughput Genotyping with SNPs, Including Human Genome-Wide Scans

Karen Maresso and Ulrich Broeckel
Departments of Pediatrics, Medicine and Physiology
Children's Hospital of Wisconsin, Human and Molecular Genetics Center,
Medical College of Wisconsin, Milwaukee, Wisconsin 53226

Advances in Genetics, Vol. 60

0065-2660/08 $35.00
DOI: 10.1016/S0065-2660(07)00405-1

ABSTRACT

The completion of the Human Genome Project (HGP) in 2003 brought the scientific community one step closer to identifying the genes underlying common, polygenic diseases. Prior to this achievement, the goal of identifying the genetic factors responsible for diseases presenting substantial public health burdens was elusive. Although the theoretical foundation for disease association studies had been discussed before the completion of the HGP, obstacles remained at that time before such studies could be considered feasible. One of these obstacles was the identification and mapping of numerous polymorphisms that could be easily and inexpensively typed. However, this challenge was overcome with the sequencing of the human genome and the subsequent cataloging of single-nucleotide polymorphisms (SNPs). The challenge then became how to rapidly and cost-effectively assay a dense set of these SNPs in the large number of samples required for disease association studies of complex traits. This challenge has been recently met as well, with the commercial offering of mass-throughput oligonucleotide array-based genotyping platforms at affordable prices. These platforms have made genome-wide association scans a reality and bring us closer than ever to elucidating the genetic mechanisms of complex disease. Here, we discuss the need for mass-throughput genotyping and then review and evaluate various platforms now available to investigators wishing to undertake high-throughput genotyping projects with SNPs, particularly genome-wide association scans.

I. INTRODUCTION: EXPLOSIVE NEED FOR MASS-THROUGHPUT GENOTYPING

Beginning in the 1980s with the use of primarily diallelic restriction fragment length polymorphisms (RFLPs), whole-genome linkage analysis became a standard tool for the identification of genes responsible for Mendelian diseases. A number of disease genes were mapped during this early time, including those for Duchenne muscular dystrophy and cystic fibrosis (Monaco *et al.*, 1986; Riordan *et al.*, 1989). The discovery of microsatellites, or short tandem repeat polymorphisms (STRPs), later in the decade made linkage analysis a much higher-throughput technology. Microsatellites were much more informative than the previously used RFLPs and were much simpler to genotype, requiring only a polymerase chain reaction (PCR) step followed by fragment sizing on polyacrylamide gels. Consequently, whole-genome linkage analysis using STRPs became the preferred method and much success resulted from this gain in throughput, with the identification of hundreds of Mendelian disease genes during the 1990s. Building on this success, there was much optimism for the application of linkage methods to the common, multifactorial diseases that resulted in substantial public health burdens, such as cancer and cardiovascular disease. However, while linkage analysis could narrow the region containing disease genes down to megabase-size intervals, association testing using single-nucleotide polymorphisms (SNPs) would be subsequently required to fine-map such regions to ultimately target the causative allele. While association analysis was being used to fine-map linkage peaks, a paper published by Risch and Merikangas (1996) described that genome-wide association analysis could be more powerful than linkage methods for the identification of genes involved in common disease. However, the main obstacle to performing genome-wide association studies at that time was the lack of a dense set of uniform DNA markers spanning the entire genome that could be typed easily, cheaply, and rapidly. Until the technology existed to identify, accurately map and then genotype such large numbers of markers in great numbers of individuals, linkage analysis followed by fine-mapping remained the best option for uncovering complex disease genes. Nevertheless, while linkage analysis proved to be a valuable tool for the mapping of numerous genes in monogenic disorders during the 1980s and 1990s, its later application to multifactorial diseases did not meet with the same success. The small effect-size of alleles and the locus heterogeneity believed to be behind complex diseases resulted in less than significant linkage findings and results that could not be replicated. Only a handful of genes contributing to common diseases have been identified to date through linkage and subsequent positional cloning methods (Helgadottir *et al.*, 2004; Horikawa *et al.*, 2000; Hugot *et al.*, 2001; Ogura *et al.*, 2001).

The explosive need for mass-throughput genotyping the field has recently witnessed stems from technological breakthroughs occurring in the last few years that are associated with the Human Genome Project (HGP). The HGP demonstrated the abundance and uniformity of SNPs and laid the foundation for their subsequent systematic discovery and cataloging in various human populations that occurred with such groups as the SNP Consortium (TSC) and the International HapMap Project. The HGP worked to develop technologies for the rapid, large-scale identification and scoring of SNPs, to identify SNPs in the coding regions of a majority of the approximately 30,000 genes annotated and to create the intellectual resources needed to study sequence variation. Groups like TSC furthered the work of the HGP with their commitment to SNP discovery and the International HapMap Project built upon these discovery efforts by characterizing the patterns of linkage disequilibrium (LD) and haplotype structure across the human genome. The work of these groups, along with others, has resulted in the identification, mapping, characterization, and the public availability of over four million validated SNPs to date (www.hapmap.org). With the advent of mass-throughput genotyping platforms, such data has the potential to contribute extensively to our basic understanding of human biology and disease.

II. MICROSATELLITES VERSUS SNPs

A. Microsatellites

Microsatellites, or STRPs, are tandem repeats of a simple DNA sequence, such as $(CA)_n$, which are usually less than 100 bp long. They can consist of a single repetitive nucleotide or be di,- tri-, tetra-, or pentanucleotide repeats. Microsatellites were first reported by Weber and May (1989) and quickly became the standard DNA marker for mapping monogenic disease genes through linkage analysis in the 1990s, replacing diallelic RFLPs. STRPs were more abundant, more informative, and technically easier to genotype than RFLPs. The increased abundancy of microsatellites over RFLPs allowed for the construction of denser maps of markers that greatly aided disease-mapping efforts. Approximately 10,000 STRPs were identified and mapped to the genome during the 1990s (Weber and Broman, 2001). Along with their increased frequency, microsatellites were also more informative than RFLPs due to their multiallelic nature. The maximum heterozygosity of any diallelic polymorphism, including RFLPs, is 50%. However, with more alleles, STRPs had average heterozygosities of 70–90%, indicating that, on average, 70–90% of a population will be heterozygous for a given STRP. The heterozygosity of a marker is important because it reflects the amount of inheritance information that can be extracted from a pedigree with that marker. As linkage analysis relies heavily on such information to precisely localize a disease gene, the increased heterozygosity of microsatellites represented a great advantage over the previously used RFLP markers. A final advantage of STRPs over RFLPs is that their analysis was much more straightforward, requiring only a PCR amplification step followed by a sizing step on a denaturing polyacrylamide gel. This represented a great simplification over the blotting and hybridization techniques needed for analysis of RFLPs. Consequently, this improvement in genotyping greatly increased the throughput of linkage methods.

Throughout the 1990s, as more and more STRPs were mapped, linkage analysis with microsatellites became the standard method for complex disease–gene identification. The typical whole-genome scan consisted of 350–400 microsatellites with an average spacing of 10 centimorgans (cM), although this marker spacing was not uniform across the genome. The high-throughput capability of microsatellite scans spurred a number of dedicated genotyping centers. The Marshfield Mammalian Genotyping Service, funded by the National Heart, Lung and Blood Institute (NHLBI), was instrumental in the identification and mapping of STRPs, assembling them into screening sets, and performing the genotyping free of charge for those investigators whose scans had been approved by NHLBI.

Hundreds of whole-genome linkage scans with microsatellites were performed during this time, both for monogenic and for multifactorial diseases. While this new method resulted in the relatively rapid discovery of numerous genes for various Mendelian disorders, it did not meet with the same success for the more complex traits. Reasons cited for this include the low heritability of multifactorial, polygenic diseases, the imprecise definition of the often complex phenotypes of common diseases, and inadequately powered studies (Hirschhorn *et al.*, 2002). However, linkage analyses with denser marker sets may be more productive. With the recent cataloging and public availability of millions of SNPs and the new mass genotyping platforms, whole-genome linkage scans with dense SNP sets are now feasible and offer another route in addition to whole-genome association scans for mapping complex disease genes.

B. SNPs

SNPs occur when a single base of DNA is substituted with another or is either inserted or deleted. SNPs can be of two types: transitions or transversions. Transitions are a purine to purine or a pyrimidine to pyrimidine substitution, for example, an adenosine to a guanine or a cytosine to a thymine. Two-thirds of all SNPs are transitions. This excess of transitions over transversions is at least partly due to the high frequency of cytosine to thymine changes, resulting from the methylation of cytosines in CpG dinucleotides followed by their spontaneous deamination to thymine. Transversions are a substitution of a purine for a pyrimidine and vice versa. Because the effects of transversions are generally more severe, resulting in a change of the chemical structure of the DNA, they are less common than transitions. While not all SNPs will predispose to disease, some of these common DNA variants are believed to be responsible for at least some of the phenotypic variations observed within and between populations, including various diseases.

SNPs can occur in coding and noncoding regions of genes as well as in the intergenic regions of the genome. Initially, SNPs falling in the coding regions of genes were of particular interest and importance because of their potential to directly impact protein function. However, due to these possible serious consequences, coding variants are generally less common than SNPs in other areas of the genome. In more recent years, as our understanding of the role of noncoding and intergenic DNA has increased, SNPs lying in such regions have been given more attention. SNPs that are outside protein-coding regions may still exert effects by impacting various epigenetic events, gene splicing, transcription factor binding, promoter regulation, and/or noncoding RNAs.

Because SNPs occur on average 1 of every 500–1000 bp, making them the most abundant and most uniformly distributed variation within the genome, they are ideal DNA markers for use in whole-genome scans; and because they are diallelic in nature, they are generally easier, quicker, and less expensive to assay than microsatellites.

Perhaps the most important aspect of SNPs with regard to their use in whole-genome scans is that they can be inherited together in haplotype or LD blocks (Daly *et al.*, 2001; Gabriel *et al.*, 2002). These blocks consist of SNPs in

LD with each other where little historical recombination has occurred. Therefore, as genotypes of SNPs in the same block tend to be correlated, not all SNPs in a block need to be directly assayed. Depending upon the size of the block and the amount of LD present within it, only one or a portion of the SNPs within a block need to be typed. However, in regions of the genome with low LD, or where haplotype blocks do not exist, a proportionately higher number of variants will need to be assayed. Given this property of SNPs, the number of variants which need to be typed in a whole-genome scan in order to comprehensively cover the genome is drastically reduced. Considerable effort has been made by the International HapMap Project to determine the haplotype block structure of the genome across various populations. This has resulted in the deposition of millions of genotypes and other SNP data into the public domain for use in planning and executing whole-genome linkage and association scans. While there is some questioning regarding the validity of the haplotype-block approach (Terwilliger and Hiekkalinna, 2006), it has become a widely accepted strategy in pursuing whole-genome scans and has proven itself a valuable tool. Current estimates as to the number of SNPs that need to be typed to provide information on the majority of common SNPs existing in the genome are between 300,000 and 500,000 (Daly *et al.*, 2001; Gabriel *et al.*, 2002; Patil *et al.*, 2001).

One aspect of SNPs that does not measure up to microsatellites is their information content (IC). Due to their diallelic nature, they are generally less informative than STRPs. Information content is a function of marker heterozygosity, density, and the number of meioses in a genetic study. As mentioned above, microsatellites have very high heterozygosities that contribute to their high IC. While SNPs generally have lower heterozygosities, often in the range of 20–30%, this is compensated by their abundance and uniform density. In addition, SNPs can be combined into haplotypes, which can increase the amount of information extracted from a given region of DNA. Haplotypes can be analyzed as multiallelic markers, much the same way as microsatellites. Finally, the lower IC of SNPs is offset by the ease with which they can be typed.

C. SNPs in whole-genome association studies

The advantages of SNPs described in the preceding section make them the ideal genetic marker for whole-genome association scans. Their identification, mapping, and cataloging in various human populations have helped make such studies a reality. There are two types of association studies: the candidate-gene approach and the genome-wide approach. While candidate-gene association studies rely on a specific hypothesis and are often limited by our current knowledge of biological and/or disease mechanisms, the genome-wide strategy is hypothesis-free and makes no assumptions regarding the mechanisms behind the phenotype being studied. Whole-genome association studies are ideally suited for common complex diseases, where the physiological mechanisms and genetic factors that contribute to them are often not well delineated. In addition, whole-genome scans are often more cost-effective in cases where large numbers of genes, or a great number of genomic loci, would otherwise be worked up in candidate-gene studies. The candidate-gene strategy has been well-tested within the literature. While a handful of genes contributing to common diseases have been identified with this approach, replicable results have been difficult to achieve for various reasons (Hirschhorn *et al.*, 2002). As the whole-genome approach has just become possible, and therefore has yet to be tested for its ability to identify the variants involved in complex traits, there are certain factors known to facilitate their successful execution. One of the most important factors relates to marker selection. SNPs for a whole-genome scan may be chosen based on a number of criteria. The two primary approaches currently in use employ an LD-based selection strategy, as exemplified by the Illumina Human-Hap-300 and -500 sets of variants, and the random approach used by Affymetrix in its 100 K and 500 K sets. Briefly, LD-based strategies exploit patterns of genomic LD to select a minimum number of SNPs to be assayed that captures the maximum amount of information across the genome, whereas random approaches ignore these LD patterns. Another type of strategy aims to ascertain SNPs based on functional criteria, such as all known nonsynonymous SNPs, as offered through the 10 K and 20 K cSNP sets by Affymetrix, or all SNPs that lie in genes, such as Illumina's Human-1 BeadChip. However, these systems are not necessarily "genome-wide," as they are designed to test only those variants directly assayed. These products test the specific hypothesis that coding variants underlie common disease and ignore other types of SNP variants existing in the genome. Although there are no current estimates as to how much of the genome is covered by these types of products, it is generally believed to be very low. These strategies and platforms will be discussed in detail in Sections III and IV.

D. SNPs in whole-genome linkage scans

Although genome-wide association studies incorporating SNPs are now feasible in the large number of individuals needed to identify the genes of small to modest effect contributing to complex disease, genome-wide linkage analysis with SNPs offers a new and improved method over previous linkage strategies with micro-satellites. Krugylak first estimated in 1997 that approximately 3000 SNPs would have more power than a 10 cM map of a few hundred single-locus multiallelic markers (Kruglyak, 1997). With today's mass-throughput genotyping systems, SNPs and STRPs can finally be directly compared. Over the past few years, a number of groups have examined the feasibility and benefits of genome-wide linkage scans with SNPs in comparison to microsatellites (Evans and Cardon, 2004; John *et al.*, 2004; Middleton *et al.*, 2004; Nsengimana *et al.*, 2005;

Schaid *et al.*, 2004; Tayo *et al.*, 2005; Xing *et al.*, 2005). Results to date demonstrate that whole-genome linkage scans employing dense SNP sets (i.e., the Affymetrix Mapping 10 K array) offer significantly higher IC across the genome as compared to the standard 300–400 ten-cM-spaced STRP scan (John *et al.*, 2004; Middleton *et al.*, 2004; Schaid *et al.*, 2004). Consequently, not only are the linkage peaks detected with SNPs more precisely defined with higher *lo*garithm of an *odds* ratio (LOD) scores and narrower 1-LOD unit support intervals than the corresponding linkage peaks in STRP scans, but also the SNP scans are able to detect linked loci that are undetectable with the microsatellites. For example, in a genome-wide linkage scan for rheumatoid arthritis comparing a typical STRP scan with an SNP scan incorporating 11,245 SNPs, John *et al.* demonstrate that while both methods detected linkage to the main rheumatoid arthritis locus of HLA*DRB1 on chromosome 6, the LOD interval in the microsatellite scan was 50 cM, but just 31 cM with the SNP scan (John *et al.*, 2004). In addition, in a genome-wide linkage study for prostate cancer comparing SNPs to STRPs, Schaid *et al.* (2004) found evidence of linkage to chromosomes 2, 6, 8, and 12; however, these loci were not identified in the STRP scan. These findings may seem counterintuitive given that SNPs are less polymorphic than multiallelic STRPs, and therefore, have less IC; however, local clusters or haplotypes of SNPs can result in more genetic information being extracted from a particular genomic region than with STRPs. The study by Schaid *et al.* (2004) demonstrated that despite lower IC per SNP compared to STRPs, the average IC across all chromosomes for SNPs was 61%, while it was just 41% for the microsatellites. The studies by John *et al.* (2004) and Middleton *et al.* (2004) demonstrated still higher ICs for the SNPs, ranging from 75% to 84%. All three studies used versions of the Affymetrix 10 K mapping array and the differences in the IC among these reports are likely due to differences in their family structures (Schaid *et al.*, 2004). Nevertheless, all studies reported substantially higher IC in the SNP scans than in the STRP scans, loci identified by the SNPs that were not identified by the microsatellite scans, and more precisely defined linkage peaks.

Many of the issues that still need resolution regarding the use of SNPs for whole-genome linkage scans relate to the impact of LD on linkage findings and optimal marker spacing. While LD can be a substantial benefit in association mapping by allowing for the indirect assaying of variants, it can be more problematic for SNP linkage studies. LD may lead to an upward bias in identity-by-descent (IBD) estimates through underestimation of haplotype frequencies. However, the impact of LD, if any, on linkage findings from empirical studies is still unclear. The study by Schaid *et al.* demonstrated inflated LOD scores resulting from the presence of high LD ($|D'| > 0.70$) among the markers used in their study. In contrast, John *et al.* concluded that a moderate amount of LD ($r^2 \geq 0.4$) had no effect on their linkage results, with similar linkage findings when SNPs in LD were either retained or removed. As part of Genetic Analysis

Workshop (GAW) 14, Murray (2005) and Peralta *et al.* (2005) used data from the Collaborative Study on the Genetics of Alcoholism to assess the impact of LD on linkage results. Neither report demonstrated a substantial effect of LD on their respective linkage findings. However, simulation studies from GAW 14 show that LD may in fact be capable of inflating LOD scores. Both Goode *et al.* (2005) and Huang *et al.* (2005) documented inflated LOD scores and even false linkage signals when dense markers in LD were used. The disagreement in findings among these studies may be due to the different family structures involved, as studies with many nongenotyped founders would be more susceptible to the effects of intermarker LD. Additionally, the different LOD statistics used and the different methods of assessing and correcting for LD among the studies may also have impacted the results.

An additional complication of high-density SNP linkage scans is the increased analytic burden they present in comparison to microsatellites. Large pedigrees with a significant amount of missing data present the greatest computational challenge. Not only are enhancements needed to current software to be able to effectively analyze SNP scans but also novel algorithms are needed that can more efficiently handle genome-wide data, particularly in the case of extended pedigrees with missing data. Any such software should be able to account for LD among markers. Much work is still needed in this area before SNP linkage scans can be efficiently analyzed, especially in the case of large pedigrees.

Overall, the advantages of a genome-wide SNP linkage scan over microsatellite-based scans include the relative ease and speed with which genotyping data can be generated, better resolution of quantitative-trait locis (QTLs), and additional detection of suggestive and significant linkage findings. The better resolution of loci and the presence of SNPs, which can serve as a partial framework for subsequent association studies, can result in considerable downstream savings when it comes to fine-mapping a linkage peak. In addition, whole-genome linkage scans require a much smaller number of SNP markers to be genotyped than whole-genome association scans. Consequently, these studies may offer an economical approach to mapping complex disease genes in family designs.

III. GENOTYPING PLATFORMS AND GOALS

A. Challenges to mass-throughput genotyping

With the completion of the HGP, the only limitation to performing genome-wide scans and other high throughput genotyping projects has been the availability of a cost-efficient, rapid, scalable genotyping technology. The successful genome-wide platform must address the issue of mass throughput while maintaining quality control, requiring a minimal amount of input DNA, ensuring high conversion rates, and costing less than a cent per genotype. Various approaches to genotyping include restriction-enzyme analysis, allele-specific amplification, single-base extension, ligation-based assays, allele-specific oligonucleotide hybridization, pyrosequencing, and a number of electrophoresis-based methods. These approaches have been incorporated into a wide variety of platforms with differing chemistries and variations on the general approaches have been described (Mamotte, 2006). However, until very recently, no platform had proven itself capable of massively parallel throughput at the scale of the whole genome while remaining cost-effective in the sample sizes needed for complex disease–gene identification; although some platforms have proven themselves extremely efficient for higher-throughput, targeted genotyping applications such as fine-mapping a QTL. These particular platforms are described in detail in Section III.C.

Challenges to highly parallel genotyping include a highly parallel array-based readout and a scalable sample preparation method amenable to high orders of multiplexing (Fan *et al.*, 2006). The high-density oligonucleotide microarray approach has been able to meet these challenges and has recently become widely adopted by those interested in pursuing whole-genome association scans. Recent advances in microarray methods have allowed high-density arrays to expand from gene expression profiling to the analysis of DNA sequence variation. While microarrays for gene-expression profiling have been in use for over a decade, their application to genotyping has been more technically challenging. This is mainly due to the specificity required when genotyping SNPs, as a single nucleotide must be sampled from the entire genome at relatively low concentrations (Fan *et al.*, 2006).

PCR has generally been used as part of the sample preparation process in order to achieve amplification of a specific locus. Multiplex PCR can be performed where a number of loci are simultaneously amplified in one reaction. However, multiplex PCR is generally limited to several dozen loci, is subject to high rates of primer–dimer formation, and requires careful and often time-consuming optimization to ensure equal amplification of all involved loci. While there are now methods existing that increase the number of loci that can be simultaneously analyzed in a single tube and that have reduced primer–dimer formation (Adessi *et al.*, 2000; Brownie *et al.*, 1997; Lin *et al.*, 1996), none of these approaches have demonstrated the scalability required to remain cost-effective at the whole-genome level. Two recent approaches have been described that incorporate established techniques with oligonucleotide arrays and avoid the problems associated with high-levels of multiplexing. These approaches are described in the following section. It is important to recognize that the technology in this area is in flux, given the rapid pace of technological development, and new products and improvements to existing ones may soon be introduced.

B. Genome-Wide platforms

1. The Affymetrix Genechip® platform

Affymetrix first offered a higher-throughput genotyping platform with their GeneChip® human mapping 100 K product. This product consists of two arrays, each capable of genotyping over 50,000 SNPs, for a total of more than 100,000 SNPs. They now offer a 500 K product that allows for the genotyping of 500,000 SNPs across two arrays. This product builds on the 100 K set but offers twice the power to map disease-susceptibility loci. The SNPs comprising the 500 K set were selected from both the public domain and the Perlegen SNP database based on accuracy, call rate, and LD analysis in Caucasians, African–Americans, and Asians. The average heterozygosity of the SNPs is 0.30 and the average distance between SNPs is 5.8 kb [(2006). GeneChip® Human Mapping 500 K Array Set datasheet. www.affymetrix.com products/arrays/specific/500k.affx]. Eighty-five percent of the genome is within 10 kb of a SNP, which represents a substantial upgrade from the 100 K GeneChip, where just 40% of the genome was within 10 kb of an SNP [(2006). GeneChip® Human Mapping 500 K Array Set and Human Mapping 100 K Array Set datasheets. www.affymetrix.com/products/arrays/specific/500k.affx and www.affymetrix.com/products/arrays/specific/100k.affx]. The coverage offered by the 500 K chip is estimated to be approximately 65%, 66%, and 41% at an r^2 of 0.80 in Caucasians, Asians, and Yorubans, respectively (Barrett and Cardon, 2006). Genotyping data quality and SNP characteristics of both the 100 K and 500 K sets are shown in Table 5.1.

In addition to SNP genotyping, Affymetrix's GeneChip® products can be used to perform genome-wide copy number/loss of heterozygosity (LOH) analyses and for the analysis of copy neutral events such as uniparental disomy. (See Section VI for a discussion of such variants as novel markers.)

All GeneChip products are based on reducing genomic complexity. In the 500 K product, this is achieved by the individual use of two restriction enzymes, Nsp I and Sty I, to digest total genomic DNA, resulting in 4 bp overhangs of all fragments. An adaptor sequence that recognizes these 4 bp overhangs is then ligated to each fragment. A single primer that recognizes the adaptor sequence is then used to amplify the DNA fragments. PCR conditions are optimized such that only those fragments falling in the size range of 200–1100 bp are amplified. The amplified DNA is then fragmented, labeled, and hybridized to one of the two arrays that constitute the GeneChip® 500 K human mapping set (Fig. 5.1).

Specificity of this assay is achieved through the use of multiple 25-mer probe pairs. Each SNP is represented on the array by a probe set consisting of multiple probe pairs (Fig. 5.2). In the probe set for each SNP, there are perfect match (PM) and mismatch (MM) probes containing the SNP in different

Table 5.1. Performance Metrics, SNP Characteristics, and Coverage Estimates of the Affymetrix 500 K and 300 K Arrays[a]

Metric	500 K array	100 K array
Number of SNPs	500,568	116,204
QC call rate	≥93%[b]	≥95%
Average MAF	0.22	0.22
Average heterozygosity	0.30	0.30
Genome within 10 kb of an SNP	85%	40%
Median physical distance between SNPs	2.5 kb	8.5 kb
Average physical distance between SNPs	5.8 kb	23.6 kb
Concordance with HapMap	99.3%[c]	99.73%
Mendelian consistency	99.9%[d]	99.97%[f]
Caucasian single-marker coverage	67%[e]	31%[g]
Yoruban single-marker coverage	41%[e]	15%[g]
Asian single-marker coverage	64%[e]	31%[g]

[a]Data taken from affymetrix's 500 K and 300 K datasheets.
[b]Based on the DM algorithm at a confidence threshold of 0.33.
[c]Estimated in 33 Caucasian and 33 Yoruban HapMap samples.
[d]Estimated in 17 HapMap trios.
[e]Estimates are for coverage of phase II HapMap SNPs at an r^2 of 0.8.
[f]Estimated in 10 trios.
[g]Estimates are for coverage of phase II HapMap data at an r^2 of 0.8 as estimated by Barrett *et al.*

locations of the oligonucleotide. For example, there is a PM and MM pair that contains the SNP at the central position of the oligonucleotide and PM and MM pairs with the SNP shifted either upstream or downstream within the oligo sequence (Fig. 5.3). Five probe pairs for the sense and five probe pairs for the antisense strand are included on the final array product. This results in a total of 20 probes interrogating each allele of an SNP or 40 total probes per SNP. Genotypes are called with the GeneChip®genotyping analysis software that uses an automated model-based algorithm to assign a confidence score to each genotype.

In addition to the probes for each SNP, Affymetrix has also incorporated a number of control probes into the array to ensure quality control. Moreover, there are 50 control SNPs located on each of the two chips. These 50 controls are cross-checked internally with the genotyping analysis software to verify that the genotypes are matching and that both arrays remain together from DNA preparation through analysis.

In addition to its whole-genome products, Affymetrix also offers a number of focused-content products and SNP kits, as well as mapping sets for the mouse, rat, and cow. These products are based on the molecular inversion probe (MIP) assay. This assay is discussed in Section III.C.

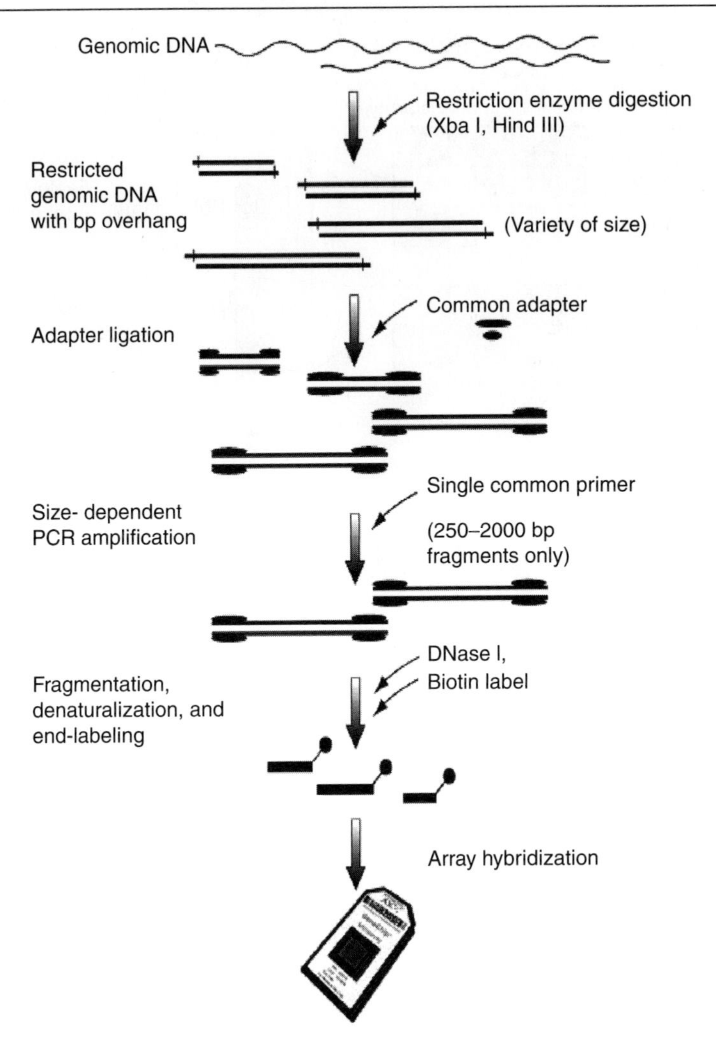

Figure 5.1. Schematic representation of the whole-genome sampling assay used by Affymetrix on their GeneChip® products. Genomic DNA is digested with restriction enzymes, yielding fragments of varying sizes. A common adapter is linked to the restriction overhangs for amplification. As only those fragments in the 250–2000 bp size range are amplified, genomic complexity is reduced. PCR products are then fragmented and labeled, resulting in a sample ready to be hybridized to the microarray. Reprinted from Dalma-Weiszhausz *et al.*, (2006), Copyright (2006), with permission from Elsevier.

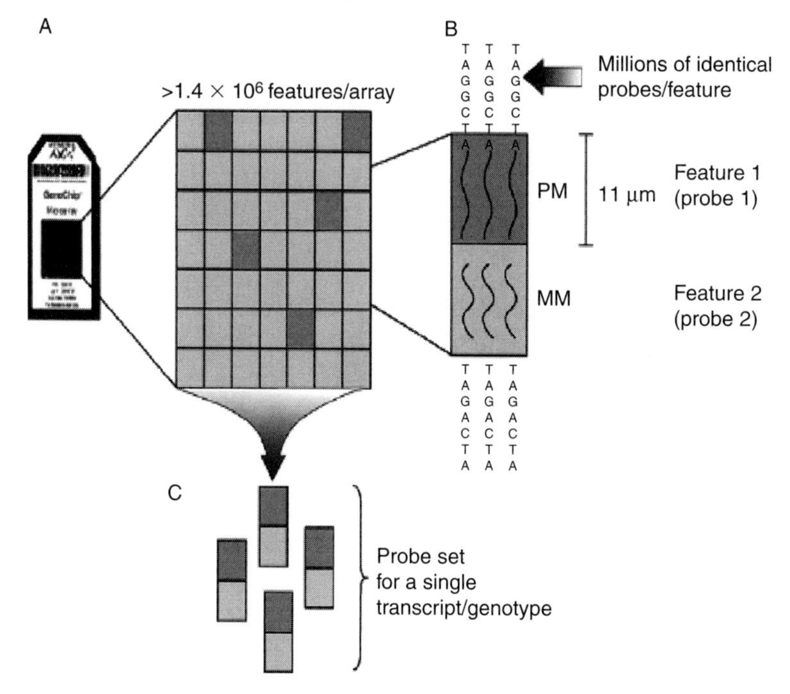

Figure 5.2. Probe array dissection. (A) A piece of quartz inside the probe array is enlarged. It contains a synthesis area of 1.28 cm^2 with more than a million different features. Each feature is composed of millions of oligonucleotide sequences. (B) For every perfect match (PM) feature, a mismatch (MM) feature is also included that is exactly the same as the PM sequence, with the exception of a nucleotide transversion at the central position. (C) PM and MM pairs that interrogate the same target sequence are referred to as a 'probe set'. Reprinted from Dalma-Weiszhausz *et al.*, (2006), Copyright (2006), with permission from Elsevier.

2. The Illumina BeadChip platform

The Illumina platforms are unique in that they make use of BeadArray technology. While conventional microarrays are manufactured by spotting or *in situ* synthesis of probes, BeadArrays rely on the random self-assembly of a bead pool onto a patterned substrate (Fan *et al.*, 2003; Michael *et al.*, 1998; Oliphant *et al.*, 2002). Each 3 μm-sized bead is coated with tens to hundreds of thousands of copies of a particular probe sequence. Batches of these beads are allowed to self-assemble into the microwells of the array. In order to determine which beads are in which wells, fluorescently labeled oligonucleotides are hybridized to a portion

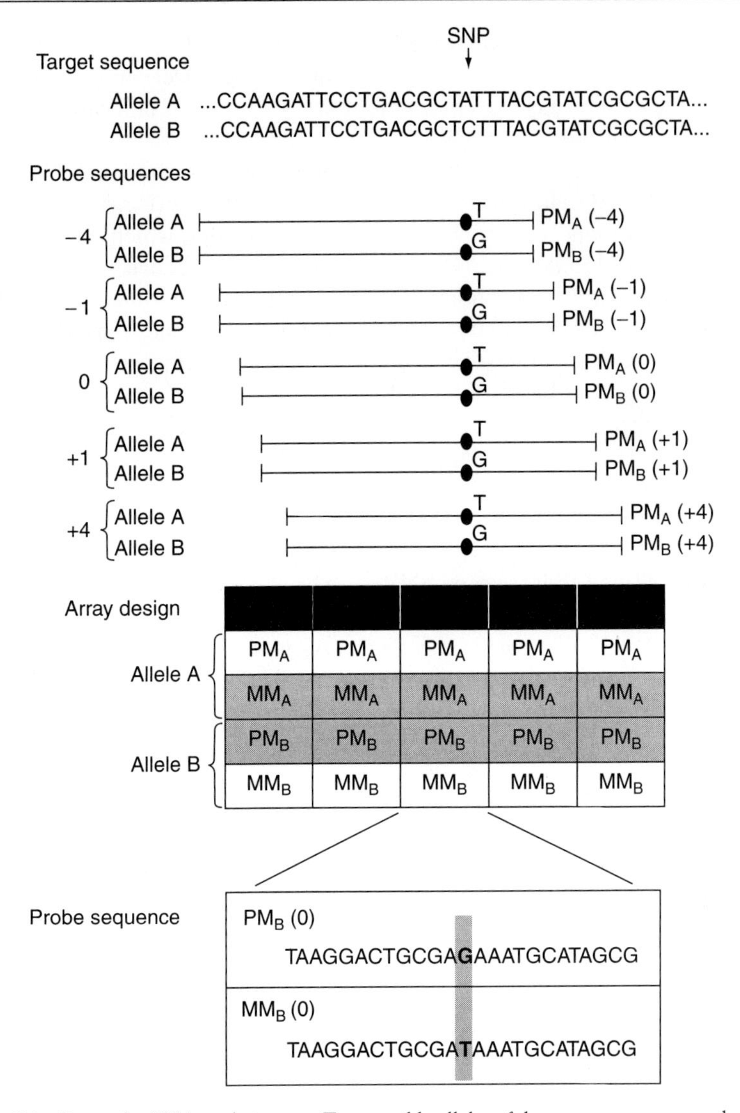

Figure 5.3. Design for DNA analysis array. Two possible alleles of the target sequence are shown at the top of the figure. The selection strategy for probe sequences includes probes that are centered at the SNP position (0), as well as probes that are shifted to the left (−4, −1) and to the right (+1, +4) of this central SNP location. Both alleles are interrogated with probes, including PM and MM probe pairs. Two features representing the B allele, the centered PM probe and its partner MM probe, are illustrated at the bottom of the figure. Reprinted from Dalma-Weiszhausz *et al.*, (2006), Copyright (2006), with permission from Elsevier.

of each probe sequence known as the "address" sequence. A proprietary decoding algorithm is then used to track which oligonucleotides successfully hybridize to their target address sequence. This platform is highly flexible and can be adapted for many different uses.

For genome-wide SNP-typing, Illumina has incorporated this BeadArray technology into its line of BeadChip product offerings (Fig. 5.4). Illumina's primary GWA products are its 300 and 550 K BeadChips, which offer access to over 317,000 and 555,000 tagSNPs, respectively. These products make use of the Infinium assay that allows genomic DNA to be genotyped directly rather than relying on a reduced-complexity representation of the genome, as in the Affymetrix protocol. The advantage of this is that specific SNPs can be chosen for genotyping. This assay has therefore allowed Illumina to offer arrays of maximally informative tagging SNPs. Tagging SNPs on the 300 K chip were selected from phase I of the HapMap Project, while the 550 K chip tag SNPs were selected from phase I + phase II HapMap data (release 20). Table 5.2 shows the tagging SNP characteristics and the performance metrics for both the 300 and 550 K BeadChips.

In addition to its tagging SNP content, the 550 K chip also offers SNPs in known copy number variable regions, nonsynonymous SNPs, as well as mitochondrial and Y chromosome SNPs. As with the Affymetrix platforms, the Illumina platforms also allow for copy number and LOH analysis in both paired and single samples, as well as the analysis of copy neutral events.

The Infinium assay can be broken into four steps: (1) whole-genome amplification of genomic DNA, (2) hybridization capture to an array of 50-mer probes, (3) array-based enzymatic SNP scoring, and (4) an antibody-based sandwich detection assay (Fig. 5.5; Gunderson *et al.*, 2006). The whole-genome amplification step enhances the sensitivity of the assay by increasing the molar concentration of each genomic loci. This amplification step has been shown to be unbiased with respect to the alleles that are amplified (Gunderson *et al.*, 2006). The amplified gDNA is then hybridized to the array using locus-specific probes consisting of a 30-mer address sequence and a 50-mer query sequence targeting the locus of interest. Following hybridization, a primer extension step is carried out on the array to score the SNP. Biotinylated nucleotides are incorporated during this step. Finally, these labels are stained and amplified in an immunohistochemistry-based sandwich assay. The BeadChip is then ready to be scanned. After scanning, the GenCall software, provided with various algorithms to automatically call genotypes, will generate reports of the results.

Aside from its GWA products, Illumina also offers a number of targeted genotyping products and custom SNP panels, including SNP linkage panels. These products use the GoldenGate® genotyping assay, described in Section III.C.

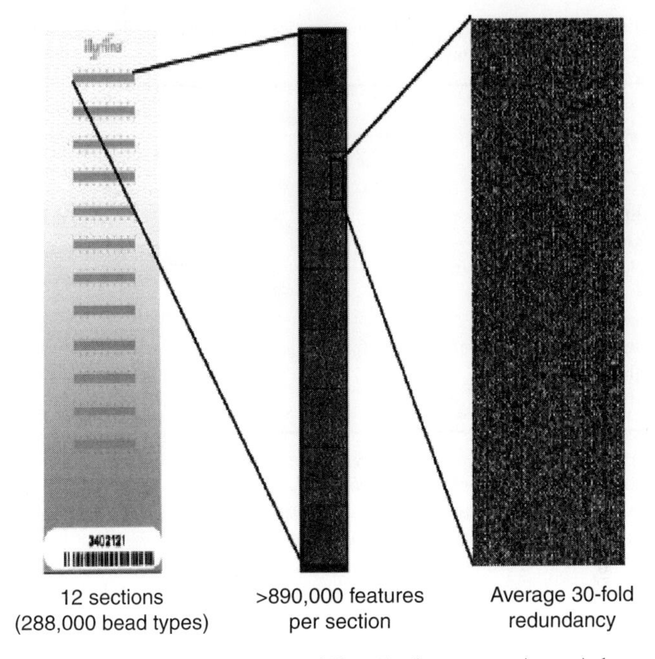

Figure 5.4. A high-density Infinium WGG BeadChip. Twelve sections (stripes) that contain over 890,000 beads each are included on each BeadChip. As many as 24,000 bead types can be loaded into each section, with average 30-fold redundancy. This platform supports about 288,000 bead types. Reprinted from Gunderson *et al.*, (2006), Copyright (2006), with permission from Elsevier.

C. Targeted genotyping platforms

For projects that do not necessitate whole-genome coverage but still require large numbers of SNPs to be typed, such as working-up a QTL, there are a number of platforms available that offer cost-effective genotyping on a smaller scale. The following three assays have been used as part of the HapMap Project, and therefore, have been extensively validated. Each assay represents a different level of multiplexing and each offers its own approach to targeted genotyping.

1. Affymetrix's molecular inversion probe technology

Affymetrix offers an assay capable of multiplexing up to 20,000 SNPs in a single reaction. The molecular inversion probe (MIP) assay (Fig. 5.6A) uses a circular probe that binds upstream and downstream of the SNP site of interest and leaves

Table 5.2. Performance Metrics, SNP Characteristics, and Coverage Estimates of the Illumina 550 K and 300 K Arrays[a]

Metric	550 K array	300 K array
Number of SNPs	>555,000	>317,000
Call rate	99.78%[b]	99.93%[e]
Reproducibility	>99.99%[b]	>99.99%[e]
Average MAF[c]	0.23, 0.21, 0.22	0.26, 0.23, 0.23
Average physical distance between SNPs	5.5, 6.5, 6.2[c]	9 kb
Concordance with HapMap	99.74%[b]	99.79%[e]
Mendelian consistency	99.97%[b]	99.97%[e]
Caucasian single-marker coverage[d]	90%	80%
Yoruban single-marker coverage[d]	57%	~70%
Asian single-marker coverage[d]	87%	~35%

[a]Data taken from illumina's 550 K and 300 K datasheets.
[b]Estimated on 120 DNA samples (8 replicates, 28 trios).
[c]Estimates are for Caucasian, Asian, and Yoruban populations, respectively.
[d]Estimates are for coverage of phase I + phase II data at an r^2 of 0.8.
[e]Estimated on 127 DNA samples (15 replicates, 25 trios).

a 1 bp gap. This gap is filled in with the base pair complementary to the SNP and is then ligated. The circularized probe is then released from the DNA through restriction enzyme digest at a consensus sequence and is PCR-amplified using common primers. The placement of the primers ensures that only circularized probes, probes that incorporated the base pair complementary to the SNP, will be amplified. The PCR product is then hybridized and read out on an array of universal-capture probes.

2. Illumina's GoldenGate assay

Illumina offers the GoldenGate assay that is capable of multiplexing up to 1536 SNPs in a single tube and can be used in conjunction with their BeadChip platform. The assay is based on an allele-specific extension reaction and universal PCR. Three oligonucleotides are designed for each SNP. Two of the oligos are specific to each allele of the SNP (allele-specific oligos, ASOs) and the third is designed to anneal several bases downstream from the SNP (locus-specific oligo, LSO). All three contain universal PCR primer sites and the LSO also contains a unique address sequence specific to a particular bead type. These three oligonucleotides hybridize to the DNA once it binds to the paramagnetic particles. Once extension of the appropriate ASO is completed, a ligation step joins this product to the downstream LSO. This couples the genotype information at the SNP locus to the address sequence of the LSO and provides a template for PCR using

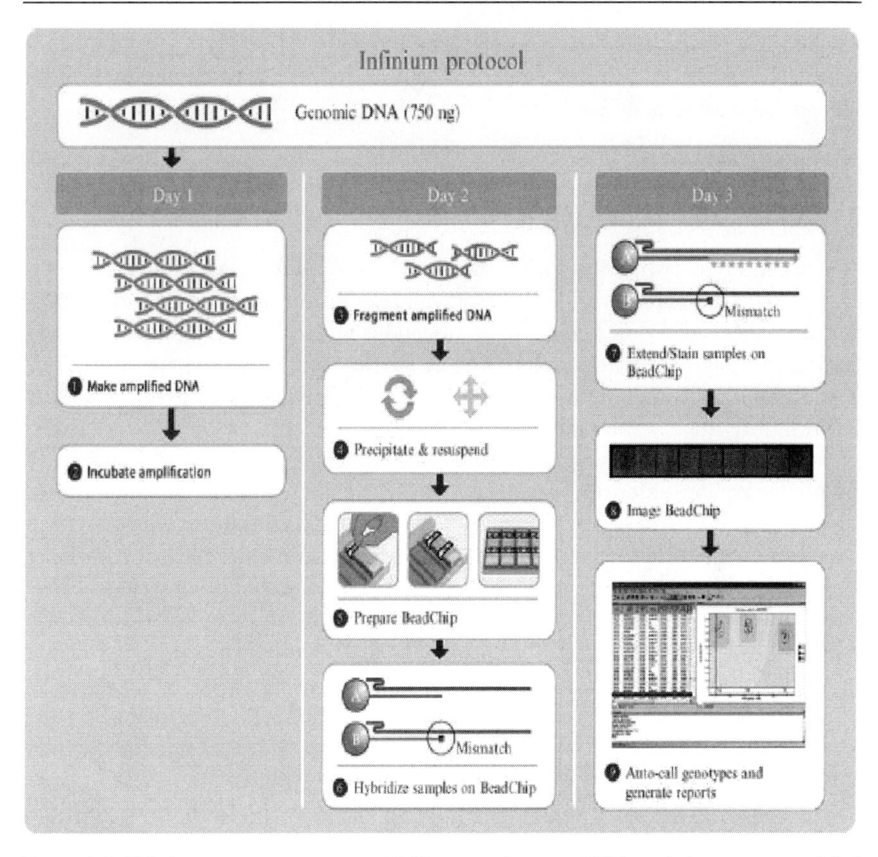

Figure 5.5. Whole-genome genotyping on DNA arrays. Genomic DNA is whole-genome amplified overnight in an isothermal reaction (steps 1 and 2). Starting with 250–750 ng of DNA, a 1000 to 2000-fold amplification is achieved. The product is fragmented enzymatically in a controlled process after amplification (step 3). After the samples are precipitated with isopropanol and resuspended (step 4), the samples are applied to the BeadChip (step 5). The samples are hybridized overnight to a BeadChip array. The SNPs are then scored using either an allele-specific primer extension (ASPE) step or a single base extension (SBE) reaction (step 7). During primer extension, biotinylated/DNP nucleotides are incorporated. These labels are stained and amplified in an immunohistochemistry-based multilayer sandwich assay. Following this, the BeadChips are read on a high-resolution confocal scanner (0.84-μm resolution). Genotype calls are made automatically by GenCall software, generating reports of the results. Reprinted from Gunderson *et al.*, (2006), Copyright (2006), with permission from Elsevier.

dye-labeled universal primers. Following PCR and some downstream processing, single-stranded, dye-labeled DNAs are then hybridized to their complement bead type through their unique address sequence. This hybridization step allows for the assay products to be separated for individual SNP readout. Genotypes are autocalled in the same manner as in Illumina's whole-genome products. The GoldenGate assay is shown in Fig. 5.6B.

3. Sequenom's MassARRAY platform

Seqeunom offers two assays for use with its MassARRAY platform. Both assays are based on matrix-assisted laser desorption/ionization time-of-flight mass spectrometry (MALDI-TOF MS). The homogenous massEXTEND (hME) assay was originally designed for the genotyping of single SNPs; however, modifications now allow it to be used in multiplexes of up to 15 SNPs. Sequenom's iPLEX assay offers slightly higher throughput with 29 SNPs.

The hME assay (Fig. 5.6C) consists of a multiplex PCR step followed by an allele-specific extension reaction. The multiplex PCR is performed with 10-mer tagged primers. These tags increase the mass of the unused primers so that they fall outside the mass range of analytical peaks. Once all loci have been amplified, a second PCR step is performed with MassEXTEND primers. These primers anneal adjacent to the SNP of interest and are extended through the polymorphic site to generate allele-specific products with unique molecular masses. These masses are then analyzed by MALDI-TOF and a genotype is assigned. All primers used in this assay are designed with Sequenom's Assay Design 2.0 software that aims to minimize the chance of overlapping spectra and to reduce unwanted inter- and intraprimer interactions.

The iPLEX assay works much the same way as the hME assay, with the most substantial difference being that a single-base extension reaction is performed in place of the hME allele-specific reaction. The smaller mass separation achieved with the SBE products contributes to the enhanced multiplexing of this assay. Modified Assay Design software is used to design iPLEX primers.

IV. SELECTION OF MARKERS AND SNPs

A. Selection of genome-wide platforms

While access to affordable whole-genome SNP-typing technology had previously been a limitation to performing GWA scans, investigators are now faced with a choice of platforms and products. In addition to overall cost, the coverage, efficiency (coverage per SNP genotyped), and redundancy of a platform must all be considered when making such a choice. In relation to the above criteria, there are two types of platforms to choose between, the random panels of SNPs offered by Affymetrix and the tagging panels offered by Illumina.

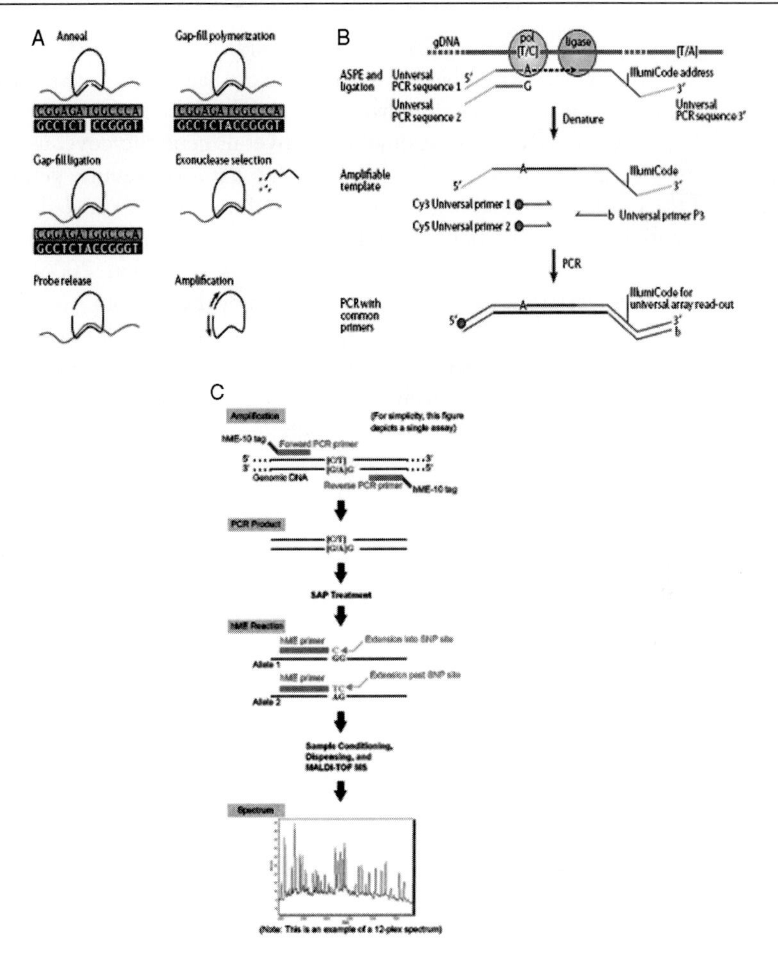

Figure 5.6. Highly parallel genotyping assays for targeted applications. (A) Molecular-inversion probe (MIP) genotyping uses probes with 50 and 30 ends capable of forming a circle that anneal up- and downstream of the SNP, leaving a 1 bp gap. dNTPs fill in the gap in a polymerase extension step. The nick is ligated, and unannealed and unligated circular probes are removed. Next, the circularized probe is released through restriction digest at a consensus sequence. The product generated is amplified using common primers to "built-in" sites on the circular probe. The orientation of the primers is such that only circularized probes are amplified. Finally, the product is hybridized and read out on an array of universal-capture probes. (B) GoldenGate genotyping is based on extension ligation between annealed locus specific oligos (LSOs) and allele-specific oligos (ASOs). An allele-specific primer extension (ASPE) reaction joins the correctly matched ASO (at the 30 end) to the 50 end of the LSO. The nick is sealed with ligation. Next, PCR amplification is used to amplify the appropriate product using common primers to universal PCR sites in both the ASO and LSO sequences.

Tagging panels consist of SNPs chosen based on LD. SNPs that are in strong LD with the greatest number of other SNPs are selected for genotyping and serve as proxies for the SNPs that are not selected. The advantage of these panels is that they offer the greatest amount of overall genomic coverage and efficiency. In a recent study using phase II HapMap data to examine genomic coverage of the currently available genome-wide platforms, Barrett *et al.* demonstrate that panels based on tagging SNPs offer, in general, greater overall coverage and efficiency than do random sets (Barrett and Cardon, 2006). The amount of common variation (MAF ≥ 0.05) captured by the random Affymetrix 500 K chip was estimated to be 65% in Caucasians at an r^2 of 0.80, while the 300 K Illumina tag set was estimated to capture 75% of such variation. Although the 550 K Illumina chip was not evaluated as part of this study, Illumina has estimated the coverage of the 550 K chip to be 90% in Caucasians at an r^2 of 0.80 [(2006). Sentrix® HumanHap 550 datasheet. www.illumina.com/Products/ArraysReagents/wgghumanhap550.ilmn]. However, the study also demonstrated that a disadvantage of tagging panels is that they are most susceptible to marker failure because all redundancies have been deliberately eliminated. If one marker fails, all SNPs in LD with that marker are lost. The authors suggest this may be a nontrivial problem given the initial reports of high failure rates, such as that reported by Klein *et al.* (2005). In addition to being vulnerable to marker failure, the study also showed that tagging sets may perform with reduced coverage and efficiency in non-Caucasian populations. The coverage of the 300 K chip drops from 75% in Caucasians to just 28% in Yorubans; and Illumina has estimated the coverage of the 550 K chip to drop from a high of 90% in Caucasians to 57% in Yorubans. To counter this problem, Illumina now offers its 550 K product with an additional 100,000 Yoruban-specific tag SNPs for enhanced coverage in African and African–American populations.

Random panels of SNPs are LD "agnostic," meaning they ignore patterns of LD. While this may not be the most efficient method of covering the genome, random panels offer the advantage of marker redundancy. Because LD is not considered when selecting SNPs for these panels, it is likely that at least some of the genotyped SNPs are correlated with each other, and the SNPs that

As in (A) above, the products generated are then hybridized and read out on an array of universal-capture probes (complementary to IllumiCodes). (C) Schematic representation of Sequenom's multiplexed massEXTEND assay. An oligonucleotide primer (hME primer) anneals adjacent to the SNP of interest. A DNA polymerase is added to the reaction along with a mixture of terminator nucleotides. This allows the extension of the hME primer through the site of the SNP and results in allele specific extension products, each having a unique molecular mass. Panels A and B are reprinted by permission from Macmillan Publishers Ltd: Fan *et al.*, (2006), copyright (2006). Panel C reprinted from Sequenom's Application Note entitled "Multiplexing the Homogenous MassEXTEND Assay," January 2004, accessed from the Sequenom website.

are not genotyped may be in LD with more than one of those that are. Therefore, if one marker fails, it is likely that information was captured by another marker. Barrett *et al.* demonstrate that the Affymetrix 500 K chip is robust to failure rates as high as 20%, with little or no loss in its coverage rate. Aside from this protective redundancy, the coverage of random sets may not vary as much among populations because their SNPs are not so carefully targeted to any one population. Coverage of the 500 K Affymetrix chip was estimated to be 66% in the Asian HapMap samples and 41% in the Yoruban samples, compared to 65% in the Caucasians (Barrett and Cardon, 2006). The drop in coverage between the Caucasian and Yoruban samples is not as great as that seen with the tagging sets. This needs to be taken into consideration when planning an association study in multiple ethnic groups.

A variation of the random approach is to increase coverage with a "fill-in" set of tag SNPs. This panel of tag SNPs could be custom made to address the specific goals of the investigation. For example, an investigator could choose to augment the 65% random coverage rate with tag SNPs designed to capture the remaining common variation, to enhance coverage in a particular genomic region(s) of interest or in a particular population, or simply to fill in known gaps in the random panel. Alternatively, investigators may use a custom "fill-in" panel to capture rare variants that are not covered well by most of the genome-wide products or to capture a particular type of variant such as non-synonymous SNPs. Final coverage would be determined by the aim of the tag SNP panel.

In summary, each platform has its own unique advantages and disadvantages, which must be taken into account when planning and executing genome-wide studies. In particular, investigators need to consider the population(s) of interest and the balance between marker efficiency and marker redundancy before selecting a genotyping platform. It is also important to keep in mind that while GWA studies overcome some of the obstacles of the past, they also present new challenges and a new set of limitations that must be considered. Perhaps one of the main challenges when performing a GWA study is addressing the large number of statistical tests that are carried out. Using a simple Bonferroni correction, which would stipulate a significance level of 10^{-7} for a study using 500,000 markers, may be overly conservative because it does not take into account LD. Two other methods include controlling the false discovery rate and permutation testing. However, a consensus has yet to be reached regarding the best method to control for multiple testing in a GWA setting. As more GWA datasets become available, this area should see further development. Other issues that must be considered include the inability of current GWA platforms to adequately capture low-frequency variants, which one can argue are underlying complex diseases, and the requirement for replication of results in different populations. Replication of results has been difficult in association studies following-up a linkage peak. It remains to be seen if this will also be the case for GWA studies.

B. Selection of targeted platforms

For projects that involve a more targeted genotyping approach, there are a number of products currently available from which investigators can choose. The choice of such a platform will be driven mainly by the specific hypotheses of the investigator. A selection of focused content products is briefly described here.

For those investigators interested in pursuing a functional SNP approach, Affymetrix currently markets 10 K and 20 K validated coding SNP panels. Both panels include double-hit nonsynonymous SNPs from the public domain, with the 20 K panel including an additional 10,000 coding SNPs validated by Affymetrix. These sets may be particularly relevant to pharmacogenomic studies, as both panels include a majority of the major drug target gene families. In addition, Affymetrix offers an immune and inflammation SNP set with over 9000 SNPs in 1000 genes related to the inflammatory response. A 3 K admixture panel is also available for those interested in performing admixture mapping to localize disease genes. These panels use the MIP assay described in Section IIIC.1.

Illumina also offers application-specific products for researchers with particular hypotheses. Currently, Illumina markets an MHC panel consisting of two sets of SNPs that can be combined or used independently. The MHC exon-centric panel has more than 1200 SNPs located with 10 kb of coding sequences of genes in the MHC region. The MHC mapping panel includes more than 1250 SNPs with an average spacing of 3.8 kb. For a candidate–gene approach to identify variants associated with cancer, there is also a cancer SNP panel with nearly 1500 validated SNPs chosen from the National Cancer Institute's SNP500 cancer database. These SNPs represent over 400 different genes. Both the MHC panels and the cancer panel employ the GoldenGate assay described in Section III.C.2.

V. NOVEL MARKERS

Variation in the genome can take many forms. While SNPs have been the primary focus due to their relative simplicity and abundance, other forms of variation, including deletions, inversions, and duplications—together forming a class of structural variants—are gaining attention. With an increasing number of studies being conducted at the whole-genome level, structural alterations, particularly copy number variants (CNVs), have been found to be more prevalent than previously thought. A CNV is defined as a duplication or deletion event involving >1 kb of DNA (Fig. 5.7). LOH is a related condition, where an allele is lost through a deletion, resulting in homozygosity or hemizygosity of the locus.

CNVs were known to be associated with various chromosomal rearrangements and genomic disorders, such as DiGeorge syndrome (Burn *et al.*, 1995), Charcot-Marie-Tooth disease (Nelis *et al.*, 1996), and Williams-Beuren syndrome (Greenberg, 1990) and also have recently been reported to contribute to HIV-1 susceptibility (Gonzalez *et al.*, 2005). However, the extent of copy number variation at the population level and its potential contribution to common disease has only begun to be examined.

CNVs have traditionally been identified through the use of array-comparative genomic hybridization (CGH) on a genome-wide scale or with quantitative PCR for screening specific regions of the genome. Array-CGH uses labeled fragments from a genome of interest and competitively hybridizes them with differentially labeled fragments from a reference genome to an array spotted with cloned DNA fragments. A number of array-CGH approaches exist that are based on the type of clones used. Arrays based on BAC clones are most commonly used due to their extensive coverage of the genome. However, the resolution provided by BAC arrays is generally low, in the range of 50 kb to just a

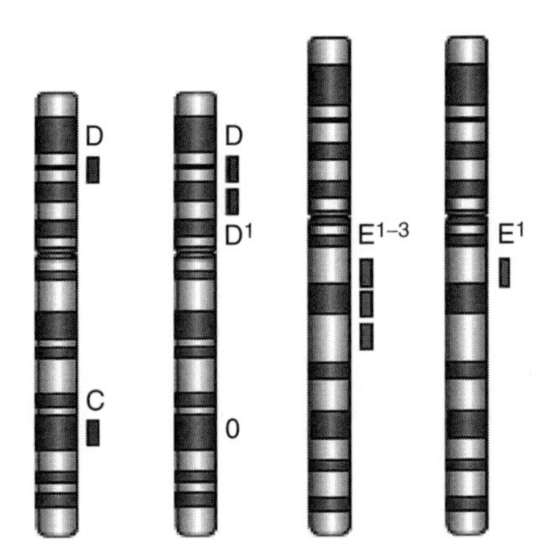

Figure 5.7. A schematic of three different types of copy number variants (CNVs) (C through E). CNV "C" represents a deletion, labeled 0, while CNV "D" represents a duplication, which is named D1. There are three copies of CNV "E" in the reference sequence, but only one on the homologous chromosome. Reprinted with permission from Macmillan Publishers Ltd.: Feuk *et al.*, (2006), Copyright (2006).

few kilobases. The use of arrays based on long oligonucleotides (~100 bp) can improve upon this resolution. Such arrays have been implemented with a technique known as representational oligonucleotide microarray analysis (ROMA). In ROMA, in order to reduce the background noise, the complexity of the input DNA is first reduced by treating it with restriction enzymes, ligating it to common adapters, and then performing amplification, similar to the assay used in the Affymetrix 100 K and 500 K arrays. The availability of the high-density SNP genotyping arrays now offers another approach to screening for CNVs on a genome-wide scale. With this method, hybridization intensities of an experimental sample are compared to average intensity values of a control sample. Deviations from these average values indicate a change in copy number. The added benefits of using the SNP arrays over the traditional clone-based approaches is their greater resolution, allowing them to identify smaller CNV variants that would be missed by the other approaches, as well as providing genotype information (Feuk *et al.*, 2006). The added advantage of SNP genotype information is that it can reveal LOH, which could act as supporting evidence for a deletion. The recent study by Redon *et al.* provides an example of using SNP genotyping arrays to identify CNVs.

CNVs may contribute to disease through gene dosage and positional affects, or in a more complicated manner, by predisposing to additional deleterious structural variation. Iafrate *et al.* (2004) and Sebat *et al.* (2004) were the first two studies to examine this type of large-scale variation across the genome. Both studies used array-based CGH in apparently healthy individuals. Each concluded that structural variation is much more common in healthy populations than initially believed and may represent an appreciable source of variation in phenotypic traits and susceptibility to disease. In the most recent and most comprehensive study to date, Redon *et al.* (2006) used both array-based CGH and SNP genotyping arrays to construct the first CNV map of the human genome. Their findings demonstrate clear variation in copy number among populations.

Given these findings, there is substantial interest in assaying these types of variants and testing for their association with disease and other phenotypic outcomes. Consequently, there has been discussion on the ability of SNP-based genome-wide association studies to capture structural variants through linkage disequilibrium. Two recent studies examined the LD pattern of common deletion polymorphisms and found that many of them are in LD with nearby SNPs and can, therefore, be indirectly assayed through SNP-based association studies (Hinds *et al.*, 2006; McCarrolll *et al.*, 2006). A third report examining CNVs in duplication-rich regions of the genome found that many of the CNVs were also in LD with surrounding SNPs, although to a lesser degree than what has been reported for deletions (Locke *et al.*, 2006). The report by Redon *et al.* also examined this issue of LD and found that just over half of the 65 CNVs they examined in their LD analysis can be tagged by neighboring SNPs. Based on this, the authors offer cautious optimism for the use of SNP-based tagging methods of assaying CNVs.

As structural variation appears to be a likely source of phenotypic variation, and as more structural variants are identified, it will become increasingly important to examine and account for such variation in disease gene association studies. While at least some CNVs may be ascertained through a tagging SNP approach, their direct screening with the use of high-density SNP genotyping arrays offers a comprehensive, economical, and robust method to CNV analysis. Both Affymetrix and Illumina's whole-genome products offer support for such analyses.

In conclusion, CNVs represent a significant source of variation that likely contributes to both normal phenotypic variation and disease susceptibility. At least some of the structural variation within the genome can be indirectly assayed through tagging SNP approaches; however, the new genome-wide SNP-typing platforms allow such variation to be directly examined at higher resolutions and with greater precision than classical array-CGH methods. The study of structural variation in complex traits is a relatively new area of research and additional work is needed to more clearly define the role of these variants within the genome, to more completely map these variants, and to determine the mechanisms by which they contribute to complex trait variation. The use of high-density SNP genotyping arrays should facilitate such work.

VI. FUTURE PROSPECTS

Novel technological developments have always preceded fundamental advances in our understanding of a given field. The development of high-density, highly parallel SNP genotyping platforms has enabled powerful whole-genome association studies that were not possible just a few years ago. Ultimately, these studies have the potential to impact on the way we diagnose and treat patients and view human phenotypic variation. To this end, mass-throughput SNP-typing platforms have the potential to usher in advances in numerous and diverse fields, but may play a particular role in the advancement of personalized medicine. Furthermore, as developments continue in the area of array-based genomic assays, multiple types of analyses will be included on one chip, allowing for comprehensive genomic analyses at the DNA, RNA, and protein levels. Such integrative analyses hold the promise to revolutionize our molecular view of health and disease.

A. SNP-typing platforms in the clinic

Currently, genetics does not play a prominent role in the evaluation of a patient in most fields of clinical medicine. The best approximation we have to account for the role of genetics in health and disease is the taking of a family history during a patient's initial evaluation. However, with the contribution of genes to chronic disease being increasingly recognized, personalized medicine based on an individual's genetic makeup is fast becoming the ultimate goal of clinical medicine. The diagnosis and treatment of patients based on knowledge of their underlying genetic variation will lead to more targeted disease management and enhanced therapeutic outcomes. Perhaps more importantly, such knowledge can lead to improved primary prevention of disease, as those known to carry particular high-risk variants can be more aggressively monitored. While this type of individualized medicine is still in the future, highly parallel, economical SNP arrays, like those offered by Affymetrix and Illumina, bring us one step closer to making it a reality. With the technology in place, the challenge now is how to interpret the massive amount of data such arrays provide and how to properly combine it with other patient data, both genomic and otherwise. This challenge will only be met, at least in part, with continued large-scale genetic epidemiological studies. Despite this, there are a few limited applications of SNP arrays already being used in the clinic.

As an example of what is currently possible, although it is not a genome-wide test, Roche Pharmaceuticals currently markets the AmpliChip®-CYP450 test, an Affymetrix array of more than 15,000 probes interrogating numerous SNPs in the CYP2D6 and CYP2C19 genes (www.amplichip.us). These genes are members of the cytochrome P450 family of genes that are responsible for metabolizing many commonly prescribed drugs, including antidepressants, antipsychotics, and antiarrhythmics. Based on variation within these two genes, patients generally fall into one of four metabolic phenotypes—ultra-rapid, extensive, intermediate, or poor. Knowledge of a patient's phenotype can assist physicians in prescribing the right dosage. Patient's can have their blood drawn and sent to one of four laboratories across the country that currently perform the test. Results are returned within a few days and the physician can then use that information to guide treatment. This test is the first of its kind to allow doctors to base drug choice and dosing for an individual on scientific data, before treatment is even begun. This test represents an early example of how mass-throughput genotyping arrays can be incorporated into the clinic and how they can directly contribute to personalized medicine. As our understanding of genomic variation grows and as risk information becomes available for more variants, more such tests can be expected to be offered. The ultimate test, of course, is to assay an individual's entire genome, either through use of the genotyping arrays or through resequencing, and then use that data to determine an individual's complete risk profile.

B. Integrative array-based genomic analyses

While SNPs can provide important information about an individual's risk for disease, they do not provide a complete picture of genomic variation and function, and hence, of human phenotypic variation. Knowledge of DNA sequence only provides a limited glimpse into the molecular mechanisms behind complex traits. It is RNA, and ultimately proteins, that carry forward the genetic message encoded in genes. Consequently, extensive research efforts are now focusing on a more integrated view of the genome, linking together all components to describe biological systems. To this end, array-based genomic analyses are quickly evolving.

Microarray technology has been in use now for over a decade for transcript profiling. While it is still a relatively new method for genotyping, it is fast on its way to becoming routine for this purpose as well. However, the potential of microarray technology exceeds its use for either of these two applications. There is a range of applications for which the use of microarrays is under development, including resequencing, epigenetic analysis, protein–DNA and protein–RNA interactions, DNA-structure analysis, as well as a number of other ideas (Hoheisel, 2006). Data from such arrays will be complementary to that offered by the genotyping arrays and the combined analysis of data from different molecular levels will offer unprecedented insights into genome variation and function, and ultimately, human phenotypic variation.

VII. CONCLUSIONS

In the past decade, the field has witnessed a dramatic change in the way in which genes underlying complex traits are identified. In the 1990s, multiallelic, highly informative microsatellites were the genetic marker of choice used in linkage studies incorporating small to moderate numbers of families. While they offered a considerable gain in throughput from the previous RFLP methods, they still were not the ideal marker for the identification of complex disease-susceptibility genes using association studies. The sequencing of the human genome laid the foundation for the next step forward, with the identification of SNPs. These variants were abundant, simple, and inexpensive to assay, and their analysis was relatively straightforward. They have since ushered in a new era of gene-mapping, enabling large-scale association studies, and with the advent of cost-effective genome-wide SNP-typing arrays, whole-genome association studies are now practical. A variety of GWA platforms, as well as a number of focused content applications, are available to investigators. Selection of the appropriate platform is influenced by a many factors and investigators must carefully consider their objectives and population of interest before choosing. Results from many of the initial association studies looking at candidate genes based on a previous linkage finding or knowledge of disease were mixed, with many unable to be replicated. It is hoped that the results of hypothesis-free genome-wide association studies unconstrained by our limited knowledge of disease mechanisms will meet with more success. We are now only just seeing the first examples of this whole-genome approach. As one of the first GWA scans to be published, Klein *et al.* (2005) identified a polymorphism in the complement factor H gene that substantially increases the risk of age-related macular degeneration. This gene is located in a region repeatedly linked to AMD in family studies. In another example, Duerr *et al.* (2006) identified strong associations for Crohn's disease with SNPs in the IL23R gene, which were subsequently replicated in additional Crohn's cohorts. Much the same way as microsatellites set in motion the explosion in Mendelian disease-gene identification in the 1990s, it is hoped that the availability of mass-throughput SNP-typing platforms will lead to a rapid growth in gene identification for polygenic diseases. The challenges are now to ensure the proper design of GWA studies, and to devise novel methods of handling and analyzing the massive amount of data these studies are sure to generate, as well as the functional evaluation of significant SNPs identified in these studies.

References

Adessi, C., Matton, G., Ayala, G., Turcatti, G., Mermod, J. J., Mayer, P., and Kawashima, E. (2000). Solid phase DNA amplification: Characterisation of primer attachment and amplification mechanisms. *Nucleic Acids Res.* **28**(20), E87.

Barrett, J. C., and Cardon, L. R. (2006). Evaluating coverage of genome-wide association scans. *Nat. Genet.* **38**(6), 659–662.

Brownie, J., Shawcross, S., Theaker, J., Whitcombe, D., Ferrie, R., Newton, C., and Little, S. (1997). The elimination of primer-dimer accumulation in PCR. *Nucleic Acids Res.* **25**(16), 3235–3241.

Burn, J., Wilson, D., and Cross, I. (1995). The clinical significance of 22q11.2 deletion. *In* "Developmental Mechanisms of Heart Disease" (E. B. Clark, R. R. Markwald, and A. Takao, eds.). Futura, New York.

Dalma-Weiszhausz, D. D., Warrington, J., Tanimoto, E. Y., and Miyada, C. G. (2006). The Affymetrix Genechip platform: An overview. *Methods Enzymol.* **410**(1), 3–28.

Daly, M. J., Rioux, J. D., Schaffner, S. F., Hudson, T. J., and Lander, E. S. (2001). High-resolution haplotype structure in the human genome. *Nat. Genet.* **29**(2), 229–232.

Duerr, R. H., Taylor, K. D., Brant, S. R., Rioux, J. D., Silverberg, M. S., Daly, M. J., Steinhart, A. H., Abraham, C., Regueiro, M., Griffiths, A., *et al.* (2006). A genome-wide association study identifies IL23R as an inflammatory bowel disease gene. *Science* **314**(5804), 1461–1463.

Evans, D. M., and Cardon, L. R. (2004). Guidelines for genotyping in genome-wide linkage studies: Single-nucleotide-polymorphism maps versus microsatellite maps. *Am. J. Hum. Genet.* **75**(4), 687–692.

Fan, J. B., Oliphant, A., Shen, R., Kermani, B. G., Garcia, F., Gunderson, K. L., Hansen, M., Steemers, F., Butler, S. L., Deloukas, P., *et al.* (2003). Highly parallel SNP genotyping. *Cold Spring Harb Symp Quant Biol.* **68,** 69–78.

Fan, J. B., Chee, M. S., and Gunderson, K. L. (2006). Highly parallel genomic assays. *Nat. Rev. Genet.* **7,** 632–644.

Feuk, L., Carson, A. R., and Scherer, S. W. (2006). Structural variation in the human genome. *Nat. Rev. Genet.* **7,** 85–97.

Gabriel, S. B., Schaffner, S. F., Nguyen, H., Moore, J. M., Roy, J., Blumenstiel, B., Higgins, J., DeFelice, M., Lochner, A., Faggart, M., *et al.* (2002). The structure of haplotype blocks in the human genome. *Science* **296**(5576), 2225–2229.

Gonzalez, E., Kulkarni, H., Bolivar, H., Mangano, A., Sanchez, R., Catano, G., Nibbs, R. J., Freedman, B. I., Quinones, M. P., Bamshad, M. J., *et al.* (2005). The influence of CCL3L1 gene-containing segmental duplications on HIV-1/AIDS susceptibility. *Science* **307**(5714), 1434–1440.

Goode, E. L., Badzioch, M. D., and Jarvik, G. P. (2005). Bias of allele-sharing linkage statistics in the presence of intermarker linkage disequilibrium. *BMC Genet.* **6**(Suppl 1), S82.

Greenberg, F. (1990). Williams syndrome professional symposium. *Am. J. Med. Genet.* **6**(Suppl), 85–88.

Gunderson, K. L., Steemers, F. J., Ren, H., Ng, P., Zhou, L., Tsan, C., Chang, W., Bullis, D., Musmacker, J., King, C., *et al.* (2006). "Whole-Genome Genotyping." *Methods Enzymol.* **410,** 359–376.

Helgadottir, A., Manolescu, A., Thorleifsson, G., Gretarsdottir, S., Jonsdottir, H., Thorsteinsdottir, U., Samani, N. J., Gudmundsson, G., Grant, S. F., Thorgeirsson, G., *et al.* (2004). The gene encoding 5-lipoxygenase activating protein confers risk of myocardial infarction and stroke. *Nat. Genet.* **36** (3), 233–239.

Hinds, D. A., Kloek, A. P., Jen, M., Chen, X., and Frazer, K. A. (2006). Common deletions and SNPs are in linkage disequilibrium in the human genome. *Nat. Genet.* **38**(1), 82–85.

Hirschhorn, J. N., Lohmueller, K., Byrne, E., and Hirschhorn, K. (2002). A comprehensive review of genetic association studies. *Genet. Med.* **4**(2), 45–61.

Hoheisel, J. D. (2006). Microarray technology: Beyond transcript profiling and genotype analysis. *Nat. Rev. Genet.* **7**(3), 200–210.

Horikawa, Y., Oda, N., Cox, N. J., Li, X., Orho-Melander, M., Hara, M., Hinokio, Y., Lindner, T. H., Mashima, H., Schwarz, P. E., *et al.* (2000). Genetic variation in the gene encoding calpain-10 is associated with type 2 diabetes mellitus. *Nat. Genet.* **26**(2), 163–175.

Huang, Q., Shete, S., Swartz, M., and Amos, C. I. (2005). Examining the effect of linkage disequilibrium on multipoint linkage analysis. *BMC Genet.* **6**(Suppl 1), S83.

Hugot, J. P., Chamaillard, M., Zouali, H., Lesage, S., Cezard, J. P., Belaiche, J., Almer, S., Tysk, C., O'Morain, C. A., Gassull, M., *et al.* (2001). Association of NOD2 leucine-rich repeat variants with susceptibility to Crohn's disease. *Nature* **411**(6837), 599–603.

Iafrate, A. J., Feuk, L., Rivera, M. N., Listewnik, M. L., Donahoe, P. K., Qi, Y., Scherer, S. W., and Lee, C. (2004). Detection of large-scale variation in the human genome. *Nat. Genet.* **36**(9), 949–951.

John, S., Shepard, N., Liu, G., Zeggini, E., Cao, M., Chen, W., Vasavda, N., Mills, T., Barton, A., Hinks, A., *et al.* (2004). Whole-genome scan, in a complex disease, using 11,245 single-nucleotide polymorphisms: Comparison to microsatellites. *Am. J. Hum. Genet.* **75,** 54–64.

Klein, R. J., Zeiss, C., Chew, E. Y., Tsai, J. Y., Sackler, R. S., Haynes, C., Henning, A. K., SanGiovanni, J. P., Mane, S. M., Mayne, S. T., *et al.* (2005). Complement factor H polymorphism in age-related macular degeneration. *Science* **308,** 385–389.

Kruglyak, L. (1997). The use of a genetic map of biallelic markers in linkage studies. *Nat. Genet.* **17** (1), 21–24.

Lin, Z., Cui, X., and Li, H. (1996). Multiplex genotype determination at a large number of gene loci. *Proc. Natl. Acad. Sci.* **93**(6), 2582–2587.

Locke, D. P., Sharp, A. J., McCarroll, S. A., McGrath, S. D., Newman, T. L., Cheng, Z., Schwartz, S., Albertson, D. G., Pinkel, D., Altshuler, D. M., *et al.* (2006). Linkage disequilibrium and heritability of copy-number polymorphisms within duplicated regions of the human genome. *Am. J. Hum. Genet.* **79**(2), 275–290.

Mamotte, C. D. (2006). Genotyping of single nucleotide substitutions. *Clin. Biochem. Rev.* **27**(1), 63–75.

McCarrolll, S. A., Hadnott, T. N., Perry, G. H., Sabeti, P. C., Zody, M. C., Barrett, J. C., Dallaire, S., Gabriel, S. B., Lee, C., Daly, M. J., *et al.* (2006). For the International HapMap Consortium. Common deletion polymorphisms in the human genome. *Nat. Genet.* **38**(1), 86–92.

Michael, K. L., Taylor, L. C., Schultz, S. L., and Walt, D. R. (1998). Randomly ordered addressable high-density optical sensor arrays. *Anal. Chem.* **70,** 1242–1248.

Middleton, F. A., Pato, M. T., Gentile, K. L., Morley, C. P., Zhao, X., Eisener, A. F., Brown, A., Petryshen, T. L., Kirby, A. N., Medeiros, H., *et al.* (2004). Genome-wide linkage analysis of bipolar disorder by use of a high-density single-nucleotide-polymorphism (SNP) genotyping assay: A comparison with microsatellite marker assays and finding of significant linkage to chromosome 6q22. *Am. J. Hum. Genet.* **74**(5), 886–897.

Monaco, A. P., Neve, R. L., Colletti-Feener, C., Bertelson, C. J., Kurnit, D. M., and Kunkel, L. M. (1986). Isolation of candidate cDNAs for portions of the Duchenne muscular dystrophy gene. *Nature* **323,** 646–650.

Murray, S. S. (2005). Evaluation of linkage disequilibrium and its effect on non-parametric multipoint linkage analysis using two high density single-nucleotide polymorphism mapping panels. *BMC Genet.* **6**(Suppl 1), S85.

Nelis, E., Van Broeckhoven, C., De Jonghe, P., Lofgren, A., Vandenberghe, A., Latour, P., Le Guern, E., Brice, A., Mostacciuolo, M. L., Schiavon, F., *et al.* (1996). Estimation of the mutation frequencies in Chracot-Marie-Tooth disease type 1 and hereditary neuropathy with liability to pressure palsies: A European Collaborative Study. *Eur. J. Hum. Genet.* **4,** 25–33.

Nsengimana, J., Renard, H., and Goldgar, D. (2005). Linkage analysis of complex diseases using microsatellites and single-nucleotide polymorphisms: Application to alcoholism. *BMC Genet.* **6** (Suppl 1), S29.

Ogura, Y., Bonen, D. K., Inohara, N., Nicolae, D. L., Chen, F. F., Ramos, R., Britton, H., Moran, T., Karaliuskas, R., Duerr, R. H., *et al.* (2001). A frameshift mutation in NOD2 associated with susceptibility to Crohn's disease. *Nature* **411**(6837), 603–606.

Oliphant, A., Barker, D. L., Stuelpnagel, J. R., and Chee, M. S. (2002). BeadArray technology: Enabling an accurate, cost-effective approach to high-throughput genotyping. *Biotechniques Suppl.* **56–58,** 60–61.

Patil, N., Berno, A. J., Hinds, D. A., Barrett, W. A., Doshi, J. M., Hacker, C. R., Kautzer, C. R., Lee, D. H., Marjoribanks, C., McDonough, D. P., *et al.* (2001). Blocks of limited haplotype diversity revealed by high-resolution scanning of human chromosome 21. *Science* **294**(5547), 1719–1723.

Peralta, J. M., Dyer, T. D., Warren, D. M., Blangero, J., and Almasy, L. (2005). Linkage disequilibrium across two different single-nucleotide polymorphism genome scans. *BMC Genet.* **6**(Suppl 1), S86.

Redon, R., Ishikawa, S., Fitch, K. R., Feuk, L., Perry, G. H., Andrews, T. D., Fiegler, H., Shapero, M. H., Carson, A. R., Chen, W., *et al.* (2006). Global variation in copy number in the human genome. *Nature* **444**(7118), 444–454.

Riordan, J. R., Rommens, J. M., Kerem, B., Alon, N., Rozmahel, R., Grzelczak, Z., Zielenski, J., Lok, S., Plavsic, N., Chou, J. L., *et al.* (1989). Identification of the cystic fibrosis gene: Cloning and characterization of complementary DNA. *Science* **245**(4922), 1066–1073.

Risch, N., and Merikangas, K. (1996). The future of genetic studies of complex human diseases. *Science* **273,** 1516–1517.

Schaid, D. J., Guenther, J. C., Christensen, G. B., Hebbring, S., Rosenow, C., Hilker, C. A., McDonnell, S. K., Cunningham, J. M., Slager, S. L., Blute, M. L., *et al.* (2004). Comparison of microsatellites versus single-nucleotide polymorphisms in a genome linkage screen for prostate cancer-susceptibility Loci. *Am. J. Hum. Genet.* **75**(6), 948–965.

Sebat, J., Lakshmi, B., Troge, J., Alexander, J., Young, J., Lundin, P., Maner, S., Massa, H., Walker, M., Chi, M., *et al.* (2004). Large-scale copy number polymorphism in the human genome. *Science* **305**(683), 525–528.

Tayo, B. O., Liang, Y., Stranges, S., and Trevisan, M. (2005). Genome-wide linkage analysis of age at onset of alcohol dependence: A comparison between microsatellites and single-nucleotide polymorphisms. *BMC Genet.* **6**(Suppl 1), S29.

Terwilliger, J. D., and Hiekkalinna, T. (2006). An utter refutation of the fundamental theorem of the HapMap. *Eur. J. Hum. Genet.* **14**(4), 426–437.

Weber, J. L., and Broman, K. W. (2001). Genotyping for human whole-genome scans: Past, present, and future. *In* "Genetic Dissection of Complex Traits" (D. C. Rao and M. A. Province, eds.). Academic Press, San Diego.

Weber, J. L., and May, P. E. (1989). Abundant class of human DNA polymorphisms which can be typed using the polymerase chain reaction. *Am. J. Hum. Genet.* **44**(3), 388–396.

Xing, C., Schumacher, F. R., Xing, G., Lu, Q., Wang, T., and Elston, R. C. (2005). Comparison of microsatellites, single-nucleotide polymorphisms (SNPs) and composite markers derived from SNPs in linkage analysis. *BMC Genet.* **6**(Suppl 1), S29.

6

Genotyping Errors and Their Impact on Genetic Analysis

Michael B. Miller,[*] Karen Schwander,[†] and D. C. Rao[†,‡]

[*]Division of Epidemiology and Community Health,
School of Public Health, and Institute of Human Genetics,
University of Minnesota, Minneapolis, Minnesota 55454
[†]Division of Biostatistics, Washington University School of Medicine,
St. Louis, Missouri 63110
[‡]Departments of Psychiatry and Genetics, Washington University in St. Louis, Missouri 63110, and Department of Mathematics, Washington University in St. Louis, Missouri 63130

Advances in Genetics, Vol. 60

0065-2660/08 $35.00
DOI: 10.1016/S0065-2660(07)00406-3

ABSTRACT

Genetic dissection of complex traits involves a series of analyses of phenotypic and genotypic datasets in samples of families and/or unrelated individuals. In particular, both linkage analysis and association analysis depend critically on the quality of phenotype and genotype data. The focus of this chapter is on certain types of genotyping errors and how marker data can help resolve problems of misspecification of familial relationships and misclassification of genotypes, and on how those kinds of errors can be detected and corrected. The impact of these types of errors on the results of genetic analyses will also be discussed. Other types of errors such as measurement error or misclassification in regard to phenotypes are not considered here.

I. INTRODUCTION

Just as phenotypes are measured with error, genotyping also involves errors although to a much lesser degree. In this chapter, the term "genotyping" will refer to the process by which individuals are classified with regard to pairs of alleles at one or more genetic loci, with misclassification being always possible. We will discuss ways of minimizing the impact of misclassification on statistical power in linkage and association studies.

In study designs that use biologic relatives (e.g., in linkage analysis), one must first ensure that estimated genotypes are assigned to the correct individuals and that the familial relationships of those individuals are correctly specified. Then, and only then, can one contend with genotype misclassification. If familial relationships are incorrectly specified, estimated genotypes might appear to be in error because individuals will then have genotypes that are inconsistent with their putative genetic relationships (e.g., alleles observed in offspring may not be seen in the purported parents). It is therefore a good practice to precede error-detection analysis of estimated genotypes with an analysis of familial genetic relationships using the uncorrected marker data. Such analyses of relationships are most effective when a large number of polymorphic markers distributed across the genome are available.

Genotyping errors tend to reduce the statistical power of both linkage and association studies. No system for genotyping, or for detecting and correcting genotyping errors, will ever be perfect. Correction of errors can improve statistical power, but the correction process is itself error prone, and if not done well it has the potential to create new errors that might unwittingly reduce statistical power even further. We will first discuss the impact of genotyping errors on genetic analysis, and then we will discuss various analytical approaches that minimize impact.

II. MISSPECIFICATION OF GENETIC RELATIONSHIPS

Every participant in a genetic study is given an identifier (ID), which is associated with that individual's data in computer records. Such data may include estimated genotypes, sex, disease affection status, age, ethnicity, height, weight, and blood pressure, etc. It may also include the IDs of the individual's parents, whether they were brought into the clinic as participants or simply recorded as founders with "dummy" IDs. Within families, IDs must be unique but it is possible to have multiple individuals with the same ID as long as they are in different families and a family ID is used. When a family ID is used, the individual can always be uniquely identified by the combination of his/her family ID with his/her individual ID. Most investigators find it convenient to use family IDs (many genetic analysis tools require them) but it is also common practice to use unique individual IDs as well, so that the family ID is not needed to identify an individual. Marker data are typically arranged for genetic analysis in a plaintext file with one record per row and space-delimited or tab-delimited fields ordered in the usual "linkage format" (see, e.g., Terwilliger and Ott, 1994).

A. Dissociation of marker data from the individual ID

One typical problem that arises during the verification of familial relationships is that the marker data may be assigned to the incorrect individual ID. This can happen when an ID associated with an individual's marker data is incorrectly entered or typed, or when the labels are switched between samples of DNA. Errors of this type can happen at various stages in the process of data collection, and the ramifications of the error depend on when/where the error occurred. For example, if the labeling error was on DNA samples, but not on blood samples, it may be possible to reassign the marker genotypes to the correct ID and fix the problem. On the other hand, if blood samples were mislabled, then any phenotypes derived from blood (e.g., lipid measurements) will have been assigned to the wrong participant. Thus, when one discovers that the marker data have been assigned to the wrong individual ID, it is not immediately clear if other phenotypes have also been incorrectly assigned to that ID. It is possible that some of the phenotypes correspond to the ID and other phenotypes correspond to the markers. In most cases, when the cause of the error is not clear, the prudent step is to remove all phenotypes derived from blood, and reassign the marker data with the correct ID.

B. Inaccurate information about familial relations

Investigators rely on family members to inform them about their genetic relationships, but this information can be inaccurate. The rate of nonpaternity is largely dependent on the population, with commonly observed percentages around 7%. The effect of nonpaternity is that some purported siblings will turn out to be maternal half siblings. Other problems that have been observed include the reporting of an adoptee as an offspring and the reporting of a grandchild as an offspring. Finally, the zygosity of twin pairs may be misdiagnosed, most often with monozygotic (MZ) pairs mistakenly identified as dizygotic (DZ). This error of zygosity diagnosis tends to increase evidence of linkage when both twins are affected, a finding that caused problems for some well-known studies involving dyslexia (Cardon *et al.*, 1995) where 4 DZ pairs were initially diagnosed as MZ, and bipolar disorder (Dick *et al.*, 2003) where 12 MZ pairs were mistakenly treated as siblings. All of these types of misspecified genetic relationships are detectable using appropriate software when enough marker data are available on pairs of purported relatives.

C. Errors in data entry

If a parental ID is entered incorrectly, genetic relationships are thereby misspecified. Such errors can create spurious cases of nonpaternity or even nonmaternity.

D. Related families specified as unrelated

In genetic analyses we assume that individuals of unknown (and presumably distant) genetic relationship are not related. This assumption is not problematic, but when we can show that families we have assumed to be unrelated are actually very closely related, we can combine those related families appropriately and thereby increase statistical power for linkage analyses. For example, when nuclear families are ascertained in a small community, it is not unusual to find that some of the parents in those families (the founders) are siblings of the parents in other families. Fortunately, with enough markers, it is possible to recognize and combine such groups of families.

III. DETECTION AND CORRECTION OF MISSPECIFICATION OF RELATIONSHIPS

A. GRR

The computer program GRR (Graphical Representation of Relationships; Abecasis *et al.*, 2000) is a model of simplicity and user-friendliness. The program converts marker data and purported family relationships into a graph where one color-coded point is plotted for every pair of genotyped individuals. The concept is simple, but the graphical output can be surprisingly informative when enough marker data are available. The program takes the pair of measured genotype vectors for every pair of subjects and compares them to see how many alleles are shared [identical by state (IBS)] at each locus. This provides an IBS count—0, 1, or 2—for every marker locus for every pair of subjects. GRR computes the mean and standard deviation of these IBS counts across marker loci within pairs of subjects. When a marker is missing for either subject, that marker is simply skipped. A scatter plot is then generated where the abscissa and ordinate (*X* and *Y* axes) of the graph are the mean and standard deviation, respectively, of the IBS counts for pairs of subjects. Because different types of relationship pairs will have different overall mean and standard deviation IBS values (e.g., the overall IBS mean for full siblings will be much higher than the overall IBS mean for unrelated subjects), the location of the different types of relative pairs tend to group together to form clusters for each type of genetic relationship. The plotted points are color coded to reflect the purported relationship of the pair of subjects. Importantly, the points that are farthest from the centroid for their purported relationship are rendered last. This means that the most interesting points are clearly visible so that misspecifications of relationships are readily detectable from the graph.

GRR has been used extensively in large studies with a few hundred microsatellite markers. In studies involving both multigenerational families and nuclear families, it has consistently revealed important aspects of the data that had previously gone unnoticed despite extensive analysis using other packages. GRR seems to be especially useful because it provides a clear view of the data and because it can quickly present results for all pairs of subjects in a data file, not just for pairs within families. When there are 3500 individuals with marker data, more than 6 million pairs are formed. It would be a massive task to compute likelihoods for every pair of individuals, but GRR's simpler approach provides a graphical view of the results in only a few minutes on an ordinary personal computer.

The ability of GRR to recognize genetic relationships across families is particularly useful in cases where the marker data has been disassociated from an individual ID because of a data entry error or sample mislabeling or mix-up. In fact, in many cases, GRR will identify subjects from two different families who have had their marker data "switched" with each other. In one real-life example of laboratory error, one sibling in each of four families was shown to be unrelated to other family members, but strangely, the four were related to each other as three full siblings and a parent. GRR demonstrated that in an entirely separate family, from a different field center, the marker data for four members of a single family had been switched with that of these four individuals. This extreme example shows how useful GRR can be in helping to recover information that otherwise would be lost. Figure 6.1 displays the GRR output for this particular case study, showing how the data appeared both before and after correction for these sample "mix-ups."

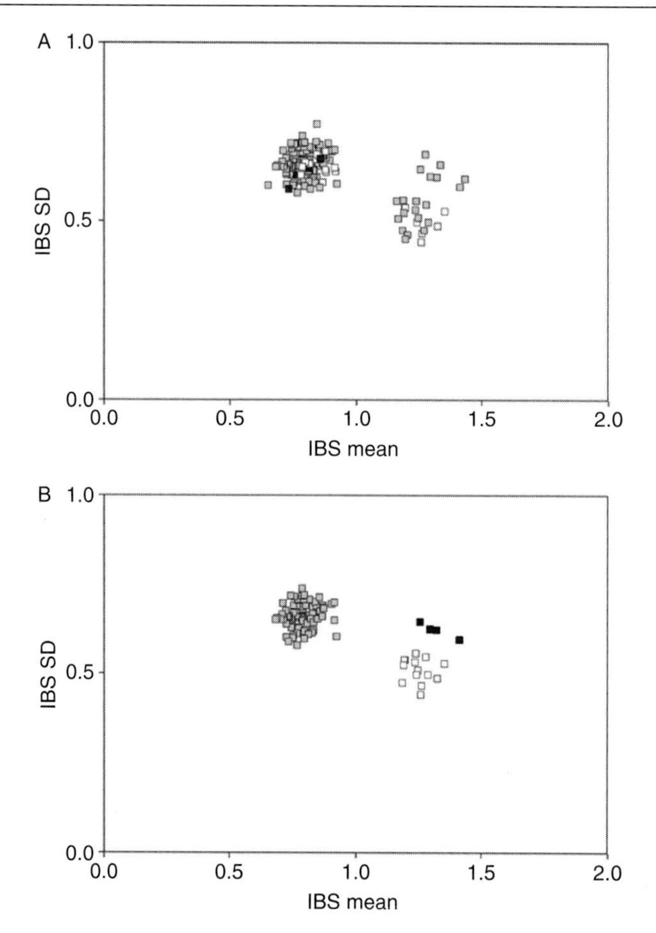

Figure 6.1. Sample graphical output from GRR using a real case study: Data for parent–offspring pairs are displayed in white, unrelated individuals in light gray, and sibling pairs in black [GRR displays them as yellow, cyan (blue-green), and red, respectively]. (A) GRR output using the raw data. (B) GRR output using the cleaned data.

GRR is not very useful if only distant relatives are genotyped. Our experience has been that we can detect problems between distant relatives only when marker data on intermediate individuals is available. For example, we can work well with cousins in GRR when the related parents (a sibling pair) are available because the cousin relationship is weak, but if the intermediate parental and sibling relationships appear to be correct, the correctness of the cousin relationship can be inferred.

B. ASPEX

The ASPEX computer program (Risch *et al.*, 1999) uses observed IBS values, like GRR, but ASPEX goes a step further by calculating, for every pair of relatives in a family, the base 10-logarithm likelihood ratios (LODs) for four relationships (parent–child, full siblings, half siblings, and MZ twins) versus unrelated. It then produces the most likely relationship based on the likelihood ratio. Because of this, ASPEX is best utilized when looking for misspecified relationships within a family, not among families. There is an obvious benefit in having these likelihood ratios, but they come at a considerable computational cost that is avoided in the GRR approach. A comparison of GRR with ASPEX when the marker data consisted of a few hundred microsatellite markers per subject indicated that the two programs agreed perfectly. This was even true for relationships deemed "indeterminate" by ASPEX: In those few cases, the points plotted by GRR would be in the region between centroids for the relationships so that we were not sure that the relationships were known accurately.

C. Eclipse

The Eclipse package (Sieberts *et al.*, 2002) includes a program called Eclipse2 that calculates likelihoods for pairs of individuals; but unlike ASPEX, these are multipoint likelihoods that use a genetic map and make use of phase information. Another program, Eclipse3, calculates likelihoods for trios of individuals. These programs are very sophisticated and not easy to use. On the other hand, they provide exactly the kind of information one would want in cases where GRR or ASPEX are unable to give a definitive answer. We used Eclipse to follow up on a suggestion from GRR that founders in certain families in a family study were siblings of founders in other families. Eclipse resoundingly confirmed the GRR finding with LODs of 100 and more. The Eclipse programs can be especially useful when only a small number of markers are available. With larger numbers of markers, GRR results tend to be very compelling on their own.

IV. GENOTYPE MISCLASSIFICATION

After detecting and correcting misspecifications of familial relationships and DNA sample mix-ups, we are in a much better position to detect genotype misclassification errors. Misclassification simply means that the estimated genotype is not the true genotype. The misclassification may be differential or nondifferential. In nondifferential misclassification, the genotyping error probability is independent of participant phenotypes such as affection status (Gordon and Finch, 2005). Such errors will introduce noise into case-control association studies or genotype–phenotype association studies of quantitative traits, which will reduce power, but they do not usually cause systematic bias and an increased rate of false-positive errors. On the other hand, in differential misclassification, the rate of error is a function of the phenotype so that a higher error rate in affected subjects, for example, could result in a false-positive association of some genotype with disease.

A. Misclassification rates

The rate of misclassification of genotypes depends on (1) the type of marker (SNP or microsatellite), (2) the specific marker being typed, (3) the quality of the DNA sample, and (4) the protocol and experience of the laboratory staff, among other factors. With current technology, SNP misclassification appears to occur in less than 1% of genotypes (Hao *et al.*, 2004) and error rates as low as 0.01% have been claimed (Murray *et al.*, 2004). Microsatellite markers show much higher rates of misclassification than do SNP markers. Murray *et al.* claim that microsatellite error rates are more than 100 times those of SNP markers, on average, but not all microsatellites have the same error rate.

B. Mendelian inconsistencies

Genotyping errors at a single locus can cause patterns of inheritance in families that are incompatible with Mendel's law of segregation. By finding such patterns in our data we are able to detect genotype misclassifications but we are rarely able to tell which of several genotypes in a family was misclassified. O'Connell and Weeks (1998) listed five patterns of incompatibility that can be observed in nuclear families and that prove that there is at least one genotyping error in the family:

(1) The alleles of a child and a parent are incompatible.
(2) The child is compatible with each parent separately but not when both parents are considered simultaneously.
(3) There are more than four alleles in a sibship.
(4) There are more than three alleles in a sibship with a homozygous child.
(5) There are more than two alleles in a sibship with two different homozygotes among the sibs.

Undetected nonpaternity can cause some of these effects at a single locus. However, GRR, ASPEX, or Eclipse makes it possible for us to detect and correct most cases of nonpaternity before searching for Mendelian inconsistencies.

O'Connell and Weeks also listed the following forms of evidence of error, but these do not require use of data from family members:

(6) An allele is out of bounds of any specified range.
(7) A male is heterozygous at an X-linked locus.
(8) An individual has only one allele defined at an autosomal locus.

In addition to these forms of inconsistency, there are other types of subtle inconsistencies often involving more distant relatives such as cousins that can be detected using computer software such as PEDCHECK (O'Connell and Weeks, 1998, 1999; also see Stringham and Boehnke, 1996; Lange and Goradia, 1987; Lange, 2002). In fact, there are a variety of programs available that will find and correct Mendelian inconsistencies within families. In most cases, one genotype (both alleles) is deleted selectively to remove inconsistencies in the family and to minimize information loss. If data from other family members is not available or informative to clarify which genotype is correct, then all inconsistent genotypes may be deleted within the family.

C. Unlikely double recombinants

Some genotyping errors will cause no Mendelian inconsistencies, but it may still be possible to detect an error using information from neighboring loci (Abecasis *et al.*, 2002; Sobel *et al.*, 2002). This is because the genotyping error may change that parental allele appears to have been transmitted in a meiosis. When compared to flanking markers this will suggest double recombination. If the flanking markers are very near, such a double recombinant should be extremely rare. If the prior probability of genotyping error is sufficiently higher than the prior probability of double recombination, it is reasonable to infer that a genotyping error probably has occurred.

Unfortunately, this method does not work well when markers are more widely spaced. The Genetic Analysis Workshop 15 simulated data including 735 autosomal microsatellite markers with an average spacing of about 5 cM and as many as 18 alleles per marker. We selected a random sample of 200 nuclear families (always two parents and two affected offsprings), with 65,600 alleles on chromosome 6 (800 subjects × 41 loci × 2 alleles per locus). We randomly selected 30 alleles and added 1 to the allele number of each of the 30 alleles, thus creating 30 errors. The program PEDSTATS from the MERLIN package (Abecasis *et al.*, 2002) was then run to identify Mendelian inconsistencies among the 735 loci. Of the 30 simulated errors, 22 were detected in this way and there were no false positive errors. At this point, all genotypes from the 22 families with detected errors were set to missing at the identified locus, and MERLIN's "error" analysis on the corrected data was run to search for unlikely double recombinants. MERLIN detected 211 likely errors (or unlikely double recombinants) in the 588,000 genotypes (800 subjects × 735 markers). Most of the 211 flagged genotypes were not errors. In fact, only 1 of 211 was an error. Thus, the remaining seven simulated errors were not detected. We therefore do not recommend use of this method with typical microsatellite maps used in genome-wide linkage studies. A denser map should provide much better results.

It is important to remember that this method of detecting errors depends strongly on the correct specification of marker order on the genetic map. If marker order is misspecified, a single recombination can look like a double recombination. In fact, a finding of many probable errors for a single marker might imply that the marker order was misspecified and not that there were many genotyping errors.

D. Allele shifting

A major cause of differential misclassification derives from allele shifting effects when DNA samples are processed in batches (Chang *et al.*, 2006). Allele shifting refers to a change in the size of the same allele from batch to batch. Microsatellite marker alleles are classified by the number of nucleotides in a repeating DNA sequence at the marker locus. For example, a trinucleotide repeat marker might have alleles 153, 156, 159, and 162 according to a certain laboratory on a certain date. But the same marker might have alleles 152, 155, 158, and 161 in a different laboratory or on a different date. The relative number of repeats is same in both of these batches but the absolute number has changed (it is possible for relative numbers also to change). A naive analysis of data pooled from these two batches would assume that there are eight different alleles instead of only four. This would then cause problems in linkage analysis because allele frequencies and information content would be misspecified. The effect of this misspecification would be lower allele frequencies and reduced linkage evidence.

Allele shifting can often be detected and corrected. Chang *et al.* (2006) detected allele shifting by comparing allele frequencies among batches within ethnic groups. The frequency of any particular allele should not change dramatically among batches unless some shifting has occurred. Chang *et al.* have proposed the following formula for detecting allele shifting:

$$S = \sqrt{\frac{\sum (x_{ij} - x_{ik})^2}{n(n-1)/2}}$$

where x_{ij} is the allele frequency of allele i in batch j, x_{ik} is the allele frequency of allele i in batch k, and n is the number of batches compared. The summation is across all batches. A large value of S indicates large changes in allele frequencies across batches and probable allele shifting. Chang *et al.* also used, as an alternative, a simple χ^2 test on the allele frequencies to see if they varied by batch. Both methods provided very similar results. They identified the same 10 markers (of ~380) as showing strong evidence of allele shifting.

These statistical tests may be used to decide whether simple and clear data corrections are indicated or whether to delete specific problematic alleles or entire markers in subsets of individuals or an entire batch. Decisions about deleting suspect genotype data should be made on a case-by-case basis; in most situations, an examination of the allelic distribution within a marker will show a clear distinction between the problematic alleles and alleles that have not shifted between batches. Decisions to "correct," instead of deleting, the genotype data should not be taken lightly. In all cases, the conservative choice is to simply delete any suspect data.

One could avoid allele shifting in principle by avoiding batch processing of DNA samples but that is not always feasible, especially in very large studies that take a long time to complete recruitment. In such studies, the calling of alleles over a long period may have changed owing to different primer designs or alterations in allele calling or there may be interindividual differences in allele calling.

V. CONCLUSION

Quality control of genotype and pedigree data is an integral part of the study process. Even though most errors are corrected by deleting some data, the quality improvement of the resulting cleaned data invariably leads to a gain in the information content and power in linkage and association studies. Therefore, in the presence of genotyping errors, more can be less and, when errors are corrected, less can be more. In addition, many commonly used genetic analysis programs will fail if they come across an error in the genotype data or pedigree file. Genetic analyses rely on genetic data that are consistent with the Mendelian rules of inheritance and in which family relationships are reported accurately. The genotyping errors discussed in this chapter will affect the outcome of data analysis, thereby impacting the power of gene discovery.

References

Abecasis, G. R., Cherny, S. S., Cookson, W. O. C., and Cardon, L. R. (2000). GRR: Graphical representation of relationship errors. *Bioinformatics* **17,** 742–743.

Abecasis, G. R., Cherny, S. S., Cookson, W. O., and Cardon, L. R. (2002). Merlin-rapid analysis of dense genetic maps using sparse gene flow trees. *Nat. Genet.* **30,** 97–101.

Cardon, L. R., Smith, S. D., Fulker, D. W., Kimberling, W. J., Pennington, B. F., and DeFries, J. C. (1995). Quantitative trait locus for reading disability: Correction. *Science* **268,** 1553.

Chang, Y. C., Kim, J. D., Schwander, K., Rao, D. C., Miller, M. B., Weder, A. B., Cooper, R. S., Schork, N. J., Province, M. A., Morrison, A. C., Kardia, S. L. R., Quertermous, T., *et al.* (2006). The impact of data quality on the identification of complex disease genes: Experience from the Family Blood Pressure Program. *Eur. J. Hum. Genet.* **4,** 469–477.

Dick, D. M., Foroud, T., Flury, L., Bowman, E. S., Miller, M. J., Rau, N. L., Moe, P. R., Samavedy, N., El-Mallakh, R., Manji, H., Glitz, D. A., Meyer, E. T., *et al.* (2003). Erratum to "Genomewide Linkage Analyses of Bipolar Disorder: A New Sample of 250 Pedigrees from the National Institute of Mental Health Genetics Initiative." *Am. J. Hum. Genet.* **73,** 979.

Gordon, D., and Finch, S. J. (2005). Factors affecting statistical power in the detection of genetic association. *J. Clin. Invest.* **115,** 1408–1418.

Hao, K., Li, C., Rosenow, C., and Hung Wong, W. (2004). Estimation of genotype error rate using samples with pedigree information—an application on the GeneChip Mapping 10K array. *Genomics* **84,** 623–630.

Lange, K. (2002). "Mathematical and Statistical Methods for Genetic Analysis," 2nd Edition, Springer, New York.

Lange, K., and Goradia, T. M. (1987). An algorithm for automatic genotype elimination. *Am. J. Hum. Genet.* **40,** 250–256.

Murray, S., Oliphant, A., Shen, R., McBride, C., Steeke, R. J., Shannon, S. G., Rubano, T., Kermani, B. G., Fan, J. B., Chee, M. S., and Hansen, M. S. (2004). A highly informative SNP linkage panel for human genetic studies. *Nat. Methods* **1,** 113–117.

O'Connell, J. R., and Weeks, D. E. (1998). PedCheck: A program for identification of genotype incompatibilities in linkage analysis. *Am. J. Hum. Genet.* **63,** 259–266.

O'Connell, J. R., and Weeks, D. E. (1999). An optimal algorithm for automatic genotype elimination. *Am. J. Hum. Genet.* **65,** 1733–1740.

Risch, N., Spiker, D., Lotspeich, L., Nouri, N., Hinds, D., Hallmayer, J., Kalaydjieva, L., McCague, P., Dimiceli, S., Pitts, T., Nguyen, L., Yang, J., *et al.* (1999). A genomic screen of autism: Evidence for a multilocus etiology. *Am. J. Hum. Genet.* **65,** 493–507.

Sieberts, S. K., Wijsman, E. M., and Thompson, E. A. (2002). Relationship inference from trios of individuals, in the presence of typing error. *Am. J. Hum. Genet.* **70,** 170–180.

Sobel, E., Papp, J. C., and Lange, K. (2002). Detection and integration of genotyping errors in statistical genetics. *Am. J. Hum. Genet.* **70,** 496–508.

Stringham, H. M., and Boehnke, M. (1996). Identifying marker typing incompatibilities in linkage analysis. *Am. J. Hum. Genet.* **59,** 946–950.

Terwilliger, J. D., and Ott, J. (1994). "Handbook of Human Genetic Linkage." Johns Hopkins University Press, Baltimore.

Part II

LINKAGE AND ASSOCIATION ANALYSIS

7

Model-Based Methods for Linkage Analysis

John P. Rice,*,† Nancy L. Saccone,† and Jonathan Corbett†
*Department of Psychiatry, Washington University School of Medicine, St. Louis, Missouri 63110
†Department of Genetics, Washington University School of Medicine, St. Louis, Missouri 63110

Advances in Genetics, Vol. 60

0065-2660/08 $35.00
DOI: 10.1016/S0065-2660(07)00407-5

ABSTRACT

The *lo*garithm of an *od*ds ratio (LOD) score method originated in a seminal article by Newton Morton in 1955. The method is broadly concerned with issues of power and the posterior probability of linkage, ensuring that a reported linkage has a high probability of being a true linkage. In addition, the method is sequential so that pedigrees or LOD curves may be combined from published reports to pool data for analysis. This approach has been remarkably successful for 50 years in identifying disease genes for Mendelian disorders.

After discussing these issues, we consider the situation for complex disorders where the maximum LOD score statistic shares some of the advantages of the traditional LOD score approach, but is limited by unknown power and the lack of sharing of the primary data needed to optimally combine analytic results. We may still learn from the LOD score method as we explore new methods in molecular biology and genetic analysis to utilize the complete human DNA sequence and the cataloging of all human genes.

I. INTRODUCTION

Recent advances in molecular biology have enabled the rapid mapping and identification of genes for single-locus disorders. These successes have relied on the LOD score method for analysis. This statistic, introduced half a century ago, has proven to have several strengths in the analysis of genetic data. The LOD score method originated in the classic 1955 paper of Newton Morton, and is broadly concerned with issues of power, false positive rates, and combining results from multiple studies.

II. THE GENERALIZED SINGLE MAJOR LOCUS MODEL DESCRIPTION

In the single major locus (SML) model, it is assumed that there is single trait locus. The trait may be either continuous or polychotomous. We consider the case of a two-allele locus for a dichotomous phenotype. Denoting the two alleles by "A" and "a," let p be the frequency of A and $q = 1 - p$ be the frequency of a. Under the assumption of random mating, the probabilities for the three genotypes AA, Aa, and aa are p^2, $2pq$, and q^2, respectively. For a continuous phenotype, it is typically assumed that the distribution within a genotype is normal, so that parameters consist of the three genotypic means and a common variance. For a dichotomous trait, the penetrances f_{AA}, f_{Aa}, and f_{aa} are defined as the probability that individuals of genotype AA, Aa, and aa, respectively, are affected. Accordingly, the SML can be described in terms of the four parameters

p, f_{AA}, f_{Aa}, and f_{aa}. We further assume that all familial resemblance is due to the single locus. Thus, the phenotype of an individual depends only on his/her genotype. That is, if $(X_1, \ldots, X_m)$ denotes the phenotypes in a family of size m, and $(g_1, \ldots, g_m)$ denotes their genotypes, then $P(X_i|g_1, \ldots, g_m) = P(X|g_i)$ and $P(X_i, X_j|g_i, g_j) = P(X_i|g_i)\, P(X_j|g_j)$, for individuals i and j, where $P()$ denotes the probability of the event in parentheses.

For a marker M, let θ denote the recombination fraction between the trait locus and M. Then the genotype denotes the joint genotype of the trait locus and marker, so that the likelihood of the family depends on p, f_{AA}, f_{Aa}, f_{aa}, and θ. These assumptions enable the efficient computation of the likelihood of large pedigrees using the algorithm by Elston and Stewart (1971) as implemented in the LINKAGE package (Lathrop *et al.*, 1984) and FASTLINK (Cottingham *et al.*, 1993) and has served as the "standard" analytic model for Mendelian traits.

III. THE LOD SCORE

A. Definition

The LOD curve $Z(\theta)$ is defined by

$$Z(\theta) = \log_{10} \frac{L(\text{data}||\theta, f_{AA}, f_{Aa}, f_{aa}, p)}{L(\text{data}||\theta = 1/2, f_{AA}, f_{Aa}, f_{aa}, p)}$$

where L denotes the likelihood.

Note that we assume that the penetrances and gene frequency are known for the trait locus so the only unknown is the recombination fraction θ. The maximum of $Z(\theta)$ is called the LOD score and denoted Z. The likelihood ratio (LR) statistic is -2 times the difference between the log of the likelihood at $\theta = 1/2$ and the log of the maximum likelihood value, where the logarithm is to the base e. Thus,

$$2(\log_e 10)Z = 4.6Z \approx \chi_1^2$$

so that a LOD score of, say 3, is equivalent to a χ^2 of 13.8 on one degree of freedom. Thus, in a one-sided test of $\theta = 1/2$, the corresponding p value is 0.0001.

B. Examples

For the sake of simplicity, assume that we are examining two loci, each of which has two alleles, and that we are able to distinguish heterozygotes from homozygotes of both types.

Let us label the alleles at the first locus by A/a and at the second locus by B/b. Consider the case that the father of our nuclear family is heterozygous at both loci whereas the mother is homozygous for the "lower case" alleles, that is, the father has genotype Aa Bb and the mother has genotype aa bb. Let us further assume that we know the phase of the father's genotype, for example, that we know he inherited both "upper case" alleles from his father and both "lower case" alleles from his mother. In this case, we know the father's haplotype is AB/ab and we can directly count the number of recombinant and nonrecombinant offspring.

Assuming that we observe k recombinants in n matings, we can test the hypothesis of free recombination as a standard hypothesis test of the parameter of a binomial distribution. If we consider the simple hypothesis test:

$$H_0 : \theta = \frac{1}{2} \tag{1}$$

$$H_1 : \theta = \theta_0 \tag{2}$$

where $0 < \theta < 1/2$, then we define the LOD score of the test as the (base 10) log of the LR

$$LR = \frac{L(k/n|\theta = \theta_0)}{L(k/n|\theta = 1/2)} = \frac{\theta_0^k(1-\theta_0)^{(n-k)}}{(1/2)^n}.$$

We then see that in the particular case that $k = 0$, the maximum likelihood estimate for θ_0 is $\theta_0 = 0$, and hence LR $= 2^n$ so that the LOD score is $\log_{10}$ LR $=$ $n \log_{10}(2)$. More specifically, in this case we would be able to reject the null hypothesis if $n \geq 10$, as a LOD score of 3 is traditionally accepted as sufficient to reject the hypothesis of free recombination and each observation adds about 0.3 to the observed LOD score.

We note that it is not necessary that all these 10 observations come from the same mating, only that they come from matings where recombinants and nonrecombinants can be distinguished without ambiguity. While such matings are the most informative for the purposes of detecting linkage, we can also make use of many other types of matings in our attempts to detect or reject linkage.

Let us now consider a nuclear family with a pair of biallelic loci labeled as above. Assume that the father is doubly heterozygous with genotype (Aa Bb) whereas the mother is doubly homozygous with genotype (aa bb), but that, unlike the above case, we do not know the phase of the father's genotype. In other words, we do not know whether the father's haplotypes are (AB/ab) (I) or (Ab/aB) (II). While it is impossible to unambiguously decide whether any given offspring of this mating is a recombinant, we can use probabilistic methods to gain linkage information. We will assume, lacking any reason to believe otherwise, that the prior probabilities of (I) and (II) are equal.

We first observe that we gain no information for linkage in the case that this mating produces only one offspring. To see why this is so, let us consider the case of one offspring and assume that this offspring has genotype (Aa bb). Let R represent the event that the offspring is the result of recombination during meiosis, and let N represent the event that the offspring is a nonrecombinant. Then observe that the event R coincides with the father's haplotype being of type (I) whereas event N coincides with the father's haplotype being of type (II). If we now let θ represent the true recombination fraction between the above loci and y the offspring's observed genotype, then we may conclude that the probability of the observed genotype in the offspring is given by:

$$P(Y) = P(y|\mathrm{I})P(\mathrm{I}) + P(y|\mathrm{II})P(\mathrm{II}) = \frac{\theta}{4} + \frac{1-\theta}{4} = \frac{1}{4}$$

It is straightforward to see that we would obtain the same result for each of the four possible genotypes that could arise from this mating. Specifically, we note that $P(y)$ is not dependent on the true value of the recombination fraction. We cannot obtain a nonzero LOD score for any of the possible genotypes of the offspring, and, thus, such a mating is not informative for linkage in the case that there is only one offspring.

However, in the case that this mating produces at least two offspring, we do gain information for linkage. Consider, for simplicity, the case where this mating produces two offspring. Because we have two offspring, each of which has 4 possible genotypes, there are 16 possible combinations of genotypes. In this biallelic system, it turns out that of these 16 possible combinations, there are really only 2 distinct states. Let us denote by y_1 the state that each of the offspring receives either both "upper case" or both "lower case" alleles from their father as well as the state that each offspring receives one "upper case" and one "lower case" allele from their father. In other words, y_1 represents the case that both offspring are either recombinants or nonrecombinants. We denote by y_2 the set of eight possibilities not contained in case y_1, that is, the case that one offspring is recombinant and the other is not. Note that in y_1 we do not know whether the offspring are recombinant or nonrecombinant, but only that their recombination status is identical. We, following convention, state that, in this case, the offspring are concordant. Similarly, in y_2, we do not know which offspring is recombinant and which offspring is nonrecombinant, but only that we have one of each. The usual term for this state is that the offspring are discordant.

Intuitively, we would believe that linkage between the loci under study is more likely under y_1 than under y_2. It turns out that our intuition is correct. Taking the same set of hypotheses as above, we calculate:

$$f(y_1|\theta_0) = (1 - \theta_0)^2 + \theta_0^2$$
$$f(y_2|\theta_0) = 2\theta_0(1 - \theta_0),$$

whereas $f(y_1 \mid \theta = 1/2) = f(y_2 \mid \theta = 1/2) = 1/2$. As $(1 - \theta_0)^2 + \theta_0^2 > 1/2$ provided that $\theta_0 \neq 1/2$, we see that we do obtain information on linkage in this mating scheme, and, furthermore, that this information provides evidence for linkage in the case that y_1 holds and that it provides evidence against linkage in the case that y_2 holds.

In the simple case that we are examining one mating with two offspring, that these offspring are concordant, and that $\theta_0 = 0$, we observe that the LR is given by:

$$\text{LR} = \frac{f(y_1|\theta_0)}{f(y_1|\theta = 1/2)} = 2$$

Thus, we obtain a LOD score of $\log_{10} \text{LR} = \log_{10} 2 = 0.30$. We note that this is the same LOD score as that obtained from one unambiguously observed recombination event, as we saw above.

C. The LOD score of 3 criterion

Let "+" denote the presence of true linkage and "−" denote nonlinkage to a particular marker. Let L denote the rejection of the null hypothesis, that is the conclusion is there is linkage.

We are interested in the quantity $P(+ \mid L)$, the probability that a linkage is correct when we conclude linkage is present. As noted by Morton (1955), "We are especially anxious to avoid the assertion that two genes are linked – Since a misleading map is worse than no map at all." The quantity $P(+ \mid L)$ depends on $P(+)$, the prior probability of linkage. Under the assumption there is a trait locus, Morton set a value of $P(+) = 0.05$, although others have argued 0.02 is more realistic.

$$\begin{aligned} P(+|L) &= P(L|+)\frac{P(+)}{P(L)} \\ &= \frac{P(L|+)P(+)}{P(L|+)P(+) + P(L|-)P(-)} \\ &= \frac{(1-\beta)P(+)}{(1-\beta)P(+) + \alpha\{1 - P(+)\}} \end{aligned}$$

This quantity increases as the significance level α becomes smaller and as the power $(1-\beta)$ becomes larger. Note that, for example, if $P(+)=0.02$, $(1-\beta)=1$, and $\alpha=0.05$, then $P(+\mid L)=0.29$. That is, even with 100% power, less than 1/3 of reported linkages would be real. This led Morton to choose $\alpha=0.001$ to reduce the "posterior type I error."

The original suggestion of Morton (1955) was to use a sequential test for linkage. In this approach, families are added one at a time to the sample and the LOD score computed. If the running LOD score goes above a cutoff log (A), we conclude linkage; if it goes below a cutoff log (B), we conclude no linkage; and if it is in-between, we continue to sample more families. In this setting, the expected sample size is lower than that of a fixed sample to achieve the same power. Wald (1947) showed that the cutoffs are given by:

$$A \approx \frac{1-\beta}{\alpha}$$

$$B \approx \frac{\beta}{1-\alpha}$$

Thus with $\alpha=0.001$, $1-\beta=0.99$, we have log (A) = 3, and log (B) = −2, the values proposed by Morton (1955).

The LOD score of 3 criterion has been used even when the analysis was not sequential. As noted above, this would correspond to $\alpha=0.0001$ for a single marker. Moreover, excluding regions with LOD scores below −2 has proven effective in exclusion mapping.

D. Genome-wide significance level

The theory of Morton applied to a single marker. In the context of a modern day genome screen, there are several hundred markers used, although the statistical tests are correlated because sets of markers are linked. Lander and Kruglyak (1995) note that for a genome-wide error rate to be 0.05, the appropriate LOD score cutoff is 3.6 for an infinitely dense map. That is, 1 in 20 genome screens will provide a LOD score of 3.6 which is false.

In this setting, the probability of having a marker linked to the trait locus is 1.0, so the real question is—What is the power to detect the trait locus using a particular LOD cutoff?

For a SML trait, the computations are straightforward. Power calculations can be done with computer programs such as SIMLINK (Boehnke, 1986) or SLINK (Weeks *et al.*, 1990) on the family material prior to genotyping and can model incomplete penetrance and incomplete marker information. In this setting, false positive high LOD scores are not problematic; areas of interest from one study can be genotyped in an independent sample.

For complex traits, the number of loci involved in the trait is unknown, so that power can not be estimated with confidence. A more subtle issue is that of replication of findings. If there were six unlinked disease loci for a disorder, and the power to detect any one of them is, say, 20%, then the power to detect at least one is $1 - (0.8)^6 = 74\%$. That is, there may be a high probability of correctly detecting a susceptibility locus, but in a replication study looking at the specific locus, the power is low (20% in our example). This presents a problem unless the replication sample is much larger than the original sample (Suarez *et al.*, 1994). This may explain, in part, the lack of replication seen in linkage studies for many disorders.

E. Affected sib pair methods and LOD scores

Popular alternatives to parametric LOD score analysis include affected sib pair (ASP) methods. Often described as a "nonparametric" or "model-free" method, ASP analyses test for excess sharing of marker alleles identical by descent (IBD) in affected-ASPs. Model-free methods are described in detail in Chapter by Almasy and Blangero, this volume.

Tests based on ASP methods do not require explicit specification of a genetic model for the disease. Hence, it might appear that ASP methods have advantages over model-based LOD score analysis for studying complex diseases of uncertain genetic etiology. However, this may be only somewhat the case. As is discussed below, it is important to be aware that equivalencies between ASP and parametric tests have been demonstrated in particular cases (Knapp *et al.*, 1994). Furthermore, implicit model assumptions exist for ASP tests, and the power of such tests is thus influenced by the appropriateness of these assumptions (Whittemore, 1996).

Knapp *et al.* (1994) have shown that commonly used ASP tests are in practice equivalent to LOD score analysis under certain assumed modes of inheritance, regardless of the true mode of inheritance. A similar equivalence was also noted in Hyer *et al.* (1991). These equivalencies imply that the power of parametric analyses under certain model assumptions remains comparable to the power of ASP tests, in such settings.

Knapp *et al.* prove these equivalencies for samples consisting of nuclear families with parents and two affected sibs. For LOD score analysis, three cases of assumed parental status are considered: both parents unaffected (I), both parents affected (II), and both parents unknown (III).

More specifically, Knapp *et al.* demonstrate the equivalence of tests based on certain IBD statistics t and tests based on the parametric LOD score, by showing that under the respective assumptions for parental status, there exists a function $f(t)$ mapping the test statistic t to the LOD score statistic Z, computed under a particular assumed mode of inheritance. The key is that this function is strictly monotone increasing over a range corresponding to significance levels (αs) of practical interest. Hence, the two statistics are equivalent over this range and describe the same test after appropriate adjustment of the test critical values.

For cases (I) and (III) (the assumption of both parents unaffected and the assumption of both parents unknown), the equivalence is proved using the sib pair mean IBD test statistic t_2, and using an assumed recessive mode of inheritance for the LOD score. The corresponding function $f(t)$ is monotone in the range $t \varepsilon [n, 2n]$. Because under the null hypothesis of no linkage ($\theta = 1/2$), $Pr_{H0} (t_2 \geq n) > 0.5$, the two statistics induce equivalent tests for any significance level $\alpha \leq 0.5$.

For case (II) where both parents are assumed affected, Knapp *et al.* prove equivalence of the sib pair "two-allele" or "proportion" test (based on the statistic $t = n_2$, the number of affected-ASPs sharing 2 alleles IBD) and the LOD score under an assumed dominant mode of inheritance. Here, the restriction for α is more stringent: the tests are equivalent if the disease gene frequency is sufficiently small and $\alpha \leq Pr_{H0} (n_2 \geq 3n/7)$. Knapp *et al.* note that this range corresponds to α values of 0.0001 or smaller for sample sizes up to $n = 88$; as n increases, Pr_{H0} $(n_2 \geq 3n/7)$ approaches 0 and the corresponding limiting α values approach 0 as well.

Further insight into the relationship between classical LOD scores and ASP IBD statistics is provided by Whittemore (1996), via a unified framework for various linkage test statistics. Whittemore introduces the efficient score statistic, obtained from the likelihood function for the observed marker data and the observed trait data. This statistic is asymptotically equivalent to the LOD score and includes affected pedigree member (and hence ASP) statistics as special cases. Writing affected pedigree member statistics as efficient score statistics, Whittemore shows that any affected pedigree member test implicitly assumes a model for the phenotypic risk ratios for a pedigree with a given IBD configuration at the trait gene, relative to an arbitrary pedigree having the same family structure. An example in the paper demonstrates that for samples of five-member nuclear families with one affected parent and three affected offspring, the mean IBD relative pair test (using the six possible affected-relative pairs) corresponds to an implicit assumption of an additive genetic model plus an additional constraint on the allele effects. Hence, the appropriateness of IBD-based ASP tests is not automatic for studies of complex disease, and such tests cannot strictly be called "model-free." Power of these tests depends upon how the underlying model assumption for the phenotypic risk ratios compares with the "true" model.

F. Maximum LOD score for ASP

As noted earlier, the SML model may be parameterized in terms of five parameters:

$$\{f_{AA}, f_{Aa}, f_{aa}, p, \text{and } \theta\}$$

Suarez *et al.* (1978) noted the IBD distribution of sib pairs, conditioned on their phenotype, does not depend on q. Risch (1990) used an alternative parameterization $\{K, \lambda_s, \lambda_O, \theta\}$ where $\lambda_R = K_R/K$ with K the population prevalence and K_R the risk to a type of relative R of an affected individual. Here, λ_s and λ_O are risk ratios to siblings and offspring, respectively. Moreover, Risch noted the IBD distribution in ASPs depends only on $\{\lambda_s, \lambda_O, \theta\}$. With multipoint marker data, all three parameters $\{\lambda_s, \lambda_O, x\}$, where x is chromosome location, can be estimated when a trait locus is present. Under the null hypothesis, the parameter space is degenerate, so that likelihood theory does not apply. However, Hauser *et al.* (1996) provide simulation results in terms of $\lambda = \lambda_O = \lambda_s$ where λ is the locus-specific recurrence risk ratio.

The important point here is that these statistics, as implemented in programs such as ASPEX (Hinds and Risch, bioweb.pasteur.fr/docs/aspex/usage.html) or MAPMAKER/SIBS (Kruglyak and Lander, 1995), are in fact the LOD score maximized over these parameters. The likelihood is P (marker data|both sibs affected and parents unknown phenotype), and maximizing the likelihood is equivalent to maximizing the LOD score (Clerget-Darpoux *et al.*, 1986).

G. Maximum LOD score versus the LOD score

The maximum LOD score (MLS) discussed above is a LOD score (maximized over λ_s, λ_O, or, alternatively, over the IBD sharing probabilities at the trait locus), and gives statistically consistent estimates of disease gene location. It is important, however, to recognize the differences between the MLS and traditional LOD score. In general, there will be more power for the traditional method when the mode of inheritance is known, and data are analyzed under the correct model.

In the case of a simple recessive model, we know that affected parents are homozygous at the disease locus, and thus noninformative for linkage analysis, whereas the unaffected parent of two affected children is heterozygous. In contrast, for a dominant disease, the affected parent is informative, and an unaffected one is not. This information is used in the traditional linkage approach, whereas parental phenotypes are not used in the MLS approach.

To illustrate this, we assume a sample of N families with ASPs for a (rare) dominant disorder. We assume that the disease gene frequency is small, so that each family will have precisely one affected parent. Assume, further, that the disease gene is completely linked to the marker. In the $2N$ parents, there will be 100% sharing with respect to the N affected parents, and 50% sharing on average, with respect to the N unaffected parents. Thus, there will be 75% sharing overall if parental diagnosis is ignored.

In the model-free approach, the resulting χ^2 would be $N/2$ if the cell counts expected under the alternative were observed. In the model-based approach, 100% sharing would be observed in the N parents, and the resulting χ^2 would be equal to N, twice that obtained when parental diagnosis is not used.

In sum, even though the approaches are equivalent under some sampling units (e.g., affected siblings with parental diagnosis unknown), the MLS and traditional LOD score methods have important differences.

IV. THE MOD SCORE METHOD

As noted above, Risch (1984) suggested the use of the likelihood of marker data conditioned on all trait phenotypes,

$$L = P(\text{marker data} \mid \text{all trait phenotypes}). \tag{3}$$

He noted that this formulation obviates the need for systematic ascertainment of pedigrees and provides information on the mode of inheritance via the linkage information. Risch viewed this as an extension of the ASP method to arbitrary pedigrees.

Clerget-Darpoux *et al.* (1986) introduced the term "MOD score," the model LOD score, computed by maximizing the LOD score under the assumption that the trait is determined by a single diallelic locus with recombination fraction θ and parameters $\phi = (p, f_1, f_2, f_3)$, where p is the gene frequency and f_1, f_2, and f_3 are the three penetrances associated with the trait locus. The MOD score is then defined as

$$\text{MOD} = \max_{\theta,\phi} \text{LOD}(\theta, \phi).$$

They note that this is equivalent to maximizing Eq. (3) where the LOD score $Z(\theta,\phi)$ is defined below (Clerget-Darpoux *et al.*, 1986; Clerget-Darpoux *et al.*, 1992).

Elston (1989) noted that,

$$Z(\theta, \phi) = \frac{L(T, M; \theta, \phi)}{L(T, M; 1/2, \phi)} = \frac{L(T, M; \theta, \phi)}{L(T; \phi)L(M)} = \frac{1}{L(M)} L(M|T; \theta, \phi).$$

Thus, the LOD score $Z(\theta,\phi)$ is given by

$$Z(\theta, \phi) = \log_{10} \frac{L(T, M; \theta, \phi)}{L(T, M; 1/2, \phi)} = \log_{10} L(M|T; \theta, \phi) + \text{constant},$$

where the constant is $\log_{10}$ L(M). This derivation assumes that the trait T and marker loci M are distinct (i.e., no pleitropy), that the alleles at the trait and marker loci are in equilibrium, and that the marker data and ascertainment scheme are independent.

Risch (1990) used the MOD score for ASPs and their parents. He parameterized the model in terms of IBD at the marker locus and notes that if IBD can be determined, the MOD score does not depend on the marker allele frequencies. Pairs where parents are missing or the IBD cannot be deduced can also be used with this method, although the MOD score is a function of the marker allele frequencies. Holmans (1993) restricted the range of the admissible parameter space in Risch's approach to yield a more powerful test. He further notes that even if the marker frequencies are known, there can be more power to use $2N$ pairs with untyped parents than to use N pairs with typed parents (with a total of $4N$ typings in each case).

MacLean *et al.* (1993) considered offspring from phase-known double backcross matings for a partially dominant trait with penetrances $(1,t,0)$. They show the MOD score is asymptotically a χ^2 with two degrees of freedom.

Hodge and Elston (1994) provide a detailed discussion of the use of the MOD score versus the "wrod" score, the LOD score computed under a wrong set of genetic parameters ϕ. They note that under the correct genetic mechanism (e.g., assuming a single locus when the true model is single locus), the MOD score will asymptotically yield the true values when there is linkage (see also Greenberg, 1989). If there is no linkage, then the MOD score contains no information on the genetic model.

V. THE LOD SCORE AND META-ANALYSIS

A. SML traits

Meta-analysis in the social sciences (Hedges and Olkin, 1985) is a technique used to integrate results from multiple studies. For SML traits, it is straightforward to combine data. If $Z_i(\theta)$ is the LOD curve for family i, then the LOD curve when N families are genotyped is

$$Z(\theta) = \sum_{i=1}^{N} Z_i(\theta).$$

Moreover, if multiple studies are used, the $Z(\theta)$ may be added from each study.

Several studies in the 1950s reported linkage between elliptocytosis and the Rh system. Morton (1956) collected the published pedigrees and derived algebraic expressions for the LOD curves to be combined in analysis. Moreover, heterogeneity was detected and explained conflicting reports of linkage.

It became standard either to display the pedigrees with marker data or to publish LOD curves in tabular form to allow pooling of independent studies. It must be emphasized that it is not proper to add MLSs, but rather the LOD curves must be added, and then the maximum taken.

For multipoint methods (i.e., analysis of more than one marker), a mapping function can be used to relate genetic distances to recombination fractions, so that the LOD curve would be related to chromosomal location.

As noted above, the LOD score is in fact a measure of statistical significance. Hedges and Olkin point out the fallacy of using outcomes of significance to interpret consistency and replication. These "vote counting" methods can give misleading conclusions. Significance/nonsignificance may reflect design or power differences. Even when two studies are significant, there may be heterogeneity—if two linkage studies give high LOD scores at markers 50 cM apart, are they consistent?

The genetic parameter of interest is x, the disease gene location, and the LOD score is a test statistic. When a single marker is analyzed, θ has been traditionally used rather than x. The ultimate validation comes from the laboratory when the disease gene is identified.

B. Complex traits

For a complex trait, if the MLS is computed in different studies, the corresponding MLS curves can not be combined like LOD curves because the MLS curves may be maximized at different λ values. Combining results from published studies represents a challenge. This is well illustrated by studies of bipolar disorder (BPD). Individual linkage results for BPD have been inconsistent. Badner and Gershon (2002) used a multiple scan probability method that combines p values across scans in regions with clusters of positive scores after adjusting for the size of the region. Segurado *et al.* (2003) applied a rank-based genome scan meta-analysis to 18 BPD genome scan datasets. The genome scan meta-analysis method divides the autosomes into bins, and ranks the results of each study within a bin. Then the average rank is computed for each bin. No bin reached genome-wide significance. The results differed from the multiple scan probability analyses, although the datasets used were not identical. These methods suffer from the tendency not to publish negative reports (publication bias), and variation of how many regions are reported in the original publications.

In contrast to single-locus traits—where data were fully shared either through publishing the original pedigrees or through LOD curves, there would appear to be inadequate sharing of data for complex traits. As noted in part A SML traits, a major strength of the LOD score method was that the published reports allowed combining of data to ultimately localize the disease gene or to reject linkage. For complex traits, published reports contain p values computed in ways that cannot be compared across studies.

The difficulties associated with combining results are well illustrated by analyses done by the Schizophrenia Linkage Collaborative Group for chromosomes 3, 6, and 8 (1996) in which data from 14 international research groups were assembled. Even when different studies reported significant LOD scores at the same marker, they were computed using different diagnostic criteria (broad vs wide); even when the same criteria were analyzed, reports used different analytic methods, and when a single method was used, the support for linkage was not always consistent with the primary analyses (cf. the commentary by Rice, 1997).

McQueen *et al.* (2005) performed a combined analysis for 11 BPD datasets by creating a single combined genetic map and found two linkage peaks that reached genome-wide significance. This approach requires having the primary data, but obviates many of the problems when only published reports are used.

More work is needed to develop methods as effective as combining LOD curves for SML traits. Besides the methodological issues, more uniform standards for reporting results, or even making genotypic data available, are needed in order to allow effective pooling of data.

VI. MULTILOCUS MODELS

Currently, significant research effort is being directed toward the genetics of complex traits. Many complex diseases are hypothesized to be due to multiple loci of varying effect size, some of which may be interacting. Yet traditional model-based linkage analysis, as described previously, is carried out assuming a single-locus model rather than a multilocus model. Note that effects of additional loci may be indirectly modeled via the penetrance functions. For example, incomplete penetrance of the disease genotype at one locus may be due in part to the modifying effects at a second locus.

In practice, it may be important to consider the effects of a particular known locus to map additional disease genes. For example, the E4 allele at the APOE gene is a well-documented risk factor for late-onset of Alzheimer's disease; nevertheless, it seems clear that additional disease-influencing loci exist (Pastor and Goate, 2004). Attempts to map these additional loci may be hampered by the strong effect of APOE allelic status in cases having the E4 allele. A useful approach to this kind of dilemma is to stratify the sample according to genotype at the known disease locus or use APOE allelic status as covariate (Wijsman *et al.*, 2005). For model-based linkage, this could mean carrying out a separate analysis of all families not segregating the known disease allele. Note that in general it is not recommended to simply drop individuals with the disease genotype while including the remaining family members in the analysis; the reason is that such exclusions would potentially introduce complex biases into the remaining sample.

Epistasis is commonly defined as the interaction (nonadditive) between (or among) two or more genetic loci. The term heterogeneity is used to describe the situation when a phenotype or disease is influenced by distinct genetic loci or independent genetic mechanisms. While it is generally acknowledged that many complex human diseases are likely to be influenced by several genetic loci (as well as environmental influences), studying two-locus models allows a first step toward unraveling this complexity. Two-locus model-based LOD score methods have been published by Lathrop and Ott (1990) and Schork *et al.* (1993); Cordell *et al.* (1995) developed a two-locus version of the MLS.

To investigate the implications of particular two-locus models of disease, both epistatic and heterogeneous, Neuman and Rice (1992) derived formulas for the recurrence risks to relatives in terms of penetrances and gene frequencies, under these models. These relationships may then be used to determine disease model parameters (gene frequencies and penetrances) that are consistent with observed population prevalences and relative recurrence risks.

The classical LOD score method requires specification of the disease model parameters. However, Neuman and Rice (1992) point out that inferring the mode of transmission in families can be problematic, as two-locus models can result in patterns of inheritance that resemble those from single-locus models. For example, consider the two-locus epistatic model in which an individual is affected if and only if a dominant disease allele is present at both loci. Then the expected proportion of affected offspring from matings of normal parents is $1/[(q_1+1)(q_2+1)]$, where q_i is the frequency of the wild-type allele at locus i. Then, as the frequency of the disease alleles (at both loci) becomes small, in the limit this proportion approaches 1/4, which is the same proportion of affected offspring from normal parents that is expected for a single-locus recessive disease.

Despite the challenges that multifactorial diseases present for gene mapping efforts, traditional linkage analysis remains an important tool. Indeed, it is important to note that linkage-based methods retain important advantages even in the face of the growing popularity of genome-wide association study designs for gene mapping. Culverhouse *et al.* (2002) demonstrated that for a large class of "purely epistatic" models with no additive or dominance variation at any of the involved loci, association methods will have no power if the loci are examined individually. However, these models do lead to increased allele sharing between affected sibs, so that model-free allele sharing linkage methods (detailed in Chapter by Almasy and Blangero, this volume) would have power to detect susceptibility loci.

VII. STRENGTHS AND WEAKNESSES OF MODEL-BASED METHODS

Model-based linkage approaches would seem the method of choice for a disease with Mendelian inheritance. These methods allow for large pedigrees (pedigree size is limited for many model-free methods), allow for meaningful power computations prior to genotyping, are relatively simple to perform, and benefit from a parametric analysis under the correct model. Moreover, a complex disease may have Mendelian subforms (such as breast cancer caused by the BRAC1 gene), and analyses of a few large pedigrees may be successful in gene identification. Many successes demonstrate the strengths of this approach.

Although these methods are often applied to complex traits, their utility compared to model-free approaches is less clear (Ott, 2001). Many investigators perform multiple analyses assuming incompletely penetrant dominant, recessive, and additive genetic single-locus models. This may lead to biased parameter estimates—a serious problem when the intention is to estimate location in a multipoint analysis. Nonetheless, this approach does seem to provide valid statistical evidence for linkage.

Strengths and weaknesses depend on study design and the genetic structure of the trait of interest. For mental retardation, there are many single-locus causes that were discovered using large pedigrees. A design sampling only ASPs would likely be heterogeneous and have little power to detect rare Mendelian subforms using model-free approaches. For an ASP design, the model-free analysis is equivalent to a MOD score analysis described above (when the parental phenotypes are coded as unknown) and by Clerget-Darpoux (2001). The MOD score is the LOD score maximized over certain parameters, so for this sampling unit, the model-based and model-free methods coalesce. For larger sampling units, model-free methods based on sib pairs do not use phenotypic information on other relatives, the MOD score conditions on this, and a model-based method can use this information.

Despite the inherent appeal in using model-free methods for complex traits, they have been surprisingly unsuccessful for disorders such as schizophrenia, BPD, and Type II diabetes; whereas case-control association designs have detected genes not suggested by linkage studies (Todd, 2006). Accordingly, the differences between linkage methods may be less important than the strengths of whole-genome association studies.

VIII. DISCUSSION

We have a complete human genome sequence, a catalog of most human genes, and a dense map of SNPs throughout the genome. However, the optimal way to utilize these technologies to identify susceptibility genes for complex disease remains unclear.

We may still learn from the LOD score method that has proven so successful in identifying genes for Mendelian traits. It must be emphasized that the detection of genetic heterogeneity and linkage of the Rh blood group to a subform of elliptocytosis by Newton Morton was quite remarkable in 1956, even though it may appear routine today.

We have highlighted two aspects of this method that, we believe, have contributed to its success. The first is the ability to share and pool data from published reports. This is consistent with the sequential nature of paper by Morton (1955). The collective LOD score will continue to rise for true linkage, and will become strongly negative for false ones. Related to this is the ability to localize a disease gene once linkage is detected. As the LOD score increases, the support interval for the region containing the gene narrows. Given that 1 cM corresponds roughly to 1 Mb (megabase) of DNA, the importance of combining all available data is evident. Until recently, positional cloning efforts usually began once the region of interest was under 2 Mb.

The other aspect of the LOD score method we highlighted is the ability to estimate the power of a linkage study and to set cutoffs to control the posterior probability of linkage. Indeed, the success of this method has been its high positive predictive value (i.e., the high probability that a reported linkage is, in fact, true). Validation of statistical linkage comes from the identification of the disease gene by the biologist. This dictates the need for methods with high predictive value.

The next decade should witness similar successes for the identification of genes that contribute to common diseases. The exact methods and study designs that affect this are as yet not clear. The MLS enjoys some of the properties of the traditional LOD score, and does provide unbiased estimates of disease gene location and the ability to do exclusion mapping for given values of the risk ratio λ.

Alternative approaches to traditional linkage analysis include (1) studying related quantitative phenotypes (endophenotypes), in addition to the dichotomous clinical diagnosis, (2) genome-wide association studies, and (3) the use of animal models. The lessons learned from the LOD score method are still timely and will have substantial impact on these new methodologies. Moreover, evidence from linkage studies may be used in the analysis of whole-genome association studies to overweight certain areas of the genome (Roeder *et al.*, 2006).

References

Badner, J. A., and Gershon, E. S. (2002). Metaanalysis of whole-genome linkage scans of bipolar disorder and schizophrenia. *Mol. Psychiatry* **7,** 405–411.

Boehnke, M. (1986). Estimating the power of a proposed linkage study: A practical computer simulation approach. *Am. J. Hum. Genet.* **39,** 513–527.

Clerget-Darpoux, F. (2001). Extension of the LOD score: The MOD score. *In* "Genetic Dissection of Complex Traits, First Edition" (D. C. Rao and M. A. Province, eds.). Academic Press, Inc., San Diego.

Clerget-Darpoux, F., and Bonaiti-Pellie, C. (1992). Strategies based on marker information for the study of human diseases. *Ann. Hum. Genet.* **56,** 145–153.

Clerget-Darpoux, F., Bonaiti-Pellie, C., and Hochez, J. (1986). Effects of misspecifying genetic parameters in lod score analysis. *Biometrics* **42,** 393–399.

Cordell, H. J., Todd, J. A., Bennett, S. T., Kawaguchi, Y., and Farrall, M. (1995). Two-locus maximum LOD score analysis of a multifactorial trait: Joint consideration of IDDM2 and IDDM4 with IDDM1 in type 1 disease. *Am. J. Hum. Genet.* **57,** 920–934.

Cottingham, R. W., Jr., Idury, R. M., and Schäffer, A. A. (1993). Faster sequential genetic linkage computations. *Am. J. Hum. Genet.* **53,** 252–263.

Culverhouse, R., Suarez, B. K., Lin, J., and Reich, T. (2002). A perspective on epistasis: Limits of models displaying no main effect. *Am. J. Hum. Genet.* **70,** 461–471.

Elston, R. C. (1989). Man bites dog? The validity of maximizing lod scores to determine mode of inheritance. *Am. J. med. Genet.* **34,** 487–488.

Elston, R. C., and Stewart, J. (1971). A general model for the analysis of pedigree data. *Hum. Hered.* **21,** 523–542.

Greenberg, D. A. (1989). Inferring mode of inheritance by comparison of lod scores. *Am. J. Med. Genet.* **34,** 480–486.

Hauser, L., Boehnke, M., Guo, S., and Risch, N. (1996). Affected-sib-pair interval mapping and exclusion for complex genetic traits. *Genet. Epidemiol.* **13**(2), 117–137.

Hedges, L. V., and Olkin, I. (1985). "Statistical Methods for Meta-Analysis." Academic Press Inc., San Diego.

Hodge, S. E., and Elston, R. C. (1994). Lods, Wrods, and Mods: The interpretation of Lod scores calculated under different models. *Genet. Epidemiol.* **11,** 329–342.

Holmans, P. (1993). Asymptotic properties of affected-sib-pair linkage analysis. *Am. J. Hum. Genet.* **52,** 362–374.

Hyer, R. N., Julier, C., Buckley, J. D., Trucco, M., Rotter, J., Spielman, R., Barnett, A., Bain, S., Boitard, C., Deschamps, I., Todd, J. A., Bell, J. I., *et al.* (1991). High-resolution linkage mapping for susceptibility genes in human polygenic disease: Insulin-dependent diabetes mellitus and chromosome 11q. *Ann. Hum. Genet.* **48,** 243–257.

Knapp, M., Seuchter, S. A., and Baur, M. P. (1994). Linkage analysis in nuclear families. 2: Relationship between affected sib-pair tests and lod score analysis. *Hum. Hered.* **44,** 44–51.

Kruglyak, L., and Lander, E. S. (1995). Complete multipoint sib-pair analysis of qualitative and quantitative traits. *Am. J. Hum. Genet.* **57,** 439–454.

Lander, E., and Kruglyak, L. (1995). Genetic dissection of complex traits: Guidelines for interpreting and reporting linkage results. *Nat. Genet.* **11,** 241–247.

Lathrop, G. M., and Ott, J. (1990). Analysis of complex diseases under oligogenic models and intrafamilial heterogeneity by the LINKAGE programs. *Am. J. Hum. Genet. Suppl.* **47,** A188.

Lathrop, G. M., Hooper, A. B., Huntsman, J. W., and Ward, R. H. (1984). Strategies for multilocus linkage analysis in humans. *Proc. Natl. Acad. Sci. USA* **81,** 3443–3446.

MacLean, C. J., Bishop, D. T., Sherman, S. L., and Diehl, S. R. (1993). Distribution of lod scores under uncertain mode of inheritance. *Am. J. Hum. Genet.* **52,** 354–361.

McQueen, M. B., Devlin, B., Faraone, S. V., Nimgaonkar, V. L., Sklar, P., Smoller, J. W., Jamra, R. A., Albus, M., Bacanu, S. A., Baron, M., Barrett, T. B., Berrettini, W., *et al.* (2005). Combined analysis from eleven linkage studies of bipolar disorder provides strong evidence of susceptibility loci on chromosomes 6q and 8q. *Am. J. Hum. Genet.* **77**(4), 582–595.

Morton, N. E. (1955). Sequential tests for the detection of linkage. *Am. J. Hum. Genet.* **7,** 277–318.

Morton, N. E. (1956). The detection and estimation of linkage between the genes for elliptocytosis and the Rh blood type. *Am. J. Hum. Genet.* **8,** 80–96.

Neuman, R. J., and Rice, J. P. (1992). Two-locus models of disease. *Genet. Epidemiol.* **9,** 347–365.

Ott, J. (2001). Major strengths and weaknesses of the lod score method. *In* "Genetic Dissection of Complex Traits" (D. C. Rao and M. A. Province, eds.). 1st Edition., Academic Press, Inc., San Diego.

Pastor, P., and Goate, A. M. (2004). Molecular genetics of Alzheimer's disease. *Curr. Psychiatry Rep.* **6,** 125–133.

Rice, J. P. (1997). The role of meta-analysis in linkage studies of complex traits. *Neuropsychiatr. Genet.* **74,** 112–114.

Risch, N. (1984). Segregation analysis incorporating linkage markers. I. Single locus models with an application to type I diabetes. *Am. J. Hum. Genet.* **36,** 363–386.

Risch, N. (1990). Linkage strategies for genetically complex traits. III. The effect of marker polymorphism on analysis of affected relative pairs. *Am. J. Hum. Genet.* **46,** 242–253.

Roeder, K., Bacanu, S. A., Wasserman, L., and Devlin, B. (2006). Using linkage genome scans to improve power of association in genome scans. *Am. J. Hum. Genet.* **78**(2), 243–252.

Schizophrenia Linkage Collaborative Group for Chromosomes 3, 6, and 8 (1996). Additional support for schizophrenia linkage on chromosomes 6 and 8: A multicenter study. *Am. J. Med. Genet. (Neuropsychiatr. Genet.)* **67,** 580–594.

Schork, N. J., Boehnke, M., Terwilliger, J. D., and Ott, J. (1993). Two-trait-locus linkage analysis: A powerful strategy for mapping complex genetic traits. *Am. J. Hum. Genet.* **53,** 1127–1136.

Segurado, R., Detera-Wadleigh, S. D., Levinson, D. F., Lewis, C. M., Gill, M., Nurnberger, J. I., Jr., Craddock, N., DePaulo, J. R., Baron, M., Gershon, E. S., Ekholm, J., Cichon, S., *et al.* (2003). Genome scan meta-analysis of schizophrenia and bipolar disorder, part III: Bipolar disorder. *Am. J. Hum. Genet.* **73,** 49–62.

Suarez, B. K., Rice, J. P., and Reich, T. (1978). The generalized sib-pair IBD distribution: Its use in the detection of linkage. *Ann. Hum. Genet.* **42,** 87–94.

Suarez, B. K., Hampe, C. L., and Van Eerdewegh, P. (1994). Problems of replicating linkage claims in psychiatry. *In* "Genetic Approaches to Mental Disorders" (E. S. Gershon and C. R. Cloninger, eds.), pp. 23–46. American Psychiatric Press, Inc., Washington DC.

Todd, J. A. (2006). Statistical false positive or true disease pathway? *Nat. Genet.* **38,** 731–733.

Wald, A. (1947). Sequential Analysis. New York: Wiley.

Whittemore, A. S. (1996). Genome scanning for linkage: An overview. *Am. J. Hum. Genet.* **59,** 704–716.

Wijsman, E. M., Daw, E. W., Yu, X., Steinbard, E. J., Nochlin, D., Bird, T. D., and Schellenberg, G. D. (2005). APOE and other loci affect age-at-onset in Alzheimer's disease in families with ps2 mutation. *Am. J. Med. Genet.* **132b,** 14–20.

Weeks, D. E., Ott, J., and Lathrop, G. M. (1990). SLINK: A general simulation program for linkage analysis. *Am. J. Hum. Genet.* **47,** A204. Abstract.

8

Contemporary Model-Free Methods for Linkage Analysis

Laura Almasy and John Blangero
Department of Genetics, Southwest Foundation for Biomedical Research, San Antonio, Texas 78245

ABSTRACT

Model-free linkage methods are based on identifying regions of the genome in which patterns of allele sharing among family members correspond to patterns of phenotype correlation among family members. Two general classes of model-free linkage methods are discussed in this chapter, relative pair methods designed

Advances in Genetics, Vol. 60

0065-2660/08 $35.00
DOI: 10.1016/S0065-2660(07)00408-7

primarily for analysis of discrete traits and variance component methods designed primarily for analysis of quantitative traits. These methods have been used to identify numerous genes influencing complex human phenotypes and remain viable approaches to gene localization in the twenty-first century.

I. INTRODUCTION

In the parlance of statistical genetics, "model-free" methods for linkage analysis refer to methods that do not require specification of the genetic model underlying the effect of an as yet undetected locus on a phenotype. That is, these methods are "model-free" when contrasted with the penetrance model-based linkage methods discussed in the Chapter by Rice *et al.*, this volume. Model-free methods for linkage analysis may require assumptions regarding the distribution of a phenotype, making them model-based from a purely statistical perspective.

There has been some pessimism recently concerning the use of model-free linkage approaches for localization of genes influencing complex traits. Generalizations are sometimes made that linkage is only useful for identifying rare loci of large effect, that is the type of loci that have been discovered through penetrance model-based linkage screens for simple Mendelian traits. In fact, there have been many successes in genome-wide linkage screens for complex diseases and quantitative traits using model-free linkage approaches, that is genes that were initially localized through model-free linkage screens with significance levels that reached genome-wide thresholds.

Some of these quantitative trait locus (QTL) localizations have progressed to identification of specific genes under a linkage peak and specific functional variants within those genes. Genome-wide linkage screens have resulted in the identification of *TCF7L2* as a QTL for type 2 diabetes (Duggirala *et al.*, 1999; Grant *et al.*, 2006; Reynisdottir *et al.*, 2003), *GABRA2* as a QTL for alcoholism and related electrophysiological endophenotypes (Edenberg *et al.*, 2004), and *IDE* as a susceptibility gene for Alzheimer's disease (Bertram *et al.*, 2000; Ertekin-Taner *et al.*, 2000). Specific functional variants have been identified within the *F7* gene that affect levels of clotting factor VII, a risk factor for thrombosis (Sabater-Lleal *et al.*, 2007; Soria *et al.*, 2005), within a member of the TAS2R bitter taste receptor family that influence ability to taste phenylthiocarbamide (PTC) (Drayna *et al.*, 2003; Kim *et al.*, 2003), and within the *SEPS1* gene that affect inflammatory response (Curran *et al.*, 2005).

Below we concentrate for the most part on methodological approaches, rather than specific software packages that implement these methods. Dudbridge (2003) provides a review of software for parametric and nonparametric linkage analysis that includes both quantitative and discrete traits and Almasy and Warren (2005) discuss software for both linkage and association analyses of quantitative traits.

II. IDENTITY BY DESCENT AND IDENTITY BY STATE

All model-free methods for linkage analysis rely on tests based on identical-by-descent (IBD) allele sharing. Two alleles are IBD if they were inherited from the same ancestral source. That is, both copies of the allele trace back to a single, specific ancestral chromosome. IBD can be contrasted with identity by state (IBS). Two alleles are IBS if they are of the same form, for example both 110 base pairs for a microsatellite or C alleles for a single nucleotide polymorphism (SNP), regardless of their ancestral origin. In the example in Fig. 8.1, the cousins 8 and 9 share their 114 base-pair alleles IBD because they each trace back to their common grandmother, individual 2. In contrast, cousins 7 and 9 only share their 118 base-pair alleles IBS. Individual 7's 118 base-pair allele traces back to her father, 3, and individual 9's 118 base-pair allele traces back to her mother, 6. Based on the pedigree information provided, individuals 3 and 6 are unrelated, making these 118 base-pair alleles IBS but not IBD, because they do not derive from a common ancestral source.

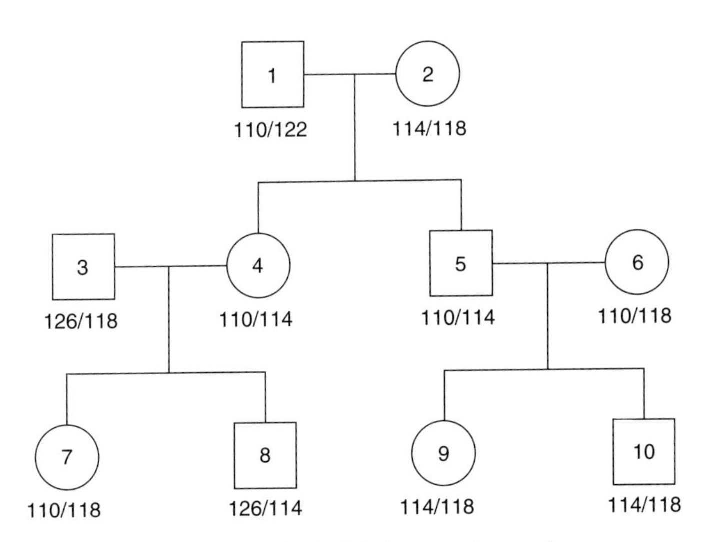

Figure 8.1. IBD and IBS allele sharing within a pedigree.

A pair of relatives may share 0, 1, or 2 alleles IBD. Only individuals who are related through both of their parents (e.g., full siblings, double first cousins) may share 2 alleles IBD. In Fig. 8.1, the siblings 9 and 10 share both of their alleles IBD, one from each heterozygous parent. In contrast, because the cousin pairs (pairs including individuals 7 and 9, 7 and 10, 8 and 9, or 8 and 10) are only related through one of their parents, they can share at most one allele IBD from a common grandparent. When inbreeding is present, an individual's two alleles may be IBD with each other and in this case the IBD sharing among a relative pair may be >2. In most cases, IBD sharing is expressed as a proportion defined as:

$$0.5 \text{ Pr (share 1 allele IBD)} + \text{Pr (share both alleles IBD)}.$$

The true proportion of IBD sharing between a noninbred relative pair is 0, 0.5, or 1, depending on whether they share 0, 1, or 2 alleles. However, estimated IBD proportions intermediate to these are produced when IBD sharing cannot be determined unambiguously. This occurs when a parent is homozygous, so it is unclear which ancestral allele was transmitted to a child, or when an ancestor is ungenotyped, in which case the IBD sharing is a function of the allele frequencies. In the example in Fig. 8.2, it is clear that the siblings inherited the same 118 base-pair allele from their heterozygous mother. This means they share at least one allele IBD. Since their father is homozygous for the 110 base-pair allele at this marker, it is unclear whether their second allele is IBD, tracing back to the same paternal grandparent, or merely IBS. In this case, their estimated IBD sharing would be 0.75, an average of the two equally likely possibilities. Similarly, if their father had been ungenotyped, the weightings for the average of

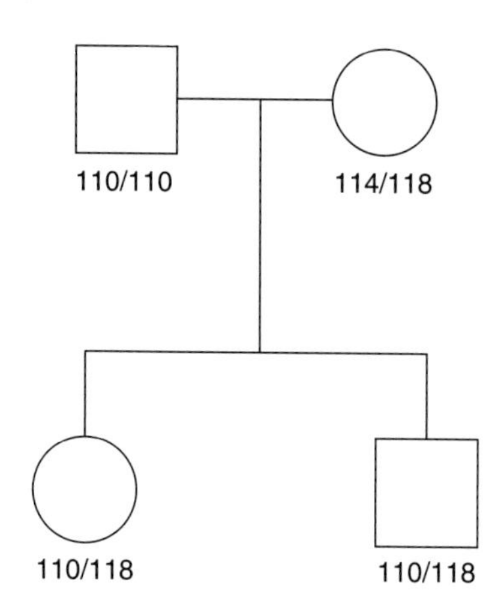

Figure 8.2. Ambiguous IBD allele sharing.

the two possibilities would depend on the frequency of the 110 base-pair allele. If the allele were rare, it would be unlikely that the father happened to be homozygous and more likely that the siblings were IBD, pushing their estimated IBD toward 1.

The expected IBD sharing between a relative pair is easily specified on the basis of their pedigree relationship. In the absence of inbreeding, parents and children obligately share exactly half their alleles IBD whereas full siblings share half their alleles on average. That is, at a particular locus, a specific sibling pair may share 0, 0.5, or 1, but the average of the IBD allele sharing for a collection of full siblings is expected to be 0.5. Alternatively, for a single full sibling pair, the average IBD allele sharing over many independent (unlinked) markers is expected to be 0.5. For each degree of relationship, this expected sharing decreases by a factor of 0.5. That is, whereas first-degree relatives (parents and children or full siblings) are expected to have an average IBD sharing of 0.5, second-degree relatives (half siblings, grandparents and grandchildren, aunts and uncles, and nieces and nephews) have an expected IBD sharing of 0.25, third-degree relatives (e.g., first cousins) have an expected IBD sharing of 0.125, and so on. The expectation of IBD allele sharing is twice the kinship coefficient for a relative pair. Affected relative pair methods of linkage analysis (discussed below) are based on identifying regions of the genome where the IBD allele sharing for a group of relative pairs differs from the expected sharing. Variance component linkage methods (discussed below) are based on identifying regions of the genome where relatives who are more alike phenotypically share more alleles IBD and relatives who are less alike phenotypically share fewer alleles, that is identifying regions where there is a good fit between patterns of phenotypic correlations among relatives and patterns of IBD allele sharing.

A. Multipoint IBD

Use of multiple markers in a chromosomal region can help to resolve ambiguities regarding which chromosome a child inherited from their parent and improve estimation of IBD allele sharing. For instance, the addition of flanking markers in our example above might help to resolve whether the siblings inherited the same chromosomal segment IBD from their homozygous or ungenotyped father. However, estimation of IBD sharing using many genotyped markers in large pedigrees can be difficult.

There are two algorithms that are commonly used for exact estimation of IBD allele sharing using multiple markers, often referred to as multipoint IBD sharing, the Elston–Stewart algorithm (Elston and Stewart, 1971), and the Lander-Green algorithm (Lander and Green, 1987). Both these approaches are limited in the complexity of the data that can be used for multipoint IBD estimation. Computing time for the Elston–Stewart algorithm is linear with the number of individuals in a family but exponential with the number of genotype markers included in the multipoint IBD estimation. Conversely, computing time for the Lander–Green algorithm is linear with increasing number of genotyped markers, but exponential for increasing family size. In practice, most genome-wide linkage screens have a large number of markers and would benefit from being able to use all of them to improve precision of the IBD estimates, making the Elston–Stewart approach less attractive. The Lander–Green algorithm can be used with nuclear families and pedigrees of moderate size, but is not viable for studies of large extended pedigrees of 40 or more individuals that provide the best power for quantitative trait linkage studies. Approximate methods of multipoint IBD estimation that use Markov Chain Monte Carlo (MCMC) approaches provide a solution for IBD estimation with a large number of genotyped markers in pedigrees of unlimited size and complexity (Heath, 1997; Sobel *et al.*, 2001).

B. IBD estimation and intermarker linkage disequilibrium

Most methods of multipoint IBD estimation assume that the genotyped markers are in linkage equilibrium, meaning that the frequency of any haplotype is the product of the allele frequencies at the individual markers. Although this assumption was generally valid for linkage screens based on a relatively sparse 5 or 10 cM map of microsatellite markers, it is clearly violated for the current SNP-based linkage panels (Goode and Jarvik, 2005). For affected sibling pair analyses, failure to account for LD between markers in a multipoint linkage scan consistently produces inflated LOD scores (Goode and Jarvik, 2005; Huang *et al.*, 2004). These effects are largest when parents or founders are untyped and the missing genotypes are inferred independently for each marker. Several solutions to this problem have been offered from clustering closely spaced markers and using their haplotypes as a multiallelic marker in IBD estimation (Abecasis and Wigginton, 2005) to the use of samples that incorporate both affected and discordant pairs as a control for general inflation of estimated IBD sharing among siblings due to disequilibrium among markers (Xing *et al.*, 2006).

III. RELATIVE PAIR LINKAGE METHODS

Relative pair linkage methods are based on selecting pairs of related individuals with particular phenotypes and then searching for chromosomal areas in which the IBD sharing for these pairs differs significantly from what is expected by chance (Penrose, 1953; Suarez *et al.*, 1978). Although sibling pairs are the most common design for this type of study, other relative pairs can be used (Risch, 1990a) and methods have been developed that utilize more than two individuals per family (Hodge, 1984; Weeks and Lange, 1988).

Sibling pairs are expected to share, on average, half of their alleles IBD. More specifically, under the rules of Mendelian segregation, at a given locus we would expect 1/4 of sibling pairs to share 0 alleles IBD, 1/2 to share 1 allele IBD, and 1/4 to share 2 alleles IBD. However, in the vicinity of a locus influencing risk of disease, *affected* sibling pairs as a group should share more alleles IBD than expected by chance. This holds true regardless of the disease model for dominant, recessive, and additive models as well as for models including gene–gene or gene–environment interactions or allelic or locus heterogeneity. Thus, affected relative pair linkage methods can be applied without knowledge of the underlying disease model that is crucial to the penetrance model-based linkage analyses described in the Chapter by Rice *et al.*, this volume.

As a simple example, imagine a rare single locus dominant model of disease. In this case, all affected sibling pairs would need to inherit the risk allele from their affected parent, guaranteeing that they share at least one allele IBD. Whether they share an allele IBD from their other parent or not is random. Their estimated IBD sharing at the disease locus, and nearby linked markers, would be 0.75, a weighted average of the two possibilities. Now imagine that the disease is a result of the interaction between two loci. An individual must carry risk alleles at both loci to be affected. By definition, both members of the affected sibling pair do carry risk alleles at both loci. So their IBD sharing in the vicinity of that dominant locus would still be expected to be 0.75. Now imagine that there is heterogeneity. You must carry a risk allele at any one of many loci to be affected. In this case, sibling pairs who carry a risk allele at our original locus are still expected to have an IBD sharing of 0.75. Sibling pairs who carry risk alleles at other loci would be expected to share, on average, half of their alleles IBD at our original locus, our expectation under the null. The actual IBD sharing among a group of affected sibling pairs in this case will depend on the degree of heterogeneity and the frequency of risk alleles at this original locus versus other loci. However, if X% of affected individuals carry a risk allele at the original locus, the expected IBD sharing among affected sibling pairs at this locus will be:

$$X0.75 + (1 - X)0.5$$

which is still >0.5 for any nonzero X.

A variety of test statistics have been developed for testing deviations of the observed IBD sharing among affected siblings pairs from the expectation. Some of these are based on the mean IBD sharing, some are based on the proportion of pairs sharing two alleles IBD, and some examine the proportions of pairs sharing 0, 1, and 2 alleles IBD. Which of these tests is most powerful will depend on the true underlying disease model. However, tests based on mean

IBD sharing are generally more powerful than those based on the proportion sharing 2 alleles IBD or on examination of the full distribution of pairs sharing 0, 1, or 2 alleles IBD over a wide range of inheritance models (Blackwelder and Elston, 1985; Knapp *et al.*, 1994). Maximum likelihood-based affected pair tests have also been developed (Risch, 1990b) and the power of these tests improved by constraining the estimated IBD sharing within a "possible triangle" that is consistent with Mendelian inheritance rules (Holmans, 1993). One advantage of these likelihood-based approaches is that they have been extended to multilocus models that may improve power to detect linkage (Cordell *et al.*, 1995, 2000).

A. Discordant pairs

A variation on the affected sibling pair approach involves the use of *discordant* sibling pairs, that is one sibling who is affected with the disorder of interest and one who is unaffected. The ideal design for this approach involves extremely discordant sibling pairs, in which siblings are drawn from opposite extremes of a trait distribution, perhaps the top and bottom 10%, 5%, or 1% (Risch and Zhang, 1995). In discordant sibling pair analyses, the expectation is that in a chromosomal region harboring a gene influencing risk of disease, the IBD sharing between discordant siblings will be <0.5. Extremely discordant and concordant sibling pairs may also be combined (Gu *et al.*, 1996).

One concern unique to discordant sibling pair studies is the possibility that selection of siblings from opposite ends of the trait distribution is enriching for nonpaternity. That is, selecting on discordant sibling pairs often increases the rate of nonpaternity in the sample and increases the proportion of pairs who are half siblings rather than full siblings. This can lead to false positive results, given that half siblings would be expected to have IBD sharing <0.5 everywhere in the genome, with the increase in false positives depending on the stringency of the selection and the heritability of the trait (Neale *et al.*, 2002). However, this concern is easily addressed either by obtaining DNA from parents, in which case half siblings are easily identified through persistent Mendelian errors of one sibling with the father, or by testing whether the observed genotypes of the siblings, genome-wide or for a large group of anonymous loci, are consistent with the stated full-sibling relationship. That is, testing whether the observed genotype sharing of a discordant sibling pair on average over all genotyped loci is more consistent with the expectation for full siblings (i.e., 0.5) or half siblings (i.e., 0.25). In general, verification of the stated pedigree relationships on the basis of the genotype data is a valuable step in any family study and there are a number of software packages for this task.

One of the most commonly used of these is PREST (Sun *et al.*, 2002), which examines consistency with a wide variety of relationship classes, ranging from nuclear family to avuncular and first cousin pairs, and includes an assessment of whether the observed allele sharing for a pair is significantly different from the expected sharing given their stated relationship. Other programs that use a similar strategy are RELCHECK (Broman and Weber, 1998) and the sib-kin module of ASPEX (Hinds and Risch, 1996). The program GRR (Abecasis *et al.*, 2001) takes a graphical approach to identifying misspecified relationships in pedigrees through plotting the mean IBS allele sharing over a group of genotyped markers against the variance in IBS allele sharing for each pair of relatives, with the point for each pair color-coded by the presumed type of relationship. Each type of relative pair in a nuclear family occupies a particular area of this space, with full siblings having high mean IBS and high variance, parent–child pairs having similar mean IBS but lower variance, half-siblings having lower mean IBS, and unrelated individuals (such as the parents paired with each other) having yet lower IBS. Misspecified pairs stand out as they are coded in a different colors than the surrounding cluster of properly specified pairs.

B. Covariates in affected sib pair analyses

A variety of techniques have been developed to incorporate information on phenotypic variability or covariates in affected relative pair linkage analyses. Most of these are oriented around the idea that there is likely to be genetic heterogeneity within the group of affected individuals, and covariates that reflect phenotypic heterogeneity among cases may be useful in selecting more genetically homogenous subgroups of families or individuals. One approach involves modeling the IBD allele sharing as a function of covariates (Glidden *et al.*, 2003; Greenwood and Bull, 1997, 1999). Alternatively, families may be stratified into separate groups or their contribution to the linkage statistic may be weighted by the covariate (Cox *et al.*, 1999; Devlin *et al.*, 2002; Hauser *et al.*, 2004; Schaid *et al.*, 2001). Recent innovations include using covariates to weight the contribution of affected individuals, rather than families, to the linkage analysis, which allows for within-family heterogeneity (Whittemore and Halpern, 2006).

C. Power

The power to detect a locus using affected or discordant sibling pair methods is a function of λ_R, the risk of recurrence to a relative of type R, and depends on the underlying trait model and also on the prevalence of the trait or the thresholds used to define affected individuals or discordant pairs (Risch, 1990a). In general, the more extreme is the threshold used for selecting concordant or discordant sibling pairs, the higher is the power. However, a more stringent threshold increases the cost of phenotyping, as it increases the number of pairs that need to be screened, and potentially limits the sample size (Risch and Zhang, 1996).

Thus, stringency of threshold for extremely discordant or concordant sibling pairs must be balanced against costs of phenotyping and sample size requirements (Gu and Rao, 1997).

IV. VARIANCE COMPONENTS LINKAGE METHODS

Whereas the relative pair methods discussed above use dichotomous yes/no categorizations of individuals into concordant and discordant groups, variance component analyses are designed for quantitative traits. The essential idea behind variance components analysis is to explain the variation in trait values among a group of individuals by partitioning the trait variance, σ_p^2, into components attributable to genetic and environmental factors. At the most basic level, these may be unidentified aggregate factors, such as additive genetic effects, σ_a^2, and residual, unspecified environmental effects, σ_e^2:

$$\sigma_p^2 = \sigma_a^2 + \sigma_e^2.$$

The matrix of phenotypic covariances among individuals, Ω, is then modeled as a function of their genetic and environmental sharing:

$$\Omega = 2\Phi\sigma_a^2 + I\sigma_e^2$$

where Φ is a matrix of kinship coefficients among the pairs of individuals and I is an identity matrix, implying an unshared environmental component that is unique to each individual. As noted above, the structuring matrix of 2Φ for each pair of individuals that is used for the additive genetic component is also the expectation of the IBD allele sharing across the genome for each relative pair. In the case of linkage analyses, the standard variance component model includes components of variance for a QTL in a specific chromosomal location, σ_q^2, as well as the aggregate additive genetic component, σ_a^2, which is now a residual due to other genes elsewhere in the genome outside the region of the QTL, and residual environmental effects, σ_e^2 (Almasy and Blangero, 1998; Amos, 1994). For the QTL-specific component, the structuring matrix is the IBD allele sharing among relatives at a specific point in the genome:

$$\Omega = \Pi\sigma_q^2 + 2\Phi\sigma_a^2 + I\sigma_e^2$$

Assuming multivariate normality, the likelihood of each pedigree can be constructed and used to obtain maximum likelihood estimates of the various model parameters, including the QTL-specific variance and the residual aggregate additive genetic variance. Likelihood ratio tests can then be used to evaluate the evidence for linkage at a specific chromosomal location, comparing the likelihood of a model in which the QTL-specific variance, σ_q^2, is estimated to one in which it is fixed at zero. Several regression-based approaches, including the revised Haseman–Elston, are essentially asymptotically equivalent to variance component models, being possibly somewhat more robust but less powerful than the maximum likelihood-based approaches (Chen *et al.*, 2004; Wang and Elston, 2005). One regression-based approach reverses the usual model, estimating IBD allele sharing as a function of covariance among relatives (Sham *et al.*, 2002). MCMC-based methods for quantitative trait linkage use this same underlying variance decomposition model, adding the number of QTLs, as well as the variance due to each, as parameters to be estimated (Daw *et al.*, 2003; Heath, 1997).

The variance component model is easily expanded to include other components, such as shared environmental effects among members of a family. Often this takes the form of a household component allowing for correlations among individuals living together at the time of sampling, a sibship or rearing environment component for traits with environmental effects that are strongest in childhood, or a component modeling a correlation among spouses (Province *et al.*, 2003). The genetic components of the model can also be expanded to incorporate dominance effects (Amos, 1994) and gene–gene (Mitchell *et al.*, 1997) or gene–environment interaction (Blangero, 1993; Diego *et al.*, 2003; Towne *et al.*, 1997).

A. Nonnormality of the trait distribution

Variance component models generally require assumptions regarding the distribution of the quantitative trait being analyzed. Typically, the standard models assume multivariate normality of the trait distribution. The method is robust to some types of nonnormality (e.g., skewness), but is sensitive to kurtosis in the trait distribution (Allison *et al.*, 1999). Specifically, leptokurtic trait distributions, with more values in the tails of the distribution than would be expected under the normal distribution, lead to an increase in the false positive rate for variance component linkage analyses performed under an assumption of multivariate normality. However, a number of approaches are available for robust variance component analyses of leptokurtic traits. Analyses may be performed using the multivariate t distribution (Blangero *et al.*, 2001) or the γ-distribution (Barber *et al.*, 2004) or robust test statistics may be calculated that take into account the observed trait distribution (Blangero *et al.*, 2000).

Nonnormality due to censoring of the trait distribution can also affect variance component analyses. Censoring occurs when the phenotype distribution is truncated, perhaps due to limitations on the instrument used to measure the trait or to inability to precisely measure the phenotype in individuals who are being treated with medications. In these cases, the observed or assumed phenotype value for an individual is actually an upper or lower bound on the possible phenotype, depending on which extreme of the distribution is censored. Tobit variance component methods have been developed to deal with censored trait distributions (Epstein *et al.*, 2003).

B. Discrete and categorical traits

Discrete yes/no traits may be analyzed in a variance component model by use of a liability threshold model (Duggirala *et al.*, 1997; Williams *et al.*, 1999; Wright, 1934a,b). The liability threshold model assumes that behind a yes/no dichotomy there is an underlying quantitative process that reflects an individual's liability to disease. A threshold is placed on this liability distribution that reflects the prevalence of the dichotomous trait. In the case of a disease trait, individuals who are affected would have liability values above the threshold and the threshold would be set such that a proportion of the liability distribution equal to the prevalence of the disease is above the threshold. Covariates may be used in the liability threshold model to indicate different thresholds (prevalences). For any given individual, their exact liability value is unknown. We only know what area of the liability distribution they are in, given their affection status and the position of the threshold for their covariate vector. The estimation of the variance components then depends on integrating over the possible liability values for each individual. Polychotomous traits, with more than two categories, may be used in this type of analysis provided the categories can be ordered with respect to each other (Gianola, 1979).

Obviously, liability threshold models provide less power for linkage than direct measures of a quantitative trait. Power for the liability threshold model for a dichotomous trait is maximized for traits with a prevalence of 50%, but even at this maximum the power is at most 40% of what it would be if the quantitative trait could be measured directly (Williams and Blangero, 2004). Given this, for discrete traits that are defined by thresholds on a measurable quantitative trait (e.g., obesity and body mass index), it is almost always preferable to analyze the quantitative trait directly.

C. Power

Power to detect linkage in variance component methods is a function of the variance due to the QTL, the total trait heritability, the sample size, and the family configuration (Williams and Blangero, 1999, 2004). The additive genetic variance due to the QTL is a function of the allele frequencies at a locus and the displacement between the genotype-specific trait means of the homozygotes. Thus, a QTL can have a large effect on the trait variance in two ways, either by having a high minor allele frequency or, if it is rarer, by causing a large displacement in the trait means. Family configuration is as important a factor in power as sample size for variance component methods. Larger families provide substantially more linkage power per person sampled (Blangero *et al.*, 2003). For fixed pedigree structures, power may be obtained analytically, but typically a sample includes families of various sizes and configuration and simulations are used to estimate power.

V. STRENGTHS AND WEAKNESSES OF MODEL-FREE LINKAGE METHODS

In general, if the correct penetrance model can be specified, model-based linkage methods are more powerful than model-free methods for dichotomous traits, though this depends on the number of penetrance models tested in the linkage analyses and the resulting penalty for multiple testing (Goldin and Weeks, 1993). However, misspecification of the penetrance model in model-based analyses not only reduces power to detect linkage but also can lead to false exclusion of a region containing a true trait locus (MacLean *et al.*, 1993; Risch and Giuffra, 1992). For quantitative traits, the linkage power of variance component analyses is similar to model-based analyses when the QTL is diallelic and better than model-based analyses in situations of allelic heterogeneity where the underlying QTL is multiallelic (Goring *et al.*, 2001).

Model-free linkage methods have significant advantages for gene localization in the context of allelic heterogeneity, both as compared to penetrance model-based linkage methods that assume a diallelic locus and as compared to association-based genome screens. A QTL may contain multiple functional variants or quantitative trait nucleotides (QTNs). For example, there is strong evidence for multiple functional promoter variants in the F7 gene in the Spanish population (Sabater-Lleal *et al.*, 2007) and evidence for multiple coding variants that affect PTC taste sensitivity in the *PTC* gene (Kim *et al.*, 2003). For linkage analyses, the effect size of this QTL is the sum of the effect size of the individual QTNs. Separately each F7 QTN accounts for 1.5–14% of the variance, but in total the F7 QTL accounts for 35.4% of the variance in clotting factor VII levels (Soria *et al.*, 2005). It is this total QTL effect size of 35.4% that determines the power to detect the locus via linkage. Conversely, to find these QTNs through an association study, the power would depend not on the total effect size of the QTL but on the individual effect sizes of the QTNs and the extent of their linkage disequilibrium with surrounding markers. The relevant effect size for localization via association would be the variance due to an individual QTN multiplied by the r^2 measure of linkage disequilibrium between that QTN and a genotyped marker.

Thus, in cases of allelic heterogeneity, where multiple QTNs underlie a single QTL, linkage studies have an advantage in terms of the effect size that is pertinent to the power to detect the locus.

Another advantage of linkage designs in general, as compared to association designs, for genome-wide screening is the lack of reliance on linkage disequilibrium. As seen above, the apparent effect size of a QTN in an association analysis will be diminished in proportion to the strength of disequilibrium between the QTN and a genotyped marker. If full sequence data is available, for example, for a candidate gene, it is possible to guarantee that the functional variants are among the genotyped markers and then concerns regarding dependence on disequilibrium do not apply. At present, complete sequencing is impractical for genome-wide scanning, though we expect that it will soon become feasible given the continuing rapid increases in the throughput and cost efficiency of the technology. Currently, genome-wide analyses must rely on a subset of genotyped markers. For linkage studies, a few hundred microsatellites or a few thousand SNPs are sufficient to resolve patterns of allele sharing within families across the whole genome with a high degree of certainty. For association studies, current genome-wide SNP sets can be shown to have acceptable levels of linkage disequilibrium ($r^2 > 0.75$) with a high proportion of common variants whose minor allele frequency is >0.05, but the coverage of variants with lower frequencies is less complete (The International HapMap Consortium, 2005). Given the issues of linkage disequilibrium and of the different effect sizes of QTNs versus QTLs, it is clear that some genes influencing complex traits will be easier to detect using linkage methods than association methods. As we cannot know ahead of time which QTLs will contain multiple QTNs or which phenotypes will be influenced by rare or common variants, the prudent strategy at this point is to maintain a diversity of approaches in our genome scans.

Of course, one limitation of the model-free linkage methods, as with any linkage method, is that they require large samples of related individuals. This may not be practical for pharmacogenetic studies that require samples of individuals treated with a specific drug or for studies of diseases that are very rare or have high mortality or late onset. On the other hand, many quantitative trait linkage studies are based on studying normal variation in disease-related risk factors. In this case, pedigrees may be readily available, particularly in population isolates or in communities that still maintain more traditional patterns of large family sizes and relatively low rates of migration. Of course, this constraint on study design for linkage is an advantage as well as a limitation. Having a pedigree-based sample permits a wide range of different types of statistical genetic analyses from estimation of heritabilities to examination of endophenotype profiles in unaffected relatives of probands, to formal segregation analyses, and to both linkage and association methods of gene localization.

VI. CONCLUSION

Even with the recent advent of genome-wide association scans with hundreds of thousands of SNPs, model-free linkage remains a viable approach for gene localization for complex and quantitative traits, as illustrated by the numerous successful model-free genome-wide linkage screens that have resulted in identification of specific genes and even specific functional variants (e.g., see Bertram *et al.*, 2000; Curran *et al.*, 2005; Drayna *et al.*, 2003; Edenberg *et al.*, 2004; Ertekin-Taner *et al.*, 2000; Grant *et al.*, 2006; Hinrichs *et al.*, 2006; Kim *et al.*, 2003; Sabater-Lleal *et al.*, 2007; Soria *et al.*, 2005). Particularly in cases where a QTL contains multiple QTNs or in cases where the QTNs have low minor allele frequencies, model-free linkage will be the most efficient method for genome-wide screening.

In the coming years, we anticipate that ways of combining information from linkage and association (Roeder *et al.*, 2006) will be a continuing area of development. Within a single family-based study, linkage and association can provide independent sources of information. Alternatively, information can be combined across studies, with prior linkage results from one study used to add power to genome-wide association studies or, eventually, to reduce the multiple testing burden in genome-wide studies using sequence data. Other likely areas for continued methodological growth in linkage analysis include models that incorporate gene–gene and gene–environment interactions and approaches for dealing with high dimensional multivariate data, such as that obtained from gene expression chips.

Acknowledgments

This research was supported in part by National Institutes of Health grants MH59490, HL45522, MH078111, MH61622, AA08403, HL070751, and GM31575.

References

The International HapMap Consortium (2005). A haplotype map of the human genome. *Nature* **437** (7063), 1299–1320.

Abecasis, G. R., and Wigginton, J. E. (2005). Handling marker-marker linkage disequilibrium: Pedigree analysis with clustered markers. *Am. J. Hum. Genet.* **77**(5), 754–767.

Abecasis, G. R., Cherny, S. S., Cookson, W. O., and Cardon, L. R. (2001). GRR: Graphical representation of relationship errors. *Bioinformatics* **17**(8), 742–743.

Allison, D. B., Neale, M. C., Zannolli, R., Schork, N. J., Amos, C. I., and Blangero, J. (1999). Testing the robustness of the likelihood-ratio test in a variance-component quantitative-trait loci-mapping procedure. *Am. J. Hum. Genet.* **65**(2), 531–544.

Almasy, L., and Blangero, J. (1998). Multipoint quantitative-trait linkage analysis in general pedigrees. *Am. J. Hum. Genet.* **62**(5), 1198–1211.

Almasy, L., and Warren, D. M. (2005). Software for quantitative trait analysis. *Hum. Genomics* **2**(3), 191–195.

Amos, C. I. (1994). Robust variance-components approach for assessing genetic linkage in pedigrees. *Am. J. Hum. Genet.* **54**(3), 535–543.

Barber, M. J., Cordell, H. J., MacGregor, A. J., and Andrew, T. (2004). Gamma regression improves Haseman-Elston and variance components linkage analysis for sib-pairs. *Genet. Epidemiol.* **26**(2), 97–107.

Bertram, L., Blacker, D., Mullin, K., Keeney, D., Jones, J., Basu, S., Yhu, S., McInnis, M. G., Go, R. C., Vekrellis, K., Selkoe, D. J., Saunders, A. J., *et al.* (2000). Evidence for genetic linkage of Alzheimer's disease to chromosome 10q. *Science* **290**(5500), 2302–2303.

Blackwelder, W. C., and Elston, R. C. (1985). A comparison of sib-pair linkage tests for disease susceptibility loci. *Genet. Epidemiol.* **2**(1), 85–97.

Blangero, J. (1993). Statistical genetic approaches to human adaptability. *Hum. Biol.* **65**(6), 941–966.

Blangero, J., Williams, J. T., and Almasy, L. (2000). Robust LOD scores for variance component-based linkage analysis. *Genet. Epidemiol.* **19**(Suppl 1), S8–S14.

Blangero, J., Williams, J. T., and Almasy, L. (2001). Variance component methods for detecting complex trait loci. *Adv. Genet.* **42,** 151–181.

Blangero, J., Williams, J. T., and Almasy, L. (2003). Novel family-based approaches to genetic risk in thrombosis. *J. Thromb. Haemost.* **1**(7), 1391–1397.

Broman, K. W., and Weber, J. L. (1998). Estimation of pairwise relationships in the presence of genotyping errors. *Am. J. Hum. Genet.* **63**(5), 1563–1564.

Chen, W. M., Broman, K. W., and Liang, K. Y. (2004). Quantitative trait linkage analysis by generalized estimating equations: Unification of variance components and Haseman-Elston regression. *Genet. Epidemiol.* **26**(4), 265–272.

Cordell, H. J., Todd, J. A., Bennett, S. T., Kawaguchi, Y., and Farrall, M. (1995). Two-locus maximum lod score analysis of a multifactorial trait: Joint consideration of IDDM2 and IDDM4 with IDDM1 in type 1 diabetes. *Am. J. Hum. Genet.* **57**(4), 920–934.

Cordell, H. J., Wedig, G. C., Jacobs, K. B., and Elston, R. C. (2000). Multilocus linkage tests based on affected relative pairs. *Am. J. Hum. Genet.* **66**(4), 1273–1286.

Cox, N. J., Frigge, M., Nicolae, D. L., Concannon, P., Hanis, C. L., Bell, G. I., and Kong, A. (1999). Loci on chromosomes 2 (NIDDM1) and 15 interact to increase susceptibility to diabetes in Mexican Americans. *Nat. Genet.* **21**(2), 213–215.

Curran, J. E., Jowett, J. B., Elliott, K. S., Gao, Y., Gluschenko, K., Wang, J., Abel Azim, D. M., Cai, G., Mahaney, M. C., Comuzzie, A. G., Dyer, T. D., Walder, K. R., *et al.* (2005). Genetic variation in selenoprotein S influences inflammatory response. *Nat. Genet.* **37**(11), 1234–1241.

Daw, E. W., Wijsman, E. M., and Thompson, E. A. (2003). A score for Bayesian genome screening. *Genet. Epidemiol.* **24**(3), 181–190.

Devlin, B., Jones, B. L., Bacanu, S. A., and Roeder, K. (2002). Mixture models for linkage analysis of affected sibling pairs and covariates. *Genet. Epidemiol.* **22**(1), 52–65.

Diego, V. P., Almasy, L., Dyer, T. D., Soler, J. M., and Blangero, J. (2003). Strategy and model building in the fourth dimension: A null model for genotype x age interaction as a Gaussian stationary stochastic process. *BMC Genet.* **4**(Suppl 1), S34.

Drayna, D., Coon, H., Kim, U. K., Elsner, T., Cromer, K., Otterud, B., Baird, L., Peiffer, A. P., and Leppert, M. (2003). Genetic analysis of a complex trait in the Utah Genetic Reference Project: A major locus for PTC taste ability on chromosome 7q and a secondary locus on chromosome 16p. *Hum. Genet.* **112**(5–6), 567–572.

Dudbridge, F. (2003). A survey of current software for linkage analysis. *Hum. Genomics* **1**(1), 63–65.

Duggirala, R., Williams, J. T., Williams-Blangero, S., and Blangero, J. (1997). A variance component approach to dichotomous trait linkage analysis using a threshold model. *Genet. Epidemiol.* **14**(6), 987–992.

Duggirala, R., Blangero, J., Almasy, L., Dyer, T. D., Williams, K. L., Leach, R. J., O'Connell, P., and Stern, M. P. (1999). Linkage of type 2 diabetes mellitus and of age at onset to a genetic location on chromosome 10q in Mexican Americans. *Am. J. Hum. Genet.* **64**(4), 1127–1140.

Edenberg, H. J., Dick, D. M., Xuei, X., Tian, H., Almasy, L., Bauer, L. O., Crowe, R. R., Goate, A., Hesselbrock, V., Jones, K., Kwon, J., Li, T.-K., *et al.* (2004). Variations in GABRA2, encoding the alpha 2 subunit of the GABA(A) receptor, are associated with alcohol dependence and with brain oscillations. *Am. J. Hum. Genet.* **74**(4), 705–714.

Elston, R. C., and Stewart, J. (1971). A general model for the genetic analysis of pedigree data. *Hum. Hered.* **21**(6), 523–542.

Epstein, M. P., Lin, X., and Boehnke, M. (2003). A tobit variance-component method for linkage analysis of censored trait data. *Am. J. Hum. Genet.* **72**(3), 611–620.

Ertekin-Taner, N., Graff-Radford, N., Younkin, L. H., Eckman, C., Baker, M., Adamson, J., Ronald, J., Blangero, J., Hutton, M., and Younkin, S. G. (2000). Linkage of plasma Abeta42 to a quantitative locus on chromosome 10 in late-onset Alzheimer's disease pedigrees. *Science* **290** (5500), 2303–2304.

Gianola, D. (1979). Heritability of polychotomous characters. *Genetics* **93**(4), 1051–1055.

Glidden, D. V., Liang, K. Y., Chiu, Y. F., and Pulver, A. E. (2003). Multipoint affected sibpair linkage methods for localizing susceptibility genes of complex diseases. *Genet. Epidemiol.* **24**(2), 107–117.

Goldin, L. R., and Weeks, D. E. (1993). Two-locus models of disease: Comparison of likelihood and nonparametric linkage methods. *Am. J. Hum. Genet.* **53**(4), 908–915.

Goode, E. L., and Jarvik, G. P. (2005). Assessment and implications of linkage disequilibrium in genome-wide single-nucleotide polymorphism and microsatellite panels. *Genet. Epidemiol.* **29** (Suppl 1), S72–S76.

Goring, H. H., Williams, J. T., and Blangero, J. (2001). Linkage analysis of quantitative traits in randomly ascertained pedigrees: Comparison of penetrance-based and variance component analysis. *Genet. Epidemiol.* **21**(Suppl 1), S783–S788.

Grant, S. F., Thorleifsson, G., Reynisdottir, I., Benediktsson, R., Manolescu, A., Sainz, J., Helgason, A., Stefansson, H., Emilsson, V., Helgadottir, A., Styrkarsdottir, U., Magnusson, K. P., *et al.* (2006). Variant of transcription factor 7-like 2 (TCF7L2) gene confers risk of type 2 diabetes. *Nat. Genet.* **38**(3), 320–323.

Greenwood, C. M., and Bull, S. B. (1997). Incorporation of covariates into genome scanning using sib-pair analysis in bipolar affective disorder. *Genet. Epidemiol.* **14**(6), 635–640.

Greenwood, C. M., and Bull, S. B. (1999). Analysis of affected sib pairs, with covariates—with and without constraints. *Am. J. Hum. Genet.* **64**(3), 871–885.

Gu, C., and Rao, D. C. (1997). A linkage strategy for detection of human quantitative-trait loci. I. Generalized relative risk ratios and power of sib pairs with extreme trait values. *Am. J. Hum. Genet.* **61**(1), 200–210.

Gu, C., Todorov, A., and Rao, D. C. (1996). Combining extremely concordant sibpairs with extremely discordant sibpairs provides a cost effective way to linkage analysis of quantitative trait loci. *Genet. Epidemiol.* **13**(6), 513–533.

Hauser, E. R., Watanabe, R. M., Duren, W. L., Bass, M. P., Langefeld, C. D., and Boehnke, M. (2004). Ordered subset analysis in genetic linkage mapping of complex traits. *Genet. Epidemiol.* **27**(1), 53–63.

Heath, S. C. (1997). Markov chain Monte Carlo segregation and linkage analysis for oligogenic models. *Am. J. Hum. Genet.* **61**(3), 748–760.

Hinds, D. A., and Risch, N. (1996). The ASPEX package: Affected sib-pair excusion mapping http://aspex.sourceforge.net/.

Hinrichs, A. L., Wang, J. C., Bufe, B., Kwon, J. M., Budde, J., Allen, R., Bertelsen, S., Evens, W., Dick, D., Rice, J., Foroud, T., Nurnberger, J., *et al.* (2006). Functional variant in a bitter-taste receptor (hTAS2R16) influences risk of alcohol dependence. *Am. J. Hum. Genet.* **78**(1), 103–111.

Hodge, S. E. (1984). The information contained in multiple sibling pairs. *Genet. Epidemiol.* **1**(2), 109–122.

Holmans, P. (1993). Asymptotic properties of affected-sib-pair linkage analysis. *Am. J. Hum. Genet.* **52**(2), 362–374.

Huang, Q., Shete, S., and Amos, C. I. (2004). Ignoring linkage disequilibrium among tightly linked markers induces false-positive evidence of linkage for affected sib pair analysis. *Am. J. Hum. Genet.* **75**(6), 1106–1112.

Kim, U. K., Jorgenson, E., Coon, H., Leppert, M., Risch, N., and Drayna, D. (2003). Positional cloning of the human quantitative trait locus underlying taste sensitivity to phenylthiocarbamide. *Science* **299**(5610), 1221–1225.

Knapp, M., Seuchter, S. A., and Baur, M. P. (1994). Linkage analysis in nuclear families. 1: Optimality criteria for affected sib-pair tests. *Hum. Hered.* **44**(1), 37–43.

Lander, E. S., and Green, P. (1987). Construction of multilocus genetic linkage maps in humans. *Proc. Natl. Acad. Sci. USA* **84**(8), 2363–2367.

MacLean, C. J., Bishop, D. T., Sherman, S. L., and Diehl, S. R. (1993). Distribution of lod scores under uncertain mode of inheritance. *Am. J. Hum. Genet.* **52**(2), 354–361.

Mitchell, B. D., Ghosh, S., Schneider, J. L., Birznieks, G., and Blangero, J. (1997). Power of variance component linkage analysis to detect epistasis. *Genet. Epidemiol.* **14**(6), 1017–1022.

Neale, M. C., Neale, B. M., and Sullivan, P. F. (2002). Nonpaternity in linkage studies of extremely discordant sib pairs. *Am. J. Hum. Genet.* **70**(2), 526–529.

Penrose, L. S. (1953). The general purpose sibpair linkage test. *Ann. Eugen.* **18**(2), 120–124.

Province, M. A., Rice, T. K., Borecki, I. B., Gu, C., Kraja, A., and Rao, D. C. (2003). Multivariate and multilocus variance components method, based on structural relationships to assess quantitative trait linkage via SEGPATH. *Genet. Epidemiol.* **24**(2), 128–138.

Reynisdottir, I., Thorleifsson, G., Benediktsson, R., Sigurdsson, G., Emilsson, V., Einarsdottir, A. S., Hjorleifsdottir, E. E., Orlygsdottir, G. T., Bjornsdottir, G. T., Saemundsdottir, J., Halldorsson, S., Hrafnkelsdottir, S., *et al.* (2003). Localization of a susceptibility gene for type 2 diabetes to chromosome 5q34-q35.2. *Am. J. Hum. Genet.* **73**(2), 323–335.

Risch, N. (1990a). Linkage strategies for genetically complex traits. II. The power of affected relative pairs. *Am. J. Hum. Genet.* **46**(2), 229–241.

Risch, N. (1990b). Linkage strategies for genetically complex traits. III. The effect of marker polymorphism on analysis of affected relative pairs. *Am. J. Hum. Genet.* **46**(2), 242–253.

Risch, N., and Giuffra, L. (1992). Model misspecification and multipoint linkage analysis. *Hum. Hered.* **42**(1), 77–92.

Risch, N., and Zhang, H. (1995). Extreme discordant sib pairs for mapping quantitative trait loci in humans. *Science* **268**(5217), 1584–1589.

Risch, N. J., and Zhang, H. (1996). Mapping quantitative trait loci with extreme discordant sib pairs: Sampling considerations. *Am. J. Hum. Genet.* **58**(4), 836–843.

Roeder, K., Bacanu, S. A., Wasserman, L., and Devlin, B. (2006). Using linkage genome scans to improve power of association in genome scans. *Am. J. Hum. Genet.* **78**(2), 243–252.

Sabater-Lleal, M., Chillon, M., Howard, T. E., Gil, E., Almasy, L., Blangero, J., Fontcuberta, J., and Soria, J. M. (2007). Functional analysis of the genetic variability in the F7 gene promoter. *Atherosclerosis* **195**(2), 262–268.

Schaid, D. J., McDonnell, S. K., and Thibodeau, S. N. (2001). Regression models for linkage heterogeneity applied to familial prostate cancer. *Am. J. Hum. Genet.* **68**(5), 1189–1196.

Sham, P. C., Purcell, S., Cherny, S. S., and Abecasis, G. R. (2002). Powerful regression-based quantitative-trait linkage analysis of general pedigrees. *Am. J. Hum. Genet.* **71**(2), 238–253.

Sobel, E., Sengul, H., and Weeks, D. E. (2001). Multipoint estimation of identity-by-descent probabilities at arbitrary positions among marker loci on general pedigrees. *Hum. Hered.* **52**(3), 121–131.

Soria, J. M., Almasy, L., Souto, J. C., Sabater-Lleal, M., Fontcuberta, J., and Blangero, J. (2005). The F7 gene and clotting factor VII levels: Dissection of a human quantitative trait locus. *Hum. Biol.* **77**(5), 561–575.

Suarez, B. K., Rice, J., and Reich, T. (1978). The generalized sib pair IBD distribution: Its use in the detection of linkage. *Ann. Hum. Genet.* **42**(1), 87–94.

Sun, L., Wilder, K., and McPeek, M. S. (2002). Enhanced pedigree error detection. *Hum. Hered.* **54**(2), 99–110.

Towne, B., Siervogel, R. M., and Blangero, J. (1997). Effects of genotype-by-sex interaction on quantitative trait linkage analysis. *Genet. Epidemiol.* **14**(6), 1053–1058.

Wang, T., and Elston, R. C. (2005). Two-level Haseman-Elston regression for general pedigree data analysis. *Genet. Epidemiol.* **29**(1), 12–22.

Weeks, D. E., and Lange, K. (1988). The affected-pedigree-member method of linkage analysis. *Am. J. Hum. Genet.* **42**(2), 315–326.

Whittemore, A. S., and Halpern, J. (2006). Nonparametric linkage analysis using person-specific covariates. *Genet. Epidemiol.* **30**(5), 369–379.

Williams, J. T., and Blangero, J. (1999). Power of variance component linkage analysis to detect quantitative trait loci. *Ann. Hum. Genet.* **63**(Pt 6), 545–563.

Williams, J. T., and Blangero, J. (2004). Power of variance component linkage analysis-II. Discrete traits. *Ann. Hum. Genet.* **68**(Pt 6), 620–632.

Williams, J. T., Van Eerdewegh, P., Almasy, L., and Blangero, J. (1999). Joint multipoint linkage analysis of multivariate qualitative and quantitative traits. I. Likelihood formulation and simulation results. *Am. J. Hum. Genet.* **65**(4), 1134–1147.

Wright, S. (1934a). An analysis of variability in number of digits in an inbred strain of Guinea pigs. *Genetics* **19**(6), 506–536.

Wright, S. (1934b). The results of crosses between inbred strains of Guinea pigs, differing in number of digits. *Genetics* **19**(6), 537–551.

Xing, C., Sinha, R., Xing, G., Lu, Q., and Elston, R. C. (2006). The affected-/discordant-sib-pair design can guarantee validity of multipoint model-free linkage analysis of incomplete pedigrees when there is marker-marker disequilibrium. *Am. J. Hum. Genet.* **79**(2), 396–401.

9

DNA Sequence-Based Phenotypic Association Analysis

Nicholas J. Schork,[*,†,‡,§,¶] **Jennifer Wessel,**[*,†,‡,¶] **and Nathalie Malo**[*,‡,¶]

*Department of Psychiatry, University of California at San Diego, La Jolla, California 92093

†Department of Family and Preventive Medicine, University of California at San Diego, La Jolla, California 92093

‡The Center for Human Genetics and Genomics, University of California at San Diego, La Jolla, California 92093

§The Stein Institute for Research on Aging, University of California at San Diego, La Jolla, California 92093

¶Scripps Genomic Medicine and Department of Molecular and Experimental Medicine, The Scripps Research Institute, La Jolla, California 92037

Advances in Genetics, Vol. 60
0065-2660/08 $35.00
DOI: 10.1016/S0065-2660(07)00409-9

ABSTRACT

The availability of cost-effective, high-throughput genotyping technologies has generated a tremendous amount of interest in genetic association studies. This interest has led to the belief that one could possibly test thousands to millions of representative polymorphic sites on the genome for association with a trait or disease in order to identify the few sites that may be of relevance to the expression of that trait or disease. The choice of which polymorphic sites are "representative" and to be interrogated in such studies is problematic and has involved considerations of the putative functional significance of the sites as well as the linkage disequilibrium relationships between variations at those sites and other neighboring sites. We consider an obvious alternative to genotyping-based strategies and settings for association studies for which decisions about which variations to interrogate are obviated. Essentially, we anticipate a time when cost-effective, high-throughput DNA sequencing technologies are available and researchers will have actual sequence information on the individuals under study rather than information about what variations they possess at a few well-chosen polymorphic genomic sites. We consider Multivariate Distance Matrix Regression analysis to evaluate associations between DNA sequence information and quantitative traits such as blood pressure and cholesterol level. We evaluate the potential of the method in a few (albeit contrived) settings via simulation studies. Ultimately, we show that the procedure has promise and argue that consideration of DNA sequence-based association data should usher in a new era in genetic association study designs and methodologies.

I. INTRODUCTION

The identification of genes that contribute to common complex traits and diseases such as hypertension, diabetes, depression, and cancer has been a focal point of a great deal of biomedical research in the last two or so decades. This interest has, in part, motivated—and been motivated by—very large-scale initiatives such as the Human Genome Project (Lander *et al.*, 2001; Venter *et al.*, 2001), the International HapMap Project (Altshuler and Clark, 2005), and, more recently, The Cancer Genome Atlas project (http://cancergenome.nih.gov/index.asp), as well as the development of very sophisticated genomic technologies such as high-throughput DNA sequencing technologies, large-scale genotyping technologies, and gene expression microarrays. The very detailed information about the human genome that has emerged from these initiatives and technologies has raised a number of questions about how they can be exploited in research focusing on the genetic basis of complex diseases.

One approach to disease gene identification that has received considerable recent interest involves genetic association studies. Genetic association studies essentially involve statistically testing naturally occurring DNA sequence variations for frequency differences between, for example, individuals with and without a particular trait or disease. If an association is found, then it can be inferred that the sequence variation in question is either likely to be causally related to the relevant trait or disease or in linkage disequilibrium (LD) with a sequence variation that is causally related to the trait or disease. Although highly problematic for a number of reasons, and known to create potential for false positive results (Hirschhorn *et al.*, 2002), there is reason to believe that many of the problems associated with genetic association studies can be overcome by testing a greater number of variations; the basic idea being that by testing more genetic variations one has a greater probability of capturing real, true positive associations despite the increase in the number of statistical tests that will be pursued. This attitude was first articulated by Risch and Merikangas (1996) but has now culminated in the initiation of large-scale "whole-genome association" (WGA) studies in which as many as 500,000 or more genetic variations are tested for association with a trait or disease (http://www.genome.gov/19518664; http://www.genome.gov/19518663).

One fundamental problem with current genetic association study approaches, whether genome-wide or not, involves the choice of variations to be tested for association. Currently genotyping technologies are too cost-prohibitive and limited to assay all known polymorphic sites (some possibly very rare) in a single individual. Therefore, variations that are either "representative" of other variations in terms of their LD relationships or known, or very likely, to alter a molecular physiological function are typically chosen for genetic association studies (see also Gu *et al.*, 2008; Chapter by Liu *et al.*, this volume). These choices can be problematic on purely biological grounds, however, if variations within and among genes and genomic regions work in combination to influence phenotypes or disease susceptibility, because the choice of a limited number of variations to be tested for association may not capture the appropriate biologically relevant effects of all the variations either in part or as a whole. There is a growing literature on precisely what variations it makes most sense to consider in genetic association studies, as it is known that LD strength varies throughout the genome (Jorgenson and Witte, 2006; Pardi *et al.*, 2005; see also Gu *et al.*, 2008; Chapter by Liu *et al.*, this volume). With this in mind, it makes sense to consider strategies for association studies that do not consider tests of a few—however well chosen—variations, but rather work with more exhaustive, individual DNA sequences obtained from research subjects in genomic regions of interest or over the entire genome.

A. DNA sequencing and association studies

There are a number of factors that motivate consideration of exhaustive DNA sequencing-based approaches to association studies. Some of these factors are technological in nature and others are biological in nature. In terms of technological factors, there is growing interest in developing fast, cost-effective DNA sequencing technologies that will rival current genotyping technologies. In fact, the National Institutes of Health has promoted the development of such technologies through its "thousand dollar" genome initiative (http://grants.nih.gov/grants/guide/rfa-files/RFA-HG-04–003.html). Most current genotyping technologies have focused on single nucleotide polymorphisms (SNPs) because of the ease with which many of them can be interrogated in single assays. However, it has become increasingly clear that other forms of variation including repeats, insertions and deletions, various forms of copy number variations, and inversions also populate the genome (Redon *et al.*, 2006). Even if SNPs are in LD with these variations, it is not clear if the strength of this LD will be sufficient to bring out relevant locus effects in association studies involving only SNPs in LD with other variations on the genome (Redon *et al.*, 2006).

In terms of the biological effects of genetic variations and the motivation to consider full-blown DNA sequencing information rather than genotyping assays that consider a subset of all variations that could be studied, the results of many recent studies suggest that combinations of variations—not necessarily working in an additive fashion—within a genomic region influence phenotypic expression. Table 9.1 provides a listing of example studies that have shown, either via *in vitro* studies, association studies, or both, that variations in small stretches of sequence associated with a gene or other DNA sequence element interact in complicated ways to influence phenotypic variation. The nature of these interactions is complicated and could reflect *cis*-acting promoter influences on gene expression or more complex diplotypic, *trans*-acting influences on protein formation and function. This raises the question as to whether consideration of a more exhaustive understanding of what actual combinations of common, rare, and possibly *de novo* variations an individual carries is necessary for association studies to work (Pritchard and Cox, 2002; Reich and Lander, 2001).

Table 9.1. Studies Suggesting That Multiple, Potential Interacting Variants Within a Gene or Specified Genomic Region Influence Phenotypic Epression

Gene	*In vitro?*	Phenotype	References
ADRB2	Yes	Bronchodilator response	Drysdale *et al.* (2000)
DRD4	No	Schizophrenia	Nakajima *et al.* (2007)
NRG1	No[a]	Schizophrenia and NRG1 mRNA levels	Law *et al.* (2006)
HTR2A	Yes	HTR2A gene expression	Myers *et al.* (2007)
ENT1	Yes	ENT1 gene expression	Myers *et al.* (2006)
CDA	Yes	CDA gene expression	Fitzgerald *et al.* (2006)
PCSK9	No	Lipoprotein levels	Kotowski *et al.* (2006)
NPC1L1	No	Lipoprotein levels	Cohen *et al.* (2006)
KRT1	Yes	KRT1 gene expression	Tao *et al.* (2006)
GH1	Yes	GH1 gene expression/ adult height	Horan *et al.* (2003)
DAT1 (SLC6A3)	Yes	DAT1 gene expression	Greenwood and Kelsoe (2003)
APOE	No	Lipid levels	Stengard *et al.* (2002)
SLC6A3	Yes	Parkinson's disease	Kelada *et al.* (2005)
CHGA	Yes	Catecholamine physiology	Wen *et al.* (2004)

[a]Note that the study of the NRG1 gene involved computational assessments of the functionality of gene variations rather than *in vitro* studies or just association studies.

B. Potential analysis methods

The analysis of exhaustive DNA sequence information in association study settings is likely to be complicated, given that some of these variations may represent structural anomalies (e.g., gross or small deletions and/or inversions) and simple base changes that exhibit complex LD relationships in the population at large. Figure 9.1 illustrates a hypothetical situation in which the genomic region associated with a particular gene is under investigation for which a total of 20 variable sites have been identified. Some of the sites (e.g., variations labeled as 1–5, 7–10, 12–14, 16, 17, and 19) are simple base changes, others represent deletions/insertions (6, 11, 15, and 18), and one represents a repeat and/or small duplication (20). DNA sequence associated with a single chromosome obtained from each of six individuals is also provided in Fig. 9.1, with three individuals possessing a disease (i.e., the individuals labeled D1, D2, and D3) and three individuals not possessing a disease (C1, C2, and C3). Also provided in the figure is a representation of the degree of cross-species conservation obtained for each site in the genomic region, as a high degree of conservation is often thought to indicate the putative functional significance of a site (Boffelli *et al.*, 2003).

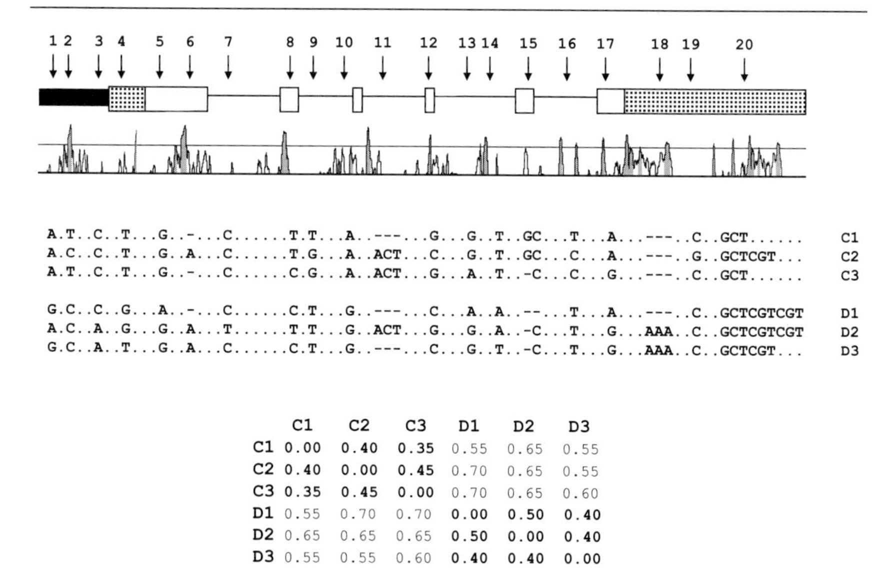

	C1	C2	C3	D1	D2	D3
C1	0.00	0.40	0.35	0.55	0.65	0.55
C2	0.40	0.00	0.45	0.70	0.65	0.55
C3	0.35	0.45	0.00	0.70	0.65	0.60
D1	0.55	0.70	0.70	0.00	0.50	0.40
D2	0.65	0.65	0.65	0.50	0.00	0.40
D3	0.55	0.55	0.60	0.40	0.40	0.00

Figure 9.1. Hypothetical diagram of polymorphic sites in and around a gene and sequence data for the genic region obtained from six individuals, three with a disease (D1, D2, and D3) and three without (C1, C2, and C3), The numbers (1–20) label the polymorphic sites (of which there are a total of 20). The black rectangle indicates the promoter region and additional regulatory sequence. The shaded rectangles indicate untranslated regions and the open rectangles indicate exons. Evolutionary conservation levels associated with different positions in the genic region are reflected in the histogram below the genic region diagram with higher peaks indicating greater cross-species conservation. The sequences below the conservation chart reflect variations in the sequence six individuals posses for three diseased (D1, D2, and D3) and three nondiseased (C1, C2, and C3) individuals. The matrix below the sequence data is a distance matrix that reflects the distance between the sequences for the 6 individuals based on a simple identify-by-state allele sharing measure over the 20 polymorphic regions. See text for more details.

To analyze such data, one could simply test each particular variation for association with the trait or disease using simple contingency table analyses, if the phenotype of interest was qualitative, or *t*-tests/ANOVAs if the phenotype was quantitative. In addition, one could create indicator variables for each locus that reflect the number of copies of an allele an individual possesses at that locus and then use multiple regression strategies to examine the combined effects of the loci on the phenotype. One could even create interaction terms involving these loci and test their significance in multiple regression contexts. These strategies have some problems in that it is not likely to be known *a priori* how the relevant alleles interact, both at a single locus (e.g., whether there are dominant, additive, or recessive allelic effects) and/or across different loci. In addition, the LD relationships between the loci will create multicollinearity problems for standard regression and analysis of variance techniques.

Analysis techniques that can potentially overcome these issues are listed in Table 9.2. Some advantages and disadvantages of these techniques are also provided in Table 9.2, although we want to emphasize that Table 9.2 is not an exhaustive listing of all possible analytic procedures, but rather a sampling of them. Thus, for example, one could use ridge regression to accommodate dependencies (i.e., LD) and therefore multicollinearity among the variations (Draper and Smith, 1981). Alternatively, one could simply obtain counts of the number of individuals with certain sequences and test them for equality between, for example, case and control groups using *sparse* contingency table analyses (Agresti, 2002; i.e., because many of the sequences will be rare, the table would have a large number of categories with few individuals in each) or via Poisson regression (Cameron and Trivedi, 1998). Another approach would involve the exploitation of the phylogeny of the sequences as described by Templeton *et al.* (2005), although this approach may be limited to effects induced at the haplotype, rather than diplotype, level. Pattern discovery and data mining approaches that could be used to test hypotheses as to whether certain sequences have features that can distinguish individuals with and without certain phenotypes could also be appropriate. Such an approach was outlined by Salamon *et al.* (1996). Logic regression, as described by Kooperberg and Ruczinski (2005), has promise since it was designed to test hypotheses about

Table 9.2. Potential Data Analysis Approaches to DNA Sequence-Based Phenotype Association Studies

Method	References	Advantages/Features	Disadvantages
Logic regression	Kooperberg and Ruczinski (2005)	Flexible; focus on interactions	Computationally intensive
Hierarchical models	Conti and Witte (2003)	Accommodates collinearity	Computationally intensive
Ridge regression	Draper and Smith (1981)	Accommodates collinearity	Requires a "ridge" parameter
Sparse contingency table	Agresti (2002)	Very general omnibus test	Works with count data
Pattern differentiation	Salamon *et al.* (1996)	Flexible; hypothesis generating	Computationally intensive
Phylogenetic/Cladistic	Templeton *et al.* (2005)	Very elegant and informative	Requires phylogeny estimation
Poisson Regression	Cameron and Trivedi (1998)	Well described and validated	Works with count data
Conditional tests	Valdes and Thomson (1997)	Separate causal from non-causal	Not much experience
MDMR	Wessel and Schork (2006)	Flexible omnibus test	Not much experience

whether combinations of predictor variables can explain a dependent variable. Other approaches such as hierarchical models (Conti and Witte, 2003) or conditional LD tests (Valdes and Thomson, 1997) also have promise.

We consider an analysis approach that is rooted in testing hypotheses about the similarity and/or "distance" between the sequences possessed by a group of individuals and the phenotypic features they have (for haplotype similarity analysis, see Gu *et al.*, 2008; Chapter by Liu *et al.*, this volume). This approach, termed Multivariate Distance Matrix Regression (MDMR) analysis, extends the work originally outlined by Wessel and Schork (2006) and exploits the fact that the assessment of sequence similarity has a great deal of precedent in modern biological research. In fact, it is arguable that the assessment of the similarity of different DNA sequences has been the *sine qua non* of evolutionary and biological research in the modern genomic era (see, e.g., standard quantitative biology and bioinformatics texts such as Durbin *et al.*, 1999). The MDMR analysis method has many advantages over traditional analysis methods in that is very flexible and subsumes both single locus and phylogeny-based association analysis techniques (Wessel and Schork, 2006), as described later.

As a simple example of the intuition behind MDMR analysis, consider the matrix below the sequence information in Fig. 9.1 that provides a distance matrix whose elements reflect the number of differences in the sequence possessed by pairs of the six individuals. It can be seen that the diseased individuals have similar sequences and the nondiseased individuals have similar sequences, but the sequences show greater differences between diseased and nondiseased individuals (i.e., the off-diagonal, nonshaded cells in the matrix are larger than the diagonal elements). Lacking in the literature are methods that can take measures of DNA similarity and directly relate them to additional information on the individuals, organisms, or species from which they were obtained, and MDMR analysis was devised to overcome this. We investigate some of the properties of MDMR analysis in some contrived settings involving quantitative traits to showcase its power and flexibility because the analysis of quantitative traits is known to be more complicated than the analysis of discrete, qualitative traits and are likely the rule, rather than the exception, in nature.

II. MULTIVARIATE DISTANCE MATRIX REGRESSION

A. Computing a distance matrix

The proposed procedure depends critically on the computation of a "distance" matrix that reflects the dissimilarity of each pair of DNA sequences obtained from the individuals in a study. There are many possible measures that could be used to construct such a matrix, and we consider a few. More detail on alternative distance measures can be found in a study by Wessel and Schork (2006). Alternatively, one could consider and possibly adapt techniques described in the vast literature on DNA sequence similarity in evolutionary biology (Felsenstein, 1988; Phillips, 2006; Templeton *et al.*, 2005). Consider a stretch of DNA sequence harboring L polymorphic sites that is evaluated for N individuals. The distance between the sequences, s^i and s^j, for individuals i and j $(i,j = 1,\ldots,N)$ can be formulated in a simple way as:

$$d_{i,j}(s^i, s^j) = \frac{1}{\sum_{l=1}^{L} w_l} \sum_{l=1}^{L} w_l \phi(p_l^i, p_l^j) \tag{1}$$

where p_l^i represents the actual sequence (i.e., nucleotide, sequence of nucleotides, absence of a nucleotide, or series of nucleotides) at position l in the sequence possessed by individual j; $\phi(p_l^i, p_l^j)$ is a measure of the dissimilarity of the nucleotides possessed by individuals i and j; and w_l is a weight assigned to position l in the sequence. The simplest measures of DNA sequence dissimilarity would have $w_l = 1$; $l = 1,\ldots, L$ and $\phi(p_l^i, p_l^j)$ as be a simple identity-by-state (IBS) measure, such that $\phi(p_l^i, p_l^j) = 0$ when $p_l^i = p_l^j$ and 1 otherwise.

There are some complications when applying DNA sequence similarity measures of the proposed type to diploid individuals, such as humans. First, the sequence data must be "phased" in the sense that the sequence associated with each of the two chromosomes an individual possesses has to be resolved. This is not necessarily trivial, although accurate phasing algorithms that work with stretches of genome harboring a small to moderate number of polymorphic sites have been developed (Salem *et al.*, 2005). For the methodology we are proposing, we assume that the phasing of chromosomes and sequence is not an issue. Second, even if one has the sequence data associated with the two chromosomes each individual possesses in a sample, there is still the problem of pairing the sequences for two individuals. We address this problem by using the minimum of the averages of the distances for the two possible pairings of the sequences from two individuals as a measure of diploid DNA sequence distance as described in Wessel and Schork (2006). Thus, let s_1^i and s_2^i represent the two sequences associated with the two chromosomes possessed by individual i and let s_1^j and s_2^j represent the two sequences associated with the two chromosomes possessed by individual j. We define the diploid sequence distance measure as:

$$d_{i,j}^D = \min\left\{\left[\frac{d_{i,j}(s_1^i, s_1^j) + d_{i,j}(s_2^i, s_2^j)}{2}\right], \left[\frac{d_{i,j}(s_1^i, s_2^j) + d_{i,j}(s_2^i, s_1^j)}{2}\right]\right\} \tag{2}$$

In terms of the weights, w_l, that can be incorporated into the construction of the distance measure, they can be based on any number of phenomena, such as cross-species DNA sequence conservation, likely functional significance based on *in silico*, *in vitro*, or model organism studies, or association studies providing statistical evidence that variations at the relevant sequence positions are found more (or less) often in, for example, diseased individuals than not (Wessel and Schork, 2006). We also note that the distance measure, $\phi(p_l^i, p_l^j)$, could be position specific in that different measures could be used to accommodate different forms of sequence variation. For example, the simple IBS sharing distance measure might be appropriate for SNPs, but something more quantitative might be more appropriate when assessing the differences in, for example, repeat or copy number variations. Obviously, if the distance measures used were position specific, then one would have to account for the different scales the measures might involve so as to not unduly weight one position more heavily than another unless that was the desired effect.

B. MDMR analysis

Once one has computed a distance matrix it can be subjected to MDMR analysis to test hypotheses about, for example, whether variation in the level of similarity/dissimilarity exhibited by pairs of individuals reflected in that matrix can be explained by other features those individuals posses (e.g., whether they have a particular disease or not, whether they have high or low values of a quantitative trait). To describe the MDMR analysis procedure, we assume that each of N individuals or study subjects has been sequenced over a genomic region harboring L variable sites and that M grouping or phenotypic variables have been collected on the N subjects. These grouping or phenotypic variables could include information about the disease status of the individuals (coded using dummy variables, such as a 1 assigned to diseased individuals and 0 assigned to nondiseased individuals), the level of a quantitative trait (e.g., blood pressure or cholesterol level), as well as important covariates such as gender, race, age.

We note that the MDMR analysis procedure is an extension of the procedure described by McArdle and Anderson (2001) and does not require that the distance matrix used have metric properties. Let the distance matrix and its elements be denoted by $D = d_{ij}$ $(i, j = 1, \ldots, N)$, for the N subjects [e.g., Eqs. (1) and (2)]. The possibility that $N \ll L$ will not pose problems in the proposed MDMR analysis procedure because the information on the L loci goes into the construction of the distance matrix, and is not analyzed in L separate analyses. Let X be an $N \times M$ matrix harboring information on the M grouping or phenotypic variables that will be modeled as predictor or regressor variables whose relationships to the values in the distance matrix are of interest. Compute the standard projection matrix, $H = X(X'X)^{-1}X'$, typically used to estimate coefficients relating the predictor variables to outcome variables in multiple regression contexts. Next, compute the matrix $A = (a_{ij}) = (-[1/2]d_{ij}^2)$ and center this matrix using the transformation discussed by Gower (1966) and denote this matrix G:

$$G = \left[I - \frac{1}{n}11'\right] A \left[I - \frac{1}{n}11'\right]$$

An F statistic can be constructed to test the hypothesis that the M regressor variables have no relationship to variation in the genomic distance or dissimilarity of the N subjects reflected in the $N \times N$ distance/dissimilarity matrix as (McArdle and Anderson, 2001):

$$F = \frac{tr(HGH)}{tr[(I - H)G(I - H)]}$$

Note that the F statistic essentially tests the hypothesis that the phenotype(s) of interest is associated with variation in the patterns of dissimilarity/similarity reflected in the distance matrix and is therefore a very general (or "omnibus") test of association between sequence variation in a genomic region and a trait. This is unlike standard approaches to association analysis that focus on individual locus (or haplotype) effects. In addition, the "sparseness" of the data that would inevitably arise with sequence-based associations—that is, the fact that very few individuals, if more than one, will have exactly the same sequences to be compared with respect to phenotypes to other sets of individuals with different sequences—is not a problem for MDMR analysis, because the distances/similarities between sequences are considered in the analysis, not discrete groupings based on individuals with exactly the same sequence.

C. Assessing significance of the *F* statistic

If the Euclidean distance is used to construct the distance matrix on a single variable (i.e., as in a univariate analysis of that variable), rather than, say, an IBS-based distance measure, and appropriate numerator and denominator degrees of freedom are accommodated in the test statistics, the F statistic above is equivalent to the standard ANOVA F statistic (McArdle and Anderson, 2001). The distributional properties of the F statistic are complicated for alternative distance measures computed for more than one variable, especially if those variables are discrete, as in genotype data. Therefore, we advocate the use of permutation tests to assess statistical significance of the pseudo-F statistic (Edgington, 1995; Good, 2000; Nievergelt *et al.*, 2007; Zapala and Schork, 2006a) when appropriate critical values either are not available or have not been estimated via prior simulation studies.

We also note that the M regressor variables can be tested individually or in a stepwise manner as in multiple regression contexts by merely fitting relevant complete and restricted models.

D. Graphical display of distance/similarity matrices

One can graphically portray the information in a distance matrix in a number of ways (e.g., heatmaps and trees) that can facilitate interpretation of MDMR analysis. Both have value, and each is constructed to showcase clustering of individuals with greater phenotypic/genomic similarity (i.e., individuals with similar phenotypes are represented as adjacent branches of the tree or adjacent cells in a heatmap). By color coding individual tree branches based on the phenotype values possessed by the individuals they represent, one can see if there are patches of a certain color on neighboring branches that would indicate that phenotype values cluster along with genetic similarity (e.g., using HyperTree v.1.0.0, http://www.kinase.com/tools/HyperTree.html). Note that if one needs a similarity matrix for graphical representations, then one can easily transform a distance matrix to a similarity matrix by subtracting the components of the matrix from 1.0 if a simple IBS distance measure is used, or subtracting them from 1.0 after each component in the matrix is divided by the theoretical or empirical maximum of the distance measure to scale the entries to lie between 0 and 1.

III. SIMULATION STUDIES

We pursued simulation studies to investigate the properties of the proposed MDMR analysis method in settings meant to capture aspects of DNA sequence-based association analyses. We acknowledge that many of the settings for which we simulated data are highly contrived (i.e., equal allele frequencies and no LD, equal allelic effect sizes in some settings) but our goal was to bring to light some of the merits of the proposed approach relative to single locus analyses and not necessarily to assess performance in many settings. As more raw DNA sequence data is collected on individuals for use in association studies, we will be in a better position to tailor simulation studies to more realistic settings. In addition, our assessment of the properties of the MDMR analysis assumed quantitative phenotypes because analyses of quantitative phenotypes have been receiving considerable interest, as they may be reflective of underlying intermediate or molecular perturbations causing disease (e.g., altered gene expression or protein levels).

A. The determination of critical values

In order to determine appropriate critical values for the MDMR procedure for use in power studies, we simulated settings for which 10, 25, 50, and 100 individuals were sampled. We assigned them phenotypic values based on random sampling from a standard normal distribution. We then assigned the hypothetical individuals two alleles at each of 1, 10, 25, 50, 100, and 250 polymorphic loci. We assumed that the alleles were equally frequent and in linkage equilibrium. Although unrealistic in genetics contexts, we wanted to consider the distribution of the pseudo-F statistic in very general situations. In all situations we examined, we considered the use of the simple unweighted IBS allele sharing sequence distance measure.

B. The assessment of the power of MDMR

To examine the power of the proposed MDMR analysis procedure, we simulated data at loci with equally frequent alleles in linkage equilibrium. We considered samples of size 50. Hypothetical alleles were assigned to each individual. One of the two alleles was assumed to increase a hypothetical quantitative trait by a factor of α. This factor was varied from values of 0.067 to 1.67 in increments of 0.067. This factor was added to an individual's phenotype value whenever the allele that increases the phenotype values was assigned to that individual via a random number generator. To create additional noise in the phenotype a standard normal variate was added to each individual's phenotype. The genotype-based phenotypic assignment process resulted in locus effects within a hypothetical genomic region that explained only a fraction of the variation. Figure 9.2 provides a plot of the percentage of variation explained (i.e., "R^2") by each locus as a function of the assumed value for the settings involving 50 individuals and 10, 25, 50, 100, and 250 loci. The values plotted in Fig. 9.2 are average R^2s obtained from regression analyses of each locus over 1000 simulations. As each locus was analyzed in isolation, the R^2s are biased upwards. We pursued this assessment of locus effects becasue the chosen sample size was small and it is known that small sample sizes can have quite pronounced influences on estimates of effect sizes. It can clearly be seen that increases in the R^2s occur with increasing α values, as expected, unless the number of loci is large. This is simply due to the fact that with such a large number of loci contributing to the expression of the phenotype, any individual locus is going to have a small effect if those loci are equal in effect size. This scenario is consistent with a truly polygenic background for the expression of the trait.

Note that in the assessment of the power of the proposed MDMR analysis method, we also considered the application of regression-based single locus analyses. We tallied the fraction of loci that resulted in significant

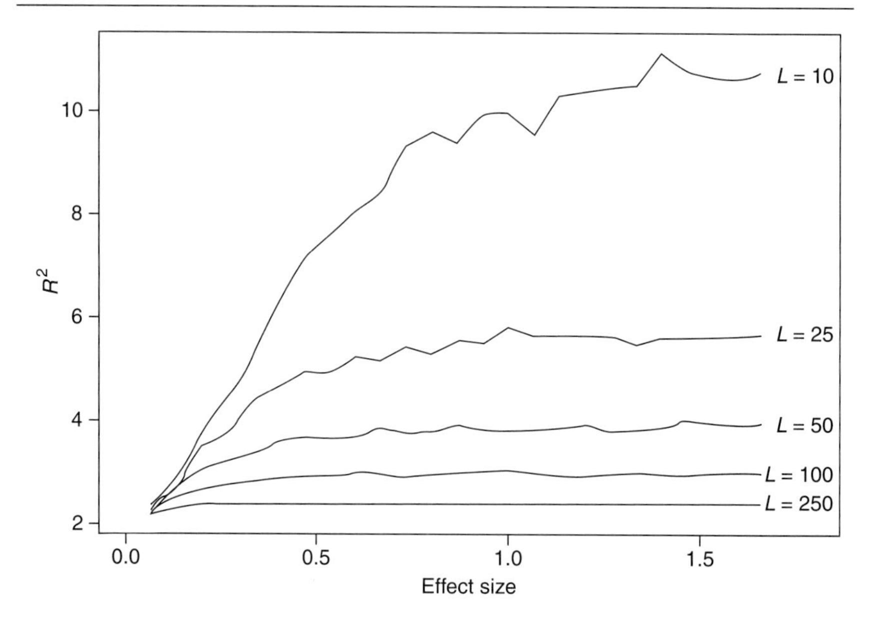

Figure 9.2. Estimated percentage of variance explained (R^2) for each locus via univariate, single locus-based, regression analysis in multilocus DNA sequence settings involving 50 hypothetical individuals for which different numbers of polymorphic loci are assumed to influence a trait. The effect size represented on the *x*-axis provides the displacement in a quantitative trait attributable to an allele at each locus. The loci were assumed to have equal effects. The R^2s were estimated from regression analysis of 1000 simulated data sets.

Bonferroni-corrected locus effects ($p < 0.05$) for each simulation and averaged this fraction over all simulations. This provides an indication of how likely it would be to detect locus effects in the region of interest if attention was confined to each locus individually.

We did consider settings in which the locus effects were assumed not to be equal, but rather had diminishing effects on the trait of interest. We generated locus effects in this scenario by assuming that the first locus had a phenotypic displacement effect of α, the effect of the second locus had an effect of $\alpha/2$, the effect of the third locus had an effect of $\alpha/3$, and so on. We also estimated the R^2s in this situation and, as expected, the loci with smaller effects only explain a small percentage of the variation in the trait. Table 9.3 provides some examples of average R^2s estimated from 1000 simulations as function of the α value for a sample size of 50 for situations in which 10 and 100 loci were studied. Table 9.3 shows only average R^2 for four of the total loci, but clearly shows that the loci have diminishing effects on the trait with the first, or leading, locus having a

Table 9.3. Estimated Contributions (Average R^2) of Each Locus Assuming a Decay in Effect Size with Each Locus Based on Regression Analyses of 1000 Simulated Datasets for Each Setting

	Number of Loci = 10					Number of Loci = 100			
Locus/α	0.07	0.60	1.07	1.67	Locus/α	0.07	0.60	1.07	1.67
1	2.34	16.46	29.86	43.00	1	2.44	15.13	31.20	42.29
4	1.96	3.71	4.13	4.79	33	1.92	1.81	2.19	2.51
7	1.71	2.00	2.80	3.38	66	2.36	1.44	2.59	2.18
10	2.11	1.85	2.16	2.28	100	2.29	2.02	1.98	2.08

pronounced effect and the additional loci having substantially reduced effects. This scenario mimics a setting in which there are a few major genes and a larger number of polygenes that influence the expression of the trait.

Finally, we considered settings in which 50 loci are tested for association in 50 individuals, but only a fraction of those loci actually have an effect on the phenotype. This setting is realistic in the sense that not all polymorphic sites within a genomic region will be functional, and tests will be performed to determine which of these loci are likely to be functional and which are not. In these settings, we again considered situations in which the loci had equal effects as well as diminishing effects on the phenotype.

IV. RESULTS

A. The determination of critical values

Table 9.4 provides estimates of critical values for sample sizes of 25, 50, and 100 assuming the number of variable sites in a genomic region of interest is 1, 10, 25, 50, 100, and 250. As noted in the discussion of the derivation of the MDMR pseudo-F statistic, the pseudo-F statistic should have properties similar to a standard F statistic, although the appropriate number of numerator and denominator degrees of freedom is not clear. We have provided F statistic percentiles in parentheses beneath the estimated critical values from the simulations that assume numerator degrees of freedom equal to the number of loci tested and denominator degrees of freedom equal to the (sample size $\times$ the number of loci) $-$ 2. These values for the degrees of freedom were shown to be reliable via simulation studies by Zapala and Schork (2006b). It can be seen that although there is some correspondence and diminishing variation in estimated F statistic critical values

Table 9.4. Estimated Pseudo-F Statistic Percentiles for MDMR Analysis of Quantitative Traits for Selected Sample Sizes and F Statistic Percentiles, Assuming Numerator Degrees of Freedom Equal to the Number of Loci and Denominator Degrees of Freedom Equal to the (Number of Loci × the Sample Size) − 2, in Parentheses

	$N = 25$			$N = 50$			$N = 100$		
	0.10	0.05	0.01	0.10	0.05	0.01	0.10	0.05	0.01
$L = 1$	2.39 (2.94)	3.12 (4.28)	5.64 (7.88)	2.32 (2.81)	3.06 (4.04)	5.33 (7.19)	2.42 (2.76)	3.19 (3.94)	5.09 (6.90)
$L = 10$	1.60 (1.62)	1.84 (1.87)	2.26 (2.39)	1.62 (1.61)	1.86 (1.85)	2.36 (2.36)	1.66 (1.61)	1.88 (1.84)	2.38 (2.34)
$L = 25$	1.37 (1.39)	1.49 (1.52)	1.73 (1.80)	1.37 (1.38)	1.50 (1.51)	1.76 (1.79)	1.40 (1.38)	1.53 (1.51)	1.79 (1.78)
$L = 50$	1.24 (1.27)	1.34 (1.36)	1.50 (1.54)	1.26 (1.27)	1.35 (1.36)	1.48 (1.53)	1.29 (1.27)	1.39 (1.35)	1.57 (1.53)
$L = 100$	1.14 (1.19)	1.21 (1.25)	1.32 (1.37)	1.18 (1.19)	1.24 (1.25)	1.35 (1.36)	1.19 (1.19)	1.26 (1.25)	1.36 (1.36)
$L = 250$	1.08 (1.12)	1.12 (1.16)	1.17 (1.23)	1.10 (1.12)	1.13 (1.15)	1.20 (1.22)	1.12 (1.12)	1.16 (1.15)	1.22 (1.22)

as a function of sample size and number of variables, consistent with the behavior of an F distribution, there is not a close correspondence with the theoretical F distribution critical values for a small number of loci although there is excellent agreement for a large number of loci. More work on the behavior of the pseudo-F statistic is being pursued (Zapala and Schork, 2006b).

B. The assessment of the power of MDMR: Equal effect sizes

The power of the MDMR analysis procedure in settings involving 50 individuals and 10, 25, 50, 100, and 250 loci with equal phenotypic effects as a function of the allelic effect size α is depicted in Fig. 9.3. It can be seen that the larger the locus effects (i.e., the smaller the number of loci assessed because the locus effects were assumed to be equal) the greater the power the MDMR analysis procedure had to detect in association between variations at the loci and the quantitative phenotype (solid lines represent MDMR analyses in Figs. 9.3 and 9.4). In addition, if one analyzed each of the loci independently using regression analyses,

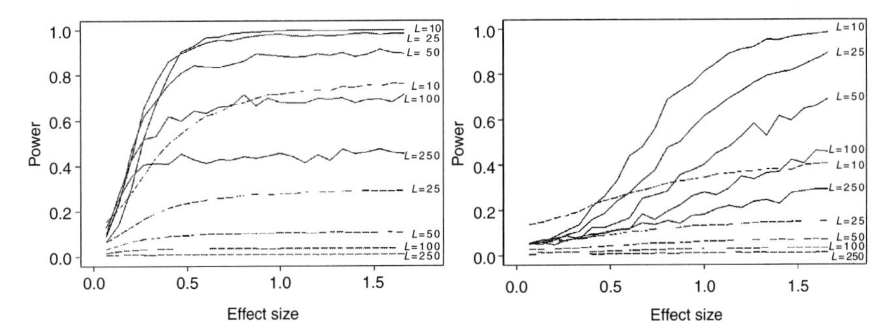

Figure 9.3. Left panel: Estimated power of MDMR DNA sequence analysis (plain lines) in settings involving different numbers of polymorphic loci (denoted "L") in a hypothetical stretch of sequence with equal effect sizes (see Fig. 2 for average locus effects in terms of estimated R^2). The x-axis provides the assumed effect size. Each power curve was estimated from 1000 simulations. The dotted lines represent the fraction of loci in each setting for which single locus regression analyses resulted in significant results assuming a Bonferroni-corrected nominal p value of 0.05. Right panel: Estimated power of MDMR DNA sequence analysis (plain lines) in settings involving different numbers of polymorphic loci (L) in a hypothetical stretch of sequence with diminishing trait effect sizes (see Table 9.3 for some representative average locus effects in terms of estimated R^2). The x-axis provides the assumed effect size for the first locus with additional loci, each having half the effect of the previous locus (i.e., the loci have diminishing effects on phenotypic expression). Each power curve was estimated from 1000 simulations. The dotted lines represent the fraction of loci in each setting for which single locus regression analyses resulted in significant results assuming a Bonferroni-corrected nominal p value of 0.05.

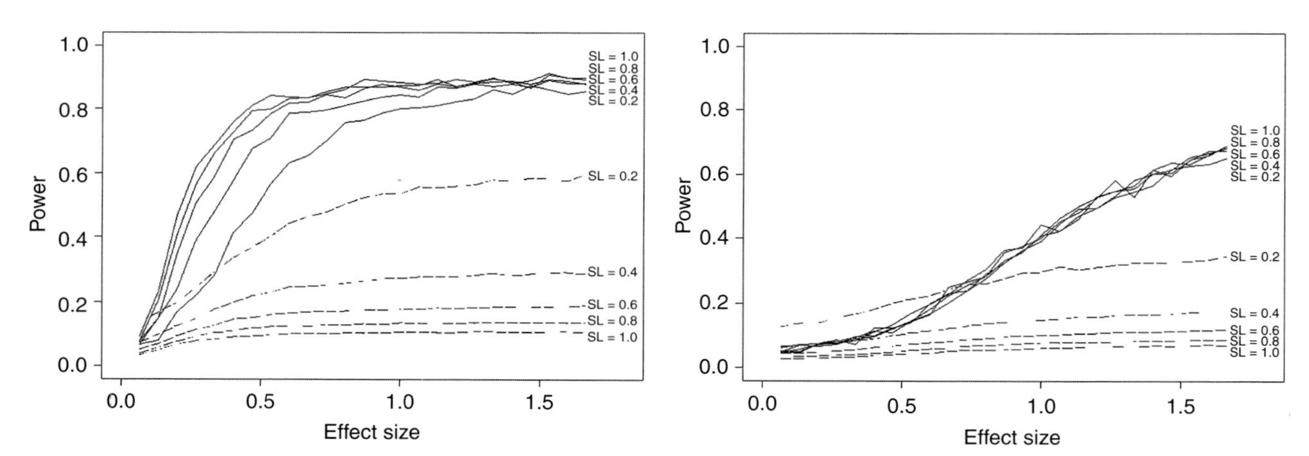

Figure 9.4. Estimated power of MDMR DNA sequence analysis (plain lines) in settings involving 50 polymorphic loci in a hypothetical stretch of sequence with only some fraction of these 50 loci (SL) having an actual effect on the hypothetic trait of interest. The loci with an effect were assumed to have equal effects on the phenotype (left panel; see Fig. 2 for average locus effects in terms of estimated R^2) or diminishing locus effects (right panel). The x-axis provides the assumed effect size for each locus (left panel) or the first locus (right panel). Each power curve was estimated from 1000 simulations. The dotted lines represent the fraction of loci in each setting for which single locus regression analyses resulted in significant results assuming a Bonferroni-corrected nominal p value of 0.05.

only a fraction of the loci would be declared significant (i.e., produce a Bonferroni-corrected p value < 0.05), with a greater fraction of the total number of loci resulting in significant results when the locus effects of each are large.

C. The assessment of the power of MDMR: Diminishing effect sizes

For situations in which the locus effects are not equal, but rather have diminishing effects on the phenotype, we find that the MDMR analysis methods have reasonable power with 50 individuals but only in situations where the loci with the largest effects do indeed have reasonably large effects (Fig. 9.3). The diminishing phenotypic influence of the loci also has a dramatic effect on the fraction of loci that can be detected as having a statistically significant association with the phenotype based on single locus analysis, as expected.

D. The assessment of the power of MDMR: The influence of neutral loci

We also considered the influence of the inclusion of loci in the analysis that did not influence the phenotype (i.e., "neutral" or nonfunctional loci). We again considered sample sizes of 50, but assumed that 50 loci were to be analyzed. We assumed that 50, 40, 30, 20, and 10 of these loci had an effect on the phenotype. We assumed settings in which the loci with a phenotypic influence had equal effects and diminishing effects. Figure 9.4 displays the results for the settings involving equal and diminishing effects. Figure 9.4 clearly shows that the greater fraction of the total loci that are functional influences the ability of single locus association studies to detect locus effects, probably due, in part, to the need to correct for the multiple comparisons involving the nonfunctional loci. However, the more "omnibus" test associated with the MDMR procedure is not influenced by these nonfunctional loci to the same degree.

V. DISCUSSION

Genetic association studies are likely to receive increasing attention in years to come as genotyping and sequencing technologies are refined and greater insight into the influences of sequence variations on biological processes is obtained. Just how association studies will be pursued, however, is likely to change considerably as these technologies emerge and new biological insights are obtained. For example, if DNA sequencing costs approach the costs of genotyping assays, then a good question will be whether it makes more sense to sequence individuals rather than ask what variations they carry at a few select sites. Obviously, a great deal of the generated sequence will not be useful as it will be monomorphic throughout the human species because this sequence has no chance of differentiating, for example, diseased and nondiseased individuals. However, this extra or redundant information may not motivate the choice not to sequence individuals in a study because the increase in information obtained about the patterns of variation each individual in a sample possesses could provide a great deal of insight not obtainable from genotype information on a few representative loci. In addition, if it is the case that it is the pattern or combination of variations that a person inherits—including rare and *de novo* variations—then exploiting LD to detect locus effects, unless it is extremely strong and one has a method that considers the relationship of multiple combinations of variations to a trait or disease, may not be appropriate or powerful enough.

The proposed MDMR approach provides a potential analytic method for analyzing associations between DNA sequences and traits or diseases. The approach is flexible and does not assume that interactions between various loci must be modeled. However, the approach is not without its problems. First, we have only considered simulation studies to investigate properties of the MDMR method that involve a small number of individuals ($n = 50$) in some highly artificial situations involving equally frequent, additive effect loci. More comprehensive investigations should be pursued with varying sample sizes and allelic effects, such as interaction effects of the type studied by Wessel and Schork (2006). We are encouraged, however, that the method appears to have power with such small sample sizes. Second, the method requires the construction of a distance measure. It is unclear which measures would be the most appropriate. Although many distance measures have been posited for cross-species sequence comparisons, it is unclear if the features they have incorporated (such as evolutionary mutation rate information) are appropriate for within-species analyses. Some measures that may show promise include measures weighted toward certain biological phenomena (Liu and Wang, 2006) or those that accommodate inversions and other phenomena, such as word and compression-based measures (Wu *et al.*, 2005). Third, although the MDMR procedure can accommodate nonfunctional variations in the formulation of the distance matrix without complete losses of power, it is clear that one aim of genetic analysis is to distinguish nonfunctional from functional loci, or to quantify the degree to which variations contribute to a phenotype. Although the MDMR analysis method can be used to identify "optimal" subsets of variables that exhibit maximal association with a particular variable (Wessel and Schork, 2006; Zapala and Schork, 2006a), it may be that other approaches, such as some of those listed in Table 9.2, are more appropriate for this task. It is important to point out that the MDMR analysis procedure can be used in genome scan contexts in which different sized "windows" of sequence are assessed for association with a trait or disease of interest (Wessel *et al.*, 2006, submitted for publication). Conceivably, one could have entire genomic sequence data on a set of individuals and move through the entire genome in an attempt to identify regions harboring variations of combinations of variations that influence a phenotype.

Ultimately, DNA sequence-based approaches to association analysis will receive greater attention as sequencing technologies become cheaper and more efficient and as experiences with sequence comparisons in other contexts, such as cancer genome analysis or evolutionary analyses, grow. It is therefore important to anticipate the data analysis needs that will inevitably arise. It is hoped that the proposed MDMR analysis procedure, if not a perfect solution, will spark extensions and the development of alternative analysis methods.

Acknowledgments

N.J.S. and his laboratory are supported in part by the following research grants: the NHLBI Family Blood Pressure Program (FBPP; U01 HL064777–06); the NIA Longevity Consortium (U19 AG023122–01); the NIMH Consortium on the Genetics of Schizophrenia (COGS; 5 R01 HLMH065571–02); NIH R01s: HL074730–02 and HL070137–01; the Scripps Genomic Medicine Initiative; and the Donald W. Reynolds Foundation (Helen Hobbs, Principal Investigator).

References

Agresti, A. (2002). "Categorical Data Analysis." John Wiley and Sons, New York.

Altshuler, D., and Clark, A. G. (2005). Genetics. Harvesting medical information from the human family tree. *Science* **307**(5712), 1052–1053.

Boffelli, D., McAuliffe, J., Ovcharenko, D., Lewis, K. D., Ovcharenko, I., Pachter, L., and Rubin, E. M. (2003). Phylogenetic shadowing of primate sequences to find functional regions of the human genome. *Science* **299,** 1391–1394.

Cameron, A. C., and Trivedi, P. K. (1998). "Regression Analysis of Count Data." Cambridge University Press, New York.

Cohen, J. C., Pertsemlidis, A., Fahmi, S., Esmail, S., Vega, G. L., Grundy, S. M., and Hobbs, H. H. (2006). Multiple rare variants in NPC1L1 associated with reduced sterol absorption and plasma low-density lipoprotein levels. *Proc. Natl. Acad. Sci. USA* **103,** 1810–1815.

Conti, D. V., and Witte, J. S. (2003). Hierarchical modeling of linkage disequilibrium: Genetic structure and spatial relations. *Am. J. Hum. Genet.* **72**(2), 351–363.

Draper, N. R., and Smith, H. (1981). "Applied Regression Analysis." John Wiley and Sons, New York.

Drysdale, C. M., McGraw, D. W., Stack, C. B., Stephens, J. C., Judson, R. S., Nandabalan, K., Arnold, K., Ruano, G., and Liggett, S. B. (2000). Complex promoter and coding region beta 2-adrenergic receptor haplotypes alter receptor expression and predict in vivo responsiveness. *Proc. Natl. Acad. Sci. USA* **97**(19), 10483–10488.

Durbin, R., Eddy, S. R., Krogh, A., and Mitchison, G. (1999). "Biological Sequence Analysis: Probabilistic Models of Proteins and Nucleic Acids." Cambridge University Press, Boston.

Edgington, E. S. (1995). "Randomization Tests," 3rd Edition. Marcel Dekker, New York.

Felsenstein, J. (1988). Phylogenies from molecular sequences: Inference and reliability. *Annu. Rev. Genet.* **22,** 521–565.

Fitzgerald, S. M., Goyal, R. K., Osborne, W. R., Roy, J. D., Wilson, J. W., and Ferrell, R. E. (2006). Identification of functional single nucleotide polymorphism haplotypes in the cytidine deaminase promoter. *Hum. Genet.* **119,** 276–283.

Good, P. I. (2000). "Permutation Tests." Springer, New York.

Gower, J. C. (1966). Some distance properties of latent root and vector methods used in multivariate analysis. *Biometrika* **53,** 325–338.

Greenwood, T. A., and Kelsoe, J. R. (2003). Promoter and intronic variants affect the transcriptional regulation of the human dopamine transporter gene. *Genomics* **82**(5), 511–520.

Gu, C. C., Yu, K., and Rao, D. C. (2008). Characterization of LD structures and the utility of HapMap in genetic association studies. *In* "Genetic Dissection of Complex Traits" (D. C. Rao and C. C. Gu, eds.), 2nd Edition, pp. 407–429. Academic Press, San Diego.

Hirschhorn, J. N., Lohmueller, K., Byrne, E., and Hirschhorn, K. (2002). A comprehensive review of genetic association studies. *Genet. Med.* **4,** 45–61.

Horan, M., Millar, D. S., Hedderich, J., Lewis, G., Newsway, V., Mo, N., Fryklund, L., Procter, A. M., Krawczak, M., and Cooper, D. N. (2003). Human growth hormone 1 (GH1) gene expression: Complex haplotype-dependent influence of polymorphic variation in the proximal promoter and locus control region. *Hum. Mutat.* **21,** 408–423.

Jorgenson, E., and Witte, J. S. (2006). A gene-centric approach to genome-wide association studies. *Nat. Rev. Genet.* **7,** 885–891.

Kelada, S. N., Costa-Mallen, P., Checkoway, H., Carlson, C. S., Weller, T. S., Swanson, P. D., Franklin, G. M., Longstreth, W. T., Afsharinejad, Z., and Costa, L. G. (2005). Dopamine transporter (SLC6A3) 5′ region haplotypes significantly affect transcriptional activity in vitro but are not associated with Parkinson's disease. *Pharmacogenet. Genomics* **15,** 659–668.

Kooperberg, C., and Ruczinski, I. (2005). Identifying interacting SNPs using Monte Carlo logic regression. *Genet. Epidemiol.* **28,** 157–170.

Kotowski, I. K., Pertsemlidis, A., Luke, A., Cooper, R. S., Vega, G. L., Cohen, J. C., and Hobbs, H. H. (2006). A spectrum of PCSK9 alleles contributes to plasma levels of low-density lipoprotein cholesterol. *Am. J. Hum. Genet.* **78,** 410–422.

Lander, ES, Linton, LM, Birren, B, Nusbaum, C, Zody, MC, Baldwin, J, Devon, K, Dewar, K, Doyle, M, FitzHugh, W, Funke, R, Gage, D, *et al.* (2001). Initial sequencing and analysis of the human genome. *Nature* **409,** 860–921.

Law, A. J., Lipska, B. K., Weickert, C. S., Hyde, T. M., Straub, R. E., Hashimoto, R., Harrison, P. J., Kleinman, J. E., and Weinberger, D. R. (2006). Neuregulin 1 transcripts are differentially expressed in schizophrenia and regulated by 5′ SNPs associated with the disease. *Proc. Natl. Acad. Sci. USA* **103**(17), 6747–6752.

Liu, N., and Wang, T. (2006). A weighted measure for the similarity analysis of DNA sequences. *J. Mol. Biol.* **12,** 897–903.

McArdle, B. H., and Anderson, M. J. (2001). Fitting multivariate models to community data: a comment on distance-based redundancy analysis. *Ecology* **82,** 290–297.

Myers, R. L., Airey, D. C., Manier, D. H., Shelton, R. C., and Sanders-Bush, E. (2007). Polymorphisms in the regulatory region of the human serotonin 5-HT(2A) receptor gene (HTR2A) influence gene expression. *Biol. Psychiatry* **61**, 167–173. 2007, [Epub ahead of print].

Myers, S. N., Goyal, R. K., Roy, J. D., Fairfull, L. D., Wilson, J. W., and Ferrell, R. E. (2006). Functional single nucleotide polymorphism haplotypes in the human equilibrative nucleoside transporter 1. *Pharmacogenet. Genomics* **16,** 315–320.

Nakajima, M., Hattori, E., Yamada, K., Iwayama, Y., Toyota, T., Iwata, Y., Tsuchiya, K. J., Sugihara, G., Hashimoto, K., Watanabe, H., Iyo, M., Hoshika, A., *et al.* (2007). Association and synergistic interaction between promoter variants of the DRD4 gene in Japanese schizophrenics. *J. Hum. Genet.* **52,** 86–91. 2007, [Epub ahead of print].

Nievergelt, C.M., Libiger, O., Schork, N. J. (2007) Generalized molecular analysis of variance. *PLOS Genet.* **3**, e51 (Epub. 2007 Feb. 22).

Pardi, F., Lewis, C. M., and Whittaker, J. C. (2005). SNP selection for association studies: Maximizing power across SNP choice and study size. *Ann. Hum. Genet.* **69,** 733–746.

Phillips, A. J. (2006). Homology assessment and molecular sequence alignment. *J. Biomed. Inform.* **39**(1), 18–33.

Pritchard, J. K., and Cox, N. J. (2002). The allelic architecture of human disease genes: Common disease-common variant…or not? *Hum. Mol. Genet.* **11,** 2417–2423.

Redon, R., Ishikawa, S., Fitch, K. R., Feuk, L., Perry, G. H., Andrews, T. D., Fiegler, H., Shapero, M. H., Carson, A. R., Chen, W., Cho, E. K., Dallaire, S., *et al.* (2006). Global variation in copy number in the human genome. *Nature* **444,** 444–454.

Reich, D. E., and Lander, E. S. (2001). On the allelic spectrum of human disease. *Trends Genet.* **17,** 502–510.

Risch, N., and Merikangas, K. (1996). The future of genetic studies of complex human diseases. *Science* **273,** 1516–1517.

Salamon, H., Tarhio, J., Ronningen, K., and Thomson, G. (1996). On distinguishing unique combinations in biological sequences. *J. Comput. Biol.* **3**(3), 407–423.

Salem, R. M., Wessel, J., and Schork, N. J. (2005). A comprehensive literature review of haplotyping software and methods for use with unrelated individuals. *Hum. Genomics* **2**(1), 39–66.

Stengard, J. H., Clark, A. G., Weiss, K. M., Kardia, S., Nickerson, D. A., Salomaa, V., Ehnholm, C., Boerwinkle, E., and Sing, C. F. (2002). Contributions of 18 additional DNA sequence variations in the gene encoding apolipoprotein E to explaining variation in quantitative measures of lipid metabolism. *Am. J. Hum. Genet.* **71,** 501–517.

Tao, H, Cox, D. R., and Frazer, K. A. (2006). Allele-specific KRT1 expression is a complex trait. *PLOS Genet.* **2,** 848–858.

Templeton, A. R., Maxwell, T., Posada, D., Stengard, J. H., Boerwinkle, E., and Sing, C. F. (2005). Tree scanning: A method for using haplotype trees in phenotype/genotype association studies. *Genetics* **169,** 441–453.

Valdes, A. M., and Thomson, G. (1997). Detecting disease-predisposing variants: The haplotype method. *Am. J. Hum. Genet.* **60**(3), 703–716.

Venter, J. C., Adams, M. D., Myers, E. W., Li, P. W., Mural, R. J., Sutton, G. G., Smith, H. O., Yandell, M., Evans, C. A., Holt, R. A., Gocayne, J. D., Amanatides, P., *et al.* (2001). The sequence of the human genome. *Science* **291**, 1304–1351. Erratum in: *Science* 2001 Jun 5; **292**(5523): 1838.

Wen, G., Mahata, S. K., Cadman, P., Mahata, M., Ghosh, S., Mahapatra, N. R., Rao, F., Stridsberg, M., Smith, D. W., Mahboubi, P., Schork, N. J., O'Connor, D. T., *et al.* (2004). Both rare and common polymorphisms contribute functional variation at CHGA, a regulator of catecholamine physiology. *Am. J. Hum. Genet.* **74,** 197–207.

Wessel, J., and Schork, N. J. (2006). Generalized genomic distance-based regression methodology for multilocus association analysis. *Am. J. Hum. Genet.* **79**(5), 792–806.

Wessel, J., Libiger, O., and Schork, N. J. (2006). Whole genome association studies using window-based Multivariate Distance Matrix Regression analysis. *Genet. Epidemiol.* (in review).

Wu, T. J., Huang, Y. H., and Li, L. A. (2005). Optimal word sizes for dissimilarity measures and estimation of the degree of dissimilarity between DNA sequences. *Bioinformatics* **21,** 4125–4132.

Zapala, M., and Schork, N. J. (2006a). Multivariate regression analysis of distance matrices for testing associations between gene expression patterns and related variables. *Proc. Natl. Acad. Sci.* **103,** 19430–19435.

Zapala, M., and Schork, N. J. (2006b). Statistical properties of multivariate distance matrix regression for high-dimensional data analysis. (in preparation).

10

Family-Based Methods for Linkage and Association Analysis

Nan M. Laird and Christoph Lange

Department of Biostatistics, Harvard School of Public Health,
Boston, Massachusetts 02115

Advances in Genetics, *Vol. 60*
0065-2660/08 $35.00
DOI: 10.1016/S0065-2660(07)00410-5

ABSTRACT

Traditional epidemiological study concepts such as case-control or cohort designs can be used in the design of genetic association studies, giving them a prominent role in genetic association analysis. A different class of designs based on related individuals, typically families, uses the concept of Mendelian transmission to achieve design-independent randomization, which permits the testing of linkage and association. Family-based designs require specialized analytic methods but they have distinct advantages: They are robust to confounding and variance inflation, which can arise in standard designs in the presence of population substructure; they test for both linkage and association; and they offer a natural solution to the multiple comparison problem. This chapter focuses on family-based designs. We describe some basic study designs as well as general approaches to analysis for qualitative, quantitative, and complex traits. Finally, we review available software.

I. INTRODUCTION

Families have dominated genetic studies, dating back to Mendel's first experiments elucidating the concepts of inheritance in plants. Later, the work of Galton, Fisher, and others on familial aggregation and segregation was built on a wealth of information about inheritance patterns derived from family studies. With the progress of the Human Genome Project, genetic markers spanning the entire human genome have enabled widespread mapping efforts based on linkage analysis using families with multiple affected individuals, leading to the discovery of many genes for Mendelian diseases and traits.

Association analysis differs fundamentally from linkage in that it is not mandatory to use families, and inferences can be made about genetic association from unrelated individuals (see Chapter by Schork *et al.*, this volume). We refer to designs that use unrelated individuals as standard designs (typically case control or cohort). Family designs for studying association based on trios (two parents and an affected offspring) were introduced by Rubenstein *et al.* (1981) and Falk and Rubinstein (1987), while the analysis of such designs was discussed by Spielman *et al.* (1993), Ott (1989), and Terwilliger and Ott (1992).

A. Hypothesis testing in family designs

In testing for genetic association with a standard design, the null and alternative hypotheses are simply given as:

H0: no association between the marker and the disease
HA: association is present between the marker and the disease

A rejection simply implies that the disease trait of interest is associated with the alleles at the marker. With a family-based test (FBAT), the null and alternative hypotheses can be phrased in terms of the underlying genetics in the population. As noted in Ott (1989), family designs have no power to detect association unless linkage is present. Thus, when testing for association or linkage with family designs, the alternative hypothesis is always HA: Both linkage and association are present between the marker and a disease susceptibility locus (DSL) underlying the trait.
There are three possibilities for the null hypothesis in a family design:

(1) H0: no linkage and no association between the marker and any DSL underlying the trait.
(2) H0: linkage but no association between the marker and any DSL underlying the trait.
(3) H0: association but no linkage between the marker and any DSL underlying the trait.

When testing candidate genes, or in a whole genome scan, the appropriate null hypothesis will ordinarily be the first null, H0: no linkage and no association. However, if we are testing for association in a study that has known linkage in the region of testing, then a more appropriate null hypothesis is the second, H0: linkage but no association. This distinction is not relevant when our sample consists of parents and one offspring. However, when the sample includes multiple offspring from the same family, with or without parents, the distribution of the test statistic under the null differs, depending on whether linkage is assumed to be present. In Section III, we will show how to construct valid tests under both types of null hypotheses.

The third null hypothesis, H0: association but no linkage was proposed by Spielman *et al.* (1993), when they introduced their transmission disequilibrium test (TDT) to test for linkage in a setting where association had been demonstrated in several population studies, but conventional linkage analysis failed to find evidence of linkage. However, as the distribution of family data under the null hypothesis is the same for the first and third null hypotheses (i.e., the null distribution depends only on whether linkage and multiple sibs are present), we usually consider only the first and second nulls. We remark that some authors (Guo *et al.*, 2007) seem to prefer the null hypothesis

(4) H0: no linkage or no association between the marker and any DSL underlying the trait.

Because the distribution under the null must consider the possibility of linkage without association, the distribution of the data under the null is the same as that for the second null hypothesis, and the two are thus equivalent, in the sense that any test valid for the second is valid for the fourth.

B. The TDT test for trios

The basic family design is the trio, consisting of two parents and one offspring. The TDT test (Spielman *et al.*, 1993) is the standard approach to the analysis when the offspring is affected with the trait of interest. The analysis is similar in principle to the alleles test in a case-control analysis, in that the number of A alleles, for example, among the cases is compared to the number expected under the null hypothesis. The main difference is how the number expected is computed under the null. With a case-control (or more generally, a standard) design, the number expected is computed by assuming the distribution of alleles is the same in cases and controls under the null, and by using that common distribution to derive an expectation for the affected group. In contrast, the trio design, and family-based designs in general, relies on using Mendel's laws to compute expectations for the offspring based on their parent's genotypes.

The basic design is diagrammed in Fig. 10.1. The analysis is very intuitive. If any of the three null hypotheses mentioned above holds, then Mendel's laws dictate the transmission of alleles from parents to offspring. The mother can only transmit the A allele and is thus not informative about association of any allele with disease in the offspring. The father transmits either A or B with probability 50/50. Thus, the child is either AA or AB, with probability 50/50. The TDT test consists of using the A alleles transmitted from heterozygous parents to their offspring (n_A). Under any H0, n_A follows a binomial distribution with $p = 0.5$ and n = the number of heterozygous parents. Because parents' transmissions are independent, each heterozygote parent has probability 0.5 of transmitting

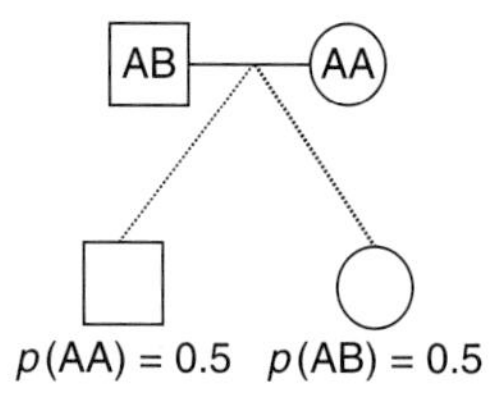

Figure 10.1. Trio Design—the TDT. Family-trios are the basis of the transmission disequilibrium test (TDT); it compares the observed number of the A alleles transmitted to the affected offspring with those expected by Mendelian transmissions. An excess of A (or B) alleles among the affected suggests that a DSL for the trait is in linkage and linkage disequilibrium (LD) with the marker locus.

the A allele. Thus, one can compute an exact test of the null hypothesis, or an asymptotic Z or χ^2 test based on this binomial distribution. The TDT usually refers to the χ^2 version of the binomial test.

As in the alleles test commonly used in standard designs (Chapter by Schork *et al.*, this volume), the potential sample size is twice the number of trios because each individual has two alleles. With the TDT test, however, that advantage is offset, as the transmissions from homozygous parents are not used. Thus, the effective sample size may be considerably less than the number of trios, depending on allele frequency. If there are multiple affected offspring, then the same test remains valid (counting n and n_A as the total number of heterozygous parent transmissions to all offspring and the number of A allele transmissions from heterozygous parents to all offspring, respectively), provided the null hypothesis tested assumes no linkage because parental transmissions to different offspring remain independent when there is no linkage between the marker and any DSL affecting the trait.

The derivation of the TDT leads to an intuitive justification for the premise that both linkage and association must be present under the alternative. If there is association, but no linkage, between the marker and the DSL in the parent population, then the marker alleles in the parents are transmitted independently of the DSL alleles, and there will be no association between the marker and any DSL in the offspring. If there is linkage of the two loci in the parents, but not association, then the two loci will be linked in the offspring, but different marker alleles will be transmitted with different DSL alleles in different families, so there will be no "population" association in the offspring. Formally, Vansteelandt *et al.* (2007) have shown that conditioning on the parents' genotypes serves to eliminate any potential confounding in the test of association, making it robust not only to population stratification and admixture but also to potential model misspecification.

C. Extensions to the TDT

Because of its great success in the analysis of trio data, there is a wealth of literature on extensions of the basic TDT. Curtis and Sham (1995a), Bickeboller and Clerget-Darpoux (1995), and Spielman and Ewens (1996) describe extensions for multiallelic tests. Schaid (1996) put the TDT test into a more general context of a score test for multinomial data, showing that the TDT is optimal for an additive alternative, and providing tests for dominant and recessive models as well. Spielman and Ewens (1998), Curtis and Sham (1995b), Schaid and Li (1997), Rabinowitz and Laird (2000), and Fulker *et al.* (1999) discuss family tests when parents are missing and/or for general pedigree designs. Martin *et al.* (2000), Horvath and Laird (1998), and Lake *et al.* (2000) describe methods for general pedigrees that are also valid when testing for association in the presence of linkage.

Fulker *et al.* (1999), Abecasis *et al.* (2000), Rabinowitz (1997), Horvath *et al.* (2001), and Laird *et al.* (2000) discuss extensions for quantitative traits. An overview of analysis methods for family designs is given by Zhao (2000). We will consider many of these extensions in Sections III and IV.

D. Design issues

There are many possible family configurations that can be used in family designs: trios, sib pairs, general nuclear families (with or without parents), and more general pedigrees. The trio design is generally the most powerful among family designs with one affected offspring per family. Although many methods have been proposed for using incomplete trios with only one parent, such methods can be biased (Curtis and Sham, 1995b) with biallelic markers, and generally it will be necessary to have at least one additional offspring to capture information from the family. Figure 10.2 shows some power comparisons for four designs: the case-control, the trio, discordant sib pairs (DSPs) (without parents), and discordant sibships (no parents, one unaffected offspring and two unaffected siblings). All four of these designs have the same number of affected cases. Panel (A) shows power for a rare disease (prevalence 0.1%) and panel (B) shows power for a common disease (14%). Both cases assume an additive disease model with allelic odds ratio of 1.3.

With rare disease, the trio design, followed by the case-control, is the most powerful relative to the number of affected offspring that need to be recruited. However, more genotyping is required (three genotypes per case, as opposed to two per case in the case-control design). In addition, it can be difficult to recruit parents; notable exceptions are childhood illnesses and when using samples originally designed for linkage analysis. With more common diseases, the case-control design is more powerful, followed closely by the parent–offspring trio and the discordant sibship trio. At both levels of prevalence, the DSP design is considerably less powerful than either the trio or the case-control (Witte *et al.*, 1999), although it requires less genotyping than either of the other family designs.

We note that unaffected siblings are most commonly used in family-based designs to compensate for missing parents. However, even when parents are present, information can be gained about association by using transmissions to unaffected offspring in the case of common disorders (Lange and Laird, 2002a; Whittaker and Lewis, 1998).

Figure 10.3A shows how the power of the basic TDT can be increased (or decreased) by using information from an additional unaffected offspring when disease prevalence is 0.3. Here the dotted horizontal line indicates power for the TDT that discards the unaffected offspring. The unaffected offspring are included by using an offset μ_Y (see Section III); when the offset is zero, unaffected offspring are not included. When the offset is close to the prevalence, the

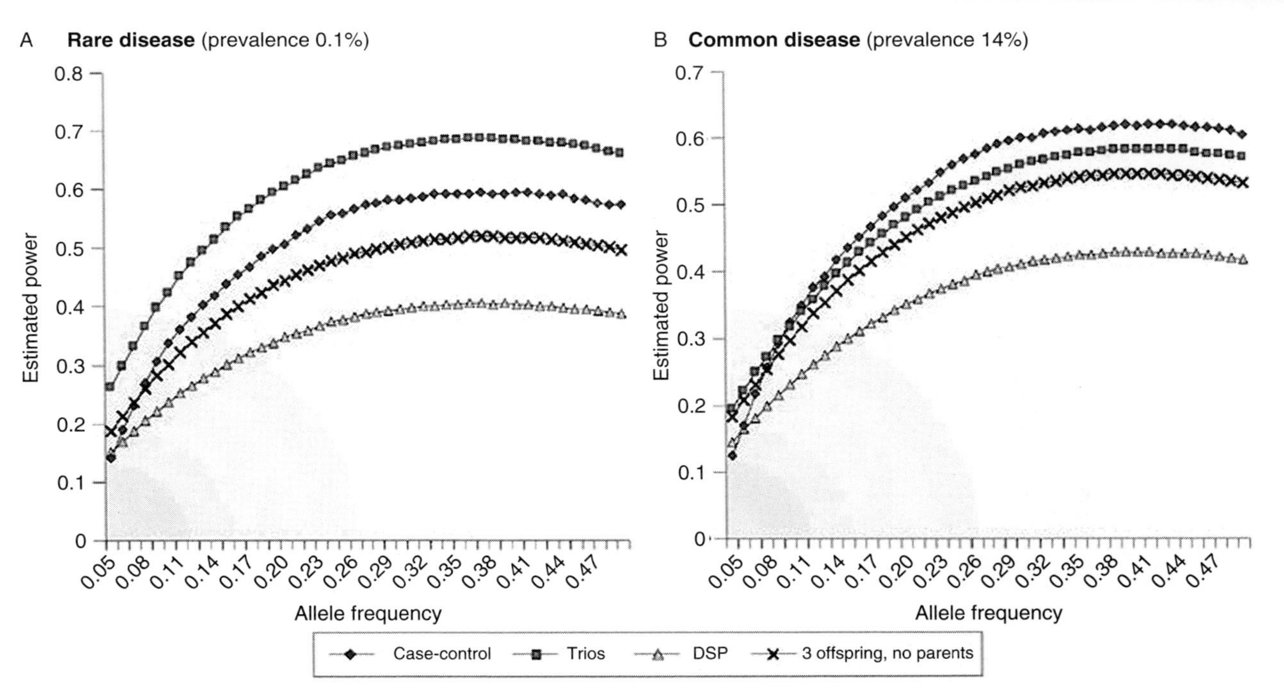

Figure 10.2. Power comparison between case-control studies and family-based designs. The estimated power levels for a case-control study with 200 cases and 200 controls are compared with those for various family-based designs: 200 trios (of an affected offspring plus parents), 200 discordant sibling (sib) pairs (DSPs; one affected and one unaffected) without parents, 200 "three discordant offspring (at least 1 affected and 1 unaffected) and no parents." Discordant sib pair designs have 50% less power than case-control designs, (Witte *et al.*, 1999). For the rare diseases (A), trio designs are more powerful than case-control designs. For common diseases (B), case-control designs are slightly more powerful than trio designs and designs with 3 discordant sibs. Although it is not shown here, for larger-effect sizes (e.g., odd ratios greater than 2), unaffected probands contain more information and the DSP design can achieve power levels that are similar to those of trios designs. The power calculations for both the family designs and the case-control designs were done in PBAT (v3.3) using Monte-Carlo simulations. These figures are reprinted from Laird and Lange (2006).

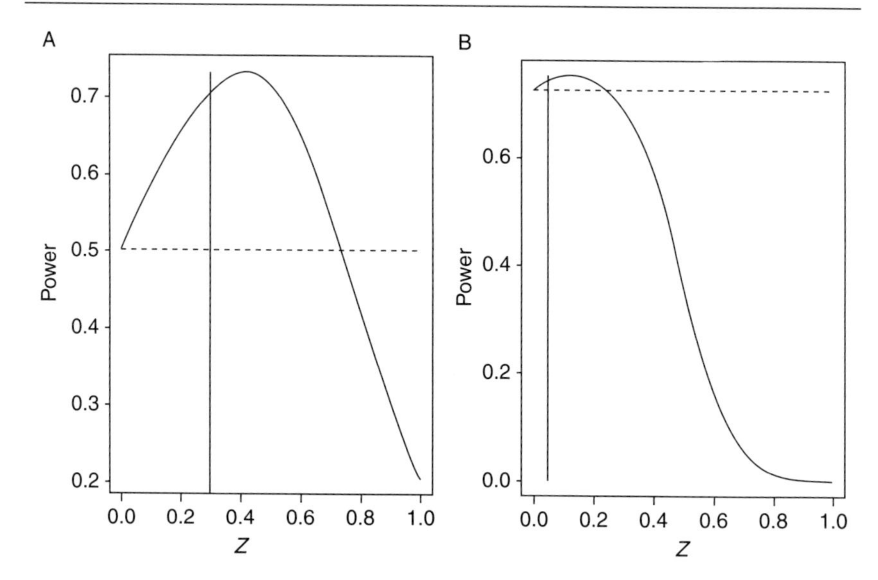

Figure 10.3. (A, B) Power increase by including unaffected offspring/dependence of the power on the offset choice. For a common and for a rare diseases, respectively, (A) and (B) show the increase in power for a sample size of 100 families if unaffected offspring are included in the FBAT statistic. The power is shown as a function of the offset μ. The solid line shows the power of the FBAT statistic and the dotted line the power of the TDT using only affected probands. The vertical line shows the prevalence of the disease. For offset $\mu = 0$, the TDT and the FBAT are identical. For a common disease including unaffected probands in the FBAT, the power increases substantially, while for rare diseases the increases is negligible. For both, the common and the rare disease scenario, the power reaches its maximum when the offset choice is approximately the disease prevalence. These figures are reprinted from Lange and Laird (2002a).

power is maximized; but if the offset is too large, too much weight is given to unaffected offspring and power is lost relative to the TDT. With rare disease, there is little to be gained from using an offset, as the maximum power is only slightly above the TDT, but again, using an offset that is too large can have negative consequences (Fig. 10.3B).

Table 10.1 shows some design and power considerations for binary traits that depend upon ascertainment conditions and family design. These power considerations assume that the optimal offset is used (see Section III) and that there is no environmental correlation between sibling phenotypes. As such, they may be slightly optimistic for designs with multiple affected offspring. For 200 families, we consider a common and a rare disease and two ascertainment conditions.

Table 10.1. Design and Power Considerations for Dichotomous Traits

No. of genotypes per family	Family type: No. of offspring	No. of parents	K = 0.3, MAF = 0.2		K = 0.05, MAF = 0.05	
			A	B	A	B
2	2	0	0.40	0.55	0.46	0.48
3	1	2	0.52	–	0.73	–
3	3	0	0.65	0.68	0.61	0.61
4	2	2	0.69	0.59	0.76	0.73
4	4	0	0.79	0.79	0.70	0.70
5	5	0	0.92	0.85	0.83	0.83

K: disease prevalence, A/B: Ascertainment condition, MAF: minor allele frequency of DSL.

Ascertainment condition A requires at least one affected offspring per family, while ascertainment condition B requires at least one affected and one unaffected offspring per family. We assume the disease prevalence is used for the offset.

For a prevalence of 30%, DSPs without parents are as powerful as trios consisting of one affected proband and parents. When the parental genotypes are missing, discordance-ascertainment conditions can more than compensate for the power loss caused by the missing parental information. When two or more additional offspring are available, there is little effect of the ascertainment condition on power, except that if parents are available, it is advantageous to have more affected offspring, making ascertainment condition A preferable.

For rare disease/rare variant, if parents are missing, it is necessary to genotype more individuals per family to attain comparable power to those cases without missing parents. As a rule of thumb, three additional siblings compensate for the power loss caused by two missing parents. Here there is little effect of ascertainment scheme on power because with a rare disease, most siblings will be unaffected.

The situation with quantitative phenotypes is somewhat different. Although it is certainly possible to ascertain individuals into a study based on their level of a quantitative trait (Risch and Zhang, 1995), such designs are difficult to implement. More likely, individuals are ascertained according to a qualitative trait, and quantitative phenotypes are also measured, for example, asthma and FEV1, or obesity and BMI. With the availability of large cohort studies with family data, we can have family designs in which there is no ascertainment with respect to trait of interest. This can be a significant advantage for the analysis of quantitative traits, although population-based samples will generally not be very useful for the analysis of rare qualitative traits. Ascertainment of subjects relative to the phenotype of interest is important because it can dictate how the analysis should be carried out (see Section III), and the power can depend quite heavily on the combination of ascertainment conditions and analytic method.

Figure 10.4 illustrates the effect on power of ascertainment conditions and analysis choices when dealing with a quantitative trait. The figure compares two strategies: random sampling from the population and selection of only those subjects whose trait is in the top 10% (considered affected). The analytic choices are to use a TDT with only affected offspring or to use the quantitative trait in the analysis (see Section III). When there is no ascertainment condition relative to the trait, then it is far preferable to analyze the quantitative trait, with an offset close to the population mean, which can be well estimated by the sample mean in this setting. With ascertainment of affected offspring only, using the basic TDT on the affected is always the best strategy; considerable power may be lost by analyzing the quantitative trait, unless the offset is carefully selected. Using the sample mean as an offset gives poor results because the sample mean is a biased estimate of the population mean.

II. ANALYSIS METHODS: FBAT AND PBAT

Here we discuss a very general approach to the analysis of family-based data. This approach permits any type of genetic model, multiallelic data, general family design, different null hypotheses, any phenotypic trait (binary, time-to-onset, measured, repeated measures, multivariate), haplotypes, and multiple markers. To motivate this approach, it is important to consider those aspects of the TDT that make it so robust and powerful. First, the test statistic is computed conditionally on the observed parental genotype. This conditioning serves to eliminate any assumptions about the distribution of alleles in the population, such as Hardy-Weinberg, or that allele frequencies are the same in cases and controls under the null. Second, the test statistic is computed conditional on the trait; this serves to eliminate assumptions about the distribution of the trait in the population. Finally, the random variable is the offspring genotype. Its distribution under the null is computed using Mendel's first law—thus the validity of the test statistic relies only on Mendel's law of random transmission of each parental allele with equal probability to each offspring.

The FBAT approach (Laird *et al.*, 2000) to the analysis of family data uses these same underlying principles in constructing a test statistic that generalizes the TDT to more complex situations. The general idea is the same: We condition on the traits (which can include any type and number of traits) and on parental genotypes (which can include multiple markers and haplotypes),

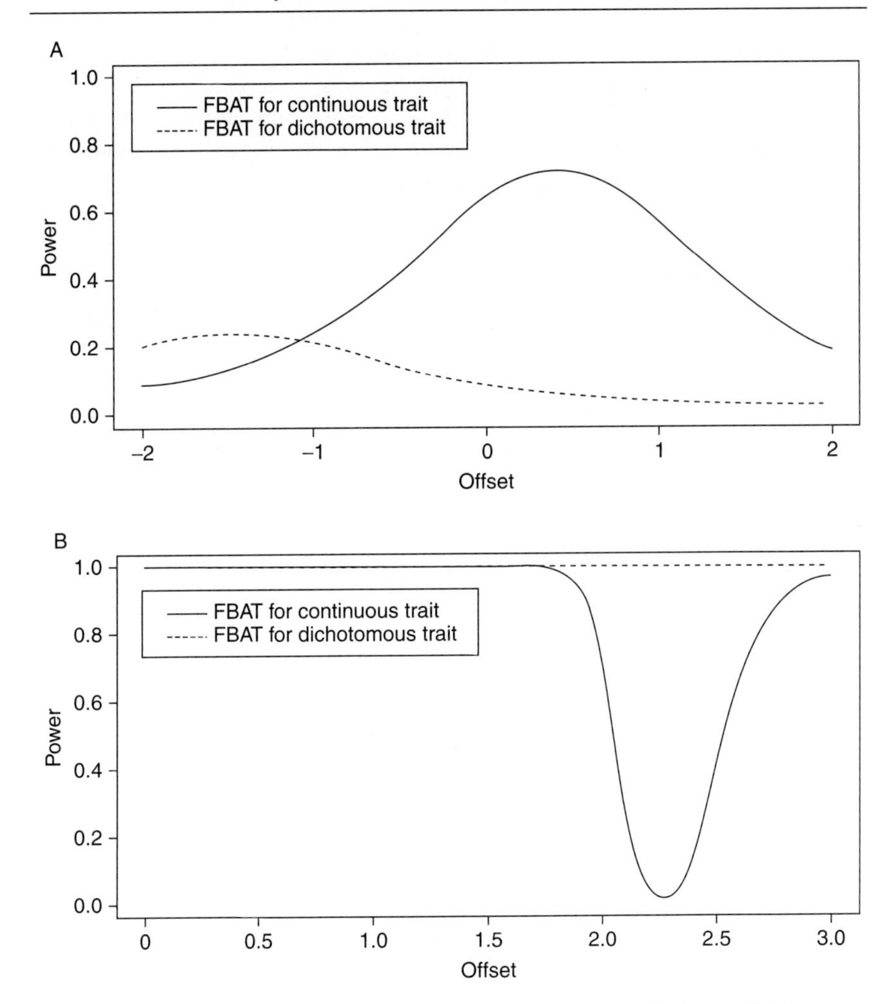

Figure 10.4. (A, B) Asymptotic power calculations for a continuous trait. The disease-allele frequency is 0.3 and the heritability, h^2, is 0.1. The dotted line shows the power of the dichotomous *FBAT* for offset choices between 0 and 1. Significance level $\alpha = 0.01$; (A) Additive model—total population sample with mean $\overline{y} = 0.39$, maximal power of FBAT 0.75, power of FBAT-o 0.74, power of PDT 0.73, and power of QTDT 0.74 ($n = 200$); (B) Additive model—affected sample with phenotypic mean $\overline{y} = 2.2$, maximal power of *FBAT* 1.00, power of FBAT-o 0.04, power of PDT 0.034, and power of QTDT 0.01 ($n = 200$). These figures are reproduced from Lange *et al.* (2002).

and we compute the distribution of the test statistic from the distribution of offspring genotypes under the null. When parents' genotypes are missing, we condition, instead, on the sufficient statistic for parental genotype (denoted S_i); see subsection on *Missing Parents or Founders* below. When analyzing haplotypes, we condition on the sufficient statistic for missing phase as well.

A. General test statistic

The test statistic uses a natural measure of association between two variables, a covariance between the traits and the genotypes. We define the covariance as

$$U = \Sigma T_{ij}(X_{ij} - E(X_{ij}|S_i)), \tag{1}$$

where i indexes family and j indexes nonfounders in the family, and summation is over all i and j. Here, T_{ij} is a coding function for the trait of interest and X_{ij} is a coding function for the genotype. The usual sample covariance centers both variables around their sample means, but with the FBAT statistic, X_{ij} is centered around its expected value, $E(X_{ij}|S_i)$, conditional on the sufficient statistic for the parental genotype, and computed under Mendel's laws. As we discuss below, T_{ij} is typically a centered phenotype. The coding function for the trait allows us to incorporate both qualitative and quantitative phenotypes, as well as time to onset. This basic formula can be used in virtually every setting; the key is the definition of the coded traits and the coded genotypes and how the distribution is computed under the null.

Note that the "centered genotype" $(X_{ij} - E(X_{ij}|S_i))$ can be thought of as the residual of the "transmission" of parental genotype to offspring. For any coded genotype, $(X_{ij} - E(X_{ij}|S_i)) = 0$ if both the parents of the ijth offspring are homozygous, regardless of what particular genotypes the parents have—that is, transmissions from homozygous parents do not contribute to the test statistic. With one homozygous parent, if we define X_{ij} as the number of A alleles, then $(X_{ij} - E(X_{ij}|S_i)) = 1/2$ if A is transmitted and $-1/2$ if the A is not transmitted (because again, transmissions from the other parent, who is homozygous, do not count). Finally, with two heterozygous parents, $(X_{ij} - E(X_{ij}|S_i))$ equals 1, 0, or -1, depending upon the number of A alleles transmitted (2, 1, or 0). Thus, in the special case where $T_{ij} = 1$ for all i and j (see below), U simply counts the total number of A transmissions from heterozygous parents, minus their expected number $(n_A - n/2)$, in the notation of Section I.B.

B. Coding the genotype

The coded genotype is chosen to reflect the selected mode of inheritance. For example, for the additive model, X_{ij} counts the number of A alleles; for the recessive, X_{ij} is 1 if the ijth offspring's genotype is AA and 0 otherwise.

In a multiallelic setting with p alleles, X_{ij} is a p-dimensional vector, each element of the vector coding for a different allele. In this case, U will also be a p-dimensional vector. Other specifications for X_{ij} will be discussed under multiple markers and haplotypes.

C. Coding the trait: Dichotomous outcomes

Consider first the case where the phenotype of interest is affection status. Setting $T_{ij} = 1$ if affected and $T_{ij} = 0$ otherwise means the test statistic will not incorporate information about transmissions to unaffected offspring. Note that this is equivalent to including only affected individuals in the test statistic. To incorporate unaffected individuals into the test statistic, we use an offset, letting $T_{ij} = (Y_{ij} - \mu)$, where Y_{ij} is the original 1/0 phenotype and μ is a user-defined offset parameter. Thus, for a dichotomous trait, with $Y_{ij} = 1$ if affected and 0 otherwise, $T_{ij} = 1 - \mu$ for affected individuals and $-\mu$ for unaffected. The optimal offset (Lange and Laird, 2002a; Whittaker and Lewis, 1998) is approximately the prevalence of the disorder, $\mu = E(Y)$. Note that the U statistic can now be thought of as a contrast between transmissions to affected offspring weighted by $(1 - \mu)$ and unaffected offspring, weighted by μ. A contrast is used because an overtransmission of the A allele to affected offspring should correspond to an undertransmission to the unaffected. As noted in Section I.D however, assigning too much weight to the unaffected (μ much larger than the prevalence) will result in a loss of power relative to using affected offspring only. When $\mu = 0$, only affected individuals are included.

Of course, a general problem is lack of knowledge about population prevalence. In addition, with dichotomous traits, ascertainment on the trait usually means that it is not possible to estimate μ from the data in hand. Fortunately, power of the test is reasonably good in a neighborhood around μ (Lange and Laird, 2002a). An alternative approach to choosing the offset is to choose μ to minimize var(U). This gives an easily computed offset (Lunetta *et al.*, 2000) that is close to the sample prevalence. However, the offset can be very large if a large number of unaffected individuals are included in the sample. Because minimizing the variance does not maximize power, we suggest limiting the offset size to a maximum of 0.5.

Quantitative and time-to-onset traits, as well as adjustment for covariates, will be discussed in Section II.H.

D. The test statistic: Large sample distribution under the null

The distribution of the FBAT statistic under the null hypothesis is obtained by treating the X_{ij} as random, but conditioning on the trait, T_{ij}, and the sufficient statistic. As $E(U) = 0$ by construction under H_0, it remains to normalize U by its standard deviation, again computed under the conditional distribution of offspring genotype, given offspring trait and S_i. For univariate X or T

$$Z = \frac{U}{\sqrt{\text{var}(U)}}, \text{ or equivalently, } \chi^2_{\text{FBAT}} = \frac{U^2}{\text{var}(U)},$$

where

$$\text{var}(U) = \sum_i \sum_{j,j'} T_{ij} T_{ij'} \text{cov}(X_{ij}, X_{ij'}, S_i, T_{ij}, T_{ij'}) \tag{2}$$

and $\text{cov}(X_{ij}, X_{ij'} S_i, T_{ij}, T_{ij'})$ is computed conditional on the traits and the sufficient statistics, assuming the null hypothesis is true. Note that this covariance only depends on S_i and not the traits when no linkage is part of the null hypothesis. For testing no association in the presence of linkage, an empirical variance can be used to estimate var(U) (Lake *et al.*, 2000). For large samples, Z is approximately distributed as $N(0,1)$, and χ^2_{FBAT} is distributed as approximately χ^2 on one degree of freedom. In the setting where U is a vector because of either multiple alleles or multiple traits, var(U) is a variance/covariance matrix, and the test statistic is the quadratic form $U^T \text{var}(U)^- U$, which is distributed as χ^2 with degrees of freedom equal to the rank of var(U) (Laird *et al.*, 2000; Lange and Laird, 2002b).

E. The TDT and χ^2_{FBAT}

When we include only affected offspring ($T_{ij} = 1$), and all parents are known, then as previously noted, $U = (n_A - n/2)$, where n_A is the number of heterozygous transmissions of A to all affected children, and n is the number of heterozygous parent–child pairs. Under a null hypothesis that includes no linkage, multiple offspring are independent, and var(U) reduces to

$$\text{var}(U) = \Sigma ij \ \text{var}(X_{ij}|S_i) = n\left(\frac{1}{2}\right)^2$$

because each transmission has variance equal to $(1/2)^2$. As a result

$$\chi^2_{\text{FBAT}} = \frac{(n_A - n/2)^2}{n(1/2)^2} = \frac{(n_A - n_B)^2}{n}$$

where n_B is the number of transmissions of the B allele from heterozygous parents to affected offspring. Thus, χ^2_{FBAT} is identical to the TDT when (1) only affected offspring are included, (2) parents' genotypes are known, and (3) the null hypothesis assumes no linkage. In other cases, χ^2_{FBAT} can be viewed as generalizing the TDT.

F. Computing the distribution with general pedigrees and/or missing founders

As defined above, the test statistic is very general. It applies to trios, parents with multiple offspring, families without parents, and general pedigrees, with or without founder genotypes. The summation over i denotes the independent pedigrees or families, and the summation over j denotes summation over all offspring in the pedigree. However, exactly how the distribution of each X_{ij} is computed depends upon the family structure and whether founder genotypes are known. For trios, it is straightforward to compute the distribution of X_{ij} given parents with known genotype, using Mendel's first law. When there are multiple offspring and no linkage, transmissions to all offspring in the family are independent, and one can treat each offspring as if it comes from a separate family. When linkage is present, the covariance between multiple offspring in the same family depends upon the unknown recombination fraction. To remove dependence of the joint distribution on the unknown recombination fraction, we can condition the distribution on patterns of identity by descent observed among the offspring (Rabinowitz and Laird, 2000). This approach to computing the conditional distribution of X_{ij} leads to discarding many families as noninformative, especially when parental genotypes are unknown. Thus, we generally use an empirical variance, as described above.

These basic ideas extend easily to general pedigrees, where the genotypes of all founders are known, and instead of conditioning on parents, we condition on the founders of the pedigrees. There is potential for a considerable gain in power in this setting when we analyze pedigrees, rather than treating all families within the pedigree as separate families (Laird and Lange, 2006; Rabinowitz and Laird, 2000).

When parents or founder's genotypes are unknown, the situation is slightly more complex, but the joint distribution of offspring outcomes can be calculated using the conditioning algorithm described in Rabinowitz and Laird (2000). The basic idea of this algorithm is to condition the distribution of observed offspring genotypes on the sufficient statistics for the unobserved parental genotypes. In this way, the distribution will not require making any assumptions about the distribution of the unobserved parental genotypes, and robustness to population substructure is maintained. To give an example, suppose we have a family with two offspring and no parents. The conditional distribution of the two offspring genotypes depends upon what genotypes are observed in the two offspring. If we observe that both genotypes are AA [or BB], then nothing can be inferred about the parents except that each has an A [or a B]. The probability of other possible outcomes for the two offspring will depend upon the unknown parental alleles. Thus, such pairs of offspring are not informative.

If instead, we observe one AA and one BB offspring, then we know that the parents are both AB. To condition on the sufficient statistic for missing parents, we require that all possible outcomes contain one AA and one BB offspring. Any other possible outcome—for example, (AA,AB)—would not allow us to infer that the parents are both AB. Thus the two possible outcomes conditioning on the sufficient statistics are (AA,BB) and (BB,AA), assuming order matters. For an additive coding with X denoting the number of A alleles, it is straightforward to show that $E(X_{i1}|S_i) = E(X_{i2}|S_i) = 1$, $\text{var}(X_{i1}|S_i) = \text{var}(X_{i2}|S_i) = 1$ and $\text{cov}(X_{i1}, X_{i2}) = -1$. Now consider this family's contribution to U and var(U). Using Eqs. (1) and (2) above, we can show that the contribution to U is $(T_{i1} - T_{i2})$ and the contribution to var(U) is $(T_{i1} - T_{i2})^2$, assuming the first child is AA and the second is BB. Thus, families in which the trait is constant (i.e., both affected or both unaffected) will make no contribution to the test statistic, but DSPs will be informative.

Finally, if we observe (AA,AB), we know that one parent must be AB and the other must have an A, but otherwise we cannot distinguish between [AA,AB] parents and [AB,AB] parents. Because transmission probabilities differ for the two possible parents, we must therefore keep fixed the set of observed genotypes, and permute them between the two offspring, as above. Again, only DSPs will be informative.

The conditioning algorithm determines the joint distribution of all offspring genotypes in a family or a pedigree; hence, it is also possible to carry out exact tests of H0 by computing the exact probability that the random χ^2_{FBAT} test statistic exceeds the observed statistic under the null (Schneiter *et al.*, 2005). The exact p value can also be estimated via Monte Carlo by drawing from the conditional distribution of offspring genotypes.

G. Haplotypes and multiple markers

A common scenario in association studies uses multiple, closely spaced markers, usually single nucleotide polymorphisms (SNPs), often within the same gene. Here, the null hypothesis of interest may be whether any marker is associated with a DSL underlying the trait. Testing each marker separately and then using FDR or Bonferonni to adjust the p values for multiple testing is one strategy, but may be quite inefficient when SNPs are in high linkage disequilibrium with each other (Chapter 10). Two approaches we discuss here use haplotypes (Chapter by Liu *et al.*, this volume) or multimarker tests.

A multilocus haplotype refers to the set of alleles, one from each marker, that are inherited from a single parent, either from the mother or from the father. There are several circumstances when using haplotypes for testing may be preferable to single-marker testing. If a disease locus is present in the region spanned by the markers, but does not correspond exactly to any of the markers tested, the DSL may not be in sufficiently high disequilibrium with any one marker to be detected by one-at-a-time testing of the markers. If we use enough markers to capture the haplotype diversity in the population, then the haplotypes should capture the variation directly at the disease locus. Alternately, if the DSL is a series of changes in base pairs at two or more of the observed markers, then using haplotypes should again be more powerful than one-at-a-time testing. However, if a single marker corresponds to only DSL in the region, or if many markers are included in the haplotype construction that are not in high LD with the putative DSL, then using haplotypes can be a poor testing strategy.

If each person's haplotype is observed, then the set of markers forming the haplotypes can be considered as a single marker with many alleles, and methods used for testing association with multiallelic markers apply. Although more can be inferred about haplotypes in the family-based setting than in population-based studies because of knowledge of parents or sibling genotypes, phase (i.e., which parent transmitted which allele) cannot always be resolved even in families, especially if parents' genotypes are missing. The principle of conditioning on the sufficient statistics for missing parental genotypes extends quite straightforwardly to handle missing parental phase (Horvath *et al.*, 2004). One now obtains a distribution for the phased offspring genotype, conditioning on the sufficient statistic for both parental genotypes and possibly missing phase. The FBAT test statistic can be computed in the same way, recognizing that the set of markers forming the haplotypes is treated as one multiallelic marker with each haplotype forming an allele. In principle, with n SNPs, there can be 2^n haplotypes, but in practice the number of haplotypes observed in a family-based analysis is usually quite a bit less than 2^n. The availability of family data enables one to eliminate many possibilities as not compatible with observed family data.

The haplotype analysis implemented in FBAT uses the principle of conditioning on the sufficient statistics for both phase and any missing parental data, and on offspring traits, and hence is not biased by population stratification and/or admixture. Either biallelic or multiallelic tests are computed in the usual way, and the empirical variance option can be used to account for the presence of linkage. Any trait can be used. A feature of the implementation of the conditioning algorithm is that it also allows one to recover information from only partially phase known families by using weights (Horvath *et al.*, 2004).

As the number of SNPs increases substantially (more than 5–10), the attractiveness of a haplotype analysis diminishes because of increasing difficulty in resolving phase. In addition, if founder information is missing, it may take considerable computer time to determine the conditional distribution. In such cases, an alternative approach uses multimarker tests. The basic idea of a multimarker approach is very straightforward; in the context of a case-control study, it is similar to a Hotelling's T^2 test where the vector of marker values is tested for equality in the cases and controls. In the family context, multimarker tests are constructed by letting X_{ij} be a vector, where each element of the vector corresponds to a different coded marker value. The full-joint distribution of the different elements of X_{ij} would require knowledge of the haplotype distribution, as discussed above. To circumvent conditioning on unknown phase, we instead use an empirical estimator of var(X_{ij}) in calculating var(U). The resulting χ^2_{FBAT} has the same quadratic form as in the multiallelic setting, with degrees of freedom again equal to the rank of var(U). The rank will be generally equal to the number of markers, unless two or more of the markers are in near perfect LD (Rakovski *et al.*, 2007). Another approach to the multimarker testing with families is discussed by Xu *et al.* (2006).

H. Coding the trait for complex phenotypes: Age-to-onset phenotypes, quantitative outcomes, and FBAT-GEE

In principle, T_{ij} can be any function of the phenotype and other individual characteristics, as long as the trait does not depend on the genotype being tested. When the phenotype of interest is time to onset, a strategy similar to the log-rank test may be used, letting T_{ij} be log-rank residuals, computed at each failure time (Lange *et al.*, 2004a).

For quantitative phenotypes, typically, a phenotypic residual is used for the coded trait, that is $T_{ij} = (Y_{ij} - \mu)$, where Y_{ij} is the original phenotype and μ is a user-defined offset parameter. Unless subjects have been ascertained into the study on the basis of their quantitative trait, the optimal choice is the phenotypic sample mean. In such situations, the FBAT statistic for quantitative phenotypes has higher statistical power than for dichotomous traits (Lange *et al.*, 2002). However, to benefit from the advantages of quantitative traits in the analysis, a few hurdles have to be overcome. Quantitative traits such as BMI or lung volume also depend on many other nongenetic factors. These can be proband characteristics but also include environmental influences. For example, BMI will depend on age and gender as well as on smoking history and dietary habits. The unadjusted, raw measurements of such traits will not accurately reflect affection status (e.g., obesity) or the severity of the disease. In such situations, it is preferable to adjust the raw measurements for all known covariates because this decreases the outcome variability, and to compute the FBAT statistic based on the phenotypic residuals.

For lung-function phenotypes, such as FEV, standard adjustment formulas have been derived (Ware and Weiss, 1996). However, such adjustment formulas typically are based on unaffected probands and, consequently, do not always incorporate all disease-specific confounding variables. For example, for FEV1, the standard adjustment includes gender and height but not smoking history, which is an important factor when looking at diseases such as chronic obstructive pulmonary disease. A detailed discussion of the limitations is given by Naylor *et al.* (2005).

In such situations, an alternative approach is to adjust the raw phenotypic measurements, by regressing the phenotype on such confounding variables, and use the phenotypic residuals in the analysis. Such an adjustment will be study specific, requires statistical model building, and might not be reproducible in other studies that may not have recorded the same confounding variables. The goal of such a within-study adjustment is to measure and incorporate all environmental factors and other covariates into the analysis. However, for many phenotypes, the confounding variables are not necessarily known prior to the study design or can be difficult to measure and model.

A second analysis issue for quantitative traits is the multiple testing problem. While affection status is defined by only one variable, the disease and its severity are often described and characterized by a set of quantitative phenotypes. Such quantitative phenotypes typically cluster together into symptom groups. For example, in asthma studies, multiple quantitative phenotypes are recorded that describe lung function (FEV, FVC) (DeMeo and Silverman, 2003). For most complex diseases, such symptom groups of quantitative phenotypes can be defined based on clinical knowledge, knowledge about the underlying biological processes, or simply based on phenotypic correlation. In such situations, it is not desirable to test all quantitative phenotypes individually and adjust for multiple testing. There are two reasons for this: first, the association tests for one symptom group will be correlated, and adjustments for multiple testing tend to be conservative in such situations. Second, if the assumption holds that the phenotypes in a symptom group are influenced by the same pathway or share similar environmental confounding, it will have more power to look at the entire symptom group and test all phenotypes jointly in a single multivariate test, without having to adjust for multiple comparisons.

In the FBAT context, such a multivariate test was introduced by Lange *et al.* (2003b), the so-called FBAT-GEE statistic. Like the original FBAT statistic, FBAT-GEE is easy to compute, tests all phenotypes simultaneously, and does not need distributional assumptions about the phenotypes that will be tested, even if the tested phenotypes are of different variable types (e.g., normally distributed phenotypes, count variables). Assuming that m traits have been recorded for each proband and they form a symptom group that we want to test simultaneously, we denote the vector containing all m observations for each proband by $Y_{ij} = (Y_{ij1}, \ldots, Y_{ijm})$ where Y_{ijk} is the kth phenotype for the jth offspring in the ith family. The multivariate FBAT-GEE statistic is then derived by defining a coding vector T_{ij}.

$$T_{ij} = Y_{ij} - \hat{Y}_{ij} = \begin{pmatrix} Y_{ij1} \\ \vdots \\ Y_{ijk} \\ \vdots \\ Y_{ijm} \end{pmatrix} - \begin{pmatrix} \hat{Y}_{ij1} \\ \vdots \\ \hat{Y}_{ijk} \\ \vdots \\ \hat{Y}_{ijm} \end{pmatrix}$$

where the $\hat{Y}_{ijk}$'s are either the observed sample means for the kth trait or the predicted trait values based on a regression model for covariates. Then the univariate coding variable T_{ij} in the FBAT statistic is replaced by the vector T_{ij} and the FBAT-GEE statistic is given by,

$$T_{\text{FBAT-GEE}} = C^T V^{-1} C.$$

Under the null hypothesis, the FBAT-GEE statistic has a χ^2 distribution with m degrees of freedom. The FBAT-GEE can also be derived based on a generalized estimating equation model with appropriated assumption about the link functions for each phenotype and the covariance structure. However, because the FBAT-GEE test is a score test, all these assumptions cancel out in the derivation of the test statistic and the FBAT-GEE statistic is obtained, making the multivariate FBAT-GEE robust against distribution assumption about the phenotype.

I. A General approach to complex phenotypes: Separating the population and family information in family data

The conditioning of the FBAT statistic on the traits and parental genotypes means that the FBAT statistic does not use all of the information about linkage and association that is available in the sample. While this means we retain robustness, the test is generally not the most efficient. Here we discuss how the extra information in the data that is not used by the FBAT statistic can be used for enhancing power of the test statistics. In particular, with the multiple comparison and model selection issues that arise in modeling complex traits and in large-scale association studies, this extra information can be used to guide the testing strategy. Here, we consider a general approach that is based on separating family data into two independent partitions corresponding to the population information, and the within family information. The population information that is subject to bias by population substructure is used for screening, or model development, and the within subject information is used for confirmatory testing. The idea is similar to cross-validation, except that the partition into the two components is not random.

Consider a simple case of offspring–parent trios. The full distribution for the data consists of a joint distribution for the offspring phenotype, Y, the offspring genotype, X, and the parental genotype, *P* (or more generally the sufficient statistics for parental or founder genotypes, S). We partition the joint distribution into two independent parts:

$$P(Y, X, S) = P(X|Y, S)P(S, Y). \qquad (3)$$

Model building, hypothesis generation, and screening can be based on S and Y, so that subsequent hypothesis testing using any test statistic whose distribution is based on $P(X|S,Y)$ will be independent of the selected model. Note that Eq. (3) simplifies further under a null hypothesis assuming no linkage because $P(X|Y,S)$ can then be replaced by $P(X|S)$.

There are numerous ways to model $P(S,Y)$, in order to obtain information about association. In general, the approach will depend on the specific design, for example, is Y quantitative or qualitative. To illustrate, consider testing a quantitative phenotype with a single marker. To utilize a population-based approach, we (Lange *et al.*, 2002, 2003a) proposed a "conditional mean model":

$$E(Y) = m + aE(X|S) \qquad (4)$$

The parameter a which determines the effect size can be fit using ordinary regression of the phenotype Y on $E(X|S)$. Note that for doubly homozygous parents, $X = E(X|S)$; otherwise we can think of X as missing if parents are informative, and $E(X|S)$ replaces the missing X. In effect, Eq. (4) defines a population regression where some Xs are imputed using parental information (or the sufficient statistics for parental information if parents are missing).

As the regression uses only (Y,S), all the information from the regression will be independent of the FBAT statistic by Eq. (3). Model (4) can be fit repeatedly for any choice of genetic model, any number of phenotypes, and any number of markers. The results of the regression can be used for generating p-values for testing the H0; $a = 0$ or by computing the conditional power of the FBAT statistic for an effect size of a. The power calculation also depends upon the observed parental genotypes and traits (Lange *et al.*, 2002, 2003a). In general, selection based on the conditional power is preferable (Van Steen *et al.*, 2005).

This basic approach has been extended to handle longitudinal and repeated measures (FBAT-PC) (Lange *et al.*, 2004b) and multivariate data (Su *et al.*, 2006) by using the screening stage to select optimal linear combinations of traits for subsequent testing and testing for multiple markers (Xu *et al.*, 2006). Jiang *et al.* (2006) proposed a method to determine the genetically relevant age range for time to onset that is particularly useful for diseases in which an early onset suggests a strong genetic component, while a late onset is mainly attributable to environmental effects, for example Alzheimer or childhood asthma.

J. Testing strategies for large-scale association studies

A major scientific obstacle in genome-wide association studies is the hundreds of thousands of SNPs and potential statistical tests that may be computed, resulting in multiple testing issues. Multistage designs have been proposed for case/control studies (Hirschhorn and Daly, 2005; Thomas *et al.*, 2004) as one way of handling this problem. The number of genotyped SNPs is reduced in each stage of the design, so that genome-wide significance is achieved step-by-step. The screening approach for family studies described above is well suited to a genome-wide association study with quantitative traits (Van Steen *et al.*, 2005). The approach is illustrated in Fig. 10.5.

With family-based designs, the screening procedure uses all families, even the "noninformative ones." Assuming moderate to low effect sizes, simulation studies suggest that if a true DSL or a SNP in LD with a DSL is included in the data set, it is sufficient to select the highest 10 or 20 SNPs for further testing and retain high power. The advantage of family-based screening is that the same data set is used for the screening step and the testing step. This means only one sample needs to be recruited, and replication in other studies serves the purpose of generalizing a significant finding to other populations. The strategy has been successful applied to a 100 k SNP scan for obesity in families from the Framingham Heart Study. Table 10.2 displays the top 10 SNPs from that study, as determined by estimated power, selected from the 100 k scan, along with their *p* values. A novel SNP for BMI was discovered (FBAT *p* value 0.0026) that would have been missed by standard approaches (e.g., the Bonferroni or Hochberg corrections for multiple testing). Using the same genetic model, the finding was replicated in four independent studies, including cohort, case-control, and family-based samples (Herbert *et al.*, 2006).

III. OTHER APPROACHES TO FAMILY-BASED ANALYSES, INCLUDING THE PDT AND THE QTDT

Many methods have been suggested to handle specific issues that arise with family designs, such as quantitative traits, multiple siblings, or missing parents. It is beyond the scope of this chapter to provide a review of all such methods, but here we mention a few of the more popular methods used to handle general family data, with either quantitative or qualitative outcomes. We first make some

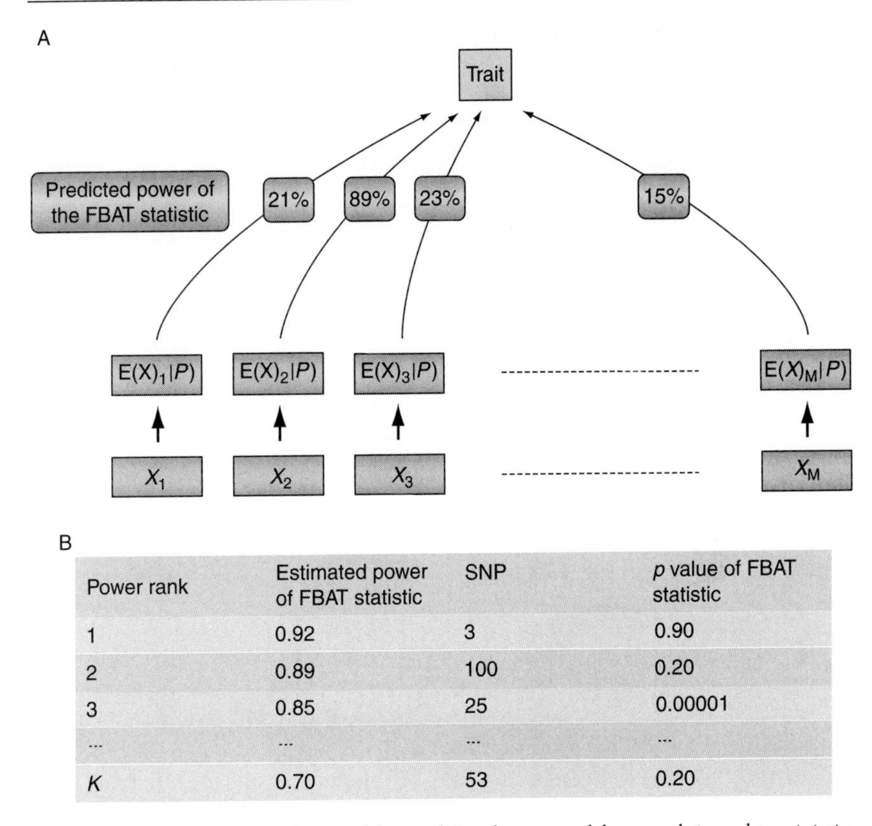

Power rank	Estimated power of FBAT statistic	SNP	p value of FBAT statistic
1	0.92	3	0.90
2	0.89	100	0.20
3	0.85	25	0.00001
...	...	...	...
K	0.70	53	0.20

Figure 10.5. The screening technique. The conditional mean model approach is used to minimize the multiple testing problem. In this example, we look at 1 quantitative trait and M SNPs. In the first step, the marker information in the offspring is assumed to be missing and imputed, using the expected markers scores conditional upon the parental genotypes/sufficient statistic. Then the conditional mean model is used to estimate the power of the FBAT statistic for each SNP. The power depends on observed parental genotypes and the effect size estimated from the conditional mean model. In the final step, the K SNPs with the highest power estimates are tested for association with the FBAT statistic at a Bonferroni-adjusted significance level of α'/K. Since only K SNPs have been selected for testing, it is only necessary to adjust for K comparisons instead of M. (A) *Step 1*: Screen SNPs with conditional mean model for testing via the FBAT statistic. (B) *Step 2:* Select the top K SNPs for testing via the FBAT statistic. The p value of the FBAT statistic has to be significant at α'/K in order to achieve overall significance. These figures are reproduced from Laird and Lange (2006).

general remarks about likelihood approaches to the analysis and how they connect with the FBAT approach. Then we consider the PDT and QTDT in more detail.

Table 10.2. Screening and Testing of SNPs for Association with BMI

Ranking from screen	SNP	Chromosome	Frequency	Informative families	*p* value FBAT
1	rs3897510	20p12.3	0.36	30	0.2934
2	rs722385	2q32.1	0.16	15	0.1520
3	rs3852352	8p12	0.33	34	0.7970
4	**rs7566605**	**2q14.1**	**0.37**	**39**	**0.0026**
5	rs4141822	13q33.3	0.29	27	0.0526
6	rs7149994	14q21.1	0.35	31	0.0695
7	rs1909459	14q21.1	0.39	38	0.2231
8	rs10520154	15q15.1	0.36	38	0.9256
9	rs440383	15q15.1	0.36	38	0.8860
10	rs9296117	6p24.1	0.40	44	0.3652

Genome-wide SNPs (86,604) were screened using parental genotypes to find those likely to affect offspring BMI. The top 10 SNPs from the screening step (ranked by power from most likely to least likely) are shown. These SNPs were tested using offspring genotypes for association with BMI using the FBAT. The rs7566605 SNP is highlighted in bold.

The general likelihood method specifies a probability density for the observed data, along with a model for how the genotype affects the phenotype. Either likelihood ratio or score tests can be used to test the hypothesis of no association. Self (1991) proposed a likelihood method for case-parent trios based on creating "pseudocontrols," using pairs of the nontransmitted alleles as unaffected siblings. Assuming a relative risk model for the genetic effect yields the conditional logistic regression likelihood; using a log-additive relative risk model yields a likelihood ratio test equivalent to the TDT. As such, this approach can be implemented using standard software packages for conditional logistic regression. This popular approach has been extended to haplotypes (Dudbridge, 2003), gene–environment interactions, and gene–gene interactions (Cordell, 2004). It can be easily generalized to multiple affected offspring in the context of testing a null hypothesis that includes no linkage. When parents are missing and unaffected siblings are available, the conditional logistic regression approach still applies, conditioning on family (Witte *et al.*, 1999).

Another approach for the analysis of trios with dichotomous phenotype data is based on multinomial likelihoods. The contribution of a case parent trio to the likelihood of the data can be factored as:

$$L_t = L_c L_p,$$

where L_c is proportional to the probability density of the child's genotype conditional on parents genotype and the child's disease status and L_p is proportional to the probability density of the parental mating type, given the child's disease status. Note that L_c depends only on Mendel's laws plus the unknown penetrance functions, that is $P(\text{disease}|X)$, whereas L_p depends on those factors as well as on the mating type frequencies. Thus, robust and efficient tests of association can be constructed from L_c alone that do not require making any assumptions about parental mating type frequencies. Likelihood ratio tests based on L_c are easily constructed for testing genetic association, using different genetic models (e.g., additive, dominant) (Clayton, 1999; Schaid and Li, 1997; Whittemore and Tu, 2000). However, score tests are generally more popular than likelihood ratio tests, partly because they yield the TDT when an additive genetic model is used; these score tests are also equivalent to FBAT tests. More importantly, score tests provide a simple way to extend the model to accommodate multiple offspring, including unaffected, without the need to specify the joint distribution of offspring under the alternative. Score tests only require specifying the distribution of the data under the null. This approach to factoring the likelihood and constructing score tests has been extended to encompass quantitative and time-to-onset phenotypes, as well as multiple offspring (Shih and Whittemore, 2002).

The FBAT statistic is a score test under more general assumptions about the distribution of the offspring phenotypes (Laird *et al.*, 2000). When parents are observed, a score test from L_c as defined above will be equivalent to the FBAT statistic for dichotomous traits. The FBAT and the likelihood approaches diverge in the treatment of missing parental data, or in the case of haplotypes, when parental phase is unknown. When parents are missing, the FBAT approach replaces conditioning on parents by conditioning on the sufficient statistics for parental genotypes, S. However, the likelihood approach estimates the probabilities of parental mating types from the likelihood, L_p, of the observed parents and averages $E(X|P)$ over the estimated distribution of parental mating types. It is thus easy to see that likelihood approaches are generally more efficient: Single cases without any parents do not have to be discarded as they are in the FBAT conditioning approach, but their inclusion relies on the assumption that their parental mating type can be estimated from the data on other parents. This is a strong assumption, and one that is unrealistic in the presence of population substructure.

A series of papers (Kistner and Weinberg, 2004; Kistner *et al.*, 2006; Umbach and Weinberg, 2000; Weinberg, 1999; Weinberg *et al.*, 1998) describes an extension of the multinomial model to the Poisson that allows the incorporation of methods for testing for parental imprinting, gene–environment interaction, and quantitative phenotypes. Rather than score tests, these authors use likelihood ratio tests. To incorporate multiple siblings, they use Wald tests computed with an empirical variance to avoid specifying the joint distribution under the alternative. Missing parents are again handled by estimating a distribution for mating types from the observed parents.

Likelihood-based approaches to the analysis of family data with quantitative traits assume that the trait follows a normal distribution, with mean depending linearly on X (Abecasis *et al.*, 2000; Fulker *et al.*, 1999; Gauderman, 2003). Unlike the setting described above, where inference about association is based on L_c, inferences about the genetic parameters are based directly on the normal likelihood for phenotype given genotype. Note that L_t can alternatively be factored as

$$L_t = L_{yx} L_x$$

Where as before, L_t is the likelihood for all the trio data, L_{yx} is the likelihood associated with the phenotype distribution, and L_x is the likelihood of the genotypes, both offspring and parents. Note that basing inferences on L_{yx} is potentially biased but fully efficient because there is no information in L_x about association. A correction for population substructure is made by incorporating $(X_{ij} - E(X_{ij}|P_i))$ into the model for the mean, as described below. We refer to this as a model-based approach because its validity will generally depend upon the correctness of the model for the distribution of the phenotypes. Likelihood ratio tests based on these models can be sensitive to distributional assumptions and ascertainment conditions.

In general, likelihood-based approaches will often be more efficient than score tests or other nonparametric approaches, and offer the possibility of testing nested models, but their validity generally depends upon the correctness of the assumed likelihood model, that is the distributional assumption about the phenotype or the appropriateness of the model for parental mating types when some parents are missing. Simulation studies are a common way of attempting to validate likelihood-based approaches in the presence of model misspecification, but in the absence of theory, simulations offer only limited assurance. In contrast, the validity of the conditioning approach depends on correctly identifying the sufficient statistics and specifying the conditional distribution under the null hypothesis.

A. The PDT and APL

Another family-based association test that is conceptionally very similar to the FBAT-approach is the PDT, the pedigree disequilibrium test (Martin *et al.*, 1997). Both the PDT and the FBAT are score tests that share a similar numerator, but instead of using $E(X|S)$ with missing parents, these tests use contrasts between affected and unaffected offspring in the same family. Additionally, the PDT approach relies on empirical variance estimators instead of variances that are analytically derived. The PDT approach is a valid test for all three null hypotheses. To obtain a more powerful test for association in the presence of linkage (second null hypothesis), the PDT approach has been extended to incorporate linkage information, the so-called "test for association in the presence of linkage" (Martin *et al.*, 2000). In contrast to the FBAT approach, which does not attempt to estimate the joint transmission probabilities to the offspring, but estimates directly the variance of the test statistic using an empirical variance/covariance estimator, the APL approach models the joint transmission probabilities to multiple offspring based on the identity by descent status. The transmission probabilities are estimated using the EM-algorithm and are used in the computation of the test statistic directly. The approach has been extended to general nuclear families with missing parental information, to haplotype analysis, to the analysis of the X-chromosome, and to handle ordinal/rank-based phenotypes such as age of onset (Chung *et al.*, 2006, 2007a,b; Martin *et al.*, 2003).

B. Quantitative traits: The QTDT

The QTDT is a widely used method for testing association with family data and quantitative traits. The essential idea relies on assuming a standard QTL model for the phenotype, that is,

$$E(Y_{ij}) = \mu + \beta X_{ij}$$

To make a correction for population substructure admixture, we may add and subtract $\beta\, E(X_{ij}|P_i)$ to the mean to obtain:

$$E(Y_{ij}) = \mu + \beta[X_{ij} - E(X_{ij}|P_i)] + \beta E(X_{ij}|P_i)$$

Subscripting the two βs by w and b, to denote within subject and between subject effects, we write:

$$E(Y_{ij}) = \mu + \beta_w[X_{ij} - E(X_{ij}|P_i)] + \beta_b E(X_{ij}|P_i) \quad (4)$$

As shown in Abecasis *et al.* (2000), OLS estimates of β_w remain unbiased for the true genetic effect in the presence of population stratification, while estimates of β_b are contaminated by population stratification. The model 4 is the basis for the QTDT test that further assumes normality of the error terms and uses a likelihood ratio test for the null hypothesis H0: $\beta_w = 0$, leaving β_b unspecified. A score test for H0: $\beta_w = 0$ that is derived from the same likelihood function will be equivalent to the FBAT statistic for quantitative traits (Lange *et al.*, 2003a).

The original model proposed by Fulker *et al.* (1999) was designed for sib pairs, or more generally sibships, with missing parental genotypes. Here, assuming an additive model, $E(X_{ij}|P_i)$ is replaced by the mean of the offspring genotypes in the *i*th family that is equivalent to $E(X_{ij}|S_i)$. In this case, the model can be expanded to include variance and covariance terms between siblings, which include components due to the genetic variance, and residual sibling resemblance. With siblings, likelihood ratio tests can also be constructed to test for linkage. However, when the observed marker is the DSL (perfect LD between the marker and the DSL), there is no power to test for linkage. As the LD between the marker and DSL decreases, the power to detect association of the trait and the marker decreases, but the power to detect linkage increases (Sham *et al.*, 2000).

An advantage of the approach is that it provides estimates of the association parameter, β_w, and the recombination parameter. However, the method does not lend itself to extension to haplotypes or multiple endpoints. Although the estimate of β_w is robust for population substructure, the contaminated estimate for the between-component β_b does not cancel out in the construction of the likelihood ratio test, causing an inflated type-1 error (Yu *et al.*, 2006).

IV. SOFTWARE

With family-based designs, there is generally a need for special software to analyze the data. Fortunately there is now a wide variety of software packages available. Most of the packages were developed by the original authors of the methods and are home-grown. Despite the lack of general support for such software packages in academia, the packages have proven to be reliable and user-friendly tools. Recently, commercial packages with professional user-support and documentation have become available that are particularly suited for less statistical-oriented users and for large-scale projects. Table 10.3 shows an overview of the most popular packages and their functions.

V. DISCUSSION

The advent of whole-genome association scans offers great promise for genetic association studies. Most projections agree that large samples of individuals will be necessary to disentangle the wheat from the chaff in these large genome scans, no matter what the design (Clayton *et al.*, 2005; Hirschhorn and Daly, 2005; Van Steen *et al.*, 2005). While it is inescapable that large samples from existing cohort or case-control studies that do not include data on relatives are generally much easier to obtain than large numbers of suitable families, we believe that the creative use of the population information contained in family data for screening and hypothesis generation, coupled with their robustness to population substructure, make these family studies competitive. In addition, with technology available for handling pedigrees with missing founders, family data already collected for linkage studies can, in many cases, be recycled for association.

Table 10.3. Software Programs for the Analysis of Family-Based Association Tests

Package	Genetic analysis capability	Phenotypic analysis capability	Special features
APL/PDT	Single marker, haplotype	Binary traits, quantitative traits, ranked traits, time-to-onset	X-chromosome
FBAT	Single marker, haplotype, multimarker	Binary traits, quantitative/multivariate traits, ranked traits, time-to-onset	X-chromosome, permutation tests
P2BAT/PBAT, PBAT GoldenHelix	Single marker, haplotype, multimarker	Binary traits, quantitative traits/ multivariate, ranked traits, time-to-onset, gene–environment interaction	Covariate adjustment, VanSteen algorithm for multiple testing, X-chromosome, permutation tests
QTDT	Single marker	Quantitative traits	Permutation tests

There are several features of family-based designs that make them less attractive than their population-based counterparts. One feature is the considerable sensitivity to genotyping errors (Gordon and Ott, 2001; Gordon *et al.*, 2001, 2004). It is clear that genotyping errors can lead to false inferences because the test distribution depends crucially on the assumption that parental genotypes are correct. In the population-based setting, nondifferential genotyping errors will only make tests conservative under the null, but with FBATs, random genotyping errors can inflate the false positive rate, sometimes substantially (Hirschhorn and Daly, 2005). This underscores the importance of validating the genotyping for any significant finding.

There is widespread belief that gene–environment (and gene–drug) interactions as well as gene–gene interactions play an important role in many complex diseases. For example, genetic interactions with smoking status and/or smoking history are believed to be determinants of the severity of chronic obstructive pulmonary disease (Celedon *et al.*, 2004; DeMeo *et al.*, 2005).

Possible unmeasured gene–environment or gene–gene interactions are often thought to be responsible for lack of reproducibility of many genetic associations. Although such hypotheses about interactions between genes and exposure variables are widely accepted in the field, adequate development of statistical methodology to test such hypotheses has lagged behind other technical developments.

While some family-based approaches have been proposed for special settings (Cordell *et al.*, 2004; Gauderman, 2002; Lake and Laird, 2004; Umbach and Weinberg, 2000; Witte *et al.*, 1999), the general problem is hampered by reliance on statistical models for main and interaction effects (which may not correspond to biological models) (Cordell *et al.*, 2004) and the difficulty of testing interactions when main effects are assumed to be present.

Much work remains to develop better statistical methods for complex disorders that are characterized by multiple, possibly interacting genes and environmental factors, and are characterized by numerous interrelated traits. The focus here has been on family-based designs for testing this is appropriate when mapping via linkage disequilibrium is the major objective. However, as we move from discovery to verification and characterization, the focus will appropriately shift to effect estimation. The development of models for complex disease phenotypes with multiple covariates, genes, and interactions will be crucial for characterizing the role of individual polymorphisms in complex disease. The challenge will be to model pathways incorporating several genes at the same time, in combination with gene–environment interactions and endophenotypes, where modest effects may add up to a substantial impact on disease.

References

Abecasis, G. R., Cardon, L. R., and Cookson, W. O. C. (2000). A general test of association for quantitative traits in nuclear families. *Am. J. Hum. Genet.* **66**(1), 279–292.

Bickeboller, H., and Clerget-Darpoux, F. (1995). Statistical properties of the allelic and genotypic transmission/disequilibrium test for multiallelic markers. *Genet. Epidemiol.* **12**(6), 865–870.

Celedon, J. C., Lange, C., Raby, B. A., Litonjua, A. A., Palmer, L. J., DeMeo, D. L., Reilly, J. J., Kwiatkowski, D. J., Chapman, H. A., Laird, N., Sylvia, J. S., Hernandez, M., *et al.* (2004). The transforming growth factor-beta1 (TGFB1) gene is associated with chronic obstructive pulmonary disease (COPD). *Hum. Mol. Genet.* **13**(15), 1649–1656.

Chung, R. H., Hauser, E. R., and Martin, E. R. (2006). The APL test: Extension to general nuclear families and haplotypes and examination of its robustness. *Hum. Hered.* **61**(4), 189–199.

Chung, R. H., Hauser, E. R., and Martin, E. R. (2007a). Interpretation of simultaneous linkage and family-based association tests in genome screens. *Genet. Epidemiol.* **31**(2), 134–142.

Chung, R. H., Morris, R. W., Zhang, L., Li, Y. J., and Martin, E. R. (2007b). X-APL: An improved family-based test of association in the presence of linkage for the X chromosome. *Am. J. Hum. Genet.* **80**(1), 59–68.

Clayton, D. (1999). A generalization of the transmission/disequilibrium test for uncertain-haplotype transmission. *Am. J. Hum. Genet.* **65**(4), 1170–1177.

Clayton, D. G., Walker, N. M., Smyth, D. J., Pask, R., Cooper, J. D., Maier, L. M., Smink, L. J., Lam, A. C., Ovington, N. R., Stevens, H. E., Nutland, S., Howson, J. M., *et al.* (2005). Population structure, differential bias and genomic control in a large-scale, case-control association study. *Nat. Genet.* **37**(11), 1243–1246.

Cordell, H. J. (2004). Properties of case/pseudocontrol analysis for genetic association studies: Effects of recombination/ascertainment, and multiple affected offspring. *Genet. Epidemiol.* **26**(3), 186–205.

Cordell, H. J., Barratt, B. J., and Clayton, D. G. (2004). Case/pseudocontrol analysis in genetic association studies: A unified framework for detection of genotype and haplotype associations, gene-gene and gene-environment interactions, and parent-of-origin effects. *Genet. Epidemiol.* **26**(3), 167–185.

Curtis, D., and Sham, P. C. (1995a). An extended transmission/disequilibrium test (TDT) for multi-allele marker loci. *Ann. Hum. Genet.* **59,** 323–336.

Curtis, D., and Sham, P. C. (1995b). A note on the application of the transmission disequilibrium test when a parent is missing. *Am. J. Hum. Genet.* **56**(3), 811–812.

DeMeo, D. L., and Silverman, E. K. (2003). Genetics of chronic obstructive pulmonary disease. *Semin. Respir. Crit. Care Med.* **24**(2), 151–160.

DeMeo, D. L., Mariani, T. J., Lange, C., Srisuma, S., Litonjua, A. A., Celedon, J. C., Lake, S. L., Reilly, J. J., Chapman, H. A., Mecham, B. H., Haley, K. J., Sylvia, J. S., *et al.* (2005). The SERPINE2 gene is associated with chronic obstructive pulmonary disease. *Am. J. Hum. Genet.* **78**(2), 253–264.

Dudbridge, F. (2003). Pedigree disequilibrium tests for multilocus haplotypes. *Genet. Epidemiol.* **25**(2), 115–121.

Falk, C. T., and Rubinstein, P. (1987). Haplotype relative risks—an easy reliable way to construct a proper control sample for risk calculations. *Ann. Hum. Genet.* **51,** 227–233.

Fulker, D. W., Cherny, S. S., Sham, P. C., and Hewitt, J. K. (1999). Combined linkage and association sib-pair analysis for quantitative traits. *Am. J. Hum. Genet.* **64**(1), 259–267.

Gauderman, W. J. (2002). Sample size requirements for matched case-control studies of gene-environment interaction. *Stat. Med.* **21**(1), 35–50.

Gauderman, W. J. (2003). Candidate gene association analysis for a quantitative trait, using parent-offspring trios. *Genet. Epidemiol.* **25**(4), 327–338.

Gordon, D., and Ott, J. (2001). Assessment and management of single nucleotide polymorphism genotype errors in genetic association analysis. *Pac. Symp. Biocomput.* 18–29.

Gordon, D., Heath, S. C., Liu, X., and Ott, J. (2001). A transmission disequilibrium test that allows for genotyping errors in the analysis of single nucleotide polymorphism data. *Am. J. Hum. Genet.* **69**(4), 507–507.

Gordon, D., Haynes, C., Johnnidis, C., Patel, S. B., Bowcock, A. M., and Ott, J. (2004). A transmission disequilibrium test for general pedigrees that is robust to the presence of random genotyping errors and any number of untyped parents. *Eur. J. Hum. Genet.* **12**(9), 752–761.

Guo, C.-Y., Lunetta, K. L., DeStefano, A. L., Ordovas, J. M., and Cupples, L. A. (2007). Informative-transimission disequilibrium test (i-TDT): Combined linkage and association mapping that includes unaffected offspring as well as affected offspring. *Genet. Epidemiol.* **31,** 115–133.

Herbert, A., Gerry, N. P., McQueen, M. B., Heid, I. M., Pfeufer, A., Illig, T., Wichmann, H. E., Meitinger, T., Hunter, D., Hu, F. B., Colditz, G., Hinney, A., *et al.* (2006). A common genetic variant is associated with adult and childhood obesity. *Science* **312**(5771), 279–283.

Hirschhorn, J. N., and Daly, M. J. (2005). Genome-wide association studies for common diseases and complex traits. *Nat. Rev. Genet.* **6**(2), 95–108.

Horvath, S., and Laird, N. M. (1998). A discordant-sibship test for disequilibrium and linkage: No need for parental data. *Am. J. Hum. Genet.* **63**(6), 1886–1897.

Horvath, S., Xu, X., and Laird, N. M. (2001). The family based association test method: Strategies for studying general genotype-phenotype associations. *Eur. J. Hum. Genet.* **9**(4), 301–306.

Horvath, S., Xu, X., Lake, S. L., Silverman, E. K., Weiss, S. T., and Laird, N. M. (2004). Family-based tests for associating haplotypes with general phenotype data: Application to asthma genetics. *Genet. Epidemiol.* **26**(1), 61–69.

Jiang, H., Harrington, D., Raby, B. A., Bertram, L., Blacker, D., Weiss, S. T., and Lange, C. (2006). Family-based association test for time-to-onset data with time-dependent differences between the hazard functions. *Genet. Epidemiol.* **30**(2), 124–132.

Kistner, E. O., and Weinberg, C. R. (2004). Method for using complete and incomplete trios to identify genes related to a quantitative trait. *Genet. Epidemiol.* **27**(1), 33–42.

Kistner, E. O., Infante-Rivard, C., and Weinberg, C. R. (2006). A method for using incomplete triads to test maternally mediated genetic effects and parent-of-origin effects in relation to a quantitative trait. *Am. J. Epidemiol.* **163**(3), 255–261.

Laird, N. M., and Lange, C. (2006). Family-based designs in the age of large-scale gene-association studies. *Nat. Rev. Genet.* **7**(5), 385–394.

Laird, N. M., Horvath, S., and Xu, X. (2000). Implementing a unified approach to family-based tests of association. *Genet. Epidemiol.* **19,** S36–S42.

Lake, S. L., and Laird, N. M. (2004). Tests of gene-environment interaction for case-parent triads with general environmental exposures. *Ann. Hum. Genet.* **68**(Pt 1), 55–64.

Lake, S. L., Blacker, D., and Laird, N. M. (2000). Family-based tests of association in the presence of linkage. *Am. J. Hum. Genet.* **67**(6), 1515–1525.

Lange, C., and Laird, N. M. (2002a). Power calculations for a general class of family-based association tests: Dichotomous traits. *Am. J. Hum. Genet.* **71**(3), 575–584.

Lange, C., and Laird, N. M. (2002b). On a general class of conditional tests for family-based association studies in genetics: The asymptotic distribution, the conditional power, and optimality considerations. *Genet. Epidemiol.* **23**(2), 165–180.

Lange, C., DeMeo, D. L., and Laird, N. M. (2002). Power and design considerations for a general class of family-based association tests: Quantitative traits. *Am. J. Hum. Genet.* **71**(6), 1330–1341.

Lange, C., DeMeo, D., Silverman, E. K., Weiss, S. T., and Laird, N. M. (2003a). Using the noninformative families in family-based association tests: A powerful new testing strategy. *Am. J. Hum. Genet.* **73**(4), 801–811.

Lange, C., Silverman, E. K., Xu, X., Weiss, S. T., and Laird, N. M. (2003b). A multivariate family-based association test using generalized estimating equations: FBAT-GEE. *Biostatistics* **4**(2), 195–206.

Lange, C., Blacker, D., and Laird, N. M. (2004a). Family-based association tests for survival and times-to-onset analysis. *Stat. Med.* **23**(2), 179–189.

Lange, C., Van Steen, K., Andrew, T., Lyon, H., DeMeo, D., Raby, B., Murphy, A., Silverman, E., MacGregor, A., Weiss, S., and Laird, N. M. (2004b). A family-based association test for repeatedly measured quantitative traits adjusting for unknown environmental and/or polygenic effects. *Stat. Appl. Genet. Mol. Biol* **3**(1), http://www.bepress.com/sagmb/vol3/iss1/art17

Lasky-Su, J., Biederman, J., Laird, N., Tsunag, M., Doyle, A. E., Smoller, J. W., Lange, C., and Faraone, S. V. (2007). Evidence for an association of the dopamine D5 receptor gene on age at onset of attention deficit hyperactivity disorder. *Am. J. Hum. Genet.* **71**(5), 648–659.

Lunetta, K. L., Faraone, S. V., Biederman, J., and Laird, N. M. (2000). Family-based tests of association and linkage that use unaffected sibs, covariates, and interactions. *Am. J. Hum. Genet.* **66**(2), 605–614.

Martin, E. R., Kaplan, N. L., and Weir, B. S. (1997). Tests for linkage and association in nuclear families. *Am. J. Hum. Genet.* **61**(2), 439–448.

Martin, E. R., Monks, S. A., Warren, L. L., and Kaplan, N. L. (2000). A test for linkage and association in general pedigrees: The pedigree disequilibrium test. *Am. J. Hum. Genet.* **67**(1), 146–154.

Martin, E. R., Bass, M. P., Hauser, E. R., and Kaplan, N. L. (2003). Accounting for linkage in family-based tests of association with missing parental genotypes. *Am. J. Hum. Genet.* **73**(5), 1016–1026.

Naylor, M., Raby, B., Weiss, S. T., and Lic, C. (2005). Generalizability of regression models for the fev1-measurements in asthmatic children. *Far East J. Theor. Stat.* **16,** 351–369.

Ott, J. (1989). Statistical properties of the haplotype relative risk. *Genet. Epidemiol.* **6**(1), 127–130.

Rabinowitz, D. (1997). A transmission disequilibrium test for quantitative trait loci. *Hum. Heredity* **47**(6), 342–350.

Rabinowitz, D., and Laird, N. (2000). A unified approach to adjusting association tests for population admixture with arbitrary pedigree structure and arbitrary missing marker information. *Hum. Heredity* **50**(4), 211–223.

Rakovski, C. S., Xu, X., Lazarus, R., Blacker, D., and Laird, N. M. (2007). A new multimarker test for family-based association studies. *Genet. Epidemiol.* **31**(1), 9–17.

Risch, N., and Zhang, H. (1995). Extreme discordant sib pairs for mapping quantitative trait loci in humans. *Science* **268**(5217), 1584–1589.

Rubenstein, P., Walker, M., Carpenter, C., Carrier, C., Krassner, J., Falk, C., and Ginsberg, F. (1981). Genetics of HLA disease associations: The use of the haplotype relative risk (HRR) and the "haplo-delta" (Dh) estimates in juvenile diabetes from three racial groups. *Hum. Immunol.* **3,** 384.

Schaid, D. J. (1996). General score tests for associations of genetic markers with disease using cases and their parents. *Genet. Epidemiol.* **13**(5), 423–449.

Schaid, D. J., and Li, H. Z. (1997). Genotype relative-risks and association tests for nuclear families with missing parental data. *Genet. Epidemiol.* **14**(6), 1113–1118.

Schneiter, K., Laird, N., and Corcoran, C. (2005). Exact family-based association tests for biallelic data. *Genet. Epidemiol.* **29**(3), 185–194.

Self, S. G. (1991). An adaptive weighted log-rank test with application to cancer prevention and screening trials. *Biometrics* **47**(3), 975–986.

Sham, P. C., Cherny, S. S., Purcell, S., and Hewitt, J. K. (2000). Power of linkage versus association analysis of quantitative traits, by use of variance-components models, for sibship data. *Am. J. Hum. Genet.* **66**(5), 1616–1630.

Shih, M. C., and Whittemore, A. S. (2002). Tests for genetic association using family data. *Genet. Epidemiol.* **22**(2), 128–145.

Spielman, R. S., and Ewens, W. J. (1996). The TDT and other family-based tests for linkage disequilibrium and association. *Am. J. Hum. Genet.* **59**(5), 983–989.

Spielman, R. S., and Ewens, W. J. (1998). A sibship test for linkage in the presence of association: The sib transmission/disequilibrium test. *Am. J. Hum. Genet.* **62,** 450–458.

Spielman, R. S., McGinnis, R. E., and Ewens, W. J. (1993). Transmission test for linkage disequilibrium: The insulin gene region and insulin-dependent diabetes mellitus (IDDM). *Am. J. Hum. Genet.* **52**(3), 506–516.

Terwilliger, J. D., and Ott, J. (1992). A haplotype-based haplotype relative risk approach to detecting allelic associations. *Hum. Heredity* **42**(6), 337–346.

Thomas, D., Xie, R., and Gebregziabher, M. (2004). Two-Stage sampling designs for gene association studies. *Genet. Epidemiol.* **27**(4), 401–414.

Umbach, D. H., and Weinberg, C. R. (2000). The use of case-parent triads to study joint meets of genotype and exposure. *Am. J. Hum. Genet.* **66**(1), 251–261.

Van Steen, K., McQueen, M. B., Herbert, A., Raby, B., Lyon, H., DeMeo, D. L., Murphy, A., Su, J., Datta, S., Rosenow, C., Christman, M., Silverman, E. K., *et al.* (2005). Genomic screening and replication using the same data set in family-based association testing. *Nat. Genet.* **37**(7), 683–691.

Vansteelandt, S., DeMeo, D. L., Su, J., Smoller, J., Murphy, A. J., McQueen, M., Schneiter, K., Celedon, J. C., Weiss, S. T., Silverman, E. K., and Lange, C. (2007). Testing and estimating gene-environment interactions with complex exposure variables. *Biometrics* (in press).

Ware, J. H., and Weiss, S. (1996). Statistical issues in longitudinal research on respiratory health. *Am. J. Respir. Crit. Care Med.* **154**(6 Pt 2), S212–S216.

Weinberg, C. R. (1999). Methods for detection of parent-of-origin effects in genetic studies of case-parents triads. *Am. J. Hum. Genet.* **65**(1), 229–235.

Weinberg, C. R., Wilcox, A. J., and Lie, R. T. (1998). A log-linear approach to case-parent-triad data: Assessing effects of disease genes that act either directly or through maternal effects and that may be subject to parental imprinting. *Am. J. Hum. Genet.* **62**(4), 969–978.

Whittaker, J. C., and Lewis, C. M. (1998). The effect of family structure on linkage tests using allelic association. *Am. J. Hum. Genet.* **63,** 889–897.

Whittemore, A. S., and Tu, I. P. (2000). Detection of disease genes by use of family data. I. Likelihood-based theory. *Am. J. Hum. Genet.* **66**(4), 1328–1340.

Witte, J. S., Gauderman, W. J., and Thomas, D. C. (1999). Asymptotic bias and efficiency in case-control studies of candidate genes and gene-environment interactions: Basic family designs. *Am. J. Epidemiol.* **149**(8), 693–705.

Xu, X., Rakovski, C., Xu, X. P., and Laird, N. (2006). An efficient family-based association test using multiple markers. *Genet. Epidemiol.* **30**(7), 620–626.

Yu, J. M., Pressoir, G., Briggs, W. H., Bi, I. V., Yamasaki, M., Doebley, J. F., McMullen, M. D., Gaut, B. S., Nielsen, D. M., Holland, J. B., Kresovich, S., and Buckler, E. S. (2006). A unified mixed-model method for association mapping that accounts for multiple levels of relatedness. *Nat. Genet.* **38**(2), 203–208.

Zhao, H. (2000). Family-based association studies. *Stat. Methods Med. Res.* **9**(6), 563–587.

11

Searching for Additional Disease Loci in a Genomic Region

Glenys Thomson,[*] Lisa F. Barcellos,[†] and Ana M. Valdes[‡]

[*]Department of Integrative Biology, University of California, Berkeley, California 94720

[†]Division of Epidemiology, School of Public Health, University of California, Berkeley, California 94720

[‡]Twin Research Unit, King's College School of Medicine, London, United Kingdom

Advances in Genetics, Vol. 60

0065-2660/08 $35.00
DOI: 10.1016/S0065-2660(07)00411-7

ABSTRACT

Our aim is to review methods to optimize detection of *all* disease genes in a genetic region. As a starting point, we assume there is sufficient evidence from linkage and/or association studies, based on significance levels or replication studies, for the involvement in disease risk of the genetic region under study. For closely linked markers, there will often be multiple associations with disease, and linkage analyses identify a region rather than the specific disease-predisposing gene. Hence, the first task is to identify the primary (major) disease-predisposing gene or genes in a genetic region, and single nucleotide polymorphisms thereof, that is, how to distinguish true associations from those that are just due to linkage disequilibrium with the actual disease-predisposing variants. Then, how do we detect additional disease genes in this genetic region? These two issues are of course very closely interrelated. No *existing* programs, either individually or in aggregate, can handle the magnitude and complexity of the analyses needed using currently available methods. Further, even with modern computers, one cannot study every possible combination of genetic markers and their haplotypes across the genome, or even within a genetic region. Although we must rely heavily on computers, in the final analysis of multiple effects in a genetic region and/or interaction or independent effects between unlinked genes, manipulation of the data by the individual investigator will play a crucial role. We recommend a multistrategy approach using a variety of complementary methods described below.

I. INTRODUCTION

A. Background

Despite extensive efforts by many groups, only a few genes and some genetic regions involved in complex diseases have been identified. The general picture is one of difficulty in locating disease genes and replication of reported linkages and associations. The major exception is the many well-established disease associations with the classical human leukocyte antigen (HLA) immune response genes [the exact number of HLA-associated diseases is not known but is in the hundreds; see, e.g., Thorsby (1997)]. Only 6 non-HLA genes were consistently replicated as associated with complex diseases in the meta-analysis by Hirschhorn *et al.* (2002), including, for example, apolipoprotein E (APOE) with Alzheimer's disease and CCR5 with AIDS, of 166 putative associations, which had been studied three or more times. However, of the remaining 160 associations, more than half were observed again one or more times, much more than expected via type 1 error. Thus, there are probably many more real associations that are hard to replicate, some presumably because of weak effects. Further, initial finding of an association tends to overestimate the odds ratio (OR) (Lohmueller *et al.*, 2003).

Since the meta-analysis in 2002, the pace of discovery of complex disease genes verified via linkage and association studies has increased due to the availability of new technologies. Whole-genome association studies are finding new disease genes, albeit with small effects (see, e.g., Sladek *et al.*, 2007; Szatmari *et al.*, 2007). As in all the above examples, complex diseases and traits may result principally from genetic variation that is relatively common in the general population. Further, all these genes are susceptibility rather than necessary loci. Similarly for the possibly large number of other genes that may be involved in a specific disease risk, many with small effects; these features make them more difficult to detect by linkage analysis (Greenberg, 1993). Pritchard (2001) has argued that rare variants with a high overall mutation rate to the susceptible class may be the main contributors to complex disease risk, as seen for many Mendelian diseases. Also, of course, a combination of these two extreme models may very well apply. Further, different individuals and families will often have disease predisposition due to a different set of disease-predisposing genes. The effects of environmental factors, age of onset effects, heterogeneity of disease which is undetected, gene–gene and gene–environment interactions of course further complicate our attempts to detect predisposing genes for complex diseases. With many HLA-associated diseases, there is evidence of the role in disease risk of additional genes in the HLA region; identification of these genes has been as difficult as study of their non-HLA disease genes, in both cases considerable heterogeneity is seen across studies. Current technology bringing the ability to quickly type large numbers of markers and large sample sizes gives us hope that our understanding of the genetics of complex diseases will rapidly increase both for primary disease genes, and for the subject of this chapter, identifying additional (secondary) disease genes in a genetic region.

There are many challenges in trying to identify appropriate techniques for the analysis of the extensive data generated by whole-genome association and/or linkage studies as well as detailed study of specific genetic regions. Even with modern computers, all possible combinations of single and multilocus haplotypes cannot be considered. Some markers that do not show a single locus association with disease may nevertheless be directly involved in disease predisposition; their effects may only become obvious once the data are stratified by the effects of a primary disease gene or interaction effects are taken into account. Many analysis programs do not work with the high degree of polymorphism found in some regions (e.g., HLA), while others only allow a limited number of loci to be evaluated. Even with sample sizes in the thousands, with stratification analyses, many cells in the data matrix will have very small or zero values, particularly for multilocus haplotypes, and more so combined with the study of age of onset and gender-specific effects. As well as lack of statistical power, another consequence is that significance levels reported from application of a software package may be spurious due to inappropriate use of a statistical test on rare alleles, haplotypes, or genotypes. Further, we face the problem that programs are not available for haplotype analyses that can handle a high level of polymorphism at some loci, for example, HLA, as well as a large number of marker loci. Some programs, for example, the population genetic analysis package PyPop (www.pypop.org) (Lancaster *et al.*, 2003) can handle the high level of polymorphism of the HLA loci or microsatellite (msat) loci, but is limited to a total of eight loci at a time, while other packages, for example, Haploview (http://www.broad.mit.edu/mpg/haploview/index.php) can handle large numbers of biallelic markers (Barrett *et al.*, 2005).

From our combined experience in analyses of data on complex diseases, we emphasize that there is a required interplay between studying individual SNP, haplotype, and genotype effects, and that all putative significant effects must be scrutinized in individual detail by the researcher. Also, fit of the data to a particular model, for example, that a particular disease gene can explain all the linkage and association data in a genetic region *does not validate* this model. It may merely mean that there is insufficient power to detect additional effects, that appropriate stratification of the data to detect additional effects was not carried out, or that additional markers need to be typed. However, *rejection* of a model, apart from type 1 error or breaking of assumptions of the model, for example, random sampling of a homogenous population, allows one to *unequivocally* state that additional factors must be incorporated into the model of disease, and guides further investigation of such factors. Our survey of methods given below particularly highlights the importance of a complementary, multistrategy array of methods to uncover all the different facets of complex genetic diseases.

In this chapter, we will not discuss the issue of correcting for multiple testing. From our experiences with HLA-associated diseases with regard to detecting the effects of non-HLA region genes as well as additional genes in the HLA region, the issue is the difficulty in finding let alone replicating effects, rather than a deluge of type 1 errors. Many initial findings of primary HLA disease associations would not have survived correction for multiple tests, yet they have been extensively replicated and verified. Although we are very sensitive to the issue of type 1 errors, our emphasis throughout will be on detecting effects to be followed up in independent studies, ranked of course by relative p values. We strongly encourage use of resampling methods in all analyses of multiple markers in a genetic region. One can then adjust for multiple testing of nonindependent results [due to linkage disequilibrium (LD)] and to take account of the fact that haplotypes are estimated rather than known as assumed in most analyses, and to give an empiric p value to guide interpretation of results.

B. Primary disease genes

Our emphasis in the description of methods in this chapter will first be on genetic features of primary (major) disease-predisposing gene(s) in a genetic region (Section II). Unless we fully understand, and stratify by, *all* heterogeneity in disease risk at the primary disease gene(s), including relatively weak effects, then application of methods to detect additional disease genes in the region may give spurious results. The definition of primary versus additional (secondary) disease genes in a region is necessarily vague. The methods we use to detect *additional* disease genes in a region (Section III) are also those that we use to distinguish a primary disease gene from markers in LD with it. How can we have it both ways? We will not attempt any formal definition, which may in fact be impossible. Instead, we will follow the rough definition that has evolved with progress in disease gene discovery: *primary disease genes* are those that "stand out" in initial association and/or linkage studies, and *for the most part*, associations of other markers in the genetic region can be explained via their LD patterns with the primary disease gene(s). *Additional (secondary) genes* are detected once we take account of these primary disease genes, usually after study of additional marker loci with an increased sample size; they are expected to have more subtle, weaker effects, and these effects may be restricted to a subset of alleles, haplotypes, or genotypes at the primary disease gene(s). We will use the generic term marker gene to refer to SNPs, msats, and classical loci, any and all of which can be used in linkage and association studies. When a primary disease gene or SNP in a region has been identified, it will be referred to by its specific nomenclature, with marker genes then referring to all additional loci typed in the region and under study. A number of genes or SNPs (not necessarily in the same gene) in a region may be included in the primary disease gene category.

HLA researchers have considered many of the issues relevant to the topic of this chapter and have developed novel methods in study of both the primary disease associations and the effects of additional disease-predisposing genes in the HLA region. The direct involvement of HLA class II, and in some cases class I, genes in the disease process has been well documented for a number of diseases, for example, class II associations: HLA DR-DQ for type 1 diabetes (T1D), HLA-DR for rheumatoid arthritis (RA) and multiple sclerosis (MS), and HLA-DQ for narcolepsy; and class I associations: HLA-B for ankylosing spondilitis. Many of the HLA-associated diseases are of autoimmune origin, but Hodgkin's disease and other cancers, infectious diseases such as malaria, tuberculosis, and AIDS, and other diseases such as narcolepsy are also included. Non-HLA region genes are also involved in all these diseases, as well as additional genes in the HLA region.

We will primarily use HLA disease associations in our examples to illustrate methods, mostly derived from T1D, but we will also discuss results from MS, RA, and narcolepsy. For a non-HLA region example, we will consider data from studies of the protein tyrosine phosphatase gene PTPN22 and RA (Begovich *et al.*, 2004; Carlton *et al.*, 2005). The R620W SNP of PTPN22 is a common susceptibility allele for autoimmunity (Siminovitch, 2004) and has been associated with increased risk for a number of autoimmune diseases, for example, T1D, RA, and Graves disease (Lee *et al.*, 2006), but not with other diseases, for example, MS (Begovich *et al.*, 2005) and psoriasis (Nistor *et al.*, 2005).

At this point, we introduce in Table 11.1 some T1D data from Noble *et al.* (1996). Consider, for the moment, the first six columns of Table 11.1: the first three columns list the haplotypes composed of the tightly linked and high LD class II HLA DRB1, DQA1, and DQB1 (abbreviated as DR-DQ) loci (haplotypes that are rare in both patients and controls are included in the "others" category and do not enter into most calculations), the next two columns are the haplotype counts (and frequencies) for the patients and controls, and the sixth column gives the OR. We can readily see (without detailed statistical analysis at this point) the breakdown of susceptibility into predisposing, "neutral," protective, and very protective categories. In actuality, there is a continuum of effects encompassing these "discrete" categories. The ORs range from 8.2 to 0.02. [Note that the LD among the HLA DR-DQ genes is so strong that in many cases there is a predominant haplotype combination; hence, we use below the shorthand DR3, DR1, and DR15 to refer to some specific haplotypes (see footnote to Table 11.1 for their definitions); in contrast, there are a number of relatively common DRB1*04XX DQB1*03XX haplotypes (worldwide there are more than shown in Table 11.1).] Additionally, with T1D and other HLA-associated diseases, for example, RA and MS, there are genotype-specific effects (e.g., Barcellos *et al.*, 2006; du Montcel *et al.*, 2005: Noble *et al.*, 1996). These will be discussed later.

C. Additional (secondary) disease genes in a genetic region

Reports of other HLA region gene and msat associations have appeared in the literature. In many of these studies, it has been difficult to determine if an additional HLA region gene is involved in disease, versus the associations reflecting LD with the antigen presenting HLA molecules directly involved in disease. However, a number of analytic strategies have been developed to remove the effects of LD with the antigen presenting HLA genes directly involved in the disease (see Section III). Controlling for the influence of class II DR-DQ haplotype and genotype effects, a role in T1D has been shown of additional HLA class II (DPB1) (Noble *et al.*, 2000; Valdes *et al.*, 2001) and class I genes (including age of onset effects and rapid disease progression) (Fujisawa *et al.*, 1995; Nakanishi *et al.*, 1995; Noble *et al.*, 2002; Steenkiste *et al.*, 2007; Tait *et al.*, 1995; Valdes *et al.*, 1999, 2005a). Also, various analyses have shown the

Table 11.1. HLA DR-DQ Haplotype Frequencies in Type 1 Diabetes

DRB1	DQA1	DQB1[a]	T1D (%)[b]	Controls (%)[c]	OR[d]	P/C ratio[e]	χ^{2f}	χ^{2g}	χ^{2h}
Predisposing									
0401	0301	0302	91 (25.3)	11 (4.0)	8.21***	6.33	29.7****	–	–
0404	0301	0302	38.5 (10.7)	5 (1.8)	6.54***	5.94	12.0***	–	–
0301	0501	0201	115 (31.9)	26 (9.4)	4.55***	3.39	20.4****	–	–
Neutral									
0101	0101	0501	27.5 (7.6)	17 (6.1)	1.27	1.24	0.0	11.2***	–
Protective									
1302	0102	0604	10.5 (2.9)	12 (4.3)	0.67	0.67	2.6	0.8	4.7*
0401	0301	0301	6 (1.7)	12 (4.3)	0.38*	0.40	6.9**	0.1	0.4
0701	0201	0201	6.5 (1.8)	30 (10.8)	0.15***	0.17	32.3****	6.1	1.4
1301	0103	0603	3 (0.8)	18 (6.5)	0.12***	0.12	21.6****	4.8*	1.6
Very protective									
1501	0102	0602	1 (0.3)	43 (15.5)	0.02***	0.02	70.3****	–	
Others			61 (17.0)	104 (37.3)	–	–	–	–	
Total			360	278			195.9****	23.0****	8.1

Data From Noble *et al.* (1996).

[a]HLA DRB1 DQA1 DQB1 haplotypes—the following abbreviations are used in the text: DR3 (DRB1*0301 DQA1*0501 DQB1*0201), DR1 (DRB1*0101 DQA1*0101 DQB1*0501), DR7 (DRB1*0701 DQA1*0201 DQB1*0201), and DR15 (DRB1*1501 DQA1*0102 DQB1*0602); for the three DR4 haplotypes, there is heterogeneity at the DRB1 and DQB1 loci, similarly for the two DR13 haplotypes, and these will be referred to by their full notation.

[b]Patient counts and (%), patient counts are the average of the two affected sibs; haplotypes listed have a frequency >4% in at least one of the patient or control haplotypes.

[c]Controls are AFBACs (Thomson, 1995a,c).

[d]Odds ratio (OR) of this haplotype versus all others: * ($p < 0.05$), ** ($p < 0.01$), *** ($p < 0.001$), **** ($p < 0.0001$).

[e]Patient/control (P/C) ratio using frequency data (see Section II.D).

[f]χ^2 contribution of the individual haplotype in the heterogeneity test of patients versus control haplotypes (1 df), and the total χ^2 (df = number of classes − 1) (the "other" category is not included in the calculations).

[g]χ^2 contribution of the individual haplotype in the RPE heterogeneity test of patients versus control haplotypes after removal of the highly protective DR15 haplotype, and the three predisposing haplotypes DRB1*0401 and 0404 with DQA1*0301 DQB1*0302 and DR3; the overall test and the individual contribution of DR1 are both highly significant.

[h]As for previous column, except DR1 is now additionally removed; the overall test is nonsignificant, with only marginal significance for one individual haplotype, so no further rounds of testing are carried out.

presence of additional disease-predisposing markers on specific high risk DR-DQ haplotypes and genotypes, that is, DR3, and DRB1*0401 and DRB1*0404 with DQB1*0302 (see, e.g., Hanifi Moghaddam *et al.*, 1998; Johansson *et al.*, 2003; Lie *et al.*, 1999a,b; Pugliese *et al.*, 2007; Steenkiste *et al.*, 2007; Valdes *et al.*, 2005c; Zavattari *et al.*, 2001). Possible heterogeneity of the DR15 haplotype has been shown, with significant reduction in the diabetes-protective effect typically associated with this haplotype (Valdes *et al.*, 2005b). The results from study of the effects of additional disease genes using different populations and even different samples from similar populations are very heterogenous and show weak effects, reminiscent of non-HLA gene effects. This is clearly demonstrated in analyses of the 13th International Histocompatibility Workshop T1D HLA data, a large worldwide collection typed for the classical HLA genes and eight msats in the HLA region (Pugliese *et al.*, 2007; Steenkiste *et al.*, 2007).

II. PRIMARY DISEASE-PREDISPOSING GENES AND THEIR GENETIC FEATURES

A. Background

The methods we discuss in Section III to detect, or not, additional disease-predisposing genes in a region are obviously the base of the formal analyses that must also be performed to detect a primary disease-predisposing gene or genes versus marker genes in LD with the primary disease gene. Detailed study of this primary disease gene, or combination of genes, as detailed in this section, would then follow. Our aim is to understand all genetic aspects of mode of inheritance and heterogeneity in risk for the primary disease gene. Differential risk between alleles, haplotypes, and genotypes at the primary disease gene will influence the categories to be considered with the stratification analyses described below in Section III to identify additional disease genes in the region.

When additional (secondary) genetic effects *are identified* in a region, one would then again apply the methods in this section to understand the genetic basis and heterogeneity of this additional effect, in combination with the primary disease gene. Again, we need to detect all heterogeneity, and determine if certain alleles, haplotypes, or genotypes can be combined as a homogenous class. This latter point is not as crucial for individual SNPs, but is crucial for highly polymorphic markers such as HLA, msats, and haplotypes of SNPs, where it becomes unrealistic to consider all variation individually (many cell counts will be extremely small), unless of course there is such heterogeneity that we must do so. We would then *repeat* the steps of Section III to detect additional disease genes, conditional now on the primary disease gene and any additional genes so far identified. We stress that all additional markers should now be *reanalyzed* under this new paradigm; markers that do not individually show an association with disease, or that appear to be explained by LD with the primary or secondary disease gene, may have their involvement masked by the primary and secondary disease genes so far identified. We continue this process until no further effects are seen. We emphasize that categories classified as homogenous with respect to risk should be continually *reevaluated* as additional genes involved in disease are discovered.

In the following text, when we talk about controls in family-based data we are referring to parental marker alleles or haplotypes *not* transmitted to an affected child in trio families [two parents and an affected child—simplex (S) pedigrees] and these form the affected family based control (AFBAC) population (Thomson, 1995a,b,c). For affected sib pair families [multiplex sib pairs (MSP) pedigrees], the parental alleles or haplotypes *never* transmitted to the affected sib pair form the AFBAC population. Even when tests are performed using the transmission disequilibrium test (TDT) (Spielman *et al.*, 1993), it is of interest to determine the AFBAC frequencies for comparison with cases, also for OR estimates and study of maternal–fetal interaction effects, etc. For trio families, *control genotypes* are formed from the two nontransmitted parental alleles or haplotypes, and are similarly approximated for affected sib pair families (MSP pedigrees) (assuming penetrance is low) (Thomson, 1995a,c). For affected sib pair families, the exact control genotypes are obtained from the two parental AFBACs in the share 2 identity by descent (IBD) category. The usual Hardy Weinberg (HW) test can be applied to these control genotypes. Alternatively, one can take allele and haplotype AFBACs and form a control population assuming HW proportions (with a sample size of half the number of AFBACs). Note that for transmitted alleles in MSP pedigrees, the two affected sibs are not independent and cannot be included as independent data points in any analyses. When we talk about a test of heterogeneity in our discussions below, for example, when testing between patient and AFBAC or case-control data, we infer (unless otherwise stated) that this could be performed as appropriate by a χ^2 test, Fisher's exact test, or resampling, with preference for the latter.

B. Linkage and association tests

The strong LD within the HLA complex and high level of polymorphism of many of the classical HLA loci have aided in the detection of genes in a large number of diseases. Initial studies of for example T1D showed class I HLA serological associations with disease. As HLA class II serological typing was developed, stronger associations with class II DR-DQ genes were found, and the class I associations could be explained by LD with the class II genes. HLA associations with a number of diseases (e.g., RA, MS, ankylosing spondilitis, hemochromatosis, and narcolepsy) were readily found, and replicated.

Age of onset effects may be important for both primary and secondary disease gene effects, both for HLA and non-HLA regions [for HLA and T1D, see, e.g., Valdes *et al.* (1999, 2005c); for Alzheimer's disease and APOE, see Blacker *et al.* (1997); for Alzheimer's disease and HLA, see Zareparsi *et al.* (2002)]. The power of the TDT and case-control association tests can be greatly affected by the age of onset of patients in the study (Li and Hsu, 2000). Further, age-related fluctuations in allele and genotype frequencies in *controls* can lead to loss of power and increase in type 1 error rates; Payami *et al.* (2005) show that APOE frequencies in controls can vary by twofold in the extreme cases due to age and gender. For cytochrome P450 2D6 (CYP2D6), the age and gender-specific fluctuations in controls were less pronounced than for APOE and non significant; however, there were significant age-dependent departures from HW, even for the youngest cohort. Further, gender differences should also be investigated.

Linkage studies using the affected sib pair method easily showed linkage of the HLA region with T1D; in fact only 15 affected sib pairs gave a p value of 0.001 [for references, see, e.g., Thomson and Valdes (2005)]. Other HLA-associated diseases such as RA and MS required much larger sample sizes, in the hundreds, to get statistically significant linkage. However, even these numbers are quite moderate. Use of msat markers *in lieu of* classical HLA typing in genome-wide linkage scans can make it more difficult to detect linkages. As for association studies, age of onset effects that are not incorporated into a linkage analysis can drastically reduce power (Hsu *et al.*, 2002; Li, 1999; Li and Hsu, 2000). The true disease model (unknown for complex diseases), especially common variant versus rare variants with extensive allelic heterogeneity, will influence whether association or linkage studies are more powerful (Pritchard, 2001; Risch and Merikangas, 1996).

Use of allele sharing IBD values can increase the power of association studies (Fingerlin *et al.*, 2004; Thomson, 1995c). Vice versa, partitioning linkage analyses by genotypes of associated alleles can also increase power (Clerget-Darpoux *et al.*, 1995; Greenberg, 1993; Greenberg and Doneshka, 1996; Hodge, 1993; Li *et al.*, 2004). With T1D IDDM2 (the VNTR 5′ to the insulin gene) and study of affected sib pair data, Dizier *et al.* (1994) show that without stratification, the IBD frequencies are 0.24, 0.53, and 0.23 and these are not significant from random expectations. However, the distribution is quite different when stratified by the genotype of the index sib: 0.27, 0.60, and 0.13 for genotype 11 and 0.07, 0.53, and 0.40 for the combined other genotypes. Li *et al.* (2004) have also shown that there can be large variability in linkage scores when families are stratified by the genotype of a single, randomly selected sib. They have developed a program (genotype IBD sharing test—GIST) in which families are weighted on the basis of the genotypes of all family members.

C. Modes of inheritance (recessive versus additive models)

1. Antigen genotype frequencies among patients method

The so-called antigen genotype frequencies among patients (AGFAP) method is informative with respect to mode of inheritance, recessive versus additive (dominant), when a marker allele is *strongly* associated with a disease-predisposing gene (Thomson, 1983, 1993, 1995a,b). For a recessive disease model, the marker locus genotype expectations are simply HW proportions based on the marker allele frequencies in patients. An early HLA A locus association with hemochromatosis (with the allele A3) correctly indicated a recessive mode of inheritance for the disease. For an additive (dominant) model, most patients are expected to be heterozygous rather than homozygous for the associated marker allele. Also, additive expectations continue to hold if we additionally allow for sporadic cases when the marker locus is itself the disease causative agent, for example, HLA-B27 and ankylosing spondilitis (Thomson, 1983). Expectations for multiple alleles are easily obtained allowing tests of mode of inheritance (Thomson, 1993, 1995a); however, the simple method of moments estimate for the additive model can give problems with multiple alleles if the mode of inheritance is not close to additive.

Original application of this model to T1D favored a recessive model; further investigation of multiallelic class II HLA DR locus associations led to discovery of heterogeneity beyond this simple model [see, e.g., Louis and Thomson (1986)]. This result highlights the fact mentioned in the Introduction (Section I) that fit to a particular model does not provide verification of that model; a more complicated model may apply but the simpler model is not rejected until more informative data is available. The data for HLA DR-DQ genotype counts and frequencies are given in Table 11.2 for the three common predisposing haplotypes and the "neutral" DR1 haplotype shown in Table 11.1. Compared to recessive expectations, there are excess heterozygotes for DR3 and the two predisposing DR4 haplotypes in this data set (DRB1*0401 and *0404 with DQA1*0301 DQB1*0302) and deficiency of the homozygous classes and the heterozygote for these two DR4 haplotypes. The base result of a more recessive like than additive (two copies vs one) mode of inheritance for the predisposing HLA DR-DQ genotypes continues to hold, with the addition of the compound heterozygote effect for these DR3/DR4 genotypes. Note that for multiallelic systems, the control frequencies for many genotypes immediately become small (see Table 11.2) unless the sample size is very large.

2. Affected sib pair ibd values and mode of inheritance

The distribution of the number of HLA haplotypes, or other marker variation, shared by affected sib pairs can also be used to obtain information on the mode of inheritance of the disease: recessive and additive as well as intermediate models

Table 11.2. HLA DR-DQ Genotype Frequencies in Type 1 Diabetes

DRB1	DQA1	DQB1	DRB1	DQA1	DQB1[a]	T1D(%)[b]	Controls (%)[c]	Recessive (%)[d]	χ^{2} [e]
0301	0501	0201	0401	0301	0302	44 (24.4)	1.0 (0.7)	29.1 (16.1)	7.6**
0301	0501	0201	0404	0301	0302	21.5 (11.9)	0.5 (0.3)	12.3 (6.8)	6.9**
0301	0501	0201	0301	0501	0201	10 (5.6)	1.2 (0.9)	18.3 (10.2)	3.8*
0101	0101	0501	0401	0301	0302	9 (5.0)	0.7 (0.5)	6.9 (3.8)	0.6
0101	0101	0501	0301	0501	0201	8 (4.4)	1.6 (1.2)	8.7 (4.9)	0.1
0401	0301	0302	0401	0301	0302	5.5 (3.1)	0.2 (0.2)	11.5 (6.4)	3.1
0401	0301	0302	0404	0301	0302	4.5 (2.5)	0.2 (0.1)	9.7 (5.4)	2.8
others						77.5	133.6	83.5 (46.4)	0.4
Total						180	139	180	

Data from Noble *et al.* (1996) and personal communication.

[a]HLA DRB1 DQA1 DQB1 genotypes.

[b]Patient genotype counts and (%), patient counts are the average of the two affected sibs; only genotypes common in patients are listed.

[c]Controls are estimated from AFBACs assuming HW proportions (Thomson, 1995a,c).

[d]Recessive expectations (Thomson, 1983) for cases under the AGFAP analysis; these are HW expectations based on the allele frequencies in the cases (from Table 11.1) (see Section II.C.1), $p \ll 0.0001$.

[e]χ^2 (df = 1) of observed versus recessive expectations for individual patient genotypes: * ($p < 0.05$), ** ($p < 0.01$), *** ($p < 0.001$), **** ($p < 0.0001$).

(Louis *et al.*, 1983; Motro and Thomson, 1985). (The test is carried out on marker genes close to, or which may include, the disease-predisposing gene.) For recessive and additive models (allowing for incomplete penetrance but not for sporadic cases of disease), the expected IBD values are functions only of the disease allele frequency, and do not involve the disease penetrance value. Ironically, as above, given we now know the extent of heterogeneity of the HLA DR-DQ contribution to T1D, the results for T1D show incredibly close fit to a recessive model, which in fact are of the form of HW proportions. Data and analyses from 538 families with 711 T1D affected sib pair comparisons show IBD share 2, 1, and 0 values of 373, 283, and 55 compared to recessive expectations of 372.3, 284.4, and 54.3, respectively (Payami *et al.*, 1985). The observed IBD frequencies of 52%, 40%, and 8% are very highly significantly different from random 25%, 50%, and 25% expectations ($p < 10^{-5}$) and reject an additive model ($p \ll 0.0001$), a hallmark of which is the share 1 class has an expectation of 50%, and of course the share 2 class is greater in frequency than the share 0 class.

3. Additional tests including modes of inheritance, maternal–fetal effects, and imprinting

To develop a complementary array of analytic methods for dissection of the genetics of complex traits, including modes of inheritance and heterogeneity effects, Thomson (1995a) considered four ascertainment schemes with nuclear families and their expectations based on a general model of disease, and also the specific expectations for recessive and additive models. The four models are based on different degrees of affectation status used in the nuclear family sampling scheme: S—simplex (child), MSP—multiplex sib pairs, MPC—multiplex parent–child, and MPS—multiplex parent–sib pairs. Note that within, for example, S pedigrees, there will be some affected parents and some affected sib pairs; these will occur in frequencies expected based on the S sampling scheme and should be retained in all analyses of these data, similarly for affected sib pair families. Each sampling scheme provides information both for detecting disease genes and determining modes of inheritance and heterogeneity effects. Data will rarely be specifically sampled for MPC and even less so for MPS pedigrees. However, these data can be extracted without bias from S and MSP data sets for additional specific analyses, including unique information. Specific tests using the different pedigree types are detailed by Thomson (1995a,b).

For family data, there are also specific tests that should be carried out on the AFBACs: independence of AFBACs from transmitted alleles [see Thomson (1995a,c) for the expectations for the different ascertainment schemes], HW proportions of AFBAC genotypes (see Section II.A), and comparison of nontransmitted alleles (AFBACs) from mothers (NIMAs—noninherited maternal antigens) versus fathers (NIPAs—noninherited paternal antigens) overall and based on specific genotypes of the proband. The exposure to NIMA via several different mechanisms may shape the immune system of the offspring and either predispose or protect against immune reactions including *in utero* exposure to NIMA, as well as postpartum exposure to NIMA mediated by breast-feeding and/or long-term persistence of maternal cells in the offspring. With HLA-associated diseases, NIMA effects should be tested for in the overall data as well as in subsets of patients with and without high-risk genotypes. NIMA effects were first reported in RA in mothers of DR4-negative patients (ten Wolde *et al.*, 1993), were not seen in the study of Silman *et al.* (1995), but were again reported by van der Horst-Bruinsma *et al.* (1998) and Harney *et al.* (2003). In T1D, the results are not consistent, with some reports of NIMA effects [e.g., see Pani *et al.* (2002)] and other studies not showing this effect (Hermann *et al.*, 2003; Lambert *et al.*, 2003).

Genomic imprinting has been implicated in susceptibility to complex diseases such as systemic lupus erythematosus and other HLA-associated autoimmune conditions. Genomic imprinting is defined as a phenomenon in which the disease phenotype depends on which parent passed on the disease-predisposing gene. While much remains to be learned about the underlying epigenetic mechanisms involved in imprinting, it has been shown to play a role in several birth defects, certain genetic diseases, and cancers, and possibly autoimmunity. Differences in maternal and paternal transmission rates of predisposing alleles have been seen in several studies of autoimmune diseases including T1D and systemic lupus erythematosus (Bennett *et al.*, 1996; Fajardy *et al.*, 2002; Sasaki *et al.*, 1999; Sekigawa *et al.*, 2003).

D. The patient/control ratio

In this section, we assume a primary disease gene has been recognized and discuss the patient/control (P/C) ratio (Thomson *et al.*, 2007). The P/C ratio is defined as follows, under the assumption, for example, that the HLA DR-DQ genes are directly involved in T1D:

$$P/C = \frac{[\text{freq}(DR-DQ)_{\text{patients}}]}{[\text{freq}(DR-DQ)_{\text{controls}}]}, \tag{1}$$

where freq(·) denotes the frequency of the haplotype or genotype under consideration (or allele or SNP or HLA amino acid as appropriate, and multiple combinations thereof), also see Clerget-Darpoux *et al.* (1988).

Under the assumption that the primary disease gene has been identified, the P/C ratios give a maximum likelihood estimate of the *relative* penetrance values for each genotype (these are a function of disease prevalence) (Thomson *et al.*, 2007). However, the ratio of two P/C ratios estimates the ratio of the *absolute* penetrance values for the two haplotypes or genotypes under consideration. Further, it is equivalent to the OR for the comparison of these two alleles, haplotypes or genotypes. The absolute penetrance values may vary between populations because they include the averaged effects of non-HLA genes, environmental factors, and other HLA genes. In the simplest case, the ratio of the absolute penetrance values for the DR-DQ genotypes and haplotypes, or other primary disease gene(s), would be the same across all studies. Further, the relative rankings based on P/C ratios for a set of haplotypes or genotypes can be compared across populations (see below).

The P/C ratios for the HLA DR-DQ haplotypes in T1D are given in Table 11.1 (column 7). In this case, the relative rankings of the HLA DR-DQ haplotypes from most predisposing through most protective are the same based on ORs and P/C ratios. This is not always the case. For haplotypes (or genotypes) with a P/C ratio < 1, the inverse of this number is used in comparison of relative strength with positively associated haplotypes (P/C ratio > 1); hence, in the data of Table 11.1, the DR15 very protective effect is much stronger than the three haplotypes listed with predisposing effects.

E. Relative predispositional effects

1. The relative predispositional effect method

With the T1D data of Table 11.1, the contributions of each haplotype to the χ^2 heterogeneity test comparing patients and controls are given in column 8 (excluding the "others" class of rare haplotypes). We see many significant positive and negative associations, similarly with the ORs in column 6. Can we determine which effects are the strongest (positive or negative associations), and which may merely reflect the consequences of these major effects (if an HLA DR-DQ haplotype is strongly positively associated with disease, others must be negatively associated, and vice versa). The very protective DR15 haplotype contributes most to the overall χ^2 value, also as noted above, it is the strongest effect based on the P/C ratio. We thus remove the DR15 haplotype and the three predisposing haplotypes [which are not significantly different in effect from each other, but are highly significantly heterogenous with the DR15 effect ($p \ll 0.0001$)] and perform another χ^2 test of heterogeneity of patient versus control values (column 9 of Table 11.1). The "neutral" DR1 haplotype, which shows roughly equal frequencies in patients and controls, now contributes most to overall significant heterogeneity ($p \ll 0.0001$). Removal of DR1 (column 10) gives a nonsignificant overall test of the remaining haplotypes (classified in the protective class).

These analyses and results demonstrate application of the relative predispositional effect (RPE) method (Payami *et al.*, 1989), where alleles, haplotypes, or genotypes with the strongest predisposing or protective effects are sequentially removed from the analysis until no further heterogeneity in risk effects is seen. TDT and pedigree disequilibrium test (PDT) values can similarly be analyzed [the PDT uses genetic data from related nuclear families and discordant sibships within extended pedigrees and can examine allele (haplotype) and genotype effects] (Martin *et al.*, 2000, 2003).

Determining the order in which haplotypes or genotypes are sequentially removed is not trivial, interplay between the contribution to the χ^2 heterogeneity test, the P/C ratio, and the control frequencies of the allele, haplotype, or genotype are needed. For the same control frequency and equivalent strength of effects, a positive association will contribute more to the overall χ^2 than a negative association. Also, less frequent classes even with stronger effects can contribute less to the overall χ^2 [this is demonstrated in Table 11.3 with comparison of the TT and TC genotype classes for PTPN22 and RA (compare the individual χ^2 contributions in column 6 vs the OR and P/C ratios in columns 4 and 5)]. Use of the OR carries the problem that using the "all others" category in the denominator may lead to considerable heterogeneity in risk in this comparison class, for example, with HLA and T1D. One solution is to always compare risk relative to one relatively "common" class, for example, the DR1

Table 11.3. PTPN22 Genotype Data for SNP R620W in Rheumatoid Arthritis

Genotype	Cases[a]	Controls[a]	OR[b]	P/C ratio[c]	χ^{2d}	Recessive[e]	Additive[f]
TT	16 (0.02)	12 (0.01)	2.30*	1.99	3.4	20.4	17.0
TC	245 (0.26)	221 (0.16)	1.91****	1.66	30.2****	236.2	242.0
CC	677 (0.72)	1168 (0.83)	–	0.87	8.9**	681.4	677.0
Total	938	1401			42.5****	938	938

Data from Begovich *et al.* (2004).

[a]Cases and controls are combined discovery and replication studies (single sib from replication study).

[b]OR is relative to CC genotype: * ($p < 0.05$), ** ($p < 0.01$), *** ($p < 0.001$), **** ($p < 0.0001$).

[c]P/C ratio using frequency data (see Section II.D).

[d]χ^2 contribution of the individual genotype in the heterogeneity test of patients versus control genotypes (1 df), and the total χ^2 (2 df).

[e]Recessive expectations for cases under the AGFAP analysis; these are HW expectations based on the allele frequencies in the cases (Thomson, 1983), see Section II.C.1, ns.

[f]Additive expectations for cases under the AGFAP analysis, using method of moments (Thomson, 1983), see Section II.C.1, ns.

haplotype or the DR3 haplotype with T1D (but these may not appear, or be rare, in some ethnic groups). Another approach is to use the P/C ratios. However, this measure involves ratios; hence, the distributional properties are less satisfactory, and the variance will also be greater the rarer the control frequencies. All these facts should be considered in any individual RPE analysis. The actual sequence of removal with RPE analysis should not dramatically alter the major effects, but will give some variation with more modest effects.

2. Application of the RPE method to T1D, MS, RA, and narcolepsy data sets

Thomson *et al.* (2007) investigated, using a meta-analysis approach, with data from 38 T1D HLA studies, whether the DR-DQ haplotypes and genotypes show the same RPEs across populations and ethnic groups using relative rankings of the P/C ratios and pairwise comparisons. Categories of consistent predisposing, intermediate ("neutral"), and protective to very protective haplotypes were identified across ethnic groups (as shown in Table 11.1 for a Caucasian population) and these correlated with disease prevalence [also see Valdes *et al.* (1997); and for HLA data, Meyer *et al.* (2006, 2007) and Single *et al.* (2007)]. Given the large number of studies, including cross-ethnic comparisons, along with the consistency of results, specific significant effects could be identified. For example, there was a statistically significant and consistent hierarchy for DRB1*04XX DQB1*0302 haplotypes from predisposing to protective effects: DRB1*0405 = *0401 = *0402 > *0404 > *0403, with the DR3 haplotype significantly between the first grouping and DRB1*0404.

This hierarchy confirms previous observations, also showing a significant effect of DRB1 on T1D predisposition. Earlier studies had readily shown the role of DQB1 on disease, with significant comparisons of the effects of haplotypes with DQB1*0302 versus DQB1*0301 with DRB1*0401 (in the predisposing vs protective categories, respectively, in Table 11.1). This is an application of the conditional haplotype method (CHM) described in Section III.F.1.

For MS, there is a predominant predisposing effect of the DRB1*1501 allele in Caucasian populations and DRB1*1501 and 1503 in African-American populations, with a predisposing DRB1*0301 effect being much weaker [see, e.g., Barcellos *et al.* (2006); Oksenberg *et al.* (2004); Yeo *et al.* (2007)]. In some instances, the DRB1*0301 effect is only observed at the genotype level (with a recessive effect) (Barcellos *et al.*, 2006). In the same study, the genotype DRB1*1501/DRB1*08 also showed a significant predisposing effect. Other significant effects of rare alleles have been observed in individual studies: DR14 protective (Barcellos *et al.*, 2006) and DRB1*0103 predisposing (Yeo *et al.*, 2007). Conditional haplotype analysis to determine which of DRB1 or DQB1 is primary in predisposition to MS cannot be carried out in Caucasian populations where the combination DRB1*1501 DQB1*0602 is rarely broken down. In African-American populations, there is some breakdown in the LD and DRB1*15 (there is no heterogeneity in the effects of *1501 and *1503) shows the primary effect independent of DQB1*0602 (Oksenberg *et al.*, 2004). Mignot *et al.* (2001) in a similar approach had shown that for narcolepsy, DQB1*0602 has the primary effect independent of DRB1*15.

Application of the AGFAP method (Section II.C.1) to the PTPN22 RA data of Begovich *et al.* (2004) (Table 11.3) cannot distinguish between recessive and additive models (assuming a gene in LD with the marker). However, conditional logistic regression (CLR) analysis (see Section III.H) significantly excludes a recessive model ($p < 0.0001$) (Begovich *et al.*, 2004). We mention this in light of our suggestion in the Introduction (Section I.A) that the data be analyzed by a variety of methods and results compared. The difference in results in this case most likely arise from the different models considered, a linked gene in LD with the marker in the AGFAP analysis, and a direct role of the marker in the CLR analysis. Considering the P/C ratios and RPE analysis, we see consistency with the CLR results with the difference in risk between the TC and CC genotypes (see Table 11.3, $p \ll 0.0001$).

In the case of a single allele showing a strong association with disease, for example, DRB1*1501 and MS and DQB1*0602 and narcolepsy, it may be beneficial to consider subsets of the data to possibly increase the power to detect

RPEs of other alleles (and also to detect other gene effects). Mignot *et al.* (2007) applied this approach to narcolepsy families where neither parent had the DQB1*0602 allele. In addition, they developed the single parent TDT: in simplex (trio) families with a DQB1*0602 allele transmitted to the proband, the TDT is applied to the other parent using only families where this other parent does not have the DQB1*0602 allele. Overall, the results indicated in addition to the well-known strong effect of DQB1*0602, additional RPEs of DQB1*0301 (susceptibility) and DQB1*0501 (resistance), replicating previously reported results in case-control studies where similar stratifications were applied (Mignot *et al.*, 2001). The rationale for this approach is to concentrate on families (or appropriate subsets of case-control data) where the effects are not overridden by the predominant risk of one or more alleles (or haplotypes or genotypes).

III. DETECTING ADDITIONAL GENES IN A GENETIC REGION

A. Background

Methods to detect genes or markers additional to a primary predisposing gene in a genetic region all rely on stratification analyses to take account of the effects of LD with the primary predisposing gene(s). Some use only a restricted set of the data. In some of these cases, power may be increased by combining data across studies and even ethnic groups; however, this gain in power may be countered by loss if there are population-specific effects. Methods to test for fit of a putative primary disease gene(s) to some or all aspects of the data are also included in this section; lack of fit implies that either the primary disease gene has not been identified or if it has, additional disease genes are involved. As mentioned previously, the identification of all primary and additional disease genes in a genetic region requires a constant interplay of application of the methods described in both this section and Section II as additional genes involved in disease are identified. With all the methods discussed below, power is an issue even with sample sizes in the hundreds or thousands, stratification approaches can quickly result in small cell sizes. All of the methods described below have been successful in detecting the role of additional disease genes in the HLA region; however, the actual genes involved have not been identified, and there is considerable heterogeneity between studies.

B. Linkage analyses of two linked disease susceptibility genes

Until relatively recently, linkage analyses assumed a single disease-predisposing gene in a region. Biernacka *et al.* (2005a), extending the work of Liang *et al.* (2001), developed a method for simultaneous localization of two susceptibility genes in one region using marker IBD sharing values in affected sib pairs. As expected, fitting of a one locus model can produce misleading localization results when there are two linked disease genes in the region. Methods to evaluate the evidence for two versus one disease loci in a region are presented in Biernacka *et al.* (2005b).

C. Linkage disequilibrium of markers with primary disease genes

If a primary disease gene in the region under study has been identified, the first step in study of additional disease genes in the region would be to plot the LD of all marker genes *with* the primary disease gene. Informal inspection would show in most cases a strong correlation of this LD pattern with the strength of association of markers with disease (before any stratification analyses have been applied and assuming the primary disease gene has been correctly identified). Obvious deviations from this pattern would indicate potential additional disease genes (or markers in strong LD with additional disease genes). (If multiple genes have a primary effect, these genes are combined together as one "super-locus" combination; haplotypes would need to be estimated to form the "alleles" at this "super-locus.")

What measure of LD should we use when studying marker associations with each other? There is no one measure that is perfect, as we are trying to depict multidimensional variables (allele frequencies and numbers of alleles at each locus and the associations (LD) between all pairwise combinations of alleles at the two loci) by one overall LD measure. For multiallelic markers, appropriate extensions of two biallelic measures are often used: the D' statistic is a weighted average of the normalized LD statistic between each pair of alleles (Hedrick, 1987), and the W_n statistic which is a reexpression of the χ^2 statistic normalized to be between zero and one (see, e.g., Meyer *et al.*, 2006); the latter measure is more informative in most cases with respect to the analyses in this section. When there are only two alleles per locus, W_n is equivalent to the correlation coefficient between the two loci, defined as $r = D/[p_A p_a p_B p_b]^{1/2}$, where p_A, p_a and p_B, p_b are the allele frequencies at the two loci (A and B) respectively and D is the standard measure of LD between the two loci.

A measure called haplotype-specific heterozygosity captures the informativeness with respect to association studies of markers on specific haplotypes of the primary disease-predisposing gene (Malkki *et al.*, 2005). The LD of all markers with each other should also be studied and will guide single association versus haplotype analyses.

D. Matched cases and controls

A modification of the classic case-control design is to match cases and controls at the genotype level for the HLA class I or class II genes known to increase risk of disease, or other primary disease-predisposing gene(s). This eliminates the effects of LD between the primary disease gene and other marker genes under study (Hanifi Moghaddam *et al.*, 1998). Significant case-control differences in the distribution of marker allele, haplotype, and genotype frequencies reflect the effect of additional disease susceptibility genes in the region. The major disadvantage of this matched approach is that it reduces the number of cases and controls available for analysis. An advantage is that effects can be summed over populations: the ratio of matched cases/controls must be the same in all data sets if they are to be combined. This approach is powerful if an effect is specific to or more easily detected in a specific subset of the data, either high risk or low risk. Note that the matched cases and controls approach is a specific case of the conditional genotype method (CGM) discussed below. This approach and the CGM can also be applied to family based data using, for example, matched genotypes from the proband (or average of affected sib pairs) and AFBACs (see Section II.A).

With matched cases and controls, the usual case, especially with T1D and its hierarchy of HLA DR-DQ disease effects, will be to restrict analyses to the genotypes common in patients. Hanifi Moghaddam *et al.* (1998) analyzed data on patients and controls matched for DR3/DRB1*0401 DQB1*0302, the most common high-risk genotype in Caucasians. Two HLA regions showed significant msat associations. With diseases with a dominant predisposing effect, for example, narcolepsy and DQB1*0602, the decision must be made whether to match specifically on the non-DQB1*0602 alleles in heterozygous individuals. As we have discussed above, it is *essential* to have performed an RPE analysis of allele, haplotype, and genotype effects at the primary disease locus before proceeding in this way. With narcolepsy, additional DQB1 effects are seen as described above and must be taken into account, similarly with MS and DRB1. The difficulty with dominant protection, for example, T1D and DR15/DRX (X = non-DR15), is that the sample size in patients will be small, but nevertheless in samples where there are sufficient numbers in this category they should be investigated. Valdes *et al.* (2005b) have found preliminary evidence of a marker that modifies the protective effect of DR15 in a Swedish population.

Additional genetic effects may be specific to a high risk category. On the other hand, they may be restricted to, or be more easily detected in, a subgroup of cases and controls or families lacking the high-risk factors at the primary disease gene(s). Removing MS trio families with DRB1*1501, and also DRB1*03 and DRB1*0103, from consideration, Yeo *et al.* (2007) found a protective effect for the allele C*05 of the HLA class I C locus. The effect could not be distinguished in the high-risk DRB1*1501 set of families.

E. Homozygous parent linkage and TDT methods

1. Homozygous parent linkage method

Using affected sib pair data, the role of genes additional to HLA DR-DQ in T1D was demonstrated by Robinson *et al.* (1993) using the homozygous parent linkage method (HPLM). Affected sib pairs (MSP pedigrees) with a parent homozygous for the DR3 haplotype were examined; a marker gene was used to distinguish between the DR3 haplotypes; in this case it was the highly polymorphic B locus, but it could be any combination of marker genes. Under the null hypothesis that no HLA region variation additional to that defined by DR3 is involved in T1D, the affected sib pairs should share the two parental DR3 haplotypes equally frequently. Significant deviation from 50% sharing was observed. Because DR3 haplotypes can for the most part be assumed to be homogenous for their DR-DQ alleles (DRB1*0301 DQA1*0501 DQB1*0201), this test implicated other HLA region genes in T1D on DR3 haplotypes. This result is consistent with more recent studies using matched case-control data (above), the homozygous parent TDT (HPTDT), and the CHM (see below).

How many parents can be expected to be homozygous for an allele (haplotype) at the primary disease-predisposing gene? This is relevant not only to the current method but also to the HPTDT described below. For S trio families, and approximately for MSP families, the expected number of homozygous parents is given by 2 (APi) (ACi), where APi is the observed frequency of the allele or haplotype (say DR3) in patients, and ACi is that in the AFBACs [determined from the results in Thomson (1995c)]. In the T1D affected sib pair families in Noble *et al.* (1996) shown in Table 11.1, the DR3 haplotype had a frequency of 0.319 in patients and 0.094 in controls, and the observed number of homozygous DR3 parents was 21 (0.058), which agrees well with the estimated expected of 0.06. Obviously, the more polymorphic the marker locus, the greater the power of this method (more informative DR3 parents); haplotypes of markers can be used to increase the power. If the sample size is sufficient, specific marker alleles and haplotypes can be studied to narrow down significant effects. The robustness of this approach is tempered by the fact that only a subset of all available data is used. Note that a significant effect with the HPLM does not mean a significant effect will be seen with the HPTDT (see below), and vice versa. As with the matched cases and controls approach, similarly with the HPTDT below, the robustness of this approach to across population analyses may be countered by the fact that only a limited subset of the data is used.

2. Homozygous parent TDT

A modification of the TDT utilizing trio family data [S pedigrees; affected sib pairs can also be used (MSP pedigrees)], with as above a parent homozygous for DR3 (the HPTDT), has also shown heterogeneity of DR3 haplotypes for T1D risk

(Johansson *et al.*, 2003; Lie *et al.*, 1999a,b). Families were examined to determine which haplotypes, defined by DR3 and an msat marker, were transmitted (T) and not transmitted (NT) from the homozygous DR3 parent to the affected child. As above, families are informative only if they are heterozygous at the marker locus.

Issues of sample size are dramatically illustrated in application of the HPTDT to the 13th IHW data on T1D (Pugliese *et al.*, 2007). Of the 307 families with a homozygous DR-DQ parent, 179, 83, and 14, respectively, were homozygous for DR3, DRB1*0401 DQB1*0302, and DRB1*0404 DQB1*0302 (these are the three predisposing haplotypes shown in Table 11.1 and common in Caucasian T1D patients). These numbers are exceedingly impressive and highlight the importance of collaboration and sharing. Nevertheless, to apply the HPTDT, the msat marker (of eight studied) had to be heterozygous in the homozygous DR parent. Further, because there are multiple msat alleles at each locus, the actual numbers in any one msat allele category is often small, even with the impressive sample size of the 13th IHW. (If one used biallelic SNPs, then many of the parents homozygous for the primary disease gene would not be heterozygous for the marker, hence haplotypes of multiple SNPs should be used.) Parents homozygous for the three predisposing haplotypes listed above were analyzed both in individual populations and combined across populations. The 13th IHW HPTDT results showed some significant effects but these were very heterogenous; different markers were significant in different populations, and the marker implicated by Lie *et al.* (1999a,b) was not replicated. However, as mentioned above, there is combined evidence from a number of studies for heterogeneity of DR3 haplotypes and T1D (references are given in Section I.C).

F. Conditional haplotype methods

1. The conditional haplotype method

The logic of the CHM is as follows: if all HLA region genes directly involved in disease susceptibility have been identified, for example, HLA DR-DQ in T1D as the null, then the *relative* frequencies of alleles at polymorphic marker loci on high-risk haplotypes containing, for example, DR3 should be the same in cases and controls; similarly, for other high-risk haplotypes, as well as neutral, and protective haplotypes (in the latter case sample size in patients may be an issue). Denote the primary disease locus by A (or the primary plus secondary loci that have been identified), that is, all putative disease-predisposing loci in the region, and alleles or haplotypes thereof by A_i, $i = 1, 2, \ldots, k_A$, and a linked marker locus by B, with alleles B_k, $k = 1, 2, \ldots, k_B$, then, under the null that the A locus defines all disease predisposition in the region:

$$\frac{f_{\text{pat}}(A_i\text{–}B_k)}{f_{\text{pat}}(A_i\text{–}B_l)} = \frac{f_{\text{con}}(A_i\text{–}B_k)}{f_{\text{con}}(A_i\text{–}B_l)}, \tag{2}$$

where $f_{\text{pat}}(\cdot)$ and $f_{\text{con}}(\cdot)$ represent patient and control frequencies. That is, although the frequencies of the haplotypes $A_i\text{–}B_k$ and $A_i\text{–}B_l$ will differ between patients and controls, the relative frequency of their ratios will be the same in patients and controls, for each A_i.

Inequality of these relative frequencies in patients and controls is expected if the allele or SNP or haplotypes thereof under study *does not include all* genes involved in the disease process and in LD with the marker loci. While fit to these expectations does not exclude the possibility that other genes in the HLA complex are involved in disease, lack of fit unequivocally shows that all disease-predisposing genes in the region have *not* been identified (provided that stratification effects have not produced spurious results). With a CETDT analysis, one similarly tests for heterogeneity of a specific haplotype using the TDT statistic (Koeleman *et al.*, 2000a,b). A hypothetical example showing application of the CHM with heterogeneity between msat marker frequencies on DR3 haplotypes in patients and controls is given in Table 11.4.

The CHM method initially showed heterogeneity of serologically defined DR3 haplotypes (Thomson *et al.*, 1988) that has also been verified with a number of other methods of analysis as described above. The method was later applied, with development of an appropriate statistical test, to amino acid sites in the HLA DR-DQ genes to detect combinations of amino acids involved in disease risk (Valdes and Thomson, 1997; Valdes *et al.*, 1997). As noted above, fit to the model does not imply that all amino acids have been identified, but lack of fit indicates that all genetic variation has not been accounted for.

2. The overall conditional haplotype method

Extending the CHM, Thomson (1984) developed a method to test for additional genetic effects over all haplotypes, henceforth referred to as the overall conditional haplotype method (OCHM). In that application of the test, haplotypes

Table 11.4. Hypothetical Application of the Conditional Haplotype Method (CHM)

DRB1	DQA1	DQB1	msat	Patients	Controls
0301	0501	0201	112	50	20
0301	0501	0201	114	30	50
0301	0501	0201	116	20	30
Total				100	100

χ^2 (2 df) = 19.87, $p < 0.00005$.

needed to be estimated. We show below (Section III.G.3) that in fact this test can be applied without resort to haplotype estimation, however, the test statistic with both these approaches is not straightforward.

One advantage of the CHM, OCHM, and CETDT, and the CGMs described below is that more of the data is used than with the matched genotype approach and the HPTDT and HPLT methods. Results across studies cannot be directly combined, although combining of, for example, *p* values can be carried out (Fisher, 1970). Care must be taken in all analyses and their interpretations with rare haplotypes and sparse cells.

3. Haplotype versus single SNP analyses

The issue of analyzing single SNP versus haplotype data is not simple. As we mentioned in the Introduction (Section I), a complex interplay and balance between these is required. We have demonstrated above with T1D data how the DRB1 and DQB1 genes are both involved in disease risk (and also probably DQA1), and thus haplotypes of these genes are considered (and genotypes thereof) in all analyses. Methods are being developed to efficiently consider single marker versus haplotype associations with disease (Browning, 2006; Browning *et al.*, 2005; Morris, 2006; Purcell *et al.*, 2007). Nonetheless, in addition, we encourage the investigator to individually examine the data because no program can take account automatically of all the nuances with respect to SNP (allele) frequencies, the haplotype configurations of the variants under consideration, and their respective frequencies; whereas the human eye can more rapidly assess possibly informative configurations, as well as those variants that cannot be distinguished from each other. This is illustrated with the PTPN22 and RA data in a study by Carlton *et al.* (2005).

In addition, we show analysis of a hypothetical example (Table 11.5). The numbers chosen are very simplistic, the sample size is small (for ease of comparison of numbers), the number of SNPs considered is small, and no correction is made for multiple testing as this example is merely to illustrate the principle of the interplay required between SNP and haplotype associations. For the first round of analyses, we consider association and RPE analyses, using OR and the P/C ratio, for the five haplotypes (Table 11.5A) and four SNPs (Table 11.5B) considered in this example. Based on the OR and the χ^2 test of heterogeneity (and the P/C ratios are in agreement with these results), haplotype 1 stands out as the strongest haplotype association with disease. In this case, the allele T of SNP 1 is uniquely carried on this haplotype, and may be a causal SNP or in LD with a causal SNP. Note that SNPs 2 and 4 show no individual association with disease; *however*, as mentioned previously, they should be considered in further analyses of the data.

Table 11.5. Haplotype Versus SNP Analyses

A. Haplotype analysis round 1

SNPs	1	2	3	4	Patients	Controls	OR	P/C	χ^2
Haplotypes									
1	T	C	A	A	40	20	2.7	2.0	6.7**
2	A	G	T	G	30	20	1.7	1.5	2.0
3	A	C	A	G	10	20	0.4	0.5	3.3
4	A	C	T	A	10	20	0.4	0.5	3.3
5	A	G	T	A	10	20	0.4	0.5	3.3
	Total				100	100			18.7***

B. SNP analysis round 1

		Patients	Controls	OR	P/C	χ^2
SNP						
1	T	40	20	2.7	2.0	6.7**
	A	60	80	0.4	0.8	2.9
Total		100	100			9.5**
2	C	60	60	1	1	–
	G	40	40	1	1	–
Total		100	100			–
3	A	50	40	1.5	1.3	1.1
	T	50	60	0.7	0.8	0.9
Total		100	100			2.0
4	A	60	60	1	1	–
	G	40	40	1	1	–
Total		100	100			–

C. Haplotype analysis round 2

SNPs	1	2	3	4	Patients	Controls	OR	P/C	χ^2
Haplotypes									
2	A	G	T	G	30	20	3.0	2.0	6.0**
3	A	C	A	G	10	20	0.6	0.7	1.1
4	A	C	T	A	10	20	0.6	0.7	1.1
5	A	G	T	A	10	20	0.6	0.7	1.1
	Total				60	80			9.3*

D. SNP analysis round 2

		Patients	Controls	OR	P/C	χ^2
SNP						
1	T	0	0	–	–	–
	A	60	80	–	–	–
Total		60	80			–
2	C	20	40	0.5	0.7	2.2
	G	40	40	2.0	1.3	1.7
Total		60	80			3.9*
3	A	10	20	0.6	0.7	1.1
	T	50	60	1.7	1.1	0.3
Total		60	80			1.4
4	A	20	40	0.5	0.7	2.2
	G	40	40	2.0	1.3	1.7
Total		60	80			3.9*

$^{*}p < 0.05$, $^{**}p < 0.01$, $^{***}p < 0.001$, $^{****}p < 0.0001$

For the second round of RPE analyses (with haplotype 1 removed) (Table 11.5C), we see that there is further heterogeneity in risk with haplotype 2 contributing the most to the overall χ^2, and also with the highest P/C ratio. (Note that the relative P/C ratios retain the same proportions between Tables 11.5A and 11.5C, that is, P/C (haplotype 2)/P/C (haplotype 3) remains the same.) The analysis in Table 11.5C is also an application of the CHM, under the hypothesis that SNP 1 is the primary disease causing SNP in this region (from Tables 11.5A and 11.5B). Is there any heterogeneity on haplotypes containing the A allele at SNP 1? The answer is yes that there is additional heterogeneity (Table 11.5C), so SNP 1 does not explain all primary disease risk in the region. Note also that now (round 2) SNPs 2 and 4 are significantly associated with disease Table 11.5D.

If we remove haplotype 2 from consideration, there are of course no more significant effects. The haplotype 2 association cannot be ascribed to a specific SNP; a combination of SNPs 2, 3, and 4 is required to explain this association, and this association may reflect LD with a single (or multiple) other SNPs that are causative. Note that in round 1 of the analyses, haplotype 1 and SNP 1 have the same strengths of associations (due to the fact that allele T of SNP 1 occurs uniquely on haplotype 1). In round 2 of the analyses, haplotype 2 has a stronger association, measured by the OR, P/C ratio, and test of heterogeneity, than any of the individual SNPs.

We emphasize again that this simplistic example in Table 11.5 is merely meant as an illustration of how one must do multiple rounds of testing of haplotype versus SNP associations until no further significant effects are detected. One can then interpret the results in terms of whether the associations can be ascribed to a few individual SNPs or haplotypes of some SNPs.

G. Conditional genotype methods

1. The conditional genotype method

Similar to the CHM, one can consider genotype frequencies (the CGM): if all disease-predisposing genes in the region have been identified and represented by the locus A, then under this null hypothesis, the genotype frequencies at a linked marker locus B (not involved in disease nor in LD with additional genes involved in disease) are expected to satisfy the relationship:

$$\frac{f_{\text{pat}}(A_iA_jB_kB_l)}{f_{\text{pat}}(A_iA_jB_mB_n)} = \frac{f_{\text{con}}(A_iA_jB_kB_l)}{f_{\text{con}}(A_iA_jB_mB_n)}. \tag{3}$$

The genotypes at the B locus include all homozygotes and heterozygotes, and similarly all homozygotes and heterozygotes can be considered at the A locus (except for those that are too rare), although each is analyzed individually with the CGM [note again the equivalence with the matched case-control method (Section III.D)].

2. The overall conditional genotype method

If the effect of an additional disease-predisposing gene in the region, for example, the B gene or one in LD with it, is not specific to a particular haplotype or genotype at the primary disease locus A, then more power should be available by considering B locus genotypes combined over all A locus genotypes [the overall conditional genotype method (OCGM)]. In this case, we use the following expectation for each genotype combination in patients (Thomson and Valdes, 2007):

$$\exp f_{\text{pat}}(A_iA_jB_kB_l) = [f_{\text{con}}(A_iA_jB_kB_l)]\left[\frac{f_{\text{pat}}(A_iA_j)}{f_{\text{con}}(A_iA_j)}\right]. \tag{4}$$

For each B locus genotype, B_kB_l, we add over the A locus effects above, and expected patient values are compared to the observed. The question of statistical testing then arises. Application of a standard test of homogeneity of the B genotype observed (obs) and expected (exp) genotype numbers does not give a χ^2 distribution, in fact the distribution is exponential. This is because of use of the ratio of the A_iA_j genotype frequencies in the estimation of expected values. Note also that the use of low frequency control genotype frequencies will be problematic notwithstanding; these genotypes can be left out of all studies. An appropriate test statistic has been developed (Thomson and Valdes, 2007); a resampling approach is also an option.

3. The overall conditional haplotype method

One can also take the observed (obs) and expected (exp) $f_{\text{pat}}(B_kB_l)$ values from above and consider the observed and expected allele frequencies at the B locus under the null, obtained simply by the method of allele (gene) counting. This is in effect the same as the OCHM of Thomson (1984) (Section III.F.2) but obviates the need to estimate LD values in controls between the A and B genes. Note also, however, that it uses ratios of A locus genotype frequencies in patients versus controls rather than ratios of allele frequencies as in Thomson (1984) that may lead to a larger variance for the estimated values. Our results re statistical testing for OCGM data also apply; further work is required to determine the appropriate test statistic, again resampling is a solution.

H. Conditional logistic regression

With CLR modeling, various aspects of the data can be analyzed: modes of inheritance of specific markers or proposed primary disease genes, dose and other heterogeneous effects of associated alleles, maximum likelihood estimates of relative penetrance values normalized to a reference genotype, and the effects of additional marker genes. Simmonds *et al.* (2005) using a stepwise logistic-regression analysis showed that the association of HLA DR-DQ with Graves disease could be explained by either DRB1 or DQA1 but not by DQB1. These data could also be analyzed using conditional haplotype and genotype methods (CHM, OCHM, CGM, and OCGM).

When a logistic regression, including CLR, is used to model the relationship between genetic factors and disease, as the distribution of data across numerous combinations of loci becomes sparse, the parameter estimates become unreasonably biased (for review see Thornton-Wells *et al.*, 2004). In other words, the analysis then suffers from the curse of dimensionality. In analyses considering a combination of loci, one or more of which have low minor allele frequencies, the number of individuals with certain multilocus genotype combinations will be so small (or perhaps equal to zero) that a reasonable estimate for that combination of genotypes cannot be derived. For HLA data on the classical loci this is particularly critical as most multilocus haplotypes are <5%.

In addition, it is quite possible that the effect of a secondary locus is due to its interaction with a major susceptibility locus and not to a "main" effect. For example, a SNP allele or genotype may only have a role on certain predisposing HLA haplotypes but not on all others. In those situations some of the traditional implementations of logistic regression models (that is, forward stepwise regression) which require significant main effects to be modeled before including interaction effects between factors represent a major methodological limitation. In practice loci with relatively small main (non-interactive) effects but more substantial interactive effects would never be even included in the analyses.

I. Combining association and IBD values

1. Background

In the final analysis of identification of all genes in a genetic region involved in disease, all aspects of the data, association and linkage and combinations thereof, must fit within these parameters of disease definition. The analyses proposed in this section, and in Section II above, for the most part look at individual aspects and not the total spectrum of information. Here we extend our investigation to include more aspects of the data. Methods combining linkage and association can greatly increase our power to detect additional significant effects; however, cell sizes can quickly become small.

2. Identification of SNPs responsible for a linkage signal

Sun *et al.* (2002) use linkage data while simultaneously taking account of association data to identify the SNP (or combination of SNPs) in a region that are directly involved in susceptibility to the disease under consideration. Their premise is that if a particular SNP is the only site in the region that influences the disease, then conditioning on the genotypes at that site for affected relatives, there should be no unexplained oversharing in the region among affected individuals. The test of Sun *et al.* can also be extended to include more polymorphic sites, and also multiple sites. Li *et al.* (2005) propose a test of whether a candidate SNP and the disease locus are in LD rather than the SNP being causal, and a second test of whether the candidate SNP can account fully for the linkage signal.

3. The marker-associated segregation χ^2 method

The marker-associated segregation χ^2 (MASC) method of Clerget-Darpoux *et al.* (1988, 1991) fits the most parsimonious model explaining the overall linkage and association observations and tests for fit to the data to test the hypothesis that a primary disease gene has been identified. The MASC method was extended (Dizier *et al.*, 1994) to take account of the role of two unlinked candidate genes in T1D, in this case HLA and IDDM2. Application of the MASC method to serological HLA DR data and T1D rejected this model (Clerget-Darpoux *et al.*, 1988); later analyses favored a complementation model (Clerget-Darpoux *et al.*, 1991; Dizier *et al.*, 1994). With molecular HLA data, fitting DR-DQ as the sole HLA susceptibility locus was strongly rejected (Valdes *et al.*, 2001); addition of HLA DPB1 gives a better fit to the data (this could be due to a locus in LD with DPB1). T1D probands were stratified into two groups: those not carrying the alleles DPB1*0301 and *0202 (which are associated with disease after accounting for the DR-DQ primary association) and those with at least one copy of either of these alleles. Interestingly, both groups have almost identical frequencies of DR-DQ haplotypes but significantly different IBD distributions in the subset of families with probands who do not carry the highly predisposing DR3/DR4 genotype. We stress again here the necessity to understand all aspects of disease heterogeneity at the primary and additional disease loci in a region.

RA is associated with HLA DRB1 alleles, and in particular, the group of alleles associated with disease has in common closely related amino acids in the third hypervariable region of the DR molecule at positions 70–74: the shared epitope (SE) hypothesis (Gregersen *et al.*, 1987). Application of the MASC method to RA HLA SE data shows lack of fit of the linkage and association data (Genin *et al.*, 1998). If the heterogeneity in risk of the SE alleles due to variation at amino acid positions 70 and 71 is taken into account, then both the linkage and association data fit the model (du Montcel *et al.*, 2005). This fit does not exclude the possibility that additional HLA region variation may modulate disease risk; a role of additional variation on HLA DR3 haplotypes has been shown with RA (Jawaheer *et al.*, 2002; Nelson *et al.*, 2007).

For PTPN22 data and RA, Bourgey *et al.* (2007) applied the MASC method and showed that the R620W variant (Begovich *et al.*, 2004) alone could not explain the observed association and linkage data; the data were compatible with three SNPs studied or possibly via the role of two untyped SNPs. This does not exclude the role of additional variants in this region.

With application of all methods described in this section, including the MASC method, it is advantageous in terms of cell sizes to pool subsets of data when appropriate. However, we must always account for all known heterogeneity, and additionally be on the lookout for different aspects of the data that highlight heterogeneity, for example, the IBD distributions described above with DPB1 and T1D (Valdes *et al.*, 2001) and RA and the SE hypothesis (du Montcel *et al.*, 2005).

IV. DISCUSSION

We are at a unique point in the study of the genetics of complex diseases in terms of the extensive genetic marker data (SNPs, msats, and classical loci), often with large sample sizes, that will increasingly be available for study. Important ongoing aspects of genetic epidemiology are investigation of methods to optimize detection of *all* disease genes in a genetic region *and* across the genome, environmental effects on disease risk, as well as gene–gene and gene–environment interactions. There is no "best practice" for the types of analyses needed for these projects. The power of different methods will vary depending on the specific genetic and environmental features of the disease under consideration; for most complex diseases, this involves many unknown factors. In this chapter, we have focused on methods to detect additional genes in a specific genetic region. Additionally, the methods are useful in detecting primary disease-predisposing genes in a region if they are not known. Also, genotype methods such as the CGM and OCGM can be used in across genome studies. The effects of unlinked disease genes may be easier to detect in samples stratified by the genotypes of a primary disease gene, except when the risk effects are strictly multiplicative.

With stratification analyses, we must take account of all SNPs, msats, or other genes that show an effect additional to the primary disease gene(s). Issues of small cell size and a large number of possible comparisons to make are then immediately apparent. This is further compounded when we consider the possibility of age- and gender-related effects. With T1D, as discussed above, HLA region effects additional to the class II DR-DQ genes have been found for the class II DPB1 gene and the class I genes, as well as a number of msats. For the latter, there is considerable heterogeneity in significant effects. Most data sets at the moment do not include typing for all these markers. In the final analysis, we must consider many combinations of these variants to determine which additional effects are due to LD with each other, and which may be the genes directly involved in these secondary effects. As continually stressed above, we need to fully understand RPE effects at all putative disease genes (primary and secondary) as we will often be obliged, due to sample size issues, to combine some alleles or haplotypes under consideration into a "homogenous" class. Note again, however, that we could miss heterogeneity with some analyses, and must constantly be on the alert for this possibility.

With a multistrategy approach, where different methods may give different results, for example, significance or not for a specific mode of inheritance, we should study the data carefully to understand differences in significance from different methods. Also, results from one analysis can guide us in confirming a result by a separate method, and specifically one that may emphasize the feature under consideration. In particular, where significant results may be found with, for example, a CLR analysis, we encourage the researcher to tabulate the data, and where possible analyze it with another method and directly show observed and expected values for the main features relevant to significant effects.

In a similar vein, along with use of programs to distinguish single SNP or gene effects versus haplotype effects of the SNPs or set of genes, we urge that the investigator individually examine the data in this context. It is probably impossible to program all the different ways of analyzing individual SNP versus haplotype effects, and as emphasized above, there is an intricate interplay required of examining single SNP (or allele) versus haplotype effects, and then determining when we can statistically differentiate between these. Haplotype analyses can lead to exclusion of a particular SNP (or gene) as directly involved in disease. On the other hand, there are many cases where no discrimination can be made between the effects of a number of SNPs if they are all very highly correlated and all fall on a haplotype associated with disease, and occur on other nondisease-associated haplotypes too infrequently to allow any disease risk distinction. We emphasize that analyses of this type should *not* be restricted to just those SNPs within a haplotype block; there can be large LD across blocks. In this context, we also again emphasize that cross-ethnic studies can be very useful in that we may see a breakdown of LD patterns.

For complex diseases, the fact that many of the genes involved have relatively small effects drastically complicates efforts to identify genetic regions involved in the disease process, and makes replication of results difficult. Additionally, each complex disease can present its own set of unique problems and statistical issues in the localization of disease genes (primary and secondary) and definition of disease-predisposing factors.

The conjunction of major advances in biocomputing and biotechnology applied to large data sets from international collaborative efforts provides a unique impetus to advance our understanding of complex diseases. New methods for data analysis, some utilizing increased computing power and others devising more powerful approaches for stratification of the data, are yielding success stories. Combining disease and population and evolutionary studies of genome and specific gene evolution is also an incisive component of current research (Harris and Meyer, 2006; Hedrick, 2006; Meyer and Thomson, 2001; Sabeti *et al.*, 2002; Voight *et al.*, 2006). Analyses are now possible that were unimaginable 20 years ago.

Acknowledgments

This work was supported by NIH grant GM35326 and NIH Contract Number: HHSN266200400076C, ADB Number: N01-AI-40076 (GT), and NIH/NIAMS grant R01A5R2300 and NIH/NIAID grant R01AI065841 (LFB).

References

Barcellos, L. F., Sawcer, S., Ramsay, P. P., Baranzini, S. E., Thomson, G., Briggs, F., Cree, B. C., Begovich, A. B., Villoslada, P., Montalban, X., Uccelli, A., Savettieri, G., *et al.* (2006). Heterogeneity at the HLA-DRB1 locus and risk for multiple sclerosis. *Hum. Mol. Genet.* **15**(18), 2813–2824.

Barrett, J. C., Fry, B., Maller, J., and Daly, M. J. (2005). Haploview: Analysis and visualization of LD and haplotype maps. *Bioinformatics* **21**(2), 263–265.

Begovich, A. B., Carlton, V. E., Honigberg, L. A., Schrodi, S. J., Chokkalingam, A. P., Alexander, H. C., Ardlie, K. G., Huang, Q., Smith, A. M., Spoerke, J. M., Conn, M. T., Chang, M., *et al.* (2004). A missense single-nucleotide polymorphism in a gene encoding a protein tyrosine phosphatase (PTPN22) is associated with rheumatoid arthritis. *Am. J. Hum. Genet.* **75** (2), 330–337.

Begovich, A. B., Caillier, S. J., Alexander, H. C., Penko, J. M., Hauser, S. L., Barcellos, L. F., and Oksenberg, J. R. (2005). The R620W polymorphism of the protein tyrosine phosphatase PTPN22 is not associated with multiple sclerosis. *Am. J. Hum. Genet.* **76**(1), 184–187.

Bennett, S. T., Wilson, A. J., Cucca, F., Nerup, J., Pociot, F., McKinney, P. A., Barnett, A. H., Bain, S. C., and Todd, J. A. (1996). IDDM2-VNTR-encoded susceptibility to type 1 diabetes: Dominant protection and parental transmission of alleles of the insulin gene-linked minisatellite locus. *J. Autoimmun.* **9**(3), 415–421.

Biernacka, J. M., Sun, L., and Bull, S. B. (2005a). Simultaneous localization of two linked disease susceptibility genes. *Genet. Epidemiol.* **28**(1), 33–47.

Biernacka, J. M., Sun, L., and Bull, S. B. (2005b). Tests for the presence of two linked disease susceptibility genes. *Genet. Epidemiol.* **29**(4), 389–401.

Blacker, D., Haines, J. L., Rodes, L., Terwedow, H., Go, R. C., Harrell, L. E., Perry, R. T., Bassett, S. S., Chase, G., Meyers, D., *et al.* (1997). ApoE-4 and age at onset of Alzheimer's disease: The NIMH genetics initiative. *Neurology* **48**(1), 139–147.

Bourgey, M., Pedry, H., and Clerget-Darpoux, F. (2007). Modeling the effect of PTPN22 in rheumatoid arthritis. *BMC Proceedings* 2007, **1**(Suppl), S37. http://www.biomedcentral.com/1753-6561/1?issue=S1.

Browning, S. R. (2006). Multilocus association mapping using variable-length Markov chains. *Am. J. Hum. Genet.* **78**(6), 903–913.

Browning, S. R., Briley, J. D., Briley, L. P., Chandra, G., Charnecki, J. H., Ehm, M. G., Johansson, K. A., Jones, B. J., Karter, A. J., Yarnall, D. P., *et al.* (2005). Case-control single-marker and haplotypic association analysis of pedigree data. *Genet. Epidemiol.* **28**(2), 110–122.

Carlton, V. E., Hu, X., Chokkalingam, A. P., Schrodi, S. J., Brandon, R., Alexander, H. C., Chang, M., Catanese, J. J., Leong, D. U., Ardlie, K. G., *et al.* (2005). PTPN22 genetic variation: Evidence for multiple variants associated with rheumatoid arthritis. *Am. J. Hum. Genet.* **77**(4), 567–581.

Clerget-Darpoux, F., Babron, M. C., Prum, B., Lathrop, G. M., Deschamps, I., and Hors, J. (1988). A new method to test genetic models in HLA associated diseases: The MASC method. *Ann. Hum. Genet.* **52**(3), 247–258.

Clerget-Darpoux, F., Babron, M. C., Deschamps, I., and Hors, J. (1991). Complementation and maternal effect in insulin-dependent diabetes. *Am. J. Hum. Genet.* **49**(1), 42–48.

Clerget-Darpoux, F., Babron, M. C., and Bickeboller, H. (1995). Comparing the power of linkage detection by the transmission disequilibrium test and the identity-by-descent test. *Genet. Epidemiol.* **12**(6), 583–588.

Dizier, M. H., Babron, M. C., and Clerget-Darpoux, F. (1994). Interactive effect of two candidate genes in a disease: Extension of the marker-association-segregation chi^2 method. *Am. J. Hum. Genet.* **55**(5), 1042–1049.

du Montcel, S. T., Michou, L., Petit-Teixeira, E., Osorio, J., Lemaire, I., Lasbleiz, S., Pierlot, C., Quillet, P., Bardin, T., Prum, B., *et al.* (2005). New classification of HLA-DRB1 alleles supports the shared epitope hypothesis of rheumatoid arthritis susceptibility. *Arthritis Rheum.* **52**(4), 1063–1068.

Fajardy, I., Vambergue, A., Stuckens, C., Weill, J., Danze, P. M., and Fontaine, P. (2002). CTLA-4 49 A/G dimorphism and type 1 diabetes susceptibility: A French case-control study and segregation analysis. Evidence of a maternal effect. *Eur. J. Immunogenet.* **29**(3), 251–257.

Fingerlin, T. E., Boehnke, M., and Abecasis, G. R. (2004). Increasing the power and efficiency of disease-marker case-control association studies through use of allele-sharing information. *Am. J. Hum. Genet.* **74**(3), 432–443.

Fisher, R. (1970). "Statistical Methods for Research Workers." Oliver and Boyd, Edinburgh.

Fujisawa, T., Ikegami, H., Kawaguchi, Y., Yamato, E., Takekawa, K., Nakagawa, Y., Hamada, Y., Ueda, H., Shima, K., and Ogihara, T. (1995). Class I HLA is associated with age-at-onset of IDDM, while class II HLA confers susceptibility to IDDM. *Diabetologia* **38**(12), 1493–1495.

Genin, E., Babron, M. C., McDermott, M. F., Mulcahy, B., Waldron-Lynch, F., Adams, C., Clegg, D. O., Ward, R. H., Shanahan, F., Molloy, M. G., *et al.* (1998). Modeling the major histocompatibility complex susceptibility to RA using the MASC method. *Genet. Epidemiol.* **15**(4), 419–430.

Greenberg, D. A. (1993). Linkage analysis of "necessary" disease loci versus "susceptibility" loci. *Am. J. Hum. Genet.* **52**(1), 135–143.

Greenberg, D. A., and Doneshka, P. (1996). Partitioned association-linkage test: Distinguishing "necessary" from "susceptibility" loci. *Genet. Epidemiol.* **13**(3), 243–252.

Gregersen, P. K., Silver, J., and Winchester, R. J. (1987). The shared epitope hypothesis: An approach to understanding the molecular genetics of susceptibility to rheumatoid arthritis. *Arthritis Rheum.* **30,** 1205–1213.

Hanifi Moghaddam, P., de Knijf, P., Roep, B. O., Van der Auwera, B., Naipal, A., Gorus, F., Schuit, F., and Giphart, M. J. (1998). Genetic structure of IDDM1: Two separate regions in the major histocompatibility complex contribute to susceptibility or protection. Belgian Diabetes Registry. *Diabetes* **47**(2), 263–269.

Harney, S., Newton, J., Milicic, A., Brown, M. A., and Wordsworth, B. P. (2003). Non-inherited maternal HLA alleles are associated with rheumatoid arthritis. *Rheumatol. (Oxford)* **42**(1), 171–174.

Harris, E. E., and Meyer, D. (2006). The molecular signature of selection underlying human adaptations. *Am. J. Phys. Anthropol. Suppl.* **43,** 89–130.

Hedrick, P. W. (1987). Gametic disequilibrium measures: Proceed with caution. *Genetics* **117**(2), 331–341.

Hedrick, P. W. (2006). Genetic polymorphism in heterogeneous environments: The age of genomics. *Annu. Rev. Ecol. Evol. Syst.* **37,** 67–93.

Hermann, R., Veijola, R., Sipila, I., Knip, M., Akerblom, H. K., Simell, O., and Ilonen, J. (2003). To: Pani MA, Van Autreve J, Van Der Auwera BJ, Gorus FK, Badenhoop K (2002). Non-transmitted maternal HLA DQ2 or DQ8 alleles and risk of type I diabetes in offspring: The importance of foetal or post partum exposure to diabetogenic molecules. *Diabetologia* **45,**1340–1343, *Diabetologia* **46**(4), 588–589; author reply 591–592.

Hirschhorn, J. N., Lohmueller, K., Byrne, E., and Hirschhorn, K. (2002). A comprehensive review of genetic association studies. *Genet. Med.* **4**(2), 45–61.

Hodge, S. E. (1993). Linkage analysis versus association analysis: Distinguishing between two models that explain disease-marker associations. *Am. J. Hum. Genet.* **53**(2), 367–384.

Hsu, L., Li, H., and Houwing-Duistermaat, J. J. (2002). A method for incorporating ages at onset in affected sibpair linkage studies. *Hum. Hered.* **54**(1), 1–12.

Jawaheer, D., Li, W., Graham, R. R., Chen, W., Damle, A., Xiao, X., Monteiro, J., Lee, A., Lundsten, R., Begovich, A., *et al.* (2002). Dissecting the genetic complexity of the association between human leukocyte antigens and rheumatoid arthritis. *Am. J. Hum. Genet.* **71**(3), 585–594.

Johansson, S., Lie, B. A., Todd, J. A., Pociot, F., Nerup, J., Cambon-Thomsen, A., Kockum, I., Akselsen, H. E., Thorsby, E., and Undlien, D. E. (2003). Evidence of at least two type 1 diabetes susceptibility genes in the HLA complex distinct from HLA-DQB1, -DQA1 and -DRB1. *Genes Immun.* **4**(1), 46–53.

Koeleman, B. P., Dudbridge, F., Cordell, H. J., and Todd, J. A. (2000a). Adaptation of the extended transmission/disequilibrium test to distinguish disease associations of multiple loci: The conditional extended transmission/disequilibrium test. *Ann. Hum. Genet.* **64**(3), 207–213.

Koeleman, B. P., Herr, M. H., Zavattari, P., Dudbridge, F., March, R., Campbell, D., Barnett, A. H., Bain, S. C., Mulargia, A. P., Loddo, M., *et al.* (2000b). Conditional ETDT analysis of the human leukocyte antigen region in type 1 diabetes. *Ann. Hum. Genet.* **64**(3), 215–221.

Lambert, A. P., Gillespie, K. M., Bingley, P. J., and Gale, E. A. (2003). To: Pani MA, Van Autreve J, Van der Auwera BJ, Gorus FK, Badenhoop K (2002). Non-transmitted maternal HLA DQ2 or DQ8 alleles and risk of type 1 diabetes in offspring: The importance of foetal or post partum exposure to diabetogenic molecules. *Diabetologia* **45,** 1340–1343. *Diabetologia* **46**(4), 590–591; author reply 591–592.

Lancaster, A., Nelson, M. P., Meyer, D., Thomson, G., and Single, R. M. (2003). PyPop: A software framework for population genomics: Analyzing large-scale multi-locus genotype data. *Pac. Symp. Biocomput.* **8,** 514–525.

Lee, Y. H., Rho, Y. H., Choi, S. J., Ji, J. D., Song, G. G., Nath, S. K., and Harley, J. B. (2006). The PTPN22 C1858T functional polymorphism and autoimmune diseases—a meta-analysis. *Rheumatology* **46**(1), 49–56.

Li, C., Scott, L. J., and Boehnke, M. (2004). Assessing whether an allele can account in part for a linkage signal: The Genotype-IBD Sharing Test (GIST). *Am. J. Hum. Genet.* **74**(3), 418–431.

Li, H. (1999). The additive genetic gamma frailty model for linkage analysis of age-of-onset variation. *Ann. Hum. Genet.* **63**(5), 455–468.

Li, H., and Hsu, L. (2000). Effects of age at onset on the power of the affected sib pair and transmission/disequilibrium tests. *Ann. Hum. Genet.* **64**(3), 239–254.

Li, M., Boehnke, M., and Abecasis, G. R. (2005). Joint modeling of linkage and association: Identifying SNPs responsible for a linkage signal. *Am. J. Hum. Genet.* **76**(6), 934–949.

Liang, K. Y., Chiu, Y. F., and Beaty, T. H. (2001). A robust identity-by-descent procedure using affected sib pairs: Multipoint mapping for complex diseases. *Hum. Hered.* **51**(1–2), 64–78.

Lie, B. A., Sollid, L. M., Ascher, H., Ek, J., Akselsen, H. E., Ronningen, K. S., Thorsby, E., and Undlien, D. E. (1999a). A gene telomeric of the HLA class I region is involved in predisposition to both type 1 diabetes and coeliac disease. *Tissue Antigens* **54**(2), 162–168.

Lie, B. A., Todd, J. A., Pociot, F., Nerup, J., Akselsen, H. E., Joner, G., Dahl-Jorgensen, K., Ronningen, K. S., Thorsby, E., and Undlien, D. E. (1999b). The predisposition to type 1 diabetes linked to the human leukocyte antigen complex includes at least one non-class II gene. *Am. J. Hum. Genet.* **64**(3), 793–800.

Lohmueller, K. E., Pearce, C. L., Pike, M., Lander, E. S., and Hirschhorn, J. N. (2003). Meta-analysis of genetic association studies supports a contribution of common variants to susceptibility to common disease. *Nat. Genet.* **33**(2), 177–182.

Louis, E. J., and Thomson, G. (1986). The three-allele synergistic mixed model for insulin-dependent diabetes mellitus. *Diabetes* **35**(9), 958–963.

Louis, E. J., Thomson, G., and Payami, H. (1983). The affected sib method. II. The intermediate model. *Ann. Hum. Genet.* **47,** 225–243.

Malkki, M., Single, R., Carrington, M., Thomson, G., and Petersdorf, E. (2005). MHC microsatellite diversity and linkage disequilibrium among common HLA-A, HLA-B, DRB1 haplotypes: Implications for unrelated donor hematopoietic transplantation and disease association studies. *Tissue Antigens* **66**(2), 114–124.

Martin, E. R., Monks, S. A., Warren, L. L., and Kaplan, N. L. (2000). A test for linkage and association in general pedigrees: The pedigree disequilibrium test. *Am. J. Hum. Genet.* **67**(1), 146–154.

Martin, E. R., Bass, M. P., Gilbert, J. R., Pericak-Vance, M. A., and Hauser, E. R. (2003). Genotype-based association test for general pedigrees: The genotype-PDT. *Genet. Epidemiol.* **25**(3), 203–213.

Meyer, D., and Thomson, G. (2001). How selection shapes variation of the human major histocompatibility complex: A review. *Ann. Hum. Genet.* **65**(1), 1–26.

Meyer, D., Single, R. M., Mack, S. J., Erlich, H. A., and Thomson, G. (2006). Signatures of demographic history and natural selection in the human major histocompatibility complex loci. *Genetics* **173**(4), 2121–2142.

Meyer, D., Single, R., Mack, S. J., Lancaster, A., Nelson, M. P., Fernandez-Vina, M., Erlich, H. A., and Thomson, G. (2007). Single locus polymorphism of classical HLA genes. *In* "Immunobiology of the Human MHC: Proceedings of the 13th International Histocompatibility Workshop and Congress, Volume I" (J. A. Hansen, ed.), pp. 653–704. IHWG Press, Seattle.

Mignot, E., Lin, L., Rogers, W., Honda, Y., Qiu, X., Lin, X., Okun, M., Hohjoh, H., Miki, T., Hsu, S., *et al.* (2001). Complex HLA-DR and -DQ interactions confer risk of narcolepsy-cataplexy in three ethnic groups. *Am. J. Hum. Genet.* **68**(3), 686–699.

Mignot, E., Lin, L., Li, H., Thomson, G., Lathrop, M., Thorsby, E., Tokunaga, K., Honda, Y., Dauvilliers, Y., Tafti, M., *et al.* (2007). HLA allele and microsatellite studies in narcolepsy. *In* "Immunobiology of the Human MHC: Proceedings of the 13th International Histocompatibility Workshop and Congress, Volume I" (J.A. Hansen, ed.), pp. 817–812. IHWG Press, Seattle.

Morris, A. P. (2006). A flexible Bayesian framework for modeling haplotype association with disease, allowing for dominance effects of the underlying causative variants. *Am. J. Hum. Genet.* **79**(4), 679–694.

Motro, U., and Thomson, G. (1985). The affected sib method. I. Statistical features of the affected sib-pair method. *Genetics* **110**(3), 525–538.

Nakanishi, K., Kobayashi, T., Inoko, H., Tsuji, K., Murase, T., and Kosaka, K. (1995). Residual beta-cell function and HLA-A24 in IDDM. Markers of glycemic control and subsequent development of diabetic retinopathy. *Diabetes* **44**(11), 1334–1339.

Nelson, J. L., Lambert, N. C., Brautbar, C., El-Gabalaway, H., Fraser, P., Gorodezky, C., Li, H., Jonas, B., Konenkov, V., Lathrop, M., *et al.* (2007). The 13th International Histocompatibility Working Group for rheumatoid arthritis joint report. *In* "Immunobiology of the Human MHC: Proceedings of the 13th International Histocompatibility Workshop and Congress, Volume I" (J. A. Hansen, ed.), pp. 797–804. IHWG Press, Seattle.

Nistor, I., Nair, R. P., Stuart, P., Hiremagalore, R., Thompson, R. A., Jenisch, S., Weichenthal, M., Abecasis, G. R., Qin, Z. S., Christophers, E., *et al.* (2005). Protein tyrosine phosphatase gene PTPN22 polymorphism in psoriasis: Lack of evidence for association. *J. Invest. Dermatol.* **125**(2), 395–396.

Noble, J. A., Valdes, A. M., Cook, M., Klitz, W., Thomson, G., and Erlich, H. A. (1996). The role of HLA class II genes in insulin-dependent diabetes mellitus: Molecular analysis of 180 Caucasian, multiplex families. *Am. J. Hum. Genet.* **59**(5), 1134–1148.

Noble, J. A., Valdes, A. M., Thomson, G., and Erlich, H. A. (2000). The HLA class II locus DPB1 can influence susceptibility to type 1 diabetes. *Diabetes* **49**(1), 121–125.

Noble, J. A., Valdes, A. M., Bugawan, T. L., Apple, R. J., Thomson, G., and Erlich, H. A. (2002). The HLA class I A locus affects susceptibility to type 1 diabetes. *Hum. Immunol.* **63**(8), 657–664.

Oksenberg, J. R., Barcellos, L. F., Cree, B. A. C., Baranzini, S. E., Bugawan, T. L., Khan, O., Lincoln, R. R., Swerdlin, A., Mignot, E., Lin, L., *et al.* (2004). Mapping multiple sclerosis susceptibility to the HLA-DR locus in African Americans. *Am. J. Hum. Genet.* **74**(1), 160–167.

Pani, M. A., Van Autreve, J., Van der Auwera, B. J., Gorus, F. K., and Badenhoop, K. (2002). Non-transmitted maternal HLA DQ2 or DQ8 alleles and risk of type I diabetes in offspring: The importance of foetal or post partum exposure to diabetogenic molecules. *Diabetologia* **45**(9), 1340–1343.

Payami, H., Thomson, G., Motro, U., Louis, E. J., and Hudes, E. (1985). The affected sib method. IV. Sib trios. *Ann. Hum. Genet.* **49**(4), 303–314.

Payami, H., Joe, S., Farid, N. R., Stenszky, V., Chan, S. H., Yeo, P. P., Cheah, J. S., and Thomson, G. (1989). Relative predispositional effects (RPEs) of marker alleles with disease: HLA-DR alleles and Graves disease. *Am. J. Hum. Genet.* **45**(4), 541–546.

Payami, H., Zhu, M., Montimurro, J., Keefe, R., McCulloch, C. C., and Moses, L. (2005). One step closer to fixing association studies: Evidence for age- and gender-specific allele frequency variations and deviations from Hardy-Weinberg expectations in controls. *Hum. Genet.* **118**(3–4), 322–330.

Pritchard, J. K. (2001). Are rare variants responsible for susceptibility to complex diseases? *Am. J. Hum. Genet.* **69**(1), 124–137.

Pugliese, A., Dorman, J. S., Steenkiste, A., Li, H., Thorsby, E., Lathrop, M., Schoch, G., Thomson, G., Caillat-Zucman, S., Awdeh, Z., *et al.* (2007). Joint Report of the 13th IHWS type 1 diabetes (T1D) component. *In* "Immunobiology of the Human MHC: Proceedings of the 13th International Histocompatibility Workshop and Congress, Volume I" (J. A. Hansen, ed.), pp. 788–796. IHWG Press, Seattle.

Purcell, S., Daly, M. J., and Sham, P. C. (2007). WHAP: Haplotype-based association analysis. *Bioinformatics* **23**(2), 255–256.

Risch, N., and Merikangas, K. (1996). The future of genetic studies of complex human diseases. *Science* **273,** 1516–1517.

Robinson, W. P., Barbosa, J., Rich, S. S., and Thomson, G. (1993). Homozygous parent affected sib pair method for detecting disease predisposing variants: Application to insulin dependent diabetes mellitus. *Genet. Epidemiol.* **10**(5), 273–288.

Sabeti, P. C., Reich, D. E., Higgins, J. M., Levine, H. Z., Richter, D. J., Schaffner, S. F., Gabriel, S. B., Platko, J. V., Patterson, N. J., McDonald, G. J., *et al.* (2002). Detecting recent positive selection in the human genome from haplotype structure. *Nature* **419**(6909), 832–837.

Sasaki, T., Nemoto, M., Yamasaki, K., and Tajima, N. (1999). Preferential transmission of maternal allele with DQA1*0301-DQB1*0302 haplotype to affected offspring in families with type 1 diabetes. *J. Hum. Genet.* **44**(5), 318–322.

Sekigawa, I., Okada, M., Ogasawara, H., Kaneko, H., Hishikawa, T., and Hashimoto, H. (2003). DNA methylation in systemic lupus erythematosus. *Lupus* **12**(2), 79–85.

Silman, A. J., Hay, E. M., Worthington, J., Thomson, W., Pepper, L., Davidson, J., Dyer, P. A., and Ollier, W. E. (1995). Lack of influence of non-inherited maternal HLA-DR alleles on susceptibility to rheumatoid arthritis. *Ann. Rheum. Dis.* **54**(4), 311–313.

Siminovitch, K. A. (2004). PTPN22 and autoimmune disease. *Nat. Genet.* **36**(12), 1248–1249.

Simmonds, M. J., Howson, J. M., Heward, J. M., Cordell, H. J., Foxall, H., Carr-Smith, J., Gibson, S. M., Walker, N., Tomer, Y., Franklyn, J. A., *et al.* (2005). Regression mapping of association between the human leukocyte antigen region and Graves disease. *Am. J. Hum. Genet.* **76**(1), 157–163.

Single, R., Meyer, D., Mack, S., Lancaster, A., Nelson, M. P., Fernandez-Vina, M., Erlich, H. A., and Thomson, G. (2007). Haplotype frequencies and linkage disequilibrium. *In* "Immunobiology of the Human MHC: Proceedings of the 13th International Histocompatibility Workshop and Congress, Volume I" (J.A. Hansen, ed.), pp. 705–746. IHWG Press, Seattle.

Sladek, R., Rocheleau, G., Rung, J., Dina, C., Shen, L., Serre, D., Boutin, P., Vincent, D., Belisle, A., Hadjadj, S., *et al.* (2007). A genome-wide association study identifies novel risk loci for type 2 diabetes. *Nature* **445**(7130), 881–885.

Spielman, R. S., McGinnis, R. E., and Ewens, W. J. (1993). Transmission test for linkage disequilibrium: The insulin gene region and insulin-dependent diabetes mellitus (IDDM). *Am. J. Hum. Genet.* **52,** 506–516.

Steenkiste, A., Valdes, A. M., Feolo, M., Hoffman, D., Concannon, P., Noble, J., Schoch, G., Hansen, J., Helmberg, W., Dorman, J. S., *et al.* (2007). The HLA component of type 1 diabetes: Report on the activities of the 13th International Histocompatibility Group and 14th International Histocompatibility Workshop. *Tissue Antigens* **69,** 214–225.

Sun, L., Cox, N. J., and McPeek, M. S. (2002). A statistical method for identification of polymorphisms that explain a linkage result. *Am. J. Hum. Genet.* **70**(2), 399–411.

Szatmari, P., Paterson, A. D., Zwaigenbaum, L., Roberts, W., Brian, J., Liu, X. Q., Vincent, J. B., Skaug, J. L., Thompson, A. P., Senman, L., *et al.* (2007). Mapping autism risk loci using genetic linkage and chromosomal rearrangements. *Nat. Genet.* **39**(3), 319–328.

Tait, B. D., Harrison, L. C., Drummond, B. P., Stewart, V., Varney, M. D., and Honeyman, M. C. (1995). HLA antigens and age at diagnosis of insulin-dependent diabetes mellitus. *Hum. Immunol.* **42**(2), 116–122.

ten Wolde, S., Breedveld, F. C., de Vries, R. R., D'Amaro, J., Rubenstein, P., Schreuder, G. M., Claas, F. H., and van Rood, J. J. (1993). Influence of non-inherited maternal HLA antigens on occurrence of rheumatoid arthritis. *Lancet* **341**(8839), 200–202.

Thomson, G. (1983). Investigation of the mode of inheritance of the HLA associated diseases by the method of antigen genotype frequencies among diseased individuals. *Tissue Antigens* **21**(2), 81–104.

Thomson, G. (1984). HLA DR antigens and susceptibility to insulin-dependent diabetes mellitus. *Am. J. Hum. Genet.* **36**(6), 1309–1317.

Thomson, G. (1993). AGFAP method: Applicability under different ascertainment schemes and a parental contributions test. *Genet. Epidemiol.* **10**(5), 289–310.

Thomson, G. (1995a). Analysis of complex human genetic traits: An ordered-notation method and new tests for mode of inheritance. *Am. J. Hum. Genet.* **57**(2), 474–486.

Thomson, G. (1995b). HLA disease associations: Models for the study of complex human genetic disorders. *Crit. Rev. Clin. Lab. Sci.* **32**(2), 183–219.

Thomson, G. (1995c). Mapping disease genes: Family-based association studies. *Am. J. Hum. Genet.* **57**(2), 487–498.

Thomson, G., and Valdes, A. M. (2005). Mapping of disease loci. *In* "Pharmacogenomics" (W. Kalow, U. A. Meyer, and R. F. Tyndale, eds.), pp. 557–587. Taylor and Francis Group, Boca Raton.

Thomson, G., and Valdes, A. M. (2007). Conditional genotype analysis: Detecting secondary disease loci in linkage disequilibrium with a primary disease locus. *BMC Proceedings* 2007, **1**(Suppl 1), S163. http://www.biomedcentral.com/1753-6561/1?issue=S1.

Thomson, G., Robinson, W. P., Kuhner, M. K., Joe, S., MacDonald, M. J., Gottschall, J. L., Barbosa, J., Rich, S. S., Bertrams, J., Baur, M. P., Partanen, J., Tait, B., *et al.* (1988). Genetic heterogeneity, modes of inheritance, and risk estimates for a joint study of Caucasians with insulin-dependent diabetes mellitus. *Am. J. Hum. Genet.* **43**(6), 799–816.

Thomson, G., Valdes, A. M., Noble, J. A., Grote, M. N., Najman, J., Erlich, H. A., Cucca, F., Pugliese, A., Steenkiste, A., Dorman, J. S., *et al.* (2007). Relative predispositional effects of HLA Class II DRB1-DQB1 haplotypes and genotypes on type 1 diabetes. *Tissue Antigens*, in press.

Thorsby, E. (1997). Invited anniversary review: HLA associated diseases. *Hum. Immunol.* **53**(1), 1–11.

Thornton-Wells, T. A., Moore, J. H., and Haines, J. L. (2004). Genetics, statistics and human disease: Analytical retooling for complexity. *Trend. Genet.* **20**(12), 640–647.

Valdes, A. M., and Thomson, G. (1997). Detecting disease-predisposing variants: The haplotype method. *Am. J. Hum. Genet.* **60**(3), 703–716.

Valdes, A. M., McWeeney, S., and Thomson, G. (1997). HLA class II DR-DQ amino acids and insulin-dependent diabetes mellitus: Application of the haplotype method. *Am. J. Hum. Genet.* **60**(3), 717–728.

Valdes, A. M., Thomson, G., Erlich, H. A., and Noble, J. A. (1999). Association between type 1 diabetes age of onset and HLA among sibling pairs. *Diabetes* **48**(8), 1658–1661.

Valdes, A. M., Noble, J. A., Genin, E., Clerget-Darpoux, F., Erlich, H. A., and Thomson, G. (2001). Modeling of HLA class II susceptibility to type I diabetes reveals an effect associated with DPB1. *Genet. Epidemiol.* **21**(3), 212–223.

Valdes, A. M., Erlich, H. A., and Noble, J. A. (2005a). Human leukocyte antigen class I B and C loci contribute to type 1 diabetes (T1D) susceptibility and age at T1D onset. *Hum. Immunol.* **66**(3), 301–313.

Valdes, A. M., Thomson, G., Graham, J., Zarghami, M., McNeney, B., Kockum, I., Smith, A., Lathrop, M., Steenkiste, A. R., Dorman, J. S., *et al.* (2005b). D6S265*15 marks a DRB1*15, DQB1*0602 haplotype associated with attenuated protection from type 1 diabetes mellitus. *Diabetologia* **48**(12), 2540–2543.

Valdes, A. M., Wapelhorst, B., Concannon, P., Erlich, H. A., Thomson, G., and Noble, J. A. (2005c). Extended DR3-D6S273-HLA-B haplotypes are associated with increased susceptibility to type 1 diabetes in US Caucasians. *Tissue Antigens* **65**(1), 115–119.

van der Horst-Bruinsma, I. E., Hazes, J. M., Schreuder, G. M., Radstake, T. R., Barrera, P., van de Putte, L. B., Mustamu, D., van Schaardenburg, D., Breedveld, F. C., and de Vries, R. (1998). Influence of non-inherited maternal HLA-DR antigens on susceptibility to rheumatoid arthritis. *Ann. Rheum. Dis.* **57**(11), 672–675.

Voight, B. F., Kudaravalli, S., Wen, X., and Pritchard, J. K. (2006). A map of recent positive selection in the human genome. *PLoS Biol* **4**(3), e72.

Yeo, T. W., De Jager, P. L., Gregory, S. G., Barcellos, L. F., Walton, A., Goris, A., Fenoglio, C., Ban, M., Taylor, C. J., Goodman, R. S., *et al.* (2007). A second major histocompatibility complex susceptibility locus for multiple sclerosis. *Ann. Neurol.* **61**(3), 228–236.

Zareparsi, S., James, D. M., Kaye, J. A., Bird, T. D., Schellenberg, G. D., and Payami, H. (2002). HLA-A2 homozygosity but not heterozygosity is associated with Alzheimer disease. *Neurology* **58**(6), 973–975.

Zavattari, P., Lampis, R., Motzo, C., Loddo, M., Mulargia, A., Whalen, M., Maioli, M., Angius, E., Todd, J. A., and Cucca, F. (2001). Conditional linkage disequilibrium analysis of a complex disease superlocus, IDDM1 in the HLA region, reveals the presence of independent modifying gene effects influencing the type 1 diabetes risk encoded by the major HLA-DQB1, -DRB1 disease loci. *Hum. Mol. Genet.* **10**(8), 881–889.

12

Methods for Handling Multiple Testing

Treva K. Rice,* Nicholas J. Schork,† and D. C. Rao*,‡

*Division of Biostatistics and Department of Psychiatry, Washington University School of Medicine, St. Louis, Missouri 63110

†Scripps Genomic Medicine and Scripps Research Institute, La Jolla, California 92069

‡Departments of Genetics and Mathematics, Washington University School of Medicine, St. Louis, Missouri 63110

ABSTRACT

The availability of high-throughput genotyping technologies and massive amounts of marker data for testing genetic associations is a dual-edged sword. On one side is the possibility that the causative gene (or a closely linked one) will be found from among those tested for association, but on the other testing,

Advances in Genetics, Vol. 60

0065-2660/08 $35.00
DOI: 10.1016/S0065-2660(07)00412-9

many loci for association creates potential false positive results and the need to accommodate the multiple testing problem. Traditional solutions involve correcting each test using adjustments such as a Bonferroni correction. This worked well in settings involving a few tests (e.g., 10–20, as is typical for candidate gene studies) and even when the number of tests was somewhat larger (e.g., a few hundred as in genome-wide microsatellite scans). However, the current dense single nucleotide polymorphism (SNP) and/or whole-genome association (WGA) studies often consider several thousand to upwards of 500,000 and 1 million SNPs. In these settings, a Bonferroni correction is not practical as it does not take into account correlations between the tests due to linkage disequilibrium and hence can be too conservative. The effect sizes of susceptibility alleles will rarely (if ever) reach the required level of significance in WGA studies if a Bonferroni correction is used, and the number of false negatives is likely to be large. Thus, one of the burning methodological issues in contemporary genetic epidemiology and statistical genetics is how to balance false positives and false negatives in large-scale association studies. This chapter reviews developments in this area from both historical and current perspectives.

I. INTRODUCTION

The problem of multiple testing in genetics studies arises when many different loci are tested for association with a particular trait or disease, as is common in genome-wide linkage and whole-genome association (WGA) scans as well as in gene expression microarray studies. The most common method for dealing with multiple testing problems involves adjusting the significance level (α) of each test to accommodate the total number of tests performed, although the need for this type of adjustment has been questioned (e.g., Rothman, 1990). Historically, the Bonferroni correction has been used to adjust significance levels (Bonferroni, 1935, 1936). The Bonferroni correction provides a simple formula for computing the required pointwise α-levels (for each test made) based on a global or experiment-wise error rate or α-level (say 0.05). This method works well when there are only a few independent tests being performed. For example, in an experiment involving 10–20 independent tests, one would expect not more than one result to be significant at the $p < 0.05$ level due to chance alone. In fact, the Bonferroni correction formed the basis for the often-cited Lander and Kruglyak (1995) lod score thresholds for mapping complex trait loci in linkage mapping contexts of 1.9 ($\alpha = 1.7 \times 10^{-3}$) for suggestive linkage and 3.3 ($\alpha = 4.9 \times 10^{-5}$) for significant linkage. These Bonferroni-based thresholds for declaring significance are based on the assumption of a dense map containing an infinite number of markers with infinitely small intermarker distances. For more realistic genome-wide linkage scans involving about 300 markers, one would expect only about 15 false positives at a pointwise α-level of 0.05, or a Bonferroni corrected α-level of 1.7×10^{-4} corresponding to a lod score of about 2.75.

Even with the use of the more realistic α-levels in linkage mapping studies, significant results involving complex, multifactorial traits have been few and far between (Altmuller *et al.*, 2001). The problems inherent to multiple testing have been compounded with the explosion of available DNA markers and high-throughput genotyping technologies for analysis, as researchers begin to undertake WGA studies that involve hundreds of thousands of tests, for example prostate cancer (Gudmundsson *et al.*, 2007) and type 2 diabetes (Sladek *et al.*, 2007). Multiple testing problems that are this pronounced have been referred to in the data mining community as the "short-fat data" or "large p, small n" problem, and currently there is no agreed upon resolution. For example, if the Bonferroni was applied to a single nucleotide polymorphism (SNP)-based WGA scan involving a half-million tests, one would expect to see a whopping 25,000 false positives if a simple $p < 0.05$ criteria for declaring significance for each SNP was applied; alternatively, one would have to adopt a pointwise α-level as small as 1.0×10^{-7} if a simple Bonferroni correction was applied to each SNP for declaring significance in order to preserve an experiment-wise error rate of 5%. Because it is unlikely that any single SNP would have an effect large enough to produce a p value that small in studies with realistic sample sizes for a complex trait, investigators are either doomed to failure at the outset or must be willing balance the risks associated with including some false positives (i.e., type I errors) in a set that might be worth pursuing in further studies against the chance of missing important loci as false negatives (i.e., type II errors).

One very important consideration in genetics mapping studies is the fact that the tests are not likely to be independent that would run counter to assumptions underlying the Bonferroni method [see, e.g., Efron (2007) for discussion]. Certain mechanisms such as linkage disequilibrium (LD) and the occurrence of multiple neighboring genes implicated in given metabolic pathways give rise to correlations among subsets of SNPs. Consequently, concurrent with the revolution in the development of molecular genetic technologies, there has been a rekindling of active study into methods of evaluating statistical significance that not only address the massive number of multiple tests that might be pursued, but also consider other mechanisms that are unique to genetic analysis settings, such as correlation structures among tests at neighboring loci due to LD and prior information about the phenotypic influence of those loci obtained from other sources such as linkage evidence and allele frequencies. In general, newer methods consider these confounding factors while attempting to strike a balance between both type I and type II errors. In later sections of this chapter, we will review aspects of the various practices that have been described for handling problems arising from multiple testing in genomics studies.

In summary, the problem inherent to many large-scale genetic linkage and association studies is that by performing multiple tests researchers increase the likelihood of obtaining false positive results (Storey and Tibshirani, 2003). Although the classical Bonferroni adjustments were developed to deal with this problem by requiring the signal to be significant at a global level across all tests, this global α-level becomes increasingly conservative when greater numbers of tests are performed because it assumes, essentially, that the tests are independent. However, practically speaking it is not likely that the effects of inherited variations for many complex traits will be overly large as there are likely multiple genes and gene interactions that influence such traits. Consequently, most researchers are willing to risk having higher false positive rates in their studies than against the risk of finding no associations or linkages at all. Ultimately, then, the problem of multiple testing in gene-finding experiments is to choose appropriate methods that will simultaneously control for both false positives and false negatives.

II. TYPES OF ERRORS IN HYPOTHESIS TESTING (TYPE I AND TYPE II)

In statistics, hypothesis testing typically involves testing a null hypothesis against an alternative hypothesis. Here we will assume that the null hypothesis is that a given gene or genetic variation has no effect on a trait of interest (i.e., it is not linked or associated with the trait), and the alternative is that there is a linkage or association. Because most contemporary genetic epidemiologic research focuses on association studies rather than linkage studies, we will consider association studies as our main examples. Unless a statistical test suggests otherwise, the null hypothesis is assumed to be true. The null hypothesis is typically articulated in a way that is opposite to what the investigator believes to be true, so that the object of conducting statistical tests for association is to see if the null hypothesis (in this case no gene-trait association) can be refuted.

The significance level for judging whether a hypothesis is true is the maximum probability that a given statistic would be observed under the null, or alternatively, the significance level is the probability that the null would be rejected when it is actually true (a type I error). The null is rejected if the p value is lower than a specified significance level (α) such as 5% or 1%. In other words, if the probability that the event specified in the null hypothesis would have occurred by chance is one in a hundred, then the significance level is 1%. A false positive (i.e., type I error or α) occurs when a test statistic suggests that the null hypothesis should be rejected even though it is true (see Table 12.1).

Table 12.1. Contingency Table Showing Type I and Type II Errors

	Alternative is true	Null is true
Reject null	True positive	False positive (type I error)
Do not reject null	False negative (type II error)	True negative

The converse is a false negative (type II error, $1 - \beta$) that occurs when a test statistic suggests that the null hypothesis should not be rejected (i.e., tentatively accepting the null) even though its alternative is true. The power of a test (β) is given by the probability of rejecting the null hypothesis in the face of the alternative being true. As emphasized, it is desirable to strike a balance between both types of errors.

The standard for judging significance at 0.05 was first suggested by Fisher (1925), who interpreted p values as providing evidence against a hypothesis. In examples involving the standard normal distribution, Fisher argued that one should reject a null hypothesis when there is only a 1 in 20 chance (5%) that it is true. To further avoid type I errors (i.e., incorrectly assuming a false positive), one can exploit criteria that are much more conservative or involve a smaller chance that the null hypothesis is true, such as 1 in 100 (1%) or 1 in 1000 (0.1%). However, when there is a decrease in the probability of making false positive inferences, there is a corresponding increase in the probability that there will be a failure to detect a real effect [type II error or $(1 - \beta)$].

Power, the complement of the type II error rate, is important in this context. For a given test, we want the type II error rate to be as low as possible (we do not want to miss a real effect or "signal") and power to be as high as possible. Power depends on the size of the effect one is investigating and the sensitivity of the data to detecting this effect. Effect size is often an index of the strength of the relationship between two variables, and for our purposes, this is taken to be the association between the marker locus and a trait or disease of interest. Sensitivity depends on the reliability of the measures and on the sample size. In general, we speak of needing at least N subjects for at least 80% power to detect an effect of a certain size at a given α-level. Under equivalent values of each of these factors, the power is greater at an α-level of 0.05 than 0.001. Power is generally calculated during the design phase of a study to ensure that an adequate sample size will be collected to test an effect or detect some minimum effect size. However, to calculate power, a researcher must often make arbitrary assumptions about effect sizes. In addition, power is often calculated assuming that single statistical tests will be pursued in order to test a single hypothesis. Thus, power calculations are often problematic when considering studies that will test many hypotheses or associations.

III. STRIKING A BALANCE BETWEEN FALSE POSITIVES AND FALSE NEGATIVES

The use of very conservative or stringent significance levels (α) to test hypotheses lead to a loss of power and an increase in the rate of false negatives. However, the use of significance levels that are too liberal lead to unacceptably high rates of false positives. Todorov and Rao (1997) demonstrated the relationship between these two errors in one linkage analysis scenario by plotting both false positives (F+) and false negatives (F−) against the pointwise significance level. The example, typical for many complex traits, uses 100 sibling pairs with the trait having a major gene allele frequency of 0.2, a residual sibling correlation of 0.2, and a total heritability of 25%. As shown in Fig. 12.1, the false negative rate decreases and the false positive rate increases with an increasing significance level. At an α-level of 0.05 (dashed vertical line at far right of figure), there are 19 false positives with almost no false negatives (1 − power). At an α-level of 0.000022, the Lander and Kruglyak (1995) criterion for genome-wide linkage scans suggest that there are only 0.05 false positives, on average,

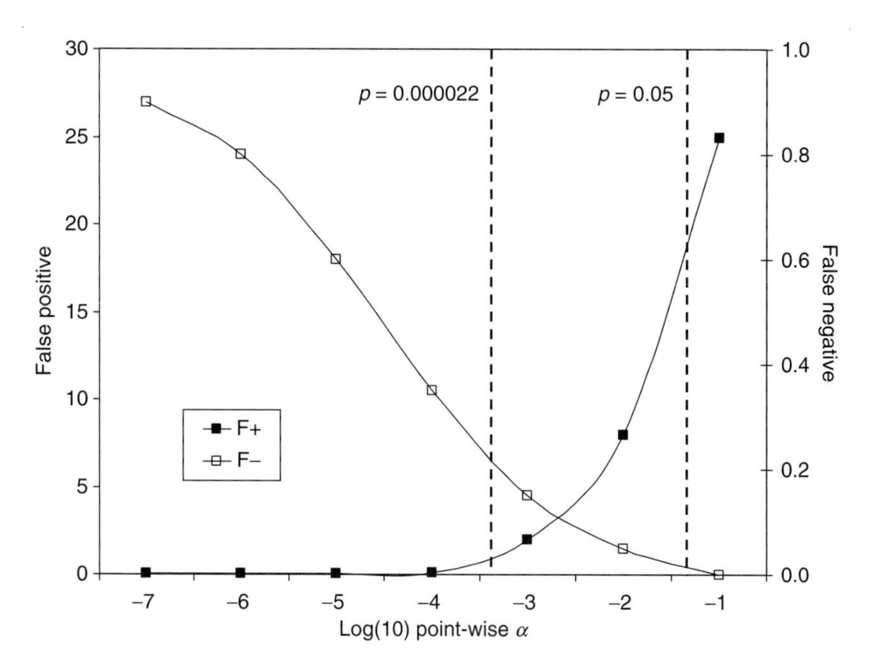

Figure 12.1. Trade-off between false positives (F+) and false negatives (F−). Expected number of F− per genomic scan is plotted on left, and F− rate (1 − power) on right, both against (the logarithm of) varying nominal significance levels. Adapted from Rao (1998).

but the false negative rate is increased. This amply demonstrates that there must be some trade-off between false negatives (to avoid missing potential signals) and false positives (to avoid chasing wrong signals).

IV. ALTERNATIVE ADJUSTMENT METHODS

Various methods attempt to control for these multiple testing issues and have been discussed in several recent articles (e.g., Balding, 2006; Carlson *et al.*, 2004; Cheverud, 2001; Elston and Spence, 2006; Guo and Elston, 2001; Lander and Kruglyak, 1995; Morton, 1998; Pounds, 2006; Province, 2001; Rao, 1998; Rao and Gu, 2001; Strug and Hodge, 2006; Thomson, 2001). At least two general types of multiple comparison procedures are used, one controlling family-wise error rates (FWERs) and the other controlling for false discovery rates (FDRs, Benjamini and Hochberg, 1995; Benjamini and Yekutieli, 2001). The family-wise approaches are based on the probability that one or more of the rejected hypotheses is true, while the FDRs control the expected number of false rejections from among the rejected hypotheses. The FWER methods tend to be more conservative and include the classical Bonferroni, sequential sampling methods, and resampling methods. Each investigator must determine which method is best for their data. Preferred methods will protect against problems associated with multiple testing by defining the most appropriate p value thresholds that have enough power to detect loci while simultaneously controlling for the number of false positives.

A. The Bonferroni correction

The Bonferroni method is a frequentist method that controls for the overall type I error rate in a study (e.g., $\alpha = 0.05$), such that the total of all tests conducted will together produce no more than a probability of α for false positive results. If n independent tests are performed, then the Bonferroni-corrected α for each individual test is $\alpha' = \alpha/n$. Thus, to attain an overall significance level of 0.05 for a small candidate gene study of, say, 10 SNPs, each individual test must reach an α-level of $(0.05/10) = 0.005$, which is not unreasonable for most complex traits in the sense that realistic effect sizes will reach this level of significance with an appropriate statistical test in settings involving realistic sample sizes. However, a Bonferroni correction for a WGA study involving, say, 500,000 SNPs corresponds to a p value of 1×10^{-7}. While these α-levels certainly will provide a safeguard against type I errors, they also will lead to an unacceptably large false negative (type II) error rate, particularly for complex traits where the locus effects are expected to be moderate at best. The major advantage of the Bonferroni correction is that is easy and straightforward to calculate. However, the major disadvantage is that it is overly conservative, particularly in WGA studies where individual tests are not independent. Consequently, less conservative methods have been devised that attempt to balance between the type I and type II errors.

B. Permutation testing

Several simulation-based resampling techniques attempt to control for multiple tests by comparing observed *p* values with *p* values calculated from simple repeated perturbations of the data. For example, the jackknife method constructs an empirical distribution of a relevant test statistic computed with the actual data though the use of subsets of the data, whereas the bootstrap method draws randomly with replacement from a set of data points to estimate the empirical distribution of a test statistic, and permutation tests estimate the distribution of a test statistic by exchanging data points among the units of observation. In genetic applications, the *permutation test* is the most widely used resampling method and grew out of the works of Fisher and Pitman in the 1930s (Fisher, 1935; Pitman, 1937a,b, 1938). In general, reference distributions for relevant test statistics calculated from the actual observed data are derived from the calculation of all possible values of some test statistic by rearranging the original data points. It is based on the assumption that if all tests are equivalent, then the test statistic should be the same even if the data points are jumbled. More specifically, this empirical method evaluates how often a given *p* value would arise by chance if the study were repeated without any true linkages or associations. Permutation testing is available within the confines of a number of genetic software packages for genome-wide linkage (e.g., SOLAR, Almasy and Blangero, 1998; Iturria *et al.*, 1999) and whole genome SNP association analysis with family-based samples (e.g., QTDT, Abecasis *et al.*, 2000). When there are too many possible orderings of the data to realistically perform permutation testing, Monte Carlo (MC) testing is asymptotically equivalent. MC methods also resample but only use a (random) subset of all possible permutations of the data to compute the reference distribution of a test statistic. As such, the accuracy of the MC-based reference distribution of a test statistic will be reflected in the number of permutations used. In general, as few as 400 rearrangements can generate accurate answers at $\alpha = 0.05$. The properties of many simulation and permutation-based tests have been described in the literature (Davison and Hinkley, 1997; Edgington, 1995; Efron, 1982; Good, 2006; Moore *et al.*, 2003; Simon, 1997).

As an example of a permutation-based test for association, a method for assessing significance using a classical TDT test has been proposed that involves randomizing the transmitted and untransmitted status of alleles from parents to offspring (e.g., Lin *et al.*, 2004). However, this exhaustive allelic TDT method is also computer intensive. The major disadvantages of resampling methods such as permutation testing are that they can be computationally intensive and may require complex programming skills and statistical knowledge. However, the availability of relatively inexpensive and fast computers and built-in permutation utilities for several available genetics programs obviates these problems to some degree. The advantages of permutation testing are that these methods are robust and useful in achieving good type I error rates and power, and they are appropriate for any test statistic regardless of whether the underlying distribution is known. Because the threshold for significance is derived directly from the data, the null distribution will be consistent with the characteristics of the original data in terms of normality, allele frequencies, and missing data patterns.

C. False discovery rate

The *FDR* (Storey and Tibshirani, 2003) is less computationally intensive than the permutation methods, and also less conservative than the Bonferroni methods. The FDR is expected proportion of positive findings that are false positives (Benjamini and Hochberg, 1995). When a well-defined statistical test is performed multiple times (testing a null against an alternative hypothesis), the FDR is the expected proportion of false rejections within the class of rejected null hypotheses. FDR can be estimated by Proc MULTTEST in SAS or the program QVALUE (freeware from http://genomine.org/qvalue/; Storey and Tibshirani, 2003), which has more flexible options that allow for exploring the behavior of the FDR in real data analysis. Various extensions to the FDR method to better accommodate specific genetic problems arise regularly in the literature. For example, one FDR approach specific to family-based association analysis incorporates Bayesian concepts by weighting the association hypothesis tests on the basis of prior information, in this case prior genome-wide linkage results (Roeder *et al.*, 2006). The association power is enhanced by upweighting the p values in relevant linkage regions and simultaneously downweighting p values in other regions. Sun *et al.* (2006) described a complementary stratified FDR approach. The variable FDRs may be applied to stratified subsamples, where the stratification may be based on allele frequencies or other genome-specific attributes such as physical location. Other extensions of the FDR procedure allow for modeling arbitrary dependencies among the test statistics (Benjamini and Yekutieli, 2001; Efron 2007 for discussion). Note that the FDR and related strategies have connections to classical graphical methods for assessing false positive rates among a large number of tests (e.g., Schweder and Spjøtvoll, 1982).

A nondiscovery rate (NDR) also has been proposed that is the proportion of false negatives from true alternatives (Craiu and Sun, 2007). This type II error rate for multiple hypothesis estimates the power as 1 − NDR), or the proportion of true positives among true alternatives. The authors propose jointly analyzing the NDR and the FDR to control both types of error.

A potentially promising avenue of research for the identification of appropriate criteria for declaring the significance of multiple hypothesis tests in FDR-like settings involves the use of receiver-operating characteristic (ROC) curves (Chen *et al.*, 2007, Tsai and Chen, 2004). ROC curves provide an intuitive statistical method for striking a balance between false negatives and false positives, given varying thresholds that differentiate two groups (e.g., true from false negatives; cases from controls). Graphical representations, much like Fig. 12.1 (see Rao, 1998), can be used to display the balance between true and false positive assignments associated with the curves as a function of appropriate significance thresholds.

D. Sequential methods

Replication is the "gold standard" for weeding out the false positives from the true signals. However, some studies are too expensive to repeat in independent samples. Consequently, methods that collect and/or analyze the data in stages (sequential methods) can be used to verify signals and/or further reduce the number of signals down to a manageable size. The underlying idea is to split the sample into at least two groups, one to explore the data and another for replication. Exactly how the sample is split remains a hot topic for discussion in the current literature.

Sequential sampling approaches were first used by Abraham Wald and Jacob Wolfowitz during the 1940s (Wald, 1945, 1947) to evaluate industrial quality control during World War II. Simultaneously, Alan Turing developed a similar procedure in order to decode German ciphers during the war, although his work was done in secret and not published until the 1980s (Hodges, 1983). The earliest application of sequential theory in genetics was by Morton (1955; see also Böddeker and Ziegler, 2001) in developing the lod score method, and since has been used in genetics and other fields quite extensively.

At the height of the multiple testing debate in genetic epidemiology, Province (2000, 2001) was the first one to develop a sequential procedure that bypasses the problem of multiple testing altogether by performing one global test at the level of the genome. This promising approach, based on the work of Bechhofer *et al.* (1968), was developed for linkage analysis using the Haseman–Elston (H–E) sibpair linkage method (Haseman and Elston, 1972). However, the method is general and is applicable to other genome-wide study designs including variance components linkage and WGA studies. In basic sequential analysis, the sample size is not fixed in advance, but rather data are evaluated in a stepwise additive manner until further sampling is stopped according to some predefined stopping rule or significance level. In general, the log likelihood of the data is a ratio of the data under alternative hypotheses over that for the null (no linkage). If there is linkage then the ratio will be positive, and this ratio should grow larger as more data points (sibpairs) are added (with increasing n). If there is no linkage then the ratio will tend to become more negative with increasing *n*. Prior to analysis, the type I and type II error rates are defined as functions of *a* and *b*, where $a^* < 0 < b^*$. Analysis of the data stops when the probability of the likelihood falls outside of these predefined upper or lower regions. Specification of the limits depends on the type of test statistic (Province, 2000 for genome-wide linkage scan; Aplenc *et al.*, 2003 for WGA).

While this method allows the investigator to set both the type I and type II error rates, there is still a problem with inflation of error rates because of the multiple testing issues. Sequential multiple decision procedures attempt to reduce this large number of hypothesis tests so that nominal significance levels can be used. Rather than a series of univariate hypothesis tests (is this the gene, or this, or this), the question is rephrased as a multiple-response procedure by assuming there is a gene and asking where is the gene. As applied to genome-wide linkage or WGA scan data, every possible subset of markers is formed and evaluated, and the one that contains only the linked or associated ones is selected. Basically, the relative evidence of linkage (or association) is compared among sets of markers rather than looking for the absolute evidence at each marker. As evidence accumulates, those signal markers will eventually rise above the level of background noise.

For linkage analysis using the H–E method and a large number of markers, the M markers are first ranked according to the linkage evidence using the sequential estimate of the error variance from the H–E regression. Once the markers are ranked, then they are divided into two subsets having the highest (t) nonsignificant linkage and the lowest ($M - t$) significant signals. The multiple decision procedure then attempts to select the best t from M populations that makes the best separation between significant and nonsignificant sets, as sampling from the sibpairs progresses. Again, preset levels for type I and type II errors are defined to regulate the stopping criterion.

These methods allow for efficient identification of true signals using the smallest possible sample size in a first stage. The excess sample of sibling pairs is then used in a subsequent stage for confirmation or replication. Presumably, there will be a smaller subset of markers to test with a corresponding deflation of the significance levels. Others (e.g., Skol *et al.*, 2006) suggest that stage 2 should consist of a joint analysis of the data from both stages because this usually results in increased power to detect genetic associations even though more stringent significance levels in the joint analysis must be used.

Other sequential methods for minimizing the effects due to multiple testing in family-based association studies incorporate both a screening and a replication stage using exactly the same data but applying different analytical methods in the two stages (Van Steen *et al.*, 2005). For example, a regression model that is statistically independent of the family based analysis can be used in a first stage to estimate a genetic effect. Here subjects are classified by genotype and a χ^2 or logistic regression model is used to test for an association. The most promising SNPs from this step are then selected for further association testing using family-based tests (FBATs) in the entire sample. The FBATs use allele-transmission rates and assess both association and linkage and the multiple testing issues are relevant only to the SNPs that were eventually selected for testing in stage 2. Simulation studies showed that successful results could be attained with this method using as many as one million SNPs, at least 1500 probands, minor allele frequencies of at least 0.01, and a heritability of at least 3%.

Another recent example of a sequential procedure used in a marker-based genetic association setting considers tests of the hypothesis that individuals exhibiting an adverse response to a particular drug have different genetic backgrounds than individuals that to do not exhibit the adverse response (Kelly *et al.*, 2006). The proposed test essentially considers all the markers simultaneously, but is applied after each individual exhibiting the adverse response is sampled. The goal of this strategy is to minimize the number of individuals exposed to the drug, who experience the adverse response and still identify the differences, if any, in genetic background they might have relative to individuals who do not experience the adverse response. Once genetic background differences have been identified, specific loci and genetic variations that cause the adverse response can be examined. The proposed procedure also points out a potential pitfall with sequential procedures to be applied in genome-wide studies in that one may not have simple analytic formulas for appropriate stopping rules and thus have to resort to computationally demanding genome-wide simulation studies to identify them (Kelly *et al.*, 2006).

E. Other methods

Another strategy to reduce problems due to multiple testing is to use haplotyping to reduce the number of tests performed. Haplotyping often results in combining the information from multiple SNPs into fewer bits of information. Haplotyping also contributes to reducing biases due to the nonindependence among linked markers. However, haplotypes typically are not known with certainty (unless expensive molecular haplotyping methods were used) and so statistical methods (with increased errors in estimation) of estimating haplotypes frequencies are used (see, e.g., Liu *et al.*, 2008; Chapter 13). At any rate, haplotyping is one method for not only reducing the number of tests performed, but according to some investigators may actually outperform single-marker tests for association, particularly with regard to detecting rare causal alleles (Akey *et al.*, 2001; de Bakker *et al.*, 2005; Fallin *et al.*, 2001; Longmate, 2001). As an interesting addition to the haplotyping approach to staving off multiple comparisons problems, it has been argued that based on known LD patterns between alleles, one may be able to draw inferences about the associations of genetic variations with a trait that have not actually been genotyped on a set of individuals, as long as variations in LD with the unobserved variations have been genotyped on those subjects (Zaitlen *et al.*, 2007).

V. CONCLUSION

In summary, molecular genetic technologies have advanced to such an extent that the sheer volume of data produced by them often overwhelms researchers' abilities to make valid inferences from those data. Therefore, what are needed are novel statistical methods and insights in order to make expedient, informed, and unbiased use of modern high-throughput genomic data. Multiple testing issues, which on the surface may seem like a simple problem, can be quite complex for genomic data for a number of reasons. In this chapter, we have sought to outline some of the available methods for dealing with this complex issue with the hope that it will spark greater interest in the subject.

Acknowledgments

T. K. R. and D.C.R. are supported in part by grants from the National Institute of General Medical Sciences (GM 28719) and from the National Heart, Lung, and Blood Institute (HL 54473), National Institutes of Health. N.J.S. is supported in part by grants from the National Heart Lung and Blood Institute Family Blood Pressure Program (HL064777), the National Institute on Aging Longevity Consortium (AG023122), the National Institute of Mental Health Consortium on the Genetics of Schizophrenia (HLMH065571), and by Scripps Genomic Medicine.

References

Abecasis, G. R., Cardon, L. R., and Cookson, W. O. (2000). A general test of association for quantitative traits in nuclear families. *Am. J. Hum. Genet.* **66,** 279–292.

Akey, J., Jin, L., and Xiong, M. (2001). Haplotypes vs single marker linkage disequilibrium tests: What do we gain? *Eur. J. Hum. Genet.* **9,** 291–300.

Almasy, L., and Blangero, J. (1998). Multipoint quantitative-trait linkage analysis in general pedigrees. *Am. J. Hum. Genet.* **62,** 1198–1211.

Altmuller, J., Palmer, L. J., Fischer, G., Scherb, H., and Wjst, M. (2001). Genomewide scans of complex human diseases: True linkage is hard to find. *Am. J. Hum. Genet.* **69,** 936–950. Erratum in: *Am. J. Hum. Genet.* **69,** 1413.

Aplenc, R., Zhao, H., Rebbeck, T. R., and Propert, K. J. (2003). Group sequential methods and sample size savings in biomarker-disease association studies. *Genetics* **163,** 1215–1219.

Balding, D. J. (2006). A tutorial on statistical methods for population association studies. *Nat. Rev. Genet.* **7,** 781–791.

Bechhofer, R. E., Kiefer, J., and Sobel, M. (1968). "Sequential Identification and Ranking Procedures." University of Chicago Press, Chicago.

Benjamini, Y., and Hochberg, Y. (1995). Controlling the false discovery rate: A practical and powerful approach to multiple testing. *J. R. Statist. Soc. B.* **57,** 289–300.

Benjamini, Y., and Yekutieli, D. (2001). The control of the false discovery rate in multiple testing under dependency. *Ann. Stat.* **29,** 1165–1188.

Böddeker, I. R., and Ziegler, A. (2001). Sequential designs for genetic epidemiological linkage or association studies: A review of the literature. *Biom. J.* **43,** 501–525.

Bonferroni, C. E. (1935). Il calcolo delle assicurazioni su gruppi di teste. *In* "Studi in Onore del Professore Salvatore Ortu Carboni." pp. 13–60. Rome, Italy.

Bonferroni, C. E. (1936). Teoria statistica delle classi e calcolo delle probabilita. *Pubblicazioni del R Instituto Superiore de Scienze Economiche e Commerciali de Firenze* **8,** 3–62.

Carlson, C. S., Eberle, M. A., Kruglyak, L., and Nickerson, D. A. (2004). Mapping complex disease loci in whole-genome association studies. *Nature* **429,** 446–452.

Chen, J. J., Wang, S. J., Tsai, C. A., and Lin, C. J. (2007). Selection of differentially expressed genes in microarray data analysis. *Pharmacogenomics J.* **7,** 212–220.

Cheverud, J. M. (2001). A simple correction for multiple comparisons in interval mapping genome scans. *Heredity Part 1* **87,** 52–58.

Craiu, R. V., and Sun, L. (2007). Choosing the lesser evil: Trade-off between false discovery rate and non-discovery rate. *Stat. Sinica (to appear).* (http://www.utstat.toronto.edu/craiu/Papers/craiu_sun_revision.pdf).

Davison, A. C., and Hinkley, D. V. (1997). "Bootstrap Methods and Their Applications" Cambridge University Press, Cambridge.

de Bakker, P. I., Yelensky, R., Pe'er, I., Gabriel, S. B., Daly, M. J., and Altshuler, D. (2005). Efficiency and power in genetic association studies. *Nat. Genet.* **37,** 1217–1223.

Edgington, E. S. (1995). "Randomization Tests." 3rd Editon, Marcel Dekker, New York.

Efron, B. (1982). The jackknife, the bootstrap, and other resampling plans. "Society of Industrial and Applied Mathematics CBMS-NSF Monographs"; Vol. 38. Capital City Press, Philadelphia.

Efron, B. (2007). Correlation and large-scale simultaneous significance testing. *J. Am. Stat. Assoc.* **102,** 93–103.

Elston, R. C., and Spence, A. M. (2006). Advances in statistical human genetics over the last 25 years. *Stat. Med.* **25,** 3049–3080.

Fallin, D., Cohen, A., Essioux, L., Chumakov, I., Blumenfeld, M., Cohen, D., and Schork, N. J. (2001). Genetic analysis of case/control data using estimated haplotype frequencies: Application to APOE locus variation and Alzheimer's disease. *Genome Res.* **11,** 143–151.

Fisher, R. A. (1925). "Statistical Methods for Research Workers." Oliver and Boyd, Edinburgh (http://psychologyyorku.ca/Fisher/Methods/).

Fisher, R. A. (1935). "The Design of Experiments." Oliver and Boyd, Edinburgh.

Good, P. (2006). "Resampling Methods: A Practical Guide to Data Analysis." 3rd Edition, Birkhäuser, Boston.

Gudmundsson, J., Sulem, P., Manolescu, A., Amundadottir, L. T., Gudbjartsson, D., Helgason, A., Rafnar, T., Bergthorsson, J. T., Agnarsson, B. A., Baker, A., Sigurdsson, A., Benediktsdottir, K. R., *et al.* (2007). Genome-wide association study identifies a second prostate cancer susceptibility variant at 8q24. *Nat. Genet.* **39,** 631–637.

Guo, X., and Elston, R. C. (2001). One-stage versus two-stage strategies for genome scans. *Adv. Genet.* **42,** 459–471.

Haseman, J. K., and Elston, R. C. (1972). The investigation of linkage between a quantitative trait and a marker locus. *Behav. Genet.* **2,** 3–19.

Hodges, A. (1983). "Alan Turing: The Enigma." Simon and Schuster, New York.

Iturria, S. J., Williams, J. T., Almasy, L., Dyer, T. D., and Blangero, J. (1999). An empirical test of the significance of an observed quantitative trait locus effect that preserves additive genetic variation. *Genet. Epidemiol.* **17**(Suppl 1), S169–S173.

Kelly, P., Stallard, N., Zhou, Y., Whitehead, J., and Bowman, C. (2006). Sequential genome-wide association studies for monitoring adverse events in the clinical evaluation of new drugs. *Stat. Med.* **25,** 3081–3092.

Lander, E., and Kruglyak, L. (1995). Genetic dissection of complex traits: Guidelines for interpreting and reporting linkage results. *Nat. Genet.* **11,** 241–247.

Longmate, J. A. (2001). Complexity and power in case-control association studies. *Am. J. Hum. Genet.* **68,** 1229–1237.

Lin, S., Chakravarti, A., and Cutler, D. J. (2004). Exhaustive allelic transmission disequilibrium tests as a new approach to genome-wide association studies. *Nat. Genet.* **36,** 1181–1188.

Liu, N., Zhang, K., and Zhao, H. (2008). Haplotype association analysis. *In* "Genetic Dissection of Complex Traits" (D. C. Rao and C. C. Gu, eds.), 2nd Edition, pp. 337–405. Academic Press, San Deigo.

Moore, D. S., McCabe, G., Duckworth, W., and Sclove, S. (2003). Bootstrap methods and permutation tests. Chapter 18. "The Practice of Business Statistics," WH Freeman, New York. (Published online athttp://bcs.whfreeman.com/pbs/).

Morton, N. E. (1955). Sequential tests for the detection of linkage. *Am. J. Hum. Genet.* **7,** 277–318.

Morton, N. E. (1998). Significance levels in complex inheritance. *Am. J. Hum. Genet.* **62,** 690–697. Erratum *Am. J. Hum. Genet.* **63,** 1252.

Pitman, E. J. G. (1937a). Significance tests which may be applied to samples from any population. *J. R. Statist. Soc.* **4,** 119–130.

Pitman, E. J. G. (1937b). Significance tests which may be applied to samples from any population. Part II. The correlation coefficient test. *J. R. Statist. Soc.* **4,** 225–232.

Pitman, E. J. G. (1938). Significance tests which may be applied to samples from any population. Part III. The analysis of variance test. *Biometrika* **29,** 322–335.

Pounds, S. B. (2006). Estimation and control of multiple testing error rates for microarray studies. *Brief. Bioinform.* **7,** 25–36.

Province, M. A. (2000). A single, sequential, genome-wide test to identify simultaneously all promising areas in a linkage scan. *Genet. Epidemiol.* **19,** 301–322.

Province, M. A. (2001). Sequential methods of analysis for genome scans. *Adv. Genet.* **42,** 499–514.

Rao, D. C. (1998). CAT scans, PET scans, and genomic scans. *Genet. Epidemiol.* **15,** 1–18.

Rao, D. C., and Gu, C. (2001). False positives and false negatives in genome scans. *Adv. Genet.* **42,** 487–498.

Roeder, K., Bacanu, S. A., Wasserman, L., and Devlin, B. (2006). Using linkage genome scans to improve power of association in genome scans. *Am. J. Hum. Genet.* **78,** 243–252.

Rothman, K. J. (1990). No adjustments are needed for multiple comparisons. *Epidemiology* **1,** 43–46.

Schweder, T., and Spjøtvoll, E. (1982). Plots of p-values to evaluate many tests simultaneously. *Biometrika* **69,** 493–502.

Simon, J. L. (1997). "Resampling: The New Statistics." 2nd Edition. (Published online at http://www.resample.com/content/text/index.shtml).

Skol, A. D., Scott, L. J., Abecasis, G. R., and Boehnke, M. (2006). Joint analysis is more efficient than replication-based analysis for two-stage genome-wide association studies. *Nat. Genet.* **38,** 209–213. Erratum *Nat. Genet.* **38,** 390.

Sladek, R., Rocheleau, G., Rung, J., Dina, C., Shen, L., Serre, D., Boutin, P., Vincent, D., Belisle, A., Hadjadj, S., Balkau, B., Heude, B., *et al.* (2007). A genome-wide association study identifies novel risk loci for type 2 diabetes. *Nature* **445,** 881–885.

Storey, J. D., and Tibshirani, R. (2003). Statistical significance for genomewide studies. *Proc. Natl. Acad. Sci. USA* **100,** 9440–9445.

Strug, L. J., and Hodge, S. E. (2006). An alternative foundation for the planning and evaluation of linkage analysis. II. Implications for multiple test adjustments. *Hum. Hered.* **61,** 200–209.

Sun, L., Craiu, R. V., Paterson, A. D., and Bull, S. B. (2006). Stratified false discovery control for large-scale hypothesis testing with application to genome-wide association studies. *Genet. Epidemiol.* **30,** 519–530.

Thomson, G. (2001). Significance levels in genome scans. *Adv. Genet.* **42,** 475–486.

Todorov, A. A., and Rao, D. C. (1997). Trade-off between false positives and false negatives in the linkage analysis of complex traits. *Genet. Epidemiol.* **14,** 453–464.

Tsai, C. A., and Chen, J. J. (2004). Significance analysis of ROC indices for comparing diagnostic markers: Applications to gene microarray data. *J. Biopharm. Stat.* **14,** 985–1003.

Van Steen, K., McQueen, M. B., Herbert, A., Raby, B., Lyon, H., Demeo, D. L., Murphy, A., Su, J., Datta, S., Rosenow, C., Christman, M., Silverman, E. K., *et al.* (2005). Genomic screening and replication using the same data set in family-based association testing. *Nat. Genet.* **37,** 683–691.

Wald, A. (1945). Sequential tests of statistical hypotheses. *Ann. Math. Stat.* **16,** 117–186.

Wald, A. (1947). "Sequential Analysis." Dover, New York.

Zaitlen, N., Kang, H. M., Eskin, E., and Halperin, E. (2007). Leveraging the HapMap correlation structure in association studies. *Am. J. Hum. Genet.* **80,** 683–691.

Part III

SPECIAL TOPICS

13

Meta-Analysis Methods

Thomas A. Trikalinos,* Georgia Salanti,† Elias Zintzaras,‡ and John P. A. Ioannidis*,†

*Department of Medicine, Tufts University School of Medicine, Boston, Massachusetts 02111

†Department of Hygiene and Epidemiology, University of Ioannina School of Medicine, Ioannina 45110, Greece

‡Department of Biomathematics, University of Thessaly School of Medicine, Larissa 41222, Greece

Advances in Genetics, Vol. 60

0065-2660/08 $35.00
DOI: 10.1016/S0065-2660(07)00413-0

ABSTRACT

Meta-analysis is the quantitative synthesis of information from several studies. It is applicable to a variety of study designs in genetics, from family-based linkage studies and population-based association studies to genome-wide scans and genome-wide association studies. By combining relevant evidence from many studies, statistical power is increased and more precise estimates may be obtained. Most importantly, meta-analysis provides a framework for the appreciation and assessment of between-study heterogeneity, that is, the methodological, epidemiological, clinical, and biological dissimilarity across the various studies. Being a retrospective research design in most cases, meta-analysis is subject to a variety of selection biases that may undermine its validity. A major challenge is to differentiate genuine between-study heterogeneity from systematic errors and biases.

I. INTRODUCTION

Integration of information on genetics of complex traits is a major and expanding challenge (Khoury and Little, 2000). Data on linkage or association of DNA sequence variations with disease phenotypes are accumulating exponentially (Anonymous, 2005; Ioannidis *et al.*, 2006). Many teams of investigators may target the same research questions, but relatively few published reports of significant findings are replicated unequivocally (Colhoun *et al.*, 2003; Dahlman *et al.*, 2002; Ioannidis, 2005; Ioannidis *et al.*, 2004; Lohmueller *et al.*, 2003; Wacholder *et al.*, 2004).

Meta-analysis, that is, the quantitative synthesis of information obtained from different studies (Lau *et al.*, 1997), can help integrate information from diverse studies; this is an important step in assessing the credibility of a proposed research finding (Ioannidis, 2003). Both family-based linkage studies and population-based association studies are amenable to meta-analysis. By combining relevant evidence across studies, one may reach more precise estimates of effect. Furthermore, meta-analysis allows the assessment of between-study heterogeneity. Heterogeneity pertains to the methodological, epidemiological, clinical, and biological dissimilarity across the meta-analyzed studies. A major challenge is to differentiate genuine between-study heterogeneity from systematic errors and biases (Lau *et al.*, 1998).

In genetics, both parametric (Hedges, 1985; Gu *et al.*, 2001) and nonparametric (Becker, 1994; Gu *et al.*, 1998, 2001) meta-analysis approaches have been used. Parametric approaches focus on quantifying the magnitude of the postulated genetic effect by combining estimates of effects across studies and are popular with association studies. Nonparametric approaches place emphasis on testing for the absence of a genetic effect by synthesizing the *p* values obtained from significance tests and are typically used to synthesize linkage signals. Moreover, besides frequentist methods, it is increasingly popular to apply Bayesian methods to genetic meta-analyses.

II. META-ANALYSIS OF POPULATION-BASED ASSOCIATION STUDIES

A. Background considerations

Assume that k studies are selected for a meta-analysis on a genetic risk factor for a biallelic locus with "A" the wild-type allele and "a" the mutant allele associated with a disease phenotype. Each study contributes two estimates d_{1i} and d_{2i}; the effect sizes comparing genotype Aa with AA, and comparing aa with AA, respectively, in study i. In general case of a multi allelic locus, we have $R + 1$ genotype risk groups that compared to a reference give R effect sizes d_{ri}, $r = 1, \ldots, R$. In practice, most investigators assume a particular genetic model that reduces the number of effect sizes into a single parameter d_i (dominant $d_{1i} = d_{2i} = d_i$, codominant $d_{1i} = 0.5 \times d_{2i} = d_i$ and recessive $d_{1i} = 0$, $d_{2i} = d_i$).

The effects of the genetic risk factor in the ith study are denoted as d_i. The Table 13.1 mentions commonly used effect size metrics along with their variances.

The individual study estimates, d_i, may deviate from the unknown common effect, d_+, because of chance or between-study heterogeneity that goes beyond chance. Depending on whether between-study heterogeneity is ignored or accounted for, *fixed* or *random genetic effects* are assessed, respectively.

1. Fixed effects

Fixed effects approaches assume that there is a single common effect size and that the observed between-study variability is entirely attributed to chance. The summary effect size d_+^F is obtained as a linear combination of k study-specific effect sizes d_i, where the weights W_i^F sum to unity (k is the number of studies). The weights that minimize the variance of the linear combination of vector d_i are inversely proportional to the variance of each study:

$$d_+^F = \sum W_i^F d_i \tag{1}$$

$$\text{where } W_i^F = \frac{w_i^F}{\sum w_i^F},\ i = 1, \ldots, k, \text{ and } w_i^F = \text{var}(d_i)^{-1}.$$

This linear estimator has the property that its asymptotic variance is smaller than that of any other linear estimator. Moreover, it has the same asymptotic distribution with the maximum likelihood estimator of the common effect size (Hedges, 1994;

Table 13.1. Examples of Commonly Used Effect Size Estimates in Parametric Meta-Analysis, and Corresponding Variances

Effect size	D	var(d)
Dichotomous outcomes		
Log odds ratio (logOR), population-based study	$\log\left(\frac{p_1}{1-p_1}\right) - \log\left(\frac{p_2}{1-p_2}\right)$	$\frac{1}{p_1(1-p_1)n_1} + \frac{1}{p_2(1-p_2)n_2}$
Continuous outcomes		
Mean difference (MD)	$m_1 - m_2$	$\frac{SD_1^2}{n_1} + \frac{SD_2^2}{n_2}$
Standardized mean difference (Hedges's g)	$\frac{(m_1 - m_2)[1 - \{3/(4(n_1 + n_2) - 9)\}]}{\sqrt{((n_1 - 1)s_1^2 + (n_2 - 1)s_2^2)/(n_1 + n_2 - 2)}}$	$\frac{n_1 + n_2}{n_1 n_2} + \frac{g^2}{2(n_1 + n_2 - 3.94)}$

The study index i has been omitted for clarity. The indices 1 and 2 refer to the compared genotypes (for genotype-based comparisons) or the compared alleles (for allele-based comparisons). p: proportion with the genetic risk factor; n: total number of people (or alleles); m: mean of the quantitative trait; s: standard deviation of the quantitative trait. A note for the odds ratio for dichotomous outcomes: When $p = 0$ or $p = 1$ in either of the compared groups, the log odds ratio or its variance are not defined. A usually employed correction adds 0.5 to all cells of the 2×2 table (Woolf, 1955).

Hedges's g is standardized version of the mean difference statistic. It expresses the mean difference relative to its standard deviation. A popular interpretation is that g expresses the average percentile standing of the average person in group 1, relative to the average person in group 2. A g of 0 indicates that the mean in group 1 is at the 50th percentile of group 2.

Zaykin *et al.*, 2002). The variance of d_+ is $\text{var}(d_+^F) = (\sum w_i^F)^{-1}$. Upper and lower boundaries of the confidence interval for d_+^F at α-level of significance are $d_+^F \pm Z_{\alpha/2} \cdot \sqrt{\text{var}(d_+^F)}$. For $\alpha = 0.05$, $Z_{0.025} \cong 1.96$.

The above-described inverse variance framework can be used for the meta-analysis of any effect size that is normally distributed, and whose variance can be calculated from the reported data (Table 13.1). Especially for dichotomous outcomes (i.e., a meta-analysis of 2×2 tables), computational alternatives include the Mantel–Haenszel method (Mantel and Haenszel, 1959) and the Peto method (not appropriate when there is not a fairly equal number of cases and controls in each study) (Yusuf *et al.*, 1985).

The Q statistic (Cochran, 1954) is the most often-used approach to assess the fit of a fixed effects model:

$$Q = \sum w_i^F (d_i - d_+^F)^2 \tag{2}$$

It has an asymptotic χ^2 distribution with $k - 1$ degrees of freedom, and is typically considered significant at the $\alpha = 0.10$ level (Cochran, 1954; Lau *et al.*, 1997).

2. Random effects

Random effect approaches allow for between-study heterogeneity and incorporate it in the calculations. It is assumed that the differences of the underlying true study-specific genetic effects from the overall summary estimates are normally distributed around zero with variance τ^2. A moment-based estimate for τ^2 (the DerSimonian and Laird estimate) is obtained from Eq. (2) by equating the observed value of Q with its expectation (DerSimonian and Laird, 1986).

$$\tau^2 = \max\left\{\frac{Q-(k-1)}{\sum w_i^{\mathrm{F}} - \left(\sum (w_i^{\mathrm{F}})^2 / \sum w_i^{\mathrm{F}}\right)}, 0\right\} \tag{3}$$

The summary meta-analysis estimate for random study effects, d_+^{R}, is obtained by a linear estimator as in Eq. (1) $d_+^{\mathrm{R}} = \sum W_i^{\mathrm{R}} d_i$, where the weights become to take heterogeneity into account. Again, the variance of the summary effect d_+^{R} is $\mathrm{var}(d_+^{\mathrm{R}}) = \left(\sum w_i^{\mathrm{R}}\right)^{-1}$, and upper and lower boundaries of the confidence interval at α-level of significance are $d_+^{\mathrm{R}} \pm Z_{\alpha/2} \cdot \sqrt{\mathrm{var}(d_+^{\mathrm{R}})}$.

3. Graphics for a parametric meta-analysis

A typical way to present a meta-analysis is a forest plot, where each study is shown by its effect size and 95% confidence interval and the summary effect and 95% confidence interval are also shown (Fig. 13.1A). Another common graphic presentation is a cumulative meta-analysis plot, where studies are placed in order (e.g., chronological) and the summary estimate and 95% confidence interval are plotted as more studies are sequentially added to the calculations (Fig. 13.1B). For other less commonly used graphical presentations, see Sutton *et al.* (2000).

4. Heterogeneity: Statistical versus other considerations

Between-study heterogeneity in the broader context may refer to differences in the design and conduct of the various combined studies, study populations, and phenotype definitions. It could also specifically reflect different linkage disequilibrium patterns across populations between the tested polymorphism and the biologically important one, variable population stratification, differential measurement errors in genotyping, and other biases. Statistical heterogeneity

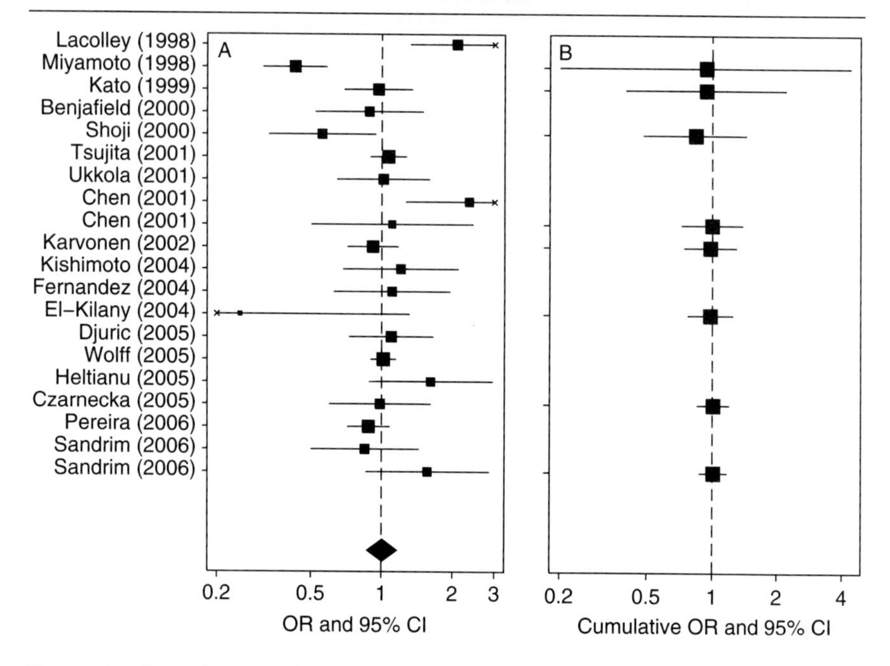

Figure 13.1. Typical meta-analysis plot. Typical forest plot (left panel, A) for a meta-analysis and typical cumulative meta-analysis of the same data (right panel, B). Based on data from Zintzaras *et al.* (2006b). CI, confidence interval; OR, odds ratio.

may point to any of these causes of diversity or combination of several causes. Moreover, like any statistical test, tests for heterogeneity are subject to false-positives and false-negatives.

Q is fairly insensitive to detect genuine heterogeneity when there are relatively few studies, and may overinterpret unimportant heterogeneity when there are many studies (Hardy and Thompson, 1998). Alternative approaches quantify the extent of heterogeneity using metrics that are less dependent on the number of synthesized studies and can be compared across meta-analyses (Higgins and Thompson, 2002). I^2 expresses the percentage of between-study variability that is attributable to heterogeneity rather than chance (Higgins and Thompson, 2002; Higgins *et al.*, 2003). I^2 is defined as

$$I^2 = \frac{\tau^2}{\tau^2 + \sigma^2}, \tag{4}$$

where σ^2 denotes the within-study variance component, and can be conveniently calculated as $1 - [k - 1)/Q]$. Expressed as a proportion, I^2 is used to quantify the extent of between-study heterogeneity. Values over 75% denote extreme heterogeneity (Higgins and Thompson, 2002). The estimates of Q, τ^2, and I^2 are unstable when only few studies are combined (Huedo-Medina *et al.*, 2006).

5. Bayesian meta-analysis

Meta-analysis can be seen as an accumulation of evidence over time and therefore Bayesian methodology seems to be its natural framework. Bayesian methods have been increasingly facilitated by advances in computing and availability of user-friendly software (Spiegelhalter *et al.*, 2003, 2004). Although often Bayesian meta-analysis methods obtain equivalent results as frequentist methods, there are some advantages, for example, it is possible to draw probabilistic conclusions for the quantities of interest (for example, about the probability that carriers of a certain allele have higher risk than the noncarriers) and to calculate predictive probabilities and effects. In contrast to frequentist approaches, both the data and the underlying parameters (such as d_+ and τ^2) are treated as unknown quantities, thus accounting for full uncertainty in all model parameters.

Prior distributions in Bayesian methods are density probability functions that reflect our beliefs regarding the unknown parameters of the model. The joint prior function for all parameters is combined with the likelihood function of the data to provide an updated belief as the joint posterior probability density function. Inferences are based on marginal posterior densities (posteriors for the parameters of interest conditional on the "nuisance" parameters). Estimations are carried out through simulations-based Markov Chain Monte Carlo and in particular Gibbs sampling (Brooks, 1998; Spiegelhalter *et al.*, 2003).

Vague (noninformative) priors are often used, that is distributions that reflect lack of prior evidence and have minimal effect on the inference relative to the data. Different priors may be applied as sensitivity analysis, particularly for the uncertainty parameters (Lambert *et al.*, 2005). On the other side, priors allow the inclusion of relevant evidence (concerning parameters for the genetic effect, the underlying genetic model, or the compliance to of Hardy–Weinberg disequilibrium) that may not necessarily come from studies eligible for the systematic review and would otherwise be ignored. Several prior elicitation methods have been described (Garthwaite *et al.*, 2005).

We describe here the distributional form of a simple Bayesian meta-analysis model as implemented in WinBUGS (Spiegelhalter *et al.*, 2003). The likelihood of the data would be normal as

$$d_i \sim \mathrm{N}(d_i^*, (w_i^{\mathrm{F}})^{-1}) \qquad (5)$$

with d_i^* the underlying study-specific random effects. Under a random effects model

$$d_i^* \sim \mathrm{N}(d_i^{\mathrm{R}}, \tau^2), \qquad (6)$$

for fixed effects $d_i^* = d_+^F$. The random effects meta-analysis model described here is a two-level hierarchical model. There is "exchangeability" between studies, which makes the study-specific estimates to shrink toward the common mean.

Common choices for prior distributions are a vague normal distribution for the location parameters (e.g., $d_i^* \sim N(0, 1000), d_+^R \sim N(0, 1000)$) and a half normal for the dispersion parameter (i.e., normal truncated at zero: $\tau \sim N(0, 1), \tau > 0$). Results using vague priors are similar to those from the frequentist approach; however, the uncertainty interval for d_+ (called "credible" interval) becomes wider, as we allow for uncertainty in between-study variance (Sutton and Abrams, 2001).

Use of Bayesian methods becomes inevitable in some more complicated but frequent situations, such as estimation of the underlying genetic model (Minelli *et al.*, 2005), generalized evidence synthesis (Ades and Sutton, 2006), incorporation of external evidence (Higgins and Whitehead, 1996), examination of baseline risk effects (Schmid *et al.*, 1998; Sharp and Thompson, 2000), and modeling of deviations from Hardy–Weinberg equilibrium (HWE) (Salanti *et al.*, 2007).

B. Special issues in genetic applications

1. Selecting the genetic model

Genetic meta-analyses have been criticized for lack of consistency in selecting genetic contrasts (Attia *et al.*, 2003; Salanti *et al.*, 2005b; Thakkinstian *et al.*, 2005). In practice, it is not easy to define *a priori* the proper, biologically meaningful genetic contrasts because the actual inheritance models are unknown. Early work in the candidate gene era has suggested that associations based on a highly selected genetic contrast are almost 10 times more likely to be refuted than other associations (Ioannidis *et al.*, 2001). In the era of genome-wide association testing, typically the primary screening of associations is based on allele-based models. One early case where a recessive model was selectively picked (Herbert *et al.*, 2006) resulted in an association that was subsequently refuted (Dina *et al.*, 2007; Loos *et al.*, 2007; Rosskopf *et al.*, 2007).

2. Hardy–Weinberg equilibrium

Disease-free groups, such as an appropriately chosen control group, should follow the HWE. Deviations from HWE may reflect attrition (Weinberg and Morris, 2003; Lee, 2003), errors in analysis or inappropriate control population (Khoury and Little, 2000; Khoury *et al.*, 1993), or genotyping error (Hosking *et al.*, 2004; Xu *et al.*, 2002), although this latter cause has been contested (Cox and Kraft, 2006; Zou and Donner, 2006). Regardless of the cause for deviation, studies not complying with HWE should be identified and treated with caution as they may lead to spurious results (Schaid and Jacobsen, 1999; Sasieni, 1997). The most commonly applied tests for HWE are the χ^2 test and the exact test (Emigh, 1980). Their results can be supplemented with a measure of the extent of deviation such as the fixation coefficient (Shoemaker *et al.*, 1998). In meta-analysis, sensitivity analysis excluding HWE-violating studies is recommended; however, given the low power of HWE tests (Salanti *et al.*, 2005a), it is difficult to distinguish compliers from violators. Adjusting for HWE deviation may affect the nominal significance for meta-analyses that are close to the nominal significance threshold (Trikalinos *et al.*, 2006b). A hierarchical meta-regression model has also been proposed for the association between the study-specific fixation coefficients and effect sizes (Salanti *et al.*, 2007).

For a biallelic locus, the observed control genotype probabilities $\tau_i^{aa}, \tau_i^{Aa}, \tau_i^{AA}$ in study i are parameterized by the fixation coefficient Φ_i and the a allele frequency λ_i:

$$\begin{aligned} \tau_i^{aa} &= \lambda_i^2 + \lambda_i(1-\lambda_i)\Phi_i \\ \tau_i^{Aa} &= 2\lambda_i(1-\lambda_i)(1-\Phi_i) \\ \tau_i^{AA} &= (1-\lambda_i)^2 + \lambda_i(1-\lambda_i)\Phi_i \end{aligned} \tag{7}$$

Then the study-specific random effects can be written as a function of the fixation coefficient $d_i^* = \delta_i^* + \beta|\Phi_i|$ with $\delta_i^* \sim N(d_+^R, \tau^2)$. The 95% posterior credible interval for β provides evidence for the dependence of effect sizes on deviations from HWE. The posterior distribution of d_+^R provides the distribution of the effect size that one would have observed if all controls were in HWE. Note that there is some confusion for the term fixation coefficient that is used with in the literature for different quantities. Here, we refer to the "system-of-mating" inbreeding coefficient that has the properties of a correlation coefficient and takes values between -1 and 1, rather than the "pedigree inbreeding coefficient" that takes values between 0 and 1.

III. META-ANALYSIS OF LINKAGE STUDIES

Several methods have been proposed for meta-analysis of linkage studies (Dempfle and Loesgen, 2004; Gu *et al.*, 1998, 1999, and 2001; Li and Rao, 1996; Minelli *et al.*, 2005; Wise *et al.*, 1999; Zintzaras and Ioannidis, 2005b). Most linkage studies typically provide LOD scores or other measures of statistical significance rather than effect sizes. Their meta-analyses are primarily nonparametric.

Nonparametric meta-analysis can accommodate studies that have collected quite diverse information on the same markers and loci (Kruglyak *et al.*, 1996). One can combine information from different study designs (affected sib pairs, extreme discordant siblings, entire sibships, affected pedigrees), regardless of the statistical analyses in the primary studies (model-free or model-based linkage, two-point or multipoint LOD score) or outcome definition (continuous or dichotomous). Drawbacks include the poor interpretation of the combined estimate, the inability to provide effect sizes, and difficulties in addressing heterogeneity.

A. Meta-analysis of significance levels

The null hypothesis for testing combined significance is that evidence for linkage is not present in any of the k assessed populations. The usual practice is to test the alternative hypothesis that evidence for linkage (either positive or negative) is present in at least one population. This is an omnibus test, and it does not address the *overall statistical significance* of linkage across the k studies. Under the null hypothesis of no linkage, the p values in the k studies are uniformly distributed. Therefore, any transformations of uniformly distributed variables with known distributional form can be used. We will describe two popular approaches.

1. Sum of logs and sum of z methods

As described by Fisher (1932), the sum

$$-2\sum \log(p_i) \tag{8}$$

is χ^2—distributed with 2*k degrees of freedom under the null hypothesis of no linkage (Becker, 1994; Fisher, 1932).

A similar method based on an inverse normal distribution transformation was proposed by Stouffer in 1949 (Becker, 1994). The sum

$$\frac{1}{\sqrt{k}}\sum \Phi^{-1}(1 - p_i) \tag{9}$$

is normally distributed (Φ^{-1} is the inverse normal distribution).

Several nonparametric linkage methods produce one-sided LOD scores that are truncated at zero. Thus, the distribution of p values under the null is a mixture of a uniform and a point mass at LOD = 0. For LOD = 0, one should choose a p value between 0.5 and 1; $p = 1/(2\log(2)) \cong 0.72$ is an approximately unbiased estimate (Province, 2001).

The above methods can be modified to weight studies. The weighted versions of Eqs. (8) and (9) are $-2\sum W_i \log(p_i)$ and $1/\sqrt{k} = \sum W_i \Phi^{-1}(1-p)$, respectively, and the sums are distributed as in the unweighted cases. Optimal weighting factors are debatable. Study size (expressed as the number of informative units, for example, trios, sib pairs, or other for different study designs) is one apparent option. Unweighted analyses and analyses weighted by size (or any other parameter that is constant for all examined loci in all studies) are suboptimal when different marker sets have been used (Dempfle and Loesgen, 2004). One would like to weight accounting for the Information Content (Kruglyak *et al.*, 1996) that varies across loci in the same study (or functions thereof). However, the locus-specific Information Content values are not routinely extractable from primary linkage studies. Finally, in a sense, p values are already weighted (Becker, 1994). The p value depends on the sample size. It is questionable whether giving larger studies even more weight is appropriate.

2. Truncated p value product method

The truncated p value product method is a generalization of the sum of log method that utilizes only p values that are smaller than a given threshold (typically 0.05) (Zaykin *et al.*, 2002). Briefly, one first estimates the product of the p values that are less than 0.05 (or any other chosen threshold). Then one evaluates the probability that this product is less than its observed value, under the null hypothesis of no linkage. Zaykin *et al.* (2002) gives an explicit formula and provides software. The truncated p value product method can also be used for meta-analysis of genome scans (Dudbridge and Koeleman, 2003; Zaykin *et al.*, 2002).

3. Methods that weight the p value and the breadth of the linkage region

Badner and Gershon (2002) proposed the multiple scan probability method that combines probabilities across regions rather than single points. This method assumes that p values on the same marker are available across the k independent studies, and is based on the sum of logs method using p values corrected for the size of the linkage region. Evidence for linkage can occur in an extended region and not in a single point; therefore, the pointwise p values across regions are combined by correcting the observed minimum pointwise p value from each study for the size of the linkage region (Feingold *et al.*, 1993). The probability of a p value being observed in a given sized region is given as:

$$p^* = C \cdot p + 2 \cdot \lambda \cdot G \cdot \Phi^{-1}(p) \cdot \Phi(p) \cdot v(p, \Delta) \tag{10}$$

where C is the number of chromosomes, p is the pointwise p value from study i, λ is the rate of crossovers per Morgan, G is the size of the region in Morgans, $\Phi^{-1}(p)$ is the standard normal inverse of p, $\Phi(p)$ is the normal density function, and $v(p,\Delta)$ is a discreteness correction for the distance Δ between markers (for continuous markers $\Delta = 0$ and $v(p,\Delta) = 1$). Then, Fisher's sum of logs method is employed.

B. Parametric meta-analysis for linkage studies

Etzel and Guerra (2002) proposed a meta-analysis method to combine evidence from independent sib pair linkage studies that report Haseman–Elston (Haseman and Elston, 1972) statistics for linkage to a quantitative-trait locus (QTL) at multiple markers on a chromosome.

Suppose there are k sib pair studies on the same QTL and each study is using m markers within the same chromosomal region of width L. We define L_q ($q = 1, \ldots, t$) as the set of analysis points, that is, putative positions of the QTL, with L_l and L_t being the start and end, respectively, of each chromosomal region. From every study, we need for each marker M_{ij} the following two statistics: the Haseman–Elston-estimated slope coefficient $\hat{\beta}_{ij}$ ($i = 1, \ldots, k$ and $j = 1, \ldots, m$) and its estimated variance of $\hat{\beta}_{ij}$, denoted as S_{ij}^2.

At L_q, for each marker M_{ij}, we calculate $\hat{\beta}_{ijq} = \hat{\beta}_{ij}/[(1-2\theta_{ijq})^2]$ and $S_{ijq}^2 = S_{ij}^2/[(1-2\theta_{ijq})^4]$, where θ_{ijq} is the recombination fraction between marker M_{ij} and the analysis point L_q.

Then, we calculate the combined estimate $\tilde{\beta}_q = \sum_{i=1}^{k}\sum_{j=1}^{n_{iq}} w_{ij}\hat{\beta}_{ijq} / \sum_{i=1}^{k}\sum_{j=1}^{n_{iq}} w_{ij}$, where n_{iq} is the number of markers from study i within D cM of L_q. The weights are $w_{ij} = \left(S_{ijq}^2 + (1/(k-1))\sum_{i=1}^{k}\sum_{j=1}^{n_{iq}}(\hat{\beta}_{ijq} - \bar{\beta}_{..q}) - 1/k\sum_{i=1}^{k}\sum_{j=1}^{n_{iq}} S_{ijq}^2\right)^{-1}$, where $\bar{\beta}_{..q}$ is the average within D cM of L_q. The variance of $\hat{\beta}_q$ is $\mathrm{var}(\tilde{\beta}_q) = 1/\sum_{i=1}^{k}\sum_{j=1}^{n_{iq}} w_{ij}$.

Significance at each analysis point L_q is tested as $t_q = \tilde{\beta}_q/\sqrt{\mathrm{var}(\tilde{\beta}_q)}$. The analysis point $L_q^{'}$ with $t_q^{'} = \min(t_q)$ is considered the point estimate (location) of the gene locus. Heterogeneity across point estimates is tested by $Q_q' = \sum_{i=1}^{k}\sum_{j=1}^{n_{iq}}(\hat{\beta}_{ijq'} - \bar{\beta}_{q'})/S_{ijq'}^2$, where $\bar{\beta}_{q'} = (\sum_{i=1}^{k}\sum_{j=1}^{n_{iq}} \hat{\beta}_{ijq}/S_{ijq}^2)/\sum_{i=1}^{k}\sum_{j=1}^{n_{iq}} 1/S_{ijq}^2$. Q_q' follows the χ^2 distribution with $(\sum_{i=1}^{k} n_{iq} - 1)$ degrees of freedom.

C. Meta-analysis of genome scans

Linkage signals in genome scans are typically low (Dempfle and Loesgen, 2004; Wise *et al.*, 1999). In addition, genome scans performed by different teams often identify linkage in different chromosomal regions. We describe here genome search meta-analysis (GSMA) (Wise *et al.*, 1999) and its extension by Zintzaras and Ioannidis (2005b) for evaluating heterogeneity between different genome-scans (HEGESMA: heterogeneity-based genome-scan meta-analysis) (Ioannidis *et al.*, 2007; Trikalinos *et al.*, 2006a; Zintzaras and Ioannidis, 2005b; Zintzaras *et al.*, 2006a).

A typical GSMA starts by splitting the chromosomes into bins of approximately equal length; a length of 30 cM gives 120 bins in total for the whole genome. For each genome scan, the most significant result of the test statistic obtained within the bin is recorded. Test statistics may include LOD score, maximum logarithm of odds score, nonparametric linkage score, z-statistics, p values, depending on the mode of analysis of the linkage data (Terwilliger and Ott, 1994). Then, for each scan, the bins are ranked in ascending order according to their significance of results and the ranks for each bin are summed across scans. The significance of the summed (or average) rank of each bin is assessed empirically against the distribution of summed (or average) ranks. Under the null hypothesis of no linkage in any chromosomal bin, the ranks are randomly assigned from each study. The probability that the ranks X_i (where $i = 1, \ldots, k$ studies) of a specific bin sum to R_{sum} across studies is:

$$P\left(\sum_{i=1}^{k} X_i = R_{\text{sum}}\right) = \begin{cases} 0, & R_{\text{sum}} < k \\ \dfrac{1}{n^k} \displaystyle\sum_{j=0}^{\text{int}[(R_{\text{sum}}-k)/n]} (-1)^j \binom{R_{\text{sum}} - jn - 1}{k-n} \binom{m}{k}, & k \le R_{\text{sum}} \le nk \\ 0, & R_{\text{sum}} < nk \end{cases} \tag{11}$$

When a bin has a high summed rank, this is considered as evidence for linkage across several studies. Equal test statistic results for bins within a scan are assigned as tied ranks. For practical application, summed ranks and average ranks are interchangeable because they are related by the number of studies. The ranks of the bins in each study can be weighted by factors such as the number of pedigrees, the number of markers, or the size of pedigrees used in each study (Wise *et al.*, 1999).

The extent of heterogeneity between studies combined in a GSMA can be quantified by estimates of the variability of the ranks for each bin across studies. The Q statistic is defined as the sum of the squared deviations of each study's bin rank from the mean of the ranks and it represents a generalization/modification of Cochran's Q statistic:

$$Q = \sum W_i^{\text{F}} (R_i - \bar{R})^2 \tag{12}$$

where R_i is the rank of the bin under investigation for study i ($i = 1$ to k studies), $\bar{R}$ is the mean rank across the available studies, and W_i^{F} is the weighting factor for study i. The Ha statistic is defined as $Ha = \sum W_i^{\text{F}} |R_i - \bar{R}|$ and the B statistic is defined as $B = \sum W_i^{\text{F}} W_j^{\text{F}} |R_i - R_j|$ where i and $j = 1$ to k studies and $i \neq j$. The B metric takes into consideration all the possible pairwise bin differences across studies.

In the typical meta-analysis application, one is interested in testing at some statistical significance level whether the observed between-study heterogeneity is too high. For GSMA, it may be important to detect both too high and too limited between-study heterogeneity. Statistically significantly limited heterogeneity may suggest that the results of different studies for the same bin are very consistent among themselves. When this is seen for a specific bin that is on the top ranks of all studies, low heterogeneity may be interpreted as further supportive evidence for the importance of this bin for linkage.

The statistical significance of the average rank and heterogeneity Q metrics is assessed using a Monte Carlo method for which software is readily available (Zintzaras and Ioannidis, 2005a; Zintzaras *et al.*, 2006a). The heterogeneity metrics are correlated with the average rank value. Very high average rank and very low average ranks are expected to have low heterogeneity because of ceiling (floor) effects (Lewis and Levinson, 2006). Restricted Monte Carlo tests have been implemented to reduce the between-metric correlation (Haseman and Elston, 1972; Lewis and Levinson, 2006; Zintzaras and Ioannidis, 2005b). In restricted tests, separate null distributions are generated for each bin, using only the distributions of bins with neighboring simulated average rank (± 2 rank places) as the bin being considered each time (Zintzaras and Ioannidis, 2005a,b). Power to detect heterogeneity with HEGESMA can be very low in many practical applications where few studies are available, especially with the restricted Monte-Carlo tests (Haseman and Elston, 1972; Lewis and Levinson, 2006; Zintzaras and Ioannidis, 2005b).

IV. SPECIAL ISSUES

A. Nonreplication of early genetic claims

Nonreplication of postulated genetic effects is common in the literature (Fig. 13.2; (Anonymous, 1999; Cardon and Bell, 2001; Colhoun *et al.*, 2003; Dahlman *et al.*, 2002; Ioannidis *et al.*, 2001; Lohmueller *et al.*, 2003). Several potential explanations have been proposed, including exaggerated early results (Ioannidis *et al.*, 2001), publication bias and time lag bias and other selection biases (see below), systematic differences between larger and smaller studies (Ioannidis *et al.*, 2003), population admixture or stratification (Cardon and Bell, 2001; Colhoun *et al.*, 2003), "racial/ethnic" heterogeneity (Ioannidis *et al.*, 2004), and Hardy–Weinberg deviations that may reflect genotyping error, stratification, or other problems (Salanti *et al.*, 2007; Trikalinos *et al.*, 2006b).

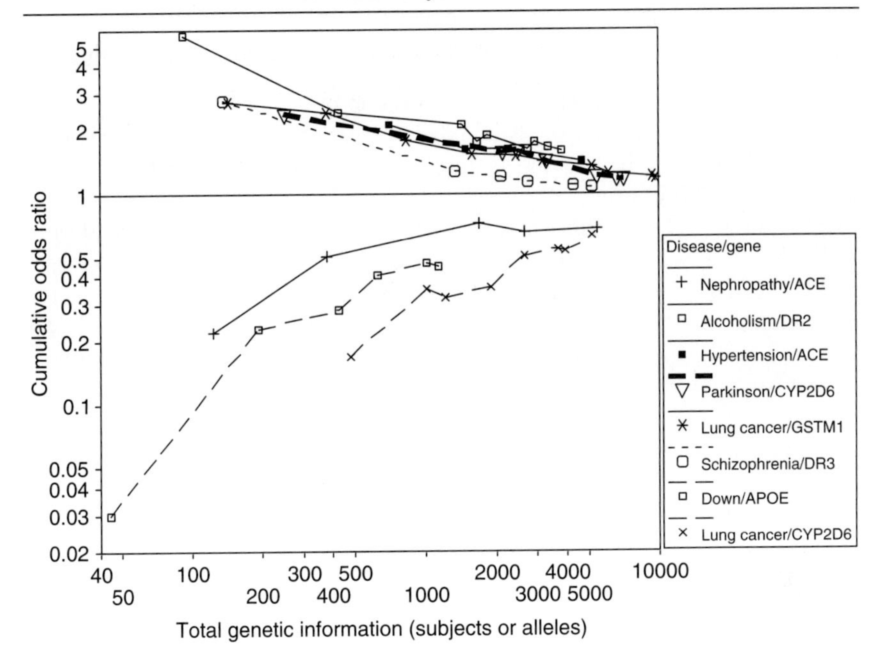

Figure 13.2. Lack of replication of proposed associations. Dissipated effect sizes over time in eight cumulative meta-analyses of proposed genetic associations. Reproduced from Ioannidis *et al.* (2001).

Empirical evidence on some of these problems is available. Most of the evidence has been generated from the candidate gene era, but some of these problems may persist in the current massive testing (e.g., genome-wide association studies).

Family-based studies are theoretically advantageous in some aspects (e.g., they are more robust to population stratification) (Laird and Lange, 2006), but otherwise they can suffer from most of the problems that can affect studies of unrelated subjects. There are several examples where linkage findings were not consistently replicated by subsequent research (Dahlman *et al.*, 2002; Zintzaras and Kitsios, 2006). Lack of power and differences in marker sets may underlie some discrepancies (Dahlman *et al.*, 2002; Dempfle and Loesgen, 2004; Zintzaras and Kitsios, 2006). Family-based studies can also be used to test for association in the presence of linkage (Laird and Lange, 2006). However, with currently available sample sizes, one cannot reliably estimate phenotypic effects of identified loci using the *same dataset* (Goring *et al.*, 2001). Phenotypic effects would be inflated (Goring *et al.*, 2001), and logically, not replicated in their magnitude by subsequent research.

B. Biases in meta-analyses, with emphasis on genetic meta-analysis

Several special biases pose threats to the validity of meta-analyses.

Publication bias operates when the publication of a study is dependent on its results: studies with "negative" (statistically nonsignificant) findings are not published, resulting in an overrepresentation of "positive" (statistically significant) findings in the literature and inflated summary results (Dickersin and Min, 1993). Many statistical tests have been introduced in the meta-analysis literature to detect publication bias and adjust for its effects (Rothstein *et al.*, 2007). These tests simply examine whether small studies give different results from larger studies. Most of the time, these tests are applied without meeting the minimal prerequisites for their application (Ioannidis and Trikalinos, 2007) and thus relying on such tests to decide on the presence or not of publication bias should be avoided (Lau *et al.*, 2006).

Time-lag bias exists when the time to completion and publication of a study is dependent on its findings. Time-lag bias has been described in clinical research where time to publication tends to be longer for "negative" compared to "positive" studies (Ioannidis, 1998). There is empirical evidence that a peculiar pattern is observed in genetic epidemiology. One often encounters in the early literature a succession of extreme opposite results. Studies with intermediate and potentially less spectacular findings may be published at a slower pace, filling the gap between the early published extremes. The meta-analysis of Fig. 13.1 shows such an example. Apparently, contradictory findings are attractive to investigators and editors, and may have an advantage for rapid publication. This special form of time-related bias has been termed the *Proteus phenomenon* (Ioannidis and Trikalinos, 2005).

Selective outcome reporting bias occurs when outcomes/analyses with "negative" results are not reported, whereas emphasis is given to "positive" outcomes/analyses in the same study. This has been systematically described in randomized trials (Chan *et al.*, 2004a,b) and selective reporting is probably larger in epidemiology (Kavvoura *et al.*, 2007). Meta-analysis may have to be aborted, if there is no concordance of reported phenotypes across the studies that one wishes to combine. In an empirical evaluation of 21 pharmacogenetic studies on a specific association, 487 different outcomes and analyses had been used (Contopoulos-Ioannidis *et al.*, 2006).

Local literature bias refers to an interplay of language bias (Egger *et al.*, 1997) and selective reporting. In a systematic comparison of the English language international literature and the local Chinese language literature (Pan *et al.*, 2005), the summary effects in meta-analyses of studies published in the local Chinese literature always implied stronger genetic associations, compared to summary effects from studies published in the English international literature. It is unlikely that this observation is attributed to "racial/ethnic heterogeneity"

(Ioannidis *et al.*, 2004; Pan *et al.*, 2005). Local literature biases are probably not specific to the local Chinese literature, but probably affect other countries' literatures as well. They may reflect a stronger prevalence of "significance-chasing" in some research environments.

C. Meta-analyses of individual participant data and consortia of investigators

Meta-analyses of individual participant data utilize participant-level information and offer advantages over the meta-analysis of aggregate data. Such initiatives are feasible though consortia and networks of investigators (Ioannidis *et al.*, 2005, 2006), and are strongly encouraged (Seminara *et al.*, 2007). Advantages include the standardization (or at least harmonization) of definitions of cases and controls, opportunities for better control of confounding, and unified and more flexible statistical analyses (Higgins *et al.*, 2001; Whitehead *et al.*, 2001). In the context of a consortium, individual teams are still free to pursue their own research enterprises, but their "discoveries" can then be tested according to a common plan in the wider consortium. Such "prospective" meta-analyses obviate the problems of selective reporting for data within the confines of the consortium. Prospective genotyping with central quality control in a consortium may also help eliminate heterogeneity due to genotyping errors in single teams. Even though meta-analyses of individual participant data are resource-intensive (Steinberg *et al.*, 1997), and they require commitment and appropriate funding, a main advantage is the enhancement of communication and collaboration between diverse teams of investigators working on the same topic. Comprehensive consortia may also be instrumental eventually for maintaining updated synopses of all evidence on genetic associations in their field (Bertram *et al.*, 2007).

D. Meta-analysis of genome-wide association studies

Genome-wide association studies with massive testing are increasingly used for dissecting complex traits (Wang *et al.*, 2005). With careful designs (Jorgenson and Witte, 2006; Laird and Lange, 2006), hypothesis-free approaches may have better success in dissecting complex traits. By mid-2007, we have already seen a flurry of major successes in identifying common gene variants through genome-wide association studies for diabetes, coronary artery disease, prostate cancer, breast cancer, Crohn's disease, age-related macular degeneration, and several other diseases and traits, and the list is growing rapidly. However, already we have examples of lack of replication of genome-wide association study findings in fields such as obesity (Dina *et al.*, 2007; Loos *et al.*, 2007; Rosskopf *et al.*, 2007) or Parkinson's disease (Elbaz *et al.*, 2006; Maraganore *et al.*, 2005). Given the large number of genotyped markers and the postulated small genetic effects of tag polymorphisms, the sample sizes of single genome-wide association studies may still be insufficient for adequate power. Placing genome-wide association data in the public domain [as advocated by the GAIN initiative and the efforts of the Wellcome Trust (Cardon, 2006)] will allow the compilation of an extensive database from diverse teams and facilitate meta-analyses. Meta-analysis of genome-wide association studies has already been explored in genome-wide datasets (Fung *et al.*, 2006; Maraganore *et al.*, 2005) in Parkinson's disease (Evangelou *et al.*, 2007). A major challenge is that overlap of polymorphisms across different genotyping platforms is often limited. Besides using similar platforms for all studies on the same phenotype, potential improvements may capitalize on the high correlation between markers that belong to the same linkage disequilibrium block. One may evaluate common and noncommon markers jointly in each block using multivariate meta-analysis (Berkey *et al.*, 1998; Gleser and Olkin, 1994) or by imputing missing polymorphisms through linked ones. Meta-analysis of genome-wide association studies and their early replication efforts from diverse teams is rapidly becoming a gold standard of discovery and presentation of newly discovered common variants. When done with appropriate methods in the setting of international collaborations and consortia, such efforts may help minimize or even eliminate some of the inherent biases of meta-analysis, such as publication and selective reporting biases.

E. Meta-analysis of gene–gene–environment data

Gene–gene–environment interactions may be very important for complex traits, but their evaluation is restricted by the low power of individual studies and the selection of the interaction's model. Sample sizes requirements depend on the design (Colhoun *et al.*, 2003; Gauderman, 2002). Although in principal meta-analysis will enhance power, in practice this would usually require individual patient data and considerable effort at harmonization because of the differential reporting of exposures across studies. To overcome this, a meta-analysis framework has been developed that combines results from studies that report information on any subset or combination of the full set of exposures (Salanti *et al.*, 2006).

Assume that we are interested in J exposure categories (genetic or environmental, e.g., $J = 4$ for the combination of carriers/noncarriers and smoking/nonsmoking) associated with a phenotype and the J corresponding odds $\pi/(1 - \pi_{j,i})$, $j = 1, \ldots, J$. Then, consider a particular "collapsed" study i, where exposures 1 to J have been collapsed into mutually exclusive groups and R_i is a subset of the J exposures (for example, R_i can be the exposure "carriers" where we have no information about smoking status). The log-odds $\pi_{R_i,i}/(1 - \pi_{R_i,i})$ for the "collapsed" exposure R_i in this study can be "decomposed" into the latent log-odds for the nested unobserved exposures (e.g., the log-odds of carriers for smokers and the log-odds of carriers for nonsmokers) as

$$\frac{\pi_{R_i,i}}{1-\pi_{R_i,i}} = \sum_{j \in R_i} \lambda_{j/R_i,i} \frac{\pi_{j,i}}{1-\pi_{j,i}} \tag{13}$$

with $\lambda_{j/R_i,i}$ being the study-specific prevalence of the exposure j in category R_i (e.g., the prevalence of smokers among carriers). Fitted within a Bayesian framework, the model requires assumptions to be made about the prevalence of the specific exposures in studies where these are unreported. These can be derived either from other studies in meta-analysis that do report all exposures or from prevalence studies.

References

Ades, A. E., and Sutton, A. J. (2006). Multi-parameter evidence synthesis in epidemiology and medical decision making: Current approaches. *JRRS* **169,** 5–35.

Anonymous (1999). Freely associating. *Nat. Genet.* **22,** 1–2.

Anonymous (2005). Framework for a fully powered risk engine. *Nat. Genet.* **37**(11), 1153.

Attia, J., Thakkinstian, A., and D'Este, C. (2003). Meta-analyses of molecular association studies: Methodologic lessons for genetic epidemiology. *J. Clin. Epidemiol.* **56,** 297–303.

Badner, J. A., and Gershon, E. S. (2002). Meta-analysis of whole-genome linkage scans of bipolar disorder and schizophrenia. *Mol. Psychiatry* **7,** 405–411.

Becker, B. J. (1994). Combining significance levels. *In* "The Handbook of Research Synthesis" (H. Cooper and L. V. Hedges, eds.), pp. 215–231. Russel Sage Foundation, New York.

Berkey, C. S., Hoaglin, D. C., Ntczak-Bouckoms, A., Mosteller, F., and Colditz, G. A. (1998). Meta-analysis of multiple outcomes by regression with random effects. *Stat. Med.* **17,** 2537–2550.

Bertram, L., McQueen, M. B., Mullin, K., Blacker, D., and Tanzi, R. E. (2007). Systematic meta-analyses of Alzheimer disease genetic association studies: The AlzGene database. *Nat. Genet.* **39,** 17–23.

Brooks, S. P. (1998). Marcov chain Monte Carlo method and its application. *Statistician* **47,** 69–100.

Cardon, L. R. (2006). Genetics. Delivering new disease genes. *Science* **314,** 1403–1405.

Cardon, L. R., and Bell, J. I. (2001). Association study designs for complex diseases. *Nat. Rev. Genet.* **2,** 91–99.

Chan, A. W., Hrobjartsson, A., Haahr, M. T., Gotzsche, P. C., and Altman, D. G. (2004a). Empirical evidence for selective reporting of outcomes in randomized trials: Comparison of protocols to published articles. *JAMA* **291,** 2457–2465.

Chan, A. W., Krleza-Jeric, K., Schmid, I., and Altman, D. G. (2004b). Outcome reporting bias in randomized trials funded by the Canadian Institutes of Health Research. *CMAJ* **171,** 735–740.

Cochran, W. G. (1954). The combination of estimates from differents expreiments. *Biometrics* **10,** 101–129.

Colhoun, H. M., McKeigue, P. M., and Davey, S. G. (2003). Problems of reporting genetic associations with complex outcomes. *Lancet* **361,** 865–872.

Contopoulos-Ioannidis, D. G., Alexiou, G. A., Gouvias, T. C., and Ioannidis, J. P. (2006). An empirical evaluation of multifarious outcomes in pharmacogenetics: Beta-2 adrenoceptor gene polymorphisms in asthma treatment. *Pharmaco Genet, Genomics* **16,** 705–711.

Cox, D. G., and Kraft, P. (2006). Quantification of the power of Hardy-Weinberg equilibrium testing to detect genotyping error. *Hum. Hered.* **61,** 10–14.

Dahlman, I., Eaves, I. A., Kosoy, R., Morrison, V. A., Heward, J., Gough, S. C., Allahabadia, A., Franklyn, J. A., Tuomilehto, J., Tuomilehto-Wolf, E., Cucca, F., Guja, C., *et al.* (2002). Parameters for reliable results in genetic association studies in common disease. *Nat. Genet.* **30,** 149–150.

Dempfle, A., and Loesgen, S. (2004). Meta-analysis of linkage studies for complex diseases: An overview of methods and a simulation study. *Ann. Hum. Genet.* **68,** 69–83.

DerSimonian, R., and Laird, N. (1986). Meta-analysis in clinical trials. *Control Clin Trials* **7,** 177–188.

Dickersin, K., and Min, Y. I. (1993). Publication bias: The problem that won't go away. *Ann. NY Acad. Sci.* **703,** 135–146.

Dina, C., Meyre, D., Samson, C., Tichet, J., Marre, M., Jouret, B., Charles, M. A., Balkau, B., and Froguel, P. (2007). Comment on "A common genetic variant is associated with adult and childhood obesity." *Science* **315,** 187.

Dudbridge, F., and Koeleman, B. P. (2003). Rank truncated product of P-values, with application to genomewide association scans. *Genet. Epidemiol.* **25,** 360–366.

Egger, M., Zellweger-Zahner, T., Schneider, M., Junker, C., Lengeler, C., and Antes, G. (1997). Language bias in randomised controlled trials published in English and German. *Lancet* **350,** 326–329.

Elbaz, A., Nelson, L. M., Payami, H., Ioannidis, J. P., Fiske, B. K., Annesi, G., Carmine, B. A., Factor, S. A., Ferrarese, C., Hadjigeorgiou, G. M., Higgins, D. S., Kawakami, H., *et al.* (2006). Lack of replication of thirteen single-nucleotide polymorphisms implicated in Parkinson's disease: A large-scale international study. *Lancet Neurol.* **5,** 917–923.

Emigh, T. (1980). A comparison of tests for Hardy-Weinberg equilibrium. *Biometrics* **36,** 627–642.

Etzel, C. J., and Guerra, R. (2002). Meta-analysis of genetic-linkage analysis of quantitative-trait loci. *Am. J. Hum. Genet.* **71,** 56–65.

Evangelou, E., Maraganore, D. M., and Ioannidis, J. P. (2007). Meta-analysis in genome-wide association datasets: Strategies and application in Parkinson's disease. *PLoS ONE* **2,** e196.

Feingold, E., Brown, P. O., and Siegmund, D. (1993). Gaussian models for genetic linkage analysis using complete high-resolution maps of identity by descent. *Am. J. Hum. Genet.* **53,** 234–251.

Fisher, R. A. (1932). "Statistical Methods for Research Workers." 4th Edition, Oliver & Boyd, London.

Fung, H. C., Scholz, S., Matarin, M., Simon-Sanchez, J., Hernandez, D., Britton, A., Gibbs, J. R., Langefeld, C., Stiegert, M. L., Schymick, J., Okun, M. S., Mandel, R. J., *et al.* (2006). Genome-wide genotyping in Parkinson's disease and neurologically normal controls: First stage analysis and public release of data. *Lancet Neurol* **5,** 911–916.

Garthwaite, P. H., Kadane, J. B., and O'Hagan, A. (2005). Statistical methods for eliciting probability distributions. *J. Am. Stat. Assoc.* **100,** 680–701.

Gauderman, W. J. (2002). Sample size requirements for association studies of gene-gene interaction. *Am. J. Epidemiol.* **155,** 478–484.

Gleser, L. J., and Olkin, I. (1994). Stochastically dependent effect sizes. *In* "The Handbook of Research Synthesis" (H. Cooper and L. V. Hedges, eds.), pp. 339–355. Rusel Sage Foundation, New York.

Goring, H. H., Terwilliger, J. D., and Blangero, J. (2001). Large upward bias in estimation of locus-specific effects from genomewide scans. *Am. J. Hum. Genet.* **69,** 1357–1369.

Gu, C., Province, M., Todorov, A., and Rao, D. C. (1998). Meta-analysis methodology for combining non-parametric sibpair linkage results: Genetic homogeneity and identical markers. *Genet. Epidemiol.* **15,** 609–626.

Gu, C., Province, M., and Rao, D. C. (1999). Meta-analysis of genetic linkage to quantitative trait loci with study-specific covariates: A mixed-effects model. *Genetic Epidemiol.* **17**(Suppl. 1), S599–S604.

Gu, C., Province, M., and Rao, D. C. (2001). Meta-analysis for model-free methods. *In* "Genetic Dissection of Complex Traits" (D. C. Rao and M. A. Provinceeds, eds.), pp. 255–272. Academic Press, San Diego, CA.

Hardy, R. J., and Thompson, S. G. (1998). Detecting and describing heterogeneity in meta-analysis. *Stat. Med.* **17,** 841–856.

Haseman, J. K., and Elston, R. C. (1972). The investigation of linkage between a quantitative trait and a marker locus. *Behav. Genet.* **2,** 3–19.

Hedges, L. V. (1985). Parametric estimation of effect size from a series of experiments. *In* "Statistical Methods for Meta-Analysis" (L. V. Hedges and I. Olkin, eds.), pp. 108–138. Academic Press, Inc., San Diego.

Hedges, L. V. (1994). Fixed effects models. *In* "The Handbook of Research Synthesis" (H. Cooper and L. V. Hedges, eds.), pp. 285–301. Rusel Sage Foundation, New York.

Herbert, A., Gerry, N. P., McQueen, M. B., Heid, I. M., Pfeufer, A., Illig, T., Wichmann, H. E., Meitinger, T., Hunter, D., Hu, F. B., *et al.* (2006). A common genetic variant is associated with adult and childhood obesity. *Science* **312,** 279–283.

Higgins, J. P., and Thompson, S. G. (2002). Quantifying heterogeneity in a meta-analysis. *Stat. Med.* **21,** 1539–1558.

Higgins, J. P., and Whitehead, A. (1996). Borrowing strength from external trials in a meta-analysis. *Stat. Med.* **15,** 2733–2749.

Higgins, J. P., Whitehead, A., Turner, R. M., Omar, R. Z., and Thompson, S. G. (2001). Meta-analysis of continuous outcome data from individual patients. *Stat. Med.* **20,** 2219–2241.

Higgins, J. P., Thompson, S. G., Deeks, J. J., and Altman, D. G. (2003). Measuring inconsistency in meta-analyses. *BMJ* **327,** 557–560.

Hosking, L., Lumsden, S., Lewis, K., Yeo, A., McCarthy, L., Bansal, A., Riley, J., Purvis, I., and Xu, C. F. (2004). Detection of genotyping errors by Hardy-Weinberg equilibrium testing. *Eur. J. Hum. Genet.* **12,** 395–399.

Huedo-Medina, T. B., Sanchez-Meca, J., Marin-Martinez, F., and Botella, J. (2006). Assessing heterogeneity in meta-analysis: Q statistic or I^2 index? *Psychol. Methods* **11,** 193–206.

Ioannidis, J. P. (1998). Effect of the statistical significance of results on the time to completion and publication of randomized efficacy trials. *JAMA* **279,** 281–286.

Ioannidis, J. P. (2003). Genetic associations: False or true?. *Trends Mol. Med.* **9,** 135–138.

Ioannidis, J. P. (2005). Why most published research findings are false. *PLoS Med.* **2,** e124.

Ioannidis, J. P., and Trikalinos, T. A. (2005). Early extreme contradictory estimates may appear in published research: The Proteus phenomenon in molecular genetics research and randomized trials. *J. Clin. Epidemiol.* **58,** 543–549.

Ioannidis, J. P., and Trikalinos, T. A. (2007). The appropriateness of asymmetry tests for publication bias in meta-analysis: A large-scale survey. CMAJ **176**(8), 1091–1096.

Ioannidis, J. P., Ntzani, E. E., Trikalinos, T. A., and Contopoulos-Ioannidis, D. G. (2001). Replication validity of genetic association studies. *Nat. Genet.* **29,** 306–309.

Ioannidis, J. P., Trikalinos, T. A., Ntzani, E. E., and Contopoulos-Ioannidis, D. G. (2003). Genetic associations in large versus small studies: An empirical assessment. *Lancet* **361,** 567–571.

Ioannidis, J. P., Ntzani, E. E., and Trikalinos, T. A. (2004). 'Racial' differences in genetic effects for complex diseases. *Nat. Genet.* **36,** 1312–1318.

Ioannidis, J. P., Bernstein, J., Boffetta, P., Danesh, J., Dolan, S., Hartge, P., Hunter, D., Inskip, P., Jarvelin, M. R., Little, J., Maraganore, D. M., Bishop, J. A., *et al.* (2005). A network of investigator networks in human genome epidemiology. *Am. J. Epidemiol.* **162,** 302–304.

Ioannidis, J. P., Gwinn, M., Little, J., Higgins, J. P., Bernstein, J. L., Boffetta, P., Bondy, M., Bray, M. S., Brenchley, P. E., Buffler, P. A., Casas, J. P., Chokkalingam, A., *et al.* (2006). A road map for efficient and reliable human genome epidemiology. *Nat. Genet.* **38,** 3–5.

Ioannidis, J. P., Ng, M. Y., Sham, P. C., Zintzaras, E., Lewis, C. M., Deng, H. W., Econs, M. J., Karasik, D., Devoto, M., Kammerer, C. M., Spector, T., Andrew, T., *et al.* (2007). Meta-analysis of genome-wide scans provides evidence for sex- and site-specific regulation of bone mass. *J. Bone Miner. Res.* **22,** 173–183.

Jorgenson, E., and Witte, J. S. (2006). A gene-centric approach to genome-wide association studies. *Nat. Rev. Genet.* **7,** 885–891.

Kavvoura, F. K., Liberopoulos, G., and Ioannidis, J. P. (2007). Selection in reported epidemiological risks: An empirical assessment. *PLoS Med.* **4,** e79.

Khoury, M. J., and Little, J. (2000). Human genome epidemiologic reviews: The beginning of something HuGE. *Am. J. Epidemiol.* **151,** 2–3.

Khoury, M. J., Beaty, T. H., and Cohen, B. H. (1993). "Fundamentals of Genetic Epidemiology." Oxford University Press, New York.

Kruglyak, L., Daly, M. J., Reeve-Daly, M. P., and Lander, E. S. (1996). Parametric and nonparametric linkage analysis: A unified multipoint approach. *Am. J. Hum. Genet.* **58,** 1347–1363.

Laird, N. M., and Lange, C. (2006). Family-based designs in the age of large-scale gene-association studies. *Nat. Rev. Genet.* **7,** 385–394.

Lambert, P. C., Sutton, A. J., Burton, P. R., Abrams, K. R., and Jones, D. R. (2005). How vague is vague? A simulation study of the impact of the use of vague prior distributions in MCMC using WinBUGS. *Stat. Med.* **24,** 2401–2428.

Lau, J., Ioannidis, J. P., and Schmid, C. H. (1997). Quantitative synthesis in systematic reviews. *Ann. Intern. Med.* **127,** 820–826.

Lau, J., Ioannidis, J. P., and Schmid, C. H. (1998). Summing up evidence: One answer is not always enough. *Lancet* **351,** 123–127.

Lau, J., Ioannidis, J. P., Terrin, N., Schmid, C. H., and Olkin, I. (2006). The case of the misleading funnel plot. *BMJ* **333,** 597–600.

Lee, W. C. (2003). Searching for disease-susceptibility loci by testing for Hardy-Weinberg disequilibrium in a gene bank of affected individuals. *Am. J. Epidemiol.* **158,** 397–400.

Lewis, C. M., and Levinson, D. E. (2006). Testing for genetic heterogeneity in the genome search meta-analysis method. *Genet. Epidemiol.* **30,** 348–355.

Li, Z., and Rao, D. C. (1996). Random effects model for meta-analysis of multiple quantitative sibpair linkage studies. *Genet Epidemiol.* **13,** 377–383.

Lohmueller, K. E., Pearce, C. L., Pike, M., Lander, E. S., and Hirschhorn, J. N. (2003). Meta-analysis of genetic association studies supports a contribution of common variants to susceptibility to common disease. *Nat Genet* **33,** 177–182.

Loos, R. J., Barroso, I., O'rahilly, S., and Wareham, N. J. (2007). Comment on "A common genetic variant is associated with adult and childhood obesity.". *Science* **315,** 187.

Mantel, N., and Haenszel, W. (1959). Statistical aspects of the analysis of data from retrospective studies of disease. *J. Natl. Cancer Inst.* **22,** 719–748.

Maraganore, D. M., de Andrade, M., Lesnick, T. G., Strain, K. J., Farrer, M. J., Rocca, W. A., Pant, P. V., Frazer, K. A., Cox, D. R., and Ballinger, D. G. (2005). High-resolution whole-genome association study of Parkinson disease. *Am. J. Hum. Genet.* **77,** 685–693.

Minelli, C., Thompson, J. R., Abrams, K. R., and Lambert, P. C. (2005). Bayesian implementation of a genetic model-free approach to the meta-analysis of genetic association studies. *Stat. Med.* **24,** 3845–3861.

Pan, Z., Trikalinos, T. A., Kavvoura, F. K., Lau, J., and Ioannidis, J. P. (2005). Local literature bias in genetic epidemiology: An empirical evaluation of the Chinese literature. *PLoS Med.* **2,** e334.

Province, M. A. (2001). The significance of not finding a gene. *Am. J. Hum. Genet.* **69,** 660–663.

Rosskopf, D., Bornhorst, A., Rimmbach, C., Schwahn, C., Kayser, A., Kruger, A., Tessmann, G., Geissler, I., Kroemer, H. K., and Volzke, H. (2007). Comment on "A common genetic variant is associated with adult and childhood obesity." *Science* **315,** 187.

Rothstein, H. R., Sutton, A. J., and Borenstein, M. (2007). "Publication Bias in Meta-Analysis. Prevention, Assessment and Adjustments." John Wiley & Sons, West Sussex.

Salanti, G., Amountza, G., Ntzani, E. E., and Ioannidis, J. P. (2005a). Hardy-Weinberg equilibrium in genetic association studies: An empirical evaluation of reporting, deviations, and power. *Eur. J. Hum. Genet.* **13,** 840–848.

Salanti, G., Sanderson, S., and Higgins, J. P. (2005b). Obstacles and opportunities in meta-analysis of genetic association studies. *Genet. Med.* **7,** 13–20.

Salanti, G., Higgins, J. P., and White, I. R. (2006). Bayesian synthesis of epidemiological evidence with different combinations of exposure groups: Application to a gene-gene-environment interaction. *Stat. Med.* **25,** 4147–4163.

Salanti, G., Higgins, J. P., Trikalinos, T. A., and Ioannidis, J. P. (2007). Bayesian meta-analysis and meta-regression for gene-disease associations and deviations from Hardy-Weinberg equilibrium. *Stat. Med.* **26,** 553–567.

Sasieni, P. D. (1997). From genotypes to genes: Doubling the sample size. *Biometrics* **53,** 1253–1261.

Schaid, D. J., and Jacobsen, S. J. (1999). Biased tests of association: Comparisons of allele frequencies when departing from Hardy-Weinberg proportions. *Am. J. Epidemiol.* **149,** 706–711.

Schmid, C. H., Lau, J., McIntosh, M. W., and Cappelleri, J. C. (1998). An empirical study of the effect of the control rate as a predictor of treatment efficacy in meta-analysis of clinical trials. *Stat. Med.* **17,** 1923–1942.

Seminara, D., Khoury, M. J., O'Brien, T. R., Manolio, T., Gwinn, M. L., Little, J., Higgins, J. P., Bernstein, J. L., Boffetta, P., Bondy, M., Bray, M. S., Brenchley, P. E., *et al.* (2007). The emergence of networks in human genome epidemiology: Challenges and opportunities. *Epidemiology* **18,** 1–8.

Sharp, S. J., and Thompson, S. G. (2000). Analysing the relationship between treatment effect and underlying risk in meta-analysis: Comparison and development of approaches. *Stat. Med.* **19,** 3251–3274.

Shoemaker, J., Painter, I., and Weir, B. S. (1998). A Bayesian characterization of Hardy-Weinberg disequilibrium. *Genetics* **149,** 2079–2088.

Spiegelhalter, D. J., Thomas, A., Best, N. G., and Lunn, D. (2003). "WinBUGS Version 1.4 Users Manual." MRC Biostatistics Unit. 2003.

Spiegelhalter, D. J., Abrams, K. R., and Myles, P. J. (2004). Eividence synthesis. *In* "Bayesian Approaches to Clinical Trials and Health-Care Evaluation" (D. J. Spiegelhalter, K. R. Abrams, and P. J. Myles, eds.). John Wiley & Sons Ltd, Chirchestor, West Sussex, England.

Steinberg, K. K., Smith, S. J., Stroup, D. F., Olkin, I., Lee, N. C., Williamson, G. D., and Thacker, S. B. (1997). Comparison of effect estimates from a meta-analysis of summary data from published studies and from a meta-analysis using individual patient data for ovarian cancer studies. *Am. J. Epidemiol.* **145,** 917–925.

Sutton, A. J., and Abrams, K. R. (2001). Bayesian methods in meta-analysis and evidence synthesis. *SMMR* **10,** 277–303.

Sutton, A. J., Abrams, K. R., Jones, D. R., Sheldon, T. A., and Song, F. (2000). "Methods for Meta-Analysis in Medical Research." John Wiley and Sons, Chirchester.

Terwilliger, J. D., and Ott, J. (1994). "Handbook of Human Genetic Linkage." Johns Hopkins University Press, Baltimore.

Thakkinstian, A., McElduff, P., D'Este, C., Duffy, D., and Attia, J. (2005). A method for meta-analysis of molecular association studies. *Stat. Med.* **24,** 1291–1306.

Trikalinos, T. A., Karvouni, A., Zintzaras, E., Ylisaukko-oja, T., Peltonen, L., Jarvela, I., and Ioannidis, J. P. (2006a). A heterogeneity-based genome search meta-analysis for autism-spectrum disorders. *Mol. Psychiatry* **11,** 29–36.

Trikalinos, T. A., Salanti, G., Khoury, M. J., and Ioannidis, J. P. (2006b). Impact of violations and deviations in Hardy-Weinberg equilibrium on postulated gene-disease associations. *Am. J. Epidemiol.* **163,** 300–309.

Wacholder, S., Chanock, S., Garcia-Closas, M., El, G. L., and Rothman, N. (2004). Assessing the probability that a positive report is false: An approach for molecular epidemiology studies. *J. Natl. Cancer Inst.* **96,** 434–442.

Wang, W. Y., Barratt, B. J., Clayton, D. G., and Todd, J. A. (2005). Genome-wide association studies: Theoretical and practical concerns. *Nat. Rev. Genet.* **6,** 109–118.

Weinberg, C. R., and Morris, R. W. (2003). Invited commentary: Testing for Hardy-Weinberg disequilibrium using a genome single-nucleotide polymorphism scan based on cases only. *Am. J. Epidemiol.* **158,** 401–403.

Whitehead, A., Omar, R. Z., Higgins, J. P., Savaluny, E., Turner, R. M., and Thompson, S. G. (2001). Meta-analysis of ordinal outcomes using individual patient data. *Stat. Med.* **20,** 2243–2260.

Wise, L. H., Lanchbury, J. S., and Lewis, C. M. (1999). Meta-analysis of genome searches. *Ann. Hum. Genet.* **63,** 263–272.

Woolf, B. (1955). On estimating the relation between blood group and disease. *Ann. Hum. Genet.* **19,** 251–253.

Xu, J., Turner, A., Little, J., Bleecker, E. R., and Meyers, D. A. (2002). Positive results in association studies are associated with departure from Hardy-Weinberg equilibrium: Hint for genotyping error?. *Hum. Genet.* **111,** 573–574.

Yusuf, S., Peto, R., Lewis, J., Collins, R., and Sleight, P. (1985). Beta blockade during and after myocardial infarction: An overview of the randomized trials. *Prog. Cardiovasc. Dis.* **27,** 335–371.

Zaykin, D. V., Zhivotovsky, L. A., Westfall, P. H., and Weir, B. S. (2002). Truncated product method for combining P-values. *Genet. Epidemiol.* **22,** 170–185.

Zintzaras, E., and Ioannidis, J. P. (2005a). HEGESMA: Genome search meta-analysis and heterogeneity testing. *Bioinformatics.* **21,** 3672–3673.

Zintzaras, E., and Ioannidis, J. P. (2005b). Heterogeneity testing in meta-analysis of genome searches. *Genet. Epidemiol.* **28,** 123–137.

Zintzaras, E., and Kitsios, G. (2006). Identification of chromosomal regions linked to premature myocardial infarction: A meta-analysis of whole-genome searches. *J. Hum. Genet.* **51,** 1015–1021.

Zintzaras, E., Kitsios, G., Harrison, G. A., Laivuori, H., Kivinen, K., Kere, J., Messinis, I., Stefanidis, I., and Ioannidis, J. P. (2006a). Heterogeneity-based genome search meta-analysis for preeclampsia. *Hum. Genet.* **120,** 360–370.

Zintzaras, E., Kitsios, G., and Stefanidis, I. (2006b). Endothelial NO synthase gene polymorphisms and hypertension: A meta-analysis. *Hypertension* **48,** 700–710.

Zou, G. Y., and Donner, A. (2006). The merits of testing Hardy-Weinberg equilibrium in the analysis of unmatched case-control data: A cautionary note. *Ann. Hum. Genet.* **70,** 923–933.

14

Haplotype-Association Analysis

Nianjun Liu,*[,1] Kui Zhang,*[,1] and Hongyu Zhao†[,2]

*Section on Statistical Genetics, Department of Biostatistics, University of Alabama at Birmingham, Birmingham, Alabama 35294

†Department of Epidemiology and Public Health and Department of Genetics, Yale University School of Medicine, New Haven, Connecticut 06520

[1]Equal contributions
[2]Corresponding author

Advances in Genetics, Vol. 60
0065-2660/08 $35.00
DOI: 10.1016/S0065-2660(07)00414-2

ABSTRACT

Association methods based on linkage disequilibrium (LD) offer a promising approach for detecting genetic variations that are responsible for complex human diseases. Although methods based on individual single nucleotide polymorphisms (SNPs) may lead to significant findings, methods based on haplotypes comprising multiple SNPs on the same inherited chromosome may provide additional power for mapping disease genes and also provide insight on factors influencing the dependency among genetic markers. Such insights may provide information essential for understanding human evolution and also for identifying *cis*-interactions between two or more causal variants. Because obtaining haplotype information directly from experiments can be cost prohibitive in most studies, especially in large scale studies, haplotype analysis presents many unique challenges. In this chapter, we focus on two main issues: haplotype inference and haplotype-association analysis. We first provide a detailed review of methods for haplotype inference using unrelated individuals as well as related individuals from pedigrees. We then cover a number of statistical methods that employ haplotype information in association analysis. In addition, we discuss the advantages and limitations of different methods.

I. INTRODUCTION

Association studies based on either genome-wide analysis or localized fine mapping through linkage disequilibrium (LD) have become increasingly popular as they represent a potentially more cost effective and powerful approach for gene mapping than linkage analysis (Botstein and Risch, 2003; Nordborg and Tavaré, 2002; Weiss and Clark, 2002). Haplotypes refer to combinations of marker alleles which are located closely together on the same chromosome and which tend to be inherited together. With the availability of high density single nucleotide polymorphism (SNP) markers, haplotypes play an important role in association studies. First, haplotypes are critical to understanding the LD pattern across the human genome, which is essential for association studies (Ardlie *et al.*, 2002; Weiss and Clark, 2002). Commonly used LD measurements based on a pair of markers, such as D' and r^2, cannot capture the higher-order dependency among markers, and often yield irregular, nonmonotonic pictures. Actually, there is no better way to understand the LD pattern than to know the haplotypes themselves. Haplotypes tell us directly how alleles are organized along the chromosome and reflect the pattern of inheritance over evolution. Daly *et al.* (2001) provided a compelling example to show that LD analysis based on underlying haplotypes can be much clearer. Second, methods based on haplotypes can be more powerful than those based on single markers in association studies of mapping complex disease genes.

The power of single marker-based methods only depends on the LD between the tested marker locus and the disease-susceptibility locus. LD information contained in flanking markers is not incorporated into such methods, which can result in potential reduction of power. In addition, even if the tested marker locus is in strong LD with the disease locus, the power can be quite low if the frequencies of the marker and disease alleles are different (e.g., Kaplan and Morris, 2001). Therefore, haplotype-based association methods are generally regarded as being more powerful than methods based on single markers (Akey *et al.*, 2001; Morris and Kaplan, 2002) since the former fully exploits LD information from multiple markers. Both simulation (Akey *et al.*, 2001; Zaykin *et al.*, 2002) and empirical studies also support this statement. Third, haplotype-based methods can potentially capture *cis*-interactions between two or more causal variants. Biologically, several mutations on a haplotype may cause a series of changes in amino acid coding and therefore lead to a larger joint effect on the trait of interest than the single amino acid change caused by a single mutation (Schaid *et al.*, 2002b). Examples include lipoprotein lipase-responsible gene in humans (Clark *et al.*, 1998) and a gene influencing initial lactase activity in humans (Hollox *et al.*, 2001). In this case, haplotypes should be more informative than individual genotypes on revealing disease-causing mechanism at a candidate gene.

When haplotypes of each individual are known, haplotypes can be considered as alleles for a single multi-allelic marker. In this situation, all association methods based on single markers can be virtually applied to analyze haplotypes. Unfortunately, the haplotypes for each individual in studies cannot be easily acquired. Current laboratory techniques (e.g., Douglas *et al.*, 2001; Yan *et al.*, 2000) can be used to determine local haplotypes experimentally, but these approaches are often too expensive and too cumbersome to be used effectively for large-scale studies. Therefore, most association studies have relied on the use of unphased genotype data, coupled with statistical and computational methods, to infer haplotypes through estimation of haplotype frequencies and resolution of haplotype pairs within individuals. In Sections II and III, we will review several commonly used methods for haplotype inference from unrelated individuals as well as related individuals from pedigrees. In Sections IV and V, we will review the association methods based on haplotypes, especially for those incorporating haplotype inference and accounting for haplotype inference uncertainty. Finally, we will discuss their applications and implications in association studies.

II. HAPLOTYPE INFERENCE FROM UNRELATED INDIVIDUALS

In the past 20 years, many methods have been developed for haplotype inference. For one individual, if there is at most one heterozygous marker, then the haplotypes can be resolved unambiguously. In general, for an individual with genotypes at k ($k > 0$) heterozygous markers, there are 2^{k-1} possible haplotype pairs that are compatible with the observed genotypes. The goals of haplotype inference include estimating haplotype frequencies from the sample and reconstructing haplotypes for each sampled individual. The existing methods for haplotype inference can be grouped into two major categories: population-based methods and family-based methods, based on the data observed. We will introduce them separately in this section and Section III.

Haplotyping based on population data is important and challenging in genetic studies. For example, because many human complex disorders are late onset, it is practically very difficult, if not impossible, and expensive to recruit the patients' parents into a family-based study (Niu, 2004). Therefore, in practice, unrelated individuals are usually recruited into the studies, not only in the retrospective case–control studies, but also in some prospective clinical trial studies, such as the Women's Health Initiative (http://www.nhlbi.nih.gov/whi/) and the Physicians' Health Study (http://phs.bwh.harvard.edu/). Haplotype analysis of genotype data from such studies needs population-based haplotype inference methods.

A. Statistical methods

The majority of the population-based haplotype inference methods are based on statistical approaches. Likelihood coupled with some population genetics assumptions, such as Hardy–Weinberg equilibrium (HWE), is the basis of these methods. As in other statistical areas, there are two primary approaches for parameter estimation: the frequentist's approach, which is usually based on the Expectation-Maximization (EM) algorithm (Dempster *et al.*, 1977), and the Bayesian approach, which is usually achieved through Gibbs sampler. These two approaches are well represented in haplotype inference.

1. The EM-based methods

The EM algorithm is widely used in haplotype inference, where the population haplotype frequencies are estimated by maximizing the likelihood function (Excoffier and Slatkin, 1995; Hawley and Kidd, 1995; Long *et al.*, 1995; Qin *et al.*, 2002; Scheet and Stephens, 2006). Let G = $(g_1, \ldots, g_N)$ denote the observed genotypes for N unrelated individuals; $S(g_i)$ the set of haplotype pairs (h,h') that are compatible with genotype g_i; $\mathbf{P} = (p_1, \ldots, p_M)$ the population frequencies of M possible haplotypes. The likelihood function of observing N genotypes conditional on the population haplotype frequencies is:

$$L(\mathbf{P}) = \Pr(G|\mathbf{P}) = \prod_{i=1}^{N} \Pr(g_i|\mathbf{P}) = \prod_{i=1}^{N} \sum_{(h,h')\in S(g_i)} \Pr(hh').$$

Under the assumption of HWE, we have $\Pr(hh') = p_h{}^2$, if $h = h'$; and $\Pr(hh') = 2p_h p_{h'}$ if $h \neq h'$. Given a set of initial values of **P**, $p_1{}^{(0)}, p_2{}^{(0)}, \ldots, p_M{}^{(0)}$, the EM procedure estimates the haplotype frequencies iteratively by repeating the following E-step and M-step until convergence:

E-step: At the $t + 1$ iteration, calculate the posterior probability for each possible haplotype pair that are compatible with the observed genotype by

$$\Pr(hh')^{(t)} = \begin{cases} [p_h^{(t)}]^2, & \text{if } h = h', \\ 2p_h^{(t)} p_{h'}^{(t)} & \text{if } h \neq h'. \end{cases}$$

M-step: At the $t + 1$ iteration, update $p_j^{(t)}$ for each j by aggregating the posterior probabilities $\Pr(hh')^{(t)}$ that involve the jth haplotype:

$$p_j^{(t)} = \frac{1}{2N} \sum_{i=1}^{N} \frac{\sum_{(h,h') \in S(g_i)} \{\Pr(hh')^{(t)} I_{(h=j)} + \Pr(hh')^{(t)} I_{(h'=j)}\}}{\sum_{(h,h') \in S(g_i)} \Pr(hh')^{(t)}} \quad \text{for } j = 1, \ldots, M,$$

where $I_{(h_1=j)}$ is an indicator variable that is 1 if the jth haplotype is involved in the haplotype pair that are compatible with g_i, and 0 otherwise.

These EM-based methods have some advantages (Niu, 2004): (1) they are based on solid statistical theory and easy to implement; (2) they are relatively robust to departures from HWE, especially when the direction of departure towards an excess of homozygosity; and (3) they are relatively easy to be incorporated into subsequent haplotype analyses. The disadvantages are (1) as pertained to many EM-based methods, the performance is sensitive to the initial values, and the EM algorithm may not converge to the global maxima; (2) the methods based on standard EM algorithm cannot handle large number of markers, since the number of possible haplotypes grows exponentially with the number of markers. In order to avoid these issues, some practical steps may be helpful, such as using the product of the allele frequencies as initial values, as suggested by Excoffier and Slatkin (1995); or start from different initial values and run the algorithm multiple times. In recent years, some statistical methods have been developed to deal with the above issues. Tregouet *et al.* (2004) proposed to use the stochastic-EM algorithm, a variant of the standard EM algorithm for haplotype inference. This algorithm is useful in avoiding convergence to local maxima (Tregouet *et al.*, 2004). Qin *et al.* (2002) and others used the partition–ligation (PL) technique to deal with large number of markers. PL is a divide-conquer-combine algorithm which was introduced to haplotype inference by Niu *et al.* (2002b). The idea of PL is as follows: First, the entire markers are broken into small "chunks" and then an algorithm is used, such as EM algorithm, to construct haplotypes within each chunk.

Afterward, two adjacent partial haplotypes are "ligated" to rebuild the phase hierarchically, through a bottom-up approach. The ligation process is repeated until the complete phase is determined. The PL technique has been employed in other haplotyping methods to deal with large number of markers (Lin *et al.*, 2002; Stephens and Donnelly, 2003; Stephens and Scheet, 2005; Zhang *et al.*, 2005). Another strategy in handling large quantity of markers is a progressive-extension technique which was first introduced and implemented in the program SNPHAP (Clayton, 2001). The basic idea of progressive-extension is to add marker one by one. This technique is also used by others (Schaid, 2005).

2. Bayesian methods

Since Bayesian approaches can incorporate prior information into statistical models, they have gained increasingly usage in genetic studies. Several Bayesian approaches have been proposed for haplotype inference (Lin *et al.*, 2002; Niu *et al.*, 2002b; Morris *et al.*, 2003; Stephens and Donnelly, 2003; Stephens and Scheet, 2005; Stephens *et al.*, 2001; Zhang *et al.*, 2006). These approaches calculate the posterior distribution of the unobserved haplotypes conditional on the observed genotypes by incorporating biological information into the priors. Markov Chain Monte Carlo (MCMC) algorithm, usually Gibbs sampler, is typically used to sample a haplotype pair to each individual genotype based on its posterior distribution. In the following, a pseudo procedure is outlined to illustrate the Gibbs sampling algorithm.

Let $H = (H_1, H_2, \ldots, H_N)$ denote the haplotype configuration for the N individuals. Given an initial haplotype reconstruction $H^{(0)} = (H_1^{(0)}, H_2^{(0)}, \ldots, H_N^{(0)})$, H is updated iteratively as following:

(1) Choose an individual, say i, from all ambiguous individuals uniformly and randomly
(2) Sample a pair of haplotypes $H_i^{(t+1)}$ to individual i based on $\Pr(H_i \mid G, H_{-i}^{(t)})$, where $H_{-i}^{(t)}$ is the set of haplotype pairs without individuals i's haplotype pair
(3) Set $H_j^{(t+1)} = H_j^{(t)}$ for $j \neq i$

Stephens *et al.* (2001) proposed to use a pseudo-Gibbs sampler, which incorporates the coalescence relationship into Gibbs sampling iterations, to reconstruct haplotypes from genotype data. The Dirichlet prior distribution they used is based on the assumption of *parent-independent-mutation*, where the type of a mutant haplotype is independent of its parental haplotypes. The pseudo-Gibbs sampler uses Gibbs sampling to obtain an approximate sample from the posterior distribution, $\Pr(H \mid G)$. They stated that the performance of the algorithm based on this Dirichlet prior is roughly comparable to the EM algorithm. They improved this algorithm by using a more practical prior distribution, an approximate coalescent prior distribution which is based on the idea that a future sampled haplotype is more likely to be the same as or similar to one of those haplotypes that have been observed. This model was later extended to estimate both haplotype frequencies and recombination parameters by including a recombination parameter and the product of approximate conditionals (PAC) likelihood into MCMC iterations (Li and Stephens, 2003; Stephens and Scheet, 2005), and allowing for decay of LD (Stephens and Scheet, 2005). Since the set of conditions do not correspond to a joint probability distribution, it is still a pseudo-MCMC sampler (Zhang *et al.*, 2006). Therefore, the interpretation of the inference results may lack clear statistical meaning (Excoffier *et al.*, 2003; Niu, 2004; Zhang *et al.*, 2006). However, it has been argued that the practical performance of this approach is good (Stephens and Donnelly, 2003; Stephens and Scheet, 2005).

Niu *et al.* (2002b) proposed a fully Bayesian model based on another Gibbs sampling algorithm and a Dirichlet prior but with no assumptions on the population evolutionary history. Their procedure first obtains an initial guess on the haplotype frequencies, denoted by $\boldsymbol{p}$, from an initial set of pseudo counts, denoted by λ, for distinct haplotypes. Then it samples each individual a pair of compatible haplotypes according to

$$\Pr[H_i^{(t+1)} = (h_g h_k) | \boldsymbol{p}, G] = \frac{p_{h_g} p_{h_k}}{\sum_{(h'_g h'_k) \in H_i} p_{h'_g} p_{h'_k}}.$$

Niu *et al.* (2002b) used a Gibbs sampler to update the assignment of haplotype pair for each individual with posterior distribution computed by

$$\Pr[H_i^{(t+1)} = (h_g h_k) | H_{-i}^{(t)}, G] \propto (N_{h_g}^{(t)} + \beta_g^{(t)})(N_{h_k}^{(t)} + \beta_k^{(t)}),$$

where N_{h_g} and N_{h_k} are the counts of haplotypes h_g and h_k in H_{-i}, respectively; $\beta = (\beta_1, \ldots, \beta_M)$ is the parameter of a Dirichlet distribution which is the priori for population haplotype frequencies $\mathbf{P} = (p_1, \ldots, p_M)$. To avoid convergence to local maxima, Niu *et al.* (2002b) applied the prior-annealing technique to shrink the pseudo count $\lambda^{(t)}$ in a fixed rate and to zero at the end of iteration. This way the pseudo counts λ have less influence on the resulting haplotype assignment.

Two new techniques were introduced in haplotype inference in the work of Niu *et al.* (2002b): PL and prior annealing. To handle a large number of loci, Niu *et al.* (2002b) firstly introduced PL in haplotype inference. The essence of PL has been introduced in Section II.A.1. In the ligation step, Niu *et al.* (2002b) proposed two strategies: progressive and hierarchical ligation. The PL strategy not only enables the algorithm to handle a larger number of markers, but also enables the algorithm to reduce the computation time. This idea was later adopted by many other algorithms, not only for dealing with population data (Lin *et al.*, 2002; Stephens and Donnelly, 2003; Stephens and Scheet, 2005), but also for dealing with family data (Lin *et al.*, 2004; Marchini *et al.*, 2006; Zhang *et al.*, 2005). The model by Niu *et al.* (2002b) was later extended to encode the coalescent information and to estimate recombination parameters (Zhang *et al.*, 2006). In the new model, a hierarchical structure is used on the prior haplotype frequency distributions in order to capture the similarities among present haplotypes attributable to their common ancestry. Therefore, the new model assigns different prior probabilities to different haplotypes according to the inferred hierarchical ancestry structure from the data. Although built on coalescence, the model does not make any assumptions about the evolution history. The coalescence relationship is only considered in the prior distribution and its influence dwindles as the sample size increases. Another advantage of this model is that the procedure is asymptotically consistent (Zhang *et al.*, 2006).

Excoffier *et al.* (2003) proposed a Bayesian algorithm for haplotype inference. The key feature of the method is that the phase of each heterozygous site in every individual is estimated based on information of two adaptive windows of neighboring markers, and the window size varies according to local level of LD (Excoffier *et al.*, 2003). This way large number of markers can be handled efficiently. Another feature of the algorithm is that mutation and recombination are explicitly incorporated into the model. As mentioned by the authors, the method is a pseudo-Gibbs sampler.

Some studies (Niu, 2004; Salem *et al.*, 2005; Stephens and Donnelly, 2003; Zhang *et al.*, 2001) have been performed to compare some of the haplotype inference methods. The results are not consistent. One possible reason may be that the performance of different methods may differ under different scenarios, and no one method seems to outperform others in all situations.

B. Combinatorial algorithms

A number of other important approaches for haplotype inference have been proposed using combinatorial methods, which often state an explicit objective function that one tries to optimize in order to obtain a solution to the inference problem (Brown and Harrower, 2006; Clark, 1990; Gusfield, 2001, 2003; Gusfield *et al.*, 2005; Halperin and Eskin, 2004; Huang *et al.*, 2005).

1. Parsimony approaches

Clark's algorithm is the first one in the literature for haplotype reconstruction from genotype data (Clark, 1990). The algorithm is based on the principle of maximum parsimony and has three steps:

(1) Form an initial set of resolved haplotypes, *R*, by identifying all unambiguous haplotypes from individuals who are homozygotes or single-site heterozygotes
(2) Determine whether each of the resolved haplotypes could resolve any unresolved genotypes from the ambiguous genotype pool (Clark's Inference Rule)
(3) Each time when an unresolved genotype is resolved, the haplotype identified which is not in *R* is added to *R*, and the genotype is removed from the ambiguous genotype pool.

Clark's Inference Rule is repeated until either all the genotypes have been resolved or no further genotype can be resolved. Unresolved genotypes are called "orphans." Clark's algorithm can produce multiple solutions, because there may be multiple resolved haplotypes that can be used to resolve a genotype, and the phasing results are dependent on the order of ambiguous genotypes that need to be resolved. However, Clark showed that the solution with the fewest orphans is the most accurate one and suggested that a solution resolving the maximum number of ambiguous genotypes is likely to be unique, that is, the solution with maximum parsimony is unique and correct (Clark, 1990). Another limitation of Clark's algorithm is that, when there is no homozygote and single-site heterozygote in the sample, the algorithm cannot start. Clark stated that as the sample size increases, the probability of failing to get the algorithm started decreases, and orphans are less likely to remain. Although Clark's algorithm does not have any explicit genetic or model assumptions, the Inference Rule in Clark's algorithm implies two major assumptions in population genetics studies (Gusfield *et al.*, 2005). First, identical genotypes will be resolved identically by Clark's algorithm. This implies the "infinite site" assumption that only one mutation can occur at any given site in the history of the sample sequences. Secondly, the Inference Rule is most applicable with the "random mating" assumption of a population (one condition of HWE). Niu *et al.* (2002b) showed that its performance is relatively sensitive to the extent of departure from HWE. The advantages of Clark's algorithm are that it is relatively simple and can handle a large number of markers.

Gusfield (2001) reformulated Clark's algorithm into a "maximum resolution" (MR) problem and proved that the MR problem is NP-hard and Max-SNP complete, which can be reduced to an integer linear programming (ILP) problem (Gusfield, 2001). To deal with the multiple-solution issue of Clark's algorithm, Gusfield and Orzach (Gusfield *et al.*, 2005) suggested a two-step approach to obtain an accurate solution. The idea of the approach is to run Clark's algorithm many times and then choose the solution that uses the smallest number of distinct haplotypes to resolve all or the most genotypes as the final solution. Simulation studies (Gusfield *et al.*, 2005) showed that the solution was almost always correct. However, the approach is very computationally intensive in nature. Gusfield (2003) proposed another approach to handling this problem, the Pure-Parsimony approach. The Pure-Parsimony problem in haplotype inference is defined as following: find a solution to minimize the total number of distinct haplotypes used to resolve all genotypes (Gusfield, 2003). Gusfield (2003) proposed to use the ILP, which is to minimize an objective function under some linear constraints, to solve this Pure-Parsimony problem. Gusfield (2003) also showed that, for a small dataset, say less than 50 individuals and less than 30 SNPs, the Pure-Parsimony approach can correctly infer 80–95% haplotype pairs. In general, the integer program uses several seconds to minutes to obtain the solution. Brown and Harrower proposed a polynomial-size ILP formulation that takes additional inequalities to reduce the running time (Brown and Harrower, 2004). Extensive experiments showed that the performance between RTIP (a reduced integer linear programming formulation for solving the haplotype inference by pure parsimony problem) and ILP formulations is comparable (Brown and Harrower, 2004, 2006).

2. Phylogeny haplotyping approaches

As can be seen from the previous description, the haplotype inference problem would be very difficult, if not impossible, without some implicit or explicit genetic assumptions about how DNA sequences evolve (Gusfield *et al.*, 2005). A coalescent model is a stochastic process that mimics the evolutionary history of a set of sampled haplotypes. Derived from coalescent model, with two assumptions, no recombination and infinite-sites, the perfect phylogeny can be formally described as following (Gusfield *et al.*, 2005): "if M is a set of binary sequences, and V is a binary sequence that will label the root, the tree displaying the evolution of the haplotypes for M and V is called a perfect phylogeny for M and V." The Perfect Phylogeny Haplotype (PPH) Problem can be stated as: given a set of genotypes, find a set of haplotypes such that each genotype is generated by a pair from this set, and this set can be derived on a perfect phylogeny. Gusfield is probably the first one who introduced and solved the PPH problem (Gusfield, 2002). The algorithm is based on reducing the PPH problem to the graph-realization problem which is linear-time solvable.

Many methods have been proposed for PPH problem since the publication of Gusfield's seminal work (Gusfield, 2002). One modification of the PPH problem is called the "imperfect", or "near-perfect", or "almost perfect" phylogeny haplotyping problem which extends the framework of PPH by allowing for both recurrent mutations and recombinations (Gusfield *et al.*, 2005; Halperin and Eskin, 2004). The imperfect phylogeny method is based on the observation that in practice, the common haplotypes fit the perfect phylogeny model, but the full set of haplotypes in the sample may not. Following the PL idea of Niu *et al.* (2002b), Halperin and Eskin (2004) used this observation to perform haplotype inference in nonoverlapping intervals of contiguous SNP sites. In each interval, perfect phylogeny method is used to resolve a subset or all of the genotypes. These haplotypes are then used to infer haplotype pairs for the remaining genotypes, which may include genotypes initially removed due to missing data (Halperin and Eskin, 2004). The imperfect phylogeny method is more robust than PPH, and can handle missing data and data with a large number of markers. In addition to the assumption of random mating, the imperfect phylogeny method has the implicit assumption that the rare haplotypes are created by recent recombinations or mutations of common haplotypes (Gusfield *et al.*, 2005).

C. Some extensions and other methods

In the above sections, we have described two main classes of methods for haplotype inference for unrelated population data. We want to emphasize here that the classification is for the convenience of description only; the methods described are the representatives of these categories and by no means the only ones. For example, there are several variants to Clark's algorithm (Orzack *et al.*, 2003). Salem *et al.* (2005) conducted a comprehensive literature review on methods and software of haplotype inference with unrelated population data, available up to June 2004. Although there are new advancements in this field since then, we agree with their conclusion (Salem *et al.*, 2005) that there is no clear evidence to suggest a distinguished algorithm, and we suggest the users to try different methods. We have compiled a list of commonly used haplotyping programs for use with unrelated population data in Table 14.1, which is an expansion of Table 1 of Chapter 6 in Feng *et al.* (2007).

Studies have shown that algorithms for inferring haplotypes from unphased genotype data provide accurate estimation (Fallin and Schork, 2000; Salem *et al.*, 2005). However, the inclusion of pedigree information may reduce haplotype ambiguity and improve the accuracy and efficiency for haplotype inference (Lin *et al.*, 2004; Liu *et al.*, 2006b), although the improvement may depend on the pedigree information and kinship added (Schouten *et al.*, 2005). Many works have been done in this field and will be introduced in next section.

Table 14.1. Haplotype Inference Methods for Samples of Unrelated Individuals

Software	Method	Web link	Platform	References
Arlequin 3.0	Bayesian	anthro.unige.ch/arlequin	Windows/MAC/LINUX	Excoffier *et al.* (2003); Schneider *et al.* (2002)
BEAGLE	Localized haplotype-cluster model	http://www.stat.auckland.ac.nz/~browning/beagle/beagle.html	Java on LINUX/MAC/UNIX/Windows	Browning and Browning (2007a)
BPPH	Imperfect phylogeny	http://wwwcsif.cs.ucdavis.edu/~gusfield/bpph.html	MAC	Chung and Gusfield (2003)
CHAPLIN	ECM	http://server2k.genetics.emory.edu/chaplin/chaplin.php	Windows	Epstein and Satten (2003)
DPPH	Perfect phylogeny	http://wwwcsif.cs.ucdavis.edu/~gusfield/dpph.html	LINUX/MAC	Gusfield (2002); Bafna *et al.* (2003)
EH	EM	http://linkage.rockefeller.edu/ott/	Windows	Xie and Ott (1993); Terwilliger and Ott (1994)
EH+	EM	http://statgen.iop.kcl.ac.uk/	Windows/UNIX	Zhao *et al.* (2000)
EM-DeCODER	EM	http://www.people.fas.harvard.edu/~junliu/em/em.html	UNIX	Niu *et al.* (2002b)
FASTEHPLUS	EM	http://www.ucl.ac.uk/~rmjdjhz/software.htm	Windows/UNIX	Zhao *et al.* (2002)
FastPHASE	EM	http://depts.washington.edu/ventures/UW_Technology/Express_Licenses/fastPHASE.php	UNIX/LINUX/MAC/Windows	Scheet and Stephens (2006); Stephens and Donnelly (2003)
GCHAP	EM	http://bioinformatics.med.utah.edu/~alun/gchap/docs/alun/gchap/GCHap.html	JRE on PC/UNIX	Thomas (2003)
Genecounting/Hap	EM	http://www.mrc-epid.cam.ac.uk/Personal/jinghua.zhao/software.htm	Windows/UNIX	Zhao *et al.* (2002); Zhao and Sham (2004)

GPPH/PPH	Perfect phylogeny	http://wwwcsif.cs.ucdavis.edu/~gusfield/pph.html	LINUX/MAC/ UNIX/Windows	Gusfield (2002); Chung and Gusfield (2003)
GS-EM	EM	http://www.people.fas.harvard.edu/~junliu/genotype/	Web-based	Kang *et al.* (2004)
HAP	Imperfect phylogeny	http://research.calit2.net/hap/	Web-based	Halperin and Eskin (2004)
HAPAR	Pasimony	http://theory.tanford.edu/~xuying/hapar/	Windows/UNIX	Wang *et al.* (2003)
HAPINFERX	Clark's	www.nslij-genetics.org/soft/hapinferx.f	UNIX	Clark (1990)
Haplo.Stats	EM	http://mayoresearch.mayo.edu/mayo/research/ biostat/schaid.cfm	R/Splus (Windows/ UNIX)	Lake *et al.* (2003); Schaid *et al.* (2002b)
HAPLO/ PERMUTE	EM	krunch.med.yale.edu/haplo/	UNIX	Hawley and Kidd (1995)
HAPLOFREQ	Imperfect phylogeny	http://www.cs.princeton.edu/haplofreq/	Web-based	Halperin and Hazan (2006)
HAPLOTYPER	Bayesian	http://www.people.fas.harvard.edu/ ~junliu/em/em.html	UNIX	Niu *et al.* (2002b)
HAPLOREC	Bayesian	http://www.cs.helsinki.fi/group/genetics/ haplotyping.html	Java virtual machine, v1.4 or newer	Eronen *et al.* (2004, 2006)
HAPLOVIEW	EM + PL	http://www.broad.mit.edu/mpg/haploview/	JRE on MAC/ UNIX/Windows	Barrett *et al.* (2005)
HAPMAX	MLE	http://www.uni-kiel.de/medinfo/mitarbeiter/krawczak/ download/index.html	Windows	Krawczak (1994)
HAPSCOPE	EM/Bayesian	http://lpg.nci.nih.gov/lpg_small/protocols/HapScope/	UNIX/Windows	Zhang *et al.* (2002)
Helix Tree Genetics Analysis	EM	http://www.goldenhelix.com/pharmhelixtreesummary. html	MAC	Excoffier and Slatkin (1995); implemented by Gold helix, INC.

(*Continues*)

Table 14.1. (*Continued*)

Software	Method	Web link	Platform	References
HPLUS	EM + EE + PL	http://qge.fhcrc.org/hplus/	MATLAB on Windows/UNIX	Li *et al.* (2003); Zhao *et al.* (2005)
LPPH	Perfect phylogeny	http://wwwcsif.cs.ucdavis.edu/~gusfield/	Windows/LINUX	Ding *et al.*(2005)
PHASE	Bayesian	www.stat.washington.edu/stephens/software.html	Windows/MAC/UNIX	Stephens *et al.* (2001); Stephens and Donnelly (2003); Stephens and Sheet (2005)
PL-EM	PL-EM	www.people.fas.harvard.edu/~junliu/plem	Windows/UNIX	Qin *et al.* (2002)
SASgenetics	EM		SAS on Windows/UNIX	Czika *et al.* (2003)
SNPEM	EM	http://polymorphism.ucsd.edu/snpem/	UNIX	Fallin *et al.* (2001)
SNPHAP	EM	www-gene.cimr.cam.ac.uk/clayton/software/	UNIX	Clayton (2001)
THESIAS	Stochastic-EM	http://genecanvas.ecgene.net/downloads.php	Windows/UNIX	Tregouet *et al.* (2004)
WHAP	EM	http://pngu.mgh.harvard.edu/~purcell/whap/	Windows/UNIX	Purcell *et al.*, (2007)
2SNP	Combinatorial	http://alla.cs.gsu.edu/~software/2SNP/	UNIX/Webbased	Brinza and Zelikovsky (2006a,b, 2007)

III. HAPLOTYPE INFERENCE FROM PEDIGREES

A. Brief introduction for haplotype inference using pedigrees

Although modern laboratory techniques, such as allele-specific long-range PCR (MichlataosBeloin *et al.*, 1996), diploid-to-haploid conversion (Douglas *et al.*, 2001; Yan *et al.*, 2000), and other methods (Tost *et al.*, 2002), have been used to determine haplotypes, these approaches are technologically demanding and cost prohibitive, which makes it extremely difficult to carry out large-scale studies. An alternative strategy to construct an individual's haplotypes is through genotyping the person's close relatives. Although this strategy may reduce haplotype ambiguity and improve the efficiency for haplotype frequency estimates (Becker and Knapp, 2003; Rohde and Fuerst, 2001; Schaid, 2002), haplotype ambiguity may still persist when the number of markers is moderately large (Hodge *et al.*, 1999), especially in the presence of missing data which is common in large pedigrees. Therefore, it is still essential to infer haplotypes for each individual from pedigrees. Actually, the availability of a large number of dense SNP markers and the presence of missing data pose daunting challenges for haplotype inference from pedigrees. Compared with haplotype inference from unrelated individuals, the complexity of haplotype inference from pedigrees not only increases with the number of markers, but also with the number of individuals within pedigrees. For haplotype inference from unrelated individuals, we only need to consider all compatible haplotype pairs for each individual. While for haplotype inference from pedigrees, we generally need to consider all compatible haplotype pairs for each individual and all compatible haplotype configurations for each pedigree.

In the past two decades, many computational and statistical methods have been developed for haplotype inference using pedigrees, including haplotype frequency estimation and haplotype reconstruction, from general pedigrees (Elston and Stewart, 1971; Gao *et al.*, 2004; Kruglyak *et al.*, 1996; Lander and Green, 1987; Li and Jiang, 2003; O'Connell, 2000; Qian and Beckmann, 2002; Sobel *et al.*, 1995; Sobel and Lange, 1996; Weeks *et al.*, 1995; Wijsman, 1987; Zhang *et al.*, 2005). Some of these methods are based on exact-likelihood computations (Du *et al.*, 1998; Lander and Green, 1987; Sobel *et al.*, 1995; Weeks *et al.*, 1995), others are based on approximate-likelihood computations (Kruglyak *et al.*, 1996; Sobel *et al.*, 1995; Weeks *et al.*, 1995), whereas others rely on rule-based strategies (Li and Jiang, 2003; O'Connell, 2000; Qian and Beckmann, 2002; Tapadar *et al.*, 2000; Wijsman, 1987; Zhang *et al.*, 2005). Many of programs have incorporated more than one type of methods (Li and Jiang, 2003; O'Connell, 2000; Zhang *et al.*, 2005).

Almost all haplotype inference methods for pedigrees assume neither mutations nor genotyping errors in the data. Under these assumptions, the Mendelian law of inheritance is used to determine the complete or partial haplotypes of an individual and further to construct all compatible haplotype configurations for a pedigree. These methods vary greatly in models and computational strategies used. Thus, they have their own strengths and weaknesses. The exact likelihood-based methods are limited to small pedigrees with a small number of markers because of extensive computations. The approximate-likelihood-based method can handle relatively large pedigree with a large number of marker loci. Additional information and assumptions, such as recombination fractions between adjacent marker loci and HWE, are generally required for likelihood-based methods. The rule-based methods generally rely on fewer assumptions and run faster than likelihood-based methods. However, the lack of statistical model makes it difficult for rule-based methods to deal with missing data, which generally exists even in moderate size pedigrees. There is also no assessment of the reliability of the results obtained from rule-based methods, which may reduce the statistical power for association studies (Morris *et al.*, 2003a; Zhao *et al.*, 2003a). In addition, rule-based methods can be computationally intensive in the presence of recombinants and missing data. In this section, we will review some of commonly used methods for haplotype inference using pedigree and discuss their advantages and disadvantages in detail.

B. Rule-based methods for haplotype inference using pedigrees

1. The importance of the Mendelian law of inheritance for rule-based methods

The Mendelian law of inheritance plays a central role in methods for haplotype inference from pedigrees. Almost all methods use the Mendelian law of inheritance to determine the complete or partial haplotypes of an individual and further to construct all compatible haplotype configurations of a pedigree. Because most rules used in rule-based methods are extensions of the Mendelian law of inheritance, let us take a simple example to see how the Mendelian law of inheritance works here. Suppose the genotypes of father, mother, and their offspring at five loci are (11,12,11,12,12), (12,11,12,22,12), and (12,12,12,12,12), respectively. Because an individual inherits one allele from its father and one allele from its mother at a marker locus, it is easy to see that the offspring has two compatible haplotype pairs (12112, 21221) and (12111, 21222). In this example, the origin of alleles of the offspring at the first four loci can be determined while the origin of alleles at the fifth locus cannot be determined because both the parents and the offspring have the same heterozygous genotypes. Although this example is quite simple, it illustrates several general principles used in haplotype inference from pedigrees. First, in the absence of missing data, the haplotypes of the offspring can be determined for those loci at which the parents and the offspring do not have the same heterozygous genotypes.

In the presence of missing data, more sophisticated rules can be developed to assign haplotypes to the offspring. Therefore, the number of compatible haplotype pairs of the offspring can be generally reduced. In this example, the number of compatible haplotype pairs for the offspring is reduced from 32 to 2 (if we consider the origin of haplotypes). Second, the haplotypes of the parents cannot be determined by the haplotypes of their offspring, unless some additional assumptions are imposed. The total number of haplotype configurations for the pedigree can be very large even for small pedigrees. In this example, the total number of compatible haplotype configurations is 128 because the father has 8 compatible haplotype pairs and the mother has 8 possible haplotype pairs. It is easy to imagine that the total number of compatible haplotype configuration for a pedigree will be greatly increased with the size of the pedigree, especially in the presence of missing data. Actually, the greatest challenge for haplotype inference using pedigrees is how to construct the most likely configuration from a large number of compatible haplotype configurations. For simple small pedigrees with a small number of marker loci, it is not difficult to enumerate all compatible haplotype configurations, calculate their likelihood function, and identify the most likely configuration for each pedigree. For large pedigrees and/or large number of marker loci, it is impractical to enumerate all compatible haplotype configurations. In this situation, most available methods limit their search and calculation on a reduced subset of configurations by imposing some additional assumptions. For example, many methods have been proposed to search the most likely configuration and calculate the likelihood function based on haplotype configurations with minimum recombinants. In this example, only 2 of 128 compatible haplotype configurations for the pedigree have zero recombinants.

2. Rule-based methods with minimum recombinants

Most rule-based methods for haplotype inference from pedigrees aim to find all compatible haplotype configurations with minimum recombinants or without recombinants. This criterion may be valid due to a shift of focus on association studies that generally involve many tightly linked markers in a small region (e.g., Cox *et al.*, 2002; Daly *et al.*, 2001; Patil *et al.*, 2001), because recombinants are unlikely events and the contribution of haplotype configurations with more recombinants in the likelihood function is much smaller than those haplotype configurations with fewer recombinants for tightly linked markers.

Nejati-Javaremi and Smith (1995) proposed a simple rule-based algorithm to find a single compatible haplotype configuration for the pedigree. Starting with nuclear families with founder parents, the partial and/or complete haplotypes of each offspring are assigned according to the Mendelian law of inheritance. Then the most common haplotypes in the offspring that are compatible with the parental genotype are assigned to the parents. Then again, the complete haplotypes with minimum recombinants are assigned to each offspring according to the parents' haplotypes and its partial haplotypes. The procedure is then repeated for all nuclear families until the haplotypes of all individuals within the pedigree have been determined. It can be seen that the haplotype configuration obtained may not have minimum recombinants because the haplotypes of parents are assigned first without considering the possible haplotypes of the offspring. In addition, this method can only be applied to genotype data without missing values.

Tapadar *et al.* (2000) developed a genetic algorithm to find haplotype configurations with minimum recombinants. In their algorithm, a pair of haplotypes is randomly assigned to each founder according to its genotype. Then each parent–offspring trio is scanned and the haplotypes of the offspring with minimum recombinants are assigned using a set of heuristic rules. If several haplotype pairs of the offspring have the same number of recombinants, one of them is randomly selected. This procedure is repeated until the haplotypes of all individuals within a pedigree are assigned. To explore possible haplotype assignments for founders without exhaustive search, they developed a genetic algorithm with modified reproduction, crossover, and mutation operations. However, reproduction, crossover, and mutation operations are only used to emulate genetic process when searching the optimal solution in a large space in the genetic algorithm. They are certainly different from the definitions of these terms in real human data (Mitchell, 1996). Since the genetic algorithm is a stochastic algorithm, Tapadar *et al.* (2000) suggested running the algorithm multiple times to find the optimal haplotype configurations. However, their algorithm can only handle genotype data without missing values, which limits the usefulness of their method.

Qian and Beckmann (2002) proposed six sets of rules to exhaustively search all haplotype confirmations with minimum recombinants within a pedigree. Basically, these rules are applied to each parents–offspring trio and nuclear family sequentially. The first two sets of rules using the Mendelian law of inheritance are applied to the trios to impute missing genotypes and assign haplotypes to the offspring. The rule sets 3, 4, and 5 are applied to each nuclear family to impute missing genotypes and assign haplotypes to the parents conditional on the haplotypes of offspring and the criterion of minimum recombinants. In general, only partial haplotypes can be determined from the first five rules. In this situation, all compatible haplotype configurations within each nuclear family in the pedigree are enumerated and only haplotype configurations with minimum recombinants are retained. This method has been implemented in the software package MRH and can only handle genotype data with no substantial missing genotypes. However, because their method only keeps haplotype configurations having minimum recombinants within nuclear families, it cannot guarantee that the identified haplotype configurations have minimum recombinants except for pedigrees with at most one recombinant.

In the absence of missing data, Li and Jiang (2003) proposed a polynomial-time exact algorithm to reconstruct all compatible haplotype configurations without recombinants. The additional extensions from them (Li and Jiang, 2003; Doi *et al.*, 2003) have been applied to small pedigrees with missing genotype data. To accommodate large pedigrees with a large portion of missing data and allow for recombinants, Li and Jiang (2005) developed an ILP method to identify the haplotype configurations with minimum recombinants. A set of rules similar to those used in Qian and Beckmann (2002) are used to formulate the constraints and guide the order of the branch-and-bound strategy. If multiple haplotype configurations with minimum recombinants are found, the configuration with the highest posterior probability is selected. Their results on simulated data based on 50 loci from a pedigree of size 29 subjects showed that the accuracy of the ILP method is more than 99.8% for data with no missingness and 98.3% even for data with 20% missing alleles (Li and Jiang, 2005). The program is also substantially faster than SIMWALK (Sobel and Lange, 1996).

3. Rule-based methods for zero recombinants

A special case of identifying haplotype configurations with minimum recombinants is to identify haplotype configurations without recombinants. In this situation, more sophisticated rules can be developed and the calculations can be substantially reduced. Wijsman (1987) proposed a set of rules to impute the missing data, check genotyping errors, and construct haplotypes under the assumption of no recombinants. These rules are then extended by Zhang *et al.* (2005). In their methods, several rules are first used to initialize the haplotypes in an individual and the other rules are sequentially applied to update each individual's haplotypes by scanning all nuclear families in the pedigree. Since the change of the haplotypes of any individual in a nuclear family could possibly affect the haplotype assignments for the other individuals of this family and the individuals of neighborhood families in the pedigree, the process will be repeated until no further changes to the haplotypes can be made. Both simulation studies and analysis from real data sets illustrated that these rules are efficient and effective for inferring haplotypes in pedigrees (Zhang *et al.*, 2005). Similar results can also be found in Baruch *et al.* (2005), in which the efficiency of several rules for haplotype inference from large pedigrees under the assumption of no recombinants was tested in both simulated and real data sets.

The genotype elimination algorithm has been proposed to accelerate the computation of likelihood in linkage analysis by eliminating those genotypes that are not consistent with the Mendelian law of inheritance (Lange and Boehnke, 1983; Lange and Goradia, 1987). Theoretically, when the haplotypes are treated as alleles at a single locus, the genotype elimination algorithm can be used to reconstruct haplotypes directly. O'Connell (2000) and Zhang *et al.* (2005) have applied this technique to identify all haplotype configurations without recombinants. To allow for a large number of markers, O'Connell (2000) implemented a divide-and-conquer strategy which essentially has the same spirit as the partition–ligation technique used by Zhang *et al.* (2005). In the divide-and-conquer approach, all of the markers are broken down into units that contain only several markers and have one or two common markers with adjacent units. The genotype elimination algorithm is first employed to reconstruct haplotypes within each unit. Then two adjacent partial haplotypes are combined using the genotype elimination algorithm again. This approach is efficient when many incompatible haplotype configurations can be removed in each unit, but will not reduce the overall complexity when there are a large number of compatible haplotype configurations (O'Connell, 2000). To further reduce the computational complexity, O'Connell (2000) employed a pruning method to discard the haplotype configurations with low probability within each unit while Zhang *et al.* (2005) removed the haplotypes with frequencies less than a threshold (e.g., 10^{-5}).

Combined with the minimum recombinant criterion, other reasonable criteria can be used to further reduce the computational complexity. For example, we may want to find the smallest set of haplotypes that can form a compatible haplotype configuration without recombinants in the pedigree, as those methods for haplotype inference from unrelated individuals that try to determine a minimum set of haplotypes that can resolve all individuals (e.g., Gusfield, 2001). Zhang *et al.* (2005) implemented a greedy algorithm in HAPLORE to accomplish this goal. They used the haplotypes carried in completely haplotyped individuals to construct the compatible haplotype configurations without recombinants. The algorithm was found to be effective when it was applied to the GAW12 data (Wijsman *et al.*, 2001) and the Oxford ACE data (Keavney *et al.*, 1998). For the GAW12 data, the identified haplotypes can provide the compatible haplotype configurations for all individuals. For the Oxford ACE data, 17 identified haplotypes are not sufficient for resolving this problem. However, 79 out of the total 82 pedigrees have compatible haplotype configurations with them. After a simple investigation, 20 haplotypes can be identified and can generate compatible haplotype configurations for all pedigrees. The success of their algorithm strongly depends on the number of complete haplotypes identified from the logic rules. The extension of this method to a general approach that can allow for recombinants is still not clear and clearly warrants further study.

4. Weaknesses of rule-based methods

Both simulation studies and analysis from real data sets illustrated that rule-based methods are efficient and effective for inferring haplotypes in pedigrees in the absence of missing data. Other than the Mendelian law of inheritance, rule-based methods do not require other parameters and assumptions, such as recombination fractions and HWE. However, rule-based methods have several inherent weaknesses. First, as we have pointed out, the criterion of minimum recombinants is only applicable for tightly linked markers. For a set of sparsely spaced markers, the probability of a haplotype configuration with more recombinants can be higher than a haplotype configuration with fewer recombinants. Second, rule-based methods are most efficient for genotype data without missing data but can be computationally intensive if recombination events are allowed among markers and there is missing data. In fact, Li and Jiang (2003) showed that the problem of finding a minimum-recombinant haplotype configuration is in general NP-complete. This is to say that there is no polynomial time algorithm that guarantees to reconstruct haplotype configurations with minimum recombinants from any input. Third, there is no assessment of the reliability of the results obtained from rule-based methods. If there is only one compatible haplotype configuration with minimum recombinants for each pedigree, the frequency of each haplotype can be calculated by counting the number of occurrences of this haplotype in founders and dividing it by two times of the total number of founders over all pedigrees. However, there are often multiple compatible haplotype configurations for each pedigree, especially in the presence of missing data. Either selecting one of them randomly or considering them equally will result in inaccurate estimates of haplotype frequencies and affect subsequent analyses. In this situation, the use of a rule-based method for haplotype inference should proceed with special cautions.

C. Likelihood-based methods for haplotype inference using pedigrees

1. Basic assumptions and notations

Since the likelihood function of pedigrees plays a central role in haplotype inference, we first introduce some basic notations used in this section and give the detailed formula for the likelihood function. Suppose there are a total of K haplotypes comprising L markers $s_1, \ldots, s_L : H = \{h_1, h_2, \ldots, h_K\}$. Their frequencies in the population are denoted as $\Pi = \{\pi_1, \pi_2, \ldots, \pi_K\}$. The recombination fraction between two adjacent SNPs, s_l and s_{l+1}, is denoted by $\theta_{l,l+1}$ $(l = 1, \ldots, L-1)$. Assume that we have a total of N pedigrees and that an individual either has both parents or no parents at all. We define an individual in a pedigree as a nonfounder if both parents are available and others are referred as founders. In a pedigree$f(1 \leq f \leq N)$, let n_f be the number of founders and m_f be the total number of individuals in the pedigree. The n_f founders are indexed as $1, n_f$ and the $m_f - n_f$ nonfounder individuals are indexed as $n_f + 1, \ldots, m_f$. The genotype and two haplotypes (unordered) of individual r in family f are denoted as G_{f_r} and $H_{f_r} = \{h_{f_r,1}, h_{f_r,2}\}$, respectively. For a nonfounder individual $r(r = n_f + 1, \ldots, m_f)$ in the pedigree, we use $H_{f_r}^F$ and $H_{f_r}^M$ to denote the haplotype pairs of its father and mother, respectively. We further assume that $h_{f_r,1}$ and $h_{f_r,2}$ are transmitted from its father and its mother, respectively, if individual r is a nonfounder. We define the function $R(H_{f_r}|H^F_{f_r}, H^M_{f_r})$ as the number of recombinants for this haplotype configuration in the trios, which is the sum of the number of recombinants for the haplotype transmitted from the father and the haplotype transmitted from the mother. Specifically, we have $R(H_{f_r}|H^F_{f_r}, H^M_{f_r}) = R(h_{f_r,1}|H^F_{f_r}) + R(h_{f_r,2}|H^M_{f_r})$. The total number of recombinants is just the summation of $R(H_{f_r}|H^F_{f_r}, H^M_{f_r})$ overall trios in the pedigree. It is likely that many haplotype pairs are compatible with the genotypes of pedigrees, especially when there are missing data. Let S_{f_r} denote all the compatible haplotype pairs for individual r in family f. If all compatible haplotype configurations have already been determined, then the likelihood of the genotypes for a pedigree f $(1 \le f \le N)$ is:

$$L_f(G_{f_1}, G_{f_2}, \ldots, G_{f_{m_f}}|\Pi) = \sum_{H_{f_1} \in S_{f_1}} \cdots \sum_{H_{f m_f} \in S_{f m_f}} P(G_{f_1}, \ldots, G_{f_{m_f}}, H_{f_1}, \ldots, H_{f_{m_f}}|\Pi)$$

$$= \sum_{H_{f_1} \in S_{f_1}} \cdots \sum_{H_{f m_f} \in S_{f m_f}} P(G_{f_1}, \ldots, G_{f_{m_f}}|H_{f_1}, \ldots, H_{f_{m_f}}, \Pi)$$

$$\times P(H_{f_1}, \ldots, H_{f_{m_f}}|\Pi) = \sum_{H_{f_1} \in S_{f_1}} \cdots \sum_{H_{f m_f} \in S_{f m_f}} \prod_{r=1}^{n_f} P(H_{f_r}|\Pi)$$

$$\times \prod_{r'=n_f+1}^{m_f} P(H_{f_{r'}}|H^F_{f_{r'}}, H^M_{f_{r'}})$$

where $P(H_{f_r}|\Pi)$ is the probability of haplotype pair H_{f_r} of the founder r in pedigree f given the haplotype frequencies, and $P(H_{f_{r'}}|H^F_{f_{r'}}, H^M_{f_{r'}}) = P(h_{f_{r'}1}|H^F_{f_{r'}})P(h_{f_{r'}2}|H^M_{f_{r'}})$ is the gamete transmission probability for a trio, where $H^F_{f_{r'}}$ and $H^M_{f_{r'}}$ are the haplotype pairs for the father and mother, respectively (Elston and Stewart, 1971).

Given the haplotype frequencies and a haplotype configuration, the likelihood function is the product of the probability of founders and the gamete transmission probability of parent–offspring trios (Elston and Stewart, 1971), which is easy to calculate. The difficulty is how to calculate the likelihood function over all possible haplotype configurations and choose one haplotype configuration with the highest probability. Different methods calculate $P(H_{f_r}|\Pi)$ and $P(H_{f_{r'}}|H^F_{f_{r'}}, H^M_{f_{r'}})$ based on different assumptions and strategies. The calculation of $P(h_{f_{r'},1}|H^F_{f_{r'}})$ and $P(h_{f_{r'},2}|H^M_{f_{r'}})$ is straightforward given $h_{f_{r'},1}$, $h_{f_{r'},2}$, $H^F_{f_{r'}}$ and $H^M_{f_{r'}}$, because the recombination events can be easily determined by them. Suppose there are $u(u \le L-1)$ recombinants between SNPs s_{l_i} and s_{l_i+1} $(i = 1, \ldots, u)$ and there are no recombinants between SNPs s_{l_i} and $s_{l_i+1}(i = u+1, \ldots, L-1)$ when the haplotype $h_{f_r,1}$ is transmitted from its father, then $P(h_{f_{r'}1}|H^F_{f_{r'}}) = \prod_{i=1}^{u} \theta_{l_i,l_i+1} \prod_{i=u+1}^{L-1}(1-\theta_{l_i,l_i+1})$ if $H^F_{f_{r'}}$ contains two identical haplotypes and $P(H_{f_{r'}1}|H^F_{f_{r'}}) = (1/2)\prod_{i=1}^{u} \theta_{l_i,l_i+1} \prod_{i=u+1}^{L-1}(1-\theta_{l_i,l_i+1})$ if $H^F_{f_{r'}}$ contains two different haplotypes. $P(h_{f_r,2}|H_{f_r}^{M})$ can be computed using the same method. The calculation of $P(H_{f_r}|\Pi)$ is also straightforward. Given the haplotype frequencies $\Pi = \{\pi_1, \pi_2, \ldots, \pi_K\}$, $P(H_{f_r}|\Pi) = 2\pi_{f_r,1}\pi_{f_r,2}$ if $h_{f_r,1} \neq h_{f_r,2}$ and $\Pr(H_{f_r}|\Pi) = \pi_{f_r,1}\pi_{f_r,2}$ if $h_{f_r,1} = h_{f_r,2}$ under the assumption of HWE. This assumption is required by almost all likelihood-based methods for haplotype inference from pedigrees as well as from unrelated individuals. However, the haplotype frequencies $\Pi = \{\pi_1, \pi_2, \ldots, \pi_K\}$ are generally unknown and need to be estimated from the data. The likelihood-based methods for haplotype inference from general pedigrees can be classified to two major categories according to the ways that $\Pi = \{\pi_1, \pi_2, \ldots, \pi_K\}$ are estimated. One group of method assumes linkage equilibrium (LE) among the markers while the other group of methods allows the presence of LD.

2. Likelihood-based methods based on LE

Most methods for haplotype inference from pedigrees used in linkage analysis assume LE among the markers. That is, the haplotype frequency is equal to the product of allele frequencies at individual markers, and this assumption ignores LD between tightly linked markers. The haplotypes inferred from these programs may not be appropriate for association studies in which the tightly linked SNP markers are generally used (Schaid *et al.*, 2002a). Thus, these methods may be only applicable for relatively sparse markers. In addition, these methods also require the input of accurate allele frequencies, which may not be readily estimated in the presence of missing data, especially when only small partially typed pedigrees are available. In this situation, it may be beneficial to use other sources of information on allele frequencies.

Because haplotype frequencies can be calculated according to the allele frequencies at each marker for methods based on LE, the only aim of these methods is to find a haplotype configuration with the highest probability. For smaller pedigrees with a small number of markers, a straightforward strategy is to enumerate all possible haplotype configurations, calculate the probability of each of them, and select one configuration with the highest probability (Sobel *et al.*, 1996). To allow for a large number of markers, Lander and Green (1987) proposed an algorithm based on inheritance vectors. For a pedigree with n_f founders and $m_f - n_f$ nonfounders, the inheritance vector at a marker locus is a vector with the length of $2(m_f - n_f)$. The inheritance vector at a marker locus completely determines which of $2n_f$ founder alleles are inherited by each nonfounder. In other words, the inheritance vector at a marker completely determines the haplotypes of each individual. To calculate the joint distribution of inheritance vectors across all markers, Lander and Green (1987) employed Hidden Markov Models, which can efficiently calculate the inheritance vector with the highest probability. In the Lander–Green algorithm, the number of possible inheritance vectors at each marker that are compatible with the pedigree and the genotype can be up to $2^{2(m_f - n_f)}$ (although the actual number of inheritance vector can be much less than this number) and all of them need to be stored and evaluated. Thus, the memory and the computational complexity required by the Lander–Green algorithm increase linearly with the number of markers but exponentially with the size of pedigrees. The original GENEHUNTER, which is the first commonly used software package that implements the Lander–Green algorithm, can only handle pedigrees with $2(m_f - n_f) - n_f \leq 16$, where n_f is the number of founders and $m_f - n_f$ is the number of nonfounders. To improve the computational efficiency of GENEHUNTER, Gudbjartsson *et al.* (2000) developed and implemented ALLEGRO, which is typically 20–100 times faster than GENHUNTER. Abecasis *et al.* (2002) observed that only a small number of inheritance vectors at each marker that are compatible with the pedigree and proposed to efficiently store and evaluate the inheritance vectors using sparse gene flow trees. They implemented it in the software package MERLIN, which can markedly reduce the computational time compared to GENEHUNTER and ALLEGRO. Recently, Gudbjartsson *et al.* (2005) implemented the 2.0 version of ALLEGRO based on multiterminal binary decision diagrams (MTBDDs). MTBDDs only store the probability distributions of all unique and nonredundant inheritance vectors. The number of unique and nonredundant inheritance vectors is much less than the number of all inheritance vectors, thus ALLEGRO 2.0 not only runs faster than GENEHUNER, ALLEGRO, and MERLIN, but also can handle larger pedigrees than that can be handled by GENEHUNTER.

Other than the Lander–Green algorithm, another commonly used algorithm for the likelihood function calculation is the Elston–Stewart algorithm (Elston and Stewart, 1971). The computational complexity of the Elston–Stewart algorithm increases linearly with the size of pedigrees but exponentially with the number of markers. Therefore, the Elston–Stewart algorithm is suitable to handle large pedigrees with a small number of marker loci. Advance techniques implemented in VITESSE (O'Connell and Weeks, 1995) and SUPERLINK (Fishelson *et al.*, 2005) allow these software packages to handle large pedigrees with a moderate number of markers. However, there is no efficient way to select the most likely haplotype configuration in the Elston–Stewart algorithm. Thus, for large pedigrees, several stochastic and deterministic algorithms have been proposed to find the most likely haplotype configurations. Sobel *et al.* (1996) and Sobel and Lange (1996) proposed a stochastic optimization algorithm, simulated annealing with the random work method, to find a nearly optimal haplotype configuration. In each step, the algorithm selects the most likely haplotype configuration with the highest probability. At the same time, the algorithm can still select other haplotype configurations with small probabilities. This strategy can avoid the algorithm to be trapped in a local optimal solution. The algorithm is implemented in a software package SIMWALK2 (Sobel and Lange, 1996) and has been successfully applied to infer haplotypes from many large pedigrees. So far SimWalk2 is still among a few software packages available that can be used for haplotype inference using very large pedigrees with a large portion of missing data. However, the computing time for SIMWLAK2 can be very long and it generally needs to be run several times to get a good solution due to its stochastic nature. Lin and Speed (1997) developed another MCMC method to identify a set of haplotype configurations with high probabilities. The multiple configurations along with their posterior probabilities can provide more information than the single configuration, especially for large pedigrees in which many haplotype configurations can have the same or very similar posterior probabilities as the most likely configuration. Other than using stochastic algorithms, Gao *et al.* (2004) proposed a deterministic method to identify a set of haplotype configurations with high posterior probabilities. In their algorithm, the haplotype pairs of each individual are assigned at each maker locus sequentially. To achieve this, the conditional probability of a haplotype pair of an individual at a marker locus is calculated by an approximation method using the information of flanking markers of this individual and its parents and offspring. Through a user defined threshold, the haplotype pairs with the lower conditional probabilities are discarded and the rest of haplotype pairs with their corresponding conditional probabilities are used for assignment of haplotype pairs of each individual at each marker and determination of the optimal order of such assignments. Gao *et al.* (2004) have tested their algorithm on real and simulated data sets. Their results showed their algorithm ran much faster than SIMWALK2 and identified haplotype configurations with higher probabilities than those obtained from SIMWALK. However, the current algorithm can only handle pedigrees with complete genotypes and extension to handle pedigrees with the missing data is warranted.

3. Likelihood-based methods based on LD

In general, there is a great deal of LD between tightly linked markers. Ignoring LD in linkage studies (Abecasis and Wigginton, 2005) as well as association studies (Schaid *et al.*, 2002b) can result in increased false positives. Due to LD, the haplotype frequencies cannot be directly calculated from the allele frequencies. In this situation, the haplotype frequencies (Π) can be estimated using the maximum-likelihood estimation (MLE) approach. Because it is difficult to directly obtain the MLE of Π, the EM algorithm, as it is used for haplotype inference from unrelated individuals, is frequently used to estimate Π. Suppose that we know that $\Pi = \Pi^{(i)}$ and we want to update $\Pi^{(i+1)}$. In the E-step, the probability of each compatible haplotype configurations is calculated. In the M-step, the expected number of occurrences of each haplotype in the sample is calculated by summing the number of occurrences of this haplotype in founders over all compatible haplotype configurations and weighting the probability of each configuration (Abecasis and Wigginton, 2005; Zhang *et al.*, 2005). Since it needs to calculate the probability of each compatible haplotype configuration and uses it in each step of iteration in the EM algorithm, it can become very computationally intensive even for a pedigree of moderate size with a small number of markers. Therefore, most EM-based methods for haplotype inference from pedigrees only use compatible haplotype configurations with zero or minimum recombinants. If one haplotype configuration has more recombinants than another haplotype configuration, then its likelihood contains more components of $\theta_{l,l+1}$ and less components of $1 - \theta_{l,l+1}$. In general, the recombination fraction between two tightly linked SNP markers, $\theta_{l,l+1}$, is small. For example, $\theta_{l,l+1}$ is in the magnitude of 10^{-4} for two SNPs with 10 kb distance. Therefore, the contribution of haplotype configurations with more recombinants in the likelihood function is much smaller than those haplotype configurations with fewer recombinants. Thus, the likelihood based on haplotype configurations with minimum recombinants can be considered a good approximation of the full likelihood for tightly linked SNPs. In addition, the haplotypes from a haplotype configuration with zero recombinants can be considered as alleles at multi-allelic marker, the more advanced techniques for the likelihood calculation can be applied to haplotype inference to further reduce the computations (e.g., Lange and Boehnke 1983; O'Connell and Weeks, 1995).

In its early stage, the EM-based methods for haplotype inference from pedigrees only considered nuclear families and parent–offspring trios (Becker and Knapp, 2004; Lin *et al.*, 2004; Rohde and Fuerst, 2001). For parent–offspring trios, compatible haplotype configurations without recombinants always exist and can be easily identified. Recently, several commonly used software packages for haplotype inference from unrelated individuals have been extended to handle parent–offspring trios (Marchini *et al.*, 2006). Cox *et al.* (2002), O'Connell (2000), and Zhang *et al.* (2005) developed the EM algorithm to estimate haplotype frequencies and assign haplotype configurations based on haplotype configurations with zero recombinants. Generally, these three methods can be divided into three steps: (1) identify the partial or complete haplotypes carried by each individual using a set of logic rules under the assumption of no recombinants; (2) identify all compatible haplotype configurations without recombinants using the genotype elimination algorithm; and (3) estimate the haplotype frequencies using the EM algorithm. To allow for a large number of markers, O'Connell (2000) implemented a branch and bound strategy in ZAPLO and Zhang *et al.* (2005) implemented a partition–ligation technique in HAPLORE. Although both methods allow for LD between adjacent markers, they do not allow for recombinants, which may be problematic for a large number of markers. To accommodate LD between tightly linked markers, Abecasis and Wigginton (2005) grouped a set of consecutive markers into clusters. Within each cluster, they assumed that there are no recombinants and developed an EM algorithm to estimate haplotype frequencies. Then haplotype frequencies across all clusters are estimated using the Lander–Green algorithm (Lander and Green, 1987). Their method does allow for recombinants between clusters but also assumes LE between clusters, which may not be true for tightly linked markers. Li and Jiang (2003, 2005) also developed the EM algorithm for haplotype frequency estimation based on the haplotype configurations with the minimum recombinants and implemented it in PEDPHASE. Rather than using all haplotype configurations with minimum recombinants in the EM algorithm, PEDPHASE only uses a single haplotype configuration with minimum recombinants in each iteration. Specifically, conditional on current haplotype frequency estimates, the haplotype configuration with the maximum probability in each pedigree is selected as its true configuration. Based on this single haplotype configuration in each pedigree, the expected number of occurrences of a haplotype is just the number of its occurrences in founders. Simulation studies have shown the performance of such EM algorithm may be compromised in the presence of missing data because there are a large number of haplotype configurations with minimum recombinants (Zhang and Zhao, 2006).

4. Weaknesses for likelihood-based methods for haplotype inference using pedigrees

The calculation of the likelihood function requires many assumptions. The departure from these assumptions may result in inaccurate haplotype inference for likelihood-based methods. Thus, these assumptions should be carefully checked before applying any haplotype inference method. The assumption of HWE is required by almost all likelihood-based methods for haplotype inference using pedigrees as well as unrelated individuals. Although simulation studies have shown that departures from HWE will affect the performance of these methods for haplotype inference (Fallin and Schork, 2000; Niu *et al.*, 2002b; Zhang and Zhao, 2006), all these studies showed that the performance of most methods compared in these studies were not strongly impacted by such departures, especially for haplotype inference using pedigrees (Zhang and Zhao, 2006). The recombination fractions are used in the calculation of the gamete transmission probability in the likelihood function. In general, it is difficult to obtain an accurate estimation of the recombination fraction between two tightly linked markers. This will potentially affect the accuracy for haplotype inference. In addition, almost all of methods assume Haldane's mapping function, in which the recombinants occurs independently on disjoint chromosome regions (Haldane, 1919). Haldane's model ignores genetic interference, which can lead to suspicious recombinants in haplotype inference (Lin *et al.*, 2003). Skrivanek *et al.* (2003) implemented a software package for haplotype inference, SIMPLE, which adopts a χ^2 model to characterize recombination events. In the presence of genetic interference, SIMPLE can reduce the number of suspicious recombinants in a short region.

D. The program packages

We have compiled a list of commonly used programs for haplotype inference using pedigrees in Table 14.2. As we have stated before, each program has its own strengths and weaknesses and is based on different assumptions, we conclude that there is no a single program than can outperform other programs in all situations. Thus, we suggest users to choose their programs upon several important factors: the pedigree size, the number and density of markers, and the missing data rate. We also suggest users to refer to Table 14.2 and their manuals for more detailed descriptions for these programs.

IV. POPULATION-BASED HAPLOTYPE-ASSOCIATION METHODS

Studies have shown that methods based on haplotypes may provide more power and accuracy in disease gene mapping than those based on single markers (Akey *et al.*, 2001; Botstein and Risch, 2003; Fallin *et al.*, 2001; Kankova *et al.*, 2005). Many methods have been proposed for haplotype-association analysis. Based on the data, these methods can be classified into population-based and family-based methods and will be introduced separately in this section and Section V.

As stated in Section II, because of the late-onset nature of many diseases and practical considerations, many studies only recruit unrelated individuals (Niu, 2004). Therefore there is a great need for population-based haplotype-association methods. In this section, we will review two main classes of such methods: methods based on score statistics and methods under the regression framework.

A. Score statistics for haplotype analysis

If haplotype information is known, many existing methods can be used for haplotype-association analysis: either to compare the frequencies of haplotypes between cases and controls, using many of the methods already developed for allele frequency comparison (Schaid, 2004), or to perform the analysis under the regression framework, where haplotypes can be treated as categorical variables. However, as stated in previous sections, haplotype phase information is usually unknown and needs to be estimated.

Table 14.2. Available Software Packages for Haplotype Inference from Pedigrees

Software	Method and assumption	Web link	Platform	References
ALLEGRO	Lander–Green algorithm, linkage equilibrium	http://www.decode.com/software/allegro		Gudbjartsson *et al.* (2000, 2005)
GENEHUNTER	Lander–Green algorithm, linkage equilibrium	http://www.broad.mit.edu/ftp/distribution/software/genehunter/	Unix	
HAPLOPED	Genetic algorithm, minimum recombinants	Contact the authors		Tapadar *et al.* (2000)
HAPLORE	Rule-based algorithm, EM algorithm, haplotype elimination algorithm, linkage disequilibrium, zero recombinants	http://bioinformatics.med.yale.edu/group/software.html	Windows, Unix, Linux	Zhang *et al.* (2005)
MERLIN	Lander–Green algorithm, Sparse Tree, EM algorithm, linkage equilibrium, linkage disequilibrium, zero recombinants	http://www.sph.umich.edu/csg/abecasis/Merlin	Unix, Linux, Windows	Abecasis *et al.* (2002); Abecasis and Wigginton (2005)
MRH	Rule-based algorithm, minimum recombinants	Contact the authors	Windows, Unix	Qian and Beckmann, 2002
PATCH	Rule-based algorithm, zero recombinants	Contact the authors	Windows	Wijsman, 1987
PedPhase	Integer linear programming, EM algorithm, linkage disequilibrium, minimum recombinants	http://vorlon.case.edu/~jxl175/haplotyping.html	Windows	Li and Jiang, 2003, 2005

(*Continues*)

Table 14.2. (*Continued*)

Software	Method and assumption	Web link	Platform	References
SIMPLE	Linkage equilibrium	http://www.stat.ohio-state.edu/~statgen/SOFTWARE/SIMPLE/	Unix	Skrivanek *et al.* (2003)
SIMWALK	Elston–Stewart algorithm, simulated annealing algorithm, linkage equilibrium	http://www.genetics.ucla.edu/software/simwalk	Unix	Sobel and Lange (1996)
ZAPLO	Haplotype elimination algorithm, EM algorithm, zero recombinants	http://www.molecular-haplotype.org/zaplo/zaplo_index.html	Unix, Linux	O'Connell (2000)

The traditional haplotype-association methods for case–control studies are usually goodness-of-fit (GOF) tests to determine whether the haplotype distributions between the cases and controls are the same. Typically a likelihood ratio statistic can be constructed as $\text{LRT} = 2(\ln L_{\text{cases}} + \ln L_{\text{controls}} - \ln L_{\text{pool}})$ which follows asymptotically a χ^2 distribution with $R - 1$ degrees of freedom, where R is the number of distinct haplotypes, under the null hypothesis. This approach has some limitations (Schaid, 2004) (1) when there are many haplotypes, there are many degrees of freedom and the power to detect association can be weak. In addition, with sparse data, the estimates for the rare haplotypes can be problematic and the null distribution may not follow a χ^2 distribution; (2) it cannot adjust for other covariates; (3) it only works for discrete outcomes; and (4) it assumes HWE for the pairs of haplotypes. Several methods have been proposed to address these limitations.

1. Haplotype sharing and clustering

Intuitively, the number of haplotypes can be reduced by clustering some "similar" haplotypes. Many statistical methods have been proposed that are based on searching for excess similarity among haplotypes from cases compared to that from controls (Beckmann *et al.*, 2005; Bourgain *et al.*, 2000, 2001; Devlin *et al.*, 2000; Tzeng, 2003, 2005; Van der Meulen and te Meerman, 1997; Yu *et al.*, 2005a,b; Zhang and Zhao, 2001; Zhang *et al.*, 2002b). The initial idea of haplotype sharing was put forth by Te Meerman and Van Der Meulen (1997) by proposing a haplotype sharing statistic (HSS) based on the variance of the shared lengths of haplotypes surrounding a possible location across all possible pairs of case haplotypes. Tzeng *et al.* (2003) showed that many of these HSSs can be formulated into the following quadratic form:

$$D = \hat{\Pi}_a^t A_a \hat{\Pi}_a - \hat{\Pi}_u^t A_u \hat{\Pi}_u, \quad T = \frac{D}{\sigma(\hat{\Pi})},$$

where $\sigma^2(\Pi) = \text{Var}(D)$; $\Pi_a = (\pi_{a1}, \ldots, \pi_{aR})$, and $\Pi_u = (\pi_{u1}, \ldots, \pi_{uR})$ are haplotype distributions for the cases and controls, respectively; A is a symmetric matrix containing the entries defined by $K(H_i, H_j)$, a symmetric kernel function of some feature in the comparison of the ith and jth haplotypes. As long as Π is bounded away from singularities in the limiting distribution, T is shown to be approximately distributed as a standard normal variable (Tzeng *et al.*, 2003).

Haplotype similarity provides a natural way to group (cluster) haplotypes, which offers one promising solution to the difficulties in the presence of many haplotypes. Haplotype grouping may enhance the efficiency of haplotype analysis by using a small number of haplotype clusters that can reduce the degrees of freedom and reduce the effects from rare haplotypes. Because haplotype sharing/clustering methods take into account the LD information between multiple markers, and usually have fewer degrees of freedom, they may have good power to detect predisposing genes (Beckmann *et al.*, 2005; Tzeng *et al.*, 2003; Yu *et al.*, 2005a).

Tzeng *et al.* (2003) showed that for common disease haplotypes, haplotype sharing tests can be more powerful than GOF tests but for rare haplotypes, the inverse is true. In addition, they showed that the power of both approaches can be enhanced by clustering rare haplotypes appropriately from the distributions prior to testing.

We can see that haplotype sharing/clustering is a technique and does not pertain to a specific test or method. We will see this again in Section IV.B.2.

2. Nonlinear test statistics

Zhao *et al.* (2005, 2006) proposed to enhance the power of the standard χ^2 statistic, which is described at the beginning of Section IV.A , through the use of nonlinear transformations to amplify the differences in haplotype frequencies between cases and controls. They stated that the standard χ^2 statistic compares haplotype frequencies—or linear transformations of haplotype frequencies—between cases and controls, although in a quadratic form. They argued that the key to increasing the power of the standard χ^2 statistic is to amplify the difference in haplotype frequencies between cases and controls.

Let $\hat{P}^A_{H_i}$ and $\hat{P}_{H_i}$ be the estimators of frequencies of haplotype H_i in cases and controls, respectively. $P^A = [P^A_{H_1}, \ldots, P^A_{H_R}]^T$, $P = [P_{H_1}, \ldots, P_{H_R}]^T$, $\Sigma^A = \text{diag}(P^A_1, \ldots, P^A_R) - P^A(P^A)^T$, and $\Sigma = \text{diag}(P_1, \ldots, P_R) - PP^T$. Let $f(x)$ be a continuously differentiable nonlinear function with a nonzero differential at x; $X_j = f(\hat{P}^A_{H_j})$, $Y_j = f(\hat{P}_{H_j})$ for $j = 1, \ldots, R$, $X = [X_1, \ldots, X_R]^T$, and $Y = [Y_1, \ldots, Y_R]^T$; $B = (b_{ij})_{R \times R}$ and $C = (c_{ij})_{R \times R}$ where $b_{ii} = \partial f(P^A_{H_i})/\partial P^A_{H_i}$, $b_{ij} = 0$, $c_{ii} = \partial f(P_{Hi})/\partial P_{Hi}$, and $c_{ij} = 0$. Define

$$\Lambda = \frac{1}{2n_A} B\Sigma^A B^T + \frac{1}{2n_U} C\Sigma C^T,$$

where n_A and n_U are the numbers of case haplotypes and control haplotypes, respectively. Let $\hat{\Lambda}$ be an estimator of the matrix Λ. The nonlinear test statistic is defined as

$$T_N = (X - Y)^T \hat{\Lambda}^- (X - Y),$$

where $\hat{\Lambda}^-$ is the generalized inverse of matrix $\hat{\Lambda}$ (Zhao *et al.*, 2006).

Zhao *et al.* (2005, 2006) evaluated the power of four nonlinear test statistics, where $f(\cdot)$ takes the form of entropy, exponential, quadratic, and reciprocal, and showed that under certain conditions, some nonlinear test statistics have much higher power than the standard χ^2 statistic. They also showed that the similarity measure of the genome region is a quadratic function of haplotype frequencies and thus the test statistics based on such similarity measures belongs to nonlinear tests. They showed through simulation that the nonlinear test did not increase false positives while maintaining high power (Zhao *et al.*, 2006).

3. Score statistics from regression models

Schaid (2004) stated that the haplotype-association methods based on a generalized linear model (GLM) framework provided a way to construct score statistics for the null hypothesis of no haplotype effects. The score statistics constructed this way can adjust for other covariates and handle many traits (such as binary traits, quantitative traits, survival traits). For the convenience of description, details will be introduced in Section IV.B.3.

B. Haplotype-association analysis using regression models

As noted earlier, haplotype phase information is usually unknown and needs to be estimated, where uncertainties are induced. If such uncertainties from haplotype inference are not taken into account, haplotype-based analyses using inferred haplotypes may be biased (Lin and Huang, 2007; Lin and Zeng, 2006; Tanck *et al.*, 2003; Zhao *et al.*, 2003b). Some methods have been proposed to incorporate haplotype inference into haplotype-association analysis using unphased genotype data. In this section, we will review haplotype analysis methods under the regression framework.

1. Regression models for unphased haplotypes

In practice, haplotypes cannot be inferred without ambiguity most of the time, no matter what method is used. A common approach is to use the most likely haplotype pair for each individual in the subsequent analysis, as if they were observed. Studies (Schaid, 2002, 2004) have shown that this two-stage approach not only may lose information substantially, but also may introduce measurement error and induce bias into the estimates of haplotype effects (Lin and Huang, 2007; Lin and Zeng, 2006; Schouten *et al.*, 2005; Tanck *et al.*, 2003; Zhao *et al.*, 2003b).

One intuitive way to deal with this problem is to use all possible haplotype pairs that are consistent with the observed genotype data (Chiano and Clayton, 1998; Lake *et al.*, 2003; Schaid, 2004; Schaid *et al.*, 2002b; Zaykin *et al.*, 2002; Zhao *et al.*, 2003b). Another more powerful way is to estimate haplotypes and their effects simultaneously (Epstein and Satten, 2003; Lin and Zeng, 2006; Spinka *et al.*, 2005). The majority of these methods are based on a prospective likelihood (Chiano and Clayton, 1998; Lake *et al.*, 2003; Lin, 2004; Mander, 2001, 2002; Schaid, 2004; Schaid *et al.*, 2002b; Stram *et al.*, 2003; Zaykin *et al.*, 2002; Zhao *et al.*, 2003b). Some methods for case–control studies are based on a retrospective likelihood to account for sample ascertainment (Epstein and Satten, 2003; Spinka *et al.*, 2005). Let G_i denote genotype information, which is directly observed, for the *i*th individual; $H_i = \{h_i, h_i'\}$ denote haplotype pair; $S(g)$ denotes the set of haplotype pairs that are consistent with $G = g$; X represents environmental covariates that may affect disease risk; $P(H)$ denotes the prior probability of haplotype pair H; Y denotes the dependent trait. In this chapter, we use the term "environmental covariates/factors" in a general way, including risk factors such as age, gender, and body mass index which are not environmental factors in a strict way. Then the prospective likelihood of the data can be expressed as (Schaid, 2004)

$$L = \prod_{i=1}^{N} \sum_{H_i \in S(G_i)} P\{Y_i | X_i, G_i(H_i), \beta\} P(H_i)$$

where N is the number of subjects in the sample and β is the vector of regression coefficients (including effects of X, H, and possibly their interactions). Usually maximum-likelihood methods are used to fit the above regression model, such as that described in Lake *et al.* (2003), where haplotype frequencies and regression parameters are estimated jointly. Zhao *et al.* (2003b) used estimating equations to estimate the logistic regression parameters based on score equations derived from the prospective likelihood, with the assumptions of rare disease and independence between haplotypes and environmental variables. They used an EM algorithm similar to that proposed in Excoffier and Slatkin (1995) to estimate haplotype frequencies that are required for evaluation of the prospective score equations using only controls. Under the assumption of independence between genes and environmental factors, it is straightforward to incorporate environmental factors in these prospective likelihood-based methods (Lake *et al.*, 2003; Zhao *et al.*, 2003b). When an exposure to external environmental factors is not directly controlled by an individual's own behavior, this independence assumption is likely to be satisfied; otherwise this assumption could be violated (Spinka *et al.*, 2005). Lin and Zeng (2006) indicated that this assumption may not hold in practice and is not statistically efficient. However, it is still not clear about the effect in the violation of this assumption.

Because the cases are over-sampled in case–control studies, as a consequence, the estimated haplotype frequencies will be biased under the alternative hypothesis if this is not handled properly (Schaid, 2004). In addition, the haplotype effects could be biased as well. This bias induced by ascertainment does not occur when phase is known, that is, for phased haplotype data (Schaid, 2004). The reason for this is that the distribution of the model covariates is fully nonparametric in this case (Lin and Zeng, 2006). The magnitude of the bias depends on the accuracy of the estimated haplotypes. The lack of consideration of the ascertainment is a limitation of the prospective likelihood methods. To deal with the ascertainment issue, Zhao *et al.* (2003b) proposed to use only controls to estimate haplotype frequencies, this may only work for rare diseases and the method could produce substantial biases for model parameter estimates when the underlying rare disease assumption is violated (Spinka *et al.*, 2005). Stram *et al.* (2003) proposed to use sampling weights based on the population disease prevalence to correct for biased estimates. Epstein and Satten (2003) proposed a retrospective likelihood approach where sampling ascertainment is taken into the model, and regression parameters and haplotype frequencies can be estimated jointly. Their retrospective likelihood can be expressed as a product of multinomial of the observed genotype data as follows:

$$L = \prod_g [\sum_{(h,h')\in S(g)} \Pr(H = (h,h')|D = 0)]^{c_g} [\sum_{(h,h')\in S(g)} \Pr(H = (h,h')|D = 1)]^{d_g},$$

where c_g and d_g are number of controls and cases with genotype g in the sample, D is the disease status ($D = 0$ indicates free of disease, $D = 1$ indicates present of disease). Note that this approach requires HWE only for the controls, yet still uses both cases and controls to estimate haplotype frequencies and haplotype odds ratio. This retrospective approach was shown to have equal or better power than some of the previously proposed prospective approaches (Satten and Epstein 2004), which may be due to the fact that the retrospective approach fully exploits the HWE assumption for the underlying population (Spinka *et al.*, 2005). However, Satten and Epstein (2004) also found that their approach was not robust to departures from HWE among the controls when haplotype effects are dominant or recessive, in contrast to some prospective methods that are robust to departures from HWE (Schaid *et al.*, 2002b; Zhao *et al.*, 2003a). They proposed to introduce an additional parameter ("fixation index") to account for the average departure from HWE in order to reduce bias. Because the retrospective likelihood involves nuisance parameters that specify the distribution of the environmental factors, incorporation of environmental factors (and haplotype–environment interactions) is complicated in this approach (Spinka *et al.* 2005). Spinka *et al.* (2005) extended the retrospective maximum-likelihood approach of Chatterjee and

Carroll (Chatterjee and Carroll, 2005) to incorporate both genetic and environmental factors and can account for the presence of missing data in the genetic factors, with a special emphasis on haplotype analysis where missing data arise due to unknown phase information. They used the following likelihood function (Spinka *et al.*, 2005)

$$L = \prod_{i=1}^{N} P(G_i, X_i | D_i) = \prod_{i=1}^{N} P(D_i | G_i, X_i) P(G_i) P(X_i) / P(D_i)$$

where

$$P(D) = \int_x \sum_{h \in H} P(D_i | X = x, H = h) P(H = h) dF(x)$$

$$P(D|X,G) = \sum_{h \in S(G)} P(D|X, H = h, G) P(H = h | X, G)$$

$$= \sum_{h \in S(G)} \frac{P(D|X, H = h) q(h; \theta)}{\sum_{h' \in S(G)} q(h'; \theta)}$$

where $q(\cdot)$ is a known function, θ is a vector of parameters, and H is the set of all possible values of genetic covariate of interest (haplotype in this context). Using a profile likelihood technique and an appropriate EM algorithm, they derived a relatively simple parameter estimation procedure and described two robust approaches that are less sensitive to the genetic–environmental independence and HWE assumptions. In addition, they showed that their retrospective maximum-likelihood procedure is equivalent to an extension of the ascertainment-corrected joint-likelihood method of Stram *et al.* (2003).

One appealing feature of GLM is that many types of responses, for example, binary and continuous, can be handled similarly with a link function. Lin and Zeng (2006) proposed a more general framework for haplotype-association analysis based on likelihood methods, where all commonly used study designs (including cross-sectional, case–control, and cohort studies) and phenotypes (including binary, quantitative, and survival traits) can be handled similarly. Their regression models can evaluate the effects of haplotypes, as well as gene–environment interactions, while accommodating a variety of genetic mechanisms (recessive, dominant, additive, and codominant models). Following Satten and Epstein (2004), they included a fixation index to allow for departure from HWE. They provided a comprehensive and mathematically rigorous way for haplotype-association analysis. Lin and Zeng (2006) proved the identifiability of the model parameters, and the consistency, asymptotic normality, and efficiency of the maximum-likelihood estimators under certain conditions. However, this method still needs to assume that environmental factors are independent of haplotypes conditional on genotypes, although this conditional independence assumption is much weaker than the gene–environment independence assumption. In addition, they only focused on SNP data and did not handle missing data on environmental covariates (Lin and Zeng, 2006).

2. Haplotype sharing and clustering

As stated in Section IV.A.1 , haplotype sharing/clustering is a technique that does not pertain to any specific method. Many statistical methods have been proposed for haplotype-association analysis using haplotype clustering information in regression models (Durrant *et al.*, 2004; Molitor *et al.*, 2003; Morris, 2005; Morris *et al.*, 2002; Seltman *et al.*, 2003; Thomas *et al.*, 2003; Tzeng, 2005; Tzeng *et al.*, 2006; Yu *et al.*, 2004, 2005a). Seltman *et al.* (2003) extended the cladistic approach proposed by Templeton and colleagues (Templeton, 1995; Templeton and Sing, 1993; Templeton *et al.*, 1987, 1988, 1992, 2000) to the GLM framework, allowing for unphased haplotypes. Durrant *et al.* (2004) proposed to use standard hierarchical clustering to create a hierarchical tree of haplotypes. A final tree is obtained by trimming the tree back towards its root, until a poor fit of the model is encountered. They used the haplotype tree together with logistic regression method for case–control data.

Tzeng *et al.* (2006) incorporated the probabilistic clustering methods of Tzeng (2005) into the GLM framework proposed by Schaid *et al.* (2002b). An evolutionary tree (ET) of haplotypes is used to cluster haplotypes (Tzeng 2005). The tree is sequentially trimmed by combining rare haplotypes into their one-step neighboring haplotypes, from leaves of the tree back to the major nodes. The final tree is determined by using an information criterion (Tzeng, 2005). The method of Tzeng *et al.* (2006) is based on clustered haplotypes that can incorporate both phase uncertainty (if phase is unknown) and clustering uncertainty, and can evaluate the overall haplotype association and the individual haplotype effects. Since it is based on GLM, the method can include covariates and model many traits (such as binary traits, quantitative traits, survival traits). The performance of this method still needs further evaluation in practice.

Using haplotype sharing/clustering can reduce the degrees of freedom and eliminate rare haplotypes. The disadvantages are (1) usually they are incapable of detecting rare variants with large effects since rare haplotypes are not retained in the clustered haplotype space (Tzeng *et al.*, 2006). Recently, a method based on haplotype clustering has been proposed that claims to work well when the disease-predisposing allele is moderately rare (Waldron *et al.*, 2006). Another way is to use a weighted penalized log-likelihood approach to deal with rare haplotypes (Souverein *et al.*, 2006); (2) most of them do not account for ascertainment and it is not clear how well they would work for case–

control studies of complex disorders (Schaid, 2004); and (3) these methods depend highly on the clustering scheme, thus the similarity measures. Further work is needed to define the best types of similarity measures. We want to emphasize that the use of haplotype clustering does not guarantee improvement. Bardel *et al.* (2006) showed that the prior grouping of haplotypes based on the algorithm of Durrant *et al.* (2004) did not increase the power of association testing compared to using the same haplotype-based logistic regression without grouping of haplotypes, except in very particular situations of LD patterns and allele frequencies.

3. Construction of score statistics

One advantage of the GLM framework is that it provides a way to construct score statistics to test the null hypothesis (Schaid, 2004). Let y_i be dependent trait, $\tilde{y}_i$ be a fitted value from a GLM with only environmental covariates, and $E[]$ be the conditional expectation over the posterior probability of haplotype pairs under the null hypothesis, given the observed genotype data. Then, the score statistics can be represented as (Schaid, 2004)

$$U = \sum_{i=1}^{N} \frac{y_i - \tilde{y}_i}{a(\phi)} E_i[X],$$

where $a(\phi)$ is used to scale the distribution, with σ^2_{mse} for normal distribution, and 1 for binomial and Poisson distributions (Schaid, 2004; Schaid *et al.*, 2002b). From the above equation, it can be seen that the score statistics measure the covariance of the residuals of a GLM model that fits only the environmental covariates with the expected haplotypes. The weights for the expected haplotypes are the posterior probabilities of the haplotype pairs given the observed genotypes (Schaid, 2004). The computation efficiency of the score statistics makes it feasible to estimate significance level by simulations, where the p-values are usually more robust for sparse haplotypes than those from asymptotic theory. Under standard regularity conditions, Xie and Stram (2005) proved that the above score test of the null hypothesis is asymptotically equivalent to the one proposed by Zaykin *et al.* (2002) where the expected haplotype scorings are used in standard regression. Xie and Stram (2005) also showed by simulations that the two methods are nearly equivalent away from the null when the sample size is small.

In this section, we have mainly introduced some representative regression-based methods in haplotype-association analysis. Table 14.3 summarizes some of the methods that fall under this general scheme. It is an

Table 14.3. Haplotype-Association Methods for Samples of Unrelated Individuals

Trait	Method	HWE	Covariates	Software	Link	Platform	References
Binary	Prospective likelihood	Pool	No				Chiano and Clayton (1998)
Binary	Prospective likelihood	Pool	Yes				Mander (2001)
Binary/ quantitative	Prospective likelihood	Pool	Yes	THESIAS	http://genecanvas.ecgene.net/downloads.php	Windows/ Linux	Tregouet *et al.* (2002) Tregouet and Garelle (2007)
Quantitative	Prospective likelihood	Pool	Yes	QHAPIPF	http://ideas.repec.org/c/boc/bocode/s425502.html	Stata	Mander (2002)
Quantitative	Prospective likelihood	Pool	Yes	weighted penalized log-likelihood maximization routine	Email Dr. M. Tanck at: m.w.tanck@amc.uva.nl	MatLab	Tanck *et al.* (2003)
Binary	Prospective likelihood	Control	Yes	HPlus	http://qge.fhcrc.org/hplus/	MatLab	Zhao *et al.* (2003b)
GLM	Prospective likelihood	Pool	Yes	Haplo Stats	http://mayoresearch.mayo.edu/mayo/research/biostat/schaid.cfm	Splus/R	Lake *et al.* (2003); Schaid *et al.* (2002b)
GLM	Prospective likelihood	Pool	Yes	eHap	http://wpicr.wpic.pitt.edu/WPICCompGen/ehap__v1.htm	Windows	Seltman, Roeder, and Devlin (2003)

(*Continues*)

Table 14.3. (*Continued*)

Trait	Method	HWE	Covariates	Software	Link	Platform	References
Binary	Prospective likelihood	Pool	Yes				Stram *et al.* (2003)
Binary	Retrospective likelihood	Control	No	CHAPLIN	http://www.genetics.emory.edu/labs/epstein/software/chaplin/	Windows	Epstein and Satten (2003)
Censored	Prospective likelihood	Pool	Yes	HA PSTAT	http://www.bios.unc.edu/~lin/hapstat	Windows	Lin (2004)
Binary	Prospective likelihood	Pool	Yes	CLADHC	Email Dr. A. Morris at: amorris@well.ox.ac.uk	Unix	Durrant *et al.* (2004)
Binary	Restrospective likelihood, modified prospective score equations	Pool	Yes	HapLogistic	Email Dr. N. Chatterjee at: chattern@mail.nih.gov	R	Spinka *et al.* (2005)
Binary, quantitative, censored	Likelihood	Pool	Yes	HAPSTAT	http://www.bios.unc.edu/~lin/hapstat	Windows	Lin and Zeng (2006)
Binary	Haplotype similarity	Pool	No	QSHS	http://www4.stat.ncsu.edu/~jytzeng/Softwares/QSHS/	R	Tzeng *et al.* (2003)
Binary	Haplotype similarity	Pool	No	Evolutionary-based haplotype clustering	http://www4.stat.ncsu.edu/~jytzeng/Softwares/Hap-Clustering/R/	R	Tzeng (2005); Tzeng *et al.* (2006)

Binary	Likelihood ratio test	Pool	No	FAMHAP	http://www.uni-bonn.de/~umt70e/becker.html	Unix	Becker and Knapp (2004); Becker *et al.* (2005a,b)
Binary	Score test	Pool	No	Nonlinear	Email Dr. M. Xiong at: Momiao.xiong@uth.tmc.edu	Windows	Zhao *et al.* (2005, 2006)
Binary	Likelihood	Pool	Yes	Weighted (penalized) logistic regression maximization routine	Email Dr. O.W. Souverein at: o.w.souverein@amc.uva.nl	MatLab	Souverein *et al.* (2006)
Binary	Localized haplotype-cluster model	Pool	No	BEAGLE	http://www.stat.auckland.ac.nz/~browning/beagle/beagle.html	Java on LINUX/MAC/UNIX/Windows	Browning and Browning (2007b)

expansion of Table II in Schaid (2004) and Table 2 of Chapter 6 in Feng *et al.* (2007). These regression approaches offer several advantages (Schaid, 2004) and constitute a very important part in haplotype analysis.

V. FAMILY-BASED ASSOCIATION METHODS USING HAPLOTYPES

A. Brief introduction

There are two different designs in association studies: family-based designs that use pedigrees and population-based designs that use unrelated individuals. Both designs are widely used in association studies and have their own strengths and weaknesses. Population-based designs are generally more powerful than family-based designs but the power difference between these two types of designs is generally small, especially when case–parent trios are used (McGinnis *et al.*, 2002; Witte *et al.*, 1999). The recruitment of families is more difficult than the recruitment of unrelated individuals, especially for late age onset diseases. However, family-based designs have two advantages compared with population-based designs. First, appropriate analyses of family-based designs are not affected by population stratification. In case–control studies, population stratification arises when samples are selected from several genetically different populations with different proportions in cases and controls. Population stratification can generate suspicious association between markers and the disease; thus, it is always a concern for association studies in heterogeneous populations, such as populations in major cities throughout the world. Second, significant findings in family-based association analysis always imply both linkage and association between the marker and the disease-susceptibility locus, which can reduce the chance of false positive findings. Therefore, family-based designs will still play important roles in association studies.

For both designs, haplotype-based association methods are generally regarded as being more powerful than methods based on single markers (Akey *et al.*, 2001; Morris and Kaplan, 2002; Zhao *et al.*, 2000) since the former fully exploits LD information from multiple markers. Both simulation and empirical studies support this conclusion (Akey *et al.*, 2001; Zaykin *et al.*, 2002). In addition, haplotype-based methods can be more powerful when multiple disease-susceptibility alleles occur within a single gene (Morris and Kaplan, 2002) and can potentially capture *cis*-interactions between two or more causal variants. It is also worth noting that the haplotype-based methods are not always more powerful than the methods based on genotypes at either a single marker locus or multiple marker loci (Chapman *et al.*, 2003; Roeder *et al.*, 2005; Xiong *et al.*, 2002). Actually, relative efficiencies of haplotype-based methods to single-locus tests and multiple-locus unphased genotype tests depend on many factors and more comprehensive evaluations will be needed to clarify these issues. Thus, we will focus on haplotype-based method in this section. We will start from a brief introduction of family-based association methods using single markers, but focus on reviewing family-based methods using haplotypes in details. The review for general methods for association studies using families can be found in the literature (e.g., Laird and Lange, 2006; Zhao, 2000) and other chapters of this book.

B. Family-based association methods using single markers

The simplest family-based design for association mapping is the case–parent design, which consists of genotypes of an affected individual and its parents. Spielman *et al.* (1993) proposed the transmission/disequilibrium test (TDT) to test the linkage between a marker locus and the disease-susceptibility locus in the presence of association. For a bi-allelic marker, the TDT compares the number of times of an allele is transmitted to the offspring from theirs parents and the number of nontransmitted alleles. Specifically, we can construct the following 2×2 transmission/nontransmission table for a bi-allelic marker with alleles A_1 and A_2 from n case–parent trios:

	Nontransmitted		
Transmitted	A_1	A_2	Total
A_1	n_{11}	n_{12}	n_{1+}
A_2	n_{21}	n_{22}	n_{2+}
Total	n_{+1}	n_{+2}	$4n$

The TDT statistic is defined by $\mathrm{TDT} = (n_{21} - n_{12})^2/(n_{21} + n_{12})$. Although the TDT is initially proposed to test linkage in the presence of LD, it is clear that the TDT is a test for both linkage and LD. Because the TDT statistic has an asymptotic χ^2 distribution with one degree of freedom under the null hypothesis of no linkage or no association, the TDT is often used as a test for the association in the presence of linkage. Actually, we can have three types of null hypothesis: no linkage and no association, no linkage in the presence of association, and no association in the presence of linkage, but only one alternative hypothesis: the maker is in both linkage and association with the disease-susceptibility locus. Thus, the rejection of null hypothesis implies that the association between the marker and the disease-susceptibility locus is due to lack of recombinants between them rather than other factors, such as population stratification. Because LD decays rapidly as the distance between two loci increases, the TDT has higher resolution than linkage analysis.

Because TDT is robust to population stratification and analytically simple, numerous extensions have been made. The TDT method has been extended to handle multi-allelic markers (Schaid, 1996; Sham, 1997; Gjessing and Lie, 2006), missing parents (Chen, 2004; Curtis and Sham, 1995; Sun *et al.*, 1999), sib pairs (Boehnke and Langefeld, 1998; Curtis and Sham, 1995; Laird *et al.*, 1998; Spielman and Ewens, 1998), general pedigrees (Abecasis *et al.*, 2000b; Gordon *et al.*, 2004; Martin *et al.*, 2000), quantitative phenotypes (Abecasis *et al.*, 2000a; Allison, 1997), and haplotypes (Clayton and Jones, 1999; Dudbridge, 2003; Yu *et al.*, 2005b; Zhang *et al.*, 2003; Zhao *et al.*, 2000). Here, we only describe the extensions of the original TDT for multi-allelic markers, because these extensions can be readily applied to data in which the haplotypes are already known or the haplotypes can be unambiguously determined according to their genotypes. Similar with the TDT statistic from a bi-allelic marker, we can construct a $k \times k$ transmission/nontransmission table for a marker with k alleles $A_1, \ldots, A_k$:

	Nontransmitted			
Transmitted	A_1	...	A_k	Total
A_1	n_{11}	...	n_{1k}	n_{1+}
...	...	...	...	...
A_k	n_{21}	...	n_{kk}	n_{k+}
Total	n_{+1}	...	n_{+k}	$4n$

Several test statistics have been constructed from this table. A direct generalization of the TDT is to test if the table of transmission/nontransmission is symmetry. The statistic can be represented as: $TDT_c = \sum_{i<j}(n_{ij} - n_{ji})^2/(n_{ij} + n_{ji})$, which has an asymptotic χ^2 distribution with $k(k-1)/2$ degrees of freedom. The statistic for the test of marginal homogeneity can be represented as: $\mathrm{TDT}_m = \sum_{i=1}^{k} (n_{i+} - n_{+i})^2/(n_{i+} + n_{+i})$. Similarly, Spielman and Ewens (1996) proposed a similar statistic to test the marginal homogeneity: $\mathrm{TDT}_{\mathrm{SE}} = ((k-1)/k) \sum_{i=1}^{k} (n_{i+} - n_{+i})^2/(n_{i+} + n_{+i} - n_{ii})$. However, as noted by Sham (1997), Schaid (1996), and Spielman and Evens (1996), TDT_m and $\mathrm{TDT}_{\mathrm{SE}}$ may not have a χ^2 distribution with $k-1$ degrees of freedom but the permutation test procedure can be used to assess their significance.

In addition to the nonparametric tests based on the contingency table, model-based methods have been developed. Sham and Curtis (1995) proposed a logistic regression model, where the logarithm of the odds of transmitting allele i, given that the parent has genotype ij, is determined by $b_i - b_j$, where b_i and b_j are parameters associated with alleles i and j. This logistic model has the following form $\log(p_{ij}/p_{ji}) = b_i - b_j$ and the likelihood function has the form $L = \sum_{i<j}(n_{ij}\ln(p_{ij}) + n_{ji}\ln(p_{ji}))$. Then the log likelihood ratio test can be constructed to test for linkage and association. The simulations showed this method is more powerful than the test of symmetry and marginal homogeneity. Schaid (1996) proposed a conditional logistic regression approach, where the offspring genotype is modeled as a function of parental genotypes and offspring disease status as follows:

$$P(g_c|g_m, g_f, D) = \frac{P(D|g_c, g_m, g_f)}{\sum_{g^* \in G} P(D|g^*, g_m, g_f)P(g^*|g_m, g_f)P(g_m, g_f)}.$$

In the above expression, D represents the affected status of the offspring, g_c, g_m, and g_f are the genotypes of the offspring, mother, and father, and g^* is one of the four possible genotypes of the offspring conditional on parental genotypes. Assuming that $P(D|g_c, g_m, g_f) = P(D|g_c)$, the above equation reduces to $P(g_c|g_m, g_f, D) = r(g_c)/\sum_{g^* \in G} r(g^*)$, where $r(g)$ is the relative risk of disease for genotype g. The above formula can be used to construct the score statistic to test the association between the marker locus and the disease-susceptibility locus under different disease models. The similar model has previously been proposed by Self *et al.* (1991).

C. Methods for phase known data

As we can see from the formula of the TDT statistic, only the parents with heterozygous genotypes will contribute to the TDT statistic. These parents are called informative parents. Thus, haplotypes across several markers have an addition advantage over single markers for family-based association studies: haplotypes can increase the heterozygosity of parents and provide more informative parents in most situations. As we have mentioned before, recombinants are unlikely events for tightly linked markers. Thus, it is reasonable to assume the haplotype configurations obtained do not contain recombinants in most situations. Then haplotypes can be considered as alleles at a multi-allelic marker and the transmission of haplotypes follows the Mendelian law of inheritance. Thus, virtually all the extended TDT methods that can handle multi-allelic markers can be directly applied to phase known haplotype data. At the same time, other methods that can handle phase known data other than treat haplotypes as alleles from a multi-allelic locus have also been developed. Finally, we would emphasize virtually all haplotype-based methods for case–control studies can also be directly applied to phase known data if the transmitted haplotypes are considered as case haplotypes and the untransmitted haplotypes as control haplotypes. But the robustness for the population stratification, which is one of the biggest advantages for family-based designs, may not be preserved in this treatment.

Van der Meulen and te Meerman (1997) constructed the HSS for association mapping using case–parent trios data. The haplotypes from the parents are assumed to be known in advance. Thus, transmitted and nontransmitted haplotypes can be determined and then are considered as case haplotypes and control haplotypes, respectively. Suppose there are n case haplotypes and n control haplotypes and that L tightly linked SNP markers are genotyped in a region of interest. The case haplotypes are denoted $h_1, \ldots, h_n$ and the control haplotypes are denoted $h_{n+1}, \ldots, h_{2n}$. For the case haplotypes, the HHS at the l th marker can be calculated as:

$$\mathrm{HSS}(l) = \sqrt{\frac{\sum_{i \neq j}(S_{h_i,h_j}(l))^2 - (\sum_{i \neq j} S_{h_i,h_j}(l))^2/(n(n-1))}{n(n-1)-1}},$$

where $S_{h_i,h_j}(l)$ is the length of the contiguous region around the lth marker over which the two haplotypes, h_i and h_j, are identical by state. Van der Meulen and te Meerman (1997) used a randomization procedure to generate the distribution of HSS under the null hypothesis: n haplotypes are generated based on allele frequencies estimated from the control haplotypes under the assumption of LE, then the HHS value is calculated. Assuming a normal distribution and after estimating its mean and variance from the randomization procedure, the p -value can be calculated for each HHS(l) from case haplotypes.

Bourgain *et al.* (2000) proposed another haplotype-sharing-based method, called the maximum identity length contrast (MILC), to search for a disease-susceptibility locus. The MILC compares the length of shared haplotypes at a marker locus in the case haplotypes with that in the controls. Specifically, if we denote the case haplotypes as $h_1, \ldots, h_n$ and the control haplotypes as $h_{n+1}, \ldots, h_{2n}$ and define

$$A(l) = \frac{2}{n(n-1)} \sum_{i=1}^{n} \sum_{j=i+1}^{n} S_{h_i,h_j}(l) \text{ and } U(l) = \frac{2}{m(m-1)} \sum_{i=n+1}^{n+m} \sum_{j=i+1}^{n+m} S_{h_i,h_j}(l),$$

then the MILC statistic proposed by Bourgain *et al.* (2000) is $\mathrm{ILCS} = \max_{1 \leq l \leq L}(A(l) - U(l))$. The idea behind haplotype-sharing-based methods is that two haplotypes containing the disease allele are more closely related than two haplotypes without the disease allele and two haplotypes with one having the disease allele and one not having the disease allele. Therefore, if the lth marker is in the region flanking the disease gene, we expect to observe an excess length of shared haplotype. That is, $A(l)$ would be greater than $U(l)$ for the marker loci in the vicinity of the disease-susceptibility locus. The significance of the observed value of MILC can be easily assessed by a simple permutation test (Bourgain *et al.*, 2000). The simulation studies (Bourgain *et al.*, 2000) has shown that the MILC method is more powerful than the TDT for markers with a higher density. Further simulation studies have shown that the power of the MILC method can greatly enhanced by the use of multiple, closely related affected individuals (Bourgain *et al.*, 2001). Bourgain *et al.* (2002) extended the MILC method to handle missing data and ambiguous haplotypes. The haplotypes of the offspring are deducted using the Mendelian law of inheritance. In situations where haplotypes cannot be unambiguously determined, alleles at the ambiguous loci are considered as missing. They proposed two models to characterize the allele distribution of missing data: missing data are unequally or equally distributed between transmitted and nontransmitted haplotypes. They also proposed two strategies to calculate $S_{h_i,h_j}(l)$ in the presence of missing data. They conducted extensive simulations to assess the impact of these models and strategies and found that the two models and strategies have similar power.

Clayton and Jones (1999) discussed the generalization of the TDT to detect association between haplotypes and the disease-susceptibility locus. They followed the framework of the score test proposed by Schaid (1996) and use a generalized haplotype risk model which includes the multiplicative model as a special case: $h(\phi_{(i,j)}) = \beta_i + \beta_j = (1/2)(h(\phi_{(i,i)}) + h(\phi_{(j,j)}))$, where $h()$ is an unspecified monotone increasing function and $\phi_{(i,j)}$ represents the haplotype pair with the haplotype h_i and h_j. If there are many marker loci, many haplotypes may exist and such an extended test tends to have low power due to the large degrees of freedom. Clayton and Jones (1999) developed a strategy to test haplotype associations for each marker in turn using the models from spatial statistics. The model assumes that the haplotype effects can be generated from a multivariate normal distribution with variance–covariance matrix υS, where S is a known matrix expressing haplotype "similarity" between pairs of haplotypes and υ is a single parameter that represents the haplotype association. The rationale behind such model is that the similar haplotypes tends to have the similar effects. Thus, the score test can be constructed to test the null hypothesis of $\upsilon = 0$. A natural way to determine the element s_{ij} in S is to classify two haplotypes h_i and h_j either as similar ($s_{ij} = 1$) or dissimilar ($s_{ij} =1$). Another nature measure of the similarity between two haplotypes is $S_{h_i,h_j}(l)$, the length of the contiguous region over which the two haplotypes, h_i and h_j, are identical by state, as was used by Van der Meulen and te Meerman (1997) and Bourgain *et al.* (2000). Clayton and Jones (1999) also suggested using simulations to evaluate the distribution of the test statistic under the null hypothesis.

To identify a subset of markers that are associated with the disease from a large number of marker loci, Lo and Zheng (2002) proposed a backward haplotype transmission association (BHTA) algorithm. In the algorithm, they defined the haplotype transmission disequilibrium (HTD) score, which is just the numerator of the test statistic TDT_m when haplotypes are treated as alleles at a multi-allelic marker. Lo and Zheng (2002) showed that the HTD score has an expectation equal to the trace of the Fisher's information matrix, parameterized by haplotype relative risks under the null hypothesis of no linkage or in the absence of LD. Thus, the HTD score can serve as a measure of association between the marker and the disease. For a set of k markers and a marker locus $i(1 \leq i \leq k)$, the difference between the HTD score based on all k markers and $k-1$ markers without the marker i is larger if the marker i is associated with the disease than if the marker i either has no linkage or no LD with the disease locus. Thus, the markers should only be kept if the HTD score is changed substantially. Based on this observation, Lo and Zheng (2002) proposed the haplotype transmission association (HTA), which is based on the change of the HTD score on all k markers and $k-1$ markers without the marker i, to eliminate those markers that either have no linkage or no LD with the disease-susceptibility locus. The results based on simulation studies showed that the BHTA method can be applied to a large number of markers and detect associations with multifactorial traits in the presence of haplotypic association.

Allen-Brady *et al.* (2006) proposed a Monte Carlo simulation method to test the association between a marker locus and a disease-susceptibility locus and implemented it in the software package PedGenie. The simulation procedure in PedGenie can be divided into the following steps. First, the haplotype frequencies are estimated from phase known data and the test statistic of interest is also calculated. Second, the haplotypes are randomly assigned to the pedigree founders in proportion to the haplotype frequencies and then assigned to the pedigree nonfounders based on the Mendelian law of inheritance. Third, the test statistic of interest is calculated using the simulated data while keeping the same missing structure as the original data. Fourth, the third step is repeated for a large number of times and the p -value is the portion of times that the values of test statistic calculated from the simulated data sets are greater than the observed test statistic. The advantages of PedGenie are that it incorporates many commonly used test statistics, including the TDT statistic, and it can handle pedigrees of arbitrary size and structure.

D. Methods for phase unknown data

In general, the haplotype phase is unknown. However, as we have mentioned before, recombinants are unlikely events for tightly linked markers. Thus, we can use methods for haplotype inference to identify the most likely haplotype configuration without recombinants and treat this data as phase known data. Once such a haplotype configuration is identified, virtually all the extended TDT methods that can handle multi-allelic markers can be directly applied. However, this strategy has several disadvantages. First, multiple haplotype configurations without recombinants will exist and the use of the most likely haplotype configuration ignores the uncertainty in haplotype inference, which can result in a loss of power. Second, the likelihood-based methods for haplotype inference involves the estimate of haplotype frequencies, which may not be robust for structured populations. Therefore, many methods that can use multiple haplotype configurations to account for the uncertainty in haplotype inference have been proposed.

Clayton (1999) generalized the TDT test to account for the uncertain haplotype transmission based on the full likelihood function conditional on disease in the offspring. Specifically, the full likelihood can be written as:

$$L^{(F)} = P(G_m, G_f, G_c|D) = P(G_m, G_f|D)P(G_c|G_m, G_f, D) = L^{(P)}L^{(C)},$$

where D represents the affected status of the offspring, G_c, G_m, and G_f are the genotypes (or paired haplotypes if they are known) of the offspring, mother, and father, respectively. The first component, $L^{(P)}$, depends not only on the haplotype relative risks, but also on the haplotype frequencies in the population. The second term, $L^{(C)}$, takes the same form as that used in Schaid (1996) and Clayton and Jones (1999) and only depends on the haplotype relative risks. Both the log of $L^{(F)}$ and $L^{(C)}$ can be used to test if the haplotype relative risks equal one. The methods based on $L^{(F)}$ are generally more powerful but they may give incorrect answers if there is population stratification. In contrast, the methods based on $L^{(C)}$ are generally less powerful but robust if there is population stratification. Therefore, $L^{(C)}$ is generally used if the haplotypes are known (Clayton and Jones, 1999; Schaid, 1996). Clayton (1999) also pointed out that the score test based on either $L^{(F)}$ or $L^{(C)}$ has a simple formula and is easy to extend to account for the uncertain haplotype transmission. Based on these arguments, Clayton (1999) proposed a score test based on the partial-likelihood to reduce the influence of population stratification as much as possible. The test statistic is derived either based on $L^{(F)}$ for those trios with the uncertain haplotype transmission or based on $L^{(C)}$ for those trios with known haplotype assignments. The full likelihood function $L^{(F)}$ has also been used by Dudbridge (2003) to construct a likelihood-ratio test. When there are ambiguous haplotype assignments, the EM algorithm is used to maximize $L^{(F)}$ under the null hypothesis as well as under the alternative hypothesis. Although the assumption of a single population with HWE is required for this method, population stratification will only lead to a conservative test (Dudbridge, 2003).

Based on a haplotype pattern mining (HPM) method developed by Toivonen *et al.* (2000), Zhang *et al.* (2002b) developed a family-based HPM (F-HPM) method that allows for the analysis of general pedigrees, qualitative and quantitative trait, haplotypes across a set of markers, and both missing marker data and ambiguous haplotype information. Both methods (HPM and F-HPM) are robust to genotyping error, missing data, and recent mutations. In the HPM, a haplotype pattern P around a candidate marker l is defined as a vector containing either alleles or symbol for missing data at the marker loci. Three parameters are used to define the haplotype pattern: the candidate marker l, the number of markers included in haplotype pattern n, and the maximum number of the markers with missing data m. For a given haplotype pattern P, the test statistic from Zhang *et al.* (2001) can be used to assess its association with the trait of interest. Thus, for a set of haplotype patterns with respect to parameters n, m, and marker location l, the haplotype patterns having significance association with the trait can be identified. Intuitively, haplotype patterns near the trait locus are likely to have stronger association than haplotype patterns further from the trait locus. Therefore, the trait locus is likely to be located at the site where there is a high proportion of strongly associated haplotype patterns (Toivern *et al.*, 2000). Zhang *et al.* (2001) first defined the proportion of strongly associated haplotype patterns around a maker as the test statistic at the marker, and then proposed to use the maximum test statistic across all markers as a test statistic to test the null hypothesis that the region being examined is not associated with the trait. To assess the significance of this test statistic, Zhang *et al.* (2002) adapted a permutation procedure proposed by Monks and Kaplan (2000) and pointed out that the simple permutation of trait values among family members is not valid. The results based on the data from the 12th Genetic Analysis Workshop GAW12 (Wijsman *et al.*, 2001) showed F-HPM can detect the association between the marker and the trait with higher power than a single marker-based test (Zhang *et al.*, 2001).

Cordell and Clayton (2002) proposed a unified stepwise regression procedure for association mapping using family data. Essentially, they used the same formula proposed by Schaid (1996) to model multilocus genotypes of the offspring:

$$P(G_c|G_m, G_f, D) = \frac{P(D|G_c, G_m, G_f)}{\sum_{G^* \in G} P(D|G^*, G_m, G_f)P(G^*|G_m, G_f)P(G_m, G_f)}.$$

In the above expression, D still represents the affected status of the offspring, G_c, G_m, and G_f are the multi-locus genotypes of the offspring, mother, and father, and G^* is one of possible genotypes of the offspring conditional on parental genotypes. The $P(G^*|G_m, G_f)$ are functions of underlying population haplotype frequencies and recombination fractions. Thus, $P(G^*|G_m, G_f)$ generally takes different values, resulting in a model that cannot be expressed as a simple conditional logistic regression. To overcome this difficulty, Cordell *et al.* (2002) proposed to use some event ξ in the family for further conditioning. Because now $P(G^*|\xi)$ does not depend on the underlying haplotype frequencies and the recombination fractions, the above formula can be conveniently expressed in a conditional logistic regression framework. They proposed several strategies to choose the event ξ. For example, we can use all families that the parental haplotypes can be unambiguously deduced without recombinants or all families that the haplotypes are "inferable" (Cordell *et al.*, 2002). This method has been extended to detect gene–gene and gene–environment interactions, and parent-of-origin effects (Cordell *et al.*, 2004). However, this method may discard many families and results in loss of power.

When the haplotypes for the parents are known, the transmission/nontransmission table can be easily constructed and the test statistics, such as TDT_{SE}, can be used to test for linkage and association between the marker and the disease. In the presence of uncertain haplotypes, Zhao *et al.* (2000) developed a method to construct the transmission/nontransmission table based on haplotype frequencies. For a case–parent trio, the contribution of a compatible haplotype assignment $\{G_m, G_f, G_c\}$ to the transmission/nontransmission table is weighted by $P(G_m)P(G_f)/\sum_{\{G_m^*, G_f^*\}} P(G_m^*)P(G_f^*)$, where $\{G_m^*, G_f^*\}$ are all possible haplotype assignments for this trios. Under the assumption of HWE, $P(G)$ is the product of haplotype frequencies. The contribution of a case–parent trio to the transmission/nontransmission table sums all possible haplotype assignments for this trio. Zhao *et al.* (2000) proved that the table is symmetric under the null hypothesis of no linkage. Thus, TDT_{SE} can be used to test such symmetry. Zhao *et al.* (2000) also proposed three methods to estimate the haplotype frequencies from the data. These methods assume all individuals from a single population with HWE. However, the symmetry of the transmission/nontransmission table is not affected by the choice of haplotype frequencies. The haplotype frequencies will only affect the power but not the validity of the test. Thus, the test proposed by Zhao *et al.* (2000) does not suffer from inflated type I error rates due to population stratification. Knapp and Becker (2003) observed that the symmetry of the transmission/nontransmission table is not essential for the validity of the test proposed by Zhao *et al.* (2000). Based on this observation, Knapp and Becker (2003) proposed several intuitive modifications for Zhao's method that can potentially increase its power but not affect its validity, and extended Zhao's method to handle general nuclear families with arbitrary number of affected and unaffected offspring.

Seltman *et al.* (2001) adapted the partial likelihood approach (Clayton, 1999) to simultaneously evaluate linkage and association between haplotypes and the disease. However, the existence of a large number of haplotypes may reduce the power because of the additional degrees of freedom required for the test. Thus, Seltman (2001) proposed to use the evolutionary relationship between haplotypes to reduce the number of degrees of freedom and to improve the power. Specifically, the "ET"–TDT developed by Seltman *et al.* (2001) can be divided into the following steps. First, the cladogram that describes the evolution relationship among haplotypes is constructed (Swofford, 1998) and the sparse nodes (rare haplotypes) are collapsed into their neighbor nodes so that no node is uncommon. Second, the cladogram is organized by clades using the algorithm developed by Templeton *et al.* (1987). Third, the score test for each clade based on the partial-likelihood (Clayton, 1999) is constructed and a "cladogram-collapsing algorithm" is used to conduct a set of tests with one degree of freedom through the cladogram. In this step, the nodes with the same parameter (haplotype relative risk) are identified. Therefore the number of different parameters will be reduced. Fourth, the p -value is evaluated using a permutation procedure. Simulation results have shown that the ET-TDT can be more powerful than other proposed methods under reasonable conditions (Seltman *et al.*, 2001).

Zhang *et al.* (2003) proposed a TDT based on haplotype sharing for tightly linked markers (HS-TDT). The MILC method proposed by Bourgain *et al.* (2000, 2001, 2002) is a special case of the HS-TDT. Like other methods based on the haplotype sharing, the degrees of freedom of the HS-TDT roughly equal the number of markers rather than the number of haplotypes. The HS-TDT for phase known data can be constructed as follows. First, for any haplotype h, the haplotype sharing score at the l th marker is defined as the average similarity between h and all of the parental $4n$ haplotypes: $X_h(l) = (1/4n)\sum_{i=1}^{4n} S_{h,h_i}(l)$, where $S_{h_i,h_j}(l)$ is the length of the contiguous region around the l th marker over which the two haplotypes, h_i and h_j, are identical by state. Second, the difference of the haplotype-sharing scores between the parental haplotypes that are transmitted and not transmitted to the k th child in the i th family at the l th marker is defined as: $x_{ik}(l) = \sum_{j=1}^{4} \delta_{ijk} X_{ij}(l)$, where $X_{ij}(l)$ is the haplotype sharing-score for four parental haplotypes in the i th family and $\delta_{ijk} = 1$ if the parental haplotype j is transmitted to the k th child and $\delta_{ijk} = -1$ if not transmitted. Third, the estimated covariance of the trait values and $x_{ik}(l)$ can be expressed as: $U_i(l) = \sum_k (y_{ik} - c)x_{ik}(l)$, where y_{ik} is the trait value for the k th child in the i th family and c is an arbitrary constant. Under the null hypothesis, the trait values are independent of $x_{ik}(l)$, thus $E(U_i(l)) = 0$. Fourth, the test statistic is defined as $U = \max_l(U(l))$, where $U(l) = \sum_i w_i U_i(l)$ and the summation is over all families and w_i is a constant. The significance of U can be assessed using a permutation procedure. If the haplotypes are unknown, the haplotype frequencies and the haplotype configurations with their corresponding posterior probabilities can be estimated using the EM algorithm with the assumption of a single population with HWE. Then $X_H(l)$ and $x_{ik}(l)$ can be defined based on all possible haplotype configurations with their posterior probabilities. Zhang *et al.* (2003) proved that the choice of c in $U_i(l)$, w_i in $U(l)$, and the haplotype frequencies will only affect the power but not the validity of HS-TDT. Thus, the HS-TDT is still valid in the presence of population stratification. The optimal values of c and w_i to achieve the maximum power is unclear. In practical, c can be set as 0 for the qualitative trait and as mean trait value over all children in all families for the quantitative trait. w_i can be set as $1/t_i$ to give equal weight to each family (Monks and Kaplan, 2000) or 1 to give larger weight to families with more children (Sun *et al.*, 1999), where t_i is the number of children in the i th family. The HS-TDT is applicable to arbitrary nuclear families and can handle both qualitative and quantitative traits. The simulation results also showed that HS-TDT is more powerful than existing single-marker TDTs and haplotype-based TDTs.

Horvath *et al.* (2004) proposed a haplotype-based approach to test for linkage and association between multiple marker loci and the disease. Their method is a direct extension of the conditional approach implemented in FABT (Horvath *et al.*, 2001; Rabinowitz and Laird, 2000) and allows for arbitrary family structures and missing parental genotypes. In the conditional approach, the conditional distribution of offspring genotypes under the null hypothesis of no linkage and no association is constructed conditioning on the sufficient statistic for any nuisance parameters. Then this conditional distribution is used to construct and evaluate the distribution of test statistics. Horvath *et al.* (2004) pointed out that the observed offspring and parental genotypes and traits can be used as a minimum sufficient statistic when the founder genotypes are incomplete. Based on this statistic, they proposed a five-step algorithm to construct an allowable set of offspring genotypes and their conditional probabilities. Then the score for the i th family can be written as: $S_i = \sum_{j=1}^{n_i} \sum_k y_{ij} X(h_{ijk,1}, h_{ijk,2}) w_{ijk}$, where y_{ij} is the trait value of j th offspring in the i th family, $X(h_{ijk,1}, h_{ijk,2})$ is a general coding of the haplotype pair $h_{ijk,1}$ and $h_{ijk,2}$, and w_{ijk} is the corresponding weight of that haplotype pair. $X(h_{ijk,1}, h_{ijk,2})$ can be a univariate function of genotypes or a vector of multiple haplotypes. The weight w_{ijk} was calculated based on the allowable set of offspring genotypes and their conditional probabilities. Horvath adapted three strategies used in Zhao *et al.* (2000) to calculate w_{ijk}.The score and its variance over all families, $U = \sum_{i=1}^{n} (S_i - E(S_i))$ and $V = \sum_{i=1}^{n} \mathrm{Var}(S_i)$, can be calculated in a similar way under the null hypothesis. The final test statistic can be a Mantel–Haenszel type test statistic: $U^T V^{-1} U$, which asymptotically follows a χ^2 distribution with degrees of freedom equal to the rank of V. Horvath *et al.* (2004) have implemented this method in FBAT and compared its performance with TRANSMIT (Clayton, 1999). They found haplotype FBAT can maintain the appropriate type I error rate while TRANSMIT had inflated type I error rate in the presence of population admixture.

Yu *et al.* (2005b) proposed the SP-TDT test that combines a sequential peeling procedure with haplotype-sharing-based TDT methods. The SP-TDT can be divided into four steps. First, the test statistic for a set of haplotype pairs F used in their peeling algorithm, $S(F)$, is constructed. As usual, $S(F)$ compares the length of shared haplotypes at a marker locus between transmitted haplotypes and untransmitted haplotypes. Different from other haplotype–sharing-based methods, Yu *et al.* (2005b) used a weighted similarity measure, where the weights are chosen to favor of identity-by-state (IBS) of rare alleles more than for common alleles. Yu *et al.* (2004) has shown that this weighted similarity measure is more powerful for case–control studies. In the second step, a sequence of nested subsets of haplotype pairs $F_0 \supset F_1 \supset \ldots \supset F_M$ is generated using a peeling procedure where F_0 is the set of all parental haplotype pairs. Due to founder heterogeneity, there are some transmitted haplotypes that may not carry disease mutations and the exclusion of such haplotypes can increase the power of haplotype-similarity-based TDT methods. In each peeling step, the subset F_{j+1} is obtained by removing haplotype pairs from F_j that maximize $S(F_{j+1})$. In the third step, the p-value, p_i, associated with $S(F_i)$, is obtained using a permutation procedure. The minimum among them, $\min_{0 \leq i \leq M} p_i$ is defined as the test statistic of the SP-TDT. In the last step, the significance of the test statistic from the third step is assessed using a permutation procedure. Yu *et al.* (2005b) showed that the SP-TDT is more powerful than some existing haplotype-similarity-based TDT methods, such as the HS-TDT method of Zhang *et al.* (2003) in the presence of founder heterogeneity.

In this subsection, we have introduced many haplotype-sharing methods. However, the haplotype similarity used in many methods (Bourgain *et al.*, 2000; Clayton and Jones, 1999; Yu *et al.*, 2005b; Zhang *et al.*, 2003) is not robust to genotyping errors, missing data, and recent marker mutations. Knapp and Becker (2004) found the haplotype similarity used in the HS-TDT (Zhang *et al.*, 2003) can result in an inflated type I error rate in the presence of genotyping errors. To accommodate data with genotyping errors, we allow a number of loci with alleles not identical by state in the vicinity of the marker considered. Given a number of mismatches d allowed in the calculation, $S_{h_1,h_2}(l)$ can be defined by the length of the longest region that contains the l th marker and has at most d loci with alleles not identical by descent for haplotypes h_1 and h_2. There are also several ways to define robust haplotype similarity in the presence of missing data. First, the missing allele on a haplotype can be considered as a different allele at one marker locus. If both haplotypes have a missing allele at a marker locus, they can be considered as difference alleles. Second, the missing allele on a haplotype can be considered as same with any other alleles. If both haplotypes have a missing allele at a marker locus, they can be considered as same alleles. Third, the missing allele on a haplotype can be imputed according to its frequencies and the haplotype similarity is the summation of haplotype similarity over all possible imputations. These ideas have been used to calculate the MILC statistic with missing data (Bourgain *et al.*, 2002), construct the haplotype pattern with gaps (Toivonen *et al.*, 2000), and calculate a robust haplotype similarity measure (Zhang *et al.*, 2004).

Table 14.4. Available Software Packages for Haplotype-Based Association Analysis Using Pedigrees

Software	Trait	Method and assumption	Link	Platform	References
ETDT	Binary	Logistic regression, TDT for multiple allelic markers	http://www.mds.qmw.ac.uk/statgen/dcurtis/software.html	MS-DOS	Sham and Curtis (1995)
ET-TDT	Binary	Evolutionary tree, haplotype, trios	http://www.stat.cmu.edu/~roeder/=ettdt/	MS-DOS	Seltman *et al.* (2001)
FBAT	Qualitative quantitative	Haplotype, general pedigrees	http://www.biostat.harvard.edu/~fbat/fbat.htm	Unix, Linux, Windows, Mac	Hovarth *et al.* (2001); Laird and Lange (2006); Rabinowitz and Laird (2000)
GASSOC	Binary	Score test, TDT for multiple allelic markers	http://mayoresearch.mayo.edu/mayo/research/schaid_lab/software.cfm		Schaid, 1996
HS-TDT	Qualitative quantitative	Haplotype sharing, nuclear families	http://www.math.mtu.edu/~shuzhang/software.html	Unix, Windows	Zhang *et al.* (2003)
MILC	Binary	Haplotype sharing, trios	Contact the authors		Bourgain *et al.* (2000, 2001, 2002)
Multiple TDT	Binary	Trios, haplotype	http://bioinformatics.med.yale.edu/group/software.html	Windows	Zhao *et al.* (2000)
PDT	Binary	General pedigrees	http://www.chg.duke.edu/software/pdt.html	Unix	Martin *et al.* (2000)
PEDGENIE	Qualitative quantitative	Simulation-based algorithm	http://bioinformatics.med.utah.edu/PedGenie/index.html	Java	Allen-Brady *et al.* (2006)
SP-TDT	Binary	Haplotype sharing	Contact the authors	Windows	Yu *et al.* (2005b)

(*Continues*)

Table 14.4. (*Continued*)

Software	Trait	Method and assumption	Link	Platform	References
TDTHAP	Binary	Haplotypes	http://www-gene.cimr.cam.ac.uk/clayton/software/		Clayton and Jones (1999)
TDT/S-TDT	Binary	Sibs, nuclear families	http://genomics.med.upenn.edu/spielman/TDT.htm	Windows	Spielman *et al.* (1993); Spielman and Ewens (1998)
TRANSMIT	Binary	Haplotype	http://www-gene.cimr.cam.ac.uk/clayton/software/	Unix	Clayton (1999)

E. The program packages

We have compiled a list of commonly used programs for haplotype-association methods using pedigrees in Table 14.4. We suggest users to choose their programs upon several important factors: the type of traits, the type of pedigrees, and the phase status. With a few exceptions, most available programs for haplotypes association methods using pedigrees can only handle trios with binary traits. Among them, ETDT (Sham and Curtis, 1995), GASSOC (Schaid, 1996), TDTHAP (Clayton and Jones, 1999), and TDT/S-TDT (Spielman *et al.*, 1993; Spielman and Ewens, 1998) are only suitable for phase-known data, while Multiple TDT (Zhao *et al.*, 2000) and TRANSMIT (Clayton, 1999) can handle phase unknown data. HS-TDT (Zhang *et al.*, 2003) implemented a haplotype-sharing-based TDT method and can handle both qualitative and quantitative traits but is only applicable to nuclear families. FBAT (Horvath *et al.*, 2001; Laird and Lange, 2006; Rabinowitz and Laird, 2000), which has been well recognized and widely used, along with PEDGENIE (Allen-Brady *et al.*, 2006) can handle general pedigree and any types of traits. We also suggest users to refer to Table 14.4 and their manuals for more detailed descriptions for these programs.

VI. DISCUSSION

Haplotypes have been playing a very important role in genetic studies and many efforts have been made to develop statistical methods of analyses (Schaid, 2004). A haplotype map of the human genome has begun to provide a wealth resource for not only practical genetic research, but also haplotype methodology development (de Bakker *et al.*, 2005; Myers *et al.*, 2005; The International HapMap Consortium, 2003, 2005).

Many methods have been developed for haplotype analysis in genetic studies, as illustrated above. In this chapter, we have only focused on haplotype-association analysis. LD mapping is another approach where many methods have been proposed using haplotype information. The major difference between the two groups of methods is that LD fine mapping methods usually take historical information into consideration and attempt to estimate the trait locus position (Duncan, 2004; Schaid, 2004). Of note, in recent years, the use of LD maps in LD units (LDU) has been proposed for association mapping (Maniatis *et al.*, 2002, 2004, 2005, 2007; Tapper *et al.*, 2005; Zhang *et al.*, 2002c). Some studies have shown that the approaches utilizing LDU perform better than haplotype analysis (Maniatis *et al.*, 2007). However, comprehensive and thorough studies are needed to have full evaluation. "DNA pooling" of individual samples is a way to reduce the genotyping cost dramatically (Amos *et al.*, 2000; Sham *et al.*, 2002). There are many haplotype methods that have been developed which are based on DNA pooling (Quade *et al.*, 2005; Wang *et al.*, 2003; Yang *et al.*, 2003; Zeng and Lin, 2005). In recent years, many studies have suggested that haplotypes tend to have block-like structures throughout the human genome and these haplotypes within the haplotype blocks can be represented by a small number of haplotype-tagging SNPs (htSNPs). Many methods have been proposed for haplotype block partitioning and for tagging SNP identification, and using this information in haplotype-association analysis (Greenspan and Geiger, 2004; Niu *et al.*, 2002a; Sai *et al.*, 2006; Xiong *et al.*, 2003; Zhang *et al.*, 2002a, 2004). Because of space limitation, these concepts are not discussed in this chapter.

Although many efforts have been made and the area of haplotype inference and analysis has been greatly advanced over the past decade, some challenging issues remain unsolved. All the methods described above depend on getting the "right size" haplotypes. If the haplotypes are too long including too many markers, the haplotypes may be composed of too many alleles, resulting in many haplotype configurations, and therefore diluting association signals with diseases (Schaid, 2004). Although methods have been proposed to deal with this issue, such as using sliding windows (Durrant *et al.*, 2004) and haplotype block information (Greenspan and Geiger, 2004), there are still no efficient approaches. Most of the methods in haplotype analysis (including haplotype inference and haplotype-association analysis) are developed under some assumptions such as HWE, no missing genotypes or missing at random, and no genotyping errors. However, these assumptions may not hold in real-world studies. Departure of the haplotype pairs from HWE can result from many sources, such as genotyping errors, gene flow, selection, and population stratification. Studies have been performed to investigate and deal with the HWE assumption (Fallin and Schork, 2000; Lin and Zeng, 2006; Liu *et al.*, 2006a,c; Niu *et al.*, 2002b; Satten and Epstein, 2004). Since human populations are usually not the result of random mating, the HWE assumption needs to be carefully evaluated and handled in haplotype analysis. Even with the advancement of high technology, missing genotypes and genotyping errors are still common in genetic studies (Abecasis *et al.*, 2001; Becker *et al.*, 2006; Cardon *et al.*, 2000; Kang *et al.*, 2004; Knapp and Becker, 2004; Liu *et al.*, 2006a; Moskvina *et al.*, 2005, 2006; Niu, 2004; Thompson *et al.*, 2005), where in addition to experimental techniques, there exist other factors that may cause missing data and genotyping errors, such as variation in DNA quality, unknown DNA variations, and the conduction of studies (e.g., in case–control studies of rare diseases, DNA samples of low quality maybe included in the studies because of the difficulty in recruiting patients). Although several studies have shown that ignoring missing genotypes and genotyping errors could significantly decrease the accuracy of estimates of haplotype frequencies (Abecasis *et al.*, 2001; Cardon *et al.*, 2000; Kang *et al.*, 2004; Knapp and Becker, 2004; Liu *et al.*, 2006a; Moskvina *et al.*, 2005, 2006; Niu, 2004), most current methods do not take these into consideration. This issue clearly needs more research in order to have valid inference. There are some other important issues in haplotype analysis, such as handling rare haplotypes. Our intention is to show that in order for haplotype methods to be more efficiently applied in genetic studies, many important methodological issues still need to be addressed.

Haplotypes are promising in the study of the genetic basis of complex traits, and the field is still evolving and vigorous which can be seen from the appearance of large quantity of new research work in the area (Allen and Satten, 2007; Carlton, 2007; Koopman *et al.*, 2007; Kwee *et al.*, 2007; Lin *et al.*, 2007; Nagy *et al.*, 2007; Purcell *et al.*, 2007; Sieh *et al.*, 2007; Zhang *et al.*, 2007; Zhao *et al.*, 2007; Zheng and McPeek, 2007). However, the relative efficiency of using haplotypes versus single markers is not always clear. We believe that no one method is universally superb to others. New methods may be needed to combine haplotype analysis and single marker analysis, such as omnibus tests. The users are advised to choose methods based on their problems, and may need to try different methods.

Acknowledgments

This work was supported in part by the grants NIH GM074913 (KZ), GM57672 and GM59507 (HZ) from the National Institute of General Medical Sciences.

References

Abecasis, G. R., and Wigginton, J. E. (2005). Handling marker-marker linkage disequilibrium: Pedigree analysis with clustered markers. *Am. J. Hum. Genet.* **77,** 754–767.

Abecasis, G. R., Cardon, L. R., and Cookson, W. O. C. (2000a). A general test of association for quantitative traits in nuclear families. *Am. J. Hum. Genet.* **66,** 279–292.

Abecasis, G. R., Cookson, W. O. C., and Cardon, L. R. (2000b). Pedigree tests of transmission disequilibrium. *Eur. J. Hum. Genet.* **8,** 545–551.

Abecasis, G. R., Cherny, S. S., and Cardon, L. R. (2001). The impact of genotyping error on family-based analysis of quantitative traits. *Eur. J. Hum. Genet.* **9,** 130–134.

Abecasis, G. R., Cherny, S. S., Cookson, W. O., and Cardon, L. R. (2002). Merlin-rapid analysis of dense genetic maps using sparse gene flow trees. *Nat. Genet.* **30,** 97–101.

Ardlie, K., Kruglyak, L., and Seielstad, M. (2002). Patterns of linkage disequilibrium in the human genome. *Nat. Rev. Genet.* **3,** 299–309.

Akey, J., Jin, L., and Xiong, M. (2001). Haplotypes vs. single marker linkage disequilibrium tests: What do we gain? *Eur. J. Hum. Genet.* **9,** 291–300.

Allen, A. S., and Satten, G. A. (2007). Inference on haplotype/disease association using parent-affected-child data: The projection conditional on parental haplotypes method. *Genet. Epidemiol.* **31,** 211–223.

Allen-Brady, K., Wong, J., and Camp, N. J. (2006). PedGenie: An analysis approach for genetic association testing in extended pedigrees and genealogies of arbitrary size. *BMC Bioinformat.* **7,** 209.

Allison, D. B. (1997). Transmission-disequilibrium tests for quantitative traits. *Am. J. Hum. Genet.* **60,** 676–690.

Amos, C. I., Frazier, M. L., and Wang, W. (2000). DNA pooling in mutation detection with reference to sequence analysis. *Am. J. Hum. Genet.* **66,** 1689–1692.

Bafna, V., Gusfield, D., Lancia, G., and Yooseph, S. (2003). Haplotyping as perfect phylogeny: A direct approach. *J. Comput. Biol.* **10,** 323–340.

Bardel, C., Darlu, P., and Genin, E. (2006). Clustering of haplotypes based on phylogeny: How good a strategy for association testing? *Eur. J. Hum. Genet.* **14,** 202–206.

Barrett, J. C., Fry, B., Maller, J., and Daly, M. J. (2005). Haploview: Analysis and visualization of LD and haplotype maps. *Bioinformatics* **21,** 263–265.

Baruch, E., Weller, J. I., Cohen-Zinder, M., Ron, M., and Seroussi, E. (2005). Efficient inference of haplotypes from genotypes on a large animal pedigree. *Genetics* **172,** 1757–1765.

Becker, T., and Knapp, M. (2003). Efficiency of haplotype frequency estimation when nuclear family information is included. *Hum. Hered.* **54,** 45–53.

Becker, T., and Knapp, M. (2004). Maximum-likelihood estimation of haplotype frequencies in nuclear families. *Genet. Epidemiol.* **27,** 21–32.

Becker, T., Cichon, S., Jonson, E., and Knapp, M. (2005a). Multiple testing in the context of haplotype analysis revisited: Application to case-control data. *Ann. Hum. Genet.* **69,** 747–756.

Becker, T., Schumacher, J., Cichon, S., Baur, M. P., and Knapp, M. (2005b). Haplotype interaction analysis of unlinked regions. *Genet. Epidemiol.* **29,** 313–322.

Becker, T., Valentonyte, R., Croucher, P. J., Strauch, K., Schreiber, S., Hampe, J., and Knapp, M. (2006). Identification of probable genotyping errors by consideration of haplotypes. *Eur. J. Hum. Genet.* **14,** 450–458.

Beckmann, L., Thomas, D. C., Fischer, C., and Chang-Claude, J. (2005). Haplotype sharing analysis using mantel statistics. *Hum. Hered.* **59,** 67–78.

Boehnke, M., and Langefeld, C. D. (1998). Genetic association mapping based on discordant sib pairs: The discordant-alleles test. *Am. J. Hum. Genet.* **62,** 950–961.

Botstein, D., and Risch, N. (2003). Discovering genotypes underlying human phenotypes: Past successes for mendelian disease, future approaches for complex disease. *Nat. Genet.* **33**(Suppl.), 228–237.

Bourgain, C., Genin, E., Quesneville, H., and Clerget-Darpoux, F. (2000). Search for multifactorial disease susceptibility genes in founder populations. *Ann. Hum. Genet.* **64,** 255–265.

Bourgain, C., Genin, E., Holopainen, P., Mustalahti, K., Maki, M., Partanen, J., and Clerget-Darpoux, F. (2001). Use of closely related affected individuals for the genetic study of complex diseases in founder populations. *Am. J. Hum. Genet.* **68,** 154–159.

Bourgain, C., Genin, E., Ober, C., and Clerget-Darproux, F. (2002). Missing data in haplotype analysis: A study on the MILC method. *Ann. Hum. Genet.* **66,** 99–108.

Brinza, D., and Zelikovsky, A. (2006a). 2SNP: Scalable Phasing Based on 2-SNP Haplotypes. *Bioinformatics* **22,** 371–373.

Brinza, D., and Zelikovsky, A. (2006b). Phasing of 2-SNP Genotypes based on Non Random Mating Model. *In* "Computational Science-ICCS 2006: Sixth International Workshop on Bioinformatics Research and Applications (IWBRA06), Part II" (V. N. Alexandrov, G. D. van Albada, P. M. A. Sloot, and J. J. Dongarra, eds.), pp. 767–774. Springer Berlin/Heidelberg, Germany.

Brinza, D., and Zelikovsky, A. (2007). 2SNP: Scalable phasing method for trios and unrelated individuals. *IEEE/ACM Transactions on Computational Biology and Bioinformatics* http://doi.ieeecomputersociety.org/10.1109/TCBB.2007.1068.

Brown, D., and Harrower, I. (2004). A new integer programming formulation for the pure parsimony problem in haplotype analysis. *In* "Workshop on Algorithms in Bioinformatics." Springer-Verlag, Berlin, Germany.

Brown, D. G., and Harrower, I. M. (2006). Integer programming approaches to haplotype inference by pure parsimony. *IEEE/ACM Trans. Comput. Biol. Bioinform.* **3,** 141–154.

Browning, S. R., and Browning, B. L. (2007a). Rapid and accurate haplotype phasing and missing data inference for whole genome association studies using localized haplotype clustering. *Am. J. Hum. Genet.* **81,** 1084–1097.

Browning, B. L., and Browning, S. R. (2007b). Efficient multifocus association mapping for whole genome association studies using localized haplotype clustering. *Genet. Epidemiol.* **31,** 365–375.

Cardon, L. R., Abecasis, G. R., and Cherny, S. S. (2000). The effect of genotype error on the power to detect linkage and association with quantitative traits. *Am. J. Hum. Genet.* **67,** 310–310.

Carlton, J. M. (2007). Toward a malaria haplotype map. *Nat. Genet.* **39,** 5–6.

Chapman, J. M., Copper, J. D., Todd, J. A., and Clayton, D. G. (2003). Detecting disease association due to linkage disequilibrium using haplotype tags: A class of tests and the determinants of statistical power. *Hum. Hered.* **56,** 18–31.

Chatterjee, N., and Carroll, R. J. (2005). Semiparametric maximum likelihood estimation exploiting gene-environment independence in case-control studies. *Biometrika* **92,** 399–418.

Chen, Y. H. (2004). New approach to association testing in case-parent designs under informative parental missingness. *Genet. Epidemiol.* **27,** 131–140.

Chiano, M. N., and Clayton, D. G. (1998). Fine genetic mapping using haplotype analysis and the missing data problem. *Ann. Hum. Genet.* **62,** 55–60.

Chung, R. H., and Gusfield, D. (2003). Perfect phylogeny haplotyper: Haplotype inferral using a tree model. *Bioinformatics* **19,** 780–781.

Clark, A. G. (1990). Inference of haplotypes from PCR-amplified samples of diploid populations. *Mol. Biol. Evol.* **7,** 111–122.

Clark, A. G., Weiss, K. M., Nickerson, D. A., Taylor, S. L., Buchanan, A., Strengård, J., Salomaa, V., Vartiainen, E., Perola, M., Boerwinkle, E., and Sing, C. F. (1998). Haplotype structure and population genetic inferences from nucleotide-sequence variation in human lipoprotein lipase. *Am. J. Hum. Genet.* **63,** 595–612.

Clayton, D. G. (1999). A generalization of the transmission/disequilibrium test for uncertain-haplotype transmission. *Am. J. Hum. Genet.* **65,** 1170–1177.

Clayton, D. (2001). SNPHAP: A program for estimating frequencies of large haplotypes of SNPs (Version 1.0) http://www.gene.cimr.cam.ac.uk/clayton/software/.

Clayton, D., and Jones, H. (1999). Transmission/disequilibrium tests for extended marker haplotypes. *Am. J. Hum. Genet.* **65,** 1161–1169.

Cordell, H. J., and Clayton, D. G. (2002). A unified stepwise regression procedure for evaluating the relative effects of polymorphisms within a gene using Case/control or family data: Application to HLA in type 1 diabetes. *Am. J. Hum. Genet.* **70,** 124–141.

Cordell, H. J., Barratt, B. J., and Clayton, D. G. (2004). Case/pseudo control analysis in genetic association studies: A unified framework for detection of genotype and haplotype associations, gene-gene and gene-environment interactions, and parent-of-origin effects. *Genet. Epidemiol.* **26,** 167–185.

Cox, R., Bouzekri, N., Martin, S., Southam, L., Hugill, A., Golamaully, M., Cooper, R., Adeyemo, A., Soubrier, F., Ward, R., Lathrop, G. M., Matsuda, F., *et al.* (2002). Angiotensin-1-converting enzyme (ACE) plasma concentration is influenced by multiple ACE-linked quantitative trait nucleotides. *Hum. Mol. Genet.* **11,** 2969–2977.

Curtis, D., and Sham, P. (1995). A note on the application of the transmission disequilibrium test when a parent is missing. *Am. J. Hum. Genet.* **56,** 811–812.

Czika, W., Yu, X., and Wolfinger, R. D. (2003). An Introduction to genetic data analysis using SAS/genetics http://support.sas.com/rnd/papers/sugi/27/genetics.pdf.

Daly, M. J., Rioux, J. D., Schaffner, S. F., Hudson, T. J., and Lander, E. S. (2001). High-resolution haplotype structure in the human genome. *Nat. Genet.* **29,** 229–232.

de Bakker, P. I., Yelensky, R., Pe'er, I., Gabriel, S. B., Daly, M. J., and Altshuler, D. (2005). Efficiency and power in genetic association studies. *Nat. Genet.* **37,** 1217–1223.

Dempster, A. P., Laird, N. M., and Rubin, D. B. (1977). Maximum likelihood estimation from incomplete data via the EM algorithm (with discussion). *J. R. Stat. Soc. B* **39,** 1–38.

Devlin, B., Roeder, K., and Wasserman, L. (2000). Genomic control for association studies: A semiparametric test to detect excess-haplotype sharing. *Biostatistics* **1,** 369–387.

Ding, Z., Filkov, V., and Gusfield, D. (2005). A linear-time algorithm for the perfect phylogeny haplotyping problem. *In* "Research in Computational Molecular Biology: Proceedings of the Ninth Annual International Conference on Computational Biology (RECOMB 2005)". (S. Miyano, J. Mesirov, S. Kasif, S. Istrail, P. Pevzner, and M. Waterman, eds.), pp. 585–600. Springer, Heidelberg, Germany.

Doi, K., Li, J., and Tao, J. (2003). Minimum recombinant haplotype configuration on the pedigrees. *In* Proc. 3rd Annual Workshop on Algorithms in Bioinformatics (WABI'03), 339–353.

Douglas, J. A., Boehnke, M., Gillanders, E., Trent, J. M., and Gruber, S. B. (2001). Experimentally derived haplotypes substantially increase the efficiency of linkage disequilibrium studies. *Nat. Genet.* **28,** 361–364.

Du, F. X., Woodward, B. W., and Denise, S. K. (1998). Haplotype construction of sires with progeny genotypes based on an exact likelihood. *J. Dairy Sci.* **81,** 1462–1468.

Dudbridge, F. (2003). Pedigree disequilibrium tests for multilocus haplotypes. *Genet. Epidemiol.* **25,** 115–121.

Duncan, C. T. (2004). "Statistical methods in Genetic Epidemiology." Oxford University Press, Inc. 198 Madison Avenue, New York.

Durrant, C., Zondervan, K. T., Cardon, L. R., Hunt, S., Deloukas, P., and Morris, A. P. (2004). Linkage disequilibrium mapping via cladistic analysis of single-nucleotide polymorphism haplotypes. *Am. J. Hum. Genet.* **75,** 35–43.

Elston, R. C., and Stewart, J. (1971). General model for genetic analysis of pedigree data. *Hum. Hered.* **21,** 523–542.

Epstein, M. P., and Satten, G. A. (2003). Inference on haplotype effects in case-control studies using unphased genotype data. *Am. J. Hum. Genet.* **73,** 1316–1329.

Eronen, L., Geerts, F., and Toivonen, H. (2006). HaploRec: Efficient and accurate large-scale reconstruction of haplotypes. *BMC Bioinformatics* **7,** 542.

Eronen, L., Geerts, F., and Toivonen, H. (2004). A Markov chain approach to reconstruction of long haplotypes. Proceedings of the 9th Pacific Symposium on Biocomputing (PSB'04), pp. 104–115. World Scientific, Singapore.

Excoffier, L., and Slatkin, M. (1995). Maximum-likelihood estimation of molecular haplotype frequencies in a diploid population. *Mol. Biol. Evol.* **12,** 921–927.

Excoffier, L., Laval, G., and Balding, D. (2003). Gametic phase estimation over large genomic regions using an adaptive window approach. *Hum. Genomics* **1,** 7–19.

Fallin, D., and Schork, N. J. (2000). Accuracy of haplotype frequency estimation for biallelic loci, via the expectation-maximization algorithm for unphased diploid genotype data. *Am. J. Hum. Genet.* **67,** 947–959.

Fallin, D., Cohen, A., Essioux, L., Chumakov, I., Blumenfeld, M., Cohen, D., and Schork, N. J. (2001). Genetic analysis of case/control data using estimated haplotype frequencies: Application to APOE locus variation and Alzheimer's disease. *Genome Res.* **11,** 143–151.

Feng, Z., Liu, N., and Zhao, H. (2007). Haplotype inference and association analysis in unrelated samples. *In* "Current Topics in Human Genetics: Studies in Complex Diseases" (H. W. Deng, H. Shen, Y. J. Liu, and H. Hu, eds.), pp. 135–176. World Scientific Publishing Company, Singapore.

Fishelson, M., Dovgolevsky, N., and Geiger, D. (2005). Maximum likelihood haplotyping for general pedigrees. *Hum. Hered.* **59,** 41–60.

Gao, G., Hoeschele, I., Sorensen, P., and Du, F. X. (2004). Conditional probability methods for haplotyping in pedigrees. *Genetics* **167,** 2055–2065.

Gjessing, H. K., and Lie, R. T. (2006). Case-parent triads: Estimating single- and double-dose effects of fetal and maternal disease gene haplotypes. *Ann. Hum. Genet.* **70,** 382–396.

Gordon, G., Haynes, C., Johnnidis, C., Patel, S. B., Bowcock, A. M., and Ott, J. (2004). A transmission disequilibrium test for general pedigrees that is robust to the presence of random genotyping errors and any number of untyped parents. *Eur. J. Hum. Genet.* **12,** 752–761.

Greenspan, G., and Geiger, D. (2004). High density linkage disequilibrium mapping using models of haplotype block variation. *Bioinformatics* **20**(Suppl. 1), I137–I144.

Gudbjartsson, D. F., Jonasson, K., Frigge, M. L., and Kong, A. (2000). Allegro, a new computer program for multipoint linkage analysis. *Nat. Genet.* **25,** 12–13.

Gudbjartsson, D. F., Thorvaldsson, T., Kong, A., Gunnarsson, G., and Ingolfsdottir, A. (2005). Allegro version 2. *Nat. Genet.* **37,** 1015–1016.

Gusfield, D. (2001). Inference of haplotypes from samples of diploid populations: Complexity and algorithms. *J. Comput. Biol.* **8,** 305–323.

Gusfield, D. (2002). Haplotyping as perfect phylogeny: Conceptual framework and efficient solutions. *In* "Sixth annual conference on Research in Computational Molecular Biology (RECOM)," pp. 166–175. ACM Press, New York.

Gusfield, D. (2003). Haplotype inference by pure parsimony. *In* "14th Annual Symposium on Combinatorial Pattern Matching (CPM'03)" (R. Baesa-Yates, E. Chavez, and M. Crochemore, eds.), pp. 144–155. Springer, Berlin/Heidelberg, Germany.

Gusfield, D., and Orzack, S. H. (2005). Haplotype inference. *In* "Handbook of Computational Molecular Biology" (S. Aluru, ed.), pp. 18.1–18.22. Chapman & Hall/CRC Press, Boca Raton, FL.

Haldane, J. B. S. (1919). The combination of linkage values, and the calculation of distance between the loci of linked factors. *J. Genet.* **8,** 299–309.

Halperin, E., and Eskin, E. (2004). Haplotype reconstruction from genotype data using imperfect phylogeny. *Bioinformatics* **20,** 1842–1849.

Halparin, E., and Hazan, E. (2006). HAPLOFREQ-Estimating haplotype frequencies efficiently. *J. Comput. Biol.* **13,** 481–500.

Hawley, M. E., and Kidd, K. K. (1995). HAPLO: A program using the EM algorithm to estimate the frequencies of multi-site haplotypes. *J. Hered.* **86,** 409–411.

Hodge, S. E., Boehnke, M., and Spence, M. A. (1999). Loss of information due to ambiguous haplotyping of SNPs. *Nat. Genet.* **21,** 360–361.

Hollox, E. J., Poulter, M., Zvarik, M., Ferak, V., Krause, A., Jenkins, T., Saha, N., Kozlov, A. I., and Swallow, D. M. (2001). Lactase haplotype diversity in the old world. *Am. J. Hum. Genet.* **68,** 160–172.

Horvath, S., Xu, X., and Laird, N. M. (2001). The family based association test method: Strategies for studying general genotype-phenotype associations. *Eur. J. Hum. Genet.* **9,** 301–306.

Horvath, S., Xu, X., Lake, S. L., Silverman, E. K., Weiss, S. T., and Laird, N. M. (2004). Family-based tests for associating haplotypes with general phenotype data: Application to asthma genetics. *Genet. Epidemiol.* **26,** 61–69.

Huang, Y. T., Chao, K. M., and Chen, T. (2005). An approximation algorithm for haplotype inference by maximum parsimony. *J. Comput. Biol.* **12,** 1261–1274.

Kang, S. J., Gordon, D., and Finch, S. J. (2004). What SNP genotyping errors are most costly for genetic association studies? *Genet. Epidemiol.* **26,** 132–141.

Kankova, K., Stejskalov, A., Hertlov, M., and Znojil, V. (2005). Haplotype analysis of RAGE gene: Identification of a haplotype marker for diabetic nephropathy in type 2 diabetes mellitus. *Nephrol. Dial. Transplant.* **20,** 1093–1102.

Kaplan, N., and Morris, R. (2001). Issues concerning association studies for fine mapping a susceptibility gene for a complex disease. *Genet. Epidemiol.* **20,** 432–457.

Keavney, B., McKenzie, C. A., Connell, J. M. C., Julier, C., Ratcliffe, P. J., Sobel, E., Lathrop, M., and Farrall, M. (1998). Measured haplotype analysis of the angiotensin-I converting enzyme gene. *Hum. Mol. Genet.* **7,** 1745–1751.

Knapp, M., and Becker, T. (2003). Family-based association analysis with tightly linked markers. *Hum. Hered.* **56,** 2–9.

Knapp, M., and Becker, T. (2004). Impact of genotyping errors on type I error rate of the haplotype-sharing transmission/disequilibrium test (HS-TDT). *Am. J. Hum. Genet.* **74,** 589–591.

Koopman, W. J., Li, Y., Coart, E., van de Weg, W. E., Vosman, B., Roldan-Ruiz, I., and Smulders, M. J. (2007). Linked vs. unlinked markers: Multilocus microsatellite haplotype-sharing as a tool to estimate gene flow and introgression. *Mol. Ecol.* **16,** 243–256.

Krawczak, M. (1994). HAPMAX documentation www.uni-kiel.de/medinfo/mitarbeiter/krawczak/download/hapmax.txt.

Kruglyak, L., Daly, M. J., Reeve-Daly, M. P., and Lander, E. S. (1996). Parametric and nonparametric linkage analysis: A unified multipoint approach. *Am. J. Hum. Genet.* **58,** 1347–1363.

Kwee, L. C., Epstein, M. P., Manatunga, A. K., Duncan, R., Allen, A. S., and Satten, G. A. (2007). Simple methods for assessing haplotype-environment interactions in case-only and case-control studies. *Genet. Epidemiol.* **31,** 75–90.

Laird, N. M., and Lange, C. (2006). Family-based designs in the age of large-scale gene-association studies. *Nat. Rev. Genet.* **7,** 385–394.

Laird, N. M., Blacker, D., and Wilcox, M. (1998). The sib transmission/disequilibrium test is a Mantel-Haenszel test. *Am. J. Hum. Genet.* **63,** 1915–1916.

Lake, S. L., Lyon, H., Tantisira, K., Silverman, E. K., Weiss, S. T., Laird, N. M., and Schaid, D. J. (2003). Estimation and tests of haplotype-environment interaction when linkage phase is ambiguous. *Hum. Hered.* **55,** 56–65.

Lander, E. S., and Green, P. (1987). Construction of multilocus genetic-linkage maps in humans. *Proc. Natl. Acad. Sci. USA* **84,** 2363–2367.

Lange, K., and Boehnke, M. (1983). Extensions to pedigree analysis V. Optimal calculation of Mendelian likelihood. *Hum. Hered.* **33,** 291–301.

Lange, K., and Goradia, T. M. (1987). An algorithm for automatic genotype elimination. *Am. J. Hum. Genet.* **40,** 250–256.

Li, J., and Jiang, T. (2003). Efficient rule-based haplotyping algorithm for pedigree data. *In* "Proceedings of the Seventh Annual International Conference on Research in Computational Molecular Biology (RECOMB 2003)." (W. Miller, M. Vingron, S. Istrail, P. Pevzner, and M. Waterman, eds.), pp. 197–206. ACM, New York.

Li, S. S., Khalid, N., Carlson, C., and Zhao, L. P. (2003). Estimating haplotype frequencies and standard errors for multiple single nucleotide polymorphisms. *Biostatistics* **4,** 513–522.

Li, J., and Jiang, T. (2005). Computing the minimum recombinant haplotype configuration from incomplete genotype data on a pedigree by integer linear programming. *J. Comput. Biol.* **12,** 719–739.

Li, N., and Stephens, M. (2003). Modeling linkage disequilibrium and identifying recombination hotspots using single-nucleotide polymorphism data. *Genetics* **165,** 2213–2233.

Lin, D. Y. (2004). Haplotype-based association analysis in cohort studies of unrelated individuals. *Genet. Epidemiol.* **26,** 255–264.

Lin, D. Y., and Huang, B. E. (2007). The use of inferred haplotypes in downstream analyses. *Am. J. Hum. Genet.* **80,** 577–579.

Lin, S., and Speed, T. P. (1997). An algorithm for haplotype analysis. *J. Comput. Biol.* **4,** 535–546.

Lin, D. Y., and Zeng, D. (2006). Likelihood-based inference on haplotype effects in genetic association studies. *J. Am. Stat. Assoc.* **101,** 89–104.

Lin, S., Cutlet, D. J., Zwick, M. E., and Chakravarti, A. (2002). Haplotype inference in random population samples. *Am. J. Hum. Genet.* **71,** 1129–1137.

Lin, S., Skrivanek, Z., and Irwin, M. (2003). Haplotyping using SIMPLE: Caution on ignoring interference. *Genet. Epidemiol.* **25,** 384–387.

Lin, S., Chakravarti, A., and Cutler, D. J. (2004). Haplotype and missing data inference in nuclear families. *Genome Res.* **14,** 1624–1632.

Lin, E., Hwang, Y., Liang, K. H., and Chen, E. Y. (2007). Pattern-recognition techniques with haplotype analysis in pharmacogenomics. *Pharmacogenomics* **8,** 75–83.

Liu, N., Beerman, I., Lifton, R., and Zhao, H. (2006a). Haplotype analysis in the presence of informatively missing genotype data. *Genet. Epidemiol.* **30,** 290–300.

Liu, P. Y., Lu, Y., and Deng, H. W. (2006b). Accurate haplotype inference for multiple linked single-nucleotide polymorphisms using sibship data. *Genetics* **174,** 499–509.

Liu, W., Zhao, W., and Chase, G. A. (2006c). The impact of missing and erroneous genotypes on tagging SNP selection and power of subsequent association tests. *Hum. Hered.* **61,** 31–44.

Lo, S., and Zheng, T. (2002). Backward haplotype transmission association (BHTA) algorithm—a fast multiple-marker screening method. *Hum. Hered.* **53,** 197–215.

Long, J. C., Williams, R. C., and Urbanek, M. (1995). An E-M algorithm and testing strategy for multiple-locus haplotypes. *Am. J. Hum. Genet.* **56,** 799–810.

Mander, P. A. (2001). Haplotype analysis in population-based association studies. *Stata J.* **1,** 58–75.

Mander, P. A. (2002). Analysis of quantitative traits using regression and log-linear modeling when phase is unknown. *Stata J.* **2,** 65–70.

Maniatis, N., Collins, A., Xu, C. F., McCarthy, L. C., Hewett, D. R., Tapper, W., Ennis, S., Ke, X., and Morton, N. E. (2002). The first linkage disequilibrium (LD) maps: Delineation of hot and cold blocks by diplotype analysis. *Proc. Natl. Acad. Sci. USA* **99,** 2228–2233.

Maniatis, N., Collins, A., Gibson, J., Zhang, W., Tapper, W., and Morton, N. E. (2004). Positional cloning by linkage disequilibrium. *Am. J. Hum. Genet.* **74,** 846–855.

Maniatis, N., Morton, N. E., Gibson, J., Xu, C. F., Hosking, L. K., and Collins, A. (2005). The optimal measure of linkage disequilibrium reduces error in association mapping of affection status. *Hum. Mol. Genet.* **14,** 145–153.

Maniatis, N., Collins, A., and Morton, N. E. (2007). Effects of single SNPs, haplotypes, and whole-genome LD maps on accuracy of association mapping. *Genet. Epidemiol.* **31,** 179–188.

Marchini, J., Cutler, D., Patterson, N., Stephens, M., Eskin, E., Halperin, E., Lin, S., Qin, Z. S., Munro, H. M., Abecasis, G. R., Donnelly, P., International HapMap Consortium, *et al.* (2006). A comparison of phasing algorithms for trios and unrelated individuals. *Am. J. Hum. Genet.* **78,** 437–450.

Martin, E. R., Monks, S. A., Warren, L. L., and Kaplan, N. L. (2000). A test for linkage and association in general pedigrees: The pedigree disequilibrium test. *Am. J. Hum. Genet.* **67,** 146–154.

McGinnis, R., Shifman, S., and Darvasi, A. (2002). Power and efficiency of the TDT and case–control design for association scans. *Behav. Genet.* **32,** 135–144.

MichlataosBeloin, S., Tishkoff, S. A., Bentley, K. L., Kidd, K. K., and Ruano, G. (1996). Molecular haplotyping of genetic markers 10 kb apart by allelic-specific long-range PCR. *Nucl. Acids Res.* **24,** 4841–4843.

Mitchell, M. (1996). "An Introduction to Genetic Algorithm" The MIT Press, Cambridge, MA.

Molitor, J., Marjoram, P., and Thomas, D. (2003). Fine-scale mapping of disease genes with multiple mutations via spatial clustering techniques. *Am. J. Hum. Genet.* **73,** 1368–1384.

Monks, S. A., and Kaplan, N. L. (2000). Removing the sampling restrictions from family-based tests of association for a quantitative-trait locus. *Am. J. Hum. Genet.* **66,** 576–592.

Morris, A. P. (2005). Direct analysis of unphased SNP genotype data in population-based association studies via Bayesian partition modelling of haplotypes. *Genet. Epidemiol.* **29,** 91–107.

Morris, R. W., and Kaplan, N. L. (2002). On the advantage of haplotype analysis in the presence of multiple disease susceptibility alleles. *Genet. Epidemiol.* **23,** 221–233.

Morris, A. P., Whittaker, J. C., and Balding, D. J. (2002). Fine-scale mapping of disease loci via shattered coalescent modeling of genealogies. *Am. J. Hum. Genet.* **70,** 686–707.

Morris, A., Pedder, A., and Ayres, K. (2003). Linkage disequilibrium assessment via log-linear modeling of SNP haplotype frequencies. *Genet. Epidemiol.* **25,** 106–114.

Moskvina, V., Craddock, N., Holmans, P., Owen, M., and O'Donovan, M. (2005). Minor genotyping error can result in substantial elevation in type I error rate in haplotype based case control analysis. *Am. J. Med. Genet. B Neuropsychiatr. Genet.* **138B,** 19–19.

Moskvina, V., Craddock, N., Holmans, P., Owen, M. J., and O'Donovan, M. C. (2006). Effects of differential genotyping error rate on the type I error probability of case-control studies. *Hum. Hered.* **61,** 55–64.

Myers, S., Bottolo, L., Freeman, C., McVean, G., and Donnelly, P. (2005). A fine-scale map of recombination rates and hotspots across the human genome. *Science* **310,** 321–324.

Nagy, M., Entz, P., Otremba, P., Schoenemann, C., Murphy, N., and Dapprich, J. (2007). Haplotype-specific extraction: A universal method to resolve ambiguous genotypes and detect new alleles—demonstrated on HLA-B. *Tissue Antigens* **69,** 176–180.

Nejati-Javaremi, A., and Smith, C. (1995). Assigning linkage haplotypes from parent and progeny genotypes. *Genetics* **142,** 1363–1367.

Niu, T. H. (2004). Algorithms for inferring haplotypes. *Genet. Epidemiol.* **27,** 334–347.

Niu, T. H., Qin, Z. S., and Liu, J. S. (2002a). Bayesian classification for genome-wide control using multiple unlinked haplotype blocks. *Am. J. Hum. Genet.* **71,** 578–578.

Niu, T. H., Qin, Z. H. S., Xu, X. P., and Liu, J. S. (2002b). Bayesian haplotype inference for multiple linked single-nucleotide polymorphisms. *Am. J. Hum. Genet.* **70,** 157–169.

Nordborg, M., and Tavaré, S. (2002). Linkage disequilibrium: What history has to tell us. *Trends Genet.* **18,** 83–90.

O'Connell, J. R. (2000). Zero-recombinant haplotyping: Applications to fine mapping using SNPs. *Genet. Epidemiol* **19,** S64–S70.

O'Connell, J. R., and Weeks, D. E. (1995). The VITESSE algorithm for rapid exact multilocus linkage analysis via genotype set-recoding and fuzzy inheritance. *Nat. Genet.* **11,** 402–408.

Orzack, S. H., Gusfield, D., Olson, J., Nesbitt, S., Subrahmanyan, L., and Stanton, V. P., Jr. (2003). Analysis and exploration of the use of rule-based algorithms and consensus methods for the inferral of haplotypes. *Genetics* **165,** 915–928.

Patil, N., Berno, A. J., Hinds, D. A., Barrett, W. A., Doshi, J. M., Hacker, C. R., Kautzer, C. R., Lee, D. H., Morjoribanks, C., McDonough, D. P., Nguyen, B. T. N., Norris, M. C., *et al.* (2001). Blocks of limited haplotype diversity revealed by high-resolution scanning of human chromosome 21. *Science* **294,** 1719–1723.

Purcell, S., Daly, M. J., and Sham, P. C. (2007). WHAP: Haplotype-based association analysis. *Bioinformatics* **23,** 255–256.

Qian, D., and Beckmann, L. (2002). Minimum-recombinant haplotyping in pedigrees. *Am. J. Hum. Genet.* **70,** 1434–1445.

Qin, Z. H. S., Niu, T. H., and Liu, J. S. (2002). Partition-ligation-expectation-maximization algorithm for haplotype inference with single-nucleotide polymorphisms. *Am. J. Hum. Genet.* **71,** 1242–1247.

Quade, S. R. E., Elston, R. C., and Goddard, K. A. B. (2005). Estimating haplotype frequencies in pooled DNA samples when there is genotyping error. *BMC Genet.* **6,** 25.

Rabinowitz, D., and Laird, N. (2000). A unified approach to adjusting association tests for population admixture with arbitrary pedigree structure and arbitrary missing marker information. *Hum. Hered.* **504,** 227–233.

Roeder, K., Bacanu, S., Sonpar, V., Zhang, X., and Devlin, B. (2005). Analysis of single-locus tests to detect gene/disease associations. *Genet. Epidemiol.* **28,** 207–219.

Rohde, K., and Fuerst, R. (2001). Haplotyping and estimation of haplotype frequencies for closely linked biallelic multilocus genetic phenotypes including nuclear family information. *Hum. Mutat.* **17,** 289–295.

Sai, K., Itoda, M., Saito, Y., Kurose, K., Katori, N., Kaniwa, N., Komamura, K., Kotake, T., Morishita, H., Tomoike, H., *et al.* (2006). Genetic variations and haplotype structures of the ABCB1 gene in a Japanese population: An expanded haplotype block covering the distal promoter region, and associated ethnic differences. *Ann. Hum. Genet.* **70,** 605–622.

Salem, R., Wessel, J., and Schork, N. (2005). A comprehensive literature review of haplotyping software and methods for use with unrelated individuals. *Hum. Genomics* **2,** 39–66.

Satten, G. A., and Epstein, M. P. (2004). Comparison of prospective and retrospective methods for haplotype inference in case-control studies. *Genet. Epidemiol.* **27,** 192–201.

Schaid, D. J. (1996). General score tests for associations of genetic markers with disease using cases and their parents. *Genet. Epidemiol.* **13,** 423–449.

Schaid, D. J. (2002). Relative efficiency of ambiguous vs. directly measured haplotype frequencies. *Genet. Epidemiol.* **23,** 426–443.

Schaid, D. J. (2004). Evaluating associations of haplotypes with traits. *Genet. Epidemiol.* **27,** 348–364.

Schaid, D. J. (2005). Haplo.Stats: http://mayoresearch.mayo.edu/mayo/research/biostat/schaid.cfm.

Schaid, D. J., McDonnell, S. K., Wang, L., Cunningham, J. M., and Thibodeau, S. N. (2002a). Caution on pedigree haplotype inference with software that assumes linkage equilibrium. *Am. J. Hum. Genet.* **71,** 992–995.

Schaid, D. J., Rowland, C. M., Tines, D. E., Jacobson, R. M., and Poland, G. A. (2002b). Score tests for association between traits and haplotypes when linkage phase is ambiguous. *Am. J. Hum. Genet.* **70,** 425–434.

Scheet, P., and Stephens, M. (2006). A fast and flexible statistical model for large-scale population genotype data: Applications to inferring missing genotypes and haplotypic phase. *Am. J. Hum. Genet.* **78,** 629–644.

Schneidler, S., Roessli, D., and Excoffier, L. (2002). Arlequin version 2.001: A software for population genetics data analysis. Genetics and Biometry Laboratory University of Geneva, Switzerland. http://lgb.unige.ch/arlequin/.

Schouten, M. T., Williams, C. K., and Haley, C. S. (2005). The impact of using related individuals for haplotype reconstruction in population studies. *Genetics* **171,** 1321–1330.

Self, S. G., Longton, G., Kopecky, K. J., and Liang, K. Y. (1991). On estimating HLA/disease association with application to a study of aplastic anemia. *Biometrics* **47,** 53–61.

Seltman, H., Roeder, K., and Devlin, B. (2001). Transmission/disequilibrium test meets measured haplotype analysis: Family-based association analysis guided by evolution of haplotypes. *Am. J. Hum. Genet.* **68,** 1250–1263.

Seltman, H., Roeder, K., and Devlin, B. (2003). Evolutionary-based association analysis using haplotype data. *Genet. Epidemiol.* **25,** 48–58.

Sham, P. (1997). The transmission/disequilibrium tests for multiallelic loci. *Am. J. Hum. Genet.* **61,** 774–778.

Sham, P., Bader, J. S., Craig, I., O'Donovan, M., and Owen, M. (2002). DNA pooling: A tool for large-scale association studies. *Nat. Rev. Genet.* **3,** 862–871.

Sieh, W., Yu, C. E., Bird, T. D., Schellenberg, G. D., and Wijsman, E. M. (2007). Accounting for linkage disequilibrium among markers in linkage analysis: Impact of haplotype frequency estimation and molecular haplotypes for a gene in a candidate region for Alzheimer's disease. *Hum. Hered.* **63,** 26–34.

Skrivanek, Z., Lin, S., and Irwin, M. (2003). Linkage analysis with sequential imputation. *Genet. Epidemiol.* **25,** 25–35.

Sobel, E., and Lange, K. (1996). Descent graphs in pedigree analysis: Application to haplotyping, location scores, and marker-sharing statistics. *Am. J. Hum. Genet.* **58,** 1323–1337.

Sobel, E., Lange, K., O'Connell, J. R., and Weeks, D. E. (1995). Haplotype algorithms. *In* "Genetic Mapping and DNA Sequencing." IMA Volumes in Mathematics and Its Applications ((edited by Frieman A, Gulliver R). T. P. Speed and M. S. Waterman, eds.), pp. 89–110. Springer, New York.

Souverein, O. W., Zwinderman, A. H., and Tanck, M. W. T. (2006). Estimating haplotype effects on dichotomous outcome for unphased genotype data using a weighted penalized log-likelihood approach. *Hum. Hered.* **61,** 104–110.

Spielman, R. S., and Ewens, W. J. (1996). The TDT and other family based tests for linkage disequilibrium and association. *Am. J. Hum. Genet.* **59,** 983–989.

Spielman, R. S., and Ewens, W. J. (1998). A sibship test for linkage in the presence of association: The sib transmission/disequilibrium test. *Am. J. Hum. Genet.* **62,** 450–458.

Spielman, R. S., McGinnis, R. E., and Ewens, W. J. (1993). Transmission test for linkage disequilibrium: The insulin gene region and insulin-dependent diabetes mellitus (IDDM). *Am. J. Hum. Genet.* **52,** 506–516.

Spinka, C., Carroll, R. J., and Chatterjee, N. (2005). Analysis of case-control studies of genetic and environmental factors with missing genetic information and haplotype-phase ambiguity. *Genet. Epidemiol.* **29,** 108–127.

Stephens, M., and Donnelly, P. (2003). A comparison of bayesian methods for haplotype reconstruction from population genotype data. *Am. J. Hum. Genet.* **73,** 1162–1169.

Stephens, M., and Scheet, P. (2005). Accounting for decay of linkage disequilibrium in haplotype inference and missing-data imputation. *Am. J. Hum. Genet.* **76,** 449–462.

Stephens, M., Smith, N. J., and Donnelly, P. (2001). A new statistical method for haplotype reconstruction from population data. *Am. J. Hum. Genet.* **68,** 978–989.

Stram, D. O., Leigh, P. C., Bretsky, P., Freedman, M., Hirschhorn, J. N., Altshuler, D., Kolonel, L. N., Henderson, B. E., and Thomas, D. C. (2003). Modeling and E-M estimation of haplotype-specific relative risks from genotype data for a case-control study of unrelated individuals. *Hum. Hered.* **55,** 179–190.

Sun, F. Z., Flanders, W. D., Yang, Q. H., and Khoury, M. J. (1999). The transmission disequilibrium test (TDT) when only one parent is available: The 1-TDT. *Am. J. Epidemiol.* **150,** 97–104.

Swofford, D. L. (1998). PAUP: Phylogenetic analysis using parsimony release 4.0b1. Sinauer Associates, Sunderland, MA.

Tanck, M. W., Klerkx, A. H., Jukema, J. W., De Knijff, P., Kastelein, J. J., and Zwinderman, A. H. (2003). Estimation of multilocus haplotype effects using weighted penalised log-likelihood: Analysis of five sequence variations at the cholesteryl ester transfer protein gene locus. *Ann. Hum. Genet.* **67,** 175–184.

Tapadar, P., Ghosh, S., and Majumder, P. P. (2000). Haplotyping in pedigrees via a genetic algorithm. *Hum. Hered.* **50,** 43–56.

Tapper, W., Collins, A., Gibson, J., Maniatis, N., Ennis, S., and Morton, N. E. (2005). A map of the human genome in linkage disequilibrium units. *Proc. Natl. Acad. Sci. USA* **102,** 11835–11839.

Templeton, A. R. (1995). A cladistic analysis of phenotypic associations with haplotypes inferred from restriction endonuclease mapping or DNA sequencing. V. Analysis of case/control sampling designs: Alzheimer's disease and the apoprotein E locus. *Genetics* **140,** 403–409.

Templeton, A. R., and Sing, C. F. (1993). A cladistic analysis of phenotypic associations with haplotypes inferred from restriction endonuclease mapping. IV. Nested analyses with cladogram uncertainty and recombination. *Genetics* **134,** 659–669.

Templeton, A. R., Boerwinkle, E., and Sing, C. F. (1987). A cladistic analysis of phenotypic associations with haplotypes inferred from restriction endonuclease mapping. I. Basic theory and an analysis of alcohol dehydrogenase activity in Drosophila. *Genetics* **117,** 343–351.

Templeton, A. R., Sing, C. F., Kessling, A., and Humphries, S. (1988). A cladistic analysis of phenotype associations with haplotypes inferred from restriction endonuclease mapping. II. The analysis of natural populations. *Genetics* **120,** 1145–1154.

Templeton, A. R., Crandall, K. A., and Sing, C. F. (1992). A cladistic analysis of phenotypic associations with haplotypes inferred from restriction endonuclease mapping and DNA sequence data. III. Cladogram estimation. *Genetics* **132,** 619–633.

Templeton, A. R., Weiss, K. M., Nickerson, D. A., Boerwinkle, E., and Sing, C. F. (2000). Cladistic structure within the human Lipoprotein lipase gene and its implications for phenotypic association studies. *Genetics* **156,** 1259–1275.

Terwilliger, J., and Ott, J. (1994). "Handbook for Human Genetic Linkage." Johns Hopkins University Press, Baltimore, MD.

The International HapMap Consortium (2003). The International HapMap Project. *Nature* **426,** 789–796.

The International HapMap Consortium (2005). A haplotype map of the human genome. *Nature* **437,** 1299–1320.

Thomas, A. (2003). GCHap: Fast MLEs for haplotype frequencies by gene counting. *Bioinformatics* **19,** 2002–2003.

Thomas, D. C., Stram, D. O., Conti, D., Molitor, J., and Marjoram, P. (2003). Bayesian spatial modeling of haplotype associations. *Hum. Hered.* **56,** 32–40.

Thompson, C. L., Baechle, D., Lu, Q., Mathew, G., Song, Y. J., Iyengar, S. K., McGuire, C. G., and Goddard, K. A. B. (2005). Effect of genotyping error in model-free linkage analysis using microsatellite or single-nucleotide polymorphism marker maps. *BMC Genet.* **6**(Suppl 1), S153.

Toivonen, H. T. T., Onkamo, P., Vasko, K., Ollikainen, V., Sevon, P., Mannila, H., Herr, M., and Kere, J. (2000). Data mining applied to linkage disequilibrium mapping. *Am. J. Hum. Genet.* **67,** 133–145.

Tost, J., Brandt, O., Boussicault, F., Derbala, D., Caloustian, C., Lechner, D., and Gut, I. G. (2002). Molecular haplotyping at high throughput. *Nucleic Acids Res.* **30,** e96.

Tregouet, D. A., Barbaux, S., Escolano, S., Tahri, N., Golmard, J. L., Tiret, L., and Cambien, F. (2002). Specific haplotypes of the P-selection gene are associated with myocardial infraction. *Hum. Mol. Genet.* **11,** 2015–2023.

Tregouet, D. A., Escolano, S., Tiret, L., Mallet, A., and Golmard, J. L. (2004). A new algorithm for haplotype-based association analysis: The Stochastic-EM algorithm. *Ann. Hum. Genet.* **68,** 165–177.

Tregouet, D. A., and Garelle, V. A. (2007). New JAVA interface implementation of THESIAS: Testing haplotypes effects in association studies. *Bioinformatics* **23,** 1038–1039.

Tzeng, J. Y. (2005). Evolutionary-based grouping of haplotypes in association analysis. *Genet. Epidemiol.* **28,** 220–231.

Tzeng, J. Y., Devlin, B., Wasserman, L., and Roeder, K. (2003). On the identification of disease mutations by the analysis of haplotype similarity and goodness of fit. *Am. J. Hum. Genet.* **72,** 891–902.

Tzeng, J. Y., Wang, C. H., Kao, J. T., and Hsiao, C. K. (2006). Regression-based association analysis with clustered haplotypes through use of genotypes. *Am. J. Hum. Genet.* **78,** 231–242.

Van der Meulen, M. A., and te Meerman, G. J. (1997). Haplotype sharing analysis in affected individuals from nuclear families with at least one affected offspring. *Genet. Epidemiol.* **14,** 915–920.

Waldron, E. R. B., Whittaker, J. C., and Balding, D. J. (2006). Fine mapping of disease genes via haplotype clustering. *Genet. Epidemiol.* **30,** 170–179.

Wang, S., Kidd, K. K., and Zhao, H. (2003). On the use of DNA pooling to estimate haplotype frequencies. *Genet. Epidemiol.* **24,** 74–82.

Weeks, D. E., Sobel, E., O'Connell, J. R., and Lange, K. (1995). Computer programs for multilocus haplotyping of general pedigrees. *Am. J. Hum. Genet.* **56,** 1506–1507.

Weiss, K. M., and Clark, A. G. (2002). Linkage disequilibrium and the mapping of complex human traits. *Trends Genet.* **18,** 19–24.

Wijsman, E. M. (1987). A deductive method of haplotype analysis in pedigrees. *Am. J. Hum. Genet.* **41,** 356–373.

Wijsman, E. M., Almasy, L., Amos, C. I., Borecki, I., Falk, C. T., King, T. M., Martinez, M. M., Meyers, D., Neuman, R., Olson, J. M., *et al.* (2001). Genetic Analysis Workshop 12: Analysis of complex genetic traits: Applications to asthma and simulated data. *Genet. Epidemiol.* **21**(Suppl. 1), S1–S853.

Witte, J. S., Gauderman, W. J., and Thomas, D. C. (1999). Asymptotic bias and efficiency in and case-control studies of candidate genes and gene-environment interactions: Basic family designs. *Am. J. Epidemiol.* **149,** 693–705.

Xie, X., and Ott, J. (1993). Testing linkage disequillibrium between a disease and marker locus. *Am. J. Hum. Genet.* **53,** 1107–1113.

Xie, R. R., and Stram, D. O. (2005). Asymptotic equivalence between two score tests for haplotype-specific risk in general linear models. *Genet. Epidemiol.* **29,** 166–170.

Xiong, M., Zhao, J., and Boerwinkle, E. (2002). Generalized T^2 test for genome association studies. *Am. J. Hum. Genet.* **70,** 1257–1268.

Xiong, M., Zhao, J., and Boerwinkle, E. (2003). Haplotype block linkage disequilibrium mapping. *Front. Biosci.* **8,** 85–93.

Yan, H., Papadopoulos, N., Marra, G., Perrera, C., Jiricny, J., Boland, C. R., Lynch, H. T., Chadwick, R. B., de la Chapelle, A., Berg, K., *et al.* (2000). Conversion of diploidy to haploidy. *Nature* **403,** 723–724.

Yang, Y., Zhang, J., Hoh, J., Matsuda, F., Xu, P., Lathrop, M., and Ott, J. (2003). Efficiency of single-nucleotide polymorphism haplotype estimation from pooled DNA. *Proc. Natl. Acad. Sci. USA* **100,** 7225–7230.

Yu, K., Gu, C. C., Province, M., Xiong, C. J., and Rao, D. C. (2004). Genetic association mapping under founder heterogeneity via weighted haplotype similarity analysis in candidate genes. *Genet. Epidemiol.* **27,** 182–191.

Yu, K., Xu, J., Rao, D. C., and Province, M. (2005a). Using tree-based recursive partitioning methods to group haplotypes for increased power in association studies. *Ann. Hum. Genet.* **69,** 577–589.

Yu, K., Zhang, S. L., Borecki, I., Kraja, A., Xiong, C. J., Myers, R., and Province, M. (2005b). A haplotype similarity based transmission/disequilibrium test under founder heterogeneity. *Ann. Hum. Genet.* **69,** 455–467.

Zaykin, D. V., Westfall, P. H., Young, S. S., Karnoub, M. A., Wagner, M. J., and Ehm, M. G. (2002). Testing association of statistically inferred haplotypes with discrete and continuous traits in samples of unrelated individuals. *Hum. Hered.* **53,** 79–91.

Zeng, D., and Lin, D. Y. (2005). Estimating haplotype-disease associations with pooled genotype data. *Genet. Epidemiol.* **28,** 70–82.

Zhang, S. L., and Zhao, H. Y. (2001). Quantitative similarity-based association tests using population samples. *Am. J. Hum. Genet.* **69,** 601–614.

Zhang, K., and Zhao, H. (2006). A comparison of several methods for haplotype frequency estimation and haplotype reconstruction for tightly linked markers from general pedigrees. *Genet. Epidemiol.* **30,** 423–437.

Zhang, S., Pakstis, A. J., Kidd, K. K., and Zhao, H. (2001). Comparisons of two methods for haplotype reconstruction and haplotype frequency estimation from population data. *Am. J. Hum. Genet.* **69,** 906–914.

Zhang, K., Calabrese, P., Nordborg, M., and Sun, F. (2002a). Haplotype block structure and its applications to association studies: Power and study designs. *Am. J. Hum. Genet.* **71,** 1386–1394.

Zhang, S. L., Kidd, K. K., and Zhao, H. Y. (2002b). Detecting genetic association in case-control studies using similarity-based association tests. *Stat. Sin.* **12,** 337–359.

Zhang, W., Collins, A., Maniatis, N., Tapper, W., and Morton, N. E. (2002c). Properties of linkage disequilibrium (LD) maps. *Proc. Natl. Acad. Sci. USA* **99,** 17004–17007.

Zhang, S., Sha, Q., Chen, H., Dong, J., and Jiang, R. (2003). Transmission/disequilibrium test based on haplotype sharing for tightly linked markers. *Am. J. Hum. Genet.* **73,** 566–579.

Zhang, K., Qin, Z. S., Liu, J. S., Chen, T., Waterman, M. S., and Sun, F. (2004). Haplotype block partitioning and tag SNP selection using genotype data and their applications to association studies. *Genome Res.* **14,** 908–916.

Zhang, K., Sun, F., and Zhao, H. (2005). HAPLORE: A program for haplotype reconstruction in general pedigrees without recombination. *Bioinformatics* **21,** 90–103.

Zhang, Y., Niu, T. H., and Liu, J. S. (2006). A coalescence-guided hierarchical Bayesian method for haplotype inference. *Am. J. Hum. Genet.* **79,** 313–322.

Zhang, H., Zhang, H., Li, Z., and Zheng, G. (2007). Statistical methods for haplotype-based matched case-control association studies. *Genet. Epidemiol* **31,** 316–326.

Zhao, H. (2000). Family-based association studies. *Stat. Methods Med. Res.* **9,** 563–587.

Zhao, H., Zhang, S., Merikangas, K. R., Trixler, M., Widenauer, D. B., Sun, F. Z., and Kidd, K. K. (2000). Transmission/disequilibrium tests using multiple tightly linked markers. *Am. J. Hum. Genet.* **67,** 936–946.

Zhao, J. H., and Sham, P. C. (2002). Faster haplotype frequency estimation using unrelated subjects. *Hum. Hered.* **53,** 36–41.

Zhao, J. H., Lissarrague, S., Essioux, L., and Sham, P. C. (2002). GENECOUNTING: Haplotype analysis with missing genotypes. *Bioinformatics* **18,** 1694–1695.

Zhao, H., Pfeiffer, R., and Gail, M. H. (2003a). Haplotype analysis in population genetics and association studies. *Pharmacogenomics* **4,** 171–178.

Zhao, L. P., Li, S. S., and Khalid, N. (2003b). A method for the assessment of disease associations with single-nucleotide polymorphism haplotypes and environmental variables in case-control studies. *Am. J. Hum. Genet.* **72,** 1231–1250.

Zhao, J., Boerwinkle, E., and Xiong, M. (2005). An entropy-based statistic for genomewide association studies. *Am. J. Hum. Genet.* **77,** 27–40.

Zhao, J., Jin, L., and Xiong, M. (2006). Nonlinear tests for genomewide association studies. *Genetics* **174,** 1529–1538.

Zhao, L. P., Li, S. S., and Shen, F. (2007). A haplotype-linkage analysis method for estimating recombination rates using dense SNP trio data. *Genet. Epidemiol.* **31,** 154–172.

Zheng, M., and McPeek, M. S. (2007). Multipoint linkage-disequilibrium mapping with haplotype-block structure. *Am. J. Hum. Genet.* **80,** 112–125.

15 Characterization of LD Structures and the Utility of HapMap in Genetic Association Studies

C. Charles Gu,* Kai Yu,‡ and D. C. Rao*,†

*Division of Biostatistics and Department of Genetics,
Washington University School of Medicine, St. Louis, Missouri 63110
†Department of Psychiatry, Washington
University School of Medicine, St. Louis, Missouri 63110
‡Division of Cancer Epidemiology and Genetics,
National Cancer Institute, Bethesda, Maryland 20892

Advances in Genetics, Vol. 60

0065-2660/08 $35.00
DOI: 10.1016/S0065-2660(07)00415-4

ABSTRACT

Observed distribution of and variation in linkage disequilibrium (LD) with respect to the evolution history and disease transmission in a population is the driving force behind the current wave of genome-wide association (GWA) studies of complex human diseases. An extensive literature covers topics from haplotype analysis that utilizes local LD structures in candidate genes and regions to genome-wide organization of LD blocks (neighborhood) that led to the development of International HapMap Project and panels of "tagSNPs" used by current GWA studies. In this chapter, we examine the scenarios where each of the major types of analysis methods may be applicable and where the current popular genotyping platforms for GWA might come short. We discuss current association analysis methods by emphasizing their reliance on the local LD structures or the global organization of the LD structures, and highlight the need to consider individual marker information content in large-scale association mapping.

I. INTRODUCTION

Genetic association mapping is becoming a popular and powerful approach for identifying genetic factors underlying common complex diseases such as hypertension and diabetes. It is because there are many causative factors (genetic and nongenetic) involved in the etiology of complex diseases, and the analytical methods exploiting linkage disequilibrium (LD) information become particularly attractive due to its sensitivity to smaller than modest genetic effects (Risch and Merikangas, 1996). In a typical association study, highly dense panels of genetic markers [e.g., single-nucleotide polymorphisms (SNPs)] are genotyped in candidate genes (or regions) or over the whole genome, and LD mapping methods are applied to "scan" for associations between the disease phenotype and the genetic variants. The numbers of SNPs employed by genome-wide association (GWA) studies grows steadily, from thousands to hundreds of thousands, and possibly to a few millions in the foreseeable future. As the number of available SNPs increases (currently over six millions SNPs across the human genome), multiple practical problems emerge. For example, how should one reduce the number of SNPs selected for a study to control genotyping cost? How to deal with the problem of inflated false positive rate and correct for multiple testing? Solutions to these questions seem to be highly related to our understanding of local LD structures and their organization at the genome level in populations.

Our understanding of local LD structures affects the efficiency of fine-mapping via haplotype (or multimarker) analysis. Haplotypes are sets of alleles at linked loci on a single chromosome inherited from a subject's parent

(Clayton *et al.*, 2004; Schaid, 2006). While single-point methods consider a single marker at a time, methods of haplotype analysis utilize information from nearby (linked) markers and tend to have enhanced power in many scenarios for detecting complex disease genes (Akey *et al.*, 2001; Morris and Kaplan, 2002; Zhang and Sun, 2005). More importantly, there is biological evidence that higher-order *cis*-interactions of multiple amino acid sites determine haplotypes as natural functional units of genes (Clark, 2004). These studies point to the potential of local LD structures and haplotype analysis in modern association mapping.

At the genome level, recent studies of genome-wide (or chromosome-wide) LD patterns in the World's populations have revealed that the human genome consists of blocks (or neighborhoods) of variable lengths where haplotype diversity is limited, and such blocks are separated by regions of recombination "hot spots" (Daly *et al.*, 2001; Reich *et al.*, 2001; Rioux *et al.*, 2001; Templeton *et al.*, 2000; Wang *et al.*, 2002). Therefore, it is hoped that a genome-wide haplotype map (HapMap) can capture empirically the architecture behind such global LD structures and enable more efficient GWA study of complex traits (Dawson *et al.*, 2002; Gabriel *et al.*, 2002; The International HapMap Consortium, 2003). For example, various strategies were proposed to select a subset of representative SNPs, so studies may focus the analysis on such "tag" SNPs (tagSNPs). The International HapMap Project (The International HapMap Consortium, 2003) has cataloged so far close to 4 million SNPs in each of four populations from different parts of the world and is expected to identify 300,000–600,000 tagSNPs that uniquely represent common haplotypes across the whole genome (The International HapMap Consortium, 2003). A great deal of recent studies have performed empirical analysis of completed HapMap data, sometimes in conjunction with data collected in local populations, to assess the practical utility of HapMap in mapping genes for complex disease. We review here some of the important findings from these studies and offer our views as to how the recent research developments in local and global LD structures can help the GWA studies.

II. HAPLOTYPE SIMILARITY IN CANDIDATE GENES AND LD MAPPING

Prior to examining the properties of global LD structures, it is helpful to first have an in-depth review of how local LD-structures in candidate genes (regions) are used in association mapping. In general, the associations between a candidate region and the trait of interest can be assessed by single-marker or haplotype analyses. Studies of power analysis showed that while both approaches can be favorable depending on the underlying disease model and factors of the study design (Akey *et al.*, 2001; Bader, 2001; Morris and Kaplan, 2002; Nielsen *et al.*, 2004), the haplotype-based approach may be more powerful for younger (hence rarer) causative variants (Schaid, 2004). More importantly, haplotype methods naturally utilize the local LD structure of the candidate region and are more informative than single-marker analyses when causative variants are themselves not (directly) assayed. This may change as new rapid high-throughput sequencing technology becomes affordable enough to allow sequence information on individuals to be used in association studies (see Schork *et al.*, 2008; Chapter IX, this volume).

A host of LD-based methods for fine-mapping are based on the premise that, over a stretch of genome covering the candidate region, "case" haplotypes harboring a disease mutation inherited from a common ancestor would exhibit excess allele sharing (compared with that in the "control" haplotypes) at markers near the mutation locus (McPeek and Strahs, 1999). Methods were developed to perform haplotype similarity analysis where similarity among haplotypes in affected individuals (cases) and that in healthy controls are compared to establish association with the disease (Bourgain *et al.*, 2000; de Vries *et al.*, 1996; te Meerman and Van der Meulen, 1997; Tzeng *et al.*, 2003). For these methods, a pair-wise similarity score was used to measure the level of similarity between two haplotypes, and the similarity of a group of haplotypes can be defined as an average of pair-wise similarity scores of the group. Association of the candidate region to a disease locus was tested by assessing whether a larger than expected score difference is present. Examples of haplotype similarities include length-based measures (physical or genetic length of the longest contiguous interval of matching alleles) and count-based measures (the number of alleles identical by state over all loci or a subinterval of consecutive markers) (Tzeng *et al.*, 2003). It is now considered essential to weight the measures according to marker allele frequencies so that artificial similarity due to frequent alleles at consecutive markers is properly discounted (sharing of rarer alleles is given more prominence) (Yu *et al.*, 2004b).

A. Founder heterogeneity

A practical difficulty arising in haplotype similarity analysis and other haplotype mapping methods is founder heterogeneity, in which case haplotypes inherit disease-causing mutations from different ancestral haplotypes. For complex traits, founder heterogeneity is likely a norm than an exception. Because case haplotypes from different founders are likely less similar than those sharing a common founder, founder heterogeneity decreases the average similarity score in cases and can reduce the power of haplotype similarity analysis (Morris and Kaplan, 2002). In the presence of founder heterogeneity, the set $\mathbf{D}$ of case haplotypes is then a union $\mathbf{D} = \mathbf{Q}_0 \cup \mathbf{Q}_1 \cup \ldots \cup \mathbf{Q}_q$ of *haplotype clusters* $\mathbf{Q}_r$, where each *cluster* $Q_r, 1 \leq r \leq q$ comprises haplotypes around a mutation of distinctive ancestral origin, and Q_0 contains the remaining noncarrier haplotypes (containing no disease mutations). While the mutation locus for each cluster may be different, it is possible to identify the substructure by clustering analysis. The key is that the clustering analysis needs to be "supervised" by the underlying haplotype–disease association so the resulted clustering of haplotypes delivers optimal power for association analysis. Such supervised clustering was successfully applied in gene expression microarray data analysis by the "gene-shaving" method (Hastie *et al.*, 2000). Similar ideas applied to haplotype similarity analysis led to the development of a peeling procedure to sequentially remove case chromosomes not likely belonging to major clusters of case haplotypes, and compare only similarity in the core clusters with that in normal controls (Yu *et al.*, 2004a,b). The peeling process starts from $\mathbf{D}_0 = \mathbf{D}$, the set of all case haplotypes, and create a sequence of nested subsets $\mathbf{D}_0 \supset \mathbf{D}_1, \ldots, \supset \mathbf{D}_M$, by deleting a proportion (e.g., 10%) of haplotypes "least similar" with others in the parent group. Association between the candidate region and the disease susceptibility locus is then repeatedly evaluated by comparing haplotype similarity among chromosomes in each identified subset with that among controls. The method can identify and remove Q_0 and clusters with small effects, and identify a *core cluster* by minimal empirical p values resulted from tests of similarity (requiring the core gives most significant difference from the control groups). A random permutation procedure suitable for large numbers of variates [e.g., the one-level permutation technique developed for microarray data analysis by Ge *et al.* (2003)] can be applied to adjust for the multiple comparisons introduced by the repeated peeling to obtain correct significance levels of the test statistic. Simulation studies showed that these modifications performed favorably when compared to existing haplotype similarity analysis methods under various genetic models (Yu *et al.*, 2004a). The power of haplotype similarity tests and the improvement of power due to the modifications will depend on the actual number of founder haplotype-carrying disease mutations (carriers), the ratio of carrier haplotype proportion in case to that in control samples, and the level of resemblance between haplotype distributions in carriers and noncarriers. It was shown that a power improvement over using simple count similarity can range from 5% to about 70% depending on the scenarios.

B. Multiple candidate genes and global test

Another practical issue in haplotype association analysis of real-world studies is the multifactorial nature of complex traits, where multiple candidate genes interact in contributing to the disease etiology. Therefore, simultaneous analyses of multiple candidate genes/regions may be essential and likely more reliable/powerful than repeated single-gene analyses. There have been methodological developments both in genotype-based and in haplotype-based approaches. Genotype-based methods are suitable when there are relatively few markers typed in each candidate genes. For example, Conti *et al.* (2003) used Bayesian model averaging to detect important genetic association in the context of complex interactions; and Schaid *et al.* (2005) proposed a generalized *U*-statistics to combine evidence of association at multiple marker loci. Improved regression models using a false-discovery rate approach were also studied to handle potential interactions of multiple candidate genes in association analysis (Devlin *et al.*, 2003). When there are a large number of SNPs genotyped in multiple candidate genes, haplotype-based approaches will be more appropriate. For example, a composite statistics may be used to study haplotype associations at multiple genes jointly by simply adding χ^2 statistics of comparing haplotype frequencies at individual regions together (Fan and Knapp, 2003). A similar idea can be applied to analyze haplotype similarities simultaneously at multiple candidate loci. One challenge is to come up with a global test that is sensitive enough in diverse situations where most or some considered genes are associated with the disease susceptibility loci. Another technical challenge is the heterogeneous effects of different genes, so similarity results collected from all genes need to be properly scaled before they can be combined. In a recent study, a nonparametric approach using signed rank-sum statistics (O'Brien, 1984) was introduced to overcome the scale problem, and a min–max composite statistic was introduced to optimize the potential power of the global association test based on haplotype similarities at multiple candidate genes (Yu *et al.*, 2005a). It first compares the structural similarity among transmitted haplotypes against nontransmitted haplotypes at individual genes, and then combines ranked haplotype similarity comparison results from all genes considered into a class of global statistics. Simulation studies are conduced to check the global tests' type I error rates in stratified populations and to compare power under various disease models. Briefly, taking 2-locus model as an example, we first constructed a rank-sum statistic $G^{(c)}, c = 1$ or 2 for each locus based on linear signed rank scores of all pair-wise similarity scores within case or control groups. We then form two other test statistics for the additive and independent effects of the two loci, respectively: $Q_{add} = G^{(1)} + G^{(2)}$ and $Q_{max} = \max(G^{(1)}, G^{(2)})$. Finally, a composite statistic is used to test association of the joint effect of both loci $Q_{comp} = \min(p(Q_{add}), p(Q_{max}))$, where $p(\cdot)$ is the empirical p value for the corresponding test resulted from a one-level permutation as before. Extensive simulation analyses showed that the composite similarity test has correct false positive rates with the best overall performance, and is recommended for initial general use when no information about the multilocus model is available.

A word of caution is that although global tests are capable of studying multiple genes collectively, they are not always superior in performance over single-gene tests adjusted for multiple comparisons. Instead, studying one gene per time with correction for multiple comparisons might be more efficient. For example, Schaid *et al.* (2005) compared their global test to single-marker tests with Bonferroni correction and found that the advantage of using their global test only occurred when the number of disease susceptibility markers was over 3 among a set of 10 markers. Therefore, it seems in situations where only relatively few genes are related to the disease, studying the entire list of genes simultaneously might not be optimal. However, it depends on the type of global tests used, and the conclusion might be reversed if the marginal effects are dismal and a test focused on interaction terms is used. In a recent study, Marchini *et al.* (2005) used a combinatorial search approach for identifying explicitly statistical interactions between loci. They showed that variations in allele frequencies among interacting loci may lead to dwindling power to detect marginal gene effects; and that the method focus on interactions can be more powerful than traditional analyses under a range of models for gene–gene interactions, even with a conservative correction for multiple testing. Because the modeling of interaction is an ongoing research field, it remains to be seen if this will become a strategy more feasible in candidate genes studies or in a GWA environment.

C. Other practical issues

1. Family-based samples

Population admixture is a potential source of spurious association findings and multiple solutions have been proposed to estimate and correct for the effect due to admixture (Devlin *et al.*, 2001; McKeigue *et al.*, 2000; Pritchard *et al.*, 2000; Tang *et al.*, 2005). It is known that regression-based methods using unrelated individuals (e.g., case-control studies) can be more efficient as compared with family-based association tests (Morton and Collins, 1998; Schaid *et al.*, 2002; Zaykin *et al.*, 2002). However, designs using family units offer many advantages, most notably in controlling undetected population stratification (Bull *et al.*, 2001). To utilize the local LD structures in candidate regions in family-based designs, methods of haplotype association analysis can be applied. Methods for single-marker analysis can be extended by constructing regression-based score tests or by estimating population-level variance structure of the haplotype effects via generalized estimation equation (Horvath *et al.*, 2001; Lange *et al.*, 2003; Schaid *et al.*, 2005). Alternatively, extensions of the transmission/disequilibrium tests may be used (Abecasis *et al.*, 2000; Spielman *et al.*, 1993). When so analyzed, haplotypes in a candidate region are viewed as “meta-alleles” of an artificial “marker” made of the multiple markers tested. This only serves to test for association with the candidate gene but provides no information about the location of disease mutations. To facilitate fine-mapping, recent developments of transmission/disequilibrium test-type methods extended to compare similarities between transmitted (to cases) and nontransmitted haplotypes (Bourgain *et al.*, 2001; Sha *et al.*, 2005; Yu *et al.*, 2005b; Zhang *et al.*, 2003). The similarity-based methods have the advantage to them the relatively small degrees of freedom, typically the number of markers (Zhang *et al.*, 2003) compared to $h - 1$ (h is the number of observed distinctive haplotypes) by other methods (Clayton, 1999; Schaid *et al.*, 2002). The same techniques of sequential peeling of haplotype clusters and min–max rank sum statistics can be used as before to combat potential problems of founder heterogeneity and interactions among multiple candidate genes (Yu *et al.*, 2005a), and facilitate fine-mapping of complex traits in family studies.

2. Unphased haplotypes and untyped markers

While most of available LD fine-mapping methods work on haplotypes with known phase, most real-world studies only have unphased genotype data available. Dealing with unphased haplotypes is statistically a missing data problem that can be computationally expensive. A popular solution is to use the most likely pair of haplotypes inferred by statistical methods (Niu, 2004). This can lead to substantial loss of information and unpredictable bias (Morris *et al.*, 2004; Schaid, 2004). Recent studies suggested that a better alternative is to directly study the unphased genotyped data and account for haplotype ambiguity by statistical modeling (Lin and Zeng, 2006; Satten and Epstein, 2004). Methods that directly testing main effects in multilocus genotype data disregard of haplotype phase, such as the genotype T^2 test (Fan *et al.*, 2005; Xiong *et al.*, 2002) and the indirect association test (Chapman *et al.*, 2003; Clayton *et al.*, 2004), have fewer degrees of freedom than the haplotype methods and can have greater power. This advantage may dissipate if interactions among candidate loci become more prominent. Nonetheless, the indirect method can be particularly useful when LD structures were carefully used to select representative tagSNPs.

III. GLOBAL ORGANIZATION OF LD STRUCTURES

Selection of a manageable core set of SNPs has been a key step in association studies of complex diseases. As the most common (~90% of total) genetic variation in humans, SNPs (variants with a minor allele frequency $\geq 1\%$) are most abundant genetic markers available across the genome. Early studies estimated that there are about 10 million such SNPs in the world's human population (i.e., ~1 SNP/300 bp) (Kruglyak and Nickerson, 2001; Reich *et al.*, 2003). It is impractical to scan all SNPs in a "genome-wide" association study. Fortunately, the redundancy in the vast number of available SNPs can be used to reduce the actual number of SNPs needed for genotyping and analysis in real studies.

The redundancy is reflected by the organization of LD structures locally in candidate regions and globally across the genome. Recent empirical studies of large genomic regions or chromosomes have observed that the human genome consists of blocks of variable lengths where pair-wise LD is high and haplotype diversity is limited, and such blocks are separated by regions of recombination "hot spots." A study of haplotype variation in 9.7 kb of the genomic sequence from the human lipoprotein lipase gene revealed distribution of recombination hot spots flanking regions with evolutionarily structured haplotype variation (Templeton *et al.*, 2000). The phenomenas were repeatedly reported by others in studies of known candidate genes [e.g., MHCII (Jeffreys *et al.*, 2001)] and genes mapped Crohn's disease (Daly *et al.*, 2001; Rioux *et al.*, 2001). At the genome level, a study by Perlegen of 36,000 SNPs on chromosome 21 in a mixed sample of African, Asian and Caucasian origin uncovered a chromosome-wide block structure where three or four common haplotypes characterized more than 80% of the population (Patil *et al.*, 2001). A later study of 51 regions across the genome in separate ethnic groups derived at similar conclusions about the block structure and estimated the average block size ranging from approximately 11 kb in Africans to approximately 22 kb in Caucasians and Asians (Gabriel *et al.*, 2002). These early studies of LD structures across the human genome proved that its certainly possible to reduce the number of SNPs needed in candidate gene studies and showed that it might be feasible to scan the whole genome by analyzing associations at a core set of SNPs. Developed and validate databases of such SNPs became an active field for research (Dawson *et al.*, 2002; Goldstein, 2001).

A. HapMap and tagSNPs

To speed the development of a public domain database of SNPs for study of genes related to common complex traits, an international research consortium combining both public and private resources (initially ~$100 million) launched a genetic variation mapping project and dubbed the "International HapMap Project." The HapMap project involves research groups from six countries and is aimed at charting genetic variation over the human genome in the world's populations. Data released from the initial phase of HapMap include a little more than 1 million SNPs across the genome in four populations with African, Asian, and European ancestry, with at least one common SNP (with minor allele frequency or MAF $\geq$ 0.05) every 5 kb (The International HapMap Consortium, 2005)..

A major goal of the HapMap was to guide the prioritization of SNP genotyping for disease association studies by identifying a core set of tagSNPs (Johnson *et al.*, 2001). As with the methods for association analysis that may be divided into genotype- or haplotype-based, the tagging methods (for selecting tagSNPs) can use a single SNP or haplotype as proxies for untyped variants. Some methods use the structure of disjoint LD blocks while others use more subtle LD structures that allow overlapping blocks (neighborhoods). The common goal of all tagging methods is to minimize the size of the core set of SNPs by reducing redundancy among SNPs while minimizing the loss of information.

All tagging methods (explicitly or indirectly) utilize both local and global organization of LD structure among SNPs across the genome. Some of the methods that can be called block-based use the information directly in various forms such as haplotype diversity, haplotype r^2 or haplotype entropy; other (block-free) methods use pairwise r^2, or multilocus R^2 (also called coefficient of determination). Besides how block information is utilized, several other major characteristics may distinguish these tagging methods. For example, genotype information may be directly used (e.g., Carlson *et al.*, 2004) or via haplotype (estimation) (Johnson *et al.*, 2001; Weale *et al.*, 2003; Zhang *et al.*, 2002). The local tagging criteria can also use different methods such as haplotype diversity (Johnson *et al.*, 2001; Zhang *et al.*, 2002), pairwise r^2 (Carlson *et al.*, 2004), haplotype r^2, or multilocus R^2 (Chapman *et al.*, 2003; Stram *et al.*, 2003; Weale *et al.*, 2003); information theoretical entropy; or mutual information (Halldorsson *et al.*, 2004; Hampe *et al.*, 2003; Sebastiani *et al.*, 2003). Finally, the global optimization algorithms may use exhaustive search (Johnson *et al.*, 2001; Weale *et al.*, 2003), greedy search or heuristics (Carlson *et al.*, 2004; Ke and Cardon, 2003; Sebastiani *et al.*, 2003), dynamic programming (Zhang *et al.*, 2002), or matrix decomposition such as PCA (Lin and Altman, 2004; Nyholt, 2004). A summary of some popular methods with their computer programs is given in Table 15.1

A great deal of literature focused on how haplotype blocks were "best" defined and constructed (i.e., "blocking" methods) and on optimal selection of tagSNPs (Abecasis *et al.*, 2001; Bafna *et al.*, 2003; Carlson *et al.*, 2004; Clayton, 2001; Halldorsson *et al.*, 2004; Johnson *et al.*, 2001; Lin and Altman, 2004; Meng *et al.*, 2003; Stram *et al.*, 2003; Zhang *et al.*, 2002). Definitions of haplotype/LD blocks can be based on haplotype diversity, pairwise LD, or recombination ("hot spots") (Wang *et al.*, 2002). Studies have shown that blocks based on these definitions share substantial similarity and robustness (Ding *et al.*, 2005; Schwartz *et al.*, 2003). Combined with the fact that multimarker aggressive tagging can be used to predict untyped markers in strong LD, the current wisdom is that blocks or neighborhoods defined by pairwise LD suffices for HapMap construction and tagSNPs selection. Because the selection of tagSNPs is sensitive to how the HapMap is constructed, the level of similarity in results from different tagging methods brings comfort on the generality of the core set of tagSNPs selected by the HapMap. The next important question is then, how well the core set of tagSNPs selected based on a population-specific HapMap can be transferable to studies of other ethnic populations?

Table 15.1. Popular Tagging Programs

Program	Block-based	Estimates blocks	Scalable	Tagging Criteria	Reference & Website
Haploview	Y	Y	Y	Gabriel criteria	(Barrett and Cardon, 2006) http://www.broad.mit.edu/mpg/haploview/
LDSelect	N	N/A	Y	pairwise r^2	(Carlson *et al.*, 2004) http://droog.gs.washington.edu/ldSelect.html
htSNP	Y	N	N	multivariate R^2, haplotype r^2	(Chapman *et al.*, 2003) http://www-gene.cimr.cam.ac.uk/clayton/software/
Hapblock	Y	Y	Y	haplotype r^2, pairwise r^2	(Zhang *et al.*, 2005) http://www-hto.usc.edu/msms/HapBlock/
tagSNPs	Y	N	Y	haplotype r^2, pairwise r^2	(Stram *et al.*, 2003) http://www-rcf.usc.edu/~stram/tagSNPs.html
TagIT	Y	N	Y	multivariate R^2, haplotype r^2	(Weale *et al.*, 2003) http://www.genome.duke.edu/tagit/
eigen2htSNP	N	N/A	unknown	pairwise r^2, PCA	(Nyholt, 2004; Lin and Altman, 2004) http://htsnp.stanford.edu/PCA/ http://gump.qimr.edu.au/general/daleN/SNPSpD/
BEST	N	N/A	unknown	mutual information	(Sebastiani *et al.*, 2003) http://genomethods.org/best/
Tagger	N	N/A	Y	pairwise r^2, haplotype r^2	(de Bakker *et al.*, 2005) http://www.broad.mit.edu/mpg/tagger/

B. Similarity of HapMaps and transferability of tagSNPs

The transferability of the core set of tagSNPs selected from one population to another is not clearly defined in the literature, although all measures are related to the extent of agreement (similarity) between the overall LD structures (HapMaps) in different populations. Earlier studies had shown that there was a great deal of variation among the LD block structures among the world's populations (Gu *et al.*, 2007a; Liu *et al.*, 2004; Mueller *et al.*, 2005; Sawyer *et al.*, 2005) (more details on this and its implications are discussed in Section IV.C). Such observations cast doubt on routine (uncritical) transferability to other populations. Alternatively, the transferability may be defined by the "genome coverage" (percentage of all SNPs captured by the tagSNPs) provided by the tagSNPs in the new population. Several recent studies adopted this notion of coverage for transferability and showed promisingly high level of transferability of the HapMap tagSNPs using completed HapMap data combined with simulation studies (de Bakker *et al.*, 2005). Gonzalez-Neira *et al.* (2006) analyzed 144 SNPs in a 1-Mb region of chromosome 22 in 1055 individuals from 38 worldwide populations in 7 continental groups. Approximating the three HapMap populations, they chose tagSNPs in three reference populations by pairwise r^2 greater than a threshold (e.g., $r^2 > 0.8$) and measured tagSNP transferability by average values of maximum pairwise r^2 with non-tagSNPs in other test populations. They found that most effective tagging was to use the Yoruba population as the reference for tagging, which resulted in an average transferability of 0.81 (range 0.69–0.98 and a little less than half of the populations with transferability < 0.8). Service *et al.* (2007) examined the transferability of HapMap Centre d'Etude du Polymorphisme Humain (CEU) chromosome 22 tagSNPs in 11 population isolates (many of which showed different LD and allele frequency distributions from the reference population) and observed coverage/transferability ranging from 82.5 to 91.8%. These empirical findings enhanced the prospects for determining a "universal" core set of tagSNPs to facilitate a feasible platform for genome-wide studies of complex diseases in multiple populations. However, in light of considerable variation in the LD patterns across populations, implications of using a universal core set of tagSNPs need further careful investigations.

C. Implications to GWA

Ideally, an optimally designed study should select the tagSNPs from a HapMap population that is most closely related to the population of interest. In reality, at least for GWA studies, the tagSNPs are predetermined by a commercial platform that likely provides a core set of tagSNPs selected primary based on a particular population (e.g., HapMap CEU). Early simulation studies estimated that it may take approximately 500,000–1,000,000 SNPs to cover the whole genome for association studies (Dunning *et al.*, 2000; Kruglyak, 1999). Empirical analysis of completed data from the HapMap (The International HapMap Consortium, 2005) showed that a "perfect" coverage of all SNPs genotyped in Phase I of the project (a little more than 1 million) would require close to half a million SNPs for each of the four populations (row 1 in Table 15.2). Estimates based on a more recent release of HapMap (Magi *et al.*, 2006) gave much higher numbers of SNPs (almost 30% higher for Yoruba) required for even a less than ideal coverage of $r^2 > 0.8$ (row 2 in Table 15.2). The numbers will grow as more new SNPs are genotyped and added to the HapMap database. Simulation analyses using data from Phase I and a representative collection of complete data (common and rare) in 5 Mb sequences over the genome (ENCODE) showed that the coverage of all common SNPs by those typed in HapMap Phase I range from 45% (YRI, Yoruba) to 74% (CEU). Therefore, a crude estimation can give an approximate upper bound of total SNPs needed for a "universal" GWA platform: over 1.7M (row 4 in Table 15.2).

Current available GWA platforms offer much smaller set of tagSNPs and therefore have markedly suboptimal transferability. Recent studies compared two representative platforms, the Affymetrix Mapping Arrays and the Illumina HumanHap BeadChip. Both platforms based tag selection on HapMap CEU SNPs but used different genotyping techniques and, more importantly, different tagging strategies. Affymetrix used randomly selected even-spaced SNPs across the genome, while Illumina used HapMap tagSNPs. While there appears to be no difference between anonymous and gene-based tagging approaches (Wiltshire *et al.*, 2006), studies showed that association analysis using tagSNPs are more powerful than those using the same number of randomly selected SNPs with a power difference as large as 20% at times (Gopalakrishnan and Qin, 2006; Zhang and Sun, 2005). Therefore, it is no surprise that earlier version of Affymetrix arrays had a relatively poor coverage in African samples compared to Illumina (see Table 15.3).

One way to improve the coverage is to apply multimarker (or haplotype) analysis (Chapman *et al.*, 2003; Clayton *et al.*, 2004). Because the power of multilocus approaches is generally higher than single point analysis, some of the

Table 15.2. Estimated Number Core Set of tagSNPs for a "Universal" GWA Platform

	CEU	CHB+JPT	YRI	Reference
$r^2 \sim 1.0$	447, 579	434, 476	604, 886	(The International HapMap Consortium, 2005)
$r^2 >= 0.8$	579,978	670,407	780,336	(Magi *et al.*, 2006)
Estimated coverage of HapMap Phase I	74%	72%	45%	(The International HapMap Consortium, 2005)
Estimated size of "universal" tagSNPs	783,754	931,121	1,734,080	

Source: Gu *et al.* (2008).

Table 15.3. Genome-Wide Coverage of Popular GWA Platforms

	CEU	CHB+JPT	YRI	Reference
Illumina HumanHap 300K	75% (86%)	63% (70%)	28% (42%)	(Pe'er *et al.*, 2006; Barrett and Cardon, 2006)
Illumina HumanHap 550K	95%	92%	68%	Illumina
Affimetrix 500K	65% (78%)	66% (72%)	41% (56%)	(Pe'er *et al.*, 2006; Barrett and Cardon, 2006)
Affymetrix 500K + 175K tag	86%	79%	49%	(Barrett and Cardon, 2006)

Source: Gu *et al.* (2008).

non-tagSNPs below the r^2 threshold will be detected with uncompromised power. This in effect increases the coverage of a given set of tagSNPs. When this "aggressive" tagging approach was used to capture 1679 common SNPs across 25 genes (a total span of 2.6 Mb with 1 SNP/1.6 kb), the genotyping burden was reduced by 15–23% compared with pairwise tagging (de Bakker *et al.*, 2006). In Table 15.3, percentages in brackets are new coverage boosted by the multimarker tagging estimated by Pe'er *et al.* (2006). A simple generalization of the results will give boosted coverage of approximately 62% to the augmented Affymetrix 500K + and of approximately 82% to Illumina 550K in Africans.

It is clear that although the tagSNPs used by both platforms provide sustentative whole-genome coverage in the usually difficult African population, the compromised transferability will severely affect the power of GWA studies using these platforms. Empirical and simulation analyses based on the completed HapMap data showed that compared to using all HapMap SNPs, there is generally a power loss of 5–10% using tagSNPs for association studies (de Bakker *et al.*, 2006; Magi *et al.*, 2006). Adding further power loss due to untyped non-HapMap SNPs [which can be as dramatic as from 81% to 63% (de Bakker *et al.*, 2006)], the result could be detrimental to some studies of complex traits where marginal signals at individual loci are weak. Improvements both in numbers of tagSNPs used by these platforms and in methods of analyses are important for success of the GWA studies.

IV. VARIATION OF LOCAL LD STRUCTURES

While HapMap transferability measured by genome-wide coverage of tagSNPs appears to be largely satisfactory, the fine structure of global organization of LD structures can vary drastically according to the genetic background of the test population. Several methods were studied for summary measure of local LD structure differences between HapMaps (constructed in similar or diverse populations). For example, Gabriel *et al.* (2002) used distribution of average block sizes; Bafna *et al.* (2003) proposed an exact test statistic of similarity between two HapMaps (LD block partitions) based on shared block boundaries; and Liu *et al.* (2004) proposed similarity measures based on concordance of SNP pairs captured by the same block. Using these measures, empirical studies showed a great deal of variability in HapMap constructions across different populations (Gu *et al.*, 2007b; Liu *et al.*, 2004; Mueller *et al.*, 2005; Sawyer *et al.*, 2005).

A result of the variability of local LD structures is the varying *information content* of individual SNPs in the reference population as compared to that in the test population. This is because that in a particular HapMap construction, the local LD structure dictates whether an SNP is captured by an LD block/neighborhood. This in turn influences the power and false positive rate for detecting the SNP as a causal variant by association test performed at tagSNPs selected based on the LD structures. There is no clear consensus as how to properly assess the information content of an SNP in the context of varying local LD structures, nor how such information may be used in disease gene association mapping. One method is to directly measure the information content of an SNP by the level of ambiguity in whether it is captured by an LD block. This was called the marker ambiguity method (Gu *et al.*, 2007a).

A. The marker ambiguity score method

We may assume that for a given population the observed LD block structures follow a probability distribution $\wp$. The distribution is affected by many factors, ranging from the genetic origin and evolutionary history of the population to sampling and LD partitioning methods. For a fixed SNP position in the genome and a particular instance of local LD structure, the SNP is either well inside a block ("captured") at the boundary of a block or being a singleton ("boundary point"). When the LD block structure varies, the SNP can alter between being captured sometimes and being a boundary point some other times. Namely, for each individual SNP, the distribution of LD structures induces a probability that measures the level of ambiguity of whether it can be tagged (captured) or not. This probability was referred to as the empirical *marker ambiguity score* (MAS) (Gu *et al.*, 2007a). Specifically, for a given pair of LD block partitionings H_1 and H_2, and a marker M that was captured by block A in H_1 and B in H_2, the block ambiguity of M is defined by $\gamma(\mathrm{M}, H_1, H_2) = 1 - |\mathrm{A} \cap \mathrm{B}|/|\mathrm{A} \cup \mathrm{B}|$, with $|\cdot|$ denoting length of a genomic region and $\cup$ and $\cap$ for union and intersection of two blocks, respectively. The definition can be modified to accommodate block definitions that allow overlapping [e.g., haplotype neighborhoods (Dawson *et al.*, 2002) or LD bins (Carlson *et al.*, 2004)]: $\gamma(\mathrm{M}, H_1, H_2) = \min_{\{\mathrm{A,B}\}}(1 - |\mathrm{A} \cap \mathrm{B}|/|\mathrm{A} \cup \mathrm{B}|)$ for all blocks A in H_1 and B in H_2 that contain M. When many instances of block partitioning are considered, the expectation of the distribution of $\gamma(\mathrm{M}, H_1, H_2)$ reflects the ambiguity of whether M falls in-between blocks (boundaries) and is defined as the $MAS\gamma(\mathrm{M})$

of SNP M : $\gamma(M) = E\{\gamma(M, H_1, H_2), (H_1, H_2) \in \wp \times \wp\}$. In practice, the sample mean $\hat{\gamma}(M)$ is used for real data analysis: $\hat{\gamma}(M) = avg_{(H_1,H_2)\in\hat{\wp}\times\hat{\wp}}\{\gamma(M, H_1, H_2)\}$, where $\hat{\wp}$ is a sample of LD block partitioning, either observed or simulated. The score is close to 0 when two global LD structures agree with each other around marker M, and close to 1 when the two are very different near M.

Gu *et al.* (2007a) applied the method to real data of 2913 SNPs in 1890 individuals (518 blacks and 1272 whites) across 56.1 Mb of one chromosome (~1 SNP/19kb) with a median gap of 8.7 kb between adjacent SNPs. Combined with Monte Carlo simulation and resampling, the study showed information content of individual SNPs (e.g., as measured by MAS) may reflect the important differences between populations. We may need more sophisticated tool and measures to gauge the difference between populations in terms of LD structures. Not only the global LD structures need to be studied, the local structure differences as measured by MAS reflect important properties critical to association analysis and should be carefully studied. Because whether an SNP is captured in an LD structure affect its potential being detected by association analysis of tagSNPs, the information content defined this way is very relevant to its power in a tagging system.

The MAS method directly measures the information content of individual SNPs in the context of varying HapMap constructions. As a result, it can reveal subtle differences in local LD structures and assess the differences between HapMap constructions at a level suitable for disease gene mapping. In addition, by studying distributional properties of the MAS, one may also compare observed LD structures at the genome level to study quality (robustness) of HapMap constructions or transferability of HapMaps between populations. For example, it was found that, within the same population, the robustness of inferred global LD structure (HapMap) is dependent on the size of samples used for the construction. Based on the study of the MAS distributions for different sample sizes, it was estimated that at least a sample of 200 or more are needed for a reasonably stable HapMap construction in both races (Gu *et al.*, 2007a). It was also found that the robustness of block boundary varies along the genome. Therefore, at some locations, larger samples may be required for a reliable inference of block boundaries.

The differences in LD structures between two ethnic groups can be seen by higher correlations of MAS scores within race than that of between races. More similarity within race is apparent as depicted in Fig. 15.1, in which scatter plots of the MAS scores estimated from simulated samples of different sample sizes (50, 100, and 150) in blacks and whites were shown. For example, correlation of MAS in the same race (whites) estimated at different sample sizes (50 vs 150) was a lot higher ($\rho = 0.69$) than that estimated in different races at same sample size (150) ($\rho = 0.22$). Across the two ethnic groups, differential LD block structures can result in quite different distributions of information content as measured by MAS. Analyses by the Mann-Whitney-Wilcoxon test and

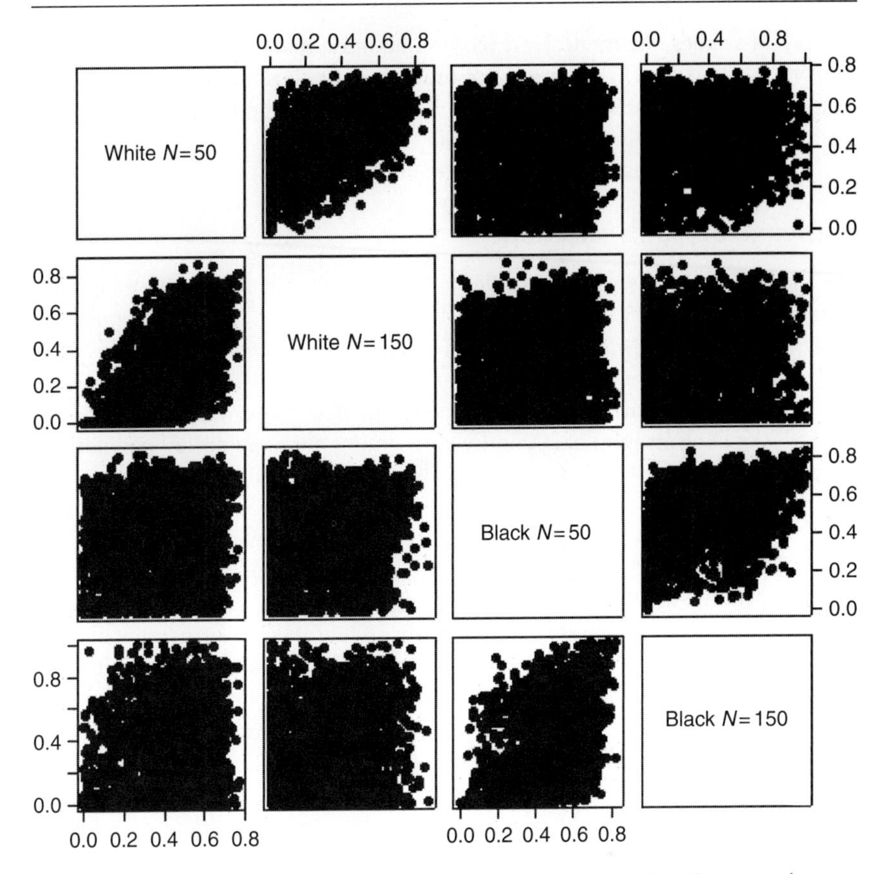

Figure 15.1. Scatter plots of MAS scores. There is a higher correlation of MAS scores within race groups than that between race groups. This is shown by the four scatter plots that are tighter around the diagonals.

quantile–quantile plots showed significant differences between the MAS distributions in blacks and that in whites. On average, MAS is slightly higher in blacks, and substantially more SNPs had very high MAS values (highly ambiguous) in blacks than in whites. This was verified using multiple sample sizes to accommodate the fact that the level of robustness of LD block boundaries vary cross the genome. More details may be found in Gu *et al.* (2007a).

B. Implications for GWA studies

In contrast to the high level of transferability of HapMap tagSNPs among populations, the substantive variability in SNP information content shown by the previous study across distinct populations has some important implications for GWA studies. At the least, it requires additional attention to (1) quantitatively assessing the information content of individual markers via MAS in HapMap constructions and (2) identifying locations where LD structures may be excessively variable in a given sample.

1. Refine tagSNPs selection

As the recent studies showed, while tagSNPs selected based on HapMap CEU data showed favorable transferability to other similar Caucasian populations (de Bakker *et al.*, 2006; Gonzalez-Neira *et al.*, 2006; Gu *et al.*, 2007b; Montpetit *et al.*, 2006), the transferability was poor when the tagSNPs were tested in more distant populations such as African–Americans (Conrad *et al.*, 2006; de Bakker *et al.*, 2006; Ding and Kullo, 2007). Even in similar but more distant populations, tagSNPs selected based on the HapMap data may have quite poor transferability in some populations or in certain genomic regions (Gu *et al.*, 2007b; Mueller *et al.*, 2005; Ribas *et al.*, 2006). In general, tagSNPs perform poorly to capture rare SNPs (e.g., MAF $< 5\%$) in test populations when such SNPs were initially excluded from a HapMap scheme (Ke *et al.*, 2005; Montpetit *et al.*, 2006; Song *et al.*, 2006). However, the poor transferability of tagSNPs in distinct populations is not entirely due to the rare SNPs. As showed by Montpetit *et al.* (2006), a substantive fraction of common SNPs were not captured by the tagSNPs in the test populations. What really make the difference in tagSNPs performance are the differential local LD structures. As we pointed out earlier, the robustness of a HapMap construction varies across the genome, so the effective sample size required for a stable HapMap construction also varies in different regions of the genome. While the current sample size of 90 subjects of current HapMap might be enough for a major part of the genome, there are regions where local LD structures are more variable and require a larger sample size or greater SNP density to achieve transferability comparable to other regions. This can be carried out by first identifying problematic regions through cross population transferability studies, followed by detailed resequencing of the regions while using individual marker information content measured by MAS as a standardized scale to control optimal choice of sample size and SNP density.

2. Prioritizing significant SNPs resulted from GWA scans

One may speculate that when studying human diseases, chromosomes in disease patients (cases) were effectively drawn from a different (sub)population than chromosomes in the controls. Therefore, there might be differences in fine LD structures attributable to the different disease status. It is reasonable to believe that the regions where LD structures differ in cases and controls may have a higher prior probability harboring disease variants. In another word, if all tested SNPs were classified into "wild-type SNPs" and "disease SNPs," one should expect the genome-wide MAS distribution being an admixture of two ("wild type" and "disease") distributions. This was exactly what we found by empirical studies [see Fig. 15.2 and more discussion in Gu *et al.* (2007a)]. In both race groups, modeling of normal mixtures on box-transformed MAS distributions indicated a mixture of two distributions centered at low MAS and at relatively high MAS values. Indeed, the subtle differences may contain location information of disease variants and can be applied to assist fine-mapping of disease variants. It should be noted that, however, the location information provided by MAS *alone* is not sufficient to capture those disease variants where LD structures were similar in cases and controls. It was also possible that some of the differences in the block structures were caused by (unmeasured) differences between cases and controls other than disease status. Therefore, one would need to further distinguish which of the regions are due to the disease (and so harboring disease variants) and which ones are not. Nonetheless, to assist in SNP selection, we can estimate the MAS values within case–control groups, and use the MAS as a weight of importance in the selection of SNPs. When faced with multiple choices of SNPs, with otherwise comparable association strengths, SNPs with higher MAS values (implying less assurance of being consistently covered by a haplotype block) should be given more consideration. This can be achieved by integrating MAS with results of conventional association tests. This way, the information content of individual SNPs can provide valuable information to prioritizing the search for important SNPs with otherwise comparable borderline *p* values.

V. DISCUSSION

Methodological developments in LD mapping and advances in empirical studies of genome-wide organization of LD structures in populations have inspired a wave of GWA studies of complex diseases. The International HapMap Project has propelled the movement by providing a growing catalog detailing the distribution of millions of SNPs and their LD structures observed in the World's populations. How to effectively use the information in designing and carrying out GWA studies that will ultimately result in meaningful gene discoveries is an active research field and deserves all the attention it can get. We discussed important recent developments in areas related to local LD structure and haplotype analysis and to global organization of LD blocks (neighborhoods). Both areas are highly relevant to the design and analysis of GWA studies.

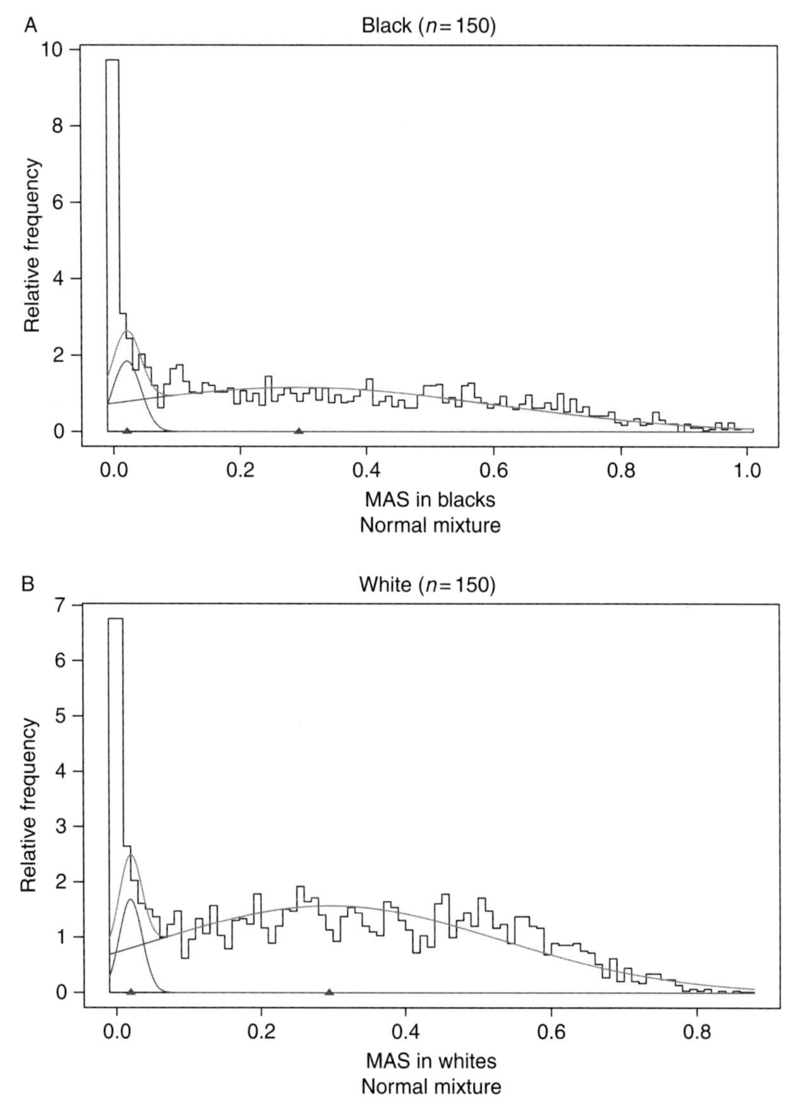

Figure 15.2. Mixed distribution of MAS scores. Mixture distribution modeling of the disease-induced MAS scores for a sample size of 150 in both races. There is strong evidence of clustering of MAS at very low and high values, implicating potential application of disease-induced MAS scores to aid selection of important SNPs.

There are several good review articles about methods of haplotype analysis and related power and sample size issues. Schaid (2004, 2006) gave a detailed account of the analytical issues involved when applying regression methods to analyzing haplotype association with complex traits. Clayton *et al.* (2004) discussed methods for "indirect association" analysis using unphased diplotype data. Clark (2004) reviewed haplotype methods in the context of functional studies and population genetics. It is commonly believed that the powers of single-point and multipoint association analyses are similar if the number of causative SNPs is less than the number of haplotypes. Some reported that, under certain models, negligible differences between methods based on pair-wise or on multilocus LD (Bader, 2001; Nielsen *et al.*, 2004), and that single marker analysis are better than haplotype analysis (North *et al.*, 2006). In addition to potential power gain, multipoint (haplotype) method also fares better in terms of capturing *cis*-interaction among variants in close range and association at untyped sites. These features of haplotype methods result in reduced dimensionality of tests in local regions. We presented in this chapter a summary of developments in methods of haplotype similarity analysis. The regression-based methods have a clear advantage in requiring minimum computational time. The indirect association methods can gracefully account for the uncertainty of inferred haplotypes and estimate association at *untyped* SNPs. Compared with the other two types of haplotype methods, the similarity-based tests can have a smaller degrees of freedom (always less than number of markers) and also give location information about potential causal variants. None of the methods has a clear advantage in terms of modeling interactions (especially higher-order interactions) among multiple variants. The computational advantage of the regression-based methods made it the method of choice of many ongoing GWA studies. Several recent studies proposed ways to scale up these local LD methods to genome-wide studies. For example, Sun *et al.* (2006) and Yang *et al.* (2006) evaluated strategies using spatial statistics or smoothing techniques (sliding window) to properly weight single-point or local haplotype tests. Browning (2006) proposed a variable-length Markov chains method for similarity analysis to obtain optimal window sizing adjusted for local LD structures, although the current implementation could not handle a large number of SNPs. Wessel and Schork (2006) proposed a multivariate distance matrix regression method for similarity analysis that has the potential to be applicable directly to sequence data for association studies (see Schork *et al.*, 2008; Chpater XX, this Volume). Further investigations are needed to develop and evaluate generalizations of these methods in the setting of GWA studies.

Recent studies on global organization of LD structures and on the utility of HapMap information in GWA have evolved from finding best ways to select tagSNPs to evaluating transferability of tagSNPs. Most relevant questions have been (1) how robust is the current construction of HapMap and

(2) transferability of HapMap-derived tagSNPs in other (similar and distant) populations. Answers to question (1) are related to sample sizes used for HapMap construction. Optimistic evaluations of empirical data recently concluded that 90–100 subjects will likely be enough for construction of a stable enough HapMap (Zeggini *et al.*, 2005), although earlier studies cautioned a lot more samples were needed for robust HapMap constructions (Liu *et al.*, 2004; Wang *et al.*, 2002). Zeggini *et al.* (2005) performed a simulation study based on empirical data over 250 kb sequence and concluded the sample size of approximately 60 is adequate for the current HapMap. Monpetit studied 1536 SNPs randomly selected in two 500 kp ENCODE regions on chromosome 2 in a Estonia population and found acceptable performance of tagSNPs at a sample size of 100. These studies considered the robustness of HapMap constructions not in terms of fine differences in LD structures but in tagSNPs performance that is directly related to the question (2) on tagSNPs transferability.

It is a clear consensus from the recent reports that transferability of the HapMap-derived tagSNPs are generally satisfactory when tested in similar populations and on HapMap SNPs (based on which the tags were selected). However, the performance of the tagSNPs detererate substantively when they applied to distant populations and/or on non-HapMap SNPs ("untyped" by the HapMap). de Bakker *et al.* (2006) reported coverage reduced from approximately 90% of HapMap SNPs to approximately 80% of "untyped" SNPs when CEU tags were tested in other Caucasian populations and from over 80% to as low as 56% when YRI tags were applied to other African samples. Another study cross-examining SNPs from the NIEHS Environment Genome Project and the HapMap also found poor performance of HapMap tagSNPs for capturing untyped variants (between 30% for YRI and 55% for CEU panels). The transferability gets worse in population isolates as showed by Service *et al.* (2007), where HapMap CEU tagSNPs were tested in 11 population isolates of European descent with resulted coverage for untyped SNPs ranging from 21% to 27%.

Given that current popular platforms for GWA studies only provide panels of about 500 K tagSNPs based largely on HapMap CEU panel, further studies are needed for better strategies to select "universal" tagSNPs. For example, one sticky issue in tagging SNPs is how to treat rare SNPs. It is well known that rare SNPs are harder to capture. But as the empirical studies showed, a substantive fraction of noncaptured SNPs were not rare SNPs (Montpetit *et al.*, 2006). On the other hand, although selection of genotyping panel based on HapMap tags (e.g., Illumina 550 K) were more efficient in capturing HapMap SNPs, a random selection strategy (e.g., Affymetrix 500 K) may perform more favorably toward rare SNPs (Zhang and Sun, 2005). Affymetrix has started β-test of a new array of 1 M SNPs as we finish this review. It will not be surprising if the new chips will adopt a somewhat hybrid selection strategy. The problem related to using a core set of "universal" tagSNPs may dissipate when individual sequence information becomes affordable for association studies. But the chapter will not entirely go away before the technology becomes truly affordable.

Caution is needed when interpreting the results from GWA scans using these genotyping panels, and more studies are needed to investigate the best strategy to reduce dimensionality while maintaining minimal information loss. New methods are sorely needed to connect global organization of LD structures in a population with local variations of LD that can be more closely linked to disease traits in a study sample. For example, to extract from the hundreds of thousands of variables (SNPs) at the initial genome-wide scanning down to a few dozens of SNPs that are more likely involved in the disease etiology requires an estimation of effective number of tests to properly adjust for multiple testing. The method of false discovery rate may be applied to derive a more realistic significance level, or a well-planned simulation scheme can be used to obtain an empirical threshold (Lin, 2006). However, none of the available methods directly combine the local variation of LD structures and the global organization of LD blocks to achieve such reduction. Moreover, although a cleverly designed combinatorial strategy has been proposed (Marchini *et al.*, 2005), most GWA strategies usually ignore interactions (be it GxE or GxG) at the initial scanning stage due to unrealistic computational demand. Some of these issues may be dealt with by studying variation of individual SNP information content in respect to changes in a person's disease status or a population's evolution history.

We discussed the MAS method as a way to approach the problem of gauging individual marker information content. Methods similar to MAS were also used by others. Gonzalez-Neira *et al.* (2006) used a frequency of capturing by bins based on a single partition that is very close to the MAS, and Bafna *et al.* (2003) used mutual information of SNP pairs. Further studies of these methods in selecting tagSNPs and in importance weighting of SNP association results seem a logical step in developing methodologies for GWA studies. These can lead to constructions of localized, disease-specific HapMaps that are not necessarily as complete as the HapMap, but with highly enriched information on particular disease induced variation in local LD structures. These information can be further integrated with functional annotations of SNPs, distribution of rare variants, and potential interaction with environmental risks, and so on, and provide a clearer idea of how marker information content varies along the genome that can ultimately facilitate more efficient GWA studies of complex diseases.

Acknowledgment

This work is partially supported by NIH grants GM028719, HL054473, HL072507, and HL071782.

References

Abecasis, G. R., Cardon, L. R., and Cookson, W. O. (2000). A general test of association for quantitative traits in nuclear families. *Am. J. Hum. Genet.* **66,** 279–292.

Abecasis, G. R., Noguchi, E., Heinzmann, A., Traherne, J. A., Bhattacharyya, S., Leaves, N. I., Anderson, G. G., Zhang, Y., Lench, N. J., Carey, A., Cardon, L. R., Moffatt, M. F., *et al.* (2001). Extent and distribution of linkage disequilibrium in three genomic regions. *Am. J. Hum. Genet.* **68,** 191–197.

Akey, J., Jin, L., and Xiong, M. (2001). Haplotypes vs single marker linkage disequilibrium tests: What do we gain?. *Eur. J. Hum. Genet.* **9,** 291–300.

Bader, J. S. (2001). The relative power of SNPs and haplotype as genetic markers for association tests. *Pharmacogenomics* **2,** 11–24.

Bafna, V., Halldórsson, B. V., Schwartz, R. S., Clark, A. G., and Istrail, S. (2003). Haplotypes and informative SNP selection algorithms: Don't block out information. *"Proceedings of the Seventh Annual International Conference on Computational Molecular Biology". (RECOMB 03).*

Barrett, J. C., and Cardon, L. R. (2006). Evaluation Coverage of genome-wide association studies. *Nat.* Genet. **38,** 659–662.

Bourgain, C., Genin, E., Quesneville, H., and Clerget-Darpoux, F. (2000). Search for multifactorial disease susceptibility genes in founder populations. *Ann. Hum. Genet.* **64,** 255–265.

Bourgain, C., Genin, E., Margaritte-Jeannin, P., and Clerget-Darpoux, F. (2001). Maximum identity length contrast: A powerful method for susceptibility gene detection in isolated populations. *Genet. Epidemiol.* **21**(Suppl 1), S560–S564.

Browning, S. R. (2006). Multilocus association mapping using variable-length Markov chains. *Am. J. Hum. Genet.* **78,** 903–913.

Bull, S. B., Darlington, G. A., Greenwood, C. M., and Shin, J. (2001). Design considerations for association studies of candidate genes in families. *Genet. Epidemiol.* **20,** 149–174.

Carlson, C. S., Eberle, M. A., Rieder, M. J., Yi, Q., Kruglyak, L., and Nickerson, D. A. (2004). Selecting a maximally informative set of single-nucleotide polymorphisms for association analyses using linkage disequilibrium. *Am. J. Hum. Genet.* **74,** 106–120.

Chapman, J. M., Cooper, J. D., Todd, J. A., and Clayton, D. G. (2003). Detecting disease associations due to linkage disequilibrium using haplotype tags: A class of tests and the determinants of statistical power. *Hum. Hered.* **56**(1–3), 18–31.

Clark, A. G. (2004). The role of haplotypes in candidate gene studies. *Genet. Epidemiol.* **27,** 321–333.

Clayton, D. (1999). A generalization of the transmission/disequilibrium test for uncertain- haplotype transmission. *Am. J. Hum. Genet.* **65,** 1170–1177.

Clayton, D. (2001). http://www.nature.com/ng/journal/v29/n2/extref/ng1001–233-S10.pdf.

Clayton, D., Chapman, J., and Cooper, J. (2004). Use of unphased multilocus genotype data in indirect association studies. *Genet. Epidemiol.* **27,** 415–428.

Conrad, D. F., Jakobsson, M., Coop, G., Wen, X., Wall, J. D., Rosenberg, N. A., and Pritchard, J. K. (2006). A worldwide survey of haplotype variation and linkage disequilibrium in the human genome. *Nat. Genet.* **38,** 1251–1260.

Conti, D. V., Cortessis, V., Molitor, J., and Thomas, D. C. (2003). Bayesian modeling of complex metabolic pathways. *Hum. Hered.* **56,** 83–93.

Daly, M. J., Rioux, J. D., Schaffner, S. F., Hudson, T. J., and Lander, E. S. (2001). High-resolution haplotype structure in the human genome. *Nat. Genet.* **29,** 229–232.

Dawson, E., Abecasis, G. R., Bumpstead, S., Chen, Y., Hunt, S., Beare, D. M., Pabial, J., Dibling, T., Tinsley, E., Kirby, S., *et al.* (2002). A first-generation linkage disequilibrium map of human chromosome 22. *Nature* **418,** 544–548.

de Bakker, P. I., Yelensky, R., Pe'er, I., Gabriel, S. B., Daly, M. J., and Altshuler, D. (2005). Efficiency and power in genetic association studies. *Nat. Genet.* **37,** 1217–1223.

de Bakker, P. I., Burtt, N. P., Graham, R. R., Guiducci, C., Yelensky, R., Drake, J. A., Bersaglieri, T., Penney, K. L., Butler, J., Young, S., *et al.* (2006). Transferability of tag SNPs in genetic association studies in multiple populations. *Nat. Genet.* **38,** 1298–1303.

de Vries, H. G., van der Meulen, M. A., Rozen, R., Halley, D. J., Scheffer, H., ten Kate, L. P., Buys, C. H., and te Meerman, G. J. (1996). Haplotype identity between individuals who share a CFTR mutation allele "identical by descent": Demonstration of the usefulness of the haplotype-sharing concept for gene mapping in real populations. *Hum. Genet.* **98,** 304–309.

Devlin, B., Roeder, K., and Bacanu, S. A. (2001). Unbiased methods for population-based association studies. *Genet. Epidemiol.* **21,** 273–284.

Devlin, B., Roeder, K., and Wasserman, L. (2003). Analysis of multilocus models of association. *Genet. Epidemiol.* **25,** 36–47.

Ding, K., and Kullo, I. J. (2007). Methods for the selection of tagging SNPs: A comparison of tagging efficiency and performance. *Eur. J. Hum. Genet.* **15,** 228–236.

Ding, K., Zhou, K., Zhang, J., Knight, J., Zhang, X., and Shen, Y. (2005). The effect of haplotype-block definitions on inference of haplotype-block structure and htSNPs selection. *Mol. Biol. Evol.* **22,** 148–159.

Dunning, A. M., Durocher, F., Healey, C. S., Teare, M. D., McBride, S. E., Carlomango, F., Xu, C. F., Dawson, E., Rhodes, S., Ueda, S., *et al.* (2000). The extent of linkage disequilibrium in four populations with distinct demographic histories. *Am. J. Hum. Genet.* **67,** 1544–1554.

Fan, R., and Knapp, M. (2003). Genome association studies of complex diseases by case-control designs. *Am. J. Hum. Genet.* **72,** 850–868.

Fan, R., Knapp, M., Wjst, M., Zhao, C., and Xiong, M. (2005). High resolution T association tests of complex diseases based on family data. *Ann. Hum. Genet.* **69,** 187–208.

Gabriel, S. B., Schaffner, S. F., Nguyen, H., Moore, J. M., Roy, J., Blumenstiel, B., Higgins, J., DeFelice, M., Lochner, A., Faggart, M., *et al.* (2002). The structure of haplotype blocks in the human genome. *Science* **296,** 2225–2229.

Ge, Y., Dudoit, S., and Speed, T. (2003). Resampling-based multiple testing for microarray data hypothesis. *Test* **12,** 1–44.

Goldstein, D. B. (2001). Islands of linkage disequilibrium. *Nat. Genet.* **29,** 109–111.

Gonzalez-Neira, A., Ke, X., Lao, O., Calafell, F., Navarro, A., Comas, D., Cann, H., Bumpstead, S., Ghori, J., Hunt, S., *et al.* (2006). The portability of tagSNPs across populations: a worldwide survey. *Genome Res.* **16,** 323–330.

Gopalakrishnan, S., and Qin, Z. S. (2006). TagSNP selection based on pairwise LD criteria and power analysis in association studies. *Pac. Symp. Biocomput.* 511–522.

Gu, C. C., Yu, K., and Boerwinkle, E. (2007a). Measuring marker information content by the ambiguity of block boundaries observed in dense SNP data. *Ann. Hum. Genet.* **71,** 127–140.

Gu, S., Pakstis, A. J., Li, H., Speed, W. C., Kidd, J. R., and Kidd, K. K. (2007b). Significant variation in haplotype block structure but conservation in tagSNP patterns among global populations. *Eur. J. Hum. Genet.* **15,** 302–312.

Gu, C. C., Yu, K., Ketkar, S., Templeton, A. R., and Rao, D. C. (2008). On transferability of genome-wide tagSNPs. *Genet. Epidemiol.* **32**(2), 89–97.

Halldorsson, B. V., Bafna, V., Lippert, R., Schwartz, R., De La Vega, F. M., Clark, A. G., and Istrail, S. (2004). Optimal haplotype block-free selection of tagging SNPs for genome-wide association studies. *Genome Res.* **14**(8), 1633–1640.

Hampe, J., Schreiber, S., and Krawczak, M. (2003). Entropy-based SNP selection for genetic association studies. *Hum. Genet.* **114,** 36–43.

Hastie, T., Tibshirani, R., Eisen, M., Brown, P., Ross, D., Scherf, U., Weinstein, J., Alizadeh, A., Staudt, L., *et al.* (2000). Gene shaving: A new class of clustering methods for expression arrays Technical Report.

Horvath, S., Xu, X., and Laird, N. M. (2001). The family based association test method: Strategies for studying general genotype–phenotype associations. *Eur. J. Hum. Genet.* **9,** 301–306.

Jeffreys, A. J., Kauppi, L., and Neumann, R. (2001). Intensely punctate meiotic recombination in the class II region of the major histocompatibility complex. *Nat. Genet.* **29,** 217–222.

Johnson, G. C., Esposito, L., Barratt, B. J., Smith, A. N., Heward, J., Di Genova, G., Veda, H., Cordell, H. J., Eaves, I. A., Dudbridge, F., *et al.* (2001). Haplotype tagging for the identification of common disease genes. *Nat. Genet.* **29,** 233–237.

Ke, X., and Cardon, L. R. (2003). Efficient selective screening of haplotype tag SNPs. *Bioinformatics* **19,** 287–288.

Ke, X., Miretti, M. M., Broxholme, J., Hunt, S., Beck, S., Bentley, D. R., Deloukas, P., and Cardon, L. R. (2005). A comparison of tagging methods and their tagging space. *Hum. Mol. Genet.* **14**(18), 2757–2767.

Kruglyak, L. (1999). Prospects for whole-genome linkage disequilibrium mapping of common disease genes. *Nat. Genet.* **22,** 139–144.

Kruglyak, L., and Nickerson, D. A. (2001). Variation is the spice of life. *Nat. Genet.* **27,** 234–236.

Lange, C., Silverman, E. K., Xu, X., Weiss, S. T., and Laird, N. M. (2003). A multivariate family-based association test using generalized estimating equations: FBAT-GEE. *Biostatistics* **4,** 195–206.

Lin, D. Y. (2006). Evaluating statistical significance in two-stage genomewide association studies. *Am. J. Hum. Genet.* **78,** 505–509.

Lin, D. Y., and Zeng, D. (2006). Likelihood-based inference on haplotype effects in genetic association studies. *J. Am. Stat. Assoc.* **101**(473), 89–104.

Lin, Z., and Altman, R. B. (2004). Finding haplotype tagging SNPs by use of principal components analysis. *Am. J. Hum. Genet.* **75,** 850–861.

Liu, N., Sawyer, S. L., Mukherjee, N., Pakstis, A. J., Kidd, J. R., Kidd, K. K., Brookes, A. J., and Zhao, H. (2004). Haplotype block structures show significant variation among populations. *Genet. Epidemiol.* **27,** 385–400.

Magi, R., Kaplinski, L., and Remm, M. (2006). The whole genome tagSNP selection and transferability among HapMap populations. *Pac. Symp. Biocomput.* **11,** 535–543.

Marchini, J., Donnelly, P., and Cardon, L. R. (2005). Genome-wide strategies for detecting multiple loci that influence complex diseases. **37**(4), 413–417.

McKeigue, P. M., Carpenter, J. R., Parra, E. J., and Shriver, M. D. (2000). Estimation of admixture and detection of linkage in admixed populations by a Bayesian approach: application to African-American populations. *Ann. Hum. Genet.* **64,** 171–186.

McPeek, M. S., and Strahs, A. (1999). Assessment of linkage disequilibrium by the decay of haplotype sharing, with application to fine-scale genetic mapping. *Am. J. Hum. Genet.* **65,** 858–875.

Meng, Z., Zaykin, D. V., Xu, C. F., Wagner, M., and Ehm, M. G. (2003). Selection of genetic markers for association analyses, using linkage disequilibrium and haplotypes. *Am. J. Hum. Genet.* **73,** 115–130.

Montpetit, A., Nelis, M., Laflamme, P., Magi, R., Ke, X., Remm, M., Cardon, L., Hudson, T. J., and Metspalu, A. (2006). An evaluation of the performance of tag SNPs derived from HapMap in a Caucasian population. *PLoS Genet* **2,** e27.

Morris, A. P., Whittaker, J. C., and Balding, D. J. (2004). Little loss of information due to unknown phase for fine-scale linkage-disequilibrium mapping with single-nucleotide-polymorphism genotype data. *Am. J. Hum. Genet.* **74,** 945–953.

Morris, R. W., and Kaplan, N. L. (2002). On the advantage of haplotype analysis in the presence of multiple disease susceptibility alleles. *Genet. Epidemiol.* **23,** 221–233.

Morton, N. E., and Collins, A. (1998). Tests and estimates of allelic association in complex inheritance. *Proc. Natl. Acad. Sci. USA* **95,** 11389–11393.

Mueller, J. C., Lohmussaar, E., Magi, R., Remm, M., Bettecken, T., Lichtner, P., Biskup, S., Illig, T., Pfeufer, A., Luedemann, J., *et al.* (2005). Linkage disequilibrium patterns and tagSNP transferability among European populations. *Am. J. Hum. Genet.* **76,** 387–398.

Nielsen, D. M., Ehm, M. G., Zaykin, D. V., and Weir, B. S. (2004). Effect of two- and three-locus linkage disequilibrium on the power to detect marker/phenotype associations. *Genetics* **168,** 1029–1040.

Niu, T. (2004). Algorithms for inferring haplotypes. *Genet. Epidemiol.* **27,** 334–347.

North, B. V., Sham, P. C., Knight, J., Martin, E. R., and Curtis, D. (2006). Investigation of the ability of haplotype association and logistic regression to identify associated susceptibility loci. *Ann. Hum. Genet.* **70,** 893–906.

Nyholt, D. R. (2004). A simple correction for multiple testing for single-nucleotide polymorphisms in linkage disequilibrium with each other. *Am. J. Hum. Genet.* **74,** 765–769.

O'Brien, P. (1984). Procedures for comparing samples with multiple endpoints. *Biometrics* **40,** 1079–1087.

Patil, N., Berno, A. J., Hinds, D. A., Barrett, W. A., Doshi, J. M., Hacker, C. R., Kautzer, C. R., Lee, D. H., Marjoribanks, C., McDonough, D. P., *et al.* (2001). Blocks of limited haplotype diversity revealed by high-resolution scanning of human chromosome 21. *Science* **294,** 1719–1723.

Pe'er, I., de Bakker, P. I., Maller, J., Yelensky, R., Altshuler, D., and Daly, M. J. (2006). Evaluating and improving power in whole-genome association studies using fixed marker sets. *Nat. Genet.* **38,** 663–667.

Pritchard, J. K., Stephens, M., Rosenberg, N. A., and Donnelly, P. (2000). Association mapping in structured populations. *Am. J. Hum. Genet.* **67,** 170–181.

Reich, D. E., Cargill, M., Bolk, S., Ireland, J., Sabeti, P. C., Richter, D. J., Lavery, T., Kouyoumjian, R., Farhadian, S. F., Ward, R., *et al.* (2001). Linkage disequilibrium in the human genome. *Nature* **411,** 199–204.

Reich, D. E., Gabriel, S. B., and Altshuler, D. (2003). Quality and completeness of SNP databases. *Nat. Genet.* **33,** 457–458.

Ribas, G., Gonzalez-Neira, A., Salas, A., Milne, R. L., Vega, A., Carracedo, B., Gonzalez, E., Barroso, E., Fernandez, L. P., Yankilevich, P., *et al.* (2006). Evaluating HapMap SNP data transferability in a large-scale genotyping project involving 175 cancer-associated genes. *Hum. Genet.* **118,** 669–679.

Rioux, J. D., Daly, M. J., Silverberg, M. S., Lindblad, K., Steinhart, H., Cohen, Z., Delmonte, T., Kocher, K., Miller, K., Guschwan, S., *et al.* (2001). Genetic variation in the 5q31 cytokine gene cluster confers susceptibility to Crohn disease. *Nat. Genet.* **29,** 223–228.

Risch, N., and Merikangas, K. (1996). The future of genetic studies of complex human diseases. *Science* **273,** 1516–1517.

Satten, G. A., and Epstein, M. P. (2004). Comparison of prospective and retrospective methods for haplotype inference in case-control studies. *Genet. Epidemiol.* **27,** 192–201.

Sawyer, S. L., Mukherjee, N., Pakstis, A. J., Feuk, L., Kidd, J. R., Brookes, A. J., and Kidd, K. K. (2005). Linkage disequilibrium patterns vary substantially among populations. *Eur. J. Hum. Genet.* **13,** 677–686.

Schaid, D. J. (2004). Evaluating associations of haplotypes with traits. *Genet. Epidemiol.* **27,** 348–364.

Schaid, D. J. (2006). Power and sample size for testing associations of haplotypes with complex traits. *Ann. Hum. Genet.* **70,** 116–130.

Schaid, D. J., Rowland, C. M., Tines, D. E., Jacobson, R. M., and Poland, G. A. (2002). Score tests for association between traits and haplotypes when linkage phase is ambiguous. *Am. J. Hum. Genet.* **70,** 425–434.

Schaid, D. J., McDonnell, S. K., Hebbring, S. J., Cunningham, J. M., and Thibodeau, S. N. (2005). Nonparametric tests of association of multiple genes with human disease. *Am. J. Hum. Genet.* **76,** 780–793.

Schork, N. J., Wessel, J., and Malo, N. (2008). DNA Sequence-Based Phenotypic Association Analysis. *In* "Genetic Dissection of Complex Traits" (D. C. Rao, and D. C. Gu, eds.), 2nd Edition, pp. 197–219. Elsevier Inc. New York.

Schwartz, R., Halldorsson, B. V., Bafna, V., Clark, A. G., and Istrail, S. (2003). Robustness of inference of haplotype block structure. *J. Comput. Biol.* **10,** 13–19.

Sebastiani, P., Lazarus, R., Weiss, S. T., Kunkel, L. M., Kohane, I. S., and Ramoni, M. F. (2003). Minimal haplotype tagging. *Proc. Natl. Acad. Sci. USA* **100,** 9900–9905.

Service, S., Sabatti, C., and Freimer, N. (2007). Tag SNPs chosen from HapMap perform well in several population isolates. *Genet. Epidemiol.* **31,** 189–194.

Sha, Q., Dong, J., Jiang, R., Chen, H. S., and Zhang, S. (2005). Haplotype sharing transmission/disequilibrium tests that allow for genotyping errors. *Genet. Epidemiol.* **28,** 341–351.

Song, C. M., Yeo, B. H., Tantoso, E., Yang, Y., Lim, Y. P., Li, K. B., and Rajagopal, G. (2006). iHAP–integrated haplotype analysis pipeline for characterizing the haplotype structure of genes. *BMC Bioinformatics* **7,** 525.

Spielman, R. S., McGinnis, R. E., and Ewens, W. J. (1993). Transmission test for linkage disequilibrium: The insulin gene region and insulin-dependent diabetes mellitus (IDDM). *Am. J. Hum. Genet.* **52,** 506–516.

Stram, D. O., Haiman, C. A., Hirschhorn, J. N., Altshuler, D., Kolonel, L. N., Henderson, B. E., and Pike, M. C. (2003). Choosing haplotype-tagging SNPS based on unphased genotype data using a preliminary sample of unrelated subjects with an example from the Multiethnic Cohort Study. *Hum. Hered.* **55,** 27–36.

Sun, Y. V., Levin, A. M., Boerwinkle, E., Robertson, H., and Kardia, S. L. (2006). A scan statistic for identifying chromosomal patterns of SNP association. *Genet. Epidemiol.* **30,** 627–635.

Tang, H., Peng, J., Wang, P., and Risch, N. J. (2005). Estimation of individual admixture: analytical and study design considerations. *Genet. Epidemiol.* **28,** 289–301.

te Meerman, G. J., and Van der Meulen, M. A. (1997). Genomic sharing surrounding alleles identical by descent: effects of genetic drift and population growth. *Genet. Epidemiol.* **14,** 1125–1130.

Templeton, A. R., Weiss, K. M., Nickerson, D. A., Boerwinkle, E., and Sing, C. F. (2000). Cladistic structure within the human lipoprotein lipase gene and its implications for phenotypic association studies. *Genetics* **156,** 1259–1275.

The International HapMap Consortium (2003). The International HapMap Project. *Nature* **426,** 789–796.

The International HapMap Consortium. http://www.hapmap.org/whatishapmap.html.en.

The International HapMap Consortium (2005). A haplotype map of the human genome. *Nature* **437,** 1299–1320.

Tzeng, J. Y., Devlin, B., Wasserman, L., and Roeder, K. (2003). On the identification of disease mutations by the analysis of haplotype similarity and goodness of fit. *Am. J. Hum. Genet.* **72,** 891–902.

Wang, N., Akey, J. M., Zhang, K., Chakraborty, R., and Jin, L. (2002). Distribution of recombination crossovers and the origin of haplotype blocks: the interplay of population history, recombination, and mutation. *Am. J. Hum. Genet.* **71,** 1227–1234.

Weale, M. E., Depondt, C., Macdonald, S. J., Smith, A., Lai, P. S., Shorvon, S. D., Wood, N. W., and Goldstein, D. B. (2003). Selection and evaluation of tagging SNPs in the neuronal-sodium-channel gene SCN1A: Implications for linkage-disequilibrium gene mapping. *Am. J. Hum. Genet.* **73,** 551–565.

Wessel, J., and Schork, N. J. (2006). Generalized genomic distance-based regression methodology for multilocus association analysis. *Am. J. Hum. Genet.* **79,** 792–806.

Wiltshire, S., de Bakker, P. I., and Daly, M. J. (2006). The value of gene-based selection of tag SNPs in genome-wide association studies. *Eur. J. Hum. Genet.* **14,** 1209–1214.

Xiong, M., Zhao, J., and Boerwinkle, E. (2002). Generalized T2 test for genome association studies. *Am. J. Hum. Genet.* **70,** 1257–1268.

Yang, H. C., Lin, C. Y., and Fann, C. S. (2006). A sliding-window weighted linkage disequilibrium test. *Genet. Epidemiol.* **30,** 531–545.

Yu, K., Gu, C. C., Province, M., Xiong, C. J., and Rao, D. C. (2004a). Genetic association mapping under founder heterogeneity via weighted haplotype similarity analysis in candidate genes. *Genet. Epidemiol.* **27,** 182–191.

Yu, K., Martin, R. B., and Whittemore, A. S. (2004b). Classifying disease chromosomes arising from multiple founders, with application to fine-scale haplotype mapping. *Genet. Epidemiol.* **27,** 173–181.

Yu, K., Gu, C. C., Xiong, C., An, P., and Province, M. A. (2005a). Global transmission/disequilibrium tests based on haplotype sharing in multiple candidate genes. *Genet. Epidemiol.* **29,** 323–335.

Yu, K., Zhang, S., Borecki, I., Kraja, A., Xiong, C., Myers, R., and Province, M. (2005b). A haplotype similarity based transmission/disequilibrium test under founder heterogeneity. *Ann. Hum. Genet.* **69,** 455–467.

Zaykin, D. V., Westfall, P. H., Young, S. S., Karnoub, M. A., Wagner, M. J., and Ehm, M. G. (2002). Testing association of statistically inferred haplotypes with discrete and continuous traits in samples of unrelated individuals. *Hum. Hered.* **53,** 79–91.

Zeggini, E., Rayner, W., Morris, A. P., Hattersley, A. T., Walker, M., Hitman, G. A., Deloukas, P., Cardon, L. R., and McCarthy, M. I. (2005). An evaluation of HapMap sample size and tagging SNP performance in large-scale empirical and simulated data sets. *Nat. Genet.* **37,** 1320–1322.

Zhang, K., Qin, Z., Chen, T., Liu, J. S., Waterman, M. S., and Sun, F. (2005). HapBlock: Haplotype block partitioning and tag SNP selection software using a set of dynamic programming algorithms. *Bioinformatics* **21,** 131–134.

Zhang, K., Deng, M., Chen, T., Waterman, M. S., and Sun, F. (2002). A dynamic programming algorithm for haplotype block partitioning. *Proc. Natl. Acad. Sci. USA* **99,** 7335–7339.

Zhang, S., Sha, Q., Chen, H. S., Dong, J., and Jiang, R. (2003). Transmission/disequilibrium test based on haplotype sharing for tightly linked markers. *Am. J. Hum. Genet.* **73,** 566–579.

16 Associations Among Multiple Markers and Complex Disease: Models, Algorithms, and Applications

Themistocles L. Assimes,* Adam B. Olshen,† Balasubramanian Narasimhan,‡ and Richard A. Olshen§

*Department of Medicine, Division of Cardiovascular Medicine, Stanford University, Stanford, California 94305–5406

†Department of Epidemiology and Biostatistics, Memorial Sloan-Kettering Cancer Center, New York, New York 10021

‡Departments of Health Research and Policy and Statistics, Stanford University, Stanford, California 94305–5405

§Departments of Health Research and Policy, Statistics, and Electrical Engineering, Stanford University, Stanford, California 94305–5405

Advances in Genetics, Vol. 60
0065-2660/08 $35.00
DOI: 10.1016/S0065-2660(07)00416-6

ABSTRACT

This chapter is a report on collaborations among its authors and others over many years. It devolves from our goal of understanding genes, their main and epistatic effects combined with interactions involving demographic and environmental features also, as together they predict genetically complex diseases. Thus, our goal is "association." Particular phenotypes of interest to us are hypertension, insulin resistance, angina, and myocardial infarction. Prediction of complex disease is notoriously difficult, though it would be made easier were we given strand-specific information on genotype. Unfortunately, with current technology, genotypic information comes to us "unphased." While obviously we have strand-specific information when genotype is homozygous, we do not have such information when genotype is heterozygous. To summarize, the ultimate goals of approaches we provide is to predict phenotype, typically untoward or not, within a specific window of time. Our approach is neither through linkage nor from finding haplotype frequencies per se.

I. INTRODUCTION

Motivation for studying the genetics of complex disease is given in the remarkable paper in *Nature* by Risch (2000). Risch observes that traits studied by the celebrated Gregor Mendel were due to changes at a single locus. However, on the one hand, the most common human diseases, those that kill the vast majority of people, are to considerable extent genetic in origin. On the other, individual differences are not attributable to abnormalities at a single or even a few variable (polymorphic) sites in the genome, and neither are the sites in the same gene. Thus, many diseases such as essential hypertension are "complex." According to Risch, "Human genetics is now at a critical juncture." (And we remain at what seems an even more critical juncture seven years later!) There are concerns about understanding the increasing volume of data whose collections are enabled by improved technologies for genotyping. It is in the spirit of these concerns that we undertook research reported in this chapter. Indeed, it is evident from this volume and much published literature and software (see e.g., Barrett *et al.*, 2005; Carlson *et al.*, 2004; Clark, 2004; Crawford and Nickerson, 2005; Schaid, 2004; Schaid *et al.*, 2002) that enormous effort has been expended producing strand-specific "haplotypes," what typically amount to strand-specific contiguous "clusters" of successive polymorphic sites. Discovering them and their frequencies of occurrence is simplified when there is information on parents. Statistical inference is simplified when subjects are assumed unrelated. As if discovering haplotypes were not difficult enough, their frequencies seem to depend upon such matters as ethnicity, gender, and other simple but telling features of data that might bear upon the phenotype of interest.

The ongoing example used for illustrative purposes here is from data generated by the GENOA network of the (U.S.) National Heart, Lung, and Blood Institute's (NHLBI's) Family Blood Pressure Program (FBPP). The FBPP is a large multicenter family-based genetics study of high blood pressure and related conditions in multiple racial groups that has been ongoing since 1995 (see The FBPP Investigators, 2002). The initial focus of the FBPP was to identify genes that contribute to essential hypertension through linkage. The strongest linkage signal to date, though hardly strong, across all racial/ethnic groups in the pooled FBPP dataset is on chromosome 2 with logarithm of the odds (LOD) score of 1.91 at location 80.22 cM (Wu *et al.*, 2006). Our example uses genotypes at multiple polymorphic sites in the IL1B (interleukin 1, beta) gene located in this region of linkage. The unique identifier for each of the eight single nucleotide polymorphisms (SNPs) is borrowed from the SeattleSNP's variation discovery resource (see Seattle SNPs, 2007). This is a NHLBI-funded program in genomic applications. It informs us both on the order and the relative distance in base pairs between SNPs on human chromosome 2 with the reference point 000000 representing the first base pair of the region in and around the gene that was re-sequenced. Corresponding rs numbers for these SNPs can be found at http://pga/gs/washington.edu/. Individuals come to us in sibships. For any fixed individual, genotypes of polymorphic sites as they vary within the individual's genome are far from independent. We lack any information regarding parents. These are important motivations of our work that follows.

There is no reason to go "through linkage" (except to suggest polymorphisms to genotype) and also no obvious reason to go "through haplotypes" to arrive at prediction. Furthermore, if predictive sites cluster so, one site can be predicted accurately enough from the others; whether the sites in a cluster are contiguous on the genome or not, we could choose a "representative" of the cluster, by strand or unphased, and entirely avoid genotyping other sites in the cluster. This is analogous to, though simpler than, the notion of "tagged SNPs."

We describe now the sections that follow. Section III is about our model for pairs of genotypes, as it applies to the putatively simple case of pairs of polymorphic sites. Genotyping is unphased, and data can be summarized in terms of counts or fractions in a simple 3×3 table. Even in the simplest case, where the sample is homogeneous with respect to theoretical distribution of pairs of genotypes, there are subtle matters regarding whether testing of the usual null hypothesis of lack of association of rows and columns is conditional on row and column marginal totals or not. Multinomial parameters figure in the well-known statistics D and D' (Hedrick, 1987; Lewontin, 1988, Mohlke *et al.*, 2001). In fact, parameterization that applies to data like our GENOA data should not be assumed simple multinomial in any case. Instead, the appropriate distribution is a mixture of distributions. Conditional on the mixing parameter, the distribution of the cited table is taken to be unconditional with respect to row and column totals. Matters are complicated for us because we cannot know the mixing distribution. We derive statistics that one can calculate and tests one can perform that are faithful to the mixture model. The tests are similar to but different from the four statistics D. Attained significance (p-values) is (are) computed by a combination of permutation testing and bootstrapping. The model for testing is that data are genuinely independent and not from sibships as they really are. We discuss the "trick" that allows us to make such an inference, and the price we pay for the trick in Section III. Section IV has focus on strand-specific models. Groundwork for sections that follow is more clear than it would be otherwise because we restrict ourselves to trivial mixtures in Section IV.

To make inferences, we compute the four-dimensional vector of p-values for our four statistics that are analogous to the four tests D. The p-values are consistent with our "random effects mixture" model and are computed for available genotype data described above. The tests are of null hypotheses of "no association" between particular, cited pairs of alleles at two fixed polymorphic sites. Our unconditional approach to modeling enables separate inferences on the four pairs of genotypes at the two sites; readers will see that the p-values can be very different for different pairs of alleles. We combine the tests of the four separate pairs of alleles per pair of SNPs into "omnibus tests" of the respective null hypotheses. Readers are directed to Section V. Our first approach to omnibus testing is an application of the Hochberg–Sime criterion (Hochberg, 1988). The second is the more lenient Benjamini–Yekutieli version of the Benjamini–Hochberg "false discovery rate (FDR)" criterion (Benjamini and Yekutieli, 2001). Omnibus testing is part of Section V, which also includes methodology for agglomerative clustering of the four-vector of p-values. Our results are compared with the output of the well-known HAPLOVIEW program (see Barrett *et al.*, 2005).

Because the ultimate goal is association of possibly genotype and other features with complex disease, Section VI has summaries of six modern approaches to association (two-class classification) that we have found useful in other contexts. Discussion involves using classification trees very simply to suggest interactions, logistic regression with l_1 (the lasso) (Park and Hastie, 2006; Tibshirani, 1996) and l_2 constraints on regression coefficients (Park and Hastie, 2007), support vector machines (Vapnik, 1999), FlexTree (Huang *et al.*, 2004), and random forests (Breiman, 2001). However, we do not apply our choice of features to methodologies for association discussed in Section VI. Definition of phenotypes is already extremely difficult. While we cannot be certain from published documentation, it appears that above and beyond availability of data on parents, construction of and inferences about frequencies of haplotypes rests upon simplifying assumptions that cannot be verified from data.

Reading Section VII requires far more mathematical background than does reading the remainder of the chapter. However, results are summarized there without proof. Material touches three topics and all supposes the model of previous sections whereby true probabilities in a contingency table we discuss in great detail are assumed fixed and not random. One topic concerns a Krein–Milman-like mixture theorem for the convex, compact set of probabilities on a discrete product space. Marginal distributions of numbers of major (alternatively minor) alleles at a polymorphic site are, separately, assumed to be binomial, albeit with separate parameters. From the point of view of genetics, this assumption means that no single allele is at all determining a phenotype. All sites appear marginally of no interest. Another part of Section VII concerns candidate joint distributions of sites with (theoretically) increasing fractions of minor alleles. This assumption brings into play Blackwell's fundamental theorem on the comparison of experiments. Possible implications for candidate joint distributions of numbers of minor alleles at any sequence of sites are given in the section. A third topic involves application of Krawtchouk polynomials, the orthogonal polynomials for the binomial distribution, to summarizing joint distributions of sites that are assumed to be in Hardy–Weinberg equilibrium as is the case for the vast majority of sites with genetically complex disease such as hypertension.

Section VIII summarizes our goals and approaches taken, and a final section has acknowledgments.

II. A MODEL FOR "DIPLOTYPES"

We begin with description of what by all rights ought to be very simple: description of a 3×3 table of unphased genotype at a pair of sites, assumed within a chosen gene. Rows refer to one site, columns to another. For simplicity, and because it is the typical case, there are two possible pairs of genotypes at each site. We designate the three columns of the table, respectively, by AA, AB, BB, and the rows by CC, CD, DD.

We will return to a study of the table from multiple points of view. In one case, it summarizes "counts" of the respective combinations of genotypes at the pair of sites. Counts can be from the entire sample, or from repeated random selections, chosen so that each family has one representative, thereby rendering the collection of "ticks" that comprise summary counts independent. As well, the summary of counts can be from sample reuse methods, that we will describe as occasions of their applications arise. Also, the table can refer to probabilities of respective pairs of genotypes. These probabilities can be described with or without reference to one of the coordinates of the four-vector of p-values associated with the tests D.

A. Independence and the multinomial model

Many analyses we report are about 3×3 tables of counts. In many statistical applications, particularly but not only regarding 2×2 tables of association and Fisher's exact test for lack of association, sampling distributions are taken to be conditional on row and column marginal totals (see e.g., Agresti, 1992). Even if we agreed on multinomial probabilities for the multinomial distribution that applies to the usual unconditional entries in a 2×2 table, once we conditioned on those row and column marginal totals, the conditional distributions of the four entries depends only on the row and column marginal totals and not on the original multinomial probabilities. Of course, the conventional conditional test in 2×2 tables does test a null hypothesis concerning the odds ratio of the table. In our case, the table is 3×3 and not 2×2. Our questions concern the unconditional probabilities of entries in the table. However, in our case because of complicated dependencies of counts induced by sibships, the multinomial distribution does not apply. Even so, by analogy with the simpler independent case and given the questions we attempt to answer, we do not condition on row and column marginal totals when we analyze tables of counts. We note aside that in the independent, identically distributed case, the disappearance of the multinomial probabilities from the conditional distributions entails that the marginal totals are "sufficient statistics" for parameters.

B. Strand-specific probabilities and resampling

One can write down entries to Table 16.1 in terms of strand-specific probabilities, respecting the four coordinates of the respective D or D' statistics that have been cited, or not. If one focuses on a specific coordinate, that is, on D for a strand-specific pair of genotypes, for example, upon the independence of genotypes A and C within a strand, then, if subjects are independent with common multinomial probabilities, each of the nine entries of the 3×3 table can be written down entirely in terms of probabilities germane to that coordinate, to that pair of site-specific genotypes. Therefore, probabilities associated with marginal (site-specific) outcomes also can be written down in coordinate-specific, if unordered,

Table 16.1. Table Summarizes Unordered Counts "true" Probabilities or Empirical Frequencies for Two Single Nucleotide Polymorphisms (SNPs)

	{AA}	{AB}	{BB}
{CC}			
{CD}			
{DD}			

terms. Here p_{AC} is the "true" probability that on one specific strand, there is an A at one specific site, while there is a C on the same strand at another specific site. It is understood that at each locus there are precisely two possible alleles; here A and B are possible at one site, and C and D at the other. More simply, p_A is the "true" probability of allele A at the site where it is a candidate to appear. It follows that, in an obvious notation, $p_A = p_{AC} + p_{AD}$.

A typical single-strand null hypothesis for either D or D' is H_{AC}: $p_{AC} = p_A p_C$. Probabilities in all nine cells of Table 16.2 can be represented in terms of these three substitutions.

- $p_{AD} = p_A - p_{AC}$
- $p_{BC} = p_C - p_{AC}$
- $p_{BD} = p_B - p_{BC} = (1 - p_A) - p_B = 1 - p_A - p_C + p_{AC}$ (1)

There are four such single-strand hypotheses.

Obviously, there are three other single-strand hypotheses analogous to H_{AC}. Given that there is only a single "degree of freedom" in a 2×2 table with fixed marginal distributions, if the four null hypotheses were studied conditional on the four empirical frequencies of alleles A, B, C, and D, then the truth of one of the four single-strand hypotheses would imply the truth of the other three. However, no conditional test statistic regarding entries in tables like Table 16.2 would make sense, since the conditional distributions in that case would not depend explicitly on the parameters at all.

We take any model by which data are generated to be family/sibship specific. Thus, we take a Bayes/empirical Bayes/random effects approach whereby parameters for a family are chosen from an admittedly complex "prior" collection of parameters. Coordinates of our approaches to studying D (or D') depend upon expectations of (squares of) probabilities with respect to their sampling distributions. Therefore, in terms of probabilities, we actually observe Table 16.3.

Table 16.2. Probabilities of Nine Possible Realizations at Two Sites for Each of Two Strands

	{AA}	{AB}	{BB}	Total
{CC}	p_{AC}^2	$2p_{AC}p_{BC}$	p_{BC}^2	p_C^2
{CD}	$2p_{AC}p_{AD}$	$2p_{AC}p_{BD} + 2p_{AD}p_{BC}$	$2p_{BC}p_{BD}$	$2p_C p_D$
{DD}	p_{AD}^2	$2p_{AD}p_{BD}$	p_{BD}^2	p_D^2
Total	p_A^2	$2p_A p_B$	p_B^2	1

The strands are distinct but are unordered in the sense that, for example, neither is known to be "the Watson strand."

Table 16.3. We Actually Observe Frequencies that are Realizations from this Table, According to Our Random Effect/Mixture Model

	{AA}	{AB}	{BB}	
{CC}	$E(p_{AC}^2)$	$2E(p_{AC}p_{BC})$	$E(p_{AC}^2)$	$E(p_C^2)$
{CD}	$2E(p_{AC}p_{AD})$	$2E(p_{AC}p_{BC}) + 2E(p_{AD}p_{BC})$	$2E(p_{BC}p_{BD})$	$2E(p_C p_D)$
{DD}	$E(p_{AD}^2)$	$2E(p_{AD}p_{BC})$	$E(p_{BD}^2)$	$E(p_D^2)$
	$E(p_A^2)$	$2E(p_A p_B)$	$E(p_B^2)$	1

Expectations are with respect to the sampling distribution of genotypes from which the sample was drawn.

If we focus on, say, H_{AC}, then while from Eq. (1) we see that there is information about that hypothesis from all nine entries of Table 16.3, it is far from clear how we might use this information other than to rephrase the hypothesis as $\{H_{AC} : E(p_{AC}^2) = E(p_A^2)E(p_C^2)\}$, which we can test on the basis of one observed entry of the table and its two corresponding observed marginal frequencies. There are three similar strand-specific analogues. These four variations of hypotheses are what we actually test.

Inference in Table 16.4 is based on a "double sampling" process of first sampling individuals within sibships, one person per sibship, followed by bootstrapping of this randomly chosen sample. The "double sampling" is repeated 200 times; each of these 200 artificially constructed "independent" samples is followed by bootstrap sampling of 200 individuals from this artificially constructed sample. The process thus involves inference whereby individuals studied in each constructed sample are independent, but the population from which they are chosen is made up of families/sibships. People in any plausible "real world" population are not independent, though for convenience often they are assumed independent. While our approach entails that inference on each artificially constructed sample necessarily concerns independent individuals, it comes at a price. At each sampling of individuals, the probability that an individual is included is inversely proportional to the size of his/her sibship. Were we to sample genuinely independent individuals in the population, assumed to consist of many sibships, this probability would not apply. There are further technical difficulties with our scheme for double sampling, partly in interpreting the sampling distribution with respect to which expectations are taken in Table 16.3. Any particular set of 3×3 tables may have insufficiently many observations in particular rows or columns to allow inferences to be made as we have outlined. Often difficulties involving sparse tables can be overcome. However, we mentioned earlier that this paper is a snapshot of ongoing research by the authors and others. At this writing, we lack information sufficient for us to comment confidently about resolution of all difficulties.

Table 16.4. This Table/Matrix has Entries Only in its Upper Triangular Part

	IL1B_012885	IL1B_009198	IL1B_008546	IL1B_005277	IL1B_004392	IL1B_003298	IL1B_002143	IL1B_001274
IL1B_012885		4×10^{-4},0,0,0	3×10^{-4},0,0,0	0,0,0,0	3×10^{-4},0,0,0	0,0,0,0	0,0,0,0	0,0,3×10^{-4},0
IL1B_009198			0,0,0,0	.01,0,0,0	.0521,0,0,0	.0665, .5599,0,0	10^{-4},0, .2973,0	0,0, .3869,0
IL1B_008546				.0102,0,0,0	.0433,0,0,0	.0631, .4696,0,0	5×10^{-4},0, .2617,0	0,0, .3504,0
IL1B_005277					0,0,0,0	0,10^{-4},0,0	9×10^{-4}, .0708, .1166,0	5×10^{-4}, .1251, .0455,0
IL1B_004392						10^{-4},0,0,0	9×10^{-4}, 6×10^{-4},0,0	4×10^{-4}, .0069, .022,0
IL1B_003298							0,0,0,0	0,0,8×10^{-4},0
IL1B_002143								0,0,0,0
IL1B_001274								

Entries in the diagonal are uninteresting, and are all 0 in any case. Further, the table is symmetric about the diagonal. The displays are subsampling/bootstrapping p-values for four respective null hypotheses. Recall that always A and C are the major alleles at the two cited single nucleotide polymorphisms (SNPs). These respective null hypotheses are $E(p_{AC}^2) = E(p_A^2)E(p_C^2)$,$E(p_{AD}^2) = E(p_A^2)E(p_D^2)$,$E(p_{BC}^2) = E(p_B^2)E(p_C^2)$,$E(p_{BD}^2) = E(p_B^2)E(p_D^2)$.

Actually computing p-values for our four hypotheses is slightly tricky. To illustrate, inference is only for the eight polymorphic sites in IL1B. Always A and C refer to the major alleles at the two respective sites—the first indexed by the row in Table 16.4 and the second by the column. B and D are the corresponding minor alleles. For each sample of individuals and each fixed hypothesis, there is a computer-generated distribution of difference between estimated expected square of joint probability minus product of estimated expectations of squares of individual probabilities. (This is possible because the computer-generated sample allows formation of an artificial table like Tables 16.2 and 16.3.) The percentile of the "true difference" in the bootstrap distribution enables computation of the p-value for the null hypothesis (of no association) for the respective sites and alleles. Of course, a p-value can be computed simultaneously from the same bootstrap sample for all four hypotheses. The 200 p-values are averaged, and these averages give rise to the four numbers in each cell of Table 16.4.

III. STRAND-SPECIFIC INFERENCE IN THE ABSENCE OF RANDOMNESS IN p

This section is like the previous one in that we assume there are two sites, each with two alleles. Data are assumed to come to us unphased, though in order that combinatorial calculations make sense there is care required in the assumed ability to identify the strands. Always, our goal is to make strand-specific inferences. As before, there are two alleles, and therefore three genotypes at each of the two sites. "Counts" can be summarized by the usual 3×3 table, that is to say, by $\{n_{ij}\}$, where i refers to row and j to column of the table. To say that there is strand-specific information from one site about the other is tantamount to saying that the 3×3 table is "close" to a permutation matrix in the sense that (at least approximately) there are positive entries for only three of the nine cells, one each per row, one each per column. Seldom will approximation be exact. However, supposing we are given a metric on distance between the matrix of counts and the "nearest permutation matrix," that is the matrix among the six candidates ($3! = 6$) with

$$\hat{n}_{ij} = \sum_{i'} \hat{n}_{i'j} + \sum_{j'} \hat{n}_{ij'}$$

Set

$$n_{..} = \sum_{i,j} n_{ij}$$

and

$$\hat{p}_{ij} = \frac{\hat{n}_{ij}}{n} \cdot$$

Table 16.5. This Table is One of Six that is Exactly of "Permutation Matrix" Form

	AA	AB	BB
CC	0	$n.\hat{p}_{12}$	0
CD	0	0	$n.\hat{p}_{23}$
DD	$n.\hat{p}_{31}$	0	0

Table 16.6. In General, the Numbers of Possible Pairs of Strands when the Strands are Identifiable

	AA	AB	BB
CC	1	2	1
CD	2	4	2
DD	1	2	1

Without loss, think of one as being "the Watson strand."

Suppose without loss of generality (for there are six cases) that the matrix is of the form shown in Table 16.5.

If we take the strands to be identifiable, then one computes that *a priori*, from Tables 16.1 and 16.2, there "might" be 16 possible pairs of strands, as is summarized in Table 16.6 that follows.

However, Table 16.5 reduces the 16 possibilities to only the 5 that are as listed.

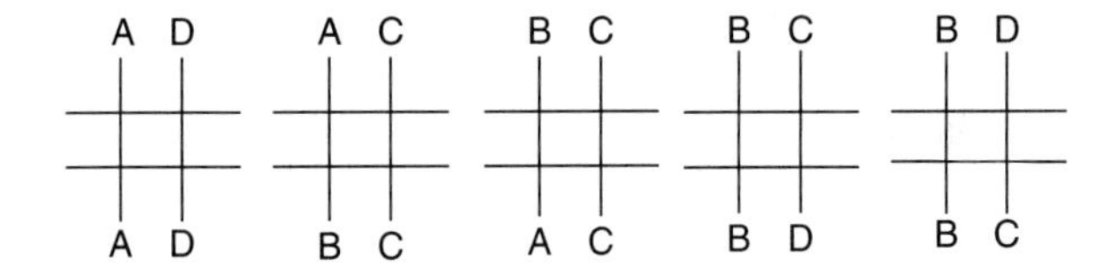

Obviously, if the strands are not identifiable, then 16 becomes 8, and 5 becomes 3 since the second and third listed strands cannot be distinguished, and neither can the fourth and fifth.

IV. HAPLOVIEW AND OUR VIEW

Haploview is a popular software package that provides computation of statistics that quantify linkage disequilibrium (LD) and population haplotype patterns. There is substantial haplotype structure in the human genome. The development of Haploview (see Barrett *et al.*, 2005) was motivated because the characterization of specific patterns has become an important tool in medical studies by association, as, for example, to a considerable extent the FBPP has become.

Haploview is written entirely in Java and can be used on any platform with Java 1.3 or later. It accepts data in many formats. The sheer scope of its output is remarkable. It produces marker quality statistics and information on LD. It also produces haplotype blocks, population haplotype frequencies, and single-marker association statistics. Of course, the goals of Haploview and related software are utterly commendable and important. It would be difficult to exaggerate current interest in finding genomic "clusters," combinations of which, along with other nongenetic predictors, may be predisposing to untoward phenotype. The HapMap project (http://www.hapmap.org) is so well known it hardly merits specific reference.

Even given virtues of the HapMap project and programs like Haploview, we note several obvious facts, to which allusion was made earlier. Since a haplotype is a set of SNPs in substantial LD with each other, necessarily the SNPs are contiguous on the genome. Thus, plausibly, predisposition to untoward phenotype might be predicted from them. That view has motivated finding "tag SNPs," what in engineering coding are termed "codewords" and in statistical clustering are termed "cluster centers." We note the obvious fact that strong covariation of genotypes in a collection of SNPs that may predispose to untoward outcome need not be contiguous. Particular SNPs may covary for reasons of viability that are distinct from their being in LD; so there is no reason that patterns of predisposing covariation necessarily are restricted to sets of SNPs that are contiguous on the genome. It is our purpose here to present the output of Haploview, and at the same time to propose models, as we have done, that are alternative to those that produce haplotypes, that are easy to understand conceptually, and also that are easy to compute. These approaches can give covarying "clusters" of SNPs that need not be contiguous on the genome.

Figure 16.1 shows one version of Haploview output for the eight polymorphic sites in IL1B computed from our GENOA data. Note that two haplotype blocks were found, one spanning 7 kb and the other a single kb. The fifth and eighth SNPs are in neither haplotype. Numbers inside each diamond are values of the statistic D' that quantifies covariability of the two SNPs that lead to the diamond, in a scale that entails a maximum of 100 (see e.g., Hedrick, 1987). Higher values mean "more covariability." Note that some SNPs not contiguous on the genome vary substantially together. Examples are the first SNP (021885) with the sixth (003298) and seventh (002143).

A. Testing: Individual pairs of alleles and omnibus tests

The reader is asked now to return to Table 16.4 and to the four allele-specific hypotheses for which p-values are reported there. It is noteworthy that the first and sixth SNPs have computed p-values of 0 for all four null hypotheses of no

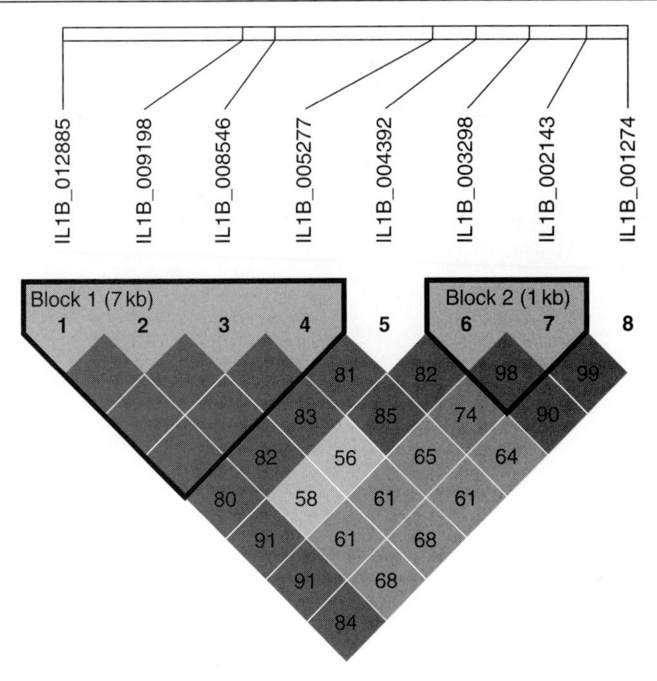

Figure 16.1. One version of Haploview output for GENOA IL1B data. Hypertensives and normotensives have been pooled. There are two haplogype blocks. Numbers inside diamonds are values of the statistic D' that quantifies covariability of the two single nucleotide polymorphisms (SNPs) that index the diamond, in a scale for which 100 is the maximum. Higher values mean "more covariability." Some SNPs that are not contiguous on the genome vary together.

association, and the same for SNPs 1 and 7. Other SNPs not at all contiguous on the genome appear close in terms of the tests of the four hypotheses. The seventh SNP (002143) is special because, unlike the other SNPs, it is far out of Hardy–Weinberg equilibrium. That our computation was done with the usual chi-square one-degree of freedom test, which is patently inappropriate given our model, does not preclude our findings that by various criteria the cited SNP is far from being in H–W equilibrium. Because nontrivial mixtures of binomial distributions cannot be binomial, it follows from our model that in principle, anyway, all sites should be out of H–W equilibrium. The one cited is the only one evident from our data. It is particularly interesting that for this site, the sample of genotyped GENOA individuals includes only 1018 people, whereas the corresponding minimum number for the other 7 sites is 1297, with maximum value 1364. We conjecture without knowing for sure that our mixture model applies to this SNP. When the unconditional distribution of alleles entails a genuine mixture, and the alleles in the components of the mixture are different, then in genotyping,

Table 16.7. This Table gives Observed Major and Minor Alleles at the Eight Polymorphic Sites in IL1B (Interleukin 1, Beta) Genotyped by the GENOA Network of the National Heart, Lung, and Blood Institute's (NHLBI's) Family Blood Pressure Program (FBPP)

SNP in IL1B	Nucleotide A	Approximate empirical frequency of nucleotide A	Nucleotide B	Hardy–Weinberg chi-square value
012885	Guanine	.671	Adenine	.067
009198	Adenine	.915	Guanine	.090
008546	Adenine	.914	Guanine	.501
005277	Adenine	.758	Guanine	3.16
004392	Guanine	.768	Adenine	1.53
003298	Guanine	.570	Adenine	.041
002143	Guanine	.655	Adenine	4.94
001274	Guanine	.598	Adenine	1.25

Also given are approximate empirical frequencies of the observed major allele and computed chi-square values for the usual respective null hypotheses that alleles are in Hardy–Weinberg equilibrium. The nominal approximate null distribution has one degree of freedom. However, it is at best very approximate here because according to our model, observations are not independent. Even without variability of allele frequencies across families, the data are from sibships. Also, we assume that allele frequencies are random by family. Note that only for IL1B_001243 is the approximate p-value small (about .013). For this single nucleotide polymorphism (SNP), the sample consists of only 1018 individuals, far fewer than the minimum of other samples (1297), not to speak of the maximum (1364). Under Hardy–Weinberg H_0, heterozygotes are over- and not underrepresented, as is typical in departures from H–W equilibrium. However, this may not be surprising in view of the possible explanation given in the text.

it may be impossible to make a primer that enables genotyping of every individual at that site. For this SNP, at which we see from Table 16.7 that guanine was the major allele and adenine the minor, heterozygotes were overrepresented relative to the best fitting single Bin$(2,p_i)$ distribution, not underrepresented for reasons of bias, as is usually the case when H–W fails. Algebra shows that when there is no systematic bias against heterozygotes, over-calling can happen if one allele is difficult to call. For reasons not reported here that owe to convexity, the problem cited is somewhat ameliorated, not exacerbated, when a mixture model applies.

We turn now to "omnibus" tests of all four hypotheses for which p-values are given in Table 16.4.

This section is about two of the many statistical approaches to multiple comparisons. We believe that they are appropriate for present purposes. The goal is to combine the four hypotheses that we study regarding each pair of SNPs within any specific, fixed gene. There should be one omnibus test of the null hypothesis that four differences are 0. The two approaches we cite here have different interpretations.

The first approach is that of Hochberg–Sime. Its basis flows from the lemma in Hochberg (1988).

Algorithmically, sort **in ascending order**, the four attained significance levels of the four distinct null hypotheses that the differences in functions of respective squares of expectations are 0. Recall that each null hypothesis is for a pair of genotypes at two fixed sites, each attained significance level resulting from a sampling followed by a bootstrapping. Refer to these sorted attained significance levels (p-values), respectively, as $P_{(1)},\ldots, P_{(4)}$. Pick a value of attained significance—call it α. For our application, we have chosen the value $\alpha = .0005$. Then for any $i = 4, 3, 2, 1$, **if** $P_{(i)} \leq \alpha/(5 - i)$, then **reject** all respective null hypotheses $H_{(i')}$ $(i' \leq i)$.

Rejection of an individual hypothesis—say H_i—means that for the pair of SNPs in question, values of the two SNPs are not independent, at least so far as the two particular genotypes are concerned. Thus, if you have the one specific genotype at the one SNP, then you can safely predict the other specific genotype at the other SNP. The way we have formulated it, this predictability is symmetric (mutual). For some pairs of SNPs, predictability depends crucially upon what the genotype at the one SNP is!

In this Hochberg–Sime context, when the omnibus test rejects, if you know the genotype at one SNP, whatever it might be, then you can safely predict the genotype at the other, no matter the specific genotype at the first.

The next approach taken here is that of Theorem 1.3 of the 2001 paper by Benjamini and Yekutieli and concerns *FDR*s, as are discussed in what follows. It may be that the PRDS hypothesis studied and used by Benjamini and Yekutieli holds in our context when it is applied to the respective absolute differences of our four test statistics. PRDS means, "Positive regression dependency on a subset." The notion is made precise by Benjamini and Yekutieli on page 1168 of their paper. However, at this writing, we lack proof that PDRS applies in our context. Further, the results of their Theorem 1.3 do not require anything of PRDS. For that result, the joint distribution of the test statistics is arbitrary.

Before giving details of Benjamini–Yekutieli, it is necessary to record two facts. First, $\sum\{1/i{:}1 \leq i \leq 4\} = 25/12 = 2.083333\ldots$ Second, since m_0 in the statement of Theorem 1.3 is unknown, but is at most m in any case, we take $m_0/m = 1$. Having said this much, let $P_{(1)},\ldots, P_{(4)}$ be as they were previously. (Note that they are written lower case in (1), page 1167 of Benjamini and Yekutieli, 2001).

Fix a number q; we suggest that here q be set equal to (25/12) (.0005). Write $k = \max\{i{:}P_{(i)} \leq (i/4)q\}$. Then, according to what Benjamini and Yekutieli call the Benjamini–Hochberg procedure, reject $H_{(1)},\ldots, H_{(k)}$. If no such i exists, then reject no null hypothesis.

Benjamini–Yekutieli's cited Theorem 1.3 says that if the Benjamini–Hochberg procedure is applied as stated, then necessarily it controls the FDR at level less than or equal to $q = .00104667\ldots$ Recall that the FDR is the expected proportion of erroneous rejections among all rejections. In our context, controlling FDR as planned would mean rejecting 1000 null hypotheses of "lack of predictability" before the expected number of false rejections was as high as 1.

The results of the two approaches to omnibus testing are given in Table 16.8. We need include results only for the four hypotheses and only for unordered pairs of SNPs. Thus, necessarily any table in matrix form that summarizes rejected hypotheses would be symmetric. Diagonal entries would always be (all, all); this is clear because a SNP is perfectly predictable from itself, so both tests would "reject" the null hypothesis of no predictability. Within each rectangle of the table, the criterion to the left of the semicolon is the stricter Hochberg–Sime criterion (Hochberg, 1988). The criterion to the right is the Benjamini–Yekutieli version of Benjamini–Hochberg procedure (Benjamini and Yekutieli, 2001), where we allow a "FDR" of about .001. Perhaps surprisingly, there are not fewer than five entries for which Benjamini and Yekutieli reject more null hypotheses than does Hochberg–Sime. Further explication is in the caption to Table 16.8.

B. Clustering

Developments earlier in the paper enable our displaying a set of polymorphic sites in a dendogram, the output of agglomerative clustering. For general references see Everitt *et al.* (2001) and Hastie *et al.* (2001). We applied a version of the hclust algorithm available in *R*. See, for example, http://sekhon.berkeley.edu/stats/html/hclust.html; hclust has been used by others very much as we have used it to find tag SNPs; see http://www.wpic.pitt.edu/WPICCompGen/hclust.htm. Of course, we have taken the approach we have in order to avoid going "through haplotypes" in general and tag SNPs in particular to find association between complex genotype and other variables as they bear upon prediction of complex human disease.

Our four-vector of *p*-values allows computation of a metric on pairwise distances of SNPs. It is not obvious what metric should apply here, and that is a topic for future research. In the application regarding IL1B that is displayed in Fig. 16.2, we use the simple four-dimensional Euclidean metric. One might use the relative frequencies of the four pairs of alleles to compute weights to the coordinates of the Euclidean metric. Alternatively, and likely better, would be to shade these empirical weights toward the equal ordinary Euclidean weights. For decision-theoretic arguments as to why the shading should help, see Bloch *et al.* (2002) and its references.

If d refers to "distance" and SNP_k and SNP_m refer to two SNPs that belong to the same cluster, then we make the obvious choice of "center" or "representative SNP" of their cluster to be an SNP that belongs to the cluster and that solves argmin max $d(\mathrm{SNP}_k, \mathrm{SNP}_m)$.

Table 16.8. This Table Displays Hypotheses Rejected According to Two Omnibus Tests, Chosen with $\alpha = .0005$

	IL1B_012885	IL1B_009198	IL1B_008546	IL1B_005277	IL1B_00592	IL1B_003298	IL1B_002143	IL1B_001274
IL1B_012885		all; all	all; all	all; all	all; all	all; all	all; all	all; all
IL1B_009198			all; all	AD,BC,BD; AD,BC,BD	AD,BC,BD; AD,BC,BD	BC,BD; BC,BD	AD,BD,AC; AD,BD,AC	AC,AD,BD; AC,AD,BD
IL1B_008546				AD,BC,BD; AD,BC,BD	AD,BC,BD; AD,BC,BD	BC,BD; BC,BD	AD,BD; AD,BD,AC	AC,AD,BD; AC,AD,BD
IL1B_005277					all; all	all; all	BD; BD	BD; BD,AD
IL1B_004392						all; all	BC,BD; all	BD; BD.AC
IL1B_003298							all; all	AC,AD,BD; all
IL1B_002143								all; all
IL1B_001274								

The criterion before the semicolon is the more strict Hochberg–Sime criterion (Hochberg, 1988). The criterion after the semicolon is the slightly more lenient Benjamini–Yekutieli version of the Benjamini–Hochberg procedure (Benjamini and Yekutieli, 2001), where we allow a "false discovery rate" of about .001. Surprisingly, there are five cells for which Benjamini–Yekutieli rejects more null hypotheses than does Hochberg–Sime. As always, A and C are the major alleles at the respective pairs of polymorphic sites of IL1B, the first given by the row and the second by the column. The omnibus test combines p-values for tests of our four respective null hypotheses $E(p_{AC}^2) = E(p_A^2)E(p_C^2), E(p_{AD}^2) = E(p_A^2)E(p_D^2), E(p_{BC}^2) = E(p_B^2)E(p_C^2), E(p_{BD}^2) = E(p_B^2)E(p_D^2)$.
Note the large number of rejections of tests of null hypotheses of no association of SNPs in HAPLOVIEW's first cluster, the four in a 7 kb contiguous block, and the sixth and seventh single nucleotide polymorphisms (SNPs), which comprise HAPLOVIEW's second cluster, in a 1 kb contiguous block.

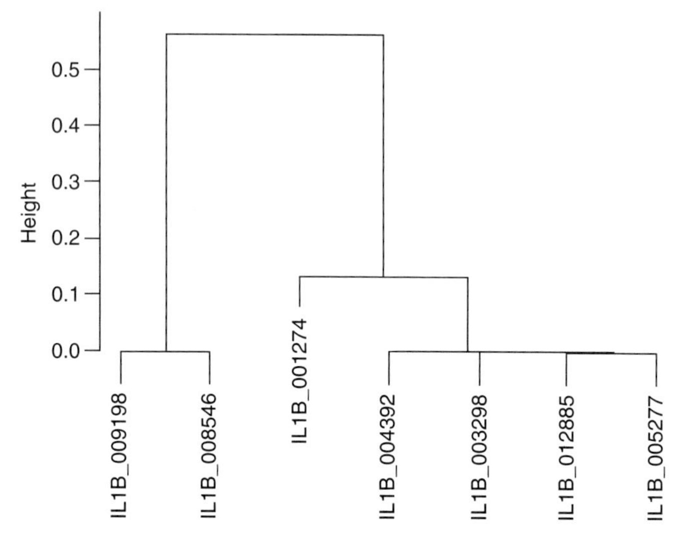

Figure 16.2. The dendogram depicted here summarizes results of agglomerative clustering of the seven polymorphic sites in ILB1 that appear to be in Hardy–Weinberg equilibrium. Clustering was based on our 4-vector of *p*-values. IL1B_002143 does not appear. That site clearly was not in H–W equilibrium for our data. Heterozygotes were over- and not underrepresented. Since nontrivial mixtures of binomials cannot be binomial, our model suggests that all sites should be out of H–W equilibrium. However, for only the one site was the departure severe enough to be detectable by simple chi-square statistics with their conventional approximate one-degree of freedom null sampling distribution (which in principle is not appropriate here). Further possible explanation is in the text. Cluster centers for the two nontrivial clusters displayed are IL1B_009198 and IL1B_005277. Explanation of the minimax criterion by which they were computed is in the text.

Figure 16.2 is labeled aptly in that it has two clear take-away messages. The first is how closely it agrees with the results of omnibus testing that were described, so SNPs that for whatever biological reasons are in the same cluster are not necessarily close on the genome. The second is the simplicity of the graphical display as opposed to listings of numbers and letters.

V. APPROACHES TO TWO-CLASS ASSOCIATION/CLASSIFICATION

This section includes discussion of six modern approaches to two-class association/ classification. While, of course, there are many classifiers that we might have included, our choices were governed by algorithms that have been applied by others and us with success to problems that are explicitly genetic. Applications of these and related methods are proceeding in parallel with research reported here. The technologies extend to more than one class, but for our main areas of application, the phenotype/class/outcome takes only two values. It seems unnecessary to complicate matters yet further in this expository context. We begin with a brief description of a sixth set of technologies, classification trees (as in CART®, Breiman *et al.*, 1984). Our focus is on applications of decision trees to discovering the synergistic impact (interactions) of predictors as they bear upon complex disease. When CART is used to classify, its performance is improved by repeated application, with emphasis on "hard to classify" cases, followed by subsequent "voting" or weighting of the results of a sequence of classifiers. This general technology goes under the name "boosting," alternatively "boosted CART" (see Friedman *et al.*, 2000 and its references). However, CART has application to finding interactions that may be more important than boosted CART in genetic contexts. In any case, boosting classifiers seems to offer little improvement in accuracy for other than tree-based classifiers.

Table 10.1 in page 313 of the fine book by Hastie *et al.* (2003) is an interesting table in which five approaches to classification are summarized in terms of nine important properties. We hesitate to include an analogous table summarizing our favorite five algorithms because once one specializes applications to large problems in genetics, SNPs, expression arrays, whatever, their properties and the wisdom of any particular application can be rather different from generic statements of the cited Table 10.1 and indeed of the entire book. Our descriptions of the algorithms that follows does include some of their respective strengths and weaknesses.

With, say, k features, there are, potentially, $O(k^2)$ second order interactions, $O(k^3)$ third order interactions, and so on. Those who analyze data argue endlessly as to whether a feature has a synergistic impact with other features if it lacks a "main effect" itself, and many preprocessors to detailed analyses search first for "main effects." However, nobody argues that in nearly all genetic applications, there are many too many candidate interactions to consider all of them, even if restricted to second order alone (which typically is unwise). Thus, imagine a CART classification tree grown so that a path from root node to terminal node splits successively on features A, B, then A again. This path suggests five "words," for candidate inclusion in a logistic or other model for prediction. They are A, B, AB, A^2, and A^2B. While two of these are "main effects," three are "interactions." Of course, every path from root node to terminal node can suggest interactions for inclusion in any model. There are exactly as many paths as there are terminal nodes to the tree. Many suggestions are redundant. And a balanced tree with T terminal nodes is of depth only $O(\log T)$. Thus, it is clear that CART can be used to narrow considerably the candidacy of interactions for inclusions in subsequent modeling. We admit that attributing attained significance or cross-validating reduction in Bayes risk to what might be an informal process of choosing interactions is difficult at best. However, if we are forced to choose between the validity of formal statistical testing and discovery of biological processes, we prefer the latter.

Of course, precise definition of phenotype is, typically, both difficult and delicate; and prediction is uninteresting without useful definition of what is being predicted. Necessarily, the present paper is limited in scope, and for present purposes we tend to think of phenotype as "disease status," untoward or not. It follows that associating phenotype with SNPs and other features, perhaps not patently genetic at all, is a crucial problem. Previous sections have had focus on selecting candidate SNPs, and earlier paragraphs have addressed choices of interactions among SNPs and/or other features. We turn now to discussion of what currently are some of our favorite methodologies for predicting/associating/classifying.

Logistic regression is a time-honored approach to classification, including in regression contexts. However, applications to the genetics of complex disease include many predictors, usually too many for inclusion in any model. This problem is extremely well known in the context of microarrays by which gene expression is quantified, and is all the more obvious when predictors are based primarily on SNPs. Therefore, even after the technologies of previous sections, there is need for further reduction in the number of features, main effects, or interactions. Thus, there is an important place for "regularized" logistic regression, that is to say, logistic regression where a linear model for the log-odds of class membership has constraints on the regression coefficients of the model. Perhaps the best known of these is "the lasso" (Hastie *et al.*, 2003; Tibshirani, 1996). There the constraint is on an l_1 norm of the vector of regression coefficients. As is well known, and reflection on the geometry involved should convince the reader, typically the lasso constraint has the effect of setting many regression coefficients to 0. This is precisely what one prefers when there are far too many SNP predictors, let alone other predictors, to enable concise biological explanation of observed phenotype. There have been many approaches to finding lasso or lasso-like solutions to problems of supervised learning (association). One important recent contribution is that of Park and Hastie (2006). Readers are encouraged to read their paper and its references.

Were constraints instead simply on the l_2 norm of the coefficients, then the desirable lasso property of setting many coefficients to 0 would disappear, though the result would be analogous to that of ridge regression, a close relative of James–Stein estimation, and known to have good decision-theoretic properties. Thus, the "elastic net" of Zou and Hastie (2005) is a plausible compromise, whereby constraints are on weighted sum of the l_1 and l_2 norms of coefficients.

The basic idea of *support vector machines* is clever and simple. Suppose that features $\mathbf{X} = (X_1, X_2, \ldots, X_k)$ are quantitative and so can be viewed as belonging to a k-dimensional Euclidean space, R^k. For each member of the n members of a *learning sample*, we observe a pair $(\mathbf{X}_j, Y_j)$, $j = 1, \ldots, n$, where the first entry is the feature vector; and the second denotes class membership. For present purposes, Y belongs to {0,1}, indicating that there are two phenotypes to the learning sample data. It may or may not be that $\{\mathbf{X}:Y = 0\}$ can be separated from $\{\mathbf{X}:Y = 1\}$ by a hyperplane in R^k. However, it is always the case that if the original features $\mathbf{X}$ are enlarged to be features $\mathbf{P} = (P_1, P_2, \ldots, P_{k'})$ for easily computed and suitable functions $P_1, \ldots, P_{k'}$, and computable $k' \geq k$, then always there is a hyperplane in $\mathrm{R}^{k'}$ by which the two cited sets can be separated. Ideally, the hyperplane can be constructed whereby it can be moved perpendicularly between the two sets so that the area between the sets intersected with large Euclidean rectangles in $\mathrm{R}^{k'}$ have as large as possible Lebesgue measure, that is to say, as large as possible Euclidean volume. See Fig. 12.1 in the book by Hastie *et al.* (2001). Now imagine that the hyperplane slides perpendicularly so as to be exactly half-way in between the two sets. And imagine further that if we are to classify a future observation for which features $\mathbf{P} = \mathbf{P}(\mathbf{X})$ are given, but Y is not, then we compute easily on which side of the hyperplane $\mathbf{P}$ lies, and we classify accordingly in the obvious way.

There remain many concerns. First, how do we choose the vector $\mathbf{P}$? Second, what if for values $k'' \geq k$ but as large a positive integer as we can tolerate, there is no perfectly separating hyperplane? Third, the motivation is for original features $\mathbf{X}$ that are quantitative, as are some non-SNP features in applications we envisage. However, features related to SNPs are discrete. Even though they can be ordered, typically they assume at most three values. We respond to these concerns briefly, one by one. As for the last of the concerns, we note that the original motivation for logistic regression is that features for the two classes are jointly Gaussian, with identical covariance matrices. They are assumed to differ only in their mean vectors. Logistic regression classifiers are then computed easily as exact likelihood ratio or Bayesian solutions to the classification problem. However, it has been found historically that logistic regression is applicable and informative when features are not at all Gaussian, indeed not ordered at all. Class conditional covariance structures need not be identical, either. Still, there are many important applications of logistic regression, including but not limited to features being indicators of particular genotypes at particular polymorphic sites. As for what to do when, practically speaking, there is no separating hyperplane for any reasonably computed set of features $\mathbf{P}$ and k'' as large as is "tolerable," we note that there is by now a truly enormous literature on this subject. It falls under the heading "support vector machines." Crucial ideas owe to Vapnik. Readers are directed to one of his books (1999), and particularly to the book by Hastie *et al.* (2001). There are important points of contact between support vector machines and kernel-based classifiers, and between support vector machines and reproducing kernel Hilbert spaces.

The last two of the short list of two-class classifiers that are particularly attractive to us both devolve from CART, that was cited (see Breiman *et al.*, 1984 and Hastie *et al.*, 2001). The first is FlexTree (Huang *et al.*, 2004).

FlexTree's motivation comes from this model for predisposition to complex disease. Such predisposition should involve untoward genotypes at many polymorphic sites. Of them, most are pure noise. Each of the values that is not the "prevalent" genotype for the minority of polymorphic sites that contribute to the signal carries a "score." Scores add. Individuals with scores above an unknown threshold are predisposed to the disease. "For the additive score problem and simulated data, FlexTree has cross-validated risk better than many cutting-edge technologies to which it was compared when small fractions of candidate" sites "carry the signal" (Huang *et al.*, 2004). FlexTree also proved of interest in separating certain groups of Chinese women into hypertensive or not, based on SNPs and other features. The idea behind FlexTree is to adapt a combination of flexible discriminant analysis and penalized discriminant analysis to a CART setting. See Sections 12.5 and 12.6 of the book by Hastie *et al.* (2001). FlexTree has many steps. The first involves transforming features so that a problem of classification is transformed to one of regression, in particular to regression regularized in the context of an l_2 constraint on the regression coefficients. Next is a node-by-node sort on features, and a methodology for "shaving" features at each node. Thus, there is risk to be evaluated, and, ultimately, as in CART, shaving a large tree to a much smaller tree based on cross-validated estimation of Bayes risk were the resulting classification tree used prospectively. As with CART, there is attention to missing data. We emphasize that FlexTree is computationally intensive and needs adapting to large sets of data as will result, for example, from 500,000+SNP chips.

The final two-class classifier that we mention is "random forests," as developed first by the late Leo Breiman but developed further by others. See the paper by Breiman (2001). At this writing, "random forests" gets more than 63,000 Google hits. Suppose that the learning sample; n, the learning sample size; and k, the dimension of the features, are the same as they were previously. Suppose further that we are given a positive integer m, which typically is much smaller than k; m denotes the number of variables to be used to determine a "split," as in CART, at each node of any tree to be grown. A learning sample is chosen from the given learning sample by bootstrapping, that is, by choosing with replacement a sample of size n from the learning sample. The remaining part of the learning sample is used to estimate the error of the resulting tree, by predicting their classes. At each node of a tree being grown, randomly choose m features on which to base the split at that node. Calculate the best split according to CART's Gini criterion (see Breiman *et al.*, 1984). Each tree is grown fully, and is ***not*** pruned, as in CART. Grow many such two-class classification trees. For a vector for which **X** is known but Y is not, classify **X** according to each of the trees grown, and vote on class membership. This, in a nutshell, is the approach of random forests. Details include assessing the importance of individual features, assessing the synergistic impact of features upon classification, and much else.

As with FlexTree, qualitative features such as SNP genotypes present no special problem. Approaches to treating "missing values" can be inherited from the approach in CART.

VI. RESTRICTIONS ON PROBABILITIES ENTAILED BY HARDY–WEINBERG EQUILIBRIUM

This section is more mathematical than other sections of the paper. At the same time, it is a bit speculative in that we make only plausibility arguments for some conclusions, and in any case, we provide no mathematical details.

For convenience, we assume there are only two possible alleles at each polymorphic site. We think of there being many such sites. From a mathematical point of view, it does not seem to matter if that number is a large integer or is taken, for convenience, to be countably infinite. Each coordinate is assumed to be the set {0,1,2}, representing the number of major alleles at the site indexed by the coordinate. We are concerned with disease that, to the extent it is genetic at all, is genetically complex. For what follows in this section, assume that the mixtures we have described are in fact trivial; that is to say, they are concentrated at a single point. We assume that no single SNP has any marginal effect on phenotype. The upshot of these assumptions is that all sites are assumed to be in Hardy–Weinberg equilibrium. Thus, at the ith site, the marginal probability distribution is $\text{Bin}(2,p_i)$, where each p_i is at least .5.

The set of probabilities on the cited product space that has each marginal probability binomial as cited is obviously convex. It is compact with respect to weak convergence, which in this instance is a metric topology. Therefore, we know from the Krein–Milman theorem that it has extreme points. Moreover, the set is precisely the closed convex hull of those extreme points, and there is a representation of every such probability in terms of some integral (average) over the extreme points. See Phelps (2001).

Plausibility arguments lead us to believe that the joint probability distribution of the set of SNPs can be described thus. There is a vector $\mathbf{p} = (p_1, p_2, \ldots)$, where each coordinate p_i has value in the interval [.5,1). (The interval being closed on the left and open on the right is deliberate, and is meant to reflect the probability of a major allele at each chromosome at the ith polymorphic site. That the interval is open on the right means that the site is polymorphic.) The vector $\mathbf{p}$ has an arbitrary distribution. Conditional on $\mathbf{p}$, the sites have binomial distributions, with respective probabilities the coordinate-wise realized values of $\mathbf{p}$. Interpretation here is subtle. For example, equality of the coordinates of $\mathbf{p}$ is to be interpreted as (random) functions on the underlying probability space, not only as realized values of these functions. If, for example, for two coordinates

i and j, $p_i = p_j$, as functions, then conditional on their common value, the realized binomials will be identical, so hardly independent. When the ps are, respectively are not, independent, then the realized binomials are, respectively are not, independent.

There is more to say about the joint distribution of SNPs. To begin, suppose once again that polymorphic sites are in H–W equilibrium. Pick any finite number of them. Sort the coordinates by the probability of the minor alleles $\{1 - p_i\}$ at the sites in the subset. Let $X_{(j)}$ be the random number of minor alleles at site j in the sorted set of polymorphic sites. Suppose that f is any convex function on $\{0,1,2\}$. Then one observes that the sequence of respective expectations, $E(X_{(j)})$, is nondecreasing in j. It follows that the joint distribution of $\{X_{(j)}\}$ is consistent with it being not only Markovian, but also a sub-martingale (that is to say, an "expectation increasing" martingale). The proof, which will be given elsewhere, depends on Blackwell's fundamental theorem on the comparison of experiments (Blackwell, 1953; Phelps, 2001). Whether joint distributions are as might be, Markovian and sub-martingale, are testable hypotheses, albeit by tests that may lack much power for reasonable alternatives.

Hardy–Weinberg equilibrium also entails other, related but at a superficial level different restrictions that owe to the orthogonal (Krawtchouk) polynomials for the joint distribution of pairs of sites. These joint probabilities are summarized by entries in a 3×3 table of probabilities, where each marginal has a binomial distribution with "$n = 2$," though the binomial parameters p might be different. See Eagleson (1969) and Lancaster (1963). These Krawtchouk polynomial expansions for pairs of sites are analogous to "canonical variables" in statistics. Because the tables have only four "degrees of freedom" anyway, and the binomial restrictions reduce that number by two, one might conjecture that the Krawtchouk expansion has only two terms that describe departure from the two sites being independent. This is in fact true (Eagleson, 1969; Lancaster, 1963). However, Krawtchouk expansions of distributions of only two sites is not terribly helpful. It would be better to consider joint Krawtchouk expansions of any finite number of sites. This approach seems to be known in mathematical physics, though not in genetics. Again, details will be pursued and published elsewhere.

VII. DISCUSSION AND SUMMARY

As we write, the results of multiple genome-wide association (GWA) studies are appearing in print (Duerr *et al.*, 2006; Frayling *et al.*, 2007; Helgadottir *et al.*, 2007; McPherson *et al.*, 2007; Rioux *et al.*, 2007; Saxena *et al.*, 2007; Scott *et al.*, 2007; Sladek *et al.*, 2007; Steinthorsdottir *et al.*, 2007; Wellcome

Trust Case-Control Consortium, 2007; Zeggini *et al.*, 2007). Data are from platforms that genotype simultaneously up to about 500,000 common SNPs (typically derived from the HapMap). These first generation GWA studies have identified a handful of new "main effects" loci, in heretofore unsuspected regions of the genome, that allow convincing association with complex traits such as diabetes mellitus (Saxena *et al.*, 2007; Scott *et al.*, 2007; Sladek *et al.*, 2007; Steinthorsdottir *et al.*, 2007; Zeggini *et al.*, 2007), coronary heart disease (Helgadottir *et al.*, 2007; McPherson *et al.*, 2007), Crohn's disease (Duerr *et al.*, 2006; Rioux *et al.*, 2007), body mass index (BMI) (Frayling *et al.*, 2007), and triglyceride levels (Saxena *et al.*, 2007). Many thousands of samples were genotyped in order to identify at most a few new variants per complex trait despite putatively high heritability of these traits. For example, the new "main effect" signals of BMI (Frayling *et al.*, 2007) and serum triglyceride levels (Saxena *et al.*, 2007) explain not more than one percent of the overall variability of these traits. Although it is too early to say with certainty given the less than optimal power in the first stage of genotyping for many of these studies, their results suggest that a large proportion of the genetic determinants of complex traits that is derived from common variants is buried in first or higher order SNP–environment or SNP–SNP interactions (epistasis). This is the context in which we have provided algorithms for the purpose of detecting such complex signals. Because of the many difficulties that attend large studies of genotype–phenotype associations, recent literature has included warnings concerning replicating such studies. One important summary is that of the NCI-NHGRI Working Group on Replication in Association Studies (2007).

We have emphasized that the ultimate goals of our exercises are to predict phenotype typically untoward or not, within a specific window of time, neither from linkage nor from finding specific haplotype frequencies. Indeed, there is no fundamental reason to go "through linkage" or "through haplotypes" to arrive at prediction. Data studied here come unphased. Furthermore, they are not independent. Dependencies owe to subjects being members of sibships. Sibships are taken to be independent. Some algorithms come from sampling dependent data in particular ways to enable analyses as if observations were independent, but we emphasize that there is a price to be paid for the sampling.

We have offered approaches to narrowing the set of SNPs that might be considered in any problem of classification. Differences between certain conditional and unconditional testing have been emphasized; we have drawn distinctions in what can be inferred in the two cases. We have made connections between problems in genetics and problems in mathematics, where mathematical models and results may be helpful in summarizing data and in thinking about relationships among polymorphic sites in the genome as together and combined with other features they bear upon complex human disease.

Acknowledgments

We gratefully acknowledge support from NIH grant 5 R37 EB002784–32 to Stanford University, NIH/NHLBI grant 2HL054527–11 to the University of Hawaii with sub-allocation to Stanford University, and to NSF grant CCR-0309701 to Stanford University. We thank Dr. Thomas Quertermous for many scientific discussions and an anonymous referee for suggestions. We also thank Jian Chu for help with the GENOA IL1B data, David Siegmund and Nancy Zhang for helpful discussions, and Bonnie Chung for her ongoing technical help. Blame for remaining mistakes rests solely with the four listed authors.

References

Agresti, A. (1992). A survey of exact inference for contingency tables. *Stat. Sci.* **7**(1), 131–177.

Barrett, J. C., Fry, B., Maller, J., and Daly, M. (2005). Haploview: Analysis and visualization of LD and hapolotype maps. *Bioinformatics* **21**(2), 263–265.

Benjamini, Y., and Yekutieli, D. (2001). The control of the false discovery rate in multiple testing under dependency. *Ann. Statist.* **29**(4), 1165–1188.

Blackwell, D. (1953). Equivalent comparisons of experiments. *Ann. Math. Statist.* **24,** 265–272.

Bloch, D., Olshen, R. A., and Walker, M. G. (2002). Probability estimation for classification trees. *J. Comp. Graphical Statist.* **11**(2), 263–288.

Breiman, L. (2001). Random forests. *Mach. Learn.* **45,** 5–32.

Breiman, L., Friedman, J. H., Olshen, R. A., and Stone, C. J. (1984). "Classification and Regression Trees." Wadsworth, Belmont, CA.

Carlson, C. S., Eberle, M. A., Rieder, M. J., Yi, Q., Kruglyak, L., and Nickerson, D. A. (2004). Selecting a maximally informative set of single-nucleotide polymorphisms for association analyses using linkage disequilibrium. *Am. J. Hum. Genet.* **74,** 106–120.

Clark, A. C. (2004). The role of haplotypes in candidate gene studies. *Genet. Epidemiol.* **27,** 321–333.

Crawford, D. C., and Nickerson, D. A. (2005). Definition and clinical importance of haplotypes. *Annu. Rev. Med.* **56,** 303–320.

Duerr, R. H., Taylor, K. D., Brant, S. R., Rioux, J. D., Silverberg, M. S., Daly, M. J., Steinhart, A. H., Abraham, C., Regueiro, M., Griffiths, A., Dassopoulos, T., Bitton, A., *et al.* (2006). A genome-wide association study identifies IL23R as an inflammatory bowel disease gene. *Science* **314**(5804), 1461–1463.

Eagleson, G. K. (1969). A characterization theorem for positive definite sequences on the Krawkchouk polynomials. *Austal. J. Statist.* **11,** 29–38.

Everitt, B. S., Landau, S., and Leese, M. (2001). "Cluster Cluster Analysis (Fourth Edition)." Oxford University Press, New York.

Frayling, T. M., Timpson, N. J., Weedon, M. N., Zeggini, E., Freathy, R. M., Lindgren, C. M., Perry, J. R. B., Elliott, K. S., Lango, H., Rayner, N. W., Shields, B., Harries, L. W., *et al.* (2007). A common variant in the FTO gene is associated with body mass index and predisposes to childhood and adult obesity. *Science* **316**(5826), 889–894.

Friedman, J., Hastie, T., and Tibshirani, R. (2000). Additive logistic regression: A statistical view of boosting (with discussion and a rejoinder by the authors). *Ann. Statist.* **28**(2), 337–407.

Hastie, T., Tibshirani, R., and Friedman, J. (2001). "Elements of Statistical Learning (corrected printing)." Springer-Verlag, New York.

Hedrick, P. W. (1987). Gametic disequilibrium measures: Proceed with caution. *Genetics* **117,** 331–341.

Helgadottir, A., Thorleifsson, G., Manolescu, A., Gretarsdottir, S., Blondal, T., Jonasdottir, A., Sigurdsson, A., Baker, A., Palsson, A., Masson, G., Gudbjartsson, D. F., Mognusson, K. P., *et al.* (2007). A common variant on chromosome 9p21 affects the risk of myocardial infarction. *Science* **316**(5830), 1491–1493.

Hochberg, Y. (1988). A sharper Bonferroni procedure for multiple tests of significance. *Biometrika* **75,** 800–802.

Huang, J., Lin, A., Narasimhan, B., Quertermous, T., Hsiung, C., Ho, L., Grove, J., Olivier, M., Ranade, K., Rishc, N., and Olshen, R. (2004). Tree-structured supervised learning and the genetics of hypertension. *Proc. Natl. Acad. Sci. U.S.A.* **101**(29), 10529–10534.

Lancaster, H. O. (1963). Correlations and canonical forms of bivariate distributions. *Ann. Math. Statist.* **34,** 532–538.

Lewontin, R. (1988). On measures of gametic disquilibrium. *Genetics* **120,** 849–852.

McPherson, R., Pertsemlidis, A., Kavaslar, N., Stewart, A., Roberts, R., Cox, D. R., Hinds, D. A., Pennachio, L. A., Tybjaerg-Hansen, A., Folsom, A. R., Boerwinkle, E., Hobbs, H. H., *et al.* (2007). A common allele on chromosome 9 associated with coronary heart disease. *Science* **316** (5830), 1488–1491.

Mohlke, K. L., Lange, E. M., Valle, T. T., Ghosh, S., Magnuson, V. L., Silander, K., Watanabe, R. M., Chines, P. S., Bergman, R. N., Tuomilehto, J., Collins, F. S., and Boehnke, M. (2001). Linkage disequilibrium between microsatellite markers extends beyond 1 cM on chromosome 20 in Finns. *Genome Res.* **11,** 1221–1226.

NCI-NHGRI Working Group on Replication in Association Studies (2007). Replicating genotype-phenotype associations. *Nature* **447**(7145), 655–660.

Park, M. Y., and Hastie, T. (2006). Regularization path algorithms for detecting gene interactions. Manuscript. http://www-stat.stanford.edu/~hastie/Papers/glasso.pdf.

Park, M. Y., and Hastie, T. (2007). L_1 regularization path algorithm for generalized linear models. *J. R. Statist. Soc.* **B 69,** 659–677. http://www-stat.stanford.edu/~hastie/Papers/glmpath.jrssb.pdf.

Phelps, R. R. (2001). "Lectures on Choquet's Theorem." Springer, Berlin/HeidelbergLecture Notes in Mathematics 1757.

Rioux, J. D., Xavier, R. J., Taylor, K. D., Silverberg, M. S., Goyette, P., Huett, A., Green, T., Kuballa, P., Barmada, M. M., Datta, L. W., Shugart, Y. Y., Griffiths, A. M., *et al.* (2007). Genome-wide association study identifies new susceptibility loci for Crohn's disease and implicates autophagy in disease pathogenesis. *Nat. Genet.* **39**(5), 596–604.

Risch, N. (2000). Searching for genetic determinants in the new millennium. *Nature* **405,** 847–856.

Saxena, R., Voight, B. F., Lyssenko, V., Burtt, N. P., de Bakker, P. I. W., Chen, H., Roix, J. J., Kathiresan, S., Hirschhorn, J. N., Daly, M. J., Hughes, T. E., Groop, L., *et al.* (2007). Genome-wide association analysis identifies loci for type 2 diabetes and triglyceride levels. *Science* **316** (5829), 1331–1336.

Schaid, D. J. (2004). Evaluating associations of haplotypes with traits. *Genet. Epidemiol.* **27,** 348–364.

Schaid, D. J., Rowland, C. M., Tines, D. E., Jacobson, R. M., and Polan, G. A. (2002). Score tests for association between traits and haplotypes when linkage phase is ambiguous. *Am. J. Hum. Genet.* **70,** 425–434.

Scott, L. J., Mohlke, K. L., Bonnycastle, L. L., Willer, C. J., Li, Y., Duren, W. L., Erdos, M. R., Stringham, H. M., Chines, P. S., Jackson, A. U., Prokunina-Olsson, L., Ding, C. J., *et al.* (2007). A genome-wide association study of type 2 diabetes in Finns detects multiple susceptibility variants. *Science* **316**(5829), 1341–1345.

Seattle SNPs (2007). NHLBI Program for Genomic Applications, SeattleSNPs, Seattle, WA (URL:http://pga.gs.washington.edu) (June, 2007).

Sladek, R., Rocheleau, G., Rung, J., Dina, C., Shen, L., Serre, D., Boutin, P., Vincent, D., Belisle, A., Hadjadj, S., Balkau, B., Heude, B., *et al.* (2007). A genome-wide association study identifies novel risk loci for type 2 diabetes. *Nature* **445**(7130), 881–885.

Steinthorsdottir, V., Thorleifsson, G., Reynisdottir, I., Benediktsson, R., Jonsdottir, T., Walters, G. B., Styrkarsdottir, U., Gretarsdottir, S., Emilsson, V., Ghosh, S., Baker, A., Snorradottir, S., *et al.* (2007). A variant in CDKAL1 influences insulin response and risk of type 2 diabetes. *Nat. Genet.* **39**(6), 770–775.

The FBPP Investigators (2002). Multi-center genetic study of hypertension: The Family Blood Pressure Program (FBPP). *Hypertension* **39**(1), 3–9.

Tibshirani, R. (1996). Regression shrinkage and selection via the lasso. *J. Royal Statist. Soc. B.* **58,** 267–288.

Vapnik, V. (1999). "The Nature of Statistical Learning Theory." Springer-Verlag, New York.

Wellcome Trust Case Control Consortium (2007). Genome-wide association study of 14,000 cases of seven common diseases and 3,000 shared controls. *Nature* **447**(7145), 661–678.

Wu, X., Kan, D., Province, M., Quertermous, T., Rao, D. C., Chang, C., Mosley, T. H., Curb, D., Boerwinkle, E., and Cooper, R. S. (2006). An updated meta-analysis of genome scans for hypertension and blood pressure in the NHLBI Family Blood Pressure Program (FBPP). *Am. J. Hypertens.* **19**(1), 122–127.

Zeggini, E., Weedon, M. N., Lindgren, C. M., Frayling, T. M., Elliott, K. S., Lango, H., Timpson, N. J., Perry, J. R. B., Rayner, N. W., Freathy, R. M., Barrett, J. C., Shields, B., *et al.* (2007). Replication of genome-wide association signals in UK samples reveals risk loci for type 2 diabetes. *Science* **316**(5829), 1336–1341.

Zou, H., and Hastie, T. (2005). Regularization and variable selection via the elastic net. *J. R. Statist. Soc. B* **67**(Part 2), 301–320.

17

Study Designs for Genome-Wide Association Studies

Peter Kraft*,† and David G. Cox*

*Program in Molecular and Genetic Epidemiology,
Harvard School of Public Health, Boston, Massachusetts 02115
†Department of Biostatistics, Harvard School of Public Health,
Boston, Massachusetts 02115

ABSTRACT

Advances in high-throughput genotyping and a flood of data on human genetic variation from the Human Genome and HapMap projects have made genome-wide association studies technically feasible. However, researchers designing

Advances in Genetics, Vol. 60

0065-2660/08 $35.00
DOI: 10.1016/S0065-2660(07)00417-8

such studies face a number of challenges, including how to avoid subtle systematic biases and how to achieve sufficient statistical power to distinguish modest association signals from chance associations. In many situations, it remains prohibitively expensive to genotype all the desired samples using a genome-wide genotyping array, so multistage designs are an attractive cost-saving measure. Here, we review some of the basic design principles for genetic association studies, discuss the properties of fixed genome-wide and custom genotyping arrays as they relate to study design, and present a theoretical framework and practical tools for power calculations. We close with a discussion of the limitations of multistage designs.

I. INTRODUCTION

More than a decade ago, Risch and Merikangas (1996) argued that genome-wide association studies could be more powerful to detect common alleles with modest effects on disease risk than family-based linkage studies. Even after a very conservative adjustment for multiple testing—they assumed one million markers would be needed to comprehensively measure genetic variation across the genome—they showed that association scans could require far fewer subjects than linkage scans. In some situations, the number of affected sib pairs needed to ensure 80% power to detect a risk allele via linkage was practically impossible (over 60,000), while the number of case-parent trios needed, although large, was quite feasible (ca. 2200).

Despite their appeal, a number of practical challenges kept genome-wide association studies in the realm of thought experiments until very recently. First, prior to the completion of the human genome project (Lander *et al.*, 2001; Sachidanandam *et al.*, 2001; International Human Genome Sequencing Consortium, 2004; Venter, 2001), the number of known polymorphic loci in the human genome fell far short of the one million envisioned by Risch and Merikangas. Nor was it clear how many markers would be required to cover the entire genome—that is, ensure that most common polymorphisms were either directly measured or in strong linkage disequilibrium with measured markers. Early estimates of the number of markers needed for genome-wide coverage ranged from 300,000 to 1.1 million (Carlson *et al.*, 2003; Gabriel *et al.*, 2002; Kruglyak, 1999; Pritchard and Przeworski, 2001), but these estimates were based on limited empirical data or simulations under simplified population genetics models. Second, genotyping technologies were too slow and far too expensive to realistically generate the billions of genotypes needed for an adequately powered genome-wide association scan.

These challenges have now been overcome.

As of March 2007 dbSNP, the publicly available database of single nucleotide polymorphisms (SNPs), contained over 5.5 million validated SNPs (http://www.ncbi.nlm.nih.gov/projects/SNP/). The International HapMap Project (http://www.hapmap.org/) has currently (release #21a) genotyped over 3.8 million SNPs on samples drawn from four sites representing three major continental populations: Europe, Africa, and Asia (Altshuler *et al.*, 2005). These data can be used to estimate patterns of linkage disequilibrium in these populations (Conrad *et al.*, 2006; de Bakker *et al.*, 2006; Need and Goldstein, 2006), which can in turn be used to design and interpret genome-wide association studies.

The Human Genome and HapMap projects also promoted technological developments. There are now several commercially available genotyping arrays that can simultaneously genotype hundreds of thousands of SNPs at a cost of a fraction of a penny per SNP (Barrett and Cardon, 2006; Maresso and Brockel, 2008; Pe'er *et al.*, 2006). To achieve such economies of scale, these arrays are restricted to fixed, predefined set of SNPs (see Gu *et al.*, 2008; Maresso and Brockel, 2008; Chapters 5 and 15, this volume). Other technologies for custom sets of SNPs exist, but are more costly on a per-SNP basis.

However, even if these genomic and technological hurdles had not existed, it was still debatable whether genome-wide association scans would be successful in discovering novel genotype–phenotype associations (Pritchard, 2001; Terwilliger and Weiss, 1998; Weiss and Terwilliger, 2000). Risch and Merikangas showed that genome-wide association scans should be able to detect common risk alleles with modest marginal effects, but it was unknown whether (or how many) such alleles existed for complex diseases and traits. If there were no such alleles—if instead complex traits were influenced by the joint effect of multiple rare alleles, say—then genome-wide association scans might fail to find any new trait loci. Results from the first genome-wide association scans (discussed below and in Table 17.1) suggest that there are detectable common alleles with modest marginal effects for complex traits.

Scans using early versions of whole-genome array technologies appeared in 2005 and 2006. As described below, these arrays covered less than half of the genome. Still, they provided important evidence that the genome-wide approach could be fruitful. One of these studies found a strong association between age-related macular degeneration (AMD) and an SNP in the Complement Factor H (*CFH*) gene (Klein *et al.*, 2005). This SNP was shown to be in strong linkage disequilibrium with a functional SNP in *CFH*, and the association has been subsequently replicated in many different studies (Thakkinstian *et al.*, 2006). On the other hand, an observed association between a SNP in *INSIG2* and body mass index (BMI) (Herbert *et al.*, 2006) has been replicated in some studies but not in others (Dina *et al.*, 2007; Loos *et al.*, 2007; Lyon *et al.*, 2007; Rosskopf *et al.*, 2007), while the strongest observed associations in a genome-wide scan for

Table 17.1. Summary of Results from Select Early Genome-Wide Association Studies

Phenotype	Study	Platform	Marker (gene)	RAF	Odds ratio	λ_s	PAF
Age-related macular degeneration	Klein *et al.* (2005)	Affymetrix GeneChip 100k	rs380390 (*CFH*)	0.48	7.4[a]	1.49	0.63
Body mass index	Herbert *et al.* (2006)	Affymetrix GeneChip 100k	rs7566605 (*INSIG2*)	0.37	1.3[b]	1.00	0.16
	Frayling *et al.* (2007)	Affymetrix GeneChip 500k	rs9939609 (*FTO*)	0.39	1.3	1.02	0.36
Cardiovascular disease	Arking *et al.* (2006)	Affymetrix GeneChip 100k	rs10494366 (*NOS1AP*)	0.36	(1.5%)[c]	na	na
	McPherson *et al.* (2007)	Custom 100k Perlegen	rs10757274 (*CDKN2A-B*)	0.49	1.7	1.06	0.52
	Wellcome Trust Case Control Consortium (2007)	Affymetrix GeneChip 500k	rs1333049[†]	0.47	1.5	1.03	0.46
	Samani *et al.* (2007)	Affymetrix GeneChip 500k	rs1333049	0.47	1.4	1.02	0.43
Prostate cancer	Gudmundsson *et al.* (2007)	Illumina HumanHap300	rs1447295[†]	0.10	1.7	1.04	0.16
	Yeager *et al.* (2007)	Illumina HumanHap550	rs1447295[†]	0.10	1.4	1.01	0.13
Diabetes	Sladek *et al.* (2007)	Illumina Human1 + HumanHap300	rs7903146 (*TCF7L2*)[†]	0.30	1.7	1.07	0.37
	Saxena *et al.* (2007)	Affymetrix Gene Chip 500k	rs7903146 (*TCF7L2*)[†]	0.26	1.4	1.03	0.28
	Scott *et al.* (2007)	Illumina HumanHap300	rs7903146 (*TCF7L2*)[†]	0.18	1.4	1.02	0.21
	Zeggini *et al.* (2007)	Affymetrix GeneChip 500k	rs8050136 (*FTO*)[†]	0.40	1.3	1.01	0.36
Breast cancer	Easton *et al.* (2007)	Custom 300k Perlegen	rs2981582 (*FGFR2*)[†]	0.38	1.3	1.01	0.37
	Hunter *et al.* (2007a)	Illumina HumanHap550	rs1219648 (*FGFR2*)	0.39	1.3	1.02	0.37

Notes: RAF = risk allele frequency; odds ratio is multiplicative increase in odds per risk-allele copy (unless otherwise noted); λ_s is sibling relative risk calculated using listed RAF (assuming Hardy–Weinberg genotype proportions π_g); and odds ratio using formula from Thomas (2004; #1221); PAF is population attributable fraction estimated as $1 - 1/(\sum_g OR_g \pi_g)$.

[a]Recessive odds ratio.

[b]Recessive odds ratio for obesity (body mass index (BMI) > 30 kg/m^2).

[c]Percent variance in quantitative trait (QT interval) explained.

[†]One of several significant associations reported.

Parkinson's disease loci (Maraganore *et al.*, 2005) have not been replicated to date, despite multiple attempts (Clarimon *et al.*, 2006; Farrer *et al.*, 2006; Goris *et al.*, 2006; Li *et al.*, 2006; Myers, 2006). These results highlight the importance and difficulty of independent replication of association findings (Chanock *et al.*, 2007; Hunter *et al.*, 2007b). An association first seen in a genome-wide scan may not be seen in other studies because of design differences (e.g., different trait definitions, control populations), differences in genetic and environmental backgrounds, differences in linkage disequilibrium patterns—or because the initial result was a false positive, perhaps due to small sample size or inadequate adjustment for multiple testing (Ioannidis, 2005; Wacholder *et al.*, 2004).

In the first five months of 2007, another wave of genome-wide association studies using arrays that covered most of the genome appeared, including four genome-wide scans for Type 2 Diabetes loci (Saxena *et al.*, 2007; Scott *et al.*, 2007; Sladek *et al.*, 2007; Zeggini *et al.*, 2007), three scans for breast cancer loci (Easton *et al.*, 2007; Hunter *et al.*, 2007a; Stacey *et al.*, 2007), three scans for Crohn disease (Hampe *et al.*, 2007; Libioulle *et al.*, 2007; Rioux *et al.*, 2007), and scans for prostate cancer (Gudmundsson *et al.*, 2007; Yeager *et al.*, 2007) and cardiovascular disease (Arking *et al.*, 2006; McPherson *et al.*, 2007) (see Table 17.1). These studies have not only replicated previously known associations (e.g., *TCF7L2* for Type 2 Diabetes) but also found novel trait loci. These early results illustrate that an agnostic approach to scanning the genome can produce unexpected findings that would have been missed taking a candidate gene approach. Before the publication of the AMD scan, few papers made the link between *CFH* and AMD, a link which provided new insights into the etiology of AMD. Genome scans have also found replicated associations between SNPs in nongenic regions and disease; these associations could lead to exciting new hypotheses about the function of these regions and their relation to disease risk (Witte, 2007).

Perhaps the biggest challenge facing investigators planning a genome-wide association study is the need to recruit and genotype enough subjects to achieve adequate power to detect the subtle associations expected for complex traits while minimizing the risk of false positives. For the most part, the associations from genome-wide association studies published to date have not achieved genome-wide significance in the initial set of samples genotyped using a genome-wide array. Instead, a small set of the most promising SNPs (chosen on the basis of statistical significance in the initial scan, plus perhaps genomic annotation such as location near a promising candidate gene) has been genotyped in additional samples. This reflects the large costs associated with genome-wide panels: even though the per-SNP genotyping costs are small, the large number of SNPs makes the per-sample costs large. Currently, it is often too expensive to genotype the requisite number of cases and controls using genome-wide arrays—as discussed in many references (Thomas, 2006; Wang *et al.*, 2005) and further below, sample sizes to reliably detect a disease locus in a population-based case–control study can top 5000 cases and 5000 controls, leading to genotyping costs from $3 to $6 million or more. This practical constraint has led to the development of multistage designs as a cost-saving measure. These designs—where a subset of samples is genotyped using a genome-wide array and then a subset of SNPs with the strongest evidence for association in this first stage are genotyped in the remaining samples—have been shown to save 50% or more on genotyping costs while retaining nearly the same power as a single-stage design where all subjects are genotyped using a genome-wide array (Kraft, 2006; Kraft *et al.*, in preparation; Satagopan and Elston, 2003; Satagopan *et al.*, 2002, 2004; Skol *et al.*, 2006; van den Oord and Sullivan, 2003; Wang *et al.*, 2006).

It should be stressed that the goal of these designs is to minimize cost while retaining power to detect a specified marginal genetic effect. The designs do not boost power by reducing the number of tests in subsequent stages, for example. The "no free lunch" axiom applies: one cannot get more information from less data. By adopting a multistage design, investigators run the risk of missing effects that exhibit heterogeneity across studies or gene–gene or gene–environment interaction effects. On the other hand, the "something is better than nothing" axiom also applies. As suggested by initial genome-wide association scan results, detectable marginal genetic associations with complex traits do exist, even if the true causal mechanism involves a gene–gene or gene–environment interaction (Pharoah, 2007). For many traits where the genetic architecture is largely unknown, it will be better to do an affordable study now that can detect some loci with less subtle effects while missing some of the more subtle effects than to wait until the ideal study becomes affordable.

The specific design that minimizes cost while achieving a target power—the proportion of samples to use in the first stage, the proportion of markers to follow up—will depend on the relative costs of the genotyping technologies used at the different stages. Currently per-genotype costs for the fixed genome-wide arrays are quickly dropping relative to the custom genotyping technologies. If this trend continues, the cost savings from a multistage design may not outweigh the logistical or opportunity costs that come from a multistage design. On the other hand, if per-genotype costs for the custom platforms were to approach the level of genome-wide costs, then the cost savings for multistage designs could be considerable, as these designs minimize the number of genotypes required.

This chapter discusses basic design principles for genome-wide association studies, with an emphasis on cost-effective multistage designs. After reviewing general design principles for genetic association studies—available study designs for binary and continuous outcomes, appropriate choice of controls, population stratification bias, etc.—we briefly discuss hypotheses regarding the genetic architecture of common complex disease, that is, the number of loci that influence a trait and the distribution of their allele frequencies and effects. This is relevant to study design because investigators need to know what kinds of effects they wish to detect. If investigators have reason to believe there is a single locus that explains much of the variation in a trait, then smaller sample sizes will suffice; for more subtle effects, larger sample sizes will be necessary. We then review properties of available genotyping technologies, including the relative costs of fixed whole-genome arrays to custom technologies, as well as the relative costs and coverage properties of several commercially available genome-wide arrays. This is followed by a discussion of principles for the design of multistage association studies, including mathematical details on power calculations for detecting marginal genetic effects and a review of the multiple testing problem. We close with a brief discussion of design and analysis challenges for studies that look beyond marginal genetic effects to the joint effects of multiple genes or genes and environmental exposures.

II. BASIC PRINCIPLES OF ASSOCIATION DESIGN

There are a number of excellent reviews of the basic principles of epidemiologic study design for testing association between an exposure and a trait in general (Breslow and Day, 1980, 1987; Rothman and Greenland, 1998; Wacholder *et al.*, 1992) and genetic association studies in particular (see Borecki and Province, 2008; Laird and Lange, 2008; Chapters 3 and 10, this volume; Thomas, 2004a; Cordell and Clayton, 2005; Newton-Cheh and Hirschhorn, 2005). Here, we will focus on three issues: choice of trait, choice of design (family-based and cross-sectional prospective, and retrospective population-based studies), and (for population-based case–control studies) choice of controls.

Choosing the right trait to study is important to maximize both the scientific and public health implications of a study and the power to detect an association (see Wojczynski and Tiwari, 2008; Chapter 4, this volume). The latter is determined in large part by the heritability of a trait: loosely, the proportion of variance in the trait due to genetic factors. This can be estimated from twin or family studies (see Rice, 2008; Chapter 2, this volume). Note that although disease incidence may be the trait of ultimate public health importance, studying correlated continuous traits such as intermediate biomarkers may be more powerful than studying the binary disease trait itself, because the heritability of an intermediate may be higher (it is "closer to the genes"; Horvath and Baur, 2000; Newton-Cheh and Hirschhorn, 2005) and there may be more power to detect an association with a continuous trait than a binary trait.

Examples of such traits include body mass index (diabetes) (Frayling *et al.*, 2007), QT length (cardiovascular disease) (Arking *et al.*, 2006), and plasma steroid hormone levels (breast cancer) (Hunter *et al.*, 2005; Missmer *et al.*, 2004).

Once researchers have decided on a trait to study, they will need to choose an appropriate design (see Table 17.2). While the strengths and weaknesses of each design should be considered, the decision of which to use will also be influenced by sample availability. Researchers who have collected DNA from families as part of a previous linkage study will justifiably want to use this existing resource in family-based association study, and researchers who have collected baseline blood samples from a large, prospectively followed cohort of unrelated individuals will want to conduct a nested case–control study. Family-based and population-based designs complement each other, so neither is absolutely pro- or prescribed. The practical advantages of using existing studies outweigh the weaknesses of any particular design. Similarly, researchers considering a *de novo* study may wish to select a design that maximizes the opportunity to study multiple related traits and possible environmental effect modifiers (Collins and Manolio, 2007)—whether the design that best accomplishes that goal is a family-based or population-based study will depend on the particulars of the trait(s) to be studied, available budget, and time frame.

The principal goals of sound design for association studies are to minimize systematic bias and maximize power. For genome-wide association studies, the primary concern is to eliminate bias under the null hypothesis of no association. If a marker is truly *unassociated* with a trait, tests of association should not reject the null hypothesis of no association any more than expected. On the other hand, if a marker is truly *associated* with a trait, it may not be essential that a design provides unbiased estimates of population risk parameters (e.g., lifetime risks of disease for subjects with different genotypes at a known susceptibility locus), considering the initial goal of genome-wide association studies is to locate regions harboring disease susceptibility loci. For example, designs that use cases with a family history of disease (Antoniou and Easton, 2003; Antoniou *et al.*, 1998; Teng and Risch, 1999) or from pedigrees showing evidence for linkage (Fingerlin *et al.*, 2004) and population-based controls have valid Type I error rates but increased power relative to designs using population-based cases and controls. This makes them attractive designs for genome-wide association studies, although they can yield biased estimates of population risk parameters (Antoniou and Easton, 2003).

A. Retrospective case–control studies

In a retrospective case–control study, researchers collect genetic and environmental exposure data on a series of subjects after they are diagnosed with the disease of interest, and then compare these data to those collected on a series of

Table 17.2. Characteristics of Common Genetic Association Study Designs

Design	Susceptible to population stratification bias?	Appropriate for late-onset diseases?	Appropriate for rare diseases?	Appropriate for intermediate biomarkers?	Environmental exposure measurement susceptible to recall bias?
Retrospective case–control	Yes, but can be minimized through appropriate design and analysis	Yes	Yes	Bias due to reverse causality is a concern	Yes
Prospective cohort (nested case–control)	Yes, but thorough ascertainment of cases from known study base further minimizes risk and magnitude of bias	Yes	May require prohibitively large cohort and follow-up time to accrue sufficient number of cases	Yes, although nested case–control sampling may induce bias	No
Family-based	No	Requires genotype data from parents or siblings, which may be difficult to collect for late-onset disease	Yes	Depends on proband ascertainment scheme	Depends on whether exposure data collected prospectively or retrospectively

disease-free subjects. This design has the advantages of being relatively inexpensive, quick, and powerful. However, it is also particularly sensitive to several sources of systematic bias, including *recall bias* and *selection bias*. *Recall bias* arises when cases report their exposure history differently than controls. This is often cited as a key drawback to retrospective designs, but should not be a problem when investigating marginal genotype–disease associations, because genotypes are measured from DNA samples, not through interviews with case and control subjects. However, differences between cases and controls in DNA collection, storage, and genotyping methods could induce systematic bias (Clayton *et al.*, 2005). Recall bias also remains a concern for retrospective studies when investigating gene–environment interactions (Garcia-Closas *et al.*, 2004). *Selection bias* occurs when the controls do not come from the same population that gave rise to the cases. For example, if cases were recruited from primary care physicians in the US midwest and controls were enrolled at a large urban hospital in southern California, then the genetic and environmental backgrounds for cases and controls may differ simply as a result of the study design. Selection bias may be a particular concern for genome-wide association studies that use controls who were recruited and genotyped for a previous study (Wellcome Trust Case Control Consortium, 2007). Multiple studies of a range of traits will be releasing individual-level genotyping data (see Table 17.3), so this approach is now quite feasible and obviously attractive from an economic standpoint—no need to genotype controls! The risk and magnitude of selection bias can be minimized by matching cases and controls on ethnicity (Thomas and Witte, 2002; Wacholder *et al.*, 2002). But considering that there are regional differences in percentage of European ancestry among African-Americans (Tang *et al.*, 2005) and north–south gradients in allele frequencies across Europe (Campbell *et al.*, 2005; Price *et al.*, 2006)—and considering above-mentioned differences in sample handling and genotyping platforms—matching cases to a control sample nominally from same ethnicity may not completely eliminate subtle biases. Methods that use a large number of markers to adjust for population stratification (discussed below) may help reduce selection bias.

Confounding bias can affect both prospective and retrospective case–control studies. It occurs when a risk factor for disease is associated with a marker under study. Several authors have argued (Clayton and McKeigue, 2001) that genetic association studies inherently minimize the risk and magnitude of confounding bias, since genotypes at a given locus are independent of most environmental factors and genotypes at other loci. Still, there are situations where genotypes are not independent of environmental or other genetic factors. For example, if disease rates and allele frequencies are correlated across ethnicity—perhaps because risk factor prevalences differ across ethnicity—an allele can appear to be associated with disease, even though within each ethnic group, it plays no causal role. This special case of confounding bias is known as

Table 17.3. Select Websites Providing Open- and Controlled-Access Data from Genome-Wide Association Scans

	Phenotype(s)	Open access	Controlled access (available to qualified investigators via registration)
Cancer genetic markers of susceptibility (http://cgems.cancer.gov)	Breast and prostate cancer	Allele frequencies, association statistics	Individual genotype, phenotype, and covariate data on ca. 4800 subjects
dbGAP (http://www.ncbi.nlm.nih.gov/sites/entrez?db=gap)	ADHD, bipolar disorder, cataract, diabetic nephropathy, macular degeneration, Parkinson disease, psoriasis, schizophrenia	Allele frequencies, association statistics	Individual genotype, phenotype, and covariate data on over 15,000 subjects
Wellcome Trust Case Control Consortium (http://www.wtccc.org.uk/)	ADHD, bipolar disorder, cataract, diabetic nephropathy, macular degeneration, Parkinson disease, psoriasis, schizophrenia		Individual genotype data from 3000 controls
Diabetes Genetics Initiative (http://www.broad.mit.edu/diabetes/scandinavs/index.html)	Type II Diabetes, fasting glucose, insulinogenic index, insulin response, anthropometric measures, cardiovascular traits (blood pressure, cholesterol, triglycerides, etc.)	Allele frequencies, association statistics	
Genomic Medicine Database (http://gmed.bu.edu/)	Body mass index, blood glucose, cardiovascular traits (blood pressure, cholesterol, triglycerides, etc.)	Allele frequencies, association statistics for “top hits”	Individual genotype, phenotype, and covariate data on original Framingham and Framingham offspring cohorts

population stratification bias. In principle, age-at-onset can also be a confounder if the genotype frequency in the source population varies with age, for example, if a locus influences disease age-at-onset, death due to competing risks or loss-to-follow up. To minimize this kind of confounding, appropriate age-stratified analyses could be performed. However, in practice, the magnitude of the bias in an age-unadjusted analysis may be quite small, as the gene–disease (or gene–loss-to-follow up) association would have to be quite large for the genotype frequencies in the population to differ appreciably with age (Thomas and Greenland, 1985). Many of the genome-wide studies published to date did not adjust for age-at-onset (Gudmundsson *et al.*, 2007; Hampe *et al.*, 2007; Klein *et al.*, 2005; Samani *et al.*, 2007; Sladek *et al.*, 2007; Wellcome Trust Case Control Consortium, 2007) or saw little difference in the distribution of the test statistics between adjusted and unadjusted analyses (Hunter *et al.*, 2007a). (Although adjusting for age-at-onset may not be necessary to eliminate confounding bias, it may increase power.) One more form of confounding bias deserves final explicit mention here: a gene may be associated with a known behavioral risk factor (e.g., smoking, alcohol use, hyperphagia) because it predisposes an individual to adopt that risky behavior. One could reduce the magnitude of this "bias" by stratifying on the risk factor when testing for genetic associations. However, an association that is due to the association between an allele and an intermediate on the causal pathway may be just as scientifically interesting as a direct association between an allele and disease risk. For example, a novel allele that influences risk of diabetes by increasing body mass index might be missed if all of the genotype–disease analyses are adjusted for body mass index (Frayling *et al.*, 2007). Again, the primary goal for genome-wide association studies is discovering genetic loci underlying variation in disease risk; characterizing these new loci—including investigating whether they act through a particular intermediate—will be the task of subsequent analyses or new studies.

B. Prospective cohort and nested case–control studies

In a prospective cohort study, DNA and exposure data are collected at baseline from a set of disease-free subjects. These subjects are then followed until they develop disease. There are a number of advantages to cohort studies relative to retrospective case–control studies (Breslow and Day, 1987; Langholz *et al.*, 1999; Rothman and Greenland, 1998). First, because the population that gave rise to the cases is explicitly defined, absolute risks can be directly estimated (assuming all cases can be ascertained). This also eliminates selection bias. Second, information about environmental exposures and intermediate biomarkers (e.g., plasma hormone levels or body mass index) are collected before diagnosis. This eliminates recall bias and lessens the risk of finding (or not finding!) a gene–biomarker association through reverse causality, where the onset of disease causes changes in biomarker levels (Thomas and Conti, 2004). Finally, multiple diseases can be studied in a cohort, while retrospective studies are generally limited to the disease used to select the cases.

The primary disadvantage to cohort studies is the large investment in both time and money they require. A large cohort—tens of thousands of subjects or more—and decades of follow-up time may be necessary to accrue the number of cases needed to detect moderate genetic effects. However, once the cohort is established, genetic association studies can be conducted rather economically using a nested case–control strategy, where cases and a small subset of appropriately matched controls are genotyped, rather than the entire cohort (Breslow and Day, 1987; Langholz *et al.*, 1999; Rothman and Greenland, 1998).

A secondary disadvantage to cohort studies is the fact that they are often restricted in terms of ethnicity, location, or socioeconomic status. This is in some respects a strength, as it may reduce the risk and magnitude of population stratification bias, but it may also limit the generalizibilty of findings to other populations with different linkage disequilibrium patterns and environmental exposures (Ambrosone *et al.*, 2007; Lin *et al.*, 2007; Pharoah, 2007). However, a robust result from one population should be a promising candidate for follow-up studies in other populations—early meta-analyses of replicated results suggest that the direction and magnitude of the odds ratios for the risk genotype(s) are similar across ethnicity (although the allele frequency varies) (Goldstein and Hirschhorn, 2004; Ioannidis *et al.*, 2004), and differences in linkage disequilibrium structure may help pinpoint the location of the causal variant (Thomas *et al.*, 2005).

C. Continuous phenotypes

Prospective studies provide numerous opportunities to study continuous phenotypes, especially if repeated measures are available: a gene might influence time-averaged values, or changes over time, or both. For example, the initial genome-wide association study reporting an association between *INSIG2* and body mass index made use of repeated body mass index measures over 24 years (Herbert *et al.*, 2006). As a cost saving measure, researchers could sample subjects from the extremes of the phenotype distribution—say subjects in the 5th percentile or lower and subjects in the 95th percentile and higher. This approach can have greater power than selecting an equivalent number of samples without regard to their phenotype values (Allison, 1997; Arking *et al.*, 2006), although for some complex multilocus penetrance functions, power might be reduced for some loci (Gu and Rao, 2002).

In practice, researchers will want to maximize the use of genome-wide retrospective and prospective case–control studies to study continuous phenotypes, including intermediate biomarkers. Sampling on case–control status may induce bias, however, even if reverse causality can be ruled out (as in many prospective studies). If the continuous phenotype is associated with disease risk, and the tested SNP is associated with disease risk, then the estimate of SNP–continuous phenotype association in pooled cases and controls can be biased—in principle, even the analysis "adjusting for" disease status by including it in a regression model or the analysis restricted to controls can be biased. This bias can be away from or towards the null. For nested case–control studies where the sampling fractions from the underlying cohort are known, an inverse-probability-of-sampling-weighted regression can provide unbiased estimates of the SNP–continuous phenotype association. (This approach is analogous to weighted complete-case analysis of missing data (Robins *et al.*, 1994) or analysis of multistage designs (Siegmund *et al.*, 1999)). The risk and magnitude of this sampling-induced bias and appropriate analytic methods require further investigation.

D. Using genome-wide SNP data to adjust for population stratification and other biases

As discussed above, sound epidemiologic design and appropriate analysis—for example, matching on and/or adjusting for self-reported ethnicity—are the first line of defense against population stratification bias. However, there may be subtle population structure within strata defined by self-reported ethnicity, and tests of gene–phenotype association may have inflated Type I error rates due to "cryptic relatedness" (Devlin and Roeder, 1999; Voight and Pritchard, 2005). If (unknown to the investigator) subjects are relatively closely related, then their genotypes will be correlated, and the usual test statistics (which assume independence) will be inflated. This is particularly a concern when only cases with a positive family history of disease are enrolled.

A number of analytic approaches that use the vast amount of genetic data available from genome-wide SNP arrays to account for latent population structure and/or cryptic relatedness have been proposed. The "structured association" approach adjusts for estimated admixture proportions (Pritchard *et al.*, 2000a,b; Satten *et al.*, 2001). "Genomic control" estimates a test-statistic inflation factor λ and uses this to correct the observed test statistics (Devlin and Roeder, 1999; Freedman *et al.*, 2004; Marchini *et al.*, 2004; Reich and Goldstein, 2001). Other approaches use genome-wide SNP data to construct principal components of genetic variation (which are then included in regression analyses as covariates) (Price *et al.*, 2006) or construct propensity scores (which are used to stratify subjects) (Epstein *et al.*, 2007). The relative performance of these methods and others (Nievergelt *et al.*, 2007; Setakis *et al.*, 2006) to adjust for population stratification, cryptic relatedness, and other forms of bias is the subject of ongoing research. Each method has its strengths and weaknesses.

Structured association is computationally intensive and makes relatively strong modeling assumptions; genomic control assumes that the magnitude of population stratification bias will be the same across the entire genome (which will generally not be the case, as allele frequencies do not noticeably differ across ethnicity for many regions)—yet the estimated admixture proportions and λ provide some evidence as to the amount of systematic bias in the data. A simple Q–Q plot can also be quite useful (Clayton *et al.*, 2005). Due diligence demands that one of these analyses adjusting for latent population structure be performed, but the best protection against false positives due to population stratification and/or cryptic relatedness will be independent replication of any promising associations in a separate study, preferably in a different population, using a different design (e.g., a family-based design) and genotyped using a different technology (Chanock *et al.*, 2007; Thomas, 2006).

E. Family-based designs

As originally proposed for case-parent trios (Self *et al.*, 1991; Schaid and Sommer, 1993; Spielman *et al.*, 1993), family-based analyses are a special case of case-only analysis (Greenland, 1999), where the observed distribution of genotypes among cases is compared to that expected under the null hypothesis, conditional on parental genotypes. The primary advantage of this design is that it is immune to population stratification bias, because each case is compared to (pseudo-)controls (Cordell *et al.*, 2004) perfectly matched on ancestry. Secondarily, family data can be used to check for genotyping errors via Mendelian inconsistencies. (Researchers conducting a case–control design could genotype HapMap trios as an additional genotyping quality-control measure.) However, case-parent trio studies can have slightly less power to detect marginal genetic effects than case–control studies with an equivalent number of subjects (Gauderman and Kraft, 2002; Risch and Teng, 1998; Witte *et al.*, 1999), are more sensitive to genotyping errors (Mitchell *et al.*, 2003), and of course require that parents are available for genotyping (making it difficult to study late-onset diseases). Analytic methods for family-based studies have been extended to include multiple cases and controls per family, continuous phenotypes, and multivariate phenotypes (see Laird and Lange, 2008; Chapter 11, this volume). New methods (Van Steen and Lange, 2005; Van Steen *et al.*, 2005) that use between-family variation ignored by standard family-based analyses to screen a large number of variants and select a small set of candidates to be tested using within-family variation are particularly promising in the context of genome-wide association scans. Such novel analytic approaches may greatly reduce or eliminate the loss in power for family-based designs relative to case–control designs. Family-based designs can be retrospective—families are enrolled and DNA samples and data on environmental exposures are collected after case diagnosis—or prospective—for example, cases are selected from a cohort with prospectively collected exposure data. Retrospective family-based designs are susceptible to some of the same problems as case–control designs, for example, increased (and perhaps differential between cases and controls where relevant) measurement error. However, in some situations (e.g., case-parent trios where the measurement error distribution does not depend on genotype, conditional on disease status), tests for gene–environment interaction should have valid Type I error rates.

III. THE GENETIC ARCHITECTURE OF COMPLEX TRAITS

By definition, complex diseases are caused by multiple genetic loci and multiple environmental factors and their interactions—as opposed to simple Mendelian disorders, which are caused by high-penetrance mutations at a single locus (Clerget-Darpoux *et al.*, 2001). The genetic architecture of complex diseases—the number of disease susceptibility loci, the distribution of their allele frequencies, and the size their effects (e.g., odds ratios for qualitative traits and differences in means across genotypes for quantitative traits)—remains mostly unknown. Until recently, what was known had to be inferred from segregation and twin studies (see e.g., Rice, 2008; Chapter 2, this volume; Rao *et al.*, 1994; Risch and Merikangas, 1996; O'Donnell *et al.*, 2002; Pharoah *et al.*, 2002; Balmain *et al.*, 2003; Fox *et al.*, 2003). The data from family studies of complex diseases are consistent with a number of genetic architectures: the disease may in fact be influenced in a simple Mendelian fashion by a single locus with a strong effect (although the failure to find such loci for most complex diseases despite great efforts argues against this); disease risk may be influenced by common alleles at a few loci, each with modest effects—or perhaps interacting in a nonadditive fashion; or the risk may be influenced by rare alleles at a large number of loci, each with small, additive effects (closer to Fisher's classic "polygenic" model (Fisher, 1918)). The ability of genome-wide association studies to detect complex disease susceptibility loci will depend upon the true genetic architecture of the disease.

Early skeptics of genome-wide association studies argued that if in fact complex diseases are principally caused by multiple rare alleles, then genome-wide association studies would have little chance of successfully identifying disease susceptibility loci (Pritchard, 2001; Terwilliger and Weiss, 1998; Weiss and Terwilliger, 2000). Even if these rare alleles were directly measured, the sample sizes required to detect small-to-modest effects would be impractical. Furthermore, rare alleles typically have low linkage disequilibrium (as measured by r^2) with surrounding loci (Altshuler *et al.*, 2005), so any approach that relies on detecting indirect effects through strong linkage disequilibrium instead of directly measuring the causal variant will fail. Arguments that there would be little useful linkage disequilibrium among common alleles have largely been refuted by empiric evidence from the HapMap and large scale resequencing efforts such as the HapMap ENCODE regions (Altshuler *et al.*, 2005) and Seattle SNPs (http://pga.gs.washington.edu/), although concerns about the relevance of linkage disequilibrium measures calculated from HapMap data in the presence of genetic heterogeneity persist (Terwilliger and Hiekkalinna, 2006).

Proponents of genome-wide association studies argued that some complex disease susceptibility alleles probably arose through either genetic drift (a.k.a. neutral evolution) or selective advantage, and hence were likely to be common (Chakravarti, 1999; Reich, and Lander 2001). (For an introduction to the basic theories behind genetic drift and genetic evolution in response to selection, see Gillespie, 1998; Hartl and Clark, 2007.) This became known as the Common Disease, Common Variant, or Common Disease Neutral Variant hypothesis. Since many complex diseases are late-onset diseases that first appear after most chances for reproduction have already occurred (e.g., most cancers, Type 2 Diabetes, coronary heart disease), they should not reduce evolutionary fitness. Hence, these alleles should not be subject to evolutionary pressures, and the frequency distribution of complex disease alleles should be determined by random drift. (Conversely, many alleles that cause simple Mendelian disorders, because they increase infant and childhood mortality, have a selective disadvantage and hence are quite rare.) Under neutral evolution, most alleles are rare, but the majority of genetic variation (heterozygosity) in the population comes from the common alleles (Hartl and Clark, 2007). Empirical data (e.g., from the HapMap ENCODE regions (Altshuler *et al.*, 2005)) conform to this distribution, although there is some evidence that there are fewer rare alleles than expected (especially in non-African populations), perhaps due to population bottlenecks (Altshuler *et al.*, 2005; Weiss, 1993). Complex disease alleles could also be common because they confer selective advantage early in life at the expense of increased risk of disease later in life, or because they represent adaptations to the environment of the past that are deleterious in the present environment. James Neel's "thrifty genotype" hypothesis for the genetic susceptibility to diabetes is an example of the latter phenomenon (Neel, 1962, 1999).

For further discussion of the genetic architecture of complex disease and its impact on the success of genome-wide association studies, see the review by Wang *et al.* (2005). As mentioned in the introduction, results from the first genome-wide studies have been promising; common variants with modest relative risks have been discovered using this approach. In fact, these results suggest that there are other susceptibility loci to be discovered even for the diseases that have been studied so far, as the proportion of familial risk explained by the new loci is still small (sibling relative risks from 1.02 to 1.06), despite their large population attributable fraction (from 15% to 60%). (For a discussion of the apparent contradiction between small sibling relative risks and large population attributable fraction, see Peto. 1980; Thomas, 2004b.)

IV. GENOTYPING TECHNOLOGIES

Table 17.4 compares different genotyping technologies in terms of scale, flexibility, and cost (see also Maresso and Brockel, 2008; Chapter 5, this volume). As mentioned above, genome-wide technologies achieve economy of scale (very low per-SNP genotyping costs) in part by fixing the set of variants to be genotyped. Custom platforms that allow users to select their own set of SNPs to be genotyped—essential for multistage designs—are more costly per SNP but less costly per subject. As discussed in the next section, the relative per-SNP genotyping costs will affect optimal design choice (how many markers to follow up after the first stage of a multistage design or whether to use a multistage design at all).

Table 17.5 compares some of the popular commercially available genome-wide arrays in terms of their design—random or tagging—and coverage (Barrett and Cardon, 2006; Pe'er *et al.*, 2006). Here "random design" denotes that the SNPs on the platform were randomly selected from the genome, without specific reference to patterns of linkage disequilibrium from the HapMap, in contrast to a "tagging design," where SNPs were explicitly chosen to serve as surrogates for common variants in the HapMap data. Coverage—often defined as the percentage of common SNPs with a pairwise r^2 above 0.8 with at least one

Table 17.4. Comparison of Fixed and Custom Content Genotyping Technologies

Platform	Number of SNPs	SNP choice	Cost per SNP	Cost per sample
Illumina HumanHap550	550,000	Fixed	$0.002	$905.00
Affymetrix GeneChip 500k	500,000	Fixed	$0.001	$405.00
Illumina Infinium	15,000	Custom	$0.013	$200.00
Illumina GoldenGate	1536	Custom	$0.059	$90.00
SNPlex	48	Custom	$0.208	$10.00
iPLEX	40	Custom	$0.125	$5.00
TaqMan	1	Custom	$0.950	$0.95

Notes: All prices estimated assuming 2000 samples to be genotyped, using information from the following websites (accessed 4 June 2007): http://www.broad.mit.edu/gen_analysis/genotyping/pricing.html, http://www.hpcgg.org/Genotyping/, and http://seq.mc.vanderbilt.edu/DNA/html/multiplexSNP.html. For illustration purposes only.

Table 17.5. Comparison of Coverage of Fixed Genome-Wide Arrays: Percent of HapMap Phase II SNPs with Maximum r^2 above Specified Thresholds

		CEU				JPT + CHB				YRI			
	Design	≥0.2	≥0.5	≥0.8	=1.0	≥0.2	≥0.5	≥0.8	=1.0	≥0.2	≥0.5	≥0.8	=1.0
Affymetrix GeneChip 100k (Hind + Xba)	Random	71	49	33	20	67	47	31	16	52	29	17	10
Affymetrix GeneChip 500k (Nsp + Sty)	Random	95	83	68	48	95	83	67	43	88	65	47	32
Illumina Human Hap300	Tagging	97	90	77	46	95	81	63	35	82	53	33	21
Illumina HumanHap550	Tagging	99	96	88	62	98	93	83	53	93	75	56	36
Illumina HumanHap650	Tagging	99	96	89	66	98	96	84	56	96	84	66	43

Note: Mark Daly and Paul De Bakker (personal communication).

SNP on the array—depends upon the patterns of linkage disequilibrium in the study population. Coverage is smaller in African samples because of the lower levels of linkage disequilibrium among Africans (Altshuler *et al.*, 2005; Gabriel *et al.*, 2002). The Illumina HumanHap550 array was designed to tag common variants in the HapMap CEPH European (CEU) samples, hence it has lower coverage in the HapMap Yoruba (YRI) samples. The additional SNPs on the Illumina HumanHap650 array were chosen to boost coverage in the YRI samples, and provide only modest gains in the CEU samples.

Table 17.5 also provides more detailed estimates of the *distribution* of maximum r^2 values for common SNPs in the Phase II HapMap. Notice that "coverage" as defined by the proportion of SNPs tagged with maximum $r^2 \geq 80\%$ only tells part of the story. Many of the loci with maximum $r^2 < 80\%$ are tagged at moderate levels ($50\% <$ maximum $r^2 < 80\%$). However, there are a few loci that are not well tagged at all (maximum $r^2 \leq 20\%$), and the proportion of these "untagged" SNPs varies by platform (random platforms have more untagged SNPs than tag platforms). Coverage also varies with allele frequency (results not shown); SNPs with higher allele frequencies are more likely to be tagged well (or moderately well) and less likely to be untagged than SNPs with lower allele frequencies (Paul de Bakker, personal communication; Altshuler *et al.*, 2005).

The calculations in Table 17.5 assume that there are no missing data—in particular, that none of the SNPs on the genome-wide array is excluded from analysis because of poor quality (e.g., high missing or low concordance rates). In practice, this is often not true, as can be seen from the completion rates observed in early genome-wide association studies (see also call rates in Tables 17.1 and 17.2 in Maresso and Brockel, 2008; Chapter 5, this volume). However, use of multimarker predictors can offset this loss of information (or where it is small, improve coverage) (Li and Abecasis, 2006; Marchini *et al.*, 2007). This approach can also help combine data from studies using different genotyping platforms. If a SNP is on the array used in the first study but not the second, it can be imputed on the second using local linkage disequilibrium patterns.

We discuss the impact of differences in the coverage of different commercially available genome-wide arrays on study power in the next section.

The discussion so far has focused on technologies that genotype each subject individually. An alternative, very cost-effective approach is to compare allele frequencies across DNA pools—say a pool of DNA from cases and a pool of DNA from controls (Prentice and Qi, 2006; Sham *et al.*, 2002). This greatly reduces the number of genotyping reactions needed, and hence cost. A genome-wide association study of 500 cases and 500 controls using individual genotyping might cost on the order of \$500,000, while the same study using pooled genotyping might cost under \$50,000. There are drawbacks to pooling, however. Pooling adds measurement variability to the usual sampling variability—there is variability in the amount of DNA each subject contributes to a pool and technical variability in measurements of allele frequencies (e.g., reading signal intensities). These extra sources of variability can reduce the efficiency (on a per-subject basis) of pooling approaches to individual genotyping approaches and could make it quite difficult to detect small differences in allele frequencies. Furthermore, not having individual genotypes makes it quite difficult to study continuous phenotypes, gene–gene interactions, gene–environment interactions, or adjust for population stratification using structured association or other techniques that rely on estimates of linkage disequilibrium. It also makes it difficult to study multiple phenotypes in the same set of samples.

V. POWER CALCULATIONS FOR MULTISTAGE DESIGN

The power and cost for a multistage design will depend on a number of factors:

- the total number of available samples
- how those samples are divided among the stages
- coverage of the genome-wide platform used at the first stage
- the number of markers genotyped in the subsequent stage(s)
- per-SNP costs of the technologies used at different stages
- stringency of false positive control (nominal Type I error rate at each locus)
- the size of the genetic effect to be detected (e.g., risk allele frequency and genotypic odds ratios)

In this section, we assume that subjects have already been recruited and phenotyped, and DNA samples have already been collected (although not genotyped). These additional costs can change optimal design parameters (Service *et al.*, 2003).

A. Appropriate significance thresholds for power calculations

Before reviewing power calculations for multistage designs, a few words about false positive control, that is, adjusting for multiple tests. For more details on this topic, see Rice *et al.* (2008) (Chapter 20, this volume). Because genome-wide association studies involve hundreds of thousands of tests, we expect to see many very large test statistics (small *p*-values) just by chance. Using standard thresholds for statistical significance such as $p < 0.05$—or for that matter, even thresholds more than an order of magnitude more stringent like $p < 0.001$—will lead to a large number of "significant" results, the vast majority of which are false positives. There are several ways to think about how to avoid false positive results. One is to control the family-wise error rate (FWER), that is, the probability of making even one mistake: rejecting the null hypothesis that a SNP is not associated with the trait under study even once when it is true. The standard Bonferroni correction (test each SNP at the $\alpha^* = \alpha/m_1$ level, where m_1 is the number of markers tested) controls the FWER, but is known to be conservative when the tests are correlated. Assuming $m_1 \approx 500{,}000$ leads to a Bonferroni-corrected threshold of $\alpha^* = 0.05/500{,}000 = 1 \times 10^{-7}$. Other methods such as estimating and adjusting for the effective number of independent tests, $m_1^* < m_1$ (Gusnato and Dudbridge, 2007; Nyholt, 2004; Patterson *et al.*, 2006), or estimating appropriate significance thresholds through permutation or simulation procedures (Dudbridge, 2006; Dudbridge and Koeleman, 2004; Lin, 2005, 2006) may increase power by accounting for test correlation.

However, because all of the above methods control the FWER—the probability of making even a single false positive mistake—they may be too conservative in principle for genome-wide association studies, where the goal is typically to screen the genome for a few very promising candidates for further study (including replication in other populations, fine mapping, and functional studies), not definitively establish the causality of novel susceptibility loci. This suggests using an alternative framework for false positive control, such as the false discovery rate (FDR) (Benjamini and Hochberg, 1995; Benjamini and Yekutieli, 2001) or false positive report probability (FPRP) (Wacholder *et al.*, 2004). These quantities attempt to balance the risk of false positives and the risk of false negatives—relevant when researchers are willing to follow up a few false leads if that greatly increases their chances of finding a true association. The FDR is defined as the expected proportion of positive tests (tests rejecting the null hypothesis) that are false positives. There are simple algebraic procedures using the ranked observed *p*-values that control the FDR when tests are independent (or "positively correlated") (Sabatti *et al.*, 2003) or regardless of the pattern of correlation among tests (Benjamini and Yekutieli, 2001). The FPRP is equivalent to the ratio of expected number of false positive tests to the expected number of positive tests. It is a Bayesian quantity, so requires investigators to specify the prior probability that a SNP is nonnull and the effect size (or distribution of effect sizes) for truly associated SNPs. The FPRP fixes these parameters (rather than estimating their posterior distribution conditional on observed data (Thomas and Clayton, 2004)), so it may not be ideal for reporting results from genome-wide association studies, as the reported FPRP will depend on readers' assumptions about the number and effect size of associated variants. Recent extensions relaxing prior assumptions on the size of genetic effects may be more appropriate (Wellcome Trust Case Control Consortium, 2007). Still, the FPRP as originally proposed can be useful when deciding on what significance threshold to use designing a study. Newton-Cheh and Hirschhorn (2005) calculate the FPRP for a range of realistic genetic architectures (monogenic, oligogenic and polygenic) and find that significance thresholds of 10^{-6} or 10^{-7} (depending on sample size) are needed to ensure FPRP $< 50\%$. For all the power calculations presented here, we assume a constant significance threshold of 10^{-7}.

All of these procedures—Bonferroni correction, FDR, and FPRP—allow researcher to "upweight" SNPs that they believe are especially promising candidates for association (e.g., nonsynonymous SNPs, SNPs under linkage peaks, SNPs in evolutionary conserved regions) by testing SNPs at different significance thresholds (Genovese *et al.*, 2006; Roeder *et al.*, 2006). To take an extreme example, researchers may very strongly believe that of all the hundreds of thousands of markers they have genotyped only one is biologically plausible, so they will test that SNP at the $\alpha = 0.05$ level while ignoring all others. This procedure controls the Type I error rate, but obviously has no power to detect any locus besides the one they tested. The trade-off between potential power gains and power losses from upweighting based on biological priors is poorly understood—and for large sample sizes (where anticipated effects will be detected with high power even using very stringent significance thresholds) changing significance thresholds may have little impact on power (although see Roeder *et al.*, 2007). For this reason, most initial genome-wide association studies have taken an "agnostic" approach and evaluated all tested variants using the same significance thresholds (Carlson, 2006; Hunter *et al.*, 2007b).

B. Power and cost calculations for multistage designs

Most of the papers on power calculations for multistage designs have focused on case–control studies using unrelated individuals (Kraft, 2006; Kraft *et al.*, in preparation; Satagopan and Elston, 2003; Satagopan *et al.*, 2002, 2004; Skol *et al.*, 2006; van den Oord and Sullivan, 2003; Wang *et al.*, 2006), and for ease of exposition so will we. However, the general framework can be extended to studies of continuous traits, family-based designs, and mixtures of family-based and population-based case–control designs (Joo *et al.*, in press).

The power of a multistage design to detect a particular marker can be calculated as the joint probability

$$\Pr(T_1 > c_1, T_2 > c_2, \ldots, T_K > c_K) \tag{1}$$

where $T_k > 0$ is the test statistic for the putative disease susceptibility allele calculated at stage $k = 1, \ldots, K$ (e.g., the Pearson's chi-squared statistic for the two-by-three cross-tabulation of disease by genotype or the absolute value of the Wald test from a regression of disease on minor allele count), c_k $(k < K)$ is the significance threshold a marker has to cross to be selected for genotyping at stage $k + 1$, and c_K is the threshold a marker has to cross in order to reach genome-wide significance. (The power could also be calculated under the restriction that the risk allele is identical across stages (Skol *et al.*, 2006).) The thresholds c_k are chosen so that under the global null (none of the tested markers are associated with disease risk) m_{k+1} markers are expected to be taken to the $k + 1$st stage (where m_{k+1} is fixed by design), and $\Pr(T_1 > c_1, T_2 > c_2, \ldots, T_K > c_K) < \alpha^*$.

The form of this calculation depends on how researchers plan to analyze the data collected at various stages. Three strategies have been proposed: a replication strategy, where data from each stage are analyzed independently of data at the other stages; a joint strategy, where evidence for association is combined across stages without assuming that the effect (e.g., trend odds ratio or case–control allele frequency differences) is the same across stages (Skol *et al.*, 2006); and a pooled strategy that assumes the effect is constant across stages (Wang *et al.*, 2006). The replication strategy has been shown to be inefficient relative to the joint and pooled analyses in most situations (one exception being a two-stage study where the effect size is greater in the second stage than in the first) (Skol *et al.*, 2006; Thomas *et al.*, 1985). The relative power of the joint versus pooled approach will depend on whether or not the true stage-specific marginal effect sizes differ. If the subjects in the different stages are randomly sampled from one large study, assuming constant effects across stage might be realistic. However, if samples at different stages are drawn from different studies—as often happens in practice, when no single study has the large number of subjects needed to ensure good power—it may not be appropriate to assume a constant effect across stages.

To illustrate, under the replication strategy expression (1) becomes

$$\Pr(T_1 > c_1) \times \Pr(T_2 > c_2) \times \ldots \times \Pr(T_K > c_K),$$

where c_k $(k < K)$ is the $(1 - m_{k+1}/m_k)$th quantile of T_k, c_K is the $(1 - \alpha/m_{K-1})$th quantile, and α is the desired FWER—in other words, all markers with $p < m_2/m_1$ are selected for the second stage, etc., and any marker that is significant in the final stage (after Bonferroni correction *for the number of markers tested in the final stage*) is declared genome-wide significant. Under the null, it can be shown by induction that the probability that a marker is declared genome-wide significant is α/m_1. For example, for two stages, this probability is (m_2/m_1) $(\alpha/m_2) = \alpha/m_1$. Under the alternative, $\Pr(T_k > c_k)$ can be calculated using standard methods (Gordon and Finch, 2005). Power calculations for joint or pooled strategies are more complicated, as the test statistics at different stages will be correlated.

Genotyping costs for a multistage design are simply $\Sigma_k m_k s_k n_k$, where s_k is the per-SNP genotyping costs for the technology used at stage k and n_k is the number of subjects in stage k.

1. Power assuming disease susceptibility loci are perfectly tagged

Table 17.6 illustrates power calculations for different two-stage designs (varying the number of subjects in the second stage n_2 and the number of markers followed up in the second stage m_2) for a study where a total of 3000 cases and 3000 controls are available. These calculations assume the disease susceptibility locus was directly measured (or tagged with $r^2 = 1$) and were performed using the very user-friendly CATS software (http://www.sph.umich.edu/csg/abecasis/CaTS/).

Table 17.6. Sample Two-Stage Power Calculations for Study with Total 3000 Cases and 3000 Controls

Design	n_2	M_2	Relative cost	Power (joint/replication) 100% at $r^2 = 1$	100% at $r^2 = 0.8$	68% at $r^2 = 0.8$
Single-stage	NA	NA	1 (ref)	0.74	0.51	0.35
Two-stage A	1500	25,000	0.61	0.74/0.35	0.51/0.20	0.35/0.14
Two-stage B	1500	1536	0.55	0.72/0.53	0.49/0.33	0.33/0.22
Two-stage C	2250	1536	0.32	0.47/0.45	0.30/0.29	0.20/0.20
Two-stage D	1758	2200	0.47	0.70/0.58	0.47/0.37	0.32/0.25
Two-stage E	1758	1536	0.47	0.68/0.58	0.46/0.37	0.32/0.25

Notes: n_2 and m_2 are number cases (= number of controls) and number of markers, respectively, in the second stage. Power calculated assuming risk allele frequency of 20% with multiplicative odds ratio of 1.3 and 500,000 markers total tested at the 1×10^{-7} level. Costs calculated using per-sample costs in Table 17.4: first stage Illumina HumanHap550; second stage Illumina Infinium (design A) or Illumina GoldenGate 1536 (designs B–E). Cost for design D is an underestimate, assuming 2200 SNPs could be genotyped at the cost of 1536.

R functions for two-stage power calculations are freely available at http://www.mskcc.org/mskcc/html/19904.cfm. Wang *et al.* (2006) present a heuristic for calculating power for a two-stage study using standard single-stage calculations, such as those implemented in Quanto (http://hydra.usc.edu/gxe/). Although much less user-friendly than other programs, the *R* function multipow (http://www.hsph.harvard.edu/faculty/kraft/soft.htm) can calculate power for general multistage studies, includes a tool for calculating power for studies of continuous outcomes, and can incorporate variability in tagging efficiency (c.f. Table 17.5).

Perhaps counter-intuitively, Table 17.6 shows that the replication analysis can have more power when fewer markers are followed up in the second stage (1536 vs 25,000) even though the number of subjects at each stage is unchanged—that is, less data can lead to greater power. This is an artifact of the replication strategy, which sets the second-stage significance threshold of $\alpha^* = \alpha/m_2$. For fixed second-stage sample size, the power of the test T_2 with $\alpha^* = \alpha/25{,}000$ will clearly be less than that of a test with $\alpha^* = \alpha/1536$. For the joint analysis, reducing the number of markers genotyped at the second stage while keeping the number of subjects in each stage fixed does reduce power, although the loss in power is slight: the joint analysis using 1500 cases and controls and 1536 markers at the second stage still achieves 97% of the power of the single-stage design using all 3000 cases and controls. Design "Two-stage D" in Table 17.6 represents the study with the smallest cost that achieves 70% power, assuming the per-SNP genotyping costs for the second stage genotyping technology are 20 times that of the first stage technology. (The "optimize" tab in CATS will find the proportion of samples that should be used in the second stage and the proportion of markers that should be followed up that minimize cost while maintaining a specified power level.) This specific design may not be practical, however, as it may be difficult to genotype the precise number of markers required using the chosen second-stage technology (e.g., the Illumina GoldenGate assay requires SNPs be submitted in batches of 1536 SNPs). The power for the slightly more realistic design in the last row of Table 17.6 has nearly identical power as the optimized design.

Note that the power reported in Table 17.6 is the power to detect a particular susceptibility locus, with a particular minor allele frequency and genetic effect. Considering that there are likely multiple susceptibility loci, and that it is unrealistic to expect a genome-wide association study to detect each locus, it may also be instructive to report the probability that *any* susceptibility locus is detected or the expected number of loci detected. For example, if there are M susceptibility loci, each of which can be detected with power P, then the probability that at least one locus is detected is $1-(1-P)^M$, and the expected number of loci detected is MP.

Figure 17.1 shows the relative costs of the least expensive one- and two-stage studies that achieve 70% power as a function of the ratio of per-SNP genotyping costs at the second stage to those at the first. As the cost of

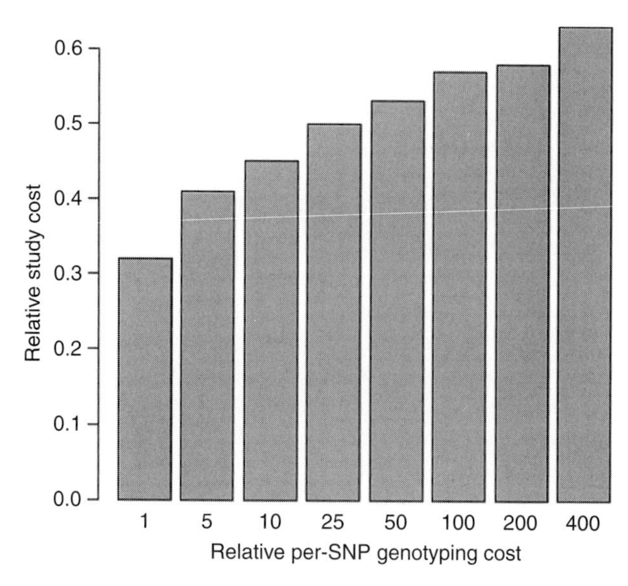

Figure 17.1. Cost of least expensive two-stage design using 3000 cases and 3000 controls total that achieves 70% power to detect a disease susceptibility locus with risk allele frequency 20% and multiplicative odds ratio of 1.3 at 1×10^{-7} level relative to the cost of a single stage study that achieves the same power (2875 cases and 2875 controls).

genome-wide technologies relative to custom technologies drops, the costs savings from a two-stage design also drop. The law of diminishing returns applies when increasing the number of stages in the design: the genotyping cost savings for a three-stage design relative to a two-stage design are smaller than those for a two-stage design relative to a single-stage design, and the cost savings for a four-stage design relative to a three-stage design are smaller yet (Kraft *et al.*, in preparation). Just as with two- versus one-stage designs, the cost savings for higher-stage designs depends on the relative per-SNP genotyping costs for the different technologies used at the different stages; the savings are the greatest when the costs are identical across stages. At some point, the logistical and scientific costs associated with multistage designs will outweigh any genotyping cost savings.

2. Power accounting for imperfect tagging (indirect association)

The calculations presented in the previous section are a best-case scenario, since they assumed the disease susceptibility locus itself or a perfect proxy for the disease susceptibility locus was genotyped. In reality, it is likely the susceptibility locus will be imperfectly correlated with a marker on the genome-wide platform. To estimate the power to detect an association between a SNP marker and disease when the marker and disease susceptibility locus are in linkage disequilibrium (measured by the squared correlation r^2), we can calculate power as if the disease susceptibility locus were directly measured, but with "effective sample size" r^2n instead of n (Pritchard and Przeworski, 2001). This approach has been criticized for oversimplification (Terwilliger and Hiekkalinna, 2006), although it should be appropriate for situations where the causal variant is a single SNP (Thomas and Stram, 2006). This approach may also be an underestimate of the true power to detect an association, because a disease susceptibility locus may be in linkage disequilibrium with multiple SNPs. Alternatively, power can be calculated via simulation, for example, using HapMap samples to estimate the correlation structure between the genotyped markers and the remaining, ungenotyped SNPs (de Bakker *et al.*, 2005).

To allow for uncertainty in the maximum r^2 between the susceptibility locus and markers in the genome-wide platform—as can be seen from Table 17.5 some loci are tagged very well and some are not tagged at all—the power (1) can be calculated by integrating over the distribution of the maximum r^2 (Jorgenson and Witte, 2006):

$$\int \Pr(T_1 > c_1, T_2 > c_2, \ldots, T_K > c_K; R)\, f(R)dR, \qquad (2)$$

where $f(R)$ is estimated using the proportion of loci in the HapMap (or HapMap ENCODE regions) with maximum $r^2 = R$, and Pr(.;R) is calculated using sample sizes $Rn_1, \ldots, Rn_K$ instead of $n_1, \ldots, n_K$. One common simplification of (2) is to assume that all loci are tagged equally well (albeit not perfectly), that is, $f(R^2)$ has a point mass some value r^2, say 80%. This is easily implemented in standard software (e.g., CATS, QUANTO) by performing calculations as usual, except with total available sample size r^2n. For example, the calculations in the 6th column of Table 17.6 were performed exactly as those in column 5, except assuming 2400 total available cases and 2400 controls, instead of 3000 cases and 3000 controls. (Functions in the *R* package dgc.genetics also calculate power given a fixed value for r^2). This approach may be unduly optimistic, however, as there will be a fraction of loci that are not tagged as well as the others. A more conservative approach assumes a certain percentage of SNPs π is tagged equally well, while the remainder are not tagged at all ($r^2 = 0$). This is illustrated in column 7 in Table 17.6, where power was calculated by multiplying column 6 by $(1 - \pi)$. The *R* function multipow generalizes this approach by summing over a user-supplied discrete distribution $f(R)$.

Figure 17.2 further illustrates the impact of different allowances for tagging performance on power calculations. The power calculated assuming all loci are tagged with $r^2 = 80\%$ is less than the power assuming all loci are perfectly tagged, although the difference decreases as the genetic effect size and sample size increase (since power asymptotes to 1 as either parameter increases). The power calculated using the distribution of r^2 in the CEU sample for the Affymetrix GeneChip 500k array (Table 17.5) can be more or less than the power calculated assuming all loci are tagged at $r^2 = 80\%$, but as the sample size increases, the power for a given effect size asymptotes to a value less than 1. This reflects the fact that there is a small fraction of common loci (5%) that are not correlated with any of the genotyped markers.

One final observation from Fig. 17.2: the sample sizes required to detect these modest genetic effects with 80% power at the 1×10^{-7} level range from 2000 cases and 2000 controls to well over 10,000 cases and controls.

Figure 17.3 compares the power of equally expensive single-stage designs using the Affymetrix GeneChip 500k and Illumina HumanHap550 arrays, using the per-subject costs listed in Table 17.4 and the tagging efficiencies from Table 17.5. For realistic budgets at today's prices (less than 10,000 samples can be genotyped using either platform), studies using the Affymetrix array are more powerful, because the reduced per-sample genotyping costs allows more subjects to be genotyped. However, as budget increases (or costs drop making larger studies feasible), the Illumina HumanHap550 platform eventually has greater power, because of its better coverage. This illustrates that power is a function of both coverage and sample size. The trade-off between these two quantities is something researchers have to consider when selecting genome-wide genotyping platforms. For example, genome-wide resequencing, which in principle has perfect coverage, may not be the best choice for genome-wide association studies when it first becomes available, because the large per-sample costs will limit feasible sample sizes.

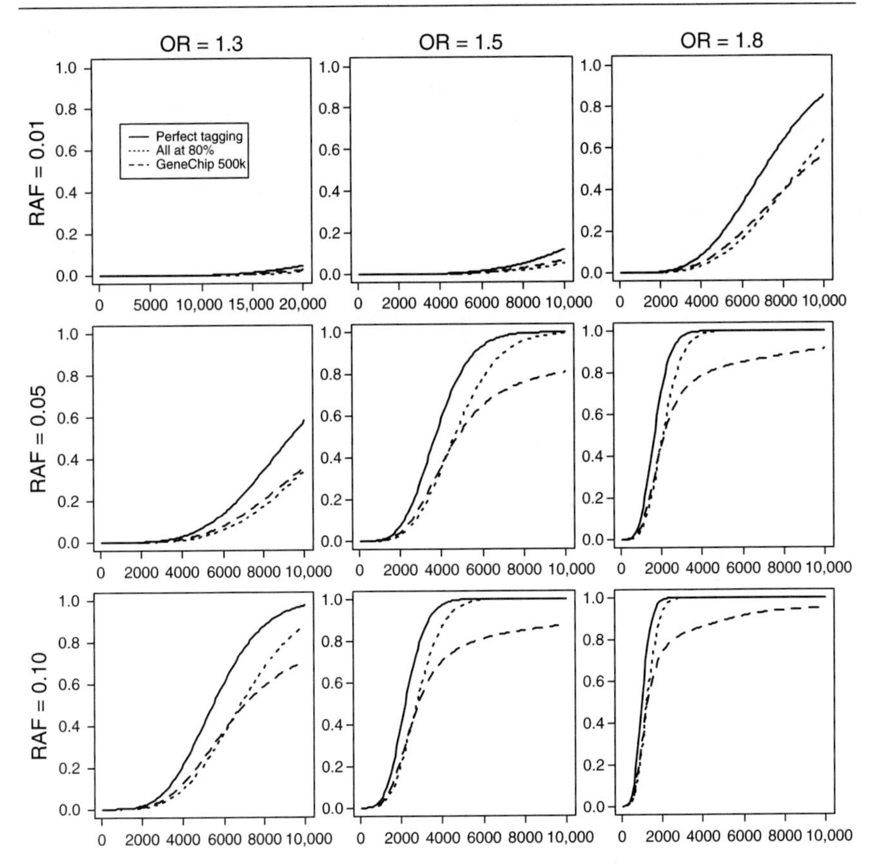

Figure 17.2. Power (*y*-axis) of a single-stage design to detect a disease susceptibility locus at the 1×10^{-7} level across a range of risk allele frequencies and dominant odds ratios, under three different assumptions about tagging efficiency: the disease locus is directly measured; the disease locus has $r^2 = 0.8$ with some marker on the genome-wide panel; and the maximum r^2 between the disease locus and markers on the genome-wide panel is unknown but follows the distribution given in Table 17.5 for the Affymetrix GeneChip 500k (i.e., max $r^2 = 0, 0.2, 0.5, 0.8$, or 1.0 with probabilities 5%, 12%, 15%, 20%, and 48%; respectively). Sample size (*x*-axis) is number of cases (assuming an equivalent number of controls).

VI. DISCUSSION

The previous section focused on choosing design parameters so that the cost to achieve power to detect a particular *marginal* genetic effect is minimized. However, there may be other genetic or environmental factors that modulate the effect of a disease susceptibility locus, such that the marginal effect of that locus

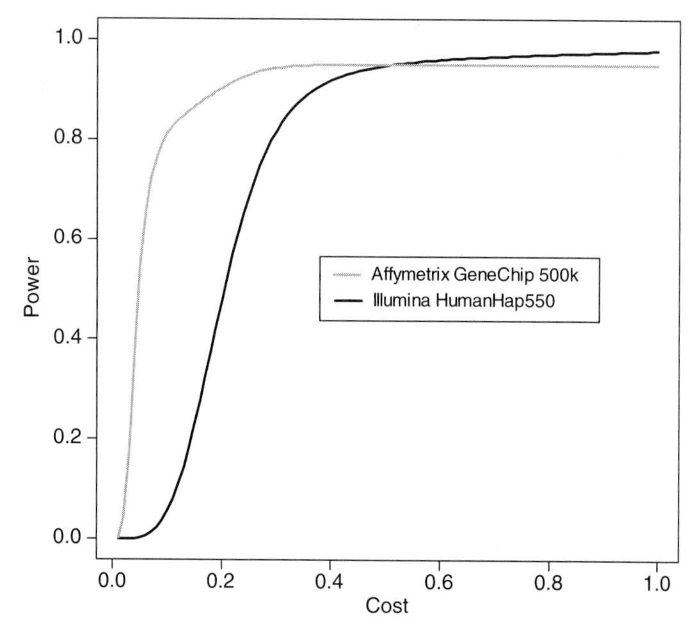

Figure 17.3. Power (y-axis) of single-stage designs to detect a disease susceptibility locus with minor allele frequency 10% and multiplicative odds ratio 1.5 at the 1×10^{-7} level as a function of cost (a function of sample size and per-sample genotyping costs). Cost scale is relative to the cost of single-stage design with 10,000 cases genotyped using the Illumina HumanHap550. All studies have equal numbers of cases and controls.

(the effect in the study sample averaged over other loci and environmental exposures) is quite small. This is one situation where a multistage study designed to detect a particular effect might be "penny-wise but pound foolish" (Hunter *et al.*, 2007b; Savage and Chanock, 2006). Methods that screen the genome for pairs of loci that jointly influence disease risk (Evans *et al.*, 2006; Marchini *et al.*, 2005) or for loci that interact with environmental exposures (Kraft *et al.*, 2007) could be applied to the first-stage genome-wide data, but the sample size judged adequate for detecting marginal effects may not be large enough to detect these more subtle interaction effects with sufficient power. Methods that search for higher-order interactions involving three or more loci—currently restricted due to computational requirements to a smaller subset of genetic variation, such as SNPs in genes with shared function or in a metabolic pathway (see Assimes *et al.*, 2008; Chapter 16, this volume; Ritchie *et al.*, 2001; Kooperberg and Ruczinski, 2005; Thomas, 2005)—will have even less power. This does not mean that loci that interact with other genes or environmental exposures will not be found by screening for marginal effects. Some such loci will have detectable marginal effects (Evans *et al.*, 2006; Kraft *et al.*, 2007; Marchini *et al.*, 2005; Pharoah, 2007).

Furthermore, when the subjects in different stages of a multistage design are drawn from different studies, a multistage approach may exacerbate the problem of effect heterogeneity. A true association may go undetected in the first-stage genome screen because of differences in study design (e.g., case and control recruitment), differences in environmental backgrounds (e.g., exposure prevalence), differences in linkage disequilibrium patterns, or sampling variability ("dumb luck"). If a disease-associated marker does not make it out of the first stage, it cannot be replicated in subsequent stages, and the entire multistage study will fail to detect it. Genome-wide data from multiple studies (drawn from multiple ethnicities) would help detect those loci with small marginal effects that may not achieve statistical significance in a few individual studies due to chance, and it would help to formally assess the existence, extent, and causes of effect heterogeneity. Ideally, investigators interested pooling their resources to study a particular disease would collaborate to choose a common protocol: genome-wide array, exposure definitions, etc. (Hunter *et al.*, 2007). The National Cancer Institute's Cancer Genetic Markers of Susceptibility (CGEMS) and pancreatic cancer genome-wide scan (PanScan) projects are examples of this approach. Alternatively, data from individual scans—ranging from precomputed summary statistics such as allele frequencies and allelic odds ratios to individual-level genotypes and phenotypes—could be released to the research community with appropriate human-subject protections (see Table 17.3).

That said, multistage designs powered to detect modest marginal genetic effects will remain important tools in the near future, as genome-wide genotyping costs (and eventually genome-wide resequencing costs (Harmon, 2007)) will prohibit performing a complete genome-wide scan in all available samples for all common complex diseases. Such studies have already found novel common variants associated with several complex traits (Table 17.1), and it remains to be seen in practice whether there are disease susceptibility loci that can be found when considering possible gene–gene or gene–environment interactions that could not be found when only considering the locus marginally. Many more successes using the multistage approach sketched here are sure to come. Making sense of these associations—fine mapping to find top candidates for the causal variant and discovering the mechanism underlying the genotype–phenotype association (Rohrwasser *et al.*, 2008; Chapter 26, this volume)—and then turning these insights into effective prevention strategies, screening tools, and treatments will keep researchers busy for years to come.

References

Allison, D. (1997). Transmission-disequilibrium tests for quantitative traits. *Am. J. Hum. Genet.* **60**, 676–690.

Altshuler, D., Brooks, L. D., Chakravarti, A., Collins, F. S., Daly, M. J., and Donnelly, P. (2005). A haplotype map of the human genome. *Nature* **437**(7063), 1299–1320.

Ambrosone, C. B., Shields, P. G., Freudenheim, J. L., and Hong, C. C. (2007). Re: Commonly studied single-nucleotide polymorphisms and breast cancer: Results from the Breast Cancer Association Consortium. *J. Natl. Cancer Inst.* **99**(6), 487; author reply 488–489.

Antoniou, A. C., and Easton, D. F. (2003). Polygenic inheritance of breast cancer: Implications for design of association studies. *Genet. Epidemiol.* **25**(3), 190–202.

Antoniou, A., Easton, D., Gayther, S., Stratton, J., and Ponder, B. (1998). Risk models for familial breast and ovarian cancer (abstract). *Genet. Epidemiol.* **15,** 522–523.

Arking, D. E., Pfeufer, A., Post, W., Kao, W. H., Newton-Cheh, C., Ikeda, M., West, K., Kashuk, C., Akyol, M., Perz, S., Jalilzadeh, S., Illing, T., *et al.* (2006). A common genetic variant in the NOS1 regulator NOS1AP modulates cardiac repolarization. *Nat. Genet.* **38**(6), 644–651.

Assimes, T., Olshen, A., and Olshen, R. (2008). Associating genotypes at one polymorphic site with those at others, or with complex disease: Models, algorithms, and applications. *In* "Genetic Dissection of Complex Traits" (D. Rao and C. Gu, eds.). Elsevier, New York.

Balmain, A., Gray, J., and Ponder, B. (2003). The genetics and genomics of cancer. *Nat. Genet.* **33** (Suppl), 238–244.

Barrett, J. C., and Cardon, L. R. (2006). Evaluating coverage of genome-wide association studies. *Nat. Genet.* **38**(6), 659–662.

Benjamini, Y., and Hochberg, Y. (1995). Controlling the false discovery rate—a practical and powerful approach to multiple testing. *J. R. Stat. Soc. B Met.* **57,** 289–300.

Benjamini, Y., and Yekutieli, D. (2001). The control of the false discovery rate in multiple testing under dependency. *Ann. Stat.* **29,** 1165–1188.

Borecki, I., and Province, M. (2008). Linkage and association: Basic concepts. *In* "Genetic Dissection of Complex Traits" (D. Rao and C. Gu, eds.). Elsevier, New York.

Breslow, N. E., and Day, N. E. (1980). "Statistical Methods in Cancer Research: I. The Analysis of Case-Control Studies." IARC Scientific publications, Lyon.

Breslow, N. E., and Day, N. E. (1987). "Statistical Methods in Cancer Research. II. The Design and Analysis of Cohort Studies." IARC Scientific Publications, Lyon.

Campbell, C. D., Ogburn, E. L., Lunetta, K. L., Lyon, H. N., Freedman, M. L., Groop, L. C., Altshuler, D., Ardlie, K. G., and Hirschhorn, J. N. (2005). Demonstrating stratification in a European American population. *Nat. Genet.* **37**(8), 868–872.

Carlson, C. S. (2006). Agnosticism and equity in genome-wide association studies. *Nat. Genet.* **38**(6), 605–606.

Carlson, C. S., Eberle, M. A., Rieder, M. J., Smith, J. D., Kruglyak, L., and Nickerson, D. A. (2003). Additional SNPs and linkage-disequilibrium analyses are necessary for whole-genome association studies in humans. *Nat. Genet.* **33**(4), 518–521.

Chakravarti, A. (1999). Population genetics—making sense out of sequence. *Nat. Genet.* **21**(1 Suppl.), 56–60.

Chanock, S. J., Manolio, T., Boehnke, M., Boerwinkle, E., Hunter, D. J., Thomas, G., Hirschhorn, J. N., Abecasis, G., Altshuler, D., Bailey-Wilson, J. E., Brooks, L. O., Cardon, L. R., *et al.* (2007). Replicating genotype-phenotype associations. *Nature* **447**(7145), 655–660.

Clarimon, J., Scholz, S., Fung, H. C., Hardy, J., Eerola, J., Hellstrom, O., Chen, C. M., Wu, Y. R., Tienari, P. J., and Singleton, A. (2006). Conflicting results regarding the semaphorin gene (SEMA5A) and the risk for Parkinson disease. *Am. J. Hum. Genet.* **78**(6), 1082–1084.

Clayton, D., and McKeigue, P. M. (2001). Epidemiological methods for studying genes and environmental factors in complex diseases. *Lancet* **358**(9290), 1356–1360.

Clayton, D. G., Walker, N. M., Smyth, D. J., Pask, R., Cooper, J. D., Maier, L. M., Smink, L. J., Lam, A. C., Ovington, N. R., Stevens, H. E., Nutland, S., Howson, J. M., *et al.* (2005). Population structure, differential bias and genomic control in a large-scale, case-control association study. *Nat. Genet.* **37**(11), 1243–1246.

Clerget-Darpoux, F., Selinger-Leneman, H., and Babron, M.-C. (2001). Why do complex traits resist DNA analysis? *IJHG* **1,** 55–63.

Collins, F. S., and Manolio, T. A. (2007). Merging and emerging cohorts: Necessary but not sufficient. *Nature* **445**(7125), 259.

Conrad, D. F., Jakobsson, M., Coop, G., Wen, X., Wall, J. D., Rosenberg, N. A., and Pritchard, J. K. (2006). A worldwide survey of haplotype variation and linkage disequilibrium in the human genome. *Nat. Genet.* **38**(11), 1251–1260.

Cordell, H. J., and Clayton, D. G. (2005). Genetic association studies. *Lancet* **366**(9491), 1121–1131.

Cordell, H. J., Barratt, B. J., and Clayton, D. G. (2004). Case/pseudocontrol analysis in genetic association studies: A unified framework for detection of genotype and haplotype associations, gene-gene and gene-environment interactions, and parent-of-origin effects. *Genet. Epidemiol.* **26**(3), 167–185.

de Bakker, P. I., Yelensky, R., Gabriel, I., Pe'er, S. B., Daly, M. J., and Altshuler, D. (2005). Efficiency and power in genetic association studies. *Nat. Genet.* **37**(11), 1217–1223.

de Bakker, P. I., Burtt, N. P., Graham, R. R., Guiducci, C., Yelensky, R., Drake, J. A., Bersaglieri, T., Penney, K. L., Butler, J., Young, S., Onofrio, R. C., Lyon, H. N., *et al.* (2006). Transferability of tag SNPs in genetic association studies in multiple populations. *Nat. Genet.* **38**(11), 1298–1303.

Devlin, B., and Roeder, K. (1999). Genomic control for association studies. *Biometrics* **55,** 997–1004.

Dina, C., Meyre, D., Samson, C., Tichet, J., Marre, M., Jouret, B., Charles, M. A., Balkau, B., and Froguel, P. (2007). Comment on "A common genetic variant is associated with adult and childhood obesity". *Science* **315**(5809), 187; author reply 187.

Dudbridge, F. (2006). A note on permutation tests in multistage association scans. *Am. J. Hum. Genet.* **78**(6), 1094–1095; author reply 196.

Dudbridge, F., and Koeleman, B. P. (2004). Efficient computation of significance levels for multiple associations in large studies of correlated data, including genomewide association studies. *Am. J. Hum. Genet.* **75**(3), 424–435.

Easton, D. F., Pooley, K. A., Dunning, A. M., Pharoah, P. D., Thompson, D., Ballinger, D. G., Struewing, J. P., Morrison, J., Field, H., Luben, R., *et al.* (2007). Genome-wide association study identifies novel breast cancer susceptibility loci. *Nature* **447**(7148), 1087–1093.

Epstein, M. P., Allen, A. S., and Satten, G. A. (2007). A simple and improved correction for population stratification in case-control studies. *Am. J. Hum. Genet.* **80**(5), 921–930.

Evans, D. M., Marchini, J., Morris, A. P., and Cardon, L. R. (2006). Two-stage two-locus models in genome-wide association. *PLoS Genet.* **2**(9), e157.

Farrer, M. J., Haugarvoll, K., Ross, O. A., Stone, J. T., Milkovic, N. M., Cobb, S. A., Whittle, A. J., Lincoln, S. J., Hulihan, M. M., Heckman, M. G., White, L. R., Aasly, J. O., *et al.* (2006). Genomewide association, Parkinson disease, and PARK10. *Am. J. Hum. Genet.* **78**(6), 1084–1088; author reply 1092–4.

Fingerlin, T. E., Boehnke, M., and Abecasis, G. R. (2004). Increasing the power and efficiency of disease-marker case-control association studies through use of allele-sharing information. *Am. J. Hum. Genet.* **74**(3), 432–443.

Fisher, R. (1918). The correlation between relatives on the supposition of Mendelian inheritance. *Trans. R. Soc. Edinburgh* **52,** 399–433.

Fox, C. S., Polak, J. F., Chazaro, I., Cupples, A., Wolf, P. A., D'Agostino, R. A., and O'Donnell, C. J. (2003). Genetic and environmental contributions to atherosclerosis phenotypes in men and women: Heritability of carotid intima-media thickness in the Framingham Heart Study. *Stroke* **34**(2), 397–401.

Frayling, T. M., Timpson, N. J., Weedon, M. N., Zeggini, E., Freathy, R. M., Lindgren, C. M., Perry, J. R., Elliott, K. S., Lango, H., Rayner, N. W., Shields, B., Harries, L. W., *et al.* (2007). A common variant in the FTO gene is associated with body mass index and predisposes to childhood and adult obesity. *Science* **316**(5826), 889–894.

Freedman, M. L., Reich, D., Penney, K. L., McDonald, G. J., Mignault, A. A., Patterson, N., Gabriel, S. B., Topol, E. J., Smoller, J. W., Pato, C. N., Pato, M. T., Petryshen, T. L., *et al.* (2004). Assessing the impact of population stratification on genetic association studies. *Nat. Genet.* **36**(4), 388–393.

Gabriel, S. B., Schaffner, S. F., Nguyen, H., Moore, J. M., Roy, J., Blumenstiel, B., Higgins, J., DeFelice, M., Lochner, A., Faggart, M., Liu-Cordero, S. N., Rotimi, C., *et al.* (2002). The structure of haplotype blocks in the human genome. *Science* **296**(5576), 2225–2229.

Garcia-Closas, M., Wacholder, S., Caporaso, N., and Rothman, N. (2004). Inference issues in cohort and case-control studies of genetic effects and gene-environment interactions. *In* "Human Genome Epidemiology: A Scientific Foundation for Using Genetic Information to Improve Health and Prevent Disease" (M. Khoury, J. Little and W. Burke, eds.). Oxford University Press, Oxford.

Gauderman, W., and Kraft, P. (2002). Family-based case-control studies. *In* "Biostatistical Genetics and Genetic Epidemiology" (R. Elston, J. Olson and L. Palmer, eds.). John Wiley and Sons, New York.

Genovese, C., Roeder, K., and Wasserman, L. (2006). False discovery control with P-value weighting. *Biometrika* **93,** 509–524.

Gillespie, J. (1998). "Population Genetics: A Concise Guide." The Johns Hopkins University Press, Baltimore.

Goldstein, D. B., and Hirschhorn, J. N. (2004). In genetic control of disease, does 'race' matter? *Nat. Genet.* **36**(12), 1243–1244.

Gordon, D., and Finch, S. J. (2005). Factors affecting statistical power in the detection of genetic association. *J. Clin. Invest.* **115**(6), 1408–1418.

Goris, A., Williams-Gray, C. H., Foltynie, T., Compston, D. A., Barker, R. A., and Sawcer, S. J. (2006). No evidence for association with Parkinson disease for 13 single-nucleotide polymorphisms identified by whole-genome association screening. *Am. J. Hum. Genet.* **78**(6), 1088–1090.

Greenland, S. (1999). A unified approach to the analysis of case-distribution (case-only) studies. *Stat. Med.* **18,** 1–15.

Gu, C., and Rao, D. C. (2002). Considerations on study designs using the extreme sibpairs methods under multilocus oligogenic models. *Genetics* **160**(4), 1733–1743.

Gu, C., Yu, K., and Rao, D. (2008). Characterization of LD structures and the Utility of HapMap in genetic association studies. *In* "Genetic Dissection of Complex Traits" (D. Rao and C. Gu, eds.). Academic Press, San Diego.

Gudmundsson, J., Sulem, P., Manolescu, A., Amundadottir, L. T., Gudbjartsson, D., Helgason, A., Rafnar, T., Bergthorsson, J. T., Agnarsson, B. A., Baker, A., Sigurdsson, A., Benediktsdottir, K. R., *et al.* (2007). Genome-wide association study identifies a second prostate cancer susceptibility variant at 8q24. *Nat. Genet.* **39**(5), 631–637.

Gusnato, A., and Dudbridge, F. (2007). Estimating genome-wide significance levels for association European Mathematical Genetics Meetings, Heidelberg.

Hampe, J., Franke, A., Rosenstiel, P., Till, A., Teuber, M., Huse, K., Albrecht, M., Mayr, G., De La Vega, F. M., Briggs, J., Gunther, S., Prescott, N. J., *et al.* (2007). A genome-wide association scan of nonsynonymous SNPs identifies a susceptibility variant for Crohn disease in ATG16L1. *Nat. Genet.* **39**(2), 207–211.

Harmon, A. (2007). 6 billion bits of data about me, me, me! *New York Times*.

Hartl, D., and Clark, A. (2007). "Principles of Population Genetics." Sinauer Associates, Inc., Sunderland, MA.

Herbert, A., Gerry, N. P., McQueen, M. B., Heid, I. M., Pfeufer, A., Illig, T., Wichmann, H. E., Meitinger, T., Hunter, D., Hu, F. B., Colditz, G., Hinney, A., *et al.* (2006). A common genetic variant is associated with adult and childhood obesity. *Science* **312**(5771), 279–283.

Horvath, S., and Baur, M. (2000). Future directions of research in statistical genetics. *Stat. Med.* **19,** 3337–3343.

Hunter, D. J., Riboli, E., Haiman, C. A., Albanes, D., Altshuler, D., Chanock, S. J., Haynes, R. B., Henderson, B. E., Kaaks, R., Stram, D. O., Thomas, G., Thun, M. J., *et al.* (2005). A candidate gene approach to searching for low-penetrance breast and prostate cancer genes. *Nat. Rev. Cancer* **5**(12), 977–985.

Hunter, D. J., Kraft, P., Jacobs, K. B., Cox, D. G., Yeager, M., Hankinson, S. E., Wacholder, S., Wang, Z., Welch, R., Hutchinson, A., Wang, J., Yu, K., *et al.* (2007a). A genome-wide association study identifies alleles in FGFR2 associated with risk of sporadic postmenopausal breast cancer. *Nat. Genet.* **39**(7), 870–874.

Hunter, D. J., Thomas, G., Hoover, R. N., and Chanock, S. J. (2007b). Scanning the horizon: What is the future of genome-wide association studies in accelerating discoveries in cancer etiology and prevention? *Cancer Causes Control* **18**(5), 479–484.

International Human Genome Sequencing Consortium (2004). Finishing the euchromatic sequence of the human genome. *Nature* **431**(7011), 931–945.

Ioannidis, J. P. (2005). Why most published research findings are false. *PLoS Med.* **2**(8), e124.

Ioannidis, J. P., Ntzani, E. E., and Trikalinos, T. A. (2004). 'Racial' differences in genetic effects for complex diseases. *Nat. Genet.* **36**(12), 1312–1318.

Joo, J., Tian, X., Zheng, G., Stylianou, M., Lin, J., and Geller, N. (in press). Joint analysis of case-parent trios and unrelated case-control designs in large scale association studies. *BMC Genet.*

Jorgenson, E., and Witte, J. S. (2006). Coverage and power in genomewide association studies. *Am. J. Hum. Genet.* **78**(5), 884–888.

Klein, R. J., Zeiss, C., Chew, E. Y., Tsai, J. Y., Sackler, R. S., Haynes, C., Henning, A. K., Sangiovanni, J. P., Mane, S. M., Mayne, S. T., Bracken, M. B., Ferris, F. L., *et al.* (2005). Complement factor H polymorphism in age-related macular degeneration. *Science* **308**(5720), 385–389.

Kooperberg, C., and Ruczinski, I. (2005). Identifying interacting SNPs using Monte Carlo logic regression. *Genet. Epidemiol.* **28**(2), 157–170.

Kraft, P. (2006). "Efficient two-stage genome-wide association designs based on false positive report probabilities." *Pac. Symp. Biocomput.* **11,** 523–534.

Kraft, P., Yen, Y. C., Stram, D. O., Morrison, J., and Gauderman, W. J. (2007). Exploiting gene-environment interaction to detect genetic associations. *Hum. Hered.* **63**(2), 111–119.

Kraft, P., Wacholder, S., Chanock, S., Hayes, R., Hunter, D., and Thomas, G. (in preparation). Cost-efficient multi-stage designs for genome-wide association studies.

Kruglyak, L. (1999). Prospects for whole-genome linkage disequilibrium mapping of common disease genes. *Nature Genet.* **22,** 139–144.

Laird, N., and Lange, C. (2008). Family-based methods for linkage and association analysis. *In* "Genetic Dissection of Complex Traits" (D. Rao and C. Gu, eds.). Elsevier, New York.

Lander, E. S., Linton, L. M., Birren, B., Nusbaum, C., Zody, M. C., Baldwin, J., Devon, K., Dewar, K., Doyle, M., FitzHugh, W., Funke, R., Goge, D., *et al.* (2001). Initial sequencing and analysis of the human genome. *Nature* **409**(6822), 860–921.

Langholz, B., Rothman, N., Wacholder, S., and Thomas, D. (1999). Cohort studies for characterizing measured genes. *Monogr. Natl. Cancer Inst.* **26,** 39–42.

Li, Y., and Abecasis, G. (2006). Mach 1.0: Rapid haplotype reconstruction and missing genotype inference. *Am. J. Hum. Genet.* **S79,** 2290.

Li, Y., Rowland, C., Schrodi, S., Laird, W., Tacey, K., Ross, D., Leong, D., Catanese, J., Sninsky, J., and Grupe, A. (2006). A case-control association study of the 12 single-nucleotide polymorphisms implicated in Parkinson disease by a recent genome scan. *Am. J. Hum. Genet.* **78**(6), 1090–1092; author reply 1092–1094.

Libioulle, C., Louis, E., Hansoul, S., Sandor, C., Farnir, F., Franchimont, D., Vermeire, S., Dewit, O., de Vos, M., Dixon, A., Demarche, B., Gut, T., *et al.* (2007). Novel crohn disease locus identified by genome-wide association maps to a gene desert on 5p13.1 and modulates expression of PTGER4. *PLoS Genet.* **3**(4), e58.

Lin, D. Y. (2005). An efficient Monte Carlo approach to assessing statistical significance in genomic studies. *Bioinformatics* **21**(6), 781–787.

Lin, D. Y. (2006). Evaluating statistical significance in two-stage genomewide association studies. *Am. J. Hum. Genet.* **78**(3), 505–509.

Lin, P. I., Vance, J. M., Pericak-Vance, M. A., and Martin, E. R. (2007). No gene is an island: The flip-flop phenomenon. *Am. J. Hum. Genet.* **80**(3), 531–538.

Loos, R. J., Barroso, I., O'Rahilly, S., and Wareham, N. J. (2007). Comment on "A common genetic variant is associated with adult and childhood obesity". *Science* **315**(5809), 187; author reply 187.

Lyon, H. N., Emilsson, V., Hinney, A., Heid, I. M., Lasky-Su, J., Zhu, X., Thorleifsson, G., Gunnarsdottir, S., Walters, G. B., Thorsteinsdottir, U., Kong, A., Gulcher, J., *et al.* (2007). The association of a SNP upstream of INSIG2 with body mass index is reproduced in several but not all cohorts. *PLoS Genet.* **3**(4), e61.

Maraganore, D. M., de Andrade, M., Lesnick, T. G., Strain, K. J., Farrer, M. J., Rocca, W. A., Pant, P. V., Frazer, K. A., Cox, D. R., and Ballinger, D. G. (2005). High-resolution whole-genome association study of Parkinson disease. *Am. J. Hum. Genet.* **77**(5), 685–693.

Marchini, J., Cardon, L. R., Phillips, M. S., and Donnelly, P. (2004). The effects of human population structure on large genetic association studies. *Nat. Genet.* **36**(5), 512–517.

Marchini, J., Donnelly, P., and Cardon, L. R. (2005). Genome-wide strategies for detecting multiple loci that influence complex diseases. *Nat. Genet.* **37**(4), 413–417.

Marchini, J., Howie, B., Myers, S., McVean, G., and Donnelly, P. (2007). A new multipoint method for genome-wide association studies by imputation of genotypes. *Nat. Genet.* **39**(7), 906–913.

Maresso, K., and Brockel, U. (2008). Genotyping platforms for mass throughput genotyping with SNPs, including human genome-wide scans. *In* "Genetic Dissection of Complex Traits" (D. Rao and C. Gu, eds.). Academic Press, San Diego.

McPherson, R., Pertsemlidis, A., Kavaslar, N., Stewart, A., Roberts, R., Cox, D. R., Hinds, D. A., Pennacchio, L. A., Tybjaerg-Hansen, A., Folsom, A. R., Boerwinkle, E., Hobbs, H. H., *et al.* (2007). A common allele on chromosome 9 associated with coronary heart disease. *Science* **316** (5830), 1488–1491.

Missmer, S. A., Eliassen, A. H., Barbieri, R. L., and Hankinson, S. E. (2004). Endogenous estrogen, androgen, and progesterone concentrations and breast cancer risk among postmenopausal women. *J. Natl. Cancer Inst.* **96**(24), 1856–1865.

Mitchell, A. A., Cutler, D. J., and Chakravarti, A. (2003). Undetected genotyping errors cause apparent overtransmission of common alleles in the transmission/disequilibrium test. *Am. J. Hum. Genet.* **72**(3), 598–610.

Myers, R. H. (2006). Considerations for genomewide association studies in Parkinson disease. *Am. J. Hum. Genet.* **78**(6), 1081–1082.

Need, A. C., and Goldstein, D. B. (2006). Genome-wide tagging for everyone. *Nat. Genet.* **38**(11), 1227–1228.

Neel, J. (1962). Diabetes mellitus: A "thrifty" genotype rendered detrimental by "progress"? *Am. J. Hum. Genet.* **14,** 353–362.

Neel, J. V. (1999). The "thrifty genotype" in 1998. *Nutr. Rev.* **57**(5 Pt. 2), S2–S9.

Newton-Cheh, C., and Hirschhorn, J. N. (2005). Genetic association studies of complex traits: Design and analysis issues. *Mutat. Res.* **573**(1–2), 54–69.

Nievergelt, C. M., Libiger, O., and Schork, N. J. (2007). Generalized analysis of molecular variance. *PLoS Genet.* **3**(4), e51.

Nyholt, D. R. (2004). A simple correction for multiple testing for single-nucleotide polymorphisms in linkage disequilibrium with each other. *Am. J. Hum. Genet.* **74**(4), 765–769.

O'Donnell, C. J., Chazaro, I., Wilson, P. W., Fox, C., Hannan, M. T., Kiel, D. P., and Cupples, L. A. (2002). Evidence for heritability of abdominal aortic calcific deposits in the Framingham Heart Study. *Circulation* **106**(3), 337–341.

Patterson, N., Price, A. L., and Reich, D. (2006). Population structure and eigenanalysis. *PLoS Genet.* **2**(12), e190.

Pe'er, I., de Bakker, P. I., Maller, J., Yelensky, R., Altshuler, D., and Daly, M. J. (2006). Evaluating and improving power in whole-genome association studies using fixed marker sets. *Nat. Genet.* **38**(6), 663–667.

Peto, J. (1980). Genetic predisposition to cancer. *In* "Cancer Incidence in Defined Populations" (J. Cairns, J. Lyon, and M. Skolnick, eds.). Cold Spring Harbor Laboratory, Cold Spring Harbor, NY.

Pharoah, P. (2007). Response: Re: Commonly studied single-nucleotide polymorphisms and breast cancer: Results from the breast cancer association consortium. *J. Natl. Cancer Inst.* **99,** 488–489.

Pharoah, P. D., Antoniou, A., Bobrow, M., Zimmern, R. L., Easton, D. F., and Ponder, B. A. (2002). Polygenic susceptibility to breast cancer and implications for prevention. *Nat. Genet.* **31**(1), 33–36.

Prentice, R. L., and Qi, L. (2006). Aspects of the design and analysis of high-dimensional SNP studies for disease risk estimation. *Biostatistics* **7**(3), 339–354.

Price, A. L., Patterson, N. J., Plenge, R. M., Weinblatt, M. E., Shadick, N. A., and Reich, D. (2006). Principal components analysis corrects for stratification in genome-wide association studies. *Nat. Genet.* **38**(8), 904–909.

Pritchard, J. (2001). Are rare variants responsible for susceptibility to complex diseases? *Am. J. Hum. Genet.* **69,** 124–137.

Pritchard, J., and Przeworski, M. (2001). Linkage disequilibrium in humans: Models and data. *Am. J. Hum. Genet.* **69,** 1–14.

Pritchard, J., Stephens, M., and Donnelly, P. (2000a). Inference of population structure using multilocus genotype data. *Genetics* **155,** 945–959.

Pritchard, J., Stephens, M., Rosenberg, N., and Donnelly, P. (2000b). Association mapping in structured populations. *Am. J. Hum. Genet.* **67,** 170–181.

Rao, V. S., van Duijn, C. M., Connor-Lacke, L., Cupples, L. A., Growdon, J. H., and Farrer, L. A. (1994). Multiple etiologies for Alzheimer disease are revealed by segregation analysis. *Am. J. Hum. Genet.* **55**(5), 991–1000.

Reich, D., and Goldstein, D. (2001). Detecting association in a case-control study while correcting for population stratification. *Genet. Epidemiol.* **20,** 4–16.

Reich, D., and Lander, E. (2001). On the allelic spectrum of human disease. *Trends Genet.* **17,** 502–510.

Rice, T. (2008). Familial resemblance and heritability. *In* "Genetic Dissection of Complex Traits" (D. Rao and C. Gu, eds.). Elsevier, New York.

Rice, T., Schork, N., and Rao, D. (2008). Methods for handling multiple testing. *In* "Genetic Dissection of Complex Traits" (D. Rao and C. Gu, eds.). Elsevier, New York.

Rioux, J. D., Xavier, R. J., Taylor, K. D., Silverberg, M. S., Goyette, P., Huett, A., Green, T., Kuballa, P., Barmada, M. M., Datta, L. W., Shugart, Y. Y., Griffiths, A. M., *et al.* (2007). Genome-wide association study identifies new susceptibility loci for Crohn disease and implicates autophagy in disease pathogenesis. *Nat. Genet.* **39**(5), 596–604.

Risch, N., and Merikangas, K. (1996). The future of genetic studies of complex human diseases. *Science* **273,** 1616–1617.

Risch, N., and Teng, J. (1998). The relative power of family-based and case-control designs for linkage disequilibrium studies of complex human diseases, I. DNA pooling. *Genome Res.* **8**(12), 1273–1278.

Ritchie, M. D., Hahn, L. W., Roodi, N., Bailey, L. R., Dupont, W. D., Parl, F. F., and Moore, J. H. (2001). Multifactor-dimensionality reduction reveals high-order interactions among estrogen-metabolism genes in sporadic breast cancer. *Am. J. Hum. Genet.* **69**(1), 138–147.

Robins, J., Rotniztky, A., and Zhao, L. (1994). Estimation of regression coefficients when some regressors are not always observed. *J. Am. Stat. Assoc.* **89,** 846–866.

Roeder, K., Bacanu, S. A., Wasserman, L., and Devlin, B. (2006). Using linkage genome scans to improve power of association in genome scans. *Am. J. Hum. Genet.* **78**(2), 243–252.

Roeder, K., Devlin, B., and Wasserman, L. (2007). Improving power in genome-wide association studies: Weights tip the scale. *Genet. Epidemiol.* **31**(7), 741–747.

Rohrwasser, A., Lott, P., Weiss, R., and Lalouel, J. (2008). From genetics to mechanism of disease liability. *In* "Genetics of Complex Disease" (D. Rao and C. Gu, eds.). Elsevier, New York.

Rosskopf, D., Bornhorst, A., Rimmbach, C., Schwahn, C., Kayser, A., Kruger, A., Tessmann, G., Geissler, I., Kroemer, H. K., and Volzke, H. (2007). Comment on "A common genetic variant is associated with adult and childhood obesity". *Science* **315**(5809), 187.

Rothman, K., and Greenland, S. (1998). "Modern Epidemiology." Lippencott-Raven, Philadelphia.

Sabatti, C., Service, S., and Freimer, N. (2003). False discovery rate in linkage and association genome screens for complex disorders. *Genetics* **164**(2), 829–833.

Sachidanandam, R., Weissman, D., Schmidt, S. C., Kakol, J. M., Stein, L. D., Marth, G., Sherry, S., Mullikin, J. C., Mortimore, B. J., Willey, D. L., Hunt, S. E., Cole, C. G., *et al.* (2001). A map of human genome sequence variation containing 1.42 million single nucleotide polymorphisms. *Nature* **409**(6822), 928–933.

Samani, N., Erdmann, J., Hall, A., Hengstenberg, C., Mangino, M., Mayer, B., Dixon, R., Meitinger, T., Braund, P., Wichmann, H., Barrett, J., König, I., Barrett, J. H., Konig, I. R., *et al.* (2007). Analysis of two genome-wide association studies identifies and validates novel gene loci for myocardial infarction. *N. Engl. J. Med.* **357,** 443–453.

Satagopan, J. M., and Elston, R. C. (2003). Optimal two-stage genotyping in population-based association studies. *Genet. Epidemiol.* **25**(2), 149–157.

Satagopan, J. M., Verbel, D. A., Venkatraman, E. S., Offit, K. E., and Begg, C. B. (2002). Two-stage designs for gene-disease association studies. *Biometrics* **58**(1), 163–170.

Satagopan, J. M., Venkatraman, E. S., and Begg, C. B. (2004). Two-stage designs for gene-disease association studies with sample size constraints. *Biometrics* **60**(3), 589–597.

Satten, G., Flanders, W., and Yang, Q. (2001). Accounting for unmeasured population substructure in case-control studies of genetic association using a novel latent-class model. *Am. J. Hum. Genet.* **68,** 466–477.

Savage, S. A., and Chanock, S. J. (2006). Genetic association studies in cancer: Good, bad or no longer ugly? *Hum. Genomics* **2**(6), 415–421.

Saxena, R., Voight, B. F., Lyssenko, V., Burtt, N. P., de Bakker, P. I., Chen, H., Roix, J. J., Kathiresan, S., Hirschhorn, J. N., Daly, M. J., Hughes, T. E., Groop, L., *et al.* (2007). Genome-wide association analysis identifies loci for type 2 diabetes and triglyceride levels. *Science* **316** (5829), 1331–1336.

Schaid, D., and Sommer, S. (1993). Genotype relative risks: Methods for design and analysis of candidate-gene association studies. *Am. J. Hum. Genet.* **53,** 1114–1126.

Scott, L. J., Mohlke, K. L., Bonnycastle, L. L., Willer, C. J., Li, Y., Duren, W. L., Erdos, M. R., Stringham, H. M., Chines, P. S., Jackson, A. U., *et al.* (2007). A genome-wide association study of type 2 diabetes in Finns detects multiple susceptibility variants. *Science* **316**(5829), 1341–1345.

Self, S., Longton, G., Kopecky, K., and Liang, K. (1991). On estimating HLA/ disease association with application to a study of aplastic anemia. *Biometrics* **47,** 53–61.

Service, S. K., Sandkuijl, L. A., and Freimer, N. B. (2003). Cost-effective designs for linkage disequilibrium mapping of complex traits. *Am. J. Hum. Genet.* **72**(5), 1213–1220.

Setakis, E., Stirnadel, H., and Balding, D. J. (2006). Logistic regression protects against population structure in genetic association studies. *Genome Res.* **16**(2), 290–296.

Sham, P., Bader, J. S., Craig, I., O'Donovan, M., and Owen, M. (2002). DNA pooling: A tool for large-scale association studies. *Nat. Rev. Genet.* **3**(11), 862–871.

Siegmund, K., Whittemore, A., and Thomas, D. (1999). Multistage sampling for disease family registries. *Monogr. Natl. Cancer Inst.* **26,** 43–48.

Skol, A. D., Scott, L. J., Abecasis, G. R., and Boehnke, M. (2006). Joint analysis is more efficient than replication-based analysis for two-stage genome-wide association studies. *Nat. Genet.* **38**(2), 209–213.

Sladek, R., Rocheleau, G., Rung, J., Dina, C., Shen, L., Serre, D., Boutin, P., Vincent, D., Belisle, A., Hadjadj, S., Balkay, B., Heude, B., *et al.* (2007). A genome-wide association study identifies novel risk loci for type 2 diabetes. *Nature* **445**(7130), 881–885.

Spielman, R., McGinnis, R., and Ewens, W. (1993). Transmission test for linkage disequilibrium: The insulin gene region and insulin-dependent diabetes mellitus (IDDM). *Am. J. Hum. Genet.* **52,** 506–516.

Stacey, S. N., Manolescu, A., Sulem, P., Rafnar, T., Gudmundsson, J., Gudjonsson, S. A., Masson, G., Jakobsdottir, M., Thorlacius, S., Helgason, A., Aben, K. K., Strobbe, L. J., *et al.* (2007). Common variants on chromosomes 2q35 and 16q12 confer susceptibility to estrogen receptor-positive breast cancer. *Nat. Genet.* **39**(7), 865–869.

Tang, H., Quertermous, T., Rodriguez, B., Kardia, S. L., Zhu, X., Brown, A., Pankow, J. S., Province, M. A., Hunt, S. C., Boerwinkle, E., Schork, N. J., and Risch, N. J. (2005). Genetic structure, self-identified race/ethnicity, and confounding in case-control association studies. *Am. J. Hum. Genet.* **76**(2), 268–275.

Teng, J., and Risch, N. (1999). The relative power of family-based and case-control designs for linkage disequilibtium studies of complex human diseases. II. Individual genotyping. *Genome Res.* **9,** 234–241.

Terwilliger, J., and Weiss, K. (1998). Linkage disequilibrium mapping of complex disease: Fantasy or reality? *Curr. Opin. Biotechnol.* **9,** 578–594.

Terwilliger, J. D., and Hiekkalinna, T. (2006). An utter refutation of the "Fundamental Theorem of the HapMap". *Eur. J. Hum. Genet.* **14**(4), 426–437.

Thakkinstian, A., Han, P., McEvoy, M., Smith, W., Hoh, J., Magnusson, K., Zhang, K., and Attia, J. (2006). Systematic review and meta-analysis of the association between complement factor H Y402H polymorphisms and age-related macular degeneration. *Hum. Mol. Genet.* **15**(18), 2784–2790.

Thomas, D. (2004a). Statistical issues in the design and analysis of gene-disease association studies. *In* "Human Genome Epidemiology" (M. Khoury, J. Little and W. Burke, eds.). Oxford University Press, New York.

Thomas, D. (2004b). "Statistical Methods in Genetic Epidemiology" Oxford University Press, Oxford.

Thomas, D., and Greenland, S. (1985). The efficiency of matching in case-control studies of risk-factor interactions. *J. Chronic Dis.* **38,** 569–574.

Thomas, D., and Witte, J. (2002). Point: Population stratification: A problem for case-control studies of candidate gene associations? *Cancer Epidemiol. Biomarkers Prev.* **11,** 505–512.

Thomas, D. C. (2005). The need for a systematic approach to complex pathways in molecular epidemiology. *Cancer Epidemiol. Biomarkers Prev.* **14**(3), 557–559.

Thomas, D. C. (2006). Are we ready for genome-wide association studies? *Cancer Epidemiol. Biomark. Prev.* **15**(4), 595–598.

Thomas, D. C., and Clayton, D. G. (2004). Betting odds and genetic associations. *J. Natl. Cancer Inst.* **96**(6), 421–423.

Thomas, D. C., and Conti, D. V. (2004). Commentary: The concept of 'Mendelian Randomization'. *Int. J. Epidemiol.* **33**(1), 21–25.

Thomas, D. C., and Stram, D. O. (2006). An utter refutation of the "Fundamental Theorem of the HapMap" by Terwilliger and Hiekkalinna. *Eur. J. Hum. Genet.* **14**(12), 1238–1239.

Thomas, D. C., Siemiatycki, J., Dewar, R., Robins, J., Goldberg, M., and Armstrong, B. G. (1985). The problem of multiple inference in studies designed to generate hypotheses. *Am. J. Epidemiol.* **122,** 1080–1095.

Thomas, D. C., Haile, R. W., and Duggan, D. (2005). Recent developments in genomewide association scans: A workshop summary and review. *Am. J. Hum. Genet.* **77**(3), 337–345.

van den Oord, E. J., and Sullivan, P. F. (2003). A framework for controlling false discovery rates and minimizing the amount of genotyping in the search for disease mutations. *Hum. Hered.* **56**(4), 188–199.

Van Steen, K., and Lange, C. (2005). PBAT: A comprehensive software package for genome-wide association analysis of complex family-based studies. *Hum. Genomics* **2**(1), 67–69.

Van Steen, K., McQueen, M. B., Herbert, A., Raby, B., Lyon, H., Demeo, D. L., Murphy, A., Su, J., Datta, S., Rosenow, C., *et al.* (2005). Genomic screening and replication using the same data set in family-based association testing. *Nat. Genet.* **37**(7), 683–691.

Voight, B. F., and Pritchard, J. K. (2005). Confounding from cryptic relatedness in case-control association studies. *PLoS Genet* **1**(3), e32.

Wacholder, S., Silverman, D., and McLaughlin, J. (1992). Selection of controls in case-control studies. II. Types of controls. *Am. J. Epidemiol.* **135,** 1029–1041.

Wacholder, S., Rothman, N., and Caporaso, N. (2002). Counterpoint: Bias from population stratification is not a major threat to the validity of conclusions from epidemiological studies of common polymorphisms and cancer. *Cancer Epidemiol. Prev. Biomark.* **11,** 513–520.

Wacholder, S., Chanock, S., Garcia-Closas, M., El Ghormli, L., and Rothman, N. (2004). Assessing the probability that a positive report is false: An approach for molecular epidemiology studies. *J. Natl. Cancer Inst.* **96**(6), 434–442.

Wang, H., Thomas, D. C., Pe'er, I., and Stram, D. O. (2006). Optimal two-stage genotyping designs for genome-wide association scans. *Genet. Epidemiol.* **30**(4), 356–368.

Wang, W. Y., Barratt, B. J., Clayton, D. G., and Todd, J. A. (2005). Genome-wide association studies: Theoretical and practical concerns. *Nat. Rev. Genet.* **6**(2), 109–118.

Weiss, K. (1993). "Genetic Human Variation and Disease." Cambridge University Press, Cambridge.

Weiss, K., and Terwilliger, J. (2000). How many diseases does it take to map a gene with SNPs? *Nat. Genet.* **26,** 151–157.

Witte, J. S. (2007). Multiple prostate cancer risk variants on 8q24. *Nat. Genet.* **39**(5), 579–580.

Witte, J. S., Gauderman, W. J., and Thomas, D. C. (1999). Asymptotic bias and efficiency in case-control studies of candidate genes and gene-environment interactions: Basic family designs. *Am. J. Epidemiol.* **148,** 693–705.

Wojczynski, M., and Tiwari, H. (2008). Definition of phenotype. *In* "Genetic Dissection of Complex Traits" (D. Rao and C. Gu, eds.). Elsevier, New York.

Wellcome Trust Case Control Consortium (2007). Genome-wide association study of 14,000 cases of seven common diseases and 3,000 controls. *Nature* **447**(7145), 661–678.

Yeager, M., Orr, N., Hayes, R. B., Jacobs, K. B., Kraft, P., Wacholder, S., Minichiello, M. J., Fearnhead, P., Yu, K., Chatterjee, N., Wang, Z., Welch, R., *et al.* (2007). Genome-wide association study of prostate cancer identifies a second risk locus at 8q24. *Nat. Genet.* **39**(5), 645–649.

Zeggini, E., Weedon, M. N., Lindgren, C. M., Frayling, T. M., Elliott, K. S., Lango, H., Timpson, N. J., Perry, J. R., Rayner, N. W., Freathy, R. M., Barrett, J. C., Shields, B., *et al.* (2007). Replication of genome-wide association signals in UK samples reveals risk loci for type 2 diabetes. *Science* **316**(5829), 1336–1341.

18

Ethical, Legal, and Social Implications of Biobanks for Genetics Research

Susanne B. Haga and Laura M. Beskow
Institute for Genome Sciences and Policy, Center for Genome Ethics, Law, and Policy, Duke University, Durham, North Carolina 27708

Advances in Genetics, Vol. 60

0065-2660/08 $35.00
DOI: 10.1016/S0065-2660(07)00418-X

ABSTRACT

The elucidation of the causes of complex diseases pivots on understanding the interaction between biological (genetic) and environmental factors that give rise to disease risk. The modest effects of genetic factors in complex diseases supports the need for large-scale studies of high-quality human biological materials, paired with detailed clinical data, to adequately detect these effects. To this end, biobanks or biorepositories have been developed around the world, by public and private entities alike, to provide researchers the opportunity to study collections of human biospecimens annotated with clinical and other health-related measurements. It has been estimated that more than 270 million tissue samples are stored in the U.S., expanding at a rate of ~20 million samples annually. In this chapter, we discuss several ethical, legal, and social issues that have been raised surrounding biobanks, including recruitment of vulnerable populations, informed consent, data disclosure to participants, intellectual property, and privacy and security. Throughout the chapter, we will highlight experiences of national biobanks in Iceland, the U.K., Sweden, and Estonia, and the proposal for a U.S. population cohort study. The dependence on public participation requires clear and transparent policies developed through inclusive processes.

I. INTRODUCTION

The elucidation of the causes of complex diseases pivots on understanding the interaction between biological (genetic) and environmental factors that give rise to disease risk. Genetic association studies have been notoriously difficult to validate due to several reasons such as imprecise or inconsistent measurement or determination of phenotype, difficulties in accurately measuring environmental exposures, population stratification, and the use of different statistical analytical methods (Hirschhorn *et al.*, 2002). In addition, some complex diseases are actually comprised of several sub-types that have recently been elucidated based on genomic expression profiling.

The modest effects of genetic factors in complex diseases supports the need for large-scale studies of high-quality human biological materials, paired with detailed clinical data, to adequately detect these effects (Ioannidis *et al.*, 2006; Lohmueller *et al.*, 2003). To this end, biobanks or biorepositories have been developed around the world, by public and private entities alike, to provide researchers the opportunity to study collections of human biospecimens annotated with clinical and other health-related measurements. As genomic technologies have advanced, the value of these biobanks has increased.

It has been estimated that more than 270 million tissue samples are stored in the U.S., expanding at a rate of ~20 million samples annually (Eiseman and Haga, 1999). Collections of human tissue samples are not new and range from small collections related to a single disease maintained by a single researcher that may be shared with no or a few researchers, to hospital collections housed by pathology departments, to institutional (public or private) or governmental biobanks of specimens and data from the general population for use in as-yet-defined studies on a range of diseases by multiple researchers.

Biobanks vary with respect to the type and number of tissues stored, the extent of clinical data, the identifiability of samples and data, and the permitted uses of the samples and data. As such, no consistent definition of "biobank" has been developed. This poses a major problem for institutions and countries alike that are considering development of legislation in this area (Kaye, 2006; Knoppers, 2005). For the purposes of this chapter, we make the distinction between *databases* that contain only genetic and clinical information and *biobanks* which contain actual samples, with or without accompanying genetic or clinical information.

In this chapter, we discuss several ethical, legal, and social issues that have been raised surrounding biobanks, including recruitment of vulnerable populations, informed consent, data disclosure to participants, intellectual property, and privacy and security (see Table 18.1). Throughout the chapter, we will highlight experiences of national biobanks in Iceland, the U.K., Sweden, and Estonia, and the proposal for a U.S. population cohort study. During this era of heightened privacy concerns, fears of commercialization and exploitation, and increased awareness of biotechnology, it is critical for the development and operation of biobanks to occur in a transparent and ethically sensitive manner (Cambon-Thomsen, 2004).

II. GOVERNANCE OF BIOBANKS

Although biobanks have been in existence for some time, they have not garnered this heightened level of public and ethical scrutiny until recently. This scrutiny is likely a result of the type and extent of information that can be gleaned from

Table 18.1. Ethical, Legal, and Social Issues Related to Biobanks

Societal/individual benefit
Recruitment of vulnerable populations
Informed consent
Privacy/confidentiality
Data access
Intellectual property/benefit-sharing

tissue samples due to the power of genetic and genomic technologies rather than the actual collection of tissue samples. Reviews of existing mechanisms for oversight have found them to be insufficient in providing uniform protections and governance for today's biobanks and large-scale genomics research projects (Bauer *et al.*, 2004; Gibbons *et al.*, 2005; Kaye, 2006; Reymond *et al.*, 2002; Winickoff, 2003). For example, although several sets of international guidelines have been developed regarding various aspects of collection, storage, and use of biological specimens (see Table 18.2), the European Union lacks an overarching regulatory system which creates challenges to the appropriate and widespread use of specimens (Kaye, 2006; Knoppers, 2005). Knoppers (2005) and Maschke (2005) provide useful overviews of the patchwork of international protections regarding the use of medical and research samples. Any legislation addressing biospecimens stored in biobanks would be severely lacking if the data generated from these samples was not also addressed (Kaye, 2006).

A number of international bodies and countries have developed guidelines and a few countries have even passed legislation governing biobanks (e.g., Iceland, Norway, Sweden; see Tables 18.2 and 18.3). In 1999, the U.S. National Bioethics Advisory Commission (NBAC) issued a report on the ethical and appropriate use of human biological materials including 23 policy recommendations. In 2005, the U.S. National Cancer Institute established the Office of Biorepositories and Biospecimen Research (OBBR) to oversee the development of standardized operating protocols, a common database, and uniform access policies for its more than 125 programs. However, coordination among all NIH institutes will be necessary to achieve harmonization of the many publicly supported biorepositories across the U.S.

In 2006, the Council of Ministers of the Council of Europe adopted a recommendation on research on biological materials of human origin. The recommendation makes the distinction between collections of human biological materials and population biobanks, the latter having the following characteristics: (1) population-based; (2) provides biological materials or data derived for multiple research projects; (3) contains human biological materials and associated personal data; and (4) follows standard protocols for the collection and distribution of samples and data. For population biobanks, oversight by an independent group is recommended to protect the welfare of research participants. The recommendation also highlights the potential risk to family members and members of the same population or group. However, the recommendation is only relevant to countries who are signatories to the Convention on Human Rights and Biomedicine; therefore, countries such as the U.K. would not be bound.

In 1998, the Act on a Health Sector Database (No. 139/1998) was passed allowing the Icelandic government to grant a private company a license to create and operate a national health database (Icelandic Health Sector

Table 18.2. International Guidelines Developed Between 1997 and 2007

CIOMS—International Ethical Guidelines for Biomedical Research Involving Human Subjects (2002)
Council of Europe, Committee of Ministers. Recommendation Rec (2006)4 of the Committee of Ministers to member states on research on biological materials of human origin. (Adopted 15 March 2006). Available at https://wcd.coe.int/ViewDoc.jsp?id = 977859&BackColorInternet = 9999CC&BackColorIntranet = FFBB55&BackColorLogged = FFAC75
ESHG—Data Storage and DNA Banking for Biomedical Research: Technical, Social and Ethical Issues (2003). Available at http://www.nature.com/ejhg/journal/v11/n2s/pdf/5201115a.pdf
HUGO—Statement on DNA Sampling Control and Access. (1998). Available at http://www.hugo-international.org/Statement_on_DNA_Sampling.htm
HUGO—Statement on Human Genomic Databases (2002) Available at http://www.hugo-international.org/PDFs/Statement%20on%20Human%20Genomic%20Databases%202002.pdf
HUGO—Statement on Benefit-Sharing (2000) http://www.gene.ucl.ac.uk/hugo/benefit.html
HUGO—Statement on the Patenting of DNA Sequences—In Particular Response to the European Biotechnology Directive (2000) Available at http://www.hugo-international.org/PDFs/Statement%20on%20Patenting%20of%20DNA%20Sequences%202000.pdf
International Society for Biological and Environmental Repositories (ISBER) (US)—Best Practices for Repositories I: Collection, Storage, and Retrieval of Human Biological Materials for Research (2005). Available at http://www.isber.org/Pubs/BestPractices.pdf
OECD—Guidance for the Operation of Biological Resource Centers (Part 4: Human-Derived Material) (2005)
UNESCO—International Declaration on Human Genetic Data (2003). Available at http://portal.unesco.org/en/ev.php-URL_ID = 17720&URL_DO – DO_TOPIC&URL_SECTION = 201.html
WHO—Proposed International Guidelines on Ethical Issues in Medical Genetics and Genetic Services (1998)
WHO—Guideline for Obtaining Informed Consent for the Procurement and Use of Human Tissues, Cells and Fluids in Research (2000)
WHO—Genetic Databases—Assessing the Benefits and the Impact on Human Rights and Patient Rights (2003)
WHO—Genomics and World Health (2002)

Database). In 2000, the company deCode Genetics was granted an exclusive license to develop such a database. Furthermore, in 2000, the Act on Biobanks granted deCode Genetics the authority to establish a genetics database. The Act also specified that a governing board would be established to oversee the biobank's activities. The Swedish Biobanks Health Act of 2002 requires that all donors must be informed about the aims or purposes for which collected samples may be used. In Estonia, in 2000, the Human Genes Research Act established the Estonian Genome Project Foundation (not-for-profit) to manage the Estonian Genome Project. In addition, a Scientific Advisory Board was established by a Supervisory Board to advise on scientific aspects of projects requesting access to biobank as well as an independent Ethics Committee to oversee the project and foundation.

Table 18.3. Regional/National Biobank Legislation and Guidelines Developed in the Between 1997 and 2007

Council of Europe—Proposal for an Instrument on the Use of Archived Human Biological Materials in Biomedical Research (2002)
Council of the European Union, European Parliament (EP)—Directive 2004/23/EC of the European Parliament and of the Council of 31 March 2004 on Setting Standards of Quality and Safety for the Donation, Procurement, Testing, Processing, Preservation, Storage and Distribution of Human Tissues and Cells. Available at http://europa.eu.int/smartapi/cgi/sga_doc?smartapi!celexapi!prod!CELEXnumdoc&lg = en&numdoc = 32004L0023&model = guicheti
Estonia—Human Genes Research Act 2000 (defines gene donors' rights)
Estonia—Personal Data Protection Act of 2003 (addresses informed consent)
EU—Council of Europe's Convention on Human Rights and Biomedicine of 1997 (otherwise known as Biomedicine Convention 1997); amendment adopted in 2005 pertaining to informed consent
EU (Council of Europe)—Recommendation Rec(2006)4 of the Committee of Ministers to member states on research on biological materials of human origin. Available at https://wcd.coe.int/ViewDoc.jsp?id = 977859&BackColorInternet = 9999CC&BackColorIntranet = FFBB55&BackColorLogged = FFAC75
EU Directive 95/46/EC—research exemption clause: informed consent not required for "further processing of already collected data for historical, statistical or scientific purposes as long as there are appropriate safeguards in place" (from Gibbons *et al.*, 2005—exemption "does not refer to medical research explicitly"—uncertain whether genetic epidemiology studies would qualify)
European Group on Ethics in Science and New Technologies (European Commission)—Opinion No.11 on the Ethical Aspects of Human Tissue Banking (1998). Available at http://europa.eu.int/comm/european_group_ethics/docs/avis11_en.pdf
European Parliament—Directive 2006/17/EC of February 2006 on implementing Directive 2004/23/EC of the European Parliament and of the Council as regards certain technical requirements for the donation, procurement and testing of human tissues and cells (2006). Available at http://europa.eu.int/eur-lex/lex/LexUriServ/site/en/oj/2006/l_038/l_03820060209en00400052.pdf
Generation Scotland—Legal and Ethical Aspects (2003) http://129.215.140.49/gs/Documents/GSlawandethicsSep03.pdf
German National Ethics Council—Biobanks for Research (2003)
German National Ethics Council—Opinion (No. 77) Ethical Issues raised by collections of biological material and associated information data: "biobanks," "biolibrairies" (2003)
Iceland Act on Biobanks No. 110/2000 (informed consent varies depending on whether sample collected from clinical care (presumed consent) or specifically donated for storage in biobank (standard informed consent process))
Generation Scotland—Legal and Ethical Aspects (2003) http://129.215.140.49/gs/Documents/GSlawandethicsSep03.pdf
German National Ethics Council—Biobanks for Research (2003)
German National Ethics Council—Opinion (No. 77) Ethical Issues raised by collections of biological material and associated information data: "biobanks," "biolibrairies" (2003)
Iceland Act on Biobanks No. 110/2000 (informed consent varies depending on whether sample collected from clinical care (presumed consent) or specifically donated for storage in biobank (standard informed consent process)
Iceland—Ministry of Health and Social Security—Regulations on the Keeping and Utilization of Biological Samples in biobanks No 134/2001
Iceland—Ministry of Health and Social Security—Regulations on the Keeping and Utilization of Biological Samples in biobanks No 134/2001 (2001)

(*Continued*)

Table 18.3. (*Continued*)

Iceland—Act on a Health Sector Database No. 139/1998. Available at http://ministryofhealth.is/laws-and-regulations/nr/659
Icelandic Act on the Protection of Privacy as Regards the Processing of Personal Data No. 77/2000 (addresses informed consent)
Icelandic National Bioethics Committee—Guidelines Concerning Informed Consent for Participation in Genetic Research Projects or other Projects Utilizing Biological Samples (2005). Available at http://www.visindasidanefnd.is/
India/Government—Guidelines for Exchange of Human Biological Material for Biomedical Research Purposes (1997)
India/Government: Department of Biotechnology—Ethical Policies on the Human Genome, Genetic Research and Services (2001)
Irish Council for Bioethics—Human Biological Material: Recommendations for Collection, Use & Storage in Research 2005. Available at http://www.bioethics.ie/pdfs/BioEthics_fin.pdf
Israel Academy of Sciences and Humanities, Bioethics Advisory Committee—Population-Based Large-Scale Collections of DNA Samples and Databases of Genetic Information (2002)
Italian Society of Human Genetics—Guidelines for Genetic Biobanks (2003)
Latvia—Human Genome Research Law (2002)
Medical Research Council (MRC)—UK biobank Ethics and Governance Framework—Background Document (2003)
Medical Research Council (UK)—Access to Collections of Data and Materials for Health Research (2006). Available at http://www.mrc.ac.uk/pdf-access_report_march_2006.pdf
Medical Research Council (UK)—MRC Operational and Ethical Guidelines: Human Tissue and Biological Samples for use in Research Clarification following passage of the Human Tissue Act 2004. Available at http://www.mrc.ac.uk/pdf-ethics_guide_human_tissue_clarification_april_2005.pdf
National Consultative Ethics Committee for Health and Life Sciences (France)—Opinion (No. 77) Ethical Issues raised by collections of biological material and associated information data: "biobanks," "biolibrairies" (2003)
Norway—Act Relating to biobanks (2003)
Norway—Medical Technology: Health Surveys and Biobanking (2004) http://www.bioethics.ntnu.no/biobanks/documents/Foresightrapport.pdf
Sweden—biobanks in Medical Care Act (2002:297) (2003)
Sweden—Biobanks in Medical Care Act (2002:297) (2003) http://www.sweden.gov.se/content/1/c6/02/31/26/f69e36fd.pdf
Sweden—Swedish Ethical Review Act of 2003
Sweden Biobanks [Health Care] Act 2002
Swedish Medical Research Council—Research Ethics Guidelines for Using Biobanks, Especially Projects Involving Genome Research (1999)
Swedish Medical Research Council—Research Ethics Guidelines for Using biobanks, Especially Projects Involving Genome Research (1999)
Swedish Personal Data Act of 1998
The Danish Council of Ethics—Health Science Information Banks—biobanks (1996)
The Danish Council of Ethics—Health Science Information Banks—biobanks (1996)
U.K. Human Tissue Act 2004

In contrast, the U.K. Biobank was not created by legislation but rather as a biomedical resource by the scientific community. In 2003, the first draft of the Ethics and Governance Framework was released and open for public comment. The development of the governance structure of the U.K. biobank was one of the most public efforts to date, permitting public comment on early drafts. The U.K. Biobank is governed by three bodies: a Board of Directors, a Science Committee, and an Ethics and Governance Council (an independent body funded by the Medical Research Council to ensure that biobank activities comply with the Framework). In addition, the U.K. biobank is governed externally by medical research ethics committees (responsible for the review of all protocols to use the database) and also by the Research Governance Framework for Health and Social Care. If the U.S. moves forward with its own biobank initiative, it will most likely follow a similar path to the U.K. biobank and not be created from legislation.

One advantage of local or national legislation over global guidelines is the ability to address unique cultural values or local norms (Kaye, 2006; Maschke and Murray, 2004; http://www.p3gconsortium.org). However, in the global market of genomic research, some general consensus or harmonization of guidelines would seem to be useful to facilitate the widespread sharing of specimens and data across borders (Blatt, 2000; Reymond *et al.*, 2002). Bauer *et al.* recommended that an international standard would need to address at least four factors: commercialization, confidentiality, informed consent, and quality of research (2004).

III. RISKS AND BENEFITS

The rise in popularity of biobanks has been spurred by genomic technology and the realization that the role of genes in complex disease is likely to be moderate and multifactorial. The opportunity to access large sample sizes with clinical and lifestyle information is expected to greatly advance understanding of the genetic etiology of many common diseases and lead to improved diagnosis, intervention, and treatment. These anticipated findings are expected to primarily benefit the population at large through improved morbidity, mortality, workforce productivity, and decreased health expenditures. On the individual level, it is expected that the practice of health care will shift to personalized care and treatment. In addition, as biobanks have been promoted to benefit the national health of future generations, participants may experience a heightened sense of patriotism or pride in fulfilling a citizen's duty or obligation by participating in a national biobank (Peterson, 2005). Table 18.4 lists several risks and potential benefits related to biobank research. The remainder of this section discusses individual and societal risks linked to biobank research.

Table 18.4. Risks and Potential Benefits Related to Biobank Research

Benefits
Improved public health
Development of new diagnostics and therapies
Movement toward individualized medicine
Risks (from Eriksson and Helgesson, 2005)
Direct physical harm (minimal or null depending on how sample is obtained)
Nonphysical harms (discrimination, stigmatization, psychological impacts)
Group harms
Moral harms (not being treated with dignity, respect for personal integrity)

A. Individual risks and potential benefits

In their report on human biological materials, NBAC found that research on human tissues "was likely to be considered of minimal risk because much of it focuses on research that is not clinically relevant to the sample source" (1999; p. 67). In contrast to mainly physical harms that may be incurred from a clinical trial, participation in biobank research often raises concerns of "informational" harms such as loss of privacy and discrimination, stigmatization, or other psychosocial risks. These risks are often difficult to quantitate, predict, and/or resolve, and may require different solutions to minimize their occurrence.

Since participation in biobanks involves substantially less risk than other types of human subjects research, it has been proposed that the oversight of this type of research should differ accordingly (Eriksson and Helgesson, 2005). Indeed, in response to the expanding use of stored human tissues in research, in 2004, the U.S. Office of Human Research Protections (OHRP) issued a guidance document on research involving coded private information or biological specimens. The guidance document aimed to clarify the distinction between more traditional human subjects research and research involving coded private information or specimens. In particular, OHRP does not consider as "human subjects research" any studies involving only private information or specimens that were not collected for the specific research study through any direct interaction with research participants and if the investigator "cannot readily ascertain the identity of the individual" of the private information or specimen. Specifically, the protection of the individual's identity may be achieved through destruction of the key or code, an agreement between the investigators and the holder of the key prohibiting release of the key under any circumstances, or an established written policy prohibiting release of the key to investigators.

Many biobanks provide an option for participants to withdraw from the biobank. However, there has been some debate on the extent and effectiveness of this option with respect to eliminating risk to the individual or group associated with the sample. Withdrawal policies may offer a range of options from anonymization to destruction of samples. For example, withdrawal by participants in the U.K. Biobank ranges from "no further contact" (but allowing continued use of information and samples, and further information to be obtained from records) and "no further use" (requiring destruction of all of the participant's information and samples) (U.K. Biobank). Similarly, the Swedish Biobanks in Medical Care Act states that upon request of withdrawal, the sample may be either anonymized or destroyed (2003). The UNESCO International Declaration on Human Genetics Data states that "when a person withdraws consent, the person's genetic data, proteomic data, and biological samples should no longer be used unless they are irretrievably unlinked to the person concerned" (2003). In contrast, for the CARTaGENE project, upon a request to withdraw, any data derived from a participant's sample but not yet analyzed with other datapoints will be withdrawn and the actual sample itself will be destroyed (2006).

The argument supporting the use of anonymization of the sample is that this action would be equivalent to destruction of the sample since any risk of harm would be eliminated with anonymization (Eriksson and Helgesson, 2005). However, some have argued that anonymization is disrespectful to participants and that the potential to re-establish the identity of the sample still exists and therefore, risk is not completely eliminated (Eriksson and Helgesson, 2005; National Bioethics Advisory Commission (NBAC), 1999). Furthermore, if general descriptive data accompanies the sample, group harms may still be possible.

B. Societal risks and benefits

Participation in population biobanks does not typically result in direct individual benefit, but more population benefit (Cambon-Thomsen, 2004). For example, the draft informed consent document for the CARTaGENE project states that the project "will benefit the Quebec population as a whole, as well as, future generations. It will improve our understanding of the genetic and non-genetic factors that affect the population's health" (2006). In a National Research Council report, the Committee on Genome Diversity concluded that "in population studies, benefit to the population has become one of the critical issues in determining the ethical justification for the study itself, and sharing benefits with the population is critical in preventing exploitation" (1997).

There is some debate about whether biobanks serve more the needs of the research community than the public (Gibbons *et al.*, 2005). In general, participants or groups do not have ownership in the sample or data and are not entitled to any financial gain from their participation in a biobank. However, this is still a lot of debate surrounding this issue. The concept of benefit-sharing arose in the early 1990s with the Convention on Biological Diversity (1992). Although not written for human genetic research per se, Article 15 of the convention is dedicated to access to genetic resources. Specifically, it states that agreement should be reached regarding the "sharing in a fair and equitable way the results of research and development and the benefits arising from the commercial and other utilization of genetic resources." Benefit-sharing would demonstrate respect for participants and community, engendering trust and support between donors and users of samples stored in biobanks. However, among the challenges of determining how best to distribute benefits obtained from genetic research of indigenous communities is the ability to define the boundaries of a community and ownership of indigenous knowledge (Schuklenk and Kleinsmidt, 2006).

If biobanks are intended to benefit the population at large, the question is raised whether *all* groups within a population will benefit. For example, will all diseases and populations benefit from data generated from resources like the HapMap project which was derived from individuals of only four populations (Foster and Sharp, 2005)? The Common Disease/Common Variant hypothesis, which serves as the premise for the HapMap project, suggests that a few common variations account for most common diseases across many populations. However, this hypothesis is not without its opponents who suggest that common diseases are more likely due to combinations of rare variants (Pritchard, 2001; Weiss and Clark, 2002). Furthermore, it is unclear whether data collected from one population can be applied to another; in other words, will the same variations have similar functional significance (Foster and Sharp, 2005)? These issues do not begin to address the issue of the uneven representation of world populations in biobank research, whether by choice (e.g., fear of group harms) or absence of opportunities to participate.

While group benefit may be one intended goal of biobank research, the potential risk of group harms looms large.

In particular, individuals identified by race or ethnicity, religious affiliation, disease type, or age—characteristics that could be easily ascertained or associated with samples or data—may be at greater risk of harm, even to members that do not participate in biobank research (Weijer and Emanuel, 2000). Even if samples are anonymous or anonymized, such general descriptive characteristics attached to the sample could increase the risk of group harms. The risk of group harm may extend from the conception of a research study to the publication and dissemination of research findings and therefore, efforts to minimize these risks should be considered throughout the research endeavor (Sharp and Foster, 2000; Weijer and Emanuel, 2000; Weijer *et al.*, 1999). Although community consultation and interactions are important in light of potential group harms, individual consent will likely remain the major requirement for research participation (see next section).

IV. RECRUITMENT OF VULNERABLE AND MINORITY POPULATIONS

Prior to the completion of the Human Genome Project, a project was proposed to survey the world's "human genetic diversity" (Cavalli-Sforza *et al.*, 1991). Known as the Human Genome Diversity Project, the goal of the project was to understand human population history through analysis of genetic variation of human specimens of anthropological interest. However, the proposal was besieged with concerns regarding the study of underserved or marginalized populations and the potential exploitation of these groups, the debate of individual versus group consent, the commercialization of genetic material, and the challenge of defining which groups people belonged to (no bright line) (Indigenous Peoples Council on Biocolonialism, 1997). As a result of these concerns and the highly publicized criticisms, plans for the project were halted, although work on the project has since commenced (Cavalli-Sforza, 2005).

Ten years later, the International HapMap Project was successful at sampling four populations for the purposes of creating a haplotype map to help identify disease genes. The DNA samples for the HapMap were collected from 270 people—90 Yorubans (Nigeria), 45 Japanese, 45 Chinese, and 90 Americans (northern and western European ancestry). In contrast to the approach taken with the Human Genome Diversity Project, the development and planning process of the HapMap Project was inclusive, seeking advice and input from bioethicists, social scientists, and community representatives from the project's conception (The International HapMap Consortium, 2004).

Many of the national biobanks have been established in countries predominantly of European or Asian ancestry. Attempts to establish a genetic database of the people of Tonga, an isolated Polynesian community, failed due to public opposition (Burton, 2002). As a result, the value of data gathered from specimens stored in current biobanks may be of limited benefit to persons of other ancestries, cultures, and lifestyles (UNESCO, 2003). A major concern identified in a survey of Swedes participating in a local biobank was that all population groups should have equal access to research results (Hoeyer *et al.*, 2005).

One unique biobank that aims to recruit only individuals of African-American ancestry is Howard University's Genomic Research in the African Diaspora (GRAD) Biobank. Initiated in 2003, the GRAD Biobank will collect tissue samples from 25,000 African-Americans over the next five years (Kaiser, 2003). This will represent the single and largest repository of samples from the African-American population. Samples and data will be accessible to Howard

University scientists and approved third parties such as drug companies. In this instance, the issue of trust is somewhat ameliorated since Howard University is a historically black university.

In many developing nations, the threat of biopiracy of indigenous populations is a major concern II. These small and typically isolated populations are of interest to geneticists given the higher degree of homogeneity of the population and/or a high prevalence of a particular condition. Several studies have been published on the genetic analysis of small island populations (Abbott *et al.*, 2006; Bonnen *et al.*, 2006; Gonzalez-Perez *et al.*, 2006) and native Indian tribes such as the Pima (Muller *et al.*, 2005) and Cherokee Indians (Stoddart *et al.*, 2002).

As much of human subjects protections currently focus on the individual participant, the shift toward the study of groups or whole populations presents different challenges that may warrant a modified approach (World Health Organization (WHO), 2003). As such, it was—and still is to some degree—unclear how investigators can effectively work with communities to address these issues. NBAC and others have recommended that investigators work with community representatives to minimize potential group harms that may incur as a result of participation in a research study (1999; Foster *et al.*, 1997). Most of the policies developed regarding community participation in biomedical research have dealt with Aboriginal and native groups (Aboriginal Health Research Ethics Committee of South Australia, 1998; Association of Canadian Universities for Northern Studies, 1982). In general, approaches to community consultation must be tailored to the specific group since a single approach is unlikely to address the cultural norms and beliefs of different groups (Foster and Sharp, 2000).

Much debate has ensued about the benefits of and approaches toward community consultation (Juengst, 1998; Reilly, 1998). Juengst (1998) recommends that participants be made aware of potential group harms that may arise from participation in population genetics research during the informed consent process (analogous to considering the family unit when deciding whether to participate in a clinical genetics study). However, he is not supportive of the practice of community consultation, arguing that rather than demonstrating respect for communities, the practice requires researchers to predefine group boundaries or reify socially defined boundaries in order to conduct community engagement. Furthermore, the practice may provide a false sense of security since true "group" consent is not possible for many groups (Juengst, 1998). Community engagement has been shown to provide useful information regarding community risks that could not be addressed through the traditional informed consent process on an individual basis (Foster *et al.*, 1998). However, public "representatives" did not appear to provide the most insight regarding community concerns or how best to address them.

V. INFORMED CONSENT

In the context of a biobank, questions surrounding informed consent arise when specimens and data are collected for research purposes; when they are collected for other purposes (e.g., clinical reasons); and when stored specimens and data are used for research (Shickle, 2006). These questions can be grouped into two broad categories: (1) when must informed consent be obtained and when can the requirement to obtain consent be waived and (2) when informed consent must be obtained, what information and options should be communicated during the consent process. With regard to the first question, federal regulations for the protection of human research subjects define a "human subject" as a living individual about whom an investigator obtains (1) data through intervention or interaction with the individual or (2) identifiable private information (45CFR46.102(f)). Because the study of biospecimens can both involve and generate private information about the persons from whom they were obtained, legal requirements to protect human subjects apply to a broader range of research than many investigators may realize. A detailed discussion of whether and how these requirements apply to research using biospecimens and data is beyond the scope of this chapter. However, several resources are available to assist researchers and institutional review boards (IRBs) in determining whether an activity using biospecimens constitutes research involving human subjects, whether it must be reviewed by an IRB, and whether informed consent must be obtained (NIH, undated; U.S. Department of Health and Human Services, Office for Human Research Protections, 2004; U.S. Department of Health and Human Services, Office for Protection from Research Risks, 1997).

With regard to the second question, U.S. federal regulations set forth general categories of information that must be provided to any prospective research participant. These include an explanation of the purposes of the research and the expected duration of the subject's participation; a description of any reasonably foreseeable risks to the subject, as well as any potential benefits to the subject or to others; a description of the extent to which confidentiality will be maintained; contact information for answers to questions about the research and research subjects' rights; and a statement that participation is voluntary and that the subject may withdraw at any time (45CFR46.116(a)). Applying these basic elements in the context of a biobank, however, presents significant challenges. Biobanks frequently involve a separation between sample collection (and thus the informed consent process) and the actual research on the sample; the research may involve hypotheses and methods that could not have been contemplated at the time of the sample collection; and the risks and potential benefits are often group-based (Rothstein, 2005). As Rothstein (2005) notes, "If informed consent is based on promoting autonomy and the rights of individuals, what ethical paradigm should govern the fundamentally social enterprise of biobank research?" [p. 91].

Thus, there is continued debate about the specific content and wording of biobank informed consent documents and processes. In addition to the information required by federal regulations, Dressler (2005) suggested the following elements for the collection of new biospecimens: (a) description of participants' rights and ownership, including financial gain or sharing of intellectual property; (b) description of the intent of the biobank for use of specimens now and in the future; (c) description of the intent of the biobank for re-contact for follow-up/participation in future research studies (e.g., epidemiologic questionnaires; family studies); and (e) description of if, when, how, and under what circumstances research data (individual and/or aggregate results) will be disclosed to the participant/family/physician.

A. Approaches to biobank consents

When human biological materials are collected—whether in a research or clinical setting—it is appropriate to ask individuals for their consent regarding the future use of their specimens (National Bioethics Advisory Commission (NBAC), 1999; see Fig. 18.1). A major ethical problem, however, is how to ensure meaningful consent when the exact research that will be conducted is unknown (Elger and Caplan, 2006; Shickle, 2006). It can be difficult (if not impossible) to anticipate all of the studies that will emerge, and attempting to describe future research in detail may be confusing rather than helpful as well as administratively burdensome (National Bioethics Advisory Commission (NBAC), 1999).

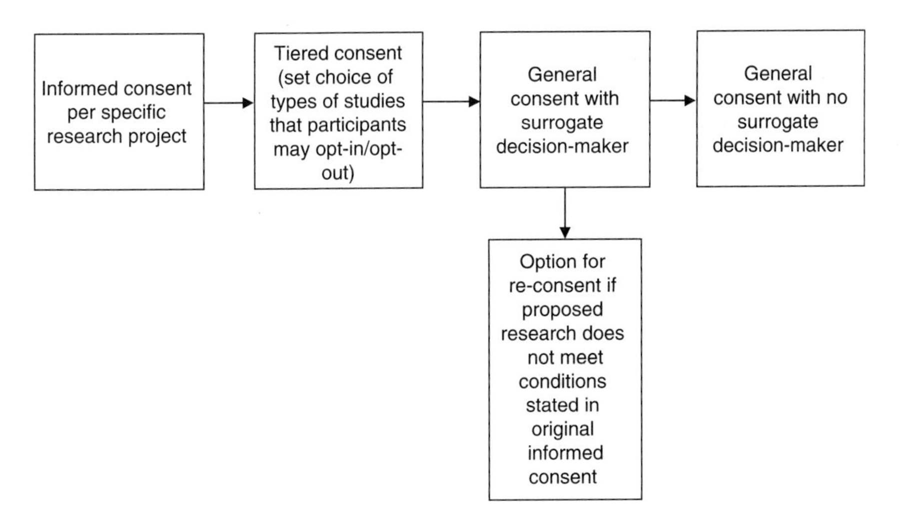

Figure 18.1. Approaches to participant consent for biobank research.

B. Tiered consent

One common approach, known as "tiered consent," is to offer participants choices about the future use of their specimens. The aim is to provide a number of options sufficient to help participants understand clearly the nature of the decision they are being asked to make (National Bioethics Advisory Commission (NBAC), 1999). Such options might include one or more of the following (Wolf and Lo, 2004):

- Permitting use of the specimen in the current study only (if collected in the context of a specific study)
- Permitting use of the specimen in the current study only, but allowing contact to request consent for future studies
- Permitting identified versus unidentified future use
- Limiting future use to certain types of research (e.g., related to a specific disease or conducted by a certain category of researcher)
- Allowing commercial use of specimens

Tiered consent has been considered by many to be a "best practice" (American Society of Human Genetics (ASHG), 1996; Clayton *et al.*, 1995; Dressler, 2005; Eiseman *et al.*, 2003; National Bioethics Advisory Commission (NBAC), 1999; National Cancer Institute, 2006; Wolf and Lo, 2004). It is believed to respect and enhance autonomy by allowing participants more control over whether, how, and with whom their specimens and data are shared (McGuire and Gibbs, 2006b). However, tiered consent has also been criticized as unwieldy or unnecessary, creating an unreasonable burden for the participant and the investigator (Grizzle *et al.*, 1999). Allowing participants to opt in or out of certain types of research requires biobanks to track the uses approved by each participant and tailor the research performed on each specimen accordingly (Shickle, 2006). This could undermine the scientific value of the biobank and greatly increase research costs (Wendler, 2006a). In addition, attempts to compartmentalize research by disease are hampered by the fact that a single gene can be associated with multiple and diverse phenotypic responses (Rothstein, 2005). Shickle (2006) further argues that tiered consent is not ethically required: "Providing that there is proper disclosure [. . .] the choice for the individual is to participate on the terms offered or not. There is a 'negative right' not to be included in the research without consent [. . .]. There is no 'positive right' for a biobank to be run in such a way just because an individual would like it to be so" [p. 516].

When a tiered consent is used, the major challenge is to balance respect for autonomy by maximizing choice with the need to ensure adequate understanding and to avoid the burdens of too many options (McGuire and Gibbs, 2006b). The options that are offered should be empirically informed and reflect potential participants' informational needs and preferred level of control over the decision-making process (McGuire and Gibbs, 2006b).

C. General consent

An alternative to tiered consent is a one-time general consent. Under this approach, consent processes are designed to describe the biobank's practices with regard to future use and then ask prospective participants whether they wish to contribute to the biobank under those conditions. If participants have concerns about the potential uses of their specimen or want certainty as a condition of their consent, they can simply decline to participate or withdraw from the study (Shickle, 2006). UNESCO's Declaration on Human Genetic Data (2003) indicates that a renewed consent on previously collected samples is not necessary if the proposed research conforms to the conditions set forth in the original consent. WHO (2003) also noted that blanket consents were acceptable as re-consent would be impractical and costly and could negatively impact research but only for samples that are anonymized. However, there is debate about the "informed" nature of the general consent process (Deschenes *et al.*, 2001), with some referring to general consent as "permission" rather than informed consent since the research to which the biospecimen will be used is not described in any detail (Greely, 1999).

In a review of the literature on individuals' views on consent for research with human biological materials, most of the data indicated that people are willing to provide samples but wanted some level of control on whether their specimens were used for research (i.e., they wanted to be asked) (Wendler, 2006a). In addition, a substantial majority were willing to provide one-time general consent and rely on ethics committees to review and approve future projects. This finding is consistent with the conclusions of others that general consent is considered ethically acceptable if at least two conditions are fulfilled: the approval of future projects by a research ethics committee (or competent body) and the participants' right to withdraw at any time (Elger and Caplan, 2006; Hansson *et al.*, 2006).

The use of "surrogate" decision-makers through ethics committees directly intersects with the issues pertaining to informed consent and the notion that the consent process is more an act of providing permission than truly "informed" consent—permitting others to make decisions on the participant's behalf under some defined general boundaries. Several national biobanks use surrogate decision-makers as an essential part of their governance framework. This mechanism serves to engender trust on the part of participants that research using their samples will be reviewed by a third party and not left to the discretion of a researcher alone.

The consistency of the data suggesting participants' support for one-time general consent led Wendler (2006a) to conclude that it should be the default approach, modified only in compelling cases. When one-time general consent is used, each future project should be assessed as to whether a reasonable individual would agree that it was covered by the information provided to potential subjects when initially asked to consent (National Bioethics Advisory Commission (NBAC), 1999; Shickle, 2006). In addition, special attention should be given research on topics that are considered sensitive or potentially objectionable to some. In other words, even when participants have given broad consent to future unspecified research, they should be afforded the additional protection provided by requiring specific consent to uses of their samples that might be considered sensitive or objectionable (e.g., "certain behavioral genetics protocols, studies differentiating traits among ethnic or racial groups, or research on stigmatizing characteristics such as addictive behavior" (National Bioethics Advisory Commission (NBAC), 1999, p. 66). A review of guidance from IRBs of top-funded NIH institutions found that 70% do not mention that re-consent may be warranted for research with new risks (Wolf and Lo, 2004).

In conclusion, the ethical requirements to respect autonomy and protect individuals from harm are consistent with several approaches to informed consent for research using biological specimens (Wendler, 2006a). Both tiered and general consent move away from the traditional model of informed consent based on a detailed description of the risks and benefits of an individual study. Therefore, it is important that both approaches be bolstered by an overarching governance framework for the biobank that promotes trust, responsibility, and accountability (Caulfield *et al.*, 2003).

VI. STORAGE OF GENETIC INFORMATION—OPTIMIZING PRIVACY

Privacy concerns have been a major issue with respect to biobanks. The risk of discrimination and stigmatization, access to samples, and dissemination of results dominate many of these concerns. Different approaches may be used to maintain the privacy of the samples as well as data obtained from or accompanying each sample. However, a combination of approaches will likely be required to address participant concerns and secure public trust.

The original holders or collectors of the samples may have the most access to the personal information stored with each sample, whereas other users of the samples may have varying degrees of access to this information, if any access at all. Therefore, the level of identity attached to each sample can vary and should be disclosed to participants during the informed consent process. In 1999, NBAC defined four types of research samples with respect to the level of identification: identified, unidentified (e.g., anonymous), unlinked (e.g., anonymized), and coded (e.g., linked or identifiable). While the "safest" of the four types of research samples is unidentified, this type of sample has the least value to researchers (Anderlik, 2003). But since even unidentified individual genotype or haplotype data may be linkable to individuals (Lin *et al.*, 2004; McGuire and Gibbs, 2006a), anonymization may not completely address concerns of privacy. In addition, the risk of group harm remains if any descriptive information is attached to the sample.

One approach to protecting privacy of participants is to limit access to specimens and/or data to approved users (discussed in next section). For some biobanks, as a condition of use of biospecimens, users must agree to abide by confidentiality and privacy provisions as set forth in a Materials Transfer Agreement or other contract. A second approach is to utilize privacy-enhancing techniques. As the translation of genomic medicine is codependent on information technology, various methods of privacy enhancement are available including "hard" de-identification, anonymization or pseudonymization, privacy risk assessment techniques, controlled database alteration, data flow segmentation, and privacy-enhancing intelligent software (Claerhout and DeMoor, 2005; De Moor *et al.*, 2003). As different databases will have different levels of risk—genetic sequence databases (no phenotypic data, e.g., GenBank) are less risky than genomic databases for medical/pharmacological research are less risky than clinical databases (direct care)—an evaluation of the most appropriate privacy-enhancing technique is required. For example, pseudonymization may be particularly useful for large research datasets where identifying and nonidentifying data are split and the identifiers are translated to pseudo-identifiers through various cryptographic techniques (generally referred to as coding) (Claerhout and DeMoor, 2005). The data may still be linked through pseudo-identifiers even if collected at different times or places. If the collectors are also the users of the biospecimens, the use of a trusted third party could also help increase the success of privacy-enhancing techniques, particularly if mistrust exists between donors, collectors, and users.

Some individuals have raised concerns about the impact of the research requirements under the Health Insurance Portability and Accountability Act (HIPAA; Public Law 104–191) and in particular, whether the general consent process is in compliance with HIPAA (Allen, 2004; Wendler, 2006b). Implemented through the Privacy Rule, HIPAA protections of individual health information apply "to research with biological samples that are associated with protected health information" (18 specified individual identifiers). However, since HIPAA "requires investigators to obtain study-specific authorization to use individuals' biological samples and associated personal information in research," it was initially unclear whether a general informed consent would be in compliance (Wendler, 2006b). After review of the regulation and the general consent process, Wendler (2006b) concluded that a general consent for the use of biobank samples is compliant with HIPAA if a two-step decision process is used, whereby an IRB assesses whether new research study is covered by individual's consent and, if so, grants a waiver of consent. In order to qualify for a waived consent, investigators must meet the following criteria: (1) the use of protected health information poses no more than minimal risk; (2) the research could not be practically conducted without a waiver; and (3) the research could not practicably be conducted without access to private health information.

Although protections provided under HIPAA limits the degree of protected health information that may be stored and shared without participant consent, other information such as diagnostic and prognostic information without identifiers may be sharable and may be cause for concern if participant identity can be discerned from the genetic sequence data or sample.

VII. BIOSPECIMEN/DATA ACCESS

Publicly accessible genetics and genomics databases employ a range of data access polices. The leaders of the Human Genome Project determined that all data generated from the project should be placed in a public repository and be accessible to all interested users. As such, rapid deposit of sequence data was required from each of the high-volume sequencing centers. Some public databases, such as GenBank, are completely open and have no specific usage policy. Other databases may require registration to track usage, but are otherwise open to all interested users. And still other databases, such as the newly proposed Genetic Association Information Network (GAIN), are limited to approved users, who must agree to a data use certification policy. Summary statistics such as allele and genotype frequency will be accessible to all users, but only approved users will be able to access individually coded phenotypes and associated genotypes and "pre-computed, simple unadjusted genotype/phenotype associations" (National Center for Biotechnology Information). In 2007, NIH issued a data-sharing policy for NIH-supported or conducted genome-wide association studies (NOT-OD-07–088).

While it is common practice for analyzed data to be publicly disseminated through professional meetings and publications, the push for the release of raw data may raise some concerns about the privacy and confidentiality of research participants, particularly as this information could be used to identify individual participants (Lin *et al.*, 2004; McGuire and Gibbs, 2006a). The NIH data-sharing policy requires that all publicly released data should not include participant identifiers or "variables that could lead to deductive disclosure of the identity of individual subjects."

With respect to biobanks and access to tissue samples or extracted components (e.g., DNA, RNA, or protein), the policies, if existent, vary. For many European biobanks, access to samples is a rather informal process, usually a result of a scientific collaboration; only a scientific purpose is requested for use of samples (Hirtzlin *et al.*, 2003). Public collections are more likely to restrict access to other public institutions whereas private collections have fewer restrictions (Hirtzlin *et al.*, 2003).

Large national or regional collections also appear to have a range of policies, though more definitive, regarding access (Hirtzlin *et al.*, 2003). For example, in Sweden and Iceland, the operators of the biobanks have the final decision regarding which scientists may have access to the biobank materials (Gibbons *et al.*, 2005). In the U.K., the requested projects must be approved by a relevant research ethics committee as well as by the U.K. Biobank's scientific committee. The U.K. Biobank's board of directors retains the final ruling over access, however, the biobank's decisions could also be overridden by a court order.

In 2006, OBBR and the Biospecimen Coordinating Committee released the First-generation Guidelines for NCI-supported Biorepositories for public comment. With respect to access to NCI-supported biorepositorites, the draft guidelines recommended the following criteria be considered: scientific validity of proposed research; investigator's agreement to abide by policies governing confidentiality, use, disposition, and security of biospecimens and associated data; investigator's written agreement in a Materials Transfer Agreement to comply with the NIH Research Tools Policy; investigator and institutional research qualifications; ethical oversight where required by Federal regulations or local institutional requirements; and adequate funding for the biorepository (Recommendation III.B.2). In addition, the guidelines recommend that ethical issues should be a primary consideration in the decision-making process and that an appeals process should be established.

Of equal interest are policies that prohibit access to biobank materials from certain groups. For instance, Estonia expressly prohibits access by police (Article 16 of the Human Genes Research Act). Interestingly, a U.K. survey found that respondents would be more willing for police (66.5%) to have access to the U.K. Biobank than private sector companies for commercial interest (50.7%) compared to public researchers (79%) (Shickle *et al.*, 2003). Respondents trusted the police (59%) more than academic scientists (38%), health/pharma companies (20%), and medical charities (19%); GPs were trusted the most (87%) followed by NHS (74%) (Shickle *et al.*, 2003).

VIII. OWNERSHIP AND INTELLECTUAL PROPERTY

There is substantial debate and concern about the ownership of participants' individual data, tissues samples, and DNA, entire databases, and downstream products. Potential owners can range from a single individual to hundreds of individuals from an institution, including researchers and participants (Gibbons *et al.*, 2005; Hirtzlin *et al.*, 2003). For example, the U.K. Biobank declares itself to be the "legal owner of the database and sample collection" (Ethics and Governance Framework), but the Swedish biobank has not established any rules or legislation pertaining to ownership. In Iceland, the law is ambiguous; the biobank licensees are not owners of the biological samples but have limited access to them. Estonia has a clear policy on ownership; Article 15 of the Human Genes Research Act of 2000 states that ownership of samples resides with the chief processor of the Estonian Genome Research Project and is nontransferable.

With respect to personal data, all data is the property of the Icelandic government. In Sweden and the U.K., information is not considered to be "property" and therefore, cannot be owned (Gibbons *et al.*, 2005). However, in Estonia, personal data can be owned as provided under Article 15 of the Human Genes Research Act; the chief processor also has ownership of all personal data in the Estonian Genome Research Project and data may not be transferred if personally identifiable.

In the U.S., two key court decisions have substantially shaped and defined the views regarding ownership of research specimens. In 1976, a patient named Moore was diagnosed with hairy-cell leukemia and sought treatment at the medical center of the University of California at Los Angeles. During his clinical care, several tissue samples were collected from Mr. Moore including blood and serum, and he underwent a splenectomy. Unknown to Mr. Moore, his tissue samples were being used in research and a cell line was derived from his T-lymphocytes. Mr. Moore alleged that his physician failed to disclose his research intentions and economic conflicts of interest prior to administering any medical treatments and filed a claim against the physicians for conversion. In 1990, the California Supreme Court ruled that the physician was obligated to disclose his fiduciary interests to his patient. However, the Court dismissed the claim of conversion since Mr. Moore did not have any property rights to the derived cell line from T-lymphocyte cells extracted from his spleen nor to any commercial profits derived from research conducted in part on that cell line (Moore v. Regents of the University of California, 1991).

In 2003, a Florida court reaffirmed the *Moore* decision by unambiguously stating that individuals do not have ownership rights in specimens donated for research purposes (Greenberg v. Miami Children's Hospital Research Institute, Inc. 2003). In this case, the plaintiffs, a group of parents of children affected with Canavan's disease and advocacy organizations, filed suit against the researcher who had isolated the Canavan disease gene from samples donated by the parents and organizations and his employer (Miami Children's Hospital Research Institute) alleging lack of informed consent, conversion, and unjust enrichment. The U.S. District Court for the Southern District of Florida found no breach of informed consent had occurred nor any basis for the claim for conversion. However, the court ruled in favor of the plaintiffs in finding that Miami Childrens did benefit unjustly from the imposed licensing fees for the Canavan genetic test.

A recent case regarding ownership of tissue samples involving another university found yet again that patients do not own their donated tissue samples. The plaintiff (a physician–researcher) argued that the donors had rights and ultimate control over the tissue samples and could either withdraw or consent to have them transferred to another location, that is, in this case, the new employer of the plaintiff (Washington University v. Catalona). The judge ruled that the patients did not have ownership of their samples since they had donated (or gifted) their tissue to the university (the university's logo was prominently displayed atop the informed consent documents) and the university supported the biorepository. The judge cited both the *Moore* and *Greenberg* decisions in his opinion.

One additional issue that the *Catalona* case brings to light is the fate of samples should the biobank become insolvent. Does the donor, the investigator, the institution, or investors own the stored tissue samples? Can a tissue collection merely be sold to the highest bidder (Anderlik, 2003; Janger, 2005)? Particularly for commercial biobanks, this issue should be addressed during the informed consent process so that donors are aware of the long-term consequences of the samples in the event of a change in biobank management.

Although the courts appear to concur on the ownership rights of human tissue specimens, ambiguity remains throughout guidelines developed by professional organizations and governmental bodies. OHRP prohibits the inclusion of language in the informed consent document regarding participants' waiver of property rights in the donated tissue specimens (45 CFR Part 46.116). State laws are mostly silent on the issue of ownership in human tissue specimens (Hakimian and Korn, 2004). The European Society of Human Genetics (2003) finds that ownership rights pivot on the identifiability and status of the data; the donor has control of anonymized and identifiable data until the data have been processed and become "research data" whereupon the investigator has ownership (2003). Anonymous data and samples are considered "abandoned" and also under the control of the investigator.

IX. DISCLOSURE OF RESEARCH RESULTS

Offering research results to participants has been the subject of many recent debates. This debate encompasses disclosure of individual results as well as aggregate or summary results. The benefits and risks of each type are discussed here.

A. Individual research results

Considerable controversy exists over the issue of disclosing individual genetic research results to research participants. This topic has been the subject of guidelines from federal entities (Bookman *et al.*, 2006; National Bioethics

Advisory Commission (NBAC), 1999) as well as numerous professional publications, and yet the debate continues to range widely between the view that all individual results should be offered and the view that such results should generally not be offered (Table 18.5; Ravitsky and Wilfond, 2006). In addition, the discussion has generally focused on results generated by individual studies, with relatively little attention to the specific context of a biobank. Here, we summarize some of the primary arguments for and against disclosure of individual research results and identify emerging areas of potential consensus that may be particularly relevant to biobank research.

1. Arguments in favor of offering individual research results

The ethical rationale most often cited as the basis for offering individual results is *respect for persons*. This principle requires that competent individuals be treated as autonomous agents, and that weight be given to their considered opinions and choices (National Commission for the Protection of Human Subjects of Biomedical and Behavioral Research, 1979). Respect for persons is said to ground the duty to offer individual results because (1) some studies show that participants want to receive such results; (2) returning results is a way of recognizing participants' contributions to the research process; (3) research results, particularly genetic results, constitute important information about oneself; and (4) participants' self-determination is enhanced when they have the choice of incorporating research results into their personal decision-making (Ossorio, 2006).

Another commonly cited rationale is the ethical principle of *beneficence*. This principle holds that persons are treated in an ethical manner not only by respecting their decisions and protecting them from harm, but also by making efforts to secure their well being (National Commission for the Protection of Human Subjects of Biomedical and Behavioral Research, 1979). Those in favor of offering individual results point to potential benefits for participants and for society at large. For participants, research results could have an impact on their clinical care, quality of life, reproductive decision-making, and/or life planning. Aside from any direct health implications, results may also have personal meaning. For example, genetic results could have ramifications for family lineage (e.g., nonpaternity), ethnic or cultural identity (e.g., tribal affiliation), and personal identity (e.g., behavioral traits) (Ravitsky and Wilfond, 2006). In other words, the value of research results depends not only on its clinical interpretation, but also on the meanings that participants bring to it (Dressler and Juengst, 2006). From a societal perspective, offering research results could affirm the key role participants play in research, raise awareness of the contribution of research to the understanding of disease and therapy, and improve trust and investment in biomedical research (Fernandez *et al.*, 2003; Shalowitz and Miller, 2005).

Finally, the offer of individual results is sometimes considered a matter of *reciprocity*—a demonstration of gratitude for participants' voluntary participation in research (Shalowitz and Miller, 2005). Some have suggested that a more intensive researcher–participant relationship creates a stronger requirement for reciprocity; thus, the imperative to communicate results that do not exceed a high threshold of utility is attenuated, for example, in a study of stored specimens that involves no personal contact with participants (Ravitsky and Wilfond, 2006).

Arguments based on all of these principles have led some to conclude that, rather than making investigators the gatekeepers of research information, participants should have the opportunity to determine what research information about themselves they wish to know (Shalowitz and Miller, 2005). Rothstein (2006), for instance, proposes that subjects should be able to select the general circumstances for disclosure of results using a tiered process during informed consent, and that investigator-developed objective criteria should be applied after the research to determine whether the findings satisfy the subject's expressed standards. Shalowitz and Miller (2005) further contend that, even when the investigator is unable to interpret results in the context of the study goals, he or she should be willing to explain this uncertainty to participants in a manner that is as understandable as possible. Ultimately, they claim, "Respect for research participants requires investigators to meet [requests for individual results] in all but a few circumstances, and the burden is on the investigators to justify nondisclosure" (Shalowitz and Miller, 2005).

2. Arguments against offering individual research results

A major argument against routinely offering research results is the definition of research itself. According to the Belmont Report (National Commission for the Protection of Human Subjects of Biomedical and Behavioral Research, 1979), "clinical practice" refers to interventions that are designed solely to enhance the well-being of an individual patient, whereas "research" is an activity designed to develop or contribute to generalizable knowledge. Thus, while treating physicians are legally and ethically obligated to act in the best interests of individual patients, researchers are obligated to conduct good science with the potential to benefit populations (Beskow *et al.*, 2001). Conducting good science includes protecting study participants, which in turn includes communicating clearly to participants that they are being asked to take part in research and, in most cases, that they should not expect individual benefit. Researchers must also avoid confusing or conflating their role with that of a treating physician (Merz *et al.*, 1997). Fulfilling these obligations is essential for minimizing participants' tendency toward therapeutic misconception, which is the mistaken perception of research as treatment designed to benefit them.

Another fundamental argument against offering individual research results is the nature of the results themselves. Finding a statistically significant association between a genetic trait and a phenotype in a single study does not imply a cause and effect relationship. Reported associations are seldom replicated and initial studies often suggest a stronger genetic effect than later studies (Hirschhorn *et al.*, 2002; Ioannidis *et al.*, 2001). Thus, conveying unsubstantiated and possibly misleading data for which the foreseeable harms could easily outweigh any benefits (Beskow, 2006).

Similar to the arguments in favor of offering individual research results, ethical principles are often cited as a rationale against routinely offering such results. The principle of *respect for persons*, typically operationalized in the requirement for informed consent (National Commission for the Protection of Human Subjects of Biomedical and Behavioral Research, 1979), requires that autonomous persons enter into research voluntarily and with adequate information. As interpreted by Clayton and Ross (2006) and echoed by others (Ossorio, 2006; Parker, 2006a):

> Showing respect for research participants requires minimizing harms so that they are not treated as mere means for scientific ends . . . Respect for persons does not mean, however, that individuals have a right to all data acquired in the research process, even if the data are about them. [p. 37]

With regard to *beneficence*, Meltzer (2006) argues that, because all research must meet the requirement of beneficence regardless of whether it returns individual research results, the principle of beneficence does not justify investigators to offer results. In its complementary expression as "do not harm," several commentators have noted that a negative duty of nonmaleficence is very different from any positive duty to promote participants' interests. Were "failure to benefit" to become a prevalent understanding of harm, it would pose intractable difficulties for IRBs in assessing risk–benefit ratios (Meltzer, 2006; Parker, 2006a).

As for reciprocity, there are many ways to acknowledge participants' contribution to research (Plummer *et al.*, 2002). Similar to concerns about payment creating inappropriate incentives to participate, the offer of learning genetic information may prompt people to enroll in research that they otherwise would not (Ossorio, 2006; Parker, 2006a). In addition, those who oppose routinely offering individual results maintain that any compensation for research participation should be available to all subjects, and not depend on the success of the study in answering the research question nor one's relationship to the findings (e.g., possession of a genetic trait or not) (Meltzer, 2006; Parker, 2006b). Further, the benefit should not be of uncertain scientific or personal value; out of fairness, it should be of established and relatively uniform value to all (Parker, 2006a).

3. Some emerging areas of potential consensus

The vigorous debate concerning all of these issues will no doubt continue and evolve. Although it is too early to identify areas of clear consensus (Table 18.5), the following points may be beginning to emerge as the majority view:

a. There will likely be an exceedingly small number of cases in which researchers are obligated to offer individual genetic research results and a large number of cases in which it may be ethically permissible to do so (Ossorio, 2006).
b. Recommendations put forth by the National Bioethics Advisory Commission (NBAC), 1999) have emerged as central to the debate. These recommendations include:
    - IRBs should develop general guidelines for the disclosure of the results of research to subjects and require investigators to address these issues explicitly in their research plans.
    - In general, these guidelines should reflect the presumption that the disclosure of research results to subjects represents an exceptional circumstance. Such disclosure should occur only when all of the following apply: (a) the findings are scientifically valid and confirmed; (b) the findings have significant implications for the subject's health concerns; and (c) a course of action to ameliorate or treat these concerns is readily available.
    - When research results are disclosed to a subject, appropriate medical advice or referral should be provided.
c. Anticipating the possible need to disclose individual results is key. Pitfalls in making an after-the-fact determination to offer results include disagreements as to whether criteria for disclosure have been met; the problem that individuals whose samples were used have no knowledge of the specific study; and questions regarding the voluntariness of consent to receive results after the results are already in hand (Beskow *et al.*, 2001). Thus, as urged in NBAC's recommendations, researchers should address the issue of offering results in their research protocol and communicate their plans during the informed process.
d. A decision about offering individual results should not be made by the investigator alone, but rather in conjunction with their IRB and other stakeholders, such as representatives from the research subject community (Bookman *et al.*, 2006; Dressler and Juengst, 2006).
e. The cost of disclosing individual results will be a major challenge in this era of restricted research resources (Fernandez and Weijer, 2006). Ethically appropriate disclosure requires resources for confirmatory testing in a Clinical Laboratory Improvement Amendments (CLIA)-certified lab, the logistics of maintaining contact with participants, and the education and counseling

Table 18.5. Offering Research Results: Examples of Consent Form Language

BIOBANK	Individual results	Aggregate results
The CARTaGENE Project[a]	*Individual benefits*: . . . you can choose to receive certain tests results from your clinical exam (blood sugar level, cholesterol level, blood pressure, percentage of body fat). These results will be provided with the appropriate explanations and can be shared with your personal doctor *Individual Research Results*: In such a population study, you will not receive any individual research results	General research results, negative or positive, will be published and publicly available in an anonymous, general, and statistical format. With [the Institute for Populations, Ethics, and Governance's] authorization, these results will be published in scientific and medical reviews and journals and posted twice a year on the CARTaGENE website
Estonian Genome Project[b]	I have the right to know my genetic data and other data about me stored in the Gene Bank, except my genealogy. I have the right to genetic counseling upon accessing my data stored in the Gene Bank. I can access my data stored in the Gene Bank free of charge	
Generation Scotland[c]	Following the clinical visit, Generation Scotland will give participants health information on some important clinical measurements such as blood pressure, cholesterol, and kidney function, but it will not be possible to give any personal genetic information	
Icelandic Biobank[d]	No results from this study regarding individual participants will ever be made available	
Marshfield Clinic Personalized Medicine Research Project[e]	No information resulting from the analysis of your DNA or about your genetic status, such as the probability of developing a specific disease, will be provided to you. Research results are often preliminary, inconclusive, and not necessarily valid for decisions concerning patient care and treatment. Preliminary information, if given to you, could create false conclusions and significant risks	Researchers will periodically send a newsletter to all project subjects. The newsletter will not contain any individual results, but will contain general information about studies. We will ask you to inform the project staff of any changes to your mailing address so that you continue to receive the newsletter.

National Institute of Environmental Health Sciences' Environmental Polymorphisms Registry[f]	To date, none of the DNA differences we will be looking at have been associated with a condition or disease. Therefore, you will not receive any results from the initial DNA testing. However, the objective of some of the follow-up studies might be to look for an association of a particular DNA difference with a certain condition or disease. If you participate in this type of follow-up study, you may or may not be given the results. That will be decided by the principal investigator of that study. All results generated from the EPR are strictly for research purposes only and cannot be used to diagnose or predict a condition or disease	
Singapore Tissue Network[g]	Neither you nor your doctor will receive the results of research done with your donated samples. This is because research can take a long time and requires samples from many people before results are known. Results from research using your samples may not be ready for many years and will not affect your care right now	
UK Biobank[h]	Apart from providing you with the results of some standard measurements made during [the initial assessment visit], none of your results will be given to you or your doctors (even if the results do not seem to be normal)	Results of research conducted on the UK Biobank resource will be made available to participants, and others who might be interested, through our website at www.ukbiobank.ac.uk

[a]CARTaGENE Information and Consent Form (2006): Available at http://www.cartagene.qc.ca/docs/InfoConsent_En.pdf.

[b]Estonian Genome Project Gene Donor Consent Form: Available http://www.geenivaramu.ee/index.php?lang = eng&sub = 74.

[c]Generation Scotland FAQ—Your Questions Answered: Available at http://129.215.140.49/gs/gFAQ.htm.

[d]DeCode Information for Participation: Available at http://www.decodegenetics.com/files/file148517.pdf.

[e]Marshfield Clinic Research Consent/Authorization Form: Available at http://www.marshfieldclinic.org/chg/pages/Proxy.aspx?Content = MCRF-Centers-CHG-Core-Units-PMRP-Consent-form_4–26–06.1.pdf.

[f]NIEHS Environmental Polymorphisms Registry Consent Form: Available at http://dir.niehs.nih.gov/direpr/assets/pdf/generalbooklet.pdf.

[g]Singapore Tissue Network FAQ For the Donor: Available at http://www.stn.org.sg/10_faq.htm.

[h]UK Biobank Participant Information Leaflet: Available at http://www.ukbiobank.ac.uk/docs/20060426%20Patient%20Info%20Leaflet%20Final.pdf.

needed to obtain consent for disclosure, explain the findings, and provide medical and psychological referrals (Beskow, 2006). Unless a solution is found for bearing these costs, we may sacrifice participants' and society's long-term interest in advancing knowledge for participants' short-term interests in obtaining particular, individual medical information (Ossorio, 2006).

f. There is a pressing need for research about offering individual research results. Participants' interest in and response to receiving such results cannot be extrapolated from those receiving results in clinical care or under research protocols designed to return predictive genetic test results for specific conditions. In both of these contexts, individuals actively seek genetic information prior to its generation and receive pretest counseling, and the tests themselves have established clinical utility (Parker, 2006a). Research is also needed on the perspectives of other stakeholders, including researchers, IRBs, and biobank directors.

In conclusion, there are several factors that suggest a prudent approach for biobanks may be to proceed with caution but consider "never saying never" with regard to offering individual results. First, although current knowledge about the exact contribution of genetic factors to health and disease is still quite rudimentary (Facio, 2006), the long-term nature of a biobank augurs for one day amassing a large amount of clinically relevant information. Second, because the process of translating research findings into clinical practice requires a convergence of political, economic, and administrative forces (Dressler and Juengst, 2006), researchers may possess data with known clinical utility well before genetic testing is available in nonresearch settings. Third, researchers may make incidental discoveries—that is, learn something about an individual that is unrelated to the research question, but may have serious, known implications—that they feel compelled to offer. Fourth, many have argued that the intensity of the researcher–participant relationship (i.e., the idea that any obligation to offer results is lower when research has been conducted on biological specimens with no personal contact) is irrelevant (Fernandez and Weijer, 2006; Fryer-Edwards and Fullerton, 2006; Ossorio, 2006). As stated by Dressler and Juengst (2006), "The lack of personal contact (visual, verbal, or otherwise) with a participant does not diminish the participant's stake in the results or the researcher's responsibility to consider the value the research finding could have for the research participant." [p. 18].

B. Aggregate research results

Offering research participants a summary of aggregate results has been encouraged as an alternative to offering individual results (Beskow *et al.*, 2001; Parker, 2006a) and as an important expression of respect and accountability in its own right (Fernandez *et al.*, 2003; Partridge and Winer, 2002; Zlotnik Shaul *et al.*, 2005).

Several studies have examined participants' views on receiving aggregate results. Richards and colleagues (2003) interviewed 21 women with breast cancer who had been enrolled in a large genetic epidemiologic study. Half of the women interviewed suggested spontaneously that they would have liked to hear more about the progress of the research and, when asked, the remaining women said they felt there ought to have been some general feedback about the outcomes of the study for all participants. A minority felt strongly about this and implied that it was a fair return for the contribution they had made to the research.

Dixon-Woods and colleagues (2006) conducted in-depth interviews with women who had participated in a randomized trial of antibiotics in pregnancy and who had requested results of the trial. Less than 20% of the women who participated in the trial indicated that they wanted to receive the results. Reactions to a leaflet summarizing the aggregate results were generally positive or neutral, although some women had difficulty in understanding the leaflet, and there was evidence of possible negative implications for women who had adverse outcomes. These findings led the authors to conclude that some caution is needed with regard to the routine provision of aggregate results, and that there should be more evaluation of how feedback of results should be handled and of the risks, benefits, and costs.

In a third example, Kaphingst and colleagues (2006) explored the views of 26 female breast cancer patients who had consented to provide blood or tissue samples for breast cancer research. Most participants initially responded that they would like to receive both aggregate and individual results. Although few had a clear preference for receiving only aggregate results, most were interested in aggregate results if individual results were not available.

For most biobank research, participants will likely have provided general consent to unspecified research based on the potential to benefit others in the future. In addition, the justification for asking participants for general consent is often based in part on the fact that they can withdraw if they object to future research. Thus, an argument in favor of providing access to aggregate results of research could be based on showing respect for research subjects and their families; maintaining and building the public's trust in the biobank; demonstrating that the benefit conveyed in the consent form is indeed being pursued; and making the option to withdraw "real" by informing participants generally about kinds of research that are being conducted.

Some points to consider with regard to the provision of aggregate results include:

1. Biobanks should consider developing a plan to make information about aggregate results available; for example, by posting such information on its public web site and having it available in hard copy upon request (Table 18.5).

2. The intent to make aggregate information available should be included in the consent document and process.
3. Any aggregate results that are provided should ideally be in a form understandable to a nonscience audience.
4. Although participants may be given aggregate information about all of the research supported by the biobank, there are significant concerns about providing information specifically about the studies in which their particular specimens happen to have been used. A primary concern is the possibility that participants may come to erroneous conclusions about why their specimens were or were not used in the study of a particular gene or condition.
5. Continued effort is needed to gather empirical data about participants' interest in and response to receiving aggregate results.

X. COMMERCIALIZATION OF BIOBANKS

The increased value of tissue samples, particularly those with substantial clinical characterization, has led to the commodification of biological tissues, genotype and sequence data, and phenotypic/environmental data (Anderlik, 2003; Rose, 2001). Government agencies such as the U.S. National Institutes of Health and the U.K. Medical Research Council also provide funding for tissue-specific biobanks (e.g., brain banks). However, given the longevity of biobanks (10–20 years for some prospective studies or undefined), maintenance costs are a major concern for the long-term outlook of biobanks. For example, the U.K. Biobank is funded jointly by the Wellcome Trust, the Medical Research Council, the Department of Health, the Scottish Executive, and the Northwest Regional Development Agency. In 2002, the biobank was allocated an initial £45 million by the Wellcome Trust (£20m), the Medical Research Council (£20m) and the Department of Health (£5m). In 2006, approval of a £61m budget was obtained following a successful 3800-person pilot project. As a result of these steep costs and efforts, academic–industry or government–industry relationships have been an important aspect to the development of several biobanks (Ashburn *et al.*, 2000; Austin *et al.*, 2003).

Several types of commercial entities exist that are involved in sample collection, storage, and/or research. For example, some companies merely serve as a warehouse for universities or companies with storage needs. Thermo Scientific serves as an offsite storage facility with 65 million stored samples at six sites in the U.S. and the U.K. Other companies may serve as a "biobroker" by facilitating the collection and exchange of tissue samples and data between those who have access to the samples (e.g., an academic medical center) and those desiring access to samples (e.g., a pharmaceutical or biotechnology company) (Anderlik, 2003).

Other companies are also involved in research in addition to sample collection and storage such as Genomics Collaborative, Inc. And finally, some companies have been able to recruit participants directly online such as DNA Sciences' Gene Trust.

The Swedish tissue bank, UmanGenomics, was created from a desire to commercialize an existing tissue collection maintained by a teaching hospital in Umea, Sweden. It was decided that the University (and the county) would own 51% of the shares and the rest would be privately held. However, little pharmaceutical interest materialized and in order for the public–private endeavor to stay afloat, it was determined that a costly shift in strategy was necessary (Rose, 2006). The involvement of a commercial stakeholder, the lack of involvement of the biobank's original director, questions regarding intellectual property and access to samples, and whether the hailed "ethical approach" (Abbott, 1999) was commercially viable besieged the partnership and contributed to the company's dissolution into a "virtual" company (Rose, 2003, 2006).

The Iceland biobank offers a slightly different experience regarding the involvement of a commercial entity. The Act on a Health Sector Database (No. 139/1998) authorized the establishment of a national medical database (Health Sector Database; HSD) that would be linked to two other databases (a genealogical database and a genetics database). An exclusive license was granted to a U.S. company, deCode Genetics, to develop the Health Sector Database. Much of the negative publicity surrounding the Iceland biobank dealt with the issues of informed consent and privacy (Arnason, 2004; Merz *et al.*, 2004). However, some of that confusion arose from the differing policies of each of the three databases (Arnason, 2004). For the Health Sector Database, the medical records of Icelanders were automatically included ("presumed consent") unless individuals wished to "opt-out." In contrast, inclusion in the deCode genetic database required explicit written informed consent. For the genealogical database, no consent was required or opt-out option available.

Based on extensive interviews with stakeholders in Iceland from government, industry, and the general public, Merz *et al.* (2004) concluded that the databases would ultimately benefit the company more so than the citizens of Iceland from which it was derived and promoted to benefit. The exclusive use and anticipated profits from research findings based on citizens' medical, genealogical, and genetic data was balanced with deCode's full support for the development of the HSD in addition to annual payments to the government and any business taxes (Merz *et al.*, 2004). Even if the company became insolvent, the country's health care infrastructure would benefit from the establishment of an electronic health records system. However, this trade off was unprecedented and perhaps unethical as other countries have shifted to an electronic system without commercial intrusion (Merz *et al.*, 2004). Withstanding the growing pains of a nascent biotechnology company with a plummeting stock price, layoffs and controversial public debate, the company has published several high profile papers (Amundadottir *et al.*, 2006; Grant *et al.*, 2006; Helgadottir *et al.* 2006) and continues to work towards the development of diagnostic and therapeutic products based on their research.

The involvement of a commercial entity in a biobank initiative can result in significantly negative public perceptions and fears of commercial exploitation. For example, while the overwhelming majority of potential donors surveyed (>90%) said they would allow for their samples to be collected and stored in the Swedish biobank, about one-third stated that their response would be different if financial gain were involved (Nilstun and Hermeren, 2006). In contrast, Jack and Womack (2003) found that only 1% of more than 3000 patients approached preoperatively to donate tissue at a British hospital that would subsequently be sold to commercial users declined to consent.

Disadvantages to the use of private DNA banking services may include limited and/or costly access to samples, limitations on data dissemination, and trust (Anderlik, 2003; Hirtzlin *et al.*, 2003). On the other hand, the use of a commercial third party or designated trustee may be advantageous in helping to protect the privacy and interests of participants (Anderlik, 2003). Governmental or academic institutions and hospitals may not possess the facilities, logistical or technical support, or commitment to collect and maintain large numbers of biospecimens over many years. While the involvement of commercial biobanks will likely continue to be viewed through polarizing lenses, the reality of today's research environment is that they will remain as a contributor, partner, and/or driver of large-scale genomics research. To prolong their involvement in the field, one of the major challenges will be to define a socially acceptable and financially viable approach to the collection, study, and marketing of data derived from biospecimens.

XI. CONCLUSION

As the creation and use of biobanks increases around the world, the need for careful consideration and analysis of ethical, legal, and social implications remains. As Knoppers and Chadwick (2005) aptly describe, "ethics does not consist of a static set of theories or principles that can unproblematically be 'applied' to new situations." As the research environment evolves and responds to the new opportunities and challenges presented by new genomic technologies and public needs, we must adapt our ethical deliberations and analyses to the changing landscape. Furthermore, while many of the issues described here are being faced by several countries, the resolutions and policies need to be based on a population's history, culture, and accepted norms rather than attempting a one-size-fits-all approach.

References

Abbott, A. (1999). Sweden sets ethical standards for use of genetic 'biobanks'. *Nature* **400,** 3.

Abbott, W. G., Winship, I. M., Gane, E. J., Finau, S. A., Munn, S. R., and Tukuitonga, C. E. (2006). Genetic diversity and linkage disequilibrium in the Polynesian population of Niue Island. *Hum. Biol.* **78**(2), 131–145.

Office of Aboriginal Health of the Health Department of Western Australia (1998). Ethical Considerations for Health-Related Research involving Aboriginal People. Available at http://www.research.murdoch.edu.au/ethics/hrec/Policies/WAAHIEC%20guidelines.pdf.

Allen, M. D. (2004). Commercial tissue repositories: HIPAA raises sponsors' fears. *IRB* **5,** 9–11.

American Society of Human Genetics (ASHG) (1996). Statement on informed consent for genetic research. *Am. J. Hum. Genet.* **59,** 471–474.

Amundadottir, L. T., Sulem, P., Gudmundsson, J., Helgason, A., Baker, A., Agnarsson, B. A., Sigurdsson, A., Benediktsdottir, K. R., Cazier, J. B., Sainz, J., Jakobsdottir, M., Kostic, J., *et al.* (2006). A common variant associated with prostate cancer in European and African populations. *Nat. Genet.* **38**(6), 652–658.

Anderlik, M. R. (2003). Commercial biobanks and genetic research: Ethical and legal issues. *Am. J. Pharmacogenom.* **3**(3), 203–215.

Arnason, V. (2004). Coding and consent: Moral challenges of the database project in Iceland. *Bioethics* **18**(1), 27–49.

Ashburn, T. T., Wilson, S. K., and Eisenstein, B. I. (2000). Human tissue research in the genomic era of medicine: Balancing individual and societal interests. *Arch. Intern. Med.* **160,** 3377–3384.

Association of Canadian Universities for Northern Studies (1982). "Ethical Principles for the Conduct of Research in the North (Ottawa)" .

Austin, M. A., Harding, S. E., and McElroy, C. E. (2003). Monitoring ethical, legal, and social issues in developing population genetic databases. *Genet. Med.* **5**(6), 451–457.

Bauer, K., Taub, S., and Parsi, K. (2004). Ethical issues in tissue banking for research: A brief review of existing organizational policies. *Theor. Med. Bioeth.* **25**(2), 113–142.

Beskow, L. M. (2006). Considering the nature of individual research results. *Am. J. Bioeth.* **6,** 38–40.

Beskow, L. M., Burke, W., Merz, J. F., Barr, P. A., Terry, S., Penchaszadeh, V. B., Gostin, L. O., Gwinn, M., and Khoury, M. J. (2001). Informed consent for population-based research involving genetics. *JAMA* **286,** 2315–2321.

Blatt, R. J. (2000). Banking biological collections: Data warehousing, data mining, and data dilemmas in genomics and global health policy. *Commun. Genet.* **3**(4), 204–211.

Bookman, E. B., Langehorne, A. A., Eckfeldt, J. H., Glass, K. C., Jarvik, G. P., Klag, M., Koski, G., Motulsky, A., Wilfond, B., Manolio, T. A., *et al.* (2006). Reporting genetic results in research studies: Summary and recommendations of an NHLBI working group. *Am. J. Med. Genet.* A **140,** 1033–1040.

Bonnen, P. E., Pe'er, I., Plenge, R. M., Salit, J., Lowe, J. K., Shapero, M. H., Lifton, R. P., Breslow, J. L., Daly, M. J., Reich, D. E., *et al.* (2006). Evaluating potential for whole-genome studies in Kosrae, an isolated population in Micronesia. *Nat. Genet.* **38**(2), 214–217.

Burton, B. (2002). Proposed genetic database on Tongans opposed. *BMJ* **324**(7335), 443.

Cambon-Thomsen, A. (2004). The social and ethical issues of post-genomic human biobanks. *Nat. Rev. Genet.* **5,** 866–873.

CARTaGENE Information and Consent Form (August 2006). Available at http://www.cartagene.qc.ca/docs/InfoConsent_En.pdf.

Caulfield, T., Upshur, R. E., and Daar, A. (2003). DNA databanks and consent: A suggested policy option involving an authorization model. *BMC Med. Ethics* **4,** E1.

Cavalli-Sforza, L. L., Wilson, A. C., Cantor, C. R., Cook-Deegan, R. M., and King, M.-C (1991). Call for a worldwide survey of human genetic diversity: A vanishing opportunity for the Human Genome Project. *Genomics* **11,** 490–491.

Cavalli-Sforza, L. L. (2005). The Human Genome Diversity Project: Past, present, and future. *Nat. Rev. Genet.* **6,** 333–340.

Claerhout, B., and DeMoor, G. J. E. (2005). Privacy protection for clinical and genomic data: The use of privacy-enhancing techniques in medicine. *Int. J. Med. Inform.* **74**(2–4), 257–265.

Clayton, E. W., and Ross, L. F. (2006). Implications of disclosing individual results of clinical research. *JAMA* **295,** 37.

Clayton, E. W., Steinberg, K. K., Khoury, M. J., Thomson, E., Andrews, L., Kahn, M. J., Kopelman, L. M., and Weiss, J. O. (1995). Informed consent for genetic research on stored tissue samples. *JAMA* **274,** 1786–1792.

Convention on Biological Diversity (1992). Available at http://www.biodiv.org/convention/default.shtml.

De Moor, G. J., Claerhout, B., and De Meyer, F. (2003). Privacy enhancing techniques—the key to secure communication and management of clinical and genomic data. *Meth. Inf. Med.* **42**(2), 148–153.

Deschenes, M., Cardinal, G., Knoppers, B. M., and Glass, K. C. (2001). Human genetic research, DNA banking and consent: A question of 'form'?. *Clin. Genet.* **59**(4), 221–239.

Dixon-Woods, M., Jackson, C., Windridge, K. C., and Kenyon, S. (2006). Receiving a summary of the results of a trial: Qualitative study of participants' views. *BMJ* **332,** 206–210.

Dressler, L. G. (2005). "Human Specimens, Cancer Research and Drug Development: How Science Policy Can Promote Progress and Protect Research Participants." Available at http://www.iom.edu/Object.File/Master/26/207/IOM_fnl.pdf.

Dressler, L. G., and Juengst, E. T. (2006). Thresholds and boundaries in the disclosure of individual genetic research results. *Am. J. Bioeth.* **6,** 18–20.

Eiseman, E., and Haga, S. B. (1999). "Handbook of Human Tissue Sources: A National Resource of Human Tissue Samples." RAND, Santa Monica.

Eiseman, E., Bloom, G., Brower, J., Clancy, N., and Olmsted, S. S. (2003). "Case Studies of Existing Human Tissue Repositories. "Best Practices" for a Biospecimen Resource for the Genomic and Proteomic Era." RAND Corporation, Santa Monica.

Elger, B. S., and Caplan, A. L. (2006). Consent and anonymization in research involving biobanks: Differing terms and norms present serious barriers to an international framework. *EMBO Rep.* **7,** 661–6.

Eriksson, S., and Helgesson, G. (2005). Potential harms, anonymization, and the right to withdraw consent to the biobank research. *Eur. J. Hum. Genet.* **13,** 1071–1076.

European Society of Human Genetics (2003). Data storage and DNA banking for biomedical research: Technical, social and ethical issues. *Eur. J. Hum. Genet.* **11**(Suppl 2), S8–S10.

Facio, F. M. (2006). One size does not fit all. *Am. J. Bioeth.* **6,** 40–42.

Fernandez, C. V., and Weijer, C. (2006). Obligations in offering to disclose genetic research results. *Am. J. Bioeth.* **6,** 44–46.

Fernandez, C. V., Kodish, E., and Weijer, C. (2003). Informing study participants of research results: An ethical imperative. *IRB* **25,** 12–19.

Foster, M. W., and Sharp, R. R. (2000). Genetic research and culturally specific risks: One size does not fit all. *Trends Genet.* **16**(2), 93–95.

Foster, M. W., and Sharp, R. R. (2005). Will investment in biobanks, prospective cohorts, and markers of common patterns of variation benefit other populations for drug response and disease susceptibility gene discovery? *Pharmacogenom. J.* **5,** 75–80.

Foster, M. W., Eisenbraun, A. J., and Carter, T. H. (1997). Communal discourse as a supplement to informed consent for genetic research. *Nat. Genet.* **17**(3), 277–279.

Foster, M. W., Bernsten, D., and Carter, T. H. (1998). A model agreement for genetic research in socially identifiable populations. *Am. J. Hum. Genet.* **63**(3), 696–702.

Fryer-Edwards, K., and Fullerton, S. M. (2006). Relationships with test-tubes: Where's the reciprocity? *Am. J. Bioeth.* **6,** 36–38.

Gibbons, S. M., Helgason, H. H., Kaye, J., Nomper, A., and Wendel, L. (2005). Lessons from European population genetic databases: Comparing the law in Estonia, Iceland, Sweden and the United Kingdom. *Eur. J. Health Law* **12**(2), 103–133.

Gonzalez-Perez, E., Esteban, E., Via, M., Garcia-Moro, C., Hernandez, M., and Moral, P. (2006). Genetic change in the polynesian population of Easter Island: Evidence from Alu insertion polymorphisms. *Ann. Hum. Genet.* **70**(Pt 6), 829–840.

Grant, S. F., Thorleifsson, G., Reynisdottir, I., Benediktsson, R., Manolescu, A., Sainz, J., Helgason, A., Stefansson, H., Emilsson, V., Helgadottir, A., *et al.* (2006). Variant of transcription factor 7-like 2 (TCF7L2) gene confers risk of type 2 diabetes. *Nat. Genet.* **38**(3), 320–323.

Greely, H. T. (1999). Breaking the stalemate: A prospective regulatory framework for unforeseen research uses of human tissue samples and health information. *Wake Forest Law Rev.* **34,** 737–66.

Greenberg, V. Miami Children's Hospital Research Institute Inc. (2003). 264 F.Supp.2d 1064 (S.D. Fla. 2003).

Grizzle, W., Grody, W. W., Noll, W. W., Sobel, M. E., Stass, S. A., Trainer, T., Travers, H., Weedn, V., and Woodruff, K. (1999). Recommended policies for uses of human tissue in research, education, and quality control. Ad Hoc Committee on Stored Tissue, College of American Pathologists. *Arch. Pathol. Lab. Med.* **123,** 296–300.

Hakimian, R., and Korn, D. (2004). Ownership and use of tissue specimens for research. *JAMA* **292,** 2500–2505.

Hansson, M. G., Gillner, J., Bartram, C. R., Carlson, J. A., and Helgesson, G. (2006). Should donors be allowed to give broad consent to future biobank research? *Lancet Oncol.* **7,** 266–269.

Helgadottir, A., Manolescu, A., Helgason, A., Thorleifsson, G., Thorsteinsdottir, U., Gudbjartsson, D. F., Gretarsdottir, S., Magnusson, K. P., Gudmundsson, G., Hicks, A., *et al.* (2006). A variant of the gene encoding leukotriene A4 hydrolase confers ethnicity-specific risk of myocardial infarction. *Nat. Genet.* **38**(1), 68–74.

Hirschhorn, J. N., Lohmueller, K., Byrne, E., and Hirschhorn, K. (2002). A comprehensive review of genetic association studies. *Genet. Med.* **4**(2), 45–61.

Hirtzlin, I., Dubreuil, C., Preaubert, N., Duchier, J., Jansen, B., Simon, J., Lobato De Faria, P., Perez-Lezaun, A., Visser, B., Williams, G. D., *et al.* (2003). An empirical survey on biobanking of human genetic material and data in six EU countries. *Eur. J. Hum. Genet.* **11**(6), 475–478.

Hoeyer, K., Olofsson, B.-O., Mjorndal, T., and Lynoe, N. (2005). The ethics of research using biobanks: Reason to question the importance attributed to informed consent. *Arch. Intern. Med.* **165,** 97–100.

Indigenous Peoples Council on Biocolonialism (1997). "Ukupseni Declaration, Kuna Yala on the Human Genome Diversity Project (HGDP)." Available at http://www.ipcb.org/resolutions/htmls/dec_ukupseni.html.

Ioannidis, J. P., Ntzani, E. E., Trikalinos, T. A., and Contopoulos-Ioannidis, D. G. (2001). Replication validity of genetic association studies. *Nat. Genet.* **29,** 306–309.

Ioannidis, J. P., Trikalinos, T. A., and Khoury, M. J. (2006). Implications of small effect sizes of individual genetic variants on the design and interpretation of genetic association studies of complex diseases. *Am. J. Epidemiol.* **164**(7), 609–614.

Jack, A. L., and Womack, C. (2003). Why surgical patients do not donate tissue for commercial research: Review of records. *BMJ* **327,** 262.

Janger, E. J. (2005). Genetic information, privacy and insolvency. *J. Law Med. Ethics* **33**(1), 79–88.

Juengst, E. T. (1998a). Group identity and human diversity: Keeping biology straight from culture. *Am. J. Hum. Genet.* **63,** 673–677.

Juengst, E. T. (1998b). Groups as gatekeepers to genomic research: Conceptually confusing, morally hazardous, and practically useless. *Kennedy Inst. Ethics J.* **8**(2), 183–200.

Kaiser, J. (2003). African-American population biobank proposed. *Science* **300**(5625), 1485.

Kaphingst, K. A., Janoff, J. M., Harris, L. N., and Emmons, K. M. (2006). Views of female breast cancer patients who donated biologic samples regarding storage and use of samples for genetic research. *Clin. Genet.* **69,** 393–398.

Kaye, J. (2006). Do we need a uniform regulatory system for biobanks across Europe? *Eur. J. Hum. Genet.* **14**(2), 245–248.

Knoppers, B. M. (2005). Biobanking: International norms. *J. Law Med. Ethics* **33**(1), 7–14.

Knoppers, B. M., and Chadwick, R. (2005). Human genetic research: Emerging trends in ethics. *Nat. Rev. Genet.* **6**(1), 75–79.

Lin, Z., Owen, A. B., and Altman, R. B. (2004). Genomic research and human subject privacy. *Science* **305,** 183.

Lohmueller, K. E., Pearce, C. L., Pike, M., Lander, E. S., and Hirschhorn, J. N. (2003). Meta-analysis of genetic association studies supports a contribution of common variants to susceptibility to common disease. *Nat. Genet.* **33**(2), 177–182.

Maschke, K. J. (2005). Navigating an ethical patchwork—human gene banks. *Nat. Biotechnol.* **23**(5), 539–545.

Maschke, K. J., and Murray, T. H. (2004). Ethical issues in tissue banking for research: The prospects and pitfalls of setting international standards. *Theor. Med. Bioeth.* **25**(2), 143–155.

McGuire, A. L., and Gibbs, R. A. (2006a). No longer de-identified. *Science* **312,** 370–371.

McGuire, A. L., and Gibbs, R. A. (2006b). Meeting the growing demands of genetic research. *J. Law. Med. Ethics* **34,** 809–812.

Meltzer, L. A. (2006). Undesirable implications of disclosing individual genetic results to research participants. *Am. J. Bioeth.* **6,** 28–30.

Merz, J. F., Sankar, P., Taube, S. E., and Livolsi, V. (1997). Use of human tissues in research: Clarifying clinician and researcher roles and information flows. *J. Investig. Med.* **45,** 252–257.

Merz, J. F., McGee, G. E., and Sankar, P. (2004). "Iceland Inc."? On the ethics of commercial population genomics. *Soc. Sci. Med.* **58**(6), 1201–1209.

Moore v. Regents of the University of California (1991). 793 P.2d 479 (Cal. 1990), cert. denied, 499 US 936.

Muller, Y. L., Infante, A. M., Hanson, R. L., Love-Gregory, L., Knowler, W., Bogardus, C., and Baier, L. J. (2005). Variants in hepatocyte nuclear factor 4alpha are modestly associated with type 2 diabetes in Pima Indians. *Diabetes* **54**(10), 3035–3039.

National Bioethics Advisory Commission (NBAC) (1999). "Research Involving Human Biological Materials: Ethical Issues and Policy Guidance (volume 1)." Available at http://www.georgetown.edu/research/nrcbl/nbac/hbm.pdf.

National Cancer Institute, Office of Biorepositories and Biospecimen Research (2006). "First-Generation Guidelines for NCI-supported Biorepositories." Available at http://biospecimens.cancer.gov/biorepositories/First%20Generation%20Guidelines%20042006.pdf.

National Center for Biotechnology Information (2006). Genetic Association Information Network. Available at http://www.fnih.org/GAIN2/home_new.shtml.

National Commission for the Protection of Human Subjects of Biomedical Behavioral Research (1979). "The Belmont Report: Ethical Principles and Guidelines for the Protection of Human Subjects of Research." US Government Printing Office, Washington DC.

National Institutes of Health (NIH; undated). "Research on Human Specimens: Are You Conducting Research Using Human Subjects?". Available at http://www.cdp.ims.nci.nith.gov/brochure.html conducting.

National Institutes of Health (2007). "Request for Information (RFI): Proposed Policy for Sharing of Data obtained in NIH supported or conducted Genome-Wide Association Studies (GWAS) (NOT-OD-07–088).". Available at http://grants.nih.gov/grants/guide/notice-files/NOT-OD-07–088.html.

National Research Council, Committee on Genome Diversity (1997). "Evaluating human genetic diversity." National Academy Press, Washington DC.

Nilstun, T., and Hermeren, G. (2006). Human tissue samples and ethics—attitudes of the general public in Sweden to biobank research. *Med. Health Care Philos* **9,** 81–86.

Ossorio, P. N. (2006). Letting the gene out of the bottle: A comment on returning individual research results to participants. *Am. J. Bioeth.* **6,** 24–25.

Public Population Project in Genomics (2006). Available at http://www.p3gconsortium.org.

Parker, L. S. (2006a). Rethinking respect for persons enrolled in research. *ASBH Exchange* **9**(1), 6–7.

Parker, L. S. (2006b). Best laid plans for offering results go awry. *Am. J. Bioeth.* **6,** 22–3.

Partridge, A. H., and Winer, E. P. (2002). Informing clinical trial participants about study results. *JAMA* **288,** 363–365.

Peterson, A. (2005). Securing our genetic health: Engendering trust in UK Biobank. *Sociol. Health Illn.* **27**(2), 271–292.

Plummer, P., Jackson, S., Konarski, J., Mahanna, E., Dunmore, C., Regan, G., Mattingly, D., Parker, B., Williams, S., Andrews, C., *et al.* (2002). Making epidemiologic studies responsive to the needs of participants and communities: The Carolina Breast Cancer Study experience. *Environ. Mol. Mutagen.* **39,** 96–101.

Pritchard, J. K. (2001). Are rare variants responsible for susceptibility to complex diseases? *Am. J. Hum. Genet.* **69,** 124–137.

Ravitsky, V., and Wilfond, B. S. (2006). Disclosing individual genetic results to research participants. *Am. J. Bioeth.* **6,** 8–17.

Reilly, P. R. (1998). Rethinking risks to human subjects in genetic research. *Am. J. Hum. Genet.* **63**(3), 682–685.

Reymond, M. A., Steinert, R., Escourrou, J., and Fourtanier, G. (2002). Ethical, legal and economic issues raised by the use of human tissue in postgenomic research. *Dig. Dis.* **20**(3–4), 257–265.

Richards, M. P., Ponder, M., Pharoah, P., Everest, S., and Mackay, J. (2003). Issues of consent and feedback in a genetic epidemiological study of women with breast cancer. *J. Med. Ethics* **29,** 93–96.

Rose, H. (2001). "The Commodification of bioinformation: The Icelandic Health Sector Database." Available at http://www.mannvernd.is/greinar/hilaryrose1_3975.pdf.

Rose, H. (2003). An ethical dilemma. *Nature* **425,** 123–124.

Rose, H. (2006). From hype to mothballs in four years: Troubles in the development of large-scale DNA biobanks in Europe. *Commun. Genet.* **9,** 184–189.

Rothstein, M. A. (2005). Expanding the ethical analysis of biobanks. *J. Law Med. Ethics* **33,** 89–101.

Rothstein, M. A. (2006). Tiered disclosure options promote the autonomy and well-being of research subjects. *Am. J. Bioeth.* **6,** 20–21.

Schuklenk, U., and Kleinsmidt, A. (2006). North-South benefit sharing arrangements in bioprospecting and genetic research: A critical ethical and legal analysis. *Develop. World Bioeth.* **6**(3), 122–134.

Shalowitz, D. I., and Miller, F. G. (2005). Disclosing individual results of clinical research: Implications of respect for participants. *JAMA* **294,** 737–740.

Sharp, R. R., and Foster, M. W. (2000). Involving study populations in the review of genetic research. *J. Law Med. Ethics* **28,** 41–51.

Shickle, D. (2006). The consent problem within DNA biobanks. *Stud. Hist. Philos. Biol. Biomed. Sci.* **37,** 503–519.

Shickle, D., Hapgood, R., Carlisle, J., Shackley, P., Morgan, A., and McCabe, C. (2003). Public attitudes to participating in UK BioBank: A DNA bank, lifestyle and morbidity database on 500,000 members of the UK public aged 45–69. *In* "Populations and Genetics: Legal and Socio-Ethical Perspectives" (B. M. Knoppers, ed.), pp. 323–344. Brill Academic Publishers, Leiden.

Stoddart, M. L., Blevins, K. S., Lee, E. T., Wang, W., Blackett, P. R., and Cherokee Diabetes Study (2002). Association of acanthosis nigricans with hyperinsulinemia compared with other selected risk factors for type 2 diabetes in Cherokee Indians: The Cherokee Diabetes Study. *Diabetes Care* **25**(6), 1009–1014.

The International HapMap Consortium (2004). Integrating ethics and science in the International HapMap Project. *Nat. Rev. Genet.* **5,** 467–475.

U.K. Biobank Limited (2007). Further information leaflet (version 27, July 2007). Available at http://www.ukbiobank.ac.uk/docs/FILB_207_Furtherinformationleafletrevision.pdf.

UNESCO (2003). "International Declaration on Human Genetic Data." Available at http://portal.unesco.org/en/ev.php-URL_ID = 17720&URL_DO = DO_TOPIC&URL_SECTION = 201.html.

U.S. Code of Federal Regulations (2000). Title 45 (Public Welfare and Human Services), Subpart A (Basic HHS Policy for Protection of Human Research Subjects), Part 46 (Protection of human subjects), Sec. 46.102 Definitions (45CFR46.102).

U.S. Code of Federal Regulations (2000). Title 45 (Public Welfare and Human Services), Subpart A (Basic HHS Policy for Protection of Human Research Subjects), Part 46 (Protection of human subjects) Sec. 46.116 General requirements for informed consent (45CFR46.116).

U.S. Department of Health and Human Services, Office for Protection from Research Risks (1997). "Issues to Consider in the Research Use of Stored Data or Tissues." Available at http://www.hhs.gov/ohrp/humansubjects/guidance/reposit.htm.

U.S. Department of Health and Human Services, Office for Human Research Protections (2004). "Guidance on Research Involving Coded Private Information or Biological Specimens." http://www.hhs.gov/ohrp/humansubjects/guidance/cdebiol.pdf.

Washington University v. Catalona, 437 F.Supp.2d 985 (E.D.Mo., 2006).

Weijer, C., and Emanuel, E. J. (2000). Protecting communities in biomedical research. *Science* **289,** 1142–1144.

Weijer, C., Goldsand, G., and Emanuel, E. J. (1999). Protecting communities in research: Current guidelines and limits of extrapolation. *Nat. Genet.* **23**(3), 275–280.

Weiss, K. M., and Clark, A. G. (2002). Linkage disequilibrium and the mapping of complex human traits. *Trends Genet.* **18**(1), 19–24.

Wendler, D. (2006a). One-time general consent for research on biological samples. *BMJ* **332,** 544–547.

Wendler, D. (2006b). One-time general consent for research on biological samples: Is it compatible with the health insurance portability and accountability act? *Arch. Intern. Med.* **166,** 1449–1452.

Winickoff, D. E. (2003). Governing population genomics: Law, bioethics, and biopolitics in three case studies. *Jurimetrics* **43**(2), 187–228.

Wolf, L. E., and Lo, B. (2004). Untapped potential: IRB guidance for the ethical research use of stored biological materials. *IRB* **26,** 1–8.

World Health Organization (WHO) (2003). "Genetic Databases—Assessing the Benefits and the Impact of Human Rights and Patient Rights." WHO, Geneva. Available at http://www.law.ed.ac.uk/ahrb/publications/online/whofinalreport.pdf.

Zlotnik Shaul, R., Reid, L., Essue, B., Gibson, J., Marzinotto, V., and Daneman, D. (2005). Dissemination to research subjects: Operationalizing investigator accountability. *Account Res.* **12,** 1–16.

Part IV

PROMISING TOPICS

19 Admixture Mapping and the Role of Population Structure for Localizing Disease Genes

Xiaofeng Zhu,* Hua Tang,† and Neil Risch‡
*Department of Epidemiology and Biostatistics,
Case Western Reserve University, Cleveland, Ohio 44106
†Department of Genetics, Stanford University, Stanford, California 94305
‡Institute for Human Genetics, University of California,
San Francisco, California 94143

Advances in Genetics, Vol. 60

0065-2660/08 $35.00
DOI: 10.1016/S0065-2660(07)00419-1

ABSTRACT

Admixture mapping, or mapping by admixture linkage disequilibrium, is a disease mapping strategy that has gained considerable popularity in recent years. It exploits the long-range linkage disequilibrium generated by admixture between genetically distinct ancestral populations. Compared to case-control association designs, admixture mapping requires fewer markers, and is more robust to allelic heterogeneity. At the same time, admixture mapping can be more powerful, and can achieve higher mapping resolution than traditional linkage studies, provided that the underlying trait variants occur at sufficiently different frequencies in the ancestral populations. In this chapter, we describe the recent methodology and software development, review successful applications, and comment on the future of this approach.

I. INTRODUCTION

Complex diseases are caused by the combination of many genes and environmental factors, as well as gene–gene and gene–environment interactions, which result in weak genotype–phenotype correlation. Linkage and association studies are currently the two main approaches for searching for genes underlying complex diseases. While linkage studies aim to identify loci that cosegregate with the trait within families, association studies seek variants associated with the trait at a population level. The linkage analysis approach has proven to be successful in studying Mendelian traits; however, studies of complex traits have been frustratingly difficult (Risch, 2000). With the development of high throughput genotyping techniques and abundant single-nucleotide polymorphisms (SNPs) available in human population, association analysis has been promoted because of its greater power and finer resolution compared with linkage studies. This is because association studies make an assumption that either the disease variant itself or a marker in strong linkage disequilibrium (LD) with the disease variant will be tested. Thus, the efficiency of association studies depends critically on the distribution of LD across the genome, which varies dramatically between populations, as well as between different genomic regions. The number of SNPs required to perform a genome-wide association study depends on the population studied (Carlson *et al.*, 2004; Gabriel *et al.*, 2002). With the completion of the International HapMap project, tagging SNPs can be selected to perform genome-wide association studies at a reduced cost (International HapMap Consortium, 2005), although such a cost may still be prohibitive for many laboratories. It has been suggested that tagging SNPs selected based on the limited samples in the HapMap project are efficient to transfer to almost all world populations with little loss in power (Conrad *et al.*, 2006, de Bakker *et al.*, 2006). Recent identification of IL23R as a gene for inflammatory bowel disease and numerous genes for T2D demonstrates the power of whole-genome association studies to uncover new disease-associated variants for complex traits (Duerr *et al.*, 2006; Saxena *et al.*, 2007; Scott *et al.*, 2007; Zeggini *et al.*, 2007).

II. POPULATION GENETIC STRUCTURE IN HUMANS

Understanding the distribution of genetic variation in human populations will help the dissection of complex diseases because most of the studies are through sampling from populations. The largest component of genetic structure in the human population falls on geographic/continental lines. For example, a study (Rosenberg *et al.*, 2002) of 377 autosomal microsatellite markers in 1056 individuals from 52 indigenous populations worldwide found significant evidence of genetic clustering along continental lines. Among six main genetic clusters identified, five of them correspond to major geographic regions: Europeans/West Asians (whites), sub-Saharan Africans, East Asians, Pacific Islanders, and Native Americans. Consistent results were obtained by using 326 autosomal microsatellite markers and 3636 subjects sampled from 15 different geographic locales within the United States and Taiwan (Tang *et al.*, 2004). Tang *et al.* (2004) found near-perfect correspondence between the self-reported racial/ethnic categories (white non-Hispanic, black non-Hispanic, Hispanic, and East Asian) and the four genetic clusters.

Population genetic structure also plays an important role in population-based case-control association design, which is one of the most commonly used designs. Concerns have been raised that the case-control designs can be thwarted by stratification bias, leading to false positive inference (Lander and Schork, 1994). Geographic ancestry is a major determinant of the genetic structure globally as well as within the United States (Rosenberg *et al.*, 2002; Tang *et al.*, 2004). Allele frequency differences range from moderate to large between continentally divided subpopulations. At the same time, environmental risk factors and disease prevalence rates also differ. In an ill-matched study, cases and controls may represent subpopulations in unequal proportion; therefore, alleles that have disparate allele frequencies among subpopulations may appear associated with disease status. Because there is no direct causal relationship between the tested loci and disease, such results are considered spurious findings.

Even seemingly homogeneous populations, such as US whites, can harbor cryptic population structure and lead to spurious association results (Campbell *et al.*, 2005; Freedman *et al.*, 2006). Cryptic population stratification is a particular concern for studies focusing on populations that have experienced recent genetic admixture, such as African-Americans and Latinos. An early example of spurious association was reported by Knowler *et al.* (1988), in which an immunoglobulin haplotype, $Gm^{3,5,13,14}$, appeared to affect the risk of type II diabetes among American Indians. The association disappeared, however, after controlling for Caucasian admixture in this population. Thus, properly adjusting for population ancestry would be important for an association test to be valid.

III. POPULATION ADMIXTURE

As a result of population genetic diversification, modern populations formed by the recent admixture of geographically diverged ancestral populations, such as African-Americans and Hispanics, can be extremely useful in admixture mapping or in detecting natural selection. Patterns of genetic variation generated by recent population admixture offer complementary information for localizing susceptibility genes (Lautenberger *et al.*, 2000; Pfaff *et al.*, 2001; Rife, 1954; Risch, 1992; Stephens *et al.*, 1994). This strategy was first proposed by Rife (1954), and its advantage and feasibility were shown in various theoretical studies (Chakraborty and Weiss, 1988; McKeigue, 1997, 1998; McKeigue *et al.*, 2000; Shriver *et al.*, 2003; Stephens *et al.*, 1994, Thomson, 1995; Zheng and Elston, 1999). Both classical likelihood-based methods (Zhang *et al.*, 2004; Zhu *et al.*, 2004) as well as Bayesian approaches (Hoggart *et al.*, 2004; Montana and Pritchard, 2004; Patterson *et al.*, 2004) have been proposed specifically for genome-wide or regional admixture mapping. According to these theoretical studies, admixture mapping is potentially more powerful than traditional linkage studies when the population risk ratio between parental populations is high; at the same time, the cost of genotyping for an admixture mapping study is likely lower than that of genome-wide association studies. Empirically, numerous studies in admixed populations have found significant correlation between individual ancestry (defined as the genome-wide average proportion of ancestry from a given population) and various phenotypes, including skin pigmentation, T2D, asthma, and blood pressure, suggesting admixture mapping is a potentially effective approach for these and other phenotypes (Burchard *et al.*, 2005; Lamason *et al.*, 2005; Martinez-Marignac *et al.*, 2007; Parra *et al.*, 2004; Shriver *et al.*, 2003; Tang *et al.*, 2006). However, applications are limited because this method requires a large number of ancestry informative markers (AIMs), which had not been available until recently.

As the first successful application of admixture mapping, Zhu *et al.* (2005) used a panel of microsatellite markers designed for traditional linkage analysis and identified two regions, on chromosomes 6 and 21, which may harbor genes influencing hypertension. More recently, Reich *et al.* (2005) used an ancestry-informative SNP panel (Smith *et al.*, 2004) and admixture mapping to identify a candidate locus for multiple sclerosis susceptibility on chromosome 1 in the centromeric region, while Freedman *et al.* (2006) reported 8q24 as the site of a prostate cancer risk locus in African-American men. These successes demonstrate that admixture mapping could be an important and promising alternative to the traditional linkage and association methods in mapping complex traits.

IV. METHODS OF ADMIXTURE MAPPING

A. The basic idea

The rationale of admixture mapping is quite simple. The key assumption is that a risk allele occurs at a substantially different frequency among ancestral populations. When this is true, we expect that, in the admixed population, affected individuals share an excess of ancestry at the disease locus from the ancestral population with the highest frequency of the risk allele compared with random locations in the genome. The same principle has been exploited in creating crosses in model organisms for QTL mapping, in which crossing of two inbred lines generates an association between strain ancestry and a QTL.

Consider a recently admixed population whose ancestors are from multiple ancestral populations. Chromosome crossovers between a pair of parental chromosomes occur during meiosis, and the resulting hybrid chromosomes are transmitted to the offspring. This process continues through subsequent generations, Thus, each chromosome of an admixed individual can be thought of as a mosaic of chromosomal blocks that originate from multiple ancestral populations. The expected length of a block in cM depends on the history of the admixed population, while the physical length also depends on the local recombination rate. Figure 19.1 llustrates such chromosomes sampled from a population admixed between two ancestral populations. We can characterize the ancestral origins of the two allele at a locus, and measure the LD between the ancestries at two loci using D' or r^2. It should be pointed out that such LD is created purely through the admixture process; hence, it is termed admixture LD in order to distinguish it from the background LD. Analytic derivation and empirical studies both show admixture LD exists at a greater genetic distance in a recently admixed population, compared with the background LD in a nonadmixed population. (Chakraborty and Weiss, 1988; Stephens *et al.*, 1994; Zhu *et al.*, 2006).

For the purpose of the illustrating the basic idea, we consider an extreme case, in which a disease allele only occurs in one of two ancestral populations (e.g., the disease allele occurs only in the African population but not in Europeans). Because of admixture LD, we would expect an overrepresentation of the ancestral blocks from the African population around the disease locus in cases, while a

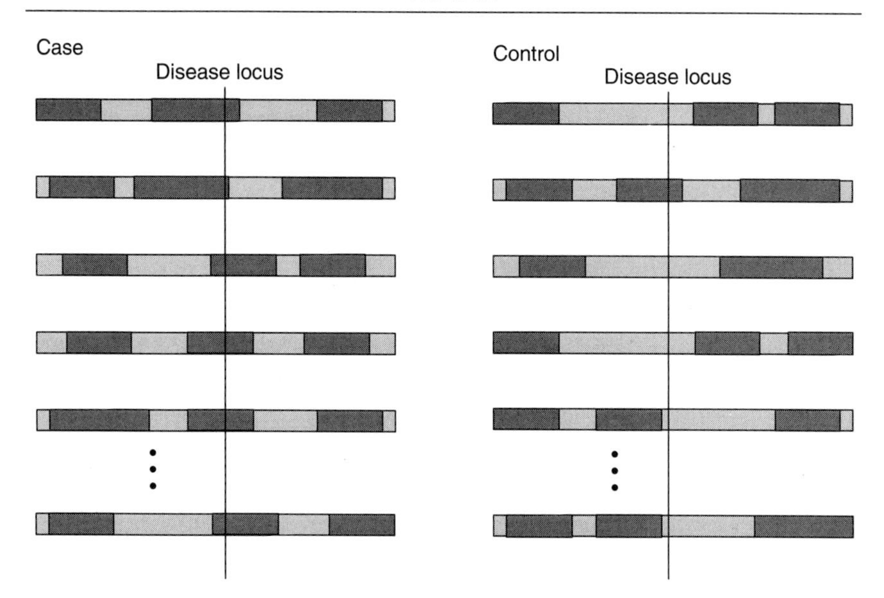

Figure 19.1. The mosaic structure of chromosomes of cases and controls sampled from an admixed population with two ancestral populations. The shaded and unshaded segments represent chromosome segments inherited from two ancestral populations. The vertical line represents the location of a disease susceptibility variant. When the disease variant has a high frequency in the shaded population and is rare in the other, more shaded segments at the disease locus in cases are observed than away from the disease locus as well as any regions in controls.

random fluctuation is expected in controls. To illustrate this property, we simulated 500 cases and 500 controls for an admixed African-European population using the method described by Zhu *et al.* (2006). Briefly, we assumed a panel of SNPs that are informative for admixture mapping across the genome as reported by Smith *et al.* (2004). The allele frequencies of the SNPs and the marker map for both the African and European ancestral populations were downloaded from www.journals.uchicago.edu. We first simulated the ancestral African and European populations based on these data by assuming Hardy–Weinberg and linkage equilibrium within each ancestral population. Admixed samples were then generated according to a continuous gene-flow model, in which the rate of admixture varied in each generation, and was drawn from a uniform distribution between 0 and 0.1 (Zhu *et al.* 2004). We assumed the disease allele frequency D in the two ancestral populations to be 0.5 and 0, respectively. We further assumed a recessive mode of inheritance with the penetrances $P(\text{disease}|\text{DD}) = 0.45$ and $P(\text{disease}|\text{Dd}) = P(\text{disease}|\text{dd}) = 0.05$. Five hundred cases and 500 controls were drawn from the 10th generation. We then calculated the average ancestry as a

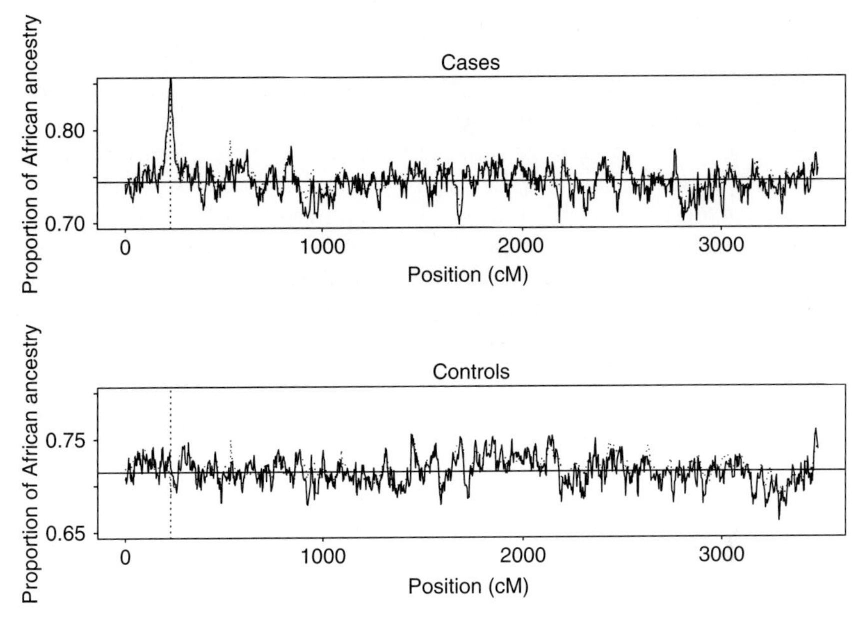

Figure 19.2. Average ancestry across the genome in cases and controls simulated according to the general continuous gene-flow model. The solid lines represent the true average ancestry and the dotted lines represent the estimated average ancestry by HMM for cases (top) and controls (bottom). The vertical line is the simulated disease susceptibility locus. Five hundred cases, 500 controls, and 1396 AIMs were used in the estimation. The risk allele frequency in one ancestral population is 0.5 and 0.0 in the other. The average ancestry from the high-risk population around the disease locus is much larger than away from the disease locus among cases. In controls, only random fluctuation is observed.

function of genome position in cases and controls, respectively. Figure 19.2 plots the average ancestry from African population among the cases and controls: clear excess ancestry from the African population is observed among the cases near the disease locus; no peak of similar magnitude was observed among the controls. It can also be observed that cases have more African ancestry on average than controls because of the admixture LD created due to the admixture process, even for unlinked loci.

B. Test statistics

Mathematical models formalizing admixture mapping as described above can be found in numerous reports (Hoggart *et al.*, 2004; McKeigue, 1998; Patterson *et al.*, 2004; Zhu *et al.*, 2004; Montana and Pritchard, 2004). Here we adopt the notation used by Zhu *et al.* (2004). Suppose we have an admixed population C resulting from two parental populations, X and Y. Let $\Pi_d(\theta)$ and $\Pi_c(\theta)$ be the proportion of alleles that are ancestry X (the "ancestral probability" from population X) among cases and controls in the current admixed population, respectively, where θ represents the genetic distance between the disease location and the candidate marker. Then both $\Pi_d(\theta)$ and $\Pi_c(\theta)$ are locally strict monotonic functions of θ (Zhu *et al.*, 2004, 2006). The null hypothesis is that the marker is unlinked to the disease allele, or $\theta = 0.5$ between the marker locus and the disease locus. This suggests two test statistics: (1) a case-only design where one tests the null hypothesis: $\Pi_d(\theta) = \Pi_d(0.5)$, where $\Pi_d(0.5)$ represents the expected proportion of alleles that are ancestry X at a locus unlinked to the disease locus in cases, (2) and a case-control design where one tests the null hypothesis: $\Pi_d(\theta) - \Pi_d(0.5) = \Pi_c(\theta) - \Pi_c(0.5)$, where $\Pi_c(0.5)$ is similarly defined in controls. These two test statistics were also discussed by Montana and Pritchard (2004). Similarly, the tests proposed by Hoggart *et al.* (2004) and Patterson *et al.* (2004) compare the locus-specific ancestry proportion with that expected under the null hypothesis. Hoggart *et al.* (2004) apply a logistic regression model, while Patterson *et al.* (2004) use a likelihood based approach by parameterizing in term of the relative risks associated with the number of alleles that are of ancestry X.

Without confusion, let $\hat{\Pi}_d(t)$ be the estimated proportion of ancestry from population X at chromosome location t, conditional on the observed marker genotypes. Both Zhu *et al.* (2004, 2006) and Montana and Pritchard (2004) proposed a test statistic

$$Z_C(t) = \frac{\hat{\Pi}_d(t) - \hat{\Pi}_d(0.5)}{\sigma(\hat{\Pi}_d(t))} \tag{1}$$

for a case-only design and

$$Z_{CC}(t) = \frac{(\hat{\Pi}_d(t) - \hat{\Pi}_d(0.5)) - (\hat{\Pi}_c(t) - \hat{\Pi}_c(0.5))}{\sigma(\hat{\Pi}_d(t) - \hat{\Pi}_c(t))} \tag{2}$$

for a case-control design, respectively. The difference between the two approaches lies in the way the variance is estimated in the denominators of the test statistics. Zhu *et al.* (2004) suggested using the empirical mean and standard error of the ancestry proportion at a set of unlinked marker loci, while Montana and Pritchard (2004) proposed a parametric bootstrap method. To illustrate in further detail, we assume an admixture mapping study consisting of n_1 unrelated cases and n_2 unrelated controls genotyped at M marker loci. Let x_{ij} be the estimated ancestry proportion from population X for the ith individual at marker j. Then the estimated ancestry proportions overall for cases and controls at marker j are given by

$$\hat{\Pi}_{\mathrm{d}}(j) = \frac{1}{n_1} \sum_{i=1}^{n_1} x_{ij}$$

and

$$\hat{\Pi}_{\mathrm{c}}(j) = \frac{1}{n_2} \sum_{i=n_1+1}^{n_1+n_2} x_{ij},$$

respectively. The ancestry proportion for cased and controls at a locus unlinked to the disease locus are denoted by $\Pi_{\mathrm{d}}(0.5)$ and $\Pi_{\mathrm{c}}(0.5)$, respectively, and can be estimated based on the ancestry at M randomly chosen and unlinked markers:

$$\hat{\Pi}_{\mathrm{d}}(0.5) = \frac{1}{Mn_1} \sum_{j=1}^{M} \sum_{i=1}^{n_1} x_{ij}$$

and

$$\hat{\Pi}_{\mathrm{c}}(0.5) = \frac{1}{Mn_2} \sum_{j=1}^{M} \sum_{i=n_1+1}^{n_1+n_2} x_{ij},$$

respectively. Then, according to Zhu *et al.* (2004), the sampling variance [i.e., the squares of the dominators in Eqs. (1) and (2)] can be estimated by

$$\sigma^2(\hat{\Pi}_{\mathrm{d}}) = \frac{1}{M} \sum_{j=1}^{M} [\hat{\Pi}_{\mathrm{d}}(j) - \hat{\Pi}_{\mathrm{d}}(0.5)]^2 \quad (3)$$

and

$$\sigma^2(\hat{\Pi}_{\mathrm{d}} - \hat{\Pi}_{\mathrm{c}}) = \frac{1}{M} \sum_{j=1}^{M} [(\hat{\Pi}_{\mathrm{d}}(j) - \hat{\Pi}_{\mathrm{c}}(j)) - (\hat{\Pi}_{\mathrm{d}}(0.5) - \hat{\Pi}_{\mathrm{c}}(0.5))]^2, \quad (4)$$

respectively. The rationale to estimate the variance by Eqs. (3) and (4) is that under the null hypothesis, the ancestral proportion at any locus comes from the same distribution when marker loci have the same ancestry informativeness. The variance estimated in this way has been shown theoretically to be asymptotically unbiased. In fact, this method even works well for linked marker loci (Sha *et al.*, 2006).

Slightly different case-only test statistics have been proposed by Hoggart *et al.* (2004) and Patterson *et al.* (2004). Let r be the ancestry risk ratio at the locus under study, defined as the ratio of risks for the trait for carrying one allele of ancestry X compared to those carrying none under the assumption of a multiplicative mode of inheritance. The risk ratio of carrying two alleles of ancestry X is r^2 compared to those with none. Let λ be the admixture proportion from the high-risk parental population (i.e., population X). Then, assuming random mating, $\Pi_d(0)$ is given by (McKeigue 1998; Zhu *et al.*, 2004):

$$\Pi_d(0) = \frac{\lambda r}{\lambda r + 1 - \lambda}.$$

Thus, the likelihood given the ancestry of an observed allele, A (A = 1 if the allele is of ancestry X and 0 otherwise), is

$$L(r; A) = \frac{(\lambda r)^A (1 - \lambda)^{1-A}}{\lambda r + 1 - \lambda}.$$

The overall likelihood can be calculated by multiplication for all alleles assuming a multiplative trait model and random mating. A standard likelihood ratio test or score test can then be carried out to test the null hypothesis $r = 1$ (Hoggart *et al.*, 2004). It should be noted that this hypothesis is slightly different from the testing null hypothesis $\Pi_d(\theta) = \Pi_d(0.5)$ (Zhu *et al.*, 2004). The reason is that r can be greater than 1 even at a locus unlinked to the disease locus because of admixture LD created in every generation of the admixture process. An alternative test was introduced by Patterson *et al.* (2004), in which the likelihood for the ith individual at marker j is defined as

$$L_{ij} = \frac{P(\text{data}|\text{disease locus})}{P(\text{data}|\text{no disease locus})} = \frac{v_{i,0}(j) + v_{i,1}(j)\varphi_1 + v_{i,2}(j)\varphi_2}{\eta_{i,0} + \eta_{i,1}\varphi_1 + \eta_{i,2}\varphi_2},$$

where $v_{i,0}(j)$, $v_{i,1}(j)$, and $v_{i,2}(j)$ are the probabilities of the ith individual having zero, one, and two alleles of ancestry X at the jth marker, respectively; $\eta_{i,0}$, $\eta_{i,1}$, and $\eta_{i,2}$ are the corresponding probabilities averaged across the genome in the same individual, and φ_1 and φ_2 are the relative risk of disease associated with having 1 and 2 alleles of ancestry X versus having no allele of ancestry X, respectively. Specific model parameters φ_1 and φ_2 are then tested: for example, $\varphi_1 = 1.3$, 1.5, 2, and 0.7, with $\varphi_2 = \varphi_1^2$ (Patterson *et al.*, 2004).

Finally, for a case-control test, logistic regression can be applied (Hoggart *et al.*, 2004) as

$$\log \frac{P(y|x)}{1 - P(y|x)} = \log \frac{\mu}{1 - \mu} + \beta(x - \lambda),$$

where y is disease status, x is the proportion of alleles that are of ancestry X at the test locus, μ is the disease prevalence in the study sample, and β is the log odds ratio of disease for individuals who carry 2 versus 0 alleles that are of ancestry X.

C. Inferring locus-specific ancestry

The discussion so far has assumed that the locus-specific ancestry is known without error. In reality, this quantity needs to be estimated. Existing methods for estimating locus-specific ancestry largely fall into two categories: Bayesian methods (Hoggart *et al.*, 2004; Montana and Pritchard, 2004; Patterson *et al.*, 2004) and likelihood-based approaches (Tang *et al.*, 2006; Zhang *et al.*, 2004; Zhu *et al.*, 2004, 2006). With the exception of the Markov-Hidden Markov Model (MHMM) proposed by Tang *et al.* (2006), all the other methods use a Hidden Markov Model (HMM) and require a set of AIMs. Let $\{g_t\}_{t=1}^{M}$ denote M ordered, observed genotypes along a chromosome and $\{v_t\}_{t=1}^{M}$ the unobservable number of alleles of ancestry X at the corresponding marker loci. We can consider $\{g_t, v_t\}_{t=1}^{M}$ in an HMM framework,

$$\begin{array}{llccccc} \text{observed} & \text{genotype} & g_1 & & g_2 & \cdots & g_M \\ & & \uparrow & & \uparrow & & \uparrow \\ \text{Hidden} & \text{States} & v_1 & \rightarrow & v_2 & \cdots \rightarrow & v_M \end{array}$$

by assuming the conditional independence given the hidden states—that is, $P(g_t|g_1, ..., g_{t-1}, v_1, ..., v_t) = P(g_t|v_t)$. This assumption may not be true when applied to dense marker sets, such as the markers used in a genome-wide association analysis, in which pairwise LD in ancestral populations is not negligible. In contrast, in MHMM (Tang *et al.*, 2006), the observed state g_t depends not only on v_t but also on the past history, as illustrated by

$$\begin{array}{llccccc} \text{observed} & \text{genotype} & g_1 & \rightarrow & g_2 & \cdots \rightarrow & g_M \\ & & \uparrow & & \uparrow & & \uparrow \\ \text{Hidden} & \text{States} & v_1 & \rightarrow & v_2 & \cdots \rightarrow & v_M \end{array}.$$

For computational tractability, Tang *et al.* (2006) consider only the first-order Markovian dependence—that is,

$$P(g_t|g_1, \ldots, g_{t-1}, v_1, \ldots, v_t) = \begin{cases} P(g_t|g_{t-1}, v_t) & \text{if } v_t = v_{t-1} \\ P(g_t|v_t) & \text{otherwise} \end{cases}.$$

Thus, MHMM is more general than HMM, and has the advantage of allowing for background LD in ancestral populations.

These existing methods also differ in their assumptions underlying the transition matrix that models the probability of the ancestral states between two adjacent markers. A transition matrix that represents a continuous gene-flow model was presented by Zhu *et al.* (2004), although this transition matrix applies only to admixed groups derived from two ancestral populations. A more flexible transition matrix is implemented in STRUCTURE (Falush *et al.*, 2003), which assumes an intermixing model; in this model, all ancestral chromosomes contributing to the admixed gene pool mixed in a single generation, followed subsequently by random mating (Long, 1991). This transition matrix has several appealing properties: it guarantees that the stationary distribution of the Markov chain coincides with the genome-average individual admixture; it allows for any number of ancestral populations; and the derived parameter can be interpreted in terms of the number of generations since admixing (Falush *et al.*, 2003). Tang *et al.* (2006) extended this transition matrix to allow for different admixing times in situations with more than two ancestral populations, which may be more realistic for some populations such as the Hispanics.

The likelihood of HMM or MHMM is calculated by the forward or backward algorithm. Directly maximizing the likelihood function works as well as computationally intensive Markov Chain Monte Carlo (MCMC) methods for estimating population parameters, such as the number of generations since the origin of admixing or the admixture rate in each generation (Zhu *et al.*, 2006). Ancestral allele frequencies can be simply counted in each surrogate ancestral population, if such information is available. However, the allele frequencies in the contemporary populations used to represent the ancestral populations may be biased, because of genetic drift or misrepresentation. To circumvent this problem, it is also possible to estimate these allele frequencies including the samples from the studied admixed population. Tang *et al.* (2005) and Zhu *et al.* (2006) suggest updating the allele frequencies via an Expectation Maximization (EM) algorithm, while others use Gibbs sampling. Simulation studies suggest both approaches performed well in general, although the EM method assuming a continuous gene-flow transition matrix outperformed Gibbs sampling when the sample was simulated according to an intermixture model (Zhu *et al.*, 2006).

Figure 19.2 demonstrates that the estimated average ancestry across the genome (dotted line) using the HMM method is reasonably good, but may be inaccurate at an individual level. This is particularly true when strong background LD exists. Ignoring background LD between markers tends to falsely infer ancestry admixture in indigenous individuals. For example, Fig. 19.3 displays the posterior mean ancestry estimates for chromosome 22 in a Han Chinese individual from Beijing genotyped in the HapMap project. When the intermarker spacing is decreased from 30 kb to 6 kb and to 3 kb, HMM mistakenly identifies increasing numbers of unexpected (and likely erroneous) ancestry switches to European or African ancestry (Tang *et al.*, 2006). In contrast, the MHMM estimated predominantly Asian ancestry.

D. Genome-wide significance level

Like genome-wide linkage and association analyses, admixture mapping also suffers from multiple comparisons issues because of the large number of loci tested. Although permutation testing is often applied in unrelated case-control

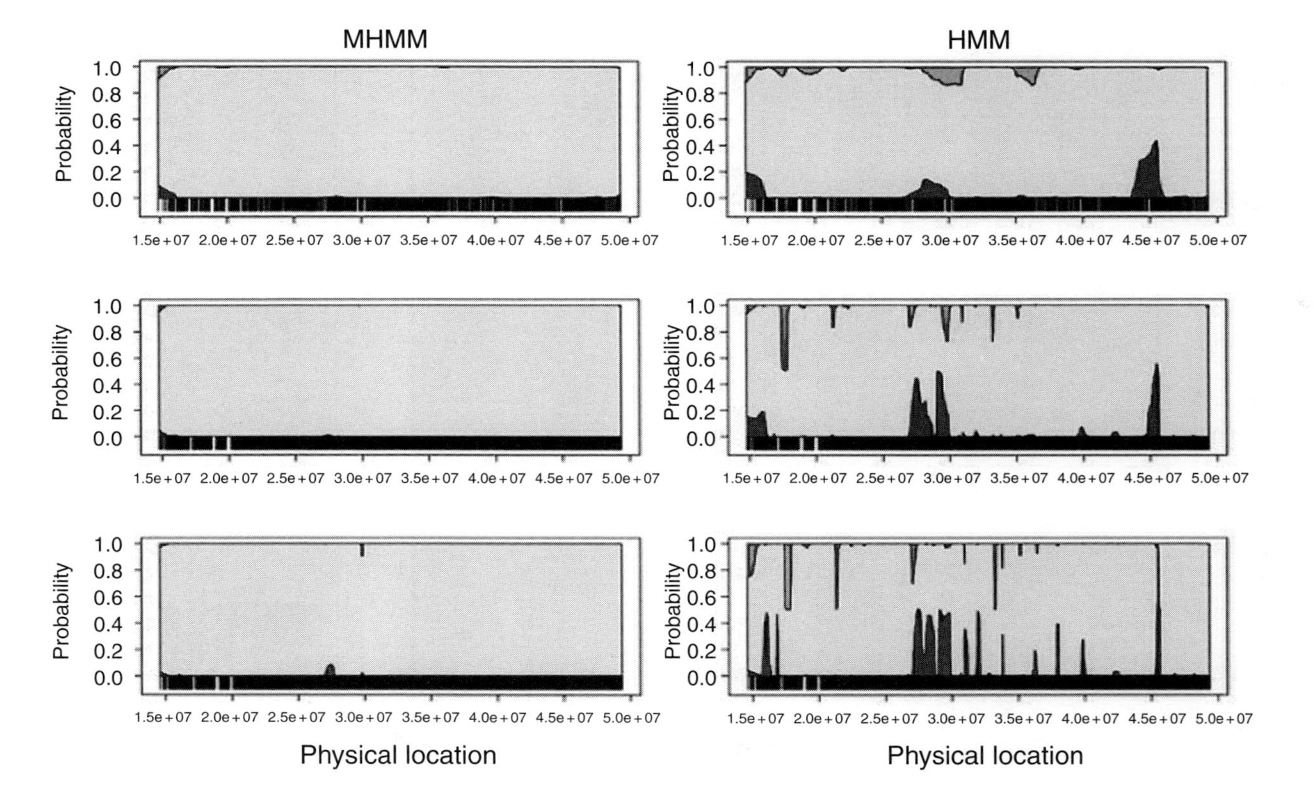

Figure 19.3. Estimated ancestry for a Han Chinese individual from Beijing. The *Y*-axis represents the posterior probability that one allele is derived from a specific ancestry and the *X*-axis indicates the physical locations of markers. Markers were sampled at an average spacing of 30 kb (top panels), 6 kb (middle panels), and 3 kb (bottom panels) that approximated the density of a 100 K SNP chip, a 500 K SNP chip, and all HapMap SNPs, respectively. Left panels, MHMM correctly infers Asian ancestry (gray) at most markers. Right panels, HMM assigns considerable probability to European ancestry (top light gray) or African ancestry (bottom dark gray) in several regions. (From Tang *et al.*, 2006. AJHG.)

studies, whether it is valid in admixture mapping is questionable, even for a case-control design. The reason is that the cases may be more related and have higher average ancestry from the ancestral population with the higher disease risk allele frequency than controls. Therefore, the exchangeability condition under null hypothesis may not be warranted. Sha *et al.* (2006) proposed a computationally efficient analytical approach for calculating the overall p value that accounts for the dependency among the tests. The method models test statistics across the genome as a Markov chain, and calculates the overall significance level using a bivariate normal distribution that represents a pair of adjacent markers. This analytic approach is much more computationally efficient than a simulation-based approach proposed by Montana and Pritchard (2004). Under the assumption of a single disease variant across the genome, an alternative approach to avoid multiple tests is the Bayesian whole-genome statistic T, which is the $\log_{10}$(average of the likelihood ratios across all the markers) (Patterson *et al.*, 2004), declaring a positive genome-wide association if $T > 2$. This statistic detects evidence for an association anywhere in the genome (Patterson *et al.*, 2004). However, it is not clear if this method is equivalent to a method correcting for multiple tests, especially when several variants contribute to the disease.

E. Software available

The available programs for carrying out admixture mapping are listed in Table 19.1. These programs differ mainly in the computational methods used for estimating the model parameters, specifically an MCMC approach versus direct likelihood maximization.

Table 19.1. Programs for Admixture Mapping

Program	References	Method	AIMs	No of ancestral populations	Background LE
STRUCTURE / MALDSOFT	Falush *et al.* (2003); Montana and Pritchard (2004)	MCMC	Microsatellite or SNP	No limit	Yes
ADMIXMAP	McKeigue (1998); Hoggart *et al.* (2004)	MCMC	Microsatellite or SNP	No limit	Yes
ANCESTRYMAP	Patterson *et al.* (2004)	MCMC	SNP	2	Yes
ADMIXPROGRAM	Zhu *et al.* (2004, 2006)	ML	SNP	2	Yes
SABER	Tang *et al.* (2006)	ML	SNP[a]	No limit	No

[a]Randomly selected single-nucleotide polymorphisms (SNPs).

V. DESIGN CONSIDERATION

A. Markers for admixture mapping

Although MHMM can use unselected dense markers, AIMs can be still useful for designing a cost-effective admixture mapping study at a reduced genotyping cost. AIMs can be either microsatellite markers (Zhu *et al.*, 2005; Smith et al., 2001) or SNPs (Reich *et al.*, 2005). Microsatellite markers can be more informative for ancestry than SNPs in general. However, far more SNPs are available than microsatellite markers across the genome, enabling the selection of highly informative SNPs for specific ancestries. Several measures have been used to evaluate the marker information content for distinguishing two ancestral populations: the absolute allele frequency difference δ, F_{ST} distance, Fisher information content, pairwise Kullback-Leibler divergence, and Informativeness for assignment (I_n) (Rosenberg *et al.*, 2003). Rosenberg *et al.* compared the performance of these measurements and concluded that all the measurements are highly correlated although I_n performs slightly better. Because HMM assumes that markers are in linkage equilibrium in ancestral populations, a preferable set of AIMs should capture as much ancestral information as possible with minimal LD in the ancestral populations. By screening 450,000 SNPs across the genome, Smith *et al.* (2004) assembled a panel of 3075 ancestry informative SNPs for populations of mixed West African and European descent. The average information content of this panel for distinguishing European from West African ancestry was 28% (measured by I_n), corresponding to $\delta = 0.56$. Zhu *et al.* (2006) studied the multiple marker information content of this panel by simulations. They found this panel can extract on average 89% of the ancestry information, although there still are some gaps that need to be filled. The Phase I of the HapMap project (Altshuler *et al.*, 2005) released over 1 million SNPs from four populations: Yoruba in Ibadan, Nigeria (YRI); Centre d'Etude du Polymorphisme Humain (CEU); Han Chinese in Beijing, China (CHB); Japanese in Tokyo, Japan (JPT). Thus, a panel of high-density ancestrally informative SNPs can be selected from HapMap. We compared the allele frequency difference among YRI, CEU, CHB, and JPT on chromosome 6 and 21 and concluded that there is an abundance of ancestral informative SNPs available across the genome (Table 19.2). We note that the δ values given in the table are likely to be exaggerated because of the relatively modest sample sizes of the groups studied ($n = 60$, 60, and 89 independent persons for YRI, CEU, and CHB-JPT, respectively). As mentioned before, the MHMM-based approach proposed by Tang *et al.* (2006) does not require preselecting AIMs. Moreover, using all available markers is likely to extract more ancestry information than selecting only a set of AIMs. It is important that the LD structure within each ancestral

Table 19.2. Distributions of AIMs in HapMap data

	CEU-YRI		YRI-CHB + JPN		CEU-CHB + JPN	
	Chr 6	Chr 21	Chr 6	Chr 21	Chr 6	Chr 21
$\delta > 0.4$						
No. of AIMs	15246	3390	21157	4270	7690	1248
Maximum distance between adjacent SNPs (kb)	3728	522.3	3248	457	7067	1007
Average distance (kb)	11.2	9.9	8.1	7.8	22.2	26.5
Average δ	0.51 ± 0.09	0.51 ± 0.09	0.52 ± 0.10	0.52 ± 0.10	0.48 ± 0.07	0.49 ± 0.08
$\delta > 0.5$						
No. of AIMs	6404	1526	10262	2000	2537	466
Maximum distance between adjacent SNPs (kb)	3980	1365	3299	616	7067	1961
Average distance (kb)	26.6	22.1	16.6	16.6	67.3	70.7
Average δ	0.59 ± 0.08	0.59 ± 0.08	0.61 ± 0.09	0.60 ± 0.08	0.57 ± 0.06	0.57 ± 0.07

δ: allele frequency difference between two populations.

population be properly incorporated in the statistical model; failure to model such LD will result in biased estimate of ancestry state, as demonstrated by Tang *et al.* (2006) and Tian *et al.* (2006).

B. Power and sample size

The power to detect linkage using admixture mapping depends on various model parameters, including parental population risk ratio at the disease locus, mode of inheritance, admixing history (both relative ancestry proportion and the time since admixing), and the distances between the testing markers and the disease locus. However, admixture mapping depends most critically on the allele frequency difference. Studies have shown that an allele frequency difference of at least 0.20 is required to have adequate power of detection even in a large sample. Furthermore, if the disease allele frequencies are sufficiently different between the ancestral populations, admixture mapping would have power even if the disease prevalence is similar in the parental populations (Smith and O'Brien, 2005).

Under an ideal situation, in which all markers are fully informative so that the ancestral state can be accurately inferred, power can be calculated analytically using a binomial distribution. When the sample is large enough,

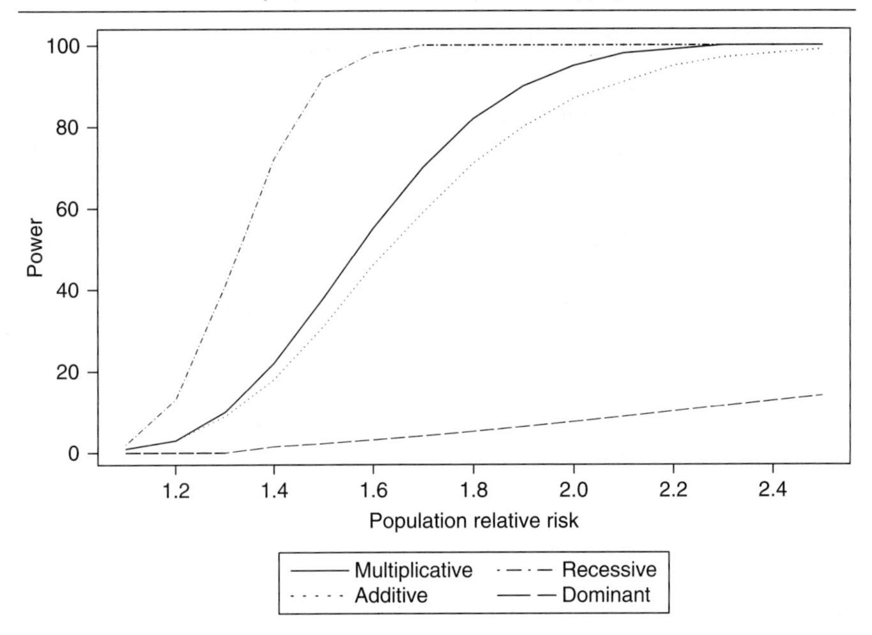

Figure 19.4. Power of admixture mapping with 800 cases for multiplative (A), additive (B), recessive (C), and dominant (D) models. Power is calculated at a significance level of 0.01 and plotted as a function of population relative risk.

we can use a normal distribution approximation. For a sample size of N cases and a two-sided type I error rate α, the power is given by

$$\text{power} = \Phi\left(\frac{\sqrt{2N}(\Pi_d(\theta) - \Pi_d(0.5)) - \sqrt{\Pi_d(0.5)(1 - \Pi_d(0.5))}Z_{1-0.5\alpha}}{\sqrt{\Pi_d(\theta)(1 - \Pi_d(\theta))}}\right),$$

where Z_α denotes the α-quantile of a normal density and Φ is the cumulative standard normal distribution function. Figure 19.4 displays the power of admixture mapping based on 800 cases from the African-American population, for whom we assumed a mixture of 18% European and 82% West African ancestries. We further assumed that markers have full ancestral information and the admixture process started 12 generations ago under the continuous gene-flow model.

VI. PROSPECTS AND CONCERNS OF ADMIXTURE MAPPING

The ideal populations for successful admixture mapping studies are those that have experienced recent admixture between genetically distinct ancestral populations, such as African-American and Hispanics populations. The first successful case of admixture mapping used the sample collected in the Family Blood Pressure Program (Province *et al.*, 2003) to identify two chromosomal regions (6q and 21q) potentially harboring genes affecting hypertension (Zhu *et al.*, 2005). Subsequently, an admixture mapping study of multiple sclerosis using a highly informative SNP panel reported a candidate locus on chromosome 1 (Reich *et al.*, 2005). More recently, an admixture mapping analysis identified a 3.8 Mb region on 8q24 that may harbor genes that influences the susceptibility for prostate cancer (Freedman *et al.*, 2006). The same region was also identified through linkage analysis, followed by fine mapping (Amundadottir *et al.*, 2006). Interestingly, follow-up association analysis revealed that the alleles described by Amundadottir *et al.* cannot account for the admixture signal, indicating that additional causal variants remain to be identified. These successes demonstrate that admixture mapping can be a powerful tool in the search for genes that influences the risk of complex diseases. In addition, admixture mapping can detect genes that are not amenable to linkage studies. Thus, evidence obtained from both linkage analysis and admixture mapping studies can be used for follow-up association studies.

An important challenge in the studies of complex diseases is the phenomenon of genetic heterogeneity, including allelic heterogeneity or locus heterogeneity that produces the same, indistinguishable phenotype (Risch, 2000). Similar to traditional linkage analysis, allelic heterogeneity is less likely to cause a problem for admixture mapping because the different mutations in a gene will likely fall in the same block created by the admixture process. However, locus heterogeneity may cause a serous problem in admixture mapping as in linkage analysis. For example, when a disease is caused by many genes, each of which contributes a small effect, then admixture mapping (as well as other mapping approaches) requires a large sample size.

Another limitation of admixture mapping is that only genetic variants with substantially different frequencies in ancestral populations can be detected. Smith and O'Brien (2005) listed 25 diseases that have different risks in African-Americans and European Americans, including various cancers, immunological pathologies, diabetes, hypertension, and multiple sclerosis. Noticeably, the risk of hypertension in African-Americans has been reported to be 2.6 times that in European Americans (Gupta *et al.*, 2003). Conversely, the risk of multiple sclerosis is two times higher in European Americans compared to African-Americans (Hogancamp *et al.*, 1997). This information provides some guidance in deciding whether admixture mapping is likely to be effective. However, as we have explained earlier, the observed ancestral risk ratio is neither a necessary nor sufficient criterion for admixture mapping.

Several authors have pointed out the limitations of admixture mapping (Smith and O'Brien, 2005, McKeigue, 2005). Most prominently, admixture mapping is likely to miss the disease variants with similar allele frequencies in the parental populations. In this case, linkage studies will be more powerful to identify these variants than admixture mapping. Success in admixture mapping also relies on having accurate information about the ancestral allele frequencies. As such information is often unknown, it has to be estimated, for example, by using the HMM and an assumed population admixture model. It has been shown that a continuous gene-flow model fits African-American population well and the corresponding allele frequencies in the parental populations can be accurately estimated (Zhu *et al.*, 2006). However, it may be more difficult to obtain accurate ancestral allele frequency estimates for Hispanic populations because of the lack of large-scale genetic data from Native American ancestral populations. In general, Hispanic populations have genetic ancestry from three populations: African, European, and Native American. There is substantial fluctuation of ancestry proportions in Hispanics depending on geographic origin. African ancestry ranges from less than 10% to over 50%, European ancestry from less than 20% to over 80%, and Native American ancestry from 5% to over 80% (Burchard *et al.* 2005; Salari *et al.*, 2005). Current HapMap data provides abundant information for identifying AIMs useful for the African-American populations. The utility of these AIMs for admixture mapping in Hispanic populations and other admixed groups remains unknown. Alternatively, Montana and Pritchard (2004) suggested that a large set of random SNPs can be equally informative as a small set of AIMs for admixture mapping studies. However, background LD among the markers in the parental populations may give rise to biased estimates of ancestral allele frequencies. The MHMM proposed by Tang *et al.* (2006) by allowing for background LD may be very useful in practice.

VII. CONCLUSIONS

Defining the genetic architecture of complex diseases remains a daunting challenge. A common feature of the underlying architecture for complex diseases is that multiple genetic variants are involved, each with a modest contribution. Nonetheless, resolving the genetics of these disorders will be a crucial step in biomedical research. In addition to the basic biological insights that will be gained, this knowledge could accelerate the development of new drugs and treatments. The rapid advance of computational and molecular technological developments offers great opportunities for studying complex diseases. Along with these exciting opportunities come new challenges, noticeably the needs for more sophisticated statistical methods. Traditional linkage-based studies are still optimal in the search for some genetic susceptibility variants. Although the genome-wide association design is a promising approach for searching susceptibility variants, the cost is still prohibitive for many laboratories. Admixture mapping combines the advantages of both linkage and association analysis, and can be especially powerful in searching for ethnic-specific variants. Admixture mapping has a resolution intermediate of linkage analysis and association studies, and the resolution depends on the admixing history and the density of the markers. Therefore, further follow-up association studies are necessary in order to identify the genetic susceptibility variants. With the advent of high-density SNPs for whole genome association studies and as the costs for genotyping decrease, a reasonable question to consider is the relative power of admixture mapping versus direct association. Analytic and simulation studies indicate that, for a large portion of the parameter space considered, direct association will be more powerful than admixture mapping; on the other hand, there are realistic scenarios in which direct association studies have low power and admixture mapping is still a better choice (Risch and Tang, 2006).

Acknowledgment

This work was supported by grant from National Human Genome Research Institute (HG003054) and National Institute of General Medicine Science (GM073059).

References

Amundadottir, L. T., Sulem, P., Gudmundsson, J., Helgason, A., Baker, A., Agnarsson, B. A., Sigurdsson, A., Benediktsdottir, K. R., Cazier, J. B., Sainz, J., Jakobsdottir, M., Kostic, J., *et al.* (2006). A common variant associated with prostate cancer in European and African populations. *Nat. Genet.* **38,** 652–658.

Burchard, G. E., Borrell, L. N., Choudhry, S., Naqvi, M., Tsai, H. J., Rodriguez-Santana, J. R., Chapela, R., Rogers, S. D., Mei, R., Rodriguez-Cintron, W., Arena, J. F., Kittle, R., *et al.* (2005). Latino populations: A unique opportunity for the study of race, genetics, and social environment in epidemiological research. *Am. J. Public Health* **95,** 2161–2168.

Campbell, C. D., Ogburn, E. L., Lunetta, K. L., Lyon, H. N., Freedman, M. L., Groop, L. C., Altshuler, D., Ardlie, K. G., and Hirschhorn, J. N. (2005). Demonstrating stratification in a European American population. *Nat. Genet.* **37,** 868–872.

Carlson, C. S., Eberle, M. A., Rieder, M. J., Yi, Q., Kruglyak, L., and Nickerson, D. A. (2004). Selecting a maximally informative set of single-nucleotide polymorphisms for association analyses using linkage disequilibrium. *Am. J. Hum. Genet.* **74,** 106–120.

Chakraborty, R., and Weiss, K. M. (1988). Admixture as a tool for finding linked genes and detecting that difference from allelic association between loci. *Proc. Natl. Acad. Sci. USA* **85,** 9119–9123.

Conrad, D. F., Jakobsson, M., Coop, G., Wen, X., Wall, J. D., Rosenberg, N. A., and Pritchard, J. K. (2006). A worldwide survey of haplotype variation and linkage disequilibrium in the human genome. *Nat. Genet.* **38,** 1251–1260.

de Bakker, P. I., Burtt, N. P., Graham, R. R., Guiducci, C., Yelensky, R., Drake, J. A., Bersaglieri, T., Penney, K. L., Butler, J., Young, S., *et al.* (2006). Transferability of tag SNPs in genetic association studies in multiple populations. *Nat. Genet.* **38,** 1298–1303.

Duerr, R. H., Taylor, K. D., Brant, S. R., Rioux, J. D., Silverberg, M. S., Daly, M. J., Steinhart, A. H., Abraham, C., Regueiro, M., Griffiths, A., *et al.* (2006). A genome-wide association study identifies IL23R as an inflammatory bowel disease gene. *Science* **314,** 1461–1463.

Falush, D., Stephens, M., and Pritchard, J. K. (2003). Inference of population structure using multilocus genotype data: linked loci and correlated allele frequencies. *Genetics* **164,** 1567–1587.

Freedman, M. L., Haiman, C. A., Patterson, N., McDonald, G. J., Tandon, A., Waliszewska, A., Penney, K., Steen, R. G., Ardlie, K., John, E. M., *et al.* (2006). Admixture mapping identifies 8q24 as a prostate cancer risk locus in African-American men. *Proc. Natl. Acad. Sci. USA* **103,** 14068–14073.

Gabriel, S. B., Schaffner, S. F., Nguyen, H., Moore, J. M., Roy, J., Blumenstiel, B., Higgins, J., DeFelice, M., Lochner, A., Faggart, M., *et al.* (2002). The structure of haplotype blocks in the human genome. *Science* **296,** 2225–2229.

Gupta, V., Nanda, N. C., Yesilbursa, D., Huang, W. Y., Gupta, V., Li, Q., and Gomez, C. (2003). Racial differences in thoracic aorta atherosclerosis among ischemic stroke patients. *Stroke* **34,** 408–412.

Hogancamp, W. E., Rodriguez, M., and Weinshenker, B. G. (1997). Identification of multiple sclerosis-associated genes. *Mayo Clin. Proc.* **72,** 965–976.

Hoggart, C. J., Shriver, M. D., Kittles, R. A., Clayton, D. G., and McKeigue, P. M. (2004). Design and analysis of admixture mapping studies. *Am. J. Hum. Genet.* **74,** 965–978.

International HapMap Consortium (2005). A haplotype map of the human genome. *Nature* **437,** 1299–1320.

Knowler, W. C., Williams, R. C., Pettitt, D. J., and Steinberg, A. G. (1988). Gm3;5,13,14 and type 2 diabetes mellitus: An association in American Indians with genetic admixture. *Am. J. Hum. Genet.* **43,** 520–526.

Lamason, R. L., Mohideen, M. A., Mest, J. R., Wong, A. C., Norton, H. L., Aros, M. C., Jurynec, M. J., Mao, X., Humphreville, V. R., Humbert, J. E., *et al.* (2005). SLC24A5, a putative cation exchanger, affects pigmentation in zebrafish and humans. *Science* **310,** 1782–1786.

Lander, E. S., and Schork, N. J. (1994). Genetic dissection of complex traits. *Science* **265,** 2037–2048.

Long, J. C. (1991). The genetic structure of admixed populations. *Genetics* **127,** 417–428.

Lautenberger, J. A., Stephens, J. C., O'Brien, S. J., and Smith, M. W. (2000). Significant admixture linkage disequilibrium across 30 cM around the FY locus in African Americans. *Am. J. Hum. Genet.* **66,** 969–978.

Kaplan, N. L., Martin, E. R., Morris, R. W., and Weir, B. S. (1998). Marker selection for the transmission/disequilibrium test, in recently admixed populations. *Am. J. Hum. Genet.* **62,** 703–712.

Martinez-Marignac, V. L., Valladares, A., Cameron, E., Chan, A., Perera, A., Globus-Goldberg, R., Wacher, N., Kumate, J., McKeigue, P., O'donnell, D., *et al.* (2007). Admixture in Mexico City: Implications for admixture mapping of Type 2 diabetes genetic risk factors. *Hum. Genet.* **120,** 807–819.

McKeigue, P. M. (1997). Mapping genes underlying ethnic differences in disease risk by linkage disequilibrium in recently admixed populations. *Am. J. Hum. Genet.* **60,** 188–196.

McKeigue, P. M. (1998). Mapping genes that underlie ethnic differences in disease risk: Methods for detecting linkage in admixed populations, by conditioning on parental admixture. *Am. J. Hum. Genet.* **63,** 241–251.

McKeigue, P. M. (2005). Prospects for admixture mapping of complex traits. *Am. J. Hum. Genet.* **76,** 1–7.

McKeigue, P. M., Carpenter, J. R., Parra, E. J., and Shriver, M. D. (2000). Estimation of admixture and detection of linkage in admixed populations by a Bayesian approach: Application to African-American populations. *Ann. Hum. Genet.* **64,** 171–186.

Montana, G., and Pritchard, J. K. (2004). Statistical tests for admixture mapping with case-control and cases-only data. *Am. J. Hum. Genet.* **75,** 771–789.

Parra, E. J., Kittles, R. A., and Shriver, M. D. (2004). Implications of correlations between skin color and genetic ancestry for biomedical research. *Nat Genet.* **36**(Suppl), S54–S60.

Patterson, N., Hattangadi, N., Lane, B., Lohmueller, K. E., Hafler, D. A., Oksenberg, J. R., Hauser, S. L., Smith, M. W., O'Brien, S. J., Altshuler, D., *et al.* (2004). Methods for high-density admixture mapping of disease genes. *Am. J. Hum. Genet.* **74,** 979–1000.

Pfaff, C. L., Parra, E. J., Bonilla, C., Hiester, K., McKeigue, P. M., Kamboh, M. I., Hutchinson, R. G., Ferrell, R. E., Boerwinkle, E., and Shriver, M. D. (2001). Population structure in admixed populations: Effect of admixture dynamics on the pattern of linkage disequilibrium. *Am. J. Hum. Genet.* **68,** 198–207.

Province, M. A., Kardia, S. L., Ranade, K., Rao, D. C., Thiel, B. A., Copper, R. S., Risch, N., Tumer, S. T., Cox, D. R., Hunt, S. C., Weder, A. B., and Boerwinkle, E. National Heart, Lung and Blood Institute Family Blood Pressure Program (2003). A meta-analysis of genome-wide linkage scans for hypertension: The National Blood Institute Family Blood Pressure Program. *Am. J. Hypertens.* **16,** 144–147.

Reich, D., Patterson, N., De Jager, P. L., McDonald, G. J., Waliszewska, A., Tandon, A., Lincoln, R. R., DeLoa, C., Fruhan, S. A., Cabre, P., *et al.* (2005). A whole-genome admixture scan finds a candidate locus for multiple sclerosis susceptibility. *Nat. Genet.* **37,** 1113–1118.

Rife, D. C. (1954). Populations of hybrid origin as source material for the detection of linkage. *Am. J. Hum. Genet.* **6,** 26–33.

Risch, N. (1992). Mapping genes for complex disease using association studies with recently admixed populations. *Am. J. Hum. Genet.* **51**(Suppl), 13.

Risch, N. J. (2000). Searching for genetic determinants in the new millennium. *Nature* **405,** 847–856.

Risch, N., and Tang, H. (2006). Whole Genome Association Studies in Admixed populations. 56th Annual Meeting, American Society of human genetics. p. 68.

Rosenberg, N. A., Pritchard, J. K., Weber, J. L., Cann, H. M., Kidd, K. K., Zhivotovsky, L. A., and Feldman, M. W. (2002). Genetic structure of human populations. *Science* **298,** 2381–2385.

Rosenberg, N. A., Li, L. M., Ward, R., and Pritchard, J. K. (2003). Informativeness of genetic markers for inference of ancestry. *Am. J. Hum. Genet.* **73,** 1402–1422.

Salari, K., Choudhry, S., Tang, H., Naqvi, M., Lind, D., Avila, P. C., Coyle, N. E., Ung, N., Nazario, S., Casal, J., *et al.* (2005). Genetic admixture and asthma-related phenotypes in Mexican American and Puerto Rican asthmatics. *Genet. Epidemiol.* **29,** 76–86.

Saxena, R., Voight, B. F., Lyssenko, V., Burtt, N. P., de Bakker, P. I., Chen, H., Roix, J. J., *et al.* Diabetes Genetics Initiative of Broad Institute of Harvard and MIT, Lund University and Novartis Institutes of BioMedical Research (2007). Genome-wide association analysis identifies loci for type 2 diabetes and triglyceride levels. *Science* **316,** 1331–1336.

Scott, L. J., Mohlke, K. L., Bonnycastle, L. L., Willer, C. J., Li, Y., Duren, W. L., Erdos, M. R., Stringham, H. M., Chines, P. S., Jackson, A. U., *et al.* (2007). A genome-wide association study of type 2 diabetes in Finns detects multiple susceptibility variants. *Science* **316,** 1341–1345.

Sha, Q., Zhang, X., Zhu, X., and Zhang, S. (2006). Analytical correction for multiple testing in admixture mapping. *Hum. Hered.* **62,** 55–63.

Shriver, M. D., Parra, E. J., Dios, S., Bonilla, C., Norton, H., Jovel, C., Pfaff, C., Jones, C., Massac, A., Cameron, N., *et al.* (2003). Skin pigmentation, biogeographical ancestry and admixture mapping. *Hum. Genet.* **112,** 387–399.

Smith, M. W., and O'Brien, S. J. (2005). Mapping by admixture linkage disequilibrium: Advances, limitations and guidelines. *Nat. Rev. Genet.* **6,** 623–632.

Smith, M. W., Lautenberger, J. A., Shin, H. D., Chretien, J. P., Shrestha, S., Gilbert, D. A., and O'Brien, S. J. (2001). Markers for mapping by admixture linkage disequilibrium in African American and Hispanic populations. *Am. J. Hum. Genet.* **69,** 1080–1094.

Smith, M. W., Patterson, N., Lautenberger, J. A., Truelove, A. L., McDonald, G. J., Waliszewska, A., Kessing, B. D., Malasky, M. J., Scafe, C., Le, E., *et al.* (2004). A high-density admixture map for disease gene discovery in African Americans. *Am. J. Hum. Genet.* **74,** 1001–1013.

Stephens, J. C., Briscoe, D., and O'Brien, S. J. (1994). Mapping by admixture linkage disequilibrium in human populations: Limits and guidelines. *Am. J. Hum. Genet.* **55,** 809–824.

Tang, H., Quertermous, T., Rodriguez, B., Kardia, S. L., Zhu, X., Brown, A., Pankow, J. S., Province, M. A., Hunt, S. C., Boerwinkle, E., *et al.* (2004). Genetic structure, self-identified race/ethnicity, and confounding in case-control association studies. *Am. J. Hum. Genet.* **76,** 268–275.

Tang, H., Peng, J., Wang, P., and Risch, N. J. (2005). Estimation of individual admixture: Analytical and study design considerations. *Genet. Epidemiol.* **28,** 289–301.

Tang, H., Coram, M., Wang, P., Zhu, X., and Risch, N. (2006). Reconstructing Genetic Ancestry Blocks in Admixed Individuals. *Am. J. Hum. Genet.* **79,** 1–12.

Thomson, G. (1995). Mapping disease genes: family-based association studies. *Am. J. Hum. Genet.* **57,** 487–498.

Tian, C., Hinds, D. A., Shigeta, R., Kittles, R., Ballinger, D. G., and Seldin, M. F. (2006). A genomewide single-nucleotide-polymorphism panel with high ancestry information for African American admixture mapping. *Am. J. Hum. Genet.* **79,** 640–649.

Zeggini, E., Weedon, M. N., Lindgren, C. M., Frayling, T. M., Elliott, K. S., Lango, H., Timpson, N. J., Perry, J. R., Rayner, N. W., Freathy, R. M., *et al.* (2007). Replication of genome-wide association signals in UK samples reveals risk loci for type 2 diabetes. *Science* **316,** 1336–1341.

Zhang, C., Chen, K., Seldin, M. F., and Li, H. (2004). A hidden Markov modeling approach for admixture mapping based on case-control data. *Genet. Epidemiol.* **27,** 225–239.

Zheng, C., and Elston, R. C. (1999). Multipoint linkage disequilibrium mapping with particular reference to the African-American population. *Genet. Epidemiol.* **17,** 79–101.

Zhu, X., Cooper, R. S., and Elston, R. C. (2004). Linkage analysis of a complex disease through use of admixed populations. *Am. J. Hum. Genet.* **74,** 1136–1153.

Zhu, X., Luke, A., Cooper, R. S., Quertermous, T., Hanis, C., Mosley, T., Gu, C. C., Tang, H., Rao, D. C., Risch, N., *et al.* (2005). Admixture mapping for hypertension loci with genome-scan markers. *Nat. Genet.* **37,** 177–181.

Zhu, X., Zhang, S., Tang, H., and Cooper, R. S. (2006). A classical likelihood based approach for admixture mapping using EM algorithm. *Hum. Genet.* **120,** 431–445.

20

Integrating Global Gene Expression Analysis and Genetics

Charles R. Farber* and Aldons J. Lusis*,†,‡

*Department of Medicine, Division of Cardiology, BH-307 CHS, University of California, Los Angeles, California 90095

†Department of Microbiology, Immunology and Molecular Genetics, University of California, Los Angeles, California 90095

‡Department of Human Genetics, University of California, Los Angeles, California 90095

Advances in Genetics, Vol. 60

0065-2660/08 $35.00
DOI: 10.1016/S0065-2660(07)00420-8

ABSTRACT

The transcriptome is defined as the collection of all RNAs produced in a cell or tissue at a defined time in development and is one of many stages that make up a biological system. It is also one of the most important; providing the critical link in the flow of information between genes and disease. Therefore, identifying gene expression changes that are reacting to or causing disease promises to significantly enhance our understanding of common disorders. However, only recently has the technology, in the form of DNA microarrays, been in place to quantitate gene expression levels on a genome-wide scale. DNA microarrays are small chips that contain arrays of DNA sequences and are capable of simultaneously quantifying the expression of thousands of genes. When applied to samples representing diseased and normal states, microarray-based expression profiling can identify differentially expressed genes that may play a role in the disease or predict progression or severity. Additionally, the integration of genetics and gene expression promises to aid in uncovering common genetic variations that control a particular disease. In animal models, this approach has already been used to identify genes correlated with disease, prioritized candidates, model causal interactions between genes and traits, and generate gene coexpression networks; all of which have shed light on novel disease mechanisms. In this chapter, we provide an overview of DNA microarray technologies and discuss ways in which microarray expression data can be combined with more traditional experimental approaches to dissect the genetic basis of disease.

I. INTRODUCTION

Biological systems can be thought of as a series of stages (commonly referred to using "-omics" nomenclature) that can be interrogated using specific technologies (Fig. 20.1). These stages include DNA (genome), RNA (transcriptome), proteins (proteome), metabolites (metabolome), and phenotypes (phenome), among many others. Although each stage can be considered individually, a great deal of cross talk between stages is required for proper cellular and physiological function. Classical genetic studies link genes (genome) to disease (phenome) without considering other stages. However, spurred by technological advances in the ability to perform bioassays in a massively parallel fashion, the sequencing of the human genome, and the development of statistical methodologies, researchers now have the capacity to leverage information from other levels of the system to better understand the role of genetic perturbations in disease. Currently, the transcriptome has proven the most accessible with regards to high-throughput analysis. The transcriptome is most commonly viewed as the full complement of mRNA species present in a given cell type or tissue at a defined time in

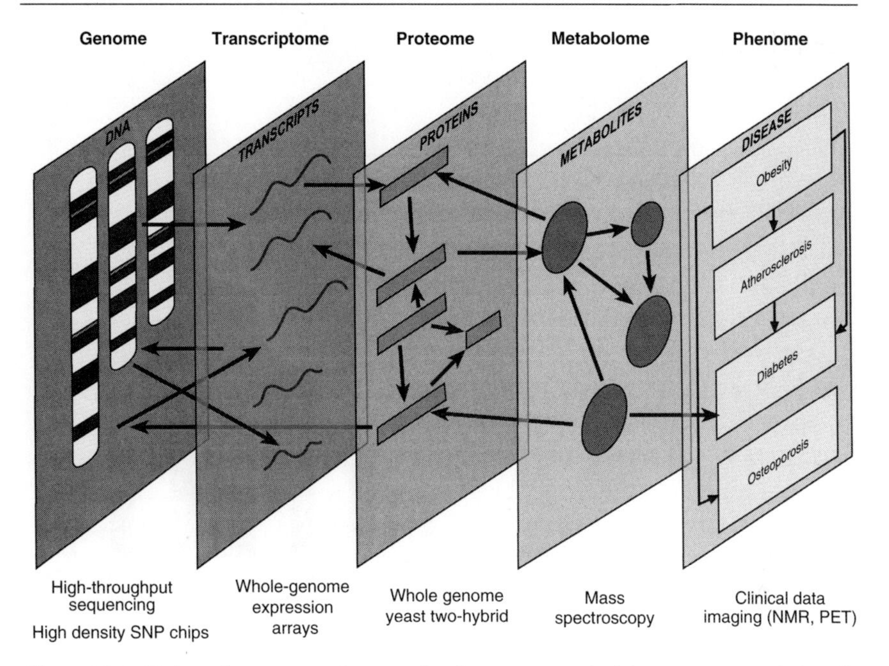

Figure 20.1. Biological systems can be viewed as being composed of discrete stages including the genome, transcriptome, proteome, metabolome, and phenome. The lines between the stages illustrate the effects of genetic perturbations and the various interactions that exist among all levels. Listed at the bottom of each panel are technologies used to generate genome-wide datasets for each stage. As discussed in the text, many limitations exist for sampling most stages, with the exception of the transcriptome due to the widespread use of DNA microarrays. PET, positron emission tomography; NMR, nuclear magnetic resonance; SNP, single nucleotide polymorphism.

development. However, recent data has proven that other RNA species such as noncoding RNAs (microRNAs, snRNAs, etc.) are important information carriers that can have profound affects on quantitative traits (Abelson *et al.*, 2005; Clop *et al.*, 2006).

The decoding of the human genome (and other model organisms such as the mouse and rat) represents an astonishing scientific achievement and has provided a comprehensive view of the first stage of the human biological system (Gibbs *et al.*, 2004; Lander *et al.*, 2001; Venter *et al.*, 2001; Waterston *et al.*, 2002). One immediate application of this "genetic parts list" was the development of DNA microarrays that are now the most widely used tool for global gene expression profiling. DNA microarrays, with the capacity to profile the entire transcriptome (at least the part we have correctly identified as transcribed), are now readily available and have been used in a plethora of applications.

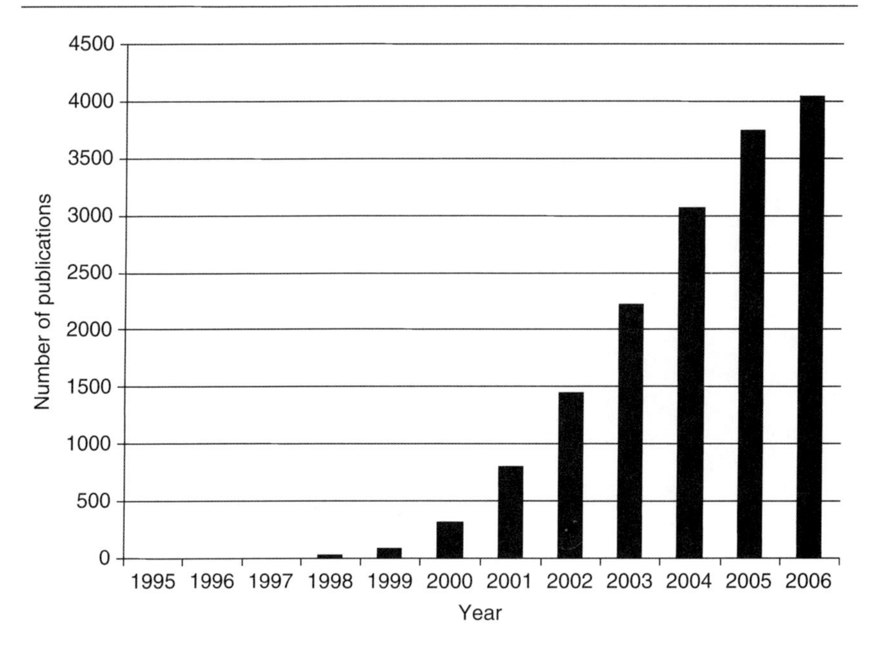

Figure 20.2. Number of DNA microarray publications per year from 1995 to 2006. Articles were identified using the search string "microarray AND gene AND expression,"

To illustrate their growing utility, a PubMed search at the National Center for Biotechnology Information (NCBI) using the search string "microarray AND gene AND expression" returned 14,331 articles, 926 (6.5%) of which were published within 90 days of this search (April 16, 2007). Figure 20.2. illustrates the growth in publications, using the same search, from 1995 to 2006.

Probably the most significant applications of gene expression microarrays to common diseases are in the cancer area. Patterns of expression in cancer tissues have been used to subdivide cancers and to predict survival and responses to specific drugs (van de Vijver *et al.*, 2002; van't Veer *et al.*, 2002). Golub *et al.* (Lamb *et al.*, 2006) recently proposed the development of a resource, they term the "connectivity map," to generalize the idea of matching patterns of expression employed by earlier studies to define disease subtypes, predict prognosis, and predict response to treatment with a particular drug. To generalize the pattern matching idea, Golub *et al.* proposed to use mRNA expression assayed on DNA microarrays to determine genomic signatures that describe all biologic states—physiologic, disease, or those induced with chemicals or genetic constructs. The connectivity map would ultimately manifest itself as a large public database of such signatures along with tools to determine pattern matching of similarities among these signatures.

The last decade has seen a paradigm shift in our ability to confront disease. The tools now exist to transition from "one gene at a time" to more global "systems-level" approaches that promise an unprecedented understanding of affected and normal states. Global snapshots of the transcriptome can now be linked to disease status and genetic polymorphisms, significantly increasing our ability to pinpoint master disease regulators. This transition will certainly lead to more creative and effective therapeutic intervention programs that are designed to confront the complexity of disease. The purpose of this chapter is to describe one aspect of this transition: the use of gene expression analysis in the context of common disease. The discussion begins with the platform for change—DNA microarrays. Our aim is to highlight technical and data analysis issues pertaining to their use in genetic studies. Our discussion then shifts to ways in which microarray technologies have and can be used to prioritize candidate genes based on potential relevance to disease. The final sections will discuss recent advances in the integration of gene expression and genetics, as well as novel analytical approaches in the development of gene coexpression networks.

II. TECHNICAL AND EXPERIMENTAL DESIGN ISSUES FOR MICROARRAYS

Approaches in systems biology rely on the collection of highly parallel information from different biological levels that can be used to infer system function in the face of genetic and environmental perturbations. The two levels that are the most amenable to comprehensive screening are the genome and transcriptome. This is due to their "relative" lack of complexity and the complementary nature of nucleic acids. In contrast, technological challenges remain for the interrogation of many of the other biological levels, such as the proteome, which is not only composed of the individual proteins but also of many regulatory relationships such as posttranslational modifications and protein–protein interactions.

Several different technologies exist for whole transcriptome profiling and detection of differentially expressed genes, including serial analysis of gene expression (Velculescu *et al.*, 1995), massively parallel signature sequencing (Brenner *et al.*, 2000), differential display (Liang and Pardee, 1992), cDNA representational difference analysis (Hubank and Schatz, 1994), and DNA microarrays (DeRisi *et al.*, 1996; Schena *et al.*, 1995). Although each is useful in certain applications, DNA microarrays are by far the most widely used. Similar to Northern blotting, the basis of microarrays is hybridization between complementary nucleic acids. In a Northern blot, a labeled probe is hybridized to a membrane containing an RNA sample (Alwine *et al.*, 1977) and the amount of probe that binds its complementary RNA is used to compare gene expression across samples. In essence, DNA microarrays simultaneously perform Northern blots for every gene in the genome.

In general terms, a DNA microarray is a collection of DNA sequences covalently attached to a stable substrate such as a glass slide, silicon wafer, or silica beads. Spots of DNA (referred to as probes and typically consisting of cDNAs or oligonucleotides) represent specific genes and are arrayed in a grid-like pattern across the solid surface. In the context of gene expression profiling, the target is composed of a population of cDNA or cRNA copies of mRNAs that are labeled and applied directly to the microarray. Once these copies are applied to the array, complementary probe–target pairs bind through hybridization. After the hybridization step, microarrays are scanned and signal intensity is quantified for each spot or feature on the array. This signal is proportional to the amount of target present in the starting RNA sample and is used as a proxy for the actual mRNA levels either in relative or absolute terms, depending on the microarray platform. Although DNA microarrays are most commonly used for gene profiling, they can also be used for a plethora of other applications such as comparative genomic hybridization, genome-wide chromatin immunoprecipitation (ChIP-chip), genomic resequencing, and single nucleotide polymorphism genotyping (Koczan and Thiesen, 2006; Stoughton, 2005).

A. DNA microarray platforms

Two general types of microarray platforms are currently in use, one- and two-color (Jaluria *et al.*, 2007). The most significant difference between one- and two-color microarrays is the type of hybridization. Two-color arrays are simultaneously hybridized using two samples, each tagged with a different label. Cyanine (Cy3 and Cy5)-labeled deoxynucleotide triphosphate incorporated into cDNA is the most popular fluorescent label used in two-color systems (Ehrenreich, 2006). The most common experimental design is hybridization of an experimental sample and control. An example would be RNA from a tissue biopsy from an affected patient hybridized simultaneously with RNA from an unaffected control. After hybridization, a scanner is used to measure the amount of fluorescent target bound to each probe. In the example above, if the ratio of affected to unaffected intensity for a gene is significantly more or less than one, the transcript level in the affected sample is up- or downregulated, respectively. In contrast, a single sample is hybridized to a one-color array, and unlike two-color systems, several different types of target and target labeling protocols exist. In general, the signal intensity for each probe is a direct readout of gene expression in absolute terms. A hypothetical experiment using one-color microarrays is illustrated in Fig. 20.3.

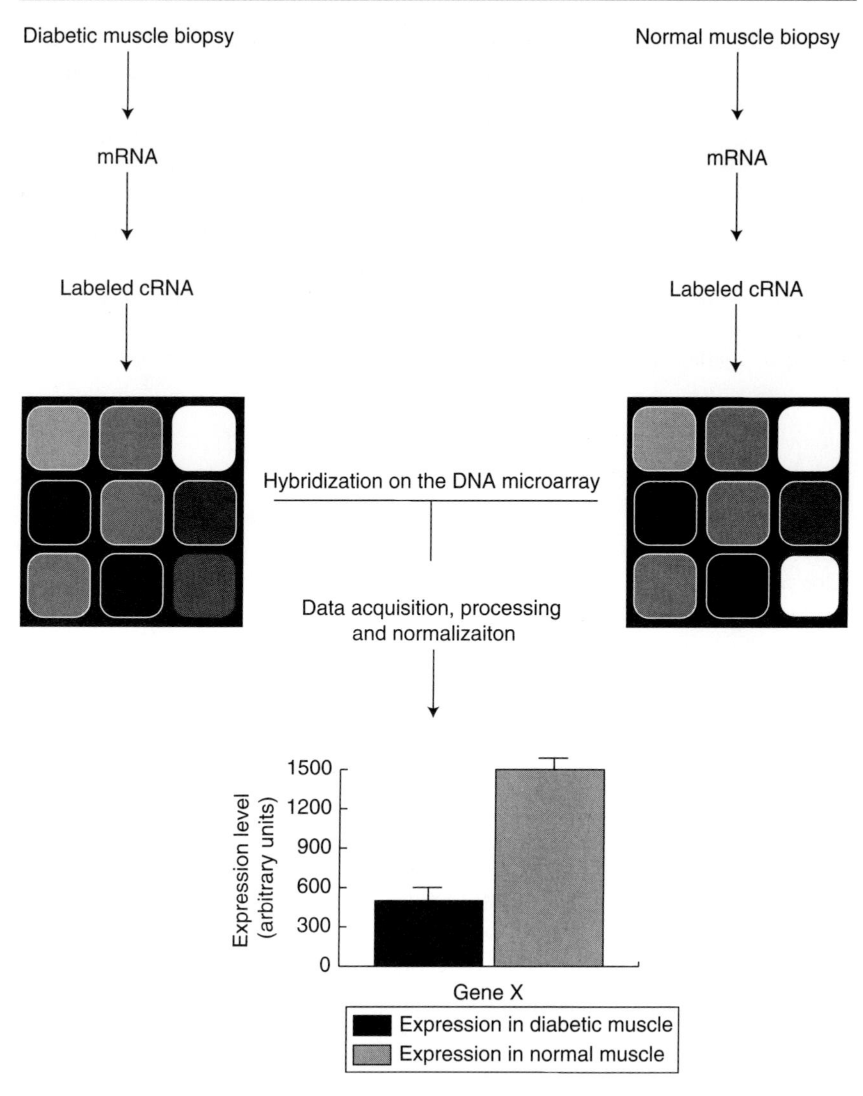

Figure 20.3. Description of a hypothetical one-color microarray analysis between affected and unaffected muscle biopsy samples. In this example, global gene profiles are generated from diabetic and normal muscle biopsies. First, mRNA is isolated from both samples and labeled cRNA (in most cases a biotin-labeled RNA copy) is hybridized to the array. A fluorescent reporter is used to determine signal intensities at each feature on the array. For simplicity, we have focused our attention on nine features, each containing a gene-specific probe sequence. In reality most one-color arrays contain millions of features. The expression of gene X is determined using the feature in the lower right-hand corner of the array. A dark spot indicates low or no signal and white represents high

In the following sections, we discuss details for the most widely used commercial platforms, such as Affymetrix, Illumina, and Agilent. It should be noted, however, that many researchers use "homemade" arrays. These are almost always of the two-color variety and are made using printing devices that directly deposit spots of DNA onto glass slides (Hager, 2006). In addition, there are technologies that are still in early stages of commercialization, but are worth noting. These include NimbleGen and CombiMatrix, both of whom have developed novel *in situ* synthesis (synthesis of the probe directly on the slide) methods: digitally controlled micromirrors and electrode-directed synthesis, respectively (Stoughton, 2005). Both platforms offer significant advantages for efficiently generating custom microarrays.

1. Affymetrix

The Affymetrix GeneChip array was one of the first commercially available whole-genome expression profiling technologies and is the market leader today. One advantage of the GeneChip array is the extremely high feature density, in excess of 1 million features/chip, relative to other platforms (Dalma-Weiszhausz *et al.*, 2006). This density is possible because of photolithography, a unique method of *in situ* synthesis (Fodor *et al.*, 1991). The process of manufacturing a GeneChip begins by adhering linker molecules with photolabile protecting groups to the surface of a silica wafer. A photolithographic mask is applied and light is introduced, removing the protecting groups at defined positions depending on the predetermined sequence and location of the oligonucleotide probes to be synthesized. Protected deoxynucleosides are added that covalently attach to the unprotected linker, and this process is repeated with new masks until oligonucleotide probe sequences are fully synthesized (Dalma-Weiszhausz *et al.*, 2006). The probe length on this platform is 25 nucleotides.

The GeneChip Human Gene 1.0 ST Array, Affymetrix's most current platform, is composed of approximately 26 probes per transcript for over 28,000 well-annotated genes. During a gene expression experiment, biotinylated cRNA is hybridized to the array and stained with a fluorescent streptavidin–phycoerythrin conjugate that binds biotin. The GeneChip is scanned and the intensity of each probe is determined. A number of software packages as well as

expression. Signal intensities are quantitiatied, processed, and normalized, yielding the level of expression of each gene within the arrayed target. The bar chart at the bottom represents the expression of gene X in both samples and indicates that it is downregulated in diabetic muscle. In a real experiment, the average of multiple biological replicates would be used to represent the expression of a gene in the affected and unaffected states.

libraries for the Bioconductor software (www.bioconductor.org) implement algorithms for calculating signal intensities from GeneChip arrays (Gentleman *et al.*, 2004; Konradi, 2005).

2. Illumina

The Illumina Universal BeadArrays represent a novel approach to genomic applications including gene expression profiling. There are two general types of BeadArrays, the Sentrix Array Matrix (SAM) and Sentrix BeadChip. The SAM is used for the analysis of specific gene sets (on the order of 1500 genes per sample) whereas BeadChips are used for whole-genome profiling. For the purpose of our discussion, we will focus on details of the BeadChip, although SAM arrays are identical in many technical aspects.

The Sentrix BeadChip consists of a silicon-coated chip with millions of microscopic wells etched in a regular pattern along its surface (Fan *et al.*, 2006). Each well is approximately 3 μm in diameter and is designed to capture and hold a signal bead. BeadChip beads are impregnated with approximately 700,000 covalently attached two-part oligonucleotide probes. The first part or sequence closest to the bead is a unique 29-mer address sequence used for array decoding. The second part is a gene-specific sequence (Kuhn *et al.*, 2004). A pool of all bead types is applied to each array and individual beads become randomly seated in microwells.

Due to the randomness of bead placement, BeadChip arrays are decoded to discern the identity of each bead type (Gunderson *et al.*, 2004). This is accomplished using the 29-mer address sequence. In the decoding process, decoder oligo pools are constructed with a set of fluorescently labeled oligonucleotides complementary to the address sequences for a subset of all bead types. Decoder pools are hybridized and the fluorescence intensity is measured for all beads across the array. In the second stage, the BeadChip is stripped and a different decoder pool is hybridized. This process is repeated for the number of stages needed to decode all possible bead types and at the end of this process a unique signature for each bead is generated. This signature provides the sequence identity of each bead and is a key quality control step ensuring the integrity of each spot on the array (Gunderson *et al.*, 2004).

One of the advantages of the BeadChip is its extremely high feature density (Kuhn *et al.*, 2004). This high density allows for the processing of six to eight unique samples per BeadChip on a substrate the size of a typical microscope slide, significantly decreasing cost per sample. For human and mouse, two different platforms are commercially available. The first quantitates the expression of over 40,000 transcripts in six samples, and the second analyzes over 20,000 genes in eight samples simultaneously. In addition, there is on average a 30-fold redundancy per bead type present on each array increasing the reliability of each gene expression measurement.

The target sample for each decoded BeadChip array is generated and labeled in a process similar to that described above for GeneChips. For data analysis, a BeadStudio analysis software package is available that is capable of data normalization and analysis. In addition, two libraries for the Bioconductor software, Beadarray and BeadExplorer (www.bioconductor.org) (Gentleman *et al.*, 2004), have been developed to assist in the analysis of BeadChip data.

3. Other platforms

A number of other commercial platforms exist, including Agilent, Applied Biosystems, and Eppendorf (Table 20.1). Recently, these platforms have been compared as part of the MicroArray Quality Control Project (Shi *et al.*, 2006). For this project, expression data on four titration pools from two distinct reference RNA samples were generated at multiple test sites using a variety of microarray-based platforms. This chapter provides a reference to an investigator by which interplatform consistency and concordance can be evaluated. For example, the study showed that in the test samples the differentially expressed genes averaged approximately 89% overlap between test sites using the same platform, and approximately 74% across one-color microarray platforms. Significant differences in various dimensions of performance between microarray platforms were noted.

Table 20.1. Commercially Available DNA Microarray Platforms Commonly Used for Expression Profiling

Platform	Highlights of each platform
Affymetrix GeneChips	Whole-genome expression profiling for a large number of species—proven platform with exceptional annotation
Illumina BeadArray	Whole-genome expression profiling in humans and rodents—ultrahigh feature density decreases cost per sample
Agilent	Whole-genome expression profiling for a large number of species
Applied Biosystems	Whole-genome expression profiling in humans and rodents—platforms provide comprehensive genome coverage
Eppendorf DualChips	Pathway-focused content
CombiMatrix CustomArray	Custom content arrays—electrode-directed synthesis allows quick developments of custom arrays
NimbleGen	Provides complete array services with standard arrays for a nearly all species with a genome sequence—platform is also flexible for designing custom content arrays

B. Microarray data analysis

Datasets generated from microarray experiments require sophisticated statistical analysis to ensure that the results are not because of technical artifacts. It is, however, useful for all researchers to have an understanding of the basic analytical steps. Below we have provided a brief overview of the most common methods. For a more detailed discussion, the reader is referred to a number of excellent reviews and volumes dealing with the subject (Butte, 2002; Quackenbush, 2001; Wit and McClure, 2004).

The first step in the analysis of microarray data involves image analysis to convert the large numbers of pixels comprising the image of the scanned microarray into expression values for each gene. Image analysis includes filtering to "clean" images, gridding, and segmentation to define the region to be quantitated, and quantification of the fluorescence intensity.

The second step involves normalizing the data, a process that attempts to remove systematic bias as a result of experimental artifacts. Generally, microarrays are analyzed using a sequential approach, in which the normalization is done before any further analysis. A set of microarrays can be normalized using a number of methods such as comparing the expression of "control" spots (spots of DNA not represented in the samples under evaluation and typically spiked in during the experiment), "housekeeping" genes or sets of probes whose expression are invariant across all samples. Following normalization, microarray data are usually scaled using a logarithmic scale. The logarithmic function can be done without loss of information regarding the original signal and it is the most natural scale to describe fold changes. Two-color data are generally reported as the logarithm of the expression ratio and one-color data as the logarithm of the intensity. Other important aspects of normalization exist, depending on the array platform, including background correction and spatial normalization.

After normalization, genes that are not expressed in the profiled sample are eliminated from the analysis. This is an important filtering step because it reduces the number of comparisons made in the next step that is the identification of genes that are differentially expressed. The ability to identify differentially expressed genes depends on the variance of the data and on the number of arrays analyzed per sample. A common preference for the analysis of differential expression is to use a *t*-test or a threshold fold change. The problem of multiple comparisons is best addressed by analysis of "false discovery rates" (Tusher *et al.*, 2001).

Another common goal is to identify genes that show similar patterns of expression, using statistical methods generally referred to as "cluster analysis." "Similarity" of expression patterns is mathematically defined using an "expression vector" for each gene that represents its location in "expression space." In such an analysis, each experiment represents a separate entry in space and the $\log_2$ intensity or $\log_2$ ratio measured for a gene represents its geometric coordinate. For instance, in a study with three experiments, the $\log_2$ expression for a given gene in Experiment 1 is its X coordinate, the $\log_2$ expression in Experiment 2 is its Y coordinate, and the $\log_2$ expression in Experiment 3 is its Z coordinate. Thus, one can represent all the information about a gene by a point in the X, Y, and Z expression space. Another gene with similar $\log_2$ expression values for each experiment will be represented by a spatially nearby point. The most widely used cluster analysis is hierarchical, with an increasing number of nested classes. Nonhierarchical clustering techniques that simply partition genes into a prespecified number of clusters can also be applied. The main advantage of all clustering procedures is their ability to organize and visualize highly dimensional datasets consisting of thousands of gene expression measurement over multiple experiments (Quackenbush, 2001; Wit and McClure, 2004). This enables the rapid extraction of biologically meaningful patterns of gene expression.

III. IDENTIFYING DISEASE CANDIDATES USING DNA MICROARRAYS

Expression array profiling can provide clues that aid in the identification of genes underlying complex traits. For example, a gene whose expression is confined to tissues affected in a certain disease may play a role in its etiology. Additionally, the differential expression of genes between diseased individuals and normal individuals can be determined. In the sections below, we outline ways in which expression profiling can be used to better understand the genetic factors that regulate disease.

A. Gene expression catalogs

Complex human disease is typically not confined to alterations in one tissue or cell type. In contrast, most disease states are the result of multiple perturbations involving an array of tissue and organ systems. However, in almost all cases, a relatively small set of relevant tissues can be considered the most likely drivers of disease. For example, a major cause of osteoporosis is an increase in the bone remodeling rate (Jilka, 2003). The bone remodeling rate is the relative activity of two cell populations, the osteoclasts that resorb old bone and osteoblasts that form new bone (Seeman and Delmas, 2006). Although changes in the bone remodeling rate are inherent to bone, gene expression changes in other tissues, such as the bone marrow and adipose, are known to influence this process. Therefore, a comprehensive catalog of gene expression of every gene in the genome, in bone, bone marrow, and adipose, would constitute a valuable filter for prioritizing candidate osteoporosis regulators.

The earliest versions of human gene expression catalogs were collections of expressed sequence tags (ESTs). ESTs are generated by sequencing clones from a complementary DNA (cDNA) library and represent a fragment of a specific mRNA. cDNA libraries are commonly generated from a single tissue and thus, relative frequency of a particular EST roughly corresponds to its expression level *in vivo*. In some cases, ESTs have been generated from both normal and diseased tissues. These datasets along with the digital differential display tool (now referred to as the digital gene expression displayer, http://cgap.nci.nih.gov/Tissues/GXS) available via the Cancer Genome Anatomy Project (CGAP) at NCBI have recently been used to identify candidate cancer genes whose EST sequences were overrepresented in malignant or normal tissue (Scheurle *et al.*, 2000). Two of the candidates were subsequently validated as true cancer genes (Narayanan, 2007). Various online sources in addition to NCBI can be used to query EST collections, such as TIGR (http://www.tigr.org/index.shtml) and RIKEN (http://read.gsc.riken.go.jp/) (Walker and Wiltshire, 2006).

The EST approach has limitations and can be biased depending on depth of sequencing. Recently, more quantitative assessments of whole-genome expression have been made in human, mouse, and rat tissues. The Genomics Institute of the Novartis Research Foundation (GNF) has generated DNA microarray expression profiles in a diverse set of human and rodent tissues (Su *et al.*, 2002). These data are publicly available through the GNF SymAtlas online database (http://symatlas.gnf.org/SymAtlas/). In addition, the Allen Institute for Brain Science has generated the Allen Brain Atlas (http://www.brain-map.org/welcome.do;jsessionid=AE1089DFC3CB220BF3CB842D84129287) that measured the expression of over 20,000 genes in the mouse brain using *in situ* hybridization (Lein *et al.*, 2007). These data provide tremendous insight into the regional expression of genes in the mouse brain.

B. Differential gene expression in disease

The use of expression profiling has been widely used in animal models to discover disease genes. In many studies, microarrays are used to interrogate regions previously found to harbor a gene(s) affecting a complex trait (referred to as quantitative trait loci or QTLs). Genes are prioritized based on differential expression dependent on QTL genotype. The resulting hypothesis is that one of the differentially expressed genes controls the phenotypic difference. If the list is short or contains biologically relevant candidates, then subsequent experiments can be used to determine which gene(s) regulates the disease. One of the first studies to demonstrate the feasibility of this approach identified the *Cd36* fatty acid translocase as the gene responsible for several metabolic defects, including insulin resistance, in the spontaneously hypertensive rat (SHR)

(Aitman *et al.*, 1999). In a cross between two rat strains (SHR and Wistar Kyoto), a QTL was identified on chromosome 4. The metabolic disturbances observed in the SHR strain were corrected in a chromosome 4 congenic strain (congenics contain a chromosomal segment from one strain introgressed onto the genetic background of a second strain). The analysis of adipose tissue gene expression between the congenic and control using two-color spotted cDNA arrays revealed a 90% reduction in the levels of *Cd36* mRNA in congenic rats. To prove the reduction in *Cd36* was causative, the authors identified multiple sequence variations in SHR *Cd36* and demonstrated that transgenic mice overexpressing *Cd36* had reduced triglycerides.

To identify genes contributing to asthma, Affymetrix GeneChip arrays were used to detect differentially expressed genes in the lungs of A/J [highly susceptible to allergen-induced airway hyperresponsiveness] and C3H/HeJ (highly resistant) mouse strains and a limited number of A/J X C3H/HeJ F1 and F1 X A/J backcross mice (Karp *et al.*, 2000). Of 21 differentially expressed genes, the complement factor 5 (C5) gene was located near the *Abhr2* (allergen-induced bronchial hyperresponsiveness 2) QTL and its expression was negatively correlated with airway hyperresponsiveness. It was previously known that A/J mice have a 2-bp deletion in exon 5 that results in a premature stop codon, while C3H possesses normal levels and activity of C5 (Wetsel *et al.*, 1990). The combination of microarray gene expression data in addition to functional studies strongly suggested a role for C5 in allergic asthma. In a subsequent study, polymorphisms in the human C5 gene were associated with bronchial asthma in a Japanese population (Hasegawa *et al.*, 2004).

Osteoporosis is one of the most common diseases associated with aging and is under strong genetic control. A QTL controlling bone mineral density, a major predictor of osteoporotic fracture risk in humans, was identified on mouse chromosome 11 between the DBA/2J and C57BL/6J strains (Klein *et al.*, 2004). The locus was captured in a congenic strain and DNA microarray analysis in kidney tissue identified a 20-fold reduction in the expression of the 12/15-lipoxygenase gene (*Alox15*) in C57BL/6J relative to congenic mice. *Alox15* knock-out mice and mice treated with a pharmacological inhibitor of 12/15 lipoxygenase had higher bone mineral density, validating the role of *Alox15* gene expression in the acquisition of bone mass.

C. Functional annotation of gene expression patterns

In many cases, the underlying biological theme of a specific group of genes altered by disease is not immediately clear and requires functional annotation. The basis for nearly all supervised annotation is the gene ontology (GO). The GO is a controlled vocabulary designed to annotate the biological process, molecular function, and cellular component of all eukaryotic genes and gene products (Ashburner *et al.*, 2000). A number of annotation tools have been developed that use GO annotation for biological interpretation of an otherwise anonymous gene set (http://www.geneontology.org/GO.tools.shtml). Of particular use is the Database for Annotation, Visualization and Integrated Discovery (DAVID) suite of annotation and visualization tools (Dennis *et al.*, 2003). DAVID allows one to identify biological themes that are enriched in a particular gene list, visualize genes in well-known biological pathways such as KEGG and BioCarta, and cluster redundant annotation terms among a group of genes. The Expression Analysis Systematic Explorer (EASE) software, developed by the DAVID bioinformatics group, also has the capacity to identify the biological theme of a gene list, and can be downloaded as a stand-alone program (Hosack *et al.*, 2003).

In many group comparisons, only a small number of genes with statistically significant changes in gene expression are identified. This is often due to low statistical power, which in most experiments is a function of small sample sizes and a large number of statistical tests. An alternative biological explanation is that some disease is caused by or elicits subtle coordinate changes in the expression of gene pathways. Small changes in pathway expression can be expected to have biologically significant effects on metabolite flux, the induction of transcriptional cascades, and, ultimately, disease. Recently, an analytical tool termed gene set enrichment analysis (GSEA) was developed to increase the statistical power of microarray experiments by identifying known biological pathways enriched for differentially coregulated genes (Mootha *et al.*, 2003; Subramanian *et al.*, 2005). GSEA takes an input set of genes, such as all genes expressed in a tissue, and ranks them based on a standard metric of differential expression between two groups. Next, a running cumulative enrichment score (ES) is calculated for each biological pathway or functionally related gene set. An example would be all genes known to be involved in atherosclerosis or inflammation. If a pathway is enriched for genes either positively or negatively correlated with disease status, a high mean ES (MES) will be assigned to that pathway. The statistical significance of the MES is assessed using permutations of the disease status label. In the seminal GSEA study, transcriptome profiles were generated from muscle biopsies collected from normal glucose tolerant, impaired glucose tolerant, and type 2 diabetic patients (Mootha *et al.*, 2003). Using traditional statistical techniques, no significant changes in gene expression were observed in any of the possible pairwise group comparisons. However, using GSEA a set of genes involved in oxidative phosphorylation possessed the highest MES. Interestingly, 89% of all genes in this pathway displayed a modest 20% reduction in expression in diabetic versus normal patients. GSEA has also been used in a mouse intercross to analyze liver gene expression profiles (Ghazalpour *et al.*, 2005). In this study, GSEA was integrated with genetics to identify metabolic pathways and regulatory loci controlling obesity. GSEA is a powerful tool to detect subtle changes in the expression of a pathway that would not be identified using the standard differential expression paradigm. However, it should be noted that this analysis relies on predefined biological pathways and will miss important changes in unannotated genes and novel pathways.

D. Identification of disease biomarkers

The discovery of disease biomarkers and prediction of disease subtypes are promising applications for expression profiling, and both are critical to the early detection and proper treatment of many diseases. In addition, biomarkers can be used to group patients in clinical trails based on observed or predicted drug responses. This may improve the clinical success of drugs with limited efficacy in the population as a whole, but which are highly efficacious for a subset of the population.

Examples of using DNA microarrays in this context have been numerous. Highlights include the work by Seo *et al.* (2004), who recently identified a set of signature genes whose expression in human aorta was predictive of atherosclerosis burden. Using the expression of this gene set, the authors were successful in classifying new aortic sections as diseased or normal over 93% of the time. Other success stories include a recent series of studies identifying distinct breast cancer subtypes using expression profiles from cancerous and normal breast samples (Kapp *et al.*, 2006; Perou *et al.*, 2000; Sorlie *et al.*, 2001).

IV. INTEGRATION OF GENETICS AND GENOMICS

Genome-wide transcript levels can be considered as intermediate phenotypes or "endophenotypes" for a disease. A powerful way to integrate genetics and genomics is to define the genetic control of transcript levels and at the same time, the genetics of disease phenotypes. In such analyses, transcript levels can be treated as quantitative traits and the loci controlling them can be mapped using classical linkage and association approaches. As summarized in Fig. 20.4, such combined genetic and genomic data can then be used to identify positional candidate genes, to identify known pathways involved in the disease, to model casual interactions involved in the disease, and to model gene networks and relate those to the disease. As yet, most studies have been performed using animal models (Doss *et al.*, 2005; Ghazalpour *et al.*, 2005, 2006; Schadt *et al.*, 2003, 2005), where the analyses are greatly simplified by the ability to control the environment, design crosses, perform invasive procedures, and sample tissues. Although likely an order of magnitude more difficult, the same approaches appear feasible in human populations.

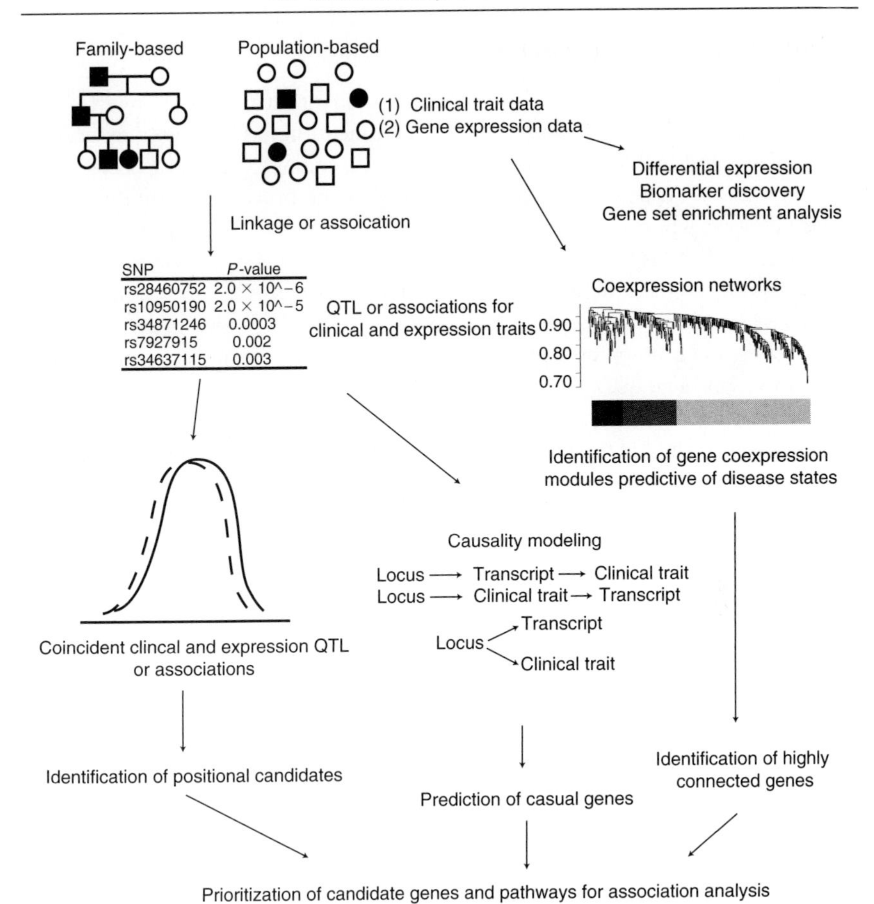

Figure 20.4. Schema for combining genetics and genomics to investigate human disease. The approach begins by collecting clinical, global gene expression, and genotype data from family- or population-based samples. The gene expression data can be used to identify differentially expressed genes, for biomarker discovery and gene set enrichment analysis by subdividing the population into groups based on disease status or genotype. Quantitative trait locus (QTL) or association analysis, depending on the population type, can be used to identify correlations between genotype and clinical/gene expression traits. These data can then be used to prioritize genes based on coincidence between gene expression and clinical QTL or associations and causality modeling. Additionally, network data on highly connected genes or genes belonging to a module correlated with a clinical trait can also be used to screen candidates. High priority genes and pathways can be validated in large populations using association analysis and/or in animal models using transgenic mice.

A. Mapping gene expression QTL

Genomic regions harboring variation affecting a quantitative trait are referred to as QTL (Falconer and Mackay, 1996). QTL identification has been used extensively in humans and model organisms to identify regions containing key disease regulators. A QTL can be composed of a single gene or as recent data indicate a cluster of genes whose cumulative effects are represented as one locus (Diament and Warden, 2004; Edderkaoui *et al.*, 2006). Statistical strategies for identifying QTL can be quite mathematically rigorous and many different types of analyses have been developed. However, correlating genotype with phenotype is the basis of all approaches. QTL mapping is crucial to any study integrating genetics and gene expression, and Fig. 20.5. illustrates a simple example for a gene expression trait. Although beyond the scope of this chapter, a more detailed description of statistical methodologies for QTL mapping is the focus of prior chapters and can be found in recent reviews (Darvasi, 1998; Doerge, 2002; Falconer and Mackay, 1996; Flint *et al.*, 2005).

The first genetical genomics experiment using global gene expression profiles was published in yeast (Brem *et al.*, 2002). In this work, the authors described two general classes of QTL controlling gene expression, *cis*- and *trans*-acting. Both terms refer to the location of the genetic variation giving rise to the eQTL. For example, if the expression of gene X is regulated by a *cis*-eQTL, the variation is located in or near the gene X locus. If the expression of gene X is controlled by a *trans*-eQTL, the variation is located outside the gene X locus. *cis*-eQTL can be due to variation such as polymorphism in a promoter or other regulatory region, whereas a *trans*-eQTL could be a polymorphism that alters the activity of a transcription factor that in turn modulates expression. Recently, Rockman and Kruglyak (2006) have proposed a more general lexicon to describe *cis* and *trans* effects. They propose using the terms local and distant linkage instead of *cis* and *trans*, respectively. Using this terminology, local linkage can be due to *cis*-acting variants affecting allele-specific transcription rates in addition to the effects of neighboring genes, autoregulation, and feedback loops. The concept of local and distant linkage is illustrated in Fig. 20.6.

One of the first applications of this approach was the investigation of the genetic architecture of gene expression. It allowed questions to be asked such as how many QTLs regulate the expression of a given gene and what fraction of variance in expression is explained by genetics? Although definitive answers are still elusive, clear trends have emerged, including the realization that expression phenotypes are relatively complex despite a direct relationship between DNA and the mRNAs it encodes. In general, many more distant linkages are observed relative to local linkages and in some cases expression phenotypes are controlled by many eQTLs. Additionally, evidence for epistasis regulating a significant fraction of gene expression traits has been reported in yeast (Brem and Kruglyak, 2005;

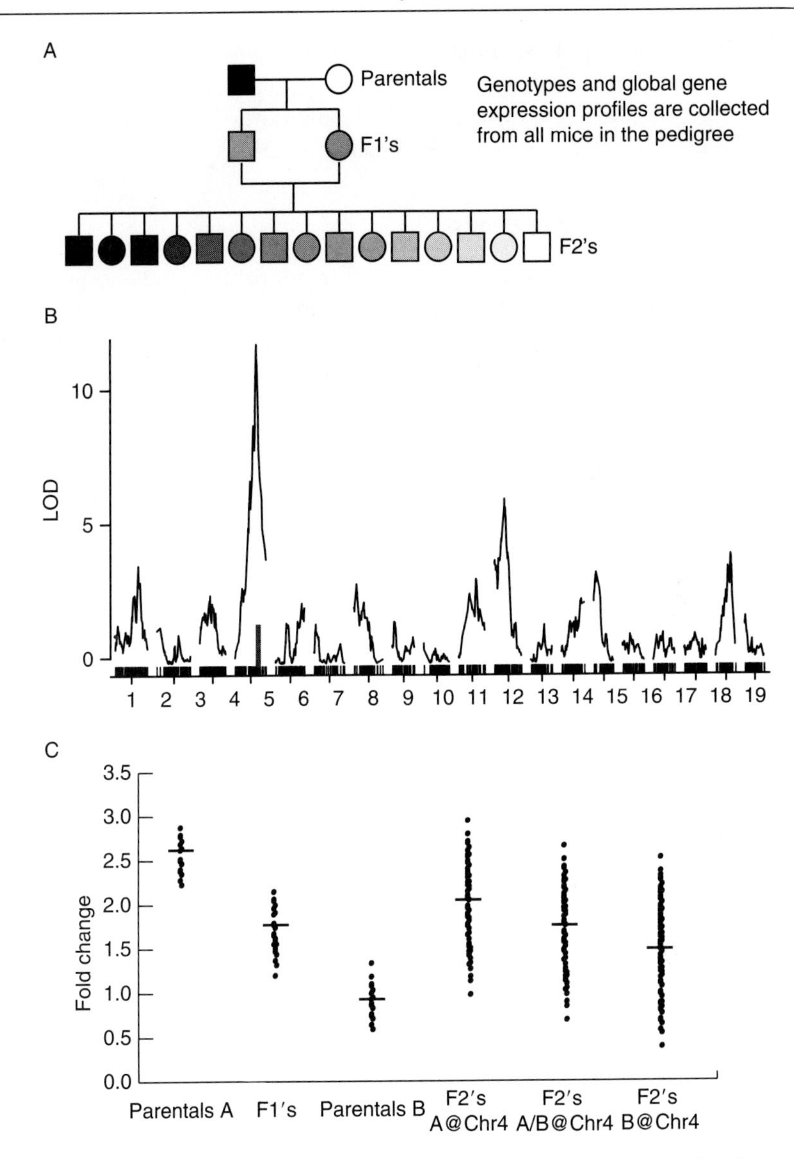

Figure 20.5. The genetics of gene expression. The example illustrates the principles of mapping expression QTL (eQTL). (A) Global gene expression profiles and genotypes are collected from a mouse F2 intercross between parental strains A and B. (B) QTL analysis is preformed by correlating the levels of an individual gene (gene X) with the presence of strain A and B alleles at markers spaced across the genome. In this example, the expression of gene X is regulated by a strong local eQTL (see text and Fig. 20.5 for description of local and distant eQTL) on chromosome 4. The position of

Brem *et al.*, 2005). Despite this surprising complexity, the average eQTLs in humans and mice, among those with statistically detectable effects, explain approximately 25% of the variation in expression, which is significantly larger than the average clinical trait QTL (Morley *et al.*, 2004; Schadt *et al.*, 2003).

B. Prioritizing candidate genes

A list of all genes with local eQTL is valuable in prioritizing candidate genes at a locus harboring clinical trait QTL and this information can then be combined with genetic fine mapping of the region. This approach has recently led to the positional cloning of the ATP-binding cassette, subfamily C (CFTR/MRP), member 6 (*Abcc6*) gene responsible for *Dyscalc1*, a major determinant of dystrophic cardiac calcification in mice (Meng *et al.*, 2007). The *Dyscalc1* QTL was first narrowed to an 840-kb region. The *Abcc6* gene, located in this region, was found to have a very strong local eQTL controlling its expression. The authors proved *Abcc6* was responsible for *Dyscalc1* using a transgenic model that recapitulated the resistance phenotype. Therefore, genes with local eQTL coincident with clinical trait QTL are excellent positional candidates and these data can be useful as a screening tool, especially when combined with additional genetic data. As yet, the list of known human eQTL is very small but this is expected to increase greatly with larger population and family studies.

Related to this, gene expression databases may help prioritize genes for diseases that display sexual dimorphism. However, until recently, it was unclear the extent of sexual differences in global gene expression. In a study by Wang *et al.* (2006), significant sex-by-QTL interactions were demonstrated for thousands of mouse liver eQTL. More importantly, obesity also differed between the sexes and many transcripts were identified that correlated with fat mass in a sex-dependent manner. A second study further demonstrated the importance of sex by showing that the expressions of thousands of genes in multiple tissues in the mouse were sexually dimorphic (Yang *et al.*, 2006). Moreover, numerous tissue-specific chromosomal hot spots were identified for eQTL, controlling the expression of sexually dimorphic genes. Together these studies indicate a strong role for gender in the control of male and female transcriptomes and the importance of sex-dependent expression in the context of disease.

---

gene X on chromosome 4 is denoted by the vertical black bar. Other distant linkages such as ones on chromosomes 11 and 18 also influence the expression of gene X. (C) A closer look at gene X expression reveals the basis of the local eQTL. Gene X is highly expressed in parental strain A, intermediate in F1 mice, and lowly expressed in strain B. F2s, homozygous for strain A alleles at markers on chromosome 4, express gene X at high levels and those inheriting strain B alleles at the same markers express gene X at a lower level, thereby explaining the strong correlation between chromosome 4 markers and the expression of gene X.

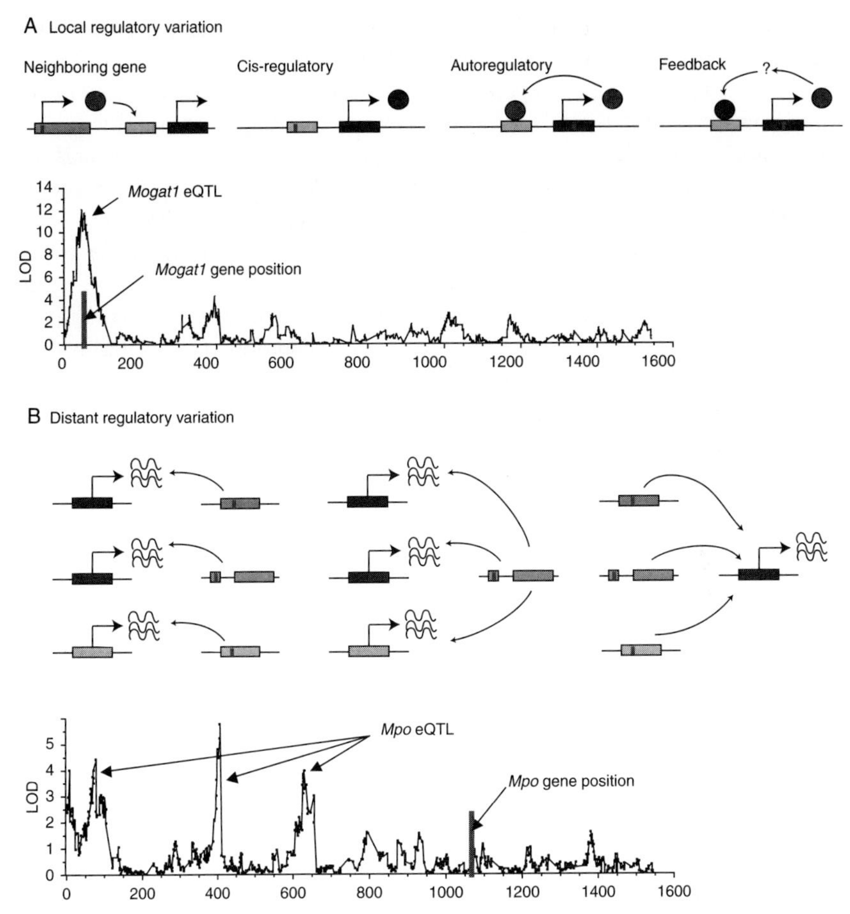

Figure 20.6. Local versus distant expression QTL (eQTL). (A) Explanation of different types of variation leading to local eQTL. For each example, darker colored boxes represent the coding region of functional genes whereas the lighter colored boxes represent upstream regulatory sequences. Variation can occur in a neighboring gene whose protein product regulates the expression of a nearby gene. *cis*-regulatory variation can also affect allele-specific transcription rates giving rise to an eQTL. Autoregulatory feedback and feedback with unknown intermediates can also produce local eQTLs. A real example is presented of a local eQTL regulating the expression of the monoacylglycerol O-acyltransferase 1 (*Mogat1*) gene in a large mouse F2 intercross. The vertical bar denotes the location of the *Mogat1* gene in the mouse. These data are not sufficient to indicate the exact nature of the local eQTL. (B) Explanation of different types of variation leading to distant eQTL. Variation in individual genes can alter the expression of a single unlinked gene, variation in one gene can alter the expression of many genes, or the expression of many genes can regulate the transcription of a single gene. A real example is presented for three distant eQTLs regulating the expression of the myeloperoxidase (*Mpo*) gene in a large mouse F2 intercross. The vertical bar denotes the location of the *Mpo* gene in the mouse. (After Drake *et al.* Mammalian Genome, 2006, 17(6), 466–479.)

Combining genetics and genomics also allows the prioritization of candidate pathways. The GSEA approach described above is an example of this. Moreover, known causal genes can be linked to known pathways by testing for significant correlations between the two. The study of dystrophic cardiac calcification discussed above is a good example. The function of *Abcc6* and how it contributed to calcification was entirely unknown and in fact, the substrate for this transporter has yet to be identified (Meng *et al.*, 2007). To examine which processes might involve *Abcc6*, correlations between *Abcc6* transcript levels and other transcripts in the mouse cross were determined. Interestingly, *Abcc6* transcripts were found to be significantly correlated with a *Wnt* signaling pathway previously proposed to contribute to calcification, suggesting testable hypotheses for the role of *Abcc6* (Meng *et al.*, 2007).

C. Modeling causal interactions

Orthogonal datasets such as genotypes, gene expression profiles, and disease status provide the data necessary to infer causality. Causality can be predicted for any gene expression—clinical trait pair by evaluating the relative likelihood of a casual, reactive, and independent model. In a causal model, a genetic variant (assayed in the population using a tightly linked genetic marker) elicits a change in gene expression that pleiotropically affects the clinical trait. In a reactive model, the genetic variant produces a change in the clinical trait that in turn alters gene expression (gene expression is reacting to the perturbed phenotype), and in an independent model, the mutation affects both the gene expression and clinical trait independently. Likelihoods for each model can be calculated based on conditional probabilities and used to assess the most probable scenario for a given gene.

Recently, Schadt *et al.* (2005) developed and applied causality modeling algorithms to a mouse intercross to predict key drivers of obesity. In that study, genes whose transcript levels correlated with adiposity were identified, and then this set was intersected with the set of genes whose eQTL overlapped with adiposity QTL (cQTL) in the cross. Several genes were predicted to be casual and have been validated using transgenic mice. Almost all the validated targets were novel obesity genes, illustrating the enormous power of this approach. A simplified example of causality modeling is presented in Fig. 20.7.

D. Gene coexpression networks

Genes do not function in isolation, but instead are members of gene groups or biological pathways that work in concert to perform particular functions. This coordinated action is due in part to transcriptional regulation. Consider the peroxisome proliferator-activated receptor family of transcription factors.

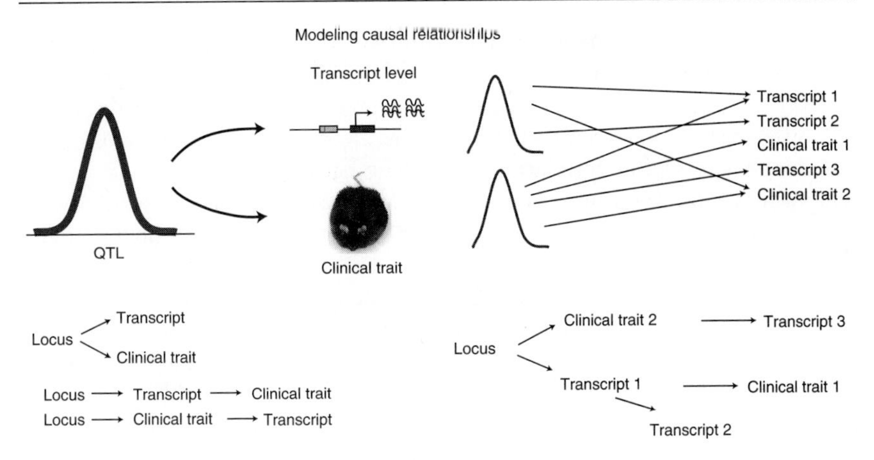

Figure 20.7. Modeling casual relationships between gene expression and clinical traits. Causality between gene expression and clinical traits can be modeled by determining the likelihoods of independent, casual, and reactive models. Additionally, information on multiple clinical trait and expression QTL can be incorporated to further strengthen casual predictions. See text for details of causality modeling.

Peroxisome proliferator-activated receptors respond to extracellular stimuli (either endogenous or exogenous) by increasing or decreasing the expression of hundreds of genes belonging to a highly diverse set of biological pathways. This concordant transcriptional regulation allows a cell to quickly respond to changing conditions. Thus, genes whose expressions are concordantly regulated over a set of differing conditions are likely to be functionally related.

Recently, much focus has been placed on developing biological networks using datasets such as gene expression, protein–protein interactions, and literature citations. A network is defined by a collection of nodes and edges, and in the case of gene coexpression networks, the nodes are genes and the edges represent a measure of expression similarity. In an unweighted coexpression network, a connection (edge) exists between two genes (nodes) only if their expression is correlated above a certain threshold. In a weighted network, all nodes are connected but the edges differ based on the strength of the relationship. Much of the theory behind the generation of biological networks comes from the work of Barabasi *et al.*, who discovered that most networks exhibit a scale-free topology. Scale-free networks consist of a small number of highly connected nodes with many edges and a large number of nodes with few edges (Barabasi and Albert, 1999).

In the context of gene expression, the purpose of network analysis is the identification of "modules" or groups of genes that share a highly similar pattern of expression. Network modules are created by grouping coregulated genes together based on a measure of similarity. An integral component of network construction is calculation of gene connectivity. In weighted gene coexpression networks, the connectivity of a gene is the sum of its connection strengths with all other genes, and connection strengths are typically measured using the absolute value of the correlation coefficient between two genes (Zhang and Horvath, 2005). If a gene is highly connected, its expression will be correlated with the expression of many other genes. Highly connected genes are referred to as network "hubs."

Gene coexpression networks have been generated in both human and mice as a tool to identify modules involved in specific cellular processes, to characterize unannotated genes, and as a tool to model the relationship between gene expression and disease. This procedure is summarized in Fig. 20.8. Gargalovic *et al.* (2006) examined a relatively small number of primary human endothelial cells for responses to oxidized phospholipids, a trait relevant to atherosclerosis. In this study, the clinical status of individuals from which the cells were derived was unknown, but the coexpression modules identified were significantly enriched in known pathways. One module was enriched for genes involved in the unfolded protein response and also contained interleukin-8, an inflammatory stimulus important in atherosclerosis. Importantly, it was shown that the unfolded protein response pathway contributed to the transcriptional regulation of interleukin-8. In the mouse, Ghazalpour *et al.* (2006) developed a weighted gene coexpression network using liver expression profiles from F2 mice. Several modules were identified, one of which contained genes highly correlated with body weight. The authors demonstrated that a model accounting for genetic information on the location of key drivers of module gene expression and network properties of module genes (namely, connectivity) was an excellent predictor of the relationship between module gene expression and adiposity.

E. Genetical genomics in human studies

Several general surveys of the genetics of gene expression in humans have now appeared (Cheung *et al.*, 2005; Monks *et al.*, 2004; Morley *et al.*, 2004). The genetical genomic studies reported in humans thus far are in their infancy and essentially represent surveys with no attempts to connect gene expression to disease. These studies have also been relatively underpowered and so a small number of clear eQTLs have been identified. Also, most of the reported studies have utilized tissue culture cells, primarily Epstein Barr virus-transformed lymphoblastoid cells, that may have significant alterations in genomic content as compared to the individuals from which they were derived. Clearly, however, the results indicate that it is possible to map loci contributing to transcript levels in humans using both linkage analysis and association. There is every reason to believe that, with larger sample numbers, databases of hundreds or thousands of

Data from a series of arrays is used to quantify gene expression changes due to a specific perturbation

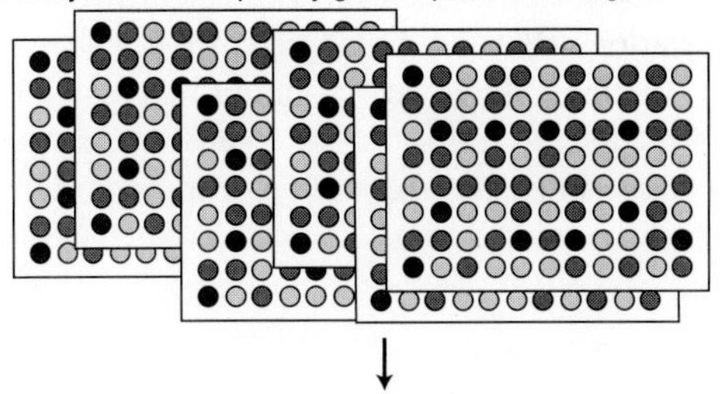

Coexpressed genes share similiar profiles of expression across samples

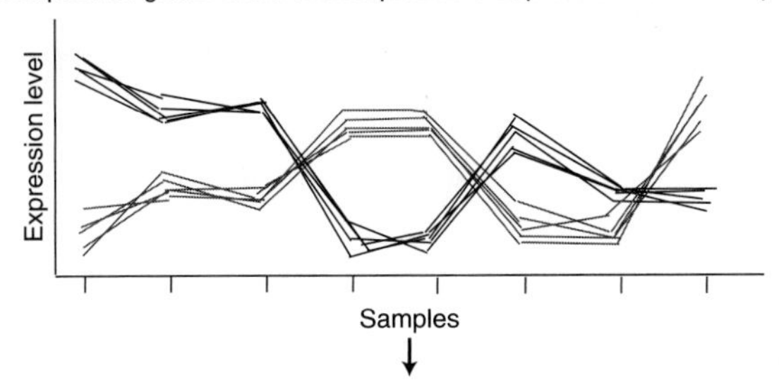

Coexpression relationships are quantified using correlation

	G1	G2	G3	G4	G5	G6	G7	G8
G1	1	0.8	0.9	0.1	0.3	0.2	0.5	0.7
G2	0.8	1	0.2	0.3	0.6	0.9	0.1	0.1
G3	0.9	0.2	1	0.9	0.8	0.2	0.1	0.6
G4	0.1	0.3	0.9	1	0.2	0.3	0.7	0.6
G5	0.3	0.6	0.8	0.2	1	0.3	0.2	0.1
G6	0.2	0.9	0.2	0.3	0.3	1	0.6	0.9
G7	0.5	0.1	0.1	0.7	0.2	0.6	1	0.8
G8	0.7	0.1	0.6	0.6	0.1	0.9	0.8	1

Gene coexpression networks are constructed by clustering correlated genes and can be visualized in a number of ways

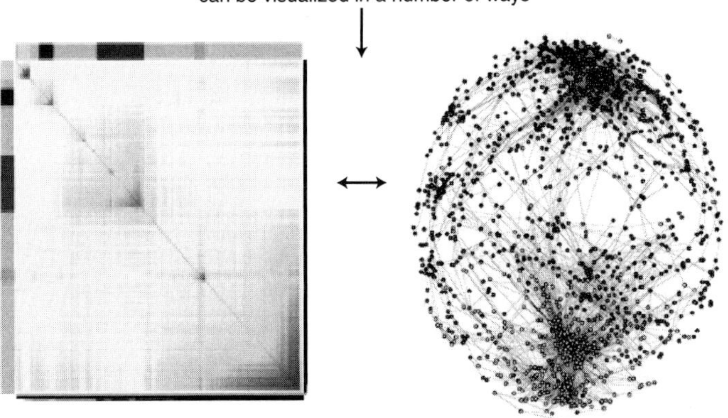

2-D heatmap representation

Multidimensional network representation

Figure 20.8. (*continued*)

genes commonly varying in transcript levels can be constructed. These will then serve to identify variations that will help prioritize the identification of genes underlying common disease. Moreover, it should be possible to correlate gene expression traits with clinical traits, as has been done in animal models, to identify potential causal genes and to begin to construct networks relevant to disease.

V. CONCLUSIONS

We have discussed a number of ways in which DNA microarray expression profiling can be used to investigate the genetic basis of disease. Our ability to predict and treat disease will only increase as novel approaches for using DNA microarrays are developed and technologies for quantifying different biological levels mature. Until recently, attempts to identify genes and pathways involved in common diseases were rarely successful. A few successful examples were primarily restricted to candidate genes that were previously identified by biochemical studies, such as apolipoprotein E and Alzheimer's disease. However, with the development of relatively inexpensive high throughput genotyping methods, including genome-wide association, and the assembly of large family-based or population-based study samples, the number of genes identified for common disease is ever increasing. The primary challenge, then, will be not to identify the underlying genes, but rather to understand pathways perturbed by genetic variation, the interactions between genes and between genes and environment, and the most suitable targets for therapeutic intervention. Global analysis of transcript levels offers an important bridge between genetic variation at the level of DNA and phenotypic variation.

Figure 20.8. Generating gene coexpression networks with global expression profiles. Coexpression networks rely on the collection of global gene expression profiles sampled across a series of perturbations such as differing genotypes. Within the collection of profiles, sets of genes will demonstrate similarity in expression patterns due to transcriptional coregulation. The coexpression relationships between genes can be quantified using correlation coefficients. Sets of correlated genes are then clustered using standard algorithms to identify "modules" of coexpressed genes. Coexpression networks can be visualized in a number of ways such as a 2-D heatmap (in which correlations between transcript levels are indicated by darker shading) or in a multidimensional spherical space consisting of nodes (genes) and edges (strength of correlation). In both plots distinct modules are labeled with different shading.

References

Abelson, J. F., Kwan, K. Y., O'Roak, B. J., Baek, D. Y., Stillman, A. A., Morgan, T. M., Mathews, C. A., Pauls, D. L., Rasin, M. R., Gunel, M., Davis, N. R., Ercan-Sencicek, A. G., *et al.* (2005). Sequence variants in SLITRK1 are associated with Tourette's syndrome. *Science* **310**(5746), 317–320.

Aitman, T. J., Glazier, A. M., Wallace, C. A., Cooper, L. D., Norsworthy, P. J., Wahid, F. N., Al-Majali, K. M., Trembling, P. M., Mann, C. J., Shoulders, C. C., Graf, D., St Lezin, E., *et al.* (1999). Identification of Cd36 (Fat) as an insulin-resistance gene causing defective fatty acid and glucose metabolism in hypertensive rats. *Nat. Genet.* **21**(1), 76–83.

Alwine, J. C., Kemp, D. J., and Stark, G. R. (1977). Method for detection of specific RNAs in agarose gels by transfer to diazobenzyloxymethyl-paper and hybridization with DNA probes. *Proc. Natl. Acad. Sci. USA* **74**(12), 5350–5354.

Ashburner, M., Ball, C. A., Blake, J. A., Botstein, D., Butler, H., Cherry, J. M., Davis, A. P., Dolinski, K., Dwight, S. S., Eppig, J. T., Harris, M. A., Hill, D. P., *et al.* (2000). Gene ontology: Tool for the unification of biology. The Gene Ontology Consortium. *Nat. Genet.* **25**(1), 25–29.

Barabasi, A. L., and Albert, R. (1999). Emergence of scaling in random networks. *Science* **286**(5439), 509–512.

Brem, R. B., and Kruglyak, L. (2005). The landscape of genetic complexity across 5,700 gene expression traits in yeast. *Proc. Natl. Acad. Sci. USA* **102**(5), 1572–1577.

Brem, R. B., Yvert, G., Clinton, R., and Kruglyak, L. (2002). Genetic dissection of transcriptional regulation in budding yeast. *Science* **296**(5568), 752–755.

Brem, R. B., Storey, J. D., Whittle, J., and Kruglyak, L. (2005). Genetic interactions between polymorphisms that affect gene expression in yeast. *Nature* **436**(7051), 701–703.

Brenner, S., Johnson, M., Bridgham, J., Golda, G., Lloyd, D. H., Johnson, D., Luo, S., McCurdy, S., Foy, M., Ewan, M., Roth, R., George, D., *et al.* (2000). Gene expression analysis by massively parallel signature sequencing (MPSS) on microbead arrays. *Nat. Biotechnol.* **18**(6), 630–634.

Butte, A. (2002). The use and analysis of microarray data. *Nat. Rev. Drug Discov.* **1**(12), 951–960.

Cheung, V. G., Spielman, R. S., Ewens, K. G., Weber, T. M., Morley, M., and Burdick, J. T. (2005). Mapping determinants of human gene expression by regional and genome-wide association. *Nature* **437**(7063), 1365–1369.

Clop, A., Marcq, F., Takeda, H., Pirottin, D., Tordoir, X., Bibe, B., Bouix, J., Caiment, F., Elsen, J. M., Eychenne, F., Meish, F., Milenkovic, D., *et al.* (2006). A mutation creating a potential illegitimate microRNA target site in the myostatin gene affects muscularity in sheep. *Nat. Genet.* **38**(7), 813–818.

Dalma-Weiszhausz, D. D., Warrington, J., Tanimoto, E. Y., and Miyada, C. G. (2006). The affymetrix GeneChip platform: An overview. *Methods Enzymol.* **410,** 3–28.

Darvasi, A. (1998). Experimental strategies for the genetic dissection of complex traits in animal models. *Nat. Genet.* **18**(1), 19–24.

Dennis, G., Jr., Sherman, B. T., Hosack, D. A., Yang, J., Gao, W., Lane, H. C., and Lempicki, R. A. (2003). DAVID: Database for annotation, visualization, and integrated discovery. *Genome Biol.* **4**(5), P3.

DeRisi, J., Penland, L., Brown, P. O., Bittner, M. L., Meltzer, P. S., Ray, M., Chen, Y., Su, Y. A., and Trent, J. M. (1996). Use of a cDNA microarray to analyse gene expression patterns in human cancer. *Nat. Genet.* **14**(4), 457–460.

Diament, A. L., and Warden, C. H. (2004). Multiple linked mouse chromosome 7 loci influence body fat mass. *Int. J. Obes. Relat. Metab. Disord.* **28**(2), 199–210.

Doerge, R. W. (2002). Mapping and analysis of quantitative trait loci in experimental populations. *Nat. Rev. Genet.* **3**(1), 43–52.

Doss, S., Schadt, E. E., Drake, T. A., and Lusis, A. J. (2005). Cis-acting expression quantitative trait loci in mice. *Genome Res.* **15**(5), 681–691.

Edderkaoui, B., Baylink, D. J., Beamer, W. G., Wergedal, J. E., Dunn, N. R., Shultz, K. L., and Mohan, S. (2006). Multiple genetic loci from CAST/EiJ chromosome 1 affect vBMD either positively or negatively in a C57BL/6J background. *J. Bone Miner. Res.* **21**(1), 97–104.

Ehrenreich, A. (2006). DNA microarray technology for the microbiologist: An overview. *Appl. Microbiol. Biotechnol.* **73**(2), 255–273.

Falconer, D. S., and Mackay, T. F. C. (1996). "Introduction to Quantitative Genetics." 4th Edition, Benjamin Cummings, Menlo Park. California.

Fan, J. B., Gunderson, K. L., Bibikova, M., Yeakley, J. M., Chen, J., Wickham Garcia, E., Lebruska, L. L., Laurent, M., Shen, R., and Barker, D. (2006). Illumina universal bead arrays. *Methods Enzymol.* **410,** 57–73.

Flint, J., Valdar, W., Shifman, S., and Mott, R. (2005). Strategies for mapping and cloning quantitative trait genes in rodents. *Nat. Rev. Genet.* **6**(4), 271–286.

Fodor, S. P., Read, J. L., Pirrung, M. C., Stryer, L., Lu, A. T., and Solas, D. (1991). Light-directed, spatially addressable parallel chemical synthesis. *Science* **251**(4995), 767–773.

Gargalovic, P. S., Imura, M., Zhang, B., Gharavi, N. M., Clark, M. J., Pagnon, J., Yang, W. P., He, A., Truong, A., Patel, S., Nelson, S. F., Horvath, S., *et al.* (2006). Identification of inflammatory gene modules based on variations of human endothelial cell responses to oxidized lipids. *Proc. Natl. Acad. Sci. USA* **103**(34), 12741–12746.

Gentleman, R. C., Carey, V. J., Bates, D. M., Bolstad, B., Dettling, M., Dudoit, S., Ellis, B., Gautier, L., Ge, Y., Gentry, J., Hornik, K., Hothorn, T., *et al.* (2004). Bioconductor: Open software development for computational biology and bioinformatics. *Genome Biol.* **5**(10), R80.

Ghazalpour, A., Doss, S., Sheth, S. S., Ingram-Drake, L. A., Schadt, E. E., Lusis, A. J., and Drake, T. A. (2005). Genomic analysis of metabolic pathway gene expression in mice. *Genome Biol.* **6**(7), R59.

Ghazalpour, A., Doss, S., Zhang, B., Wang, S., Plaisier, C., Castellanos, R., Brozell, A., Schadt, E. E., Drake, T. A., Lusis, A. J., and Horvath, S. (2006). Integrating genetic and network analysis to characterize genes related to mouse weight. *PLoS Genet.* **2**(8), e130.

Gibbs, R. A., Weinstock, G. M., Metzker, M. L., Muzny, D. M., Sodergren, E. J., Scherer, S., Scott, G., Steffen, D., Worley, K. C., Burch, P. E., Okwaonu, G., Hines, S., *et al.* (2004). Genome sequence of the Brown Norway rat yields insights into mammalian evolution. *Nature* **428**(6982), 493–521.

Gunderson, K. L., Kruglyak, S., Graige, M. S., Garcia, F., Kermani, B. G., Zhao, C., Che, D., Dickinson, T., Wickham, E., Bierle, J., Doucet, D., Milewski, M., *et al.* (2004). Decoding randomly ordered DNA arrays. *Genome Res.* **14**(5), 870–877.

Hager, J. (2006). Making and using spotted DNA microarrays in an academic core laboratory. *Methods Enzymol.* **410,** 135–168.

Hasegawa, K., Tamari, M., Shao, C., Shimizu, M., Takahashi, N., Mao, X. Q., Yamasaki, A., Kamada, F., Doi, S., Fujiwara, H., Miyatake, A., Fujita, K., *et al.* (2004). Variations in the C3, C3a receptor, and C5 genes affect susceptibility to bronchial asthma. *Hum. Genet.* **115**(4), 295–301.

Hosack, D. A., Dennis, G., Jr., Sherman, B. T., Lane, H. C., and Lempicki, R. A. (2003). Identifying biological themes within lists of genes with EASE. *Genome Biol.* **4**(10), R70.

Hubank, M., and Schatz, D. G. (1994). Identifying differences in mRNA expression by representational difference analysis of cDNA. *Nucleic Acids Res.* **22**(25), 5640–5648.

Jaluria, P., Konstantopoulos, K., Betenbaugh, M., and Shiloach, J. (2007). A perspective on microarrays: Current applications, pitfalls, and potential uses. *Microb. Cell Fact.* **6,** 4.

Jilka, R. L. (2003). Biology of the basic multicellular unit and the pathophysiology of osteoporosis. *Med. Pediatr. Oncol.* **41**(3), 182–185.

Kapp, A. V., Jeffrey, S. S., Langerod, A., Borresen-Dale, A. L., Han, W., Noh, D. Y., Bukholm, I. R., Nicolau, M., Brown, P. O., and Tibshirani, R. (2006). Discovery and validation of breast cancer subtypes. *BMC Genomics* **7,** 231.

Karp, C. L., Grupe, A., Schadt, E., Ewart, S. L., Keane-Moore, M., Cuomo, P. J., Kohl, J., Wahl, L., Kuperman, D., Germer, S., Aud, D., Peltz, G., *et al.* (2000). Identification of complement factor 5 as a susceptibility locus for experimental allergic asthma. *Nat. Immunol.* **1**(3), 221–226.

Klein, R. F., Allard, J., Avnur, Z., Nikolcheva, T., Rotstein, D., Carlos, A. S., Shea, M., Waters, R. V., Belknap, J. K., Peltz, G., *et al.* (2004). Regulation of bone mass in mice by the lipoxygenase gene Alox15. *Science* **303**(5655), 229–232.

Koczan, D., and Thiesen, H. J. (2006). Survey of microarray technologies suitable to elucidate transcriptional networks as exemplified by studying KRAB zinc finger gene families. *Proteomics* **6**(17), 4704–4715.

Konradi, C. (2005). Gene expression microarray studies in polygenic psychiatric disorders: Applications and data analysis. *Brain Res. Brain Res. Rev.* **50**(1), 142–155.

Kuhn, K., Baker, S. C., Chudin, E., Lieu, M. H., Oeser, S., Bennett, H., Rigault, P., Barker, D., McDaniel, T. K., and Chee, M. S. (2004). A novel, high-performance random array platform for quantitative gene expression profiling. *Genome Res.* **14**(11), 2347–2356.

Lamb, J., Crawford, E. D., Peck, D., Modell, J. W., Blat, I. C., Wrobel, M. J., Lerner, J., Brunet, J. P., Subramanian, A., Ross, K. N., Reich, M., Hieronymus, H., *et al.* (2006). The connectivity map: Using gene-expression signatures to connect small molecules, genes, and disease. *Science* **313** (5795), 1929–1935.

Lander, E. S., Linton, L. M., Birren, B., Nusbaum, C., Zody, M. C., Baldwin, J., Devon, K., Dewar, K., Doyle, M., FitzHugh, W., Funke, R., Gage, D., *et al.* (2001). Initial sequencing and analysis of the human genome. *Nature* **409**(6822), 860–921.

Lein, E. S., Hawrylycz, M. J., Ao, N., Ayres, M., Bensinger, A., Bernard, A., Boe, A. F., Boguski, M. S., Brockway, K. S., Byrnes, E. J., Chen, L., Chen, L., *et al.* (2007). Genome-wide atlas of gene expression in the adult mouse brain. *Nature* **445**(7124), 168–176.

Liang, P., and Pardee, A. B. (1992). Differential display of eukaryotic messenger RNA by means of the polymerase chain reaction. *Science* **257**(5072), 967–971.

Meng, H., Vera, I., Che, N., Wang, X., Wang, S. S., Ingram-Drake, L., Schadt, E. E., Drake, T. A., and Lusis, A. J. (2007). Identification of Abcc6 as the major causal gene for dystrophic cardiac calcification in mice through integrative genomics. *Proc. Natl. Acad. Sci. USA* **104**(11), 4530–4535.

Monks, S. A., Leonardson, A., Zhu, H., Cundiff, P., Pietrusiak, P., Edwards, S., Phillips, J. W., Sachs, A., and Schadt, E. E. (2004). Genetic inheritance of gene expression in human cell lines. *Am. J. Hum. Genet.* **75**(6), 1094–1105.

Mootha, V. K., Lindgren, C. M., Eriksson, K. F., Subramanian, A., Sihag, S., Lehar, J., Puigserver, P., Carlsson, E., Ridderstrale, M., Laurila, E., Houstis, N., Daly, M. J., *et al.* (2003). PGC-1alpha-responsive genes involved in oxidative phosphorylation are coordinately downregulated in human diabetes. *Nat. Genet.* **34**(3), 267–273.

Morley, M., Molony, C. M., Weber, T. M., Devlin, J. L., Ewens, K. G., Spielman, R. S., and Cheung, V. G. (2004). Genetic analysis of genome-wide variation in human gene expression. *Nature* **430**(7001), 743–747.

Narayanan, R. (2007). Bioinformatics approaches to cancer gene discovery. *Methods Mol, Biol.* **360,** 13–31.

Perou, C. M., Sorlie, T., Eisen, M. B., van de Rijn, M., Jeffrey, S. S., Rees, C. A., Pollack, J. R., Ross, D. T., Johnsen, H., Akslen, L. A., Fluge, O., Pergamenschikov, A., *et al.* (2000). Molecular portraits of human breast tumours. *Nature* **406**(6797), 747–752.

Quackenbush, J. (2001). Computational analysis of microarray data. *Nat. Rev. Genet.* **2**(6), 418–427.

Rockman, M. V., and Kruglyak, L. (2006). Genetics of global gene expression. *Nat. Rev. Genet.* **7** (11), 862–872.

Schadt, E. E., Monks, S. A., Drake, T. A., Lusis, A. J., Che, N., Colinayo, V., Ruff, T. G., Milligan, S. B., Lamb, J. R., Cavet, G., Linsley, P. S., Mao, M., *et al.* (2003). Genetics of gene expression surveyed in maize, mouse and man. *Nature* **422**(6929), 297–302.

Schadt, E. E., Lamb, J., Yang, X., Zhu, J., Edwards, S., Guhathakurta, D., Sieberts, S. K., Monks, S., Reitman, M., Zhang, C., Lum, P. Y., Leonardson, A., *et al.* (2005). An integrative genomics approach to infer causal associations between gene expression and disease. *Nat. Genet.* **37**(7), 710–717.

Schena, M., Shalon, D., Davis, R. W., and Brown, P. O. (1995). Quantitative monitoring of gene expression patterns with a complementary DNA microarray. *Science* **270**(5235), 467–470.

Scheurle, D., DeYoung, M. P., Binninger, D. M., Page, H., Jahanzeb, M., and Narayanan, R. (2000). Cancer gene discovery using digital differential display. *Cancer Res.* **60**(15), 4037–4043.

Seeman, E., and Delmas, P. D. (2006). Bone quality—the material and structural basis of bone strength and fragility. *N. Engl. J. Med.* **354**(21), 2250–2261.

Seo, D., Wang, T., Dressman, H., Herderick, E. E., Iversen, E. S., Dong, C., Vata, K., Milano, C. A., Rigat, F., Pittman, J., Nevins, J. R., West, M., *et al.* (2004). Gene expression phenotypes of atherosclerosis. *Arterioscler. Thromb. Vasc. Biol.* **24**(10), 1922–1927.

Shi, L., Reid, L. H., Jones, W. D., Shippy, R., Warrington, J. A., Baker, S. C., Collins, P. J., de Longueville, F., Kawasaki, E. S., Lee, K. Y., Luo, Y., Sun, Y. A., *et al.* (2006). The MicroArray Quality Control (MAQC) project shows inter- and intraplatform reproducibility of gene expression measurements. *Nat. Biotechnol.* **24**(9), 1151–1161.

Sorlie, T., Perou, C. M., Tibshirani, R., Aas, T., Geisler, S., Johnsen, H., Hastie, T., Eisen, M. B., van de Rijn, M., Jeffrey, S. S., Thorsen, T., Quist, H., *et al.* (2001). Gene expression patterns of breast carcinomas distinguish tumor subclasses with clinical implications. *Proc. Natl. Acad. Sci. USA* **98** (19), 10869–10874.

Stoughton, R. B. (2005). Applications of DNA microarrays in biology. *Annu. Rev. Biochem.* **74,** 53–82.

Su, A. I., Cooke, M. P., Ching, K. A., Hakak, Y., Walker, J. R., Wiltshire, T., Orth, A. P., Vega, R. G., Sapinoso, L. M., Moqrich, A., Patapoutian, A., Hampton, G. M., *et al.* (2002). Large-scale analysis of the human and mouse transcriptomes. *Proc. Natl. Acad. Sci. USA* **99**(7), 4465–4470.

Subramanian, A., Tamayo, P., Mootha, V. K., Mukherjee, S., Ebert, B. L., Gillette, M. A., Paulovich, A., Pomeroy, S. L., Golub, T. R., Lander, E. S., and Mesirov, J. P. (2005). Gene set enrichment analysis: A knowledge-based approach for interpreting genome-wide expression profiles. *Proc. Natl. Acad. Sci USA* **102**(43), 15545–15550.

Tusher, V. G., Tibshirani, R., and Chu, G. (2001). Significance analysis of microarrays applied to the ionizing radiation response. *Proc. Natl. Acad. Sci. USA* **98**(9), 5116–5121.

van de Vijver, M. J., He, Y. D., van't Veer, L. J., Dai, H., Hart, A. A., Voskuil, D. W., Schreiber, G. J., Peterse, J. L., Roberts, C., Marton, M. J., Parrish, M., Atsma, D., *et al.* (2002). A gene-expression signature as a predictor of survival in breast cancer. *N. Engl. J. Med.* **347**(25), 1999–2009.

van't Veer, L. J., Dai, H., van de Vijver, M. J., He, Y. D., Hart, A. A., Mao, M., Peterse, H. L., van der Kooy, K., Marton, M. J., Witteveen, A. T., Schreiber, G. J., Kerkhoven, R. M., *et al.* (2002). Gene expression profiling predicts clinical outcome of breast cancer. *Nature* **415**(6871), 530–536.

Velculescu, V. E., Zhang, L., Vogelstein, B., and Kinzler, K. W. (1995). Serial analysis of gene expression. *Science* **270**(5235), 484–487.

Venter, J. C., Adams, M. D., Myers, E. W., Li, P. W., Mural, R. J., Sutton, G. G., Smith, H. O., Yandell, M., Evans, C. A., Holt, R. A., Gocayne, J. D., Amanatides, P., *et al.* (2001). The sequence of the human genome. *Science* **291**(5507), 1304–1351.

Walker, J. R., and Wiltshire, T. (2006). Databases of free expression. *Mamm. Genome.* **17**(12), 1141–1146.

Wang, S., Yehya, N., Schadt, E. E., Wang, H., Drake, T. A., and Lusis, A. J. (2006). Genetic and genomic analysis of a fat mass trait with complex inheritance reveals marked sex specificity. *PLoS Genet.* **2**(2), e15.

Waterston, R. H., Lindblad-Toh, K., Birney, E., Rogers, J., Abril, J. F., Agarwal, P., Agarwala, R., Ainscough, R., Alexandersson, M., An, P., Antonarakis, S. E., Attwood, J., *et al.* (2002). Initial sequencing and comparative analysis of the mouse genome. *Nature* **420**(6915), 520–562.

Wetsel, R. A., Fleischer, D. T., and Haviland, D. L. (1990). Deficiency of the murine fifth complement component (C5). A 2-base pair gene deletion in a 5′-exon. *J. Biol. Chem.* **265**(5), 2435–2440.

Wit, E., and McClure, J. (2004). "Statistics for Microarrays: Design, Analysis and Inference." 1st Edition, Wiley, New York.

Yang, X., Schadt, E. E., Wang, S., Wang, H., Arnold, A. P., Ingram-Drake, L., Drake, T. A., and Lusis, A. J. (2006). Tissue-specific expression and regulation of sexually dimorphic genes in mice. *Genome Res.* **16**(8), 995–1004.

Zhang, B., and Horvath, S. (2005). A general framework for weighted gene co-expression network analysis. *Stat. Appl. Genet. Mol. Biol.* **4**, Article17.

21

A Systems Biology Approach to Drug Discovery

Jun Zhu, Bin Zhang, and Eric E. Schadt
Rosetta Inpharmatics, LLC, a Wholly Owned Subsidiary of Merck and Co., Inc., Seattle, Washington 98109

ABSTRACT

Common human diseases like obesity and diabetes are driven by complex networks of genes and any number of environmental factors. To understand this complexity in hopes of identifying targets and developing drugs against disease, a systematic approach is required to elucidate the genetic and

Advances in Genetics, Vol. 60 0065-2660/08 $35.00
DOI: 10.1016/S0065-2660(07)00421-X

environmental factors and interactions among and between these factors, and to establish how these factors induce changes in gene networks that in turn lead to disease. The explosion of large-scale, high-throughput technologies in the biological sciences has enabled researchers to take a more systems biology approach to study complex traits like disease. Genotyping of hundreds of thousands of DNA markers and profiling tens of thousands of molecular phenotypes simultaneously in thousands of individuals is now possible, and this scale of data is making it possible for the first time to reconstruct whole gene networks associated with disease. In the following sections, we review different approaches for integrating genetic expression and clinical data to infer causal relationships among gene expression traits and between expression and disease traits. We further review methods to integrate these data in a more comprehensive manner to identify common pathways shared by the causal factors driving disease, including the reconstruction of association and probabilistic causal networks. Particular attention is paid to integrating diverse information to refine these types of networks so that they are more predictive. To highlight these different approaches in practice, we step through an example on how *Insig2* was identified as a causal factor for plasma cholesterol levels in mice.

I. INTRODUCTION

One of the primary aims of drug discovery over the last 30 years has been to design drugs against targets that were thought to affect simple signaling pathways associated with disease. This approach to drug discovery has not met with huge success, largely because it represents an overly simplistic view of the molecular mechanisms underlying complex traits like common human diseases. The view that simple, well-ordered (often linear) pathways drive key biological processes is a consequence of biological reductionism, brought about by the need to form a basic understanding of the fundamental attributes of biological systems, driven historically by limitations in the set of tools available for analysis of biological systems. In reality, complex biological systems are best modeled as highly modular, fluid systems exhibiting a plasticity that allows them to adapt to a vast array of conditions. While this viewpoint has long represented the ideal, the tools needed to examine and describe this complexity were often lacking. However, the tools now exist to take comprehensive snapshots on many different levels of complex systems, so that a more network-oriented view of biological systems for explaining the underlying causes of disease, as well as the best ways to target disease, are now possible.

The explosion of large-scale, high-throughput technologies in the biological sciences has motivated a rapid shift away from reductionism in favor of systems of biology (Hartwell *et al.*, 1999). New tools are now available to look at interactions among tens of thousands of genes in different tissues and in different states. Genomic sequencing has been completed for a plethora of organisms, and a growing computational infrastructure has enabled integrated views of DNA, RNA, and protein interaction data to elucidate the fundamental nature of disease and living systems more generally. Thus, future successes in biomedical research will likely demand a more comprehensive view of the complex array of interactions in biological systems and how such interactions are influenced by genetic background, infection, environmental states, lifestyle choices, and social structures more generally (Barabasi and Oltvai, 2004; Zerhouni, 2003). This holistic view requires embracing complexity in its entirety, so that complex biological systems are beginning to be seen as dynamic, fluid systems able to reconfigure themselves as conditions demand (Han *et al.*, 2004; Luscombe *et al.*, 2004; Pinto *et al.*, 2004).

Systems biology is defined by Hood *et al.* (Ideker *et al.*, 2001a) as the study of "biological systems by systematically perturbing them (biologically, genetically, or chemically); monitoring the gene, protein, and informational pathway responses; integrating these data; and ultimately, formulating mathematical models that describe the structure of the system and its response to individual perturbations." This definition emphasizes systematic perturbations and high-dimensional data generation in biological systems, followed by the integration of these data to get at the processes driving complex traits. There are multiple ways to generate perturbations systematically. For example, genes can be knocked out, multiple compound treatments can be administered, and genes can be knocked down using RNAi-based technologies in biological systems. An alternative to these artificial, more simplistic perturbations are naturally occurring genetic variations that segregate in populations such as those obtained from experimental crosses or that occur naturally. Not only are these types of perturbations naturally occurring, possibly making them more relevant for identifying key nodes in networks that drive disease, but they also occur in a multifactorial context that enables the study of additive effects and epistatic interactions. Because most common human diseases are thought to manifest themselves as a result of many weak contributors rather than a few dominant factors, it is important to study complex traits in this more realistic setting. In the following sections, we review different approaches for integrating genetic, expression, and clinical data to infer causal relationships among gene expression traits and between expression and disease traits. We further review methods to integrate these data in a more comprehensive manner to identify common pathways shared by the causal factors at different levels, including association and probabilistic causal networks. Particular attention is paid to integrating diverse information to refine these types of networks so that they are more predictive. To highlight these different approaches in practice, we step through an example on how *Insig2* was identified as a causal factor for plasma cholesterol levels in mice.

II. CAUSAL INFERENCE: AN INTEGRATED APPROACH TO DRUG DISCOVERY

Establishing causal relationships between genes, gene products, protein states, metabolites and complex phenotypes like human diseases is key to understanding mechanisms of complex diseases and how to modulate diseases by targeting of specific genes. Causality is a logical relationship between two events, with one event referred to as the cause and the other, being the direct consequence of the first, referred to as the effect. Inferring causal relationships between events of interest has been well studied in economics and the neural sciences (Gourevitch *et al.*, 2006), where the Granger causality test (Granger, 1969) applied to time series data has played a prominent role. Tests like the Granger causality test have been applied to high-dimensional data like microarray data to infer transcriptional regulatory networks (Mukhopadhyay and Chatterjee, 2007). However, at present the number of time points sampled is far too small to make this approach useful in practice. Instead, additional information is needed to constrain the space of possible models to consider among multiple genes or traits of interest (Mani and Cooper, 2000). One way to constrain this space is to leverage DNA variation information.

Genetic variation contributes to almost all human diseases, such as diabetes, obesity, cardiovascular diseases, infection, neurological diseases, and cancer. A given individual's genetic background, and the interaction of this background with a diversity of environmental factors, affects disease susceptibility, disease progression, and response to drug treatment, in addition to other clinically relevant factors. Identifying the specific genetic loci associated with disease traits is one of the first steps commonly applied to understanding how changes in DNA affect the onset of disease and disease progression. Such understanding would ultimately lead to the development of therapies to treat and prevent diseases as well as to the design of diagnostic tests.

However, associating DNA variation information with disease states on its own typically does not provide the necessary functional information needed to understand the mechanisms that lead to disease states. In fact, from genetic studies it is usually not possible to even definitively declare that the gene most proximal to a DNA variant associated with disease is the actual causal gene, let alone discern from the genetic data whether activating or inactivating the gene would be required for treatment of disease. The DNA variation information, however, can be combined with molecular profiling data like gene expression to provide a functional context within which to understand a gene's function. DNA variations that affect transcriptional and other molecular, cellular, tissue, and organism networks, interactions among these networks and environment factors, have their impact on the complex physiological processes underlying different diseases via this complex array of molecular networks

(Schadt and Lum, 2006). Therefore, considering these intermediate, molecular phenotypes, together with the genetic and disease data, provides an opportunity to elucidate the key drivers of disease more fully.

Classically, causal loci for a small number of clinical traits have been identified via linkage and/or association studies. Fine mapping (positional cloning) of these genetic loci found to be associated with disease was then carried out to identify the causal genes and actual DNA variants responsible for susceptibility to disease-associated traits (Rosen *et al.*, 1993; Van Eerdewegh *et al.*, 2002). This approach worked well for Mendelian traits and has begun to show great promise for common human diseases, now that comprehensive genome-wide association scans can be carried out. However, such studies still do not elucidate the mechanism through which the causal genes affect disease traits. Recent advances in microarray technologies have now enabled researchers to monitor tens of thousands of genes for differences in expression simultaneously. Gene transcripts have been identified that are associated with complex disease phenotypes (Adams *et al.*, 2000; Schadt *et al.*, 2003) or disease subtypes (Mootha *et al.*, 2003; Schadt *et al.*, 2003; van 't Veer *et al.*, 2002). For more information on specific microarray technologies, see the chapter by Farber and Lusis (2008) in this volume. Changes in gene expression often reflect changes in a gene's activity and the impact a gene has on different phenotypes. It is now well established that gene expression is a significantly heritable trait (Brem *et al.*, 2002; Bystrykh *et al.*, 2005; Chesler *et al.*, 2005; Jansen and Nap, 2001; Monks *et al.*, 2004; Morley *et al.*, 2004; Petretto *et al.*, 2006a,b; Schadt *et al.*, 2003, 2005a). Genetic loci affecting an expression trait is referred to as an expression quantitative trait locus or eQTL (Brem *et al.*, 2002; Schadt *et al.*, 2003). Therefore, if a gene expression trait is highly correlated with a disease trait of interest, both the expression trait and the disease trait may be under the control of common genetic loci. Further, if we knew that this gene was physically located in the same region to which the expression and disease trait linked, this would provide an objective and direct path to identify candidate causal genes for the disease traits of interest (Brem *et al.*, 2002; Bystrykh *et al.*, 2005; Chesler *et al.*, 2005; Jansen and Nap, 2001; Monks *et al.*, 2004; Morley *et al.*, 2004; Petretto *et al.*, 2006a,b; Schadt *et al.*, 2003, 2005a). More important than this type of *cis*-acting effect are the *trans*-acting effects whereby many more gene expression traits are genetically linked to the same region as the disease traits, where these genes do not physically reside in this region. In such cases, the cluster of expression traits linking to the disease locus may be in the causal networks that are intermediate to the genetic loci and disease traits, so that they also lead to an increased susceptibility to disease. Since in these types of cases entire networks are identified that are causal for disease, a much broader net is cast to identify the key drivers of disease, compared to what could be achieved using classic genetic methods.

Genes that play an intermediate causal role have been directly identified and validated via directed perturbation experiments, which have shown that these intermediate phenotypes can in fact be causal for disease (Schadt *et al.*, 2005a).

In a segregating population like an F2 intercross, the flow of information, from changes in DNA to changes in RNA and protein function that in turn give rise to a disease phenotype, is unambiguous (Fig. 21.1). Given the unambiguous flow of information and a couple of simplifying assumptions, there are a limited number of ways two traits linked to the same DNA locus can be related (Schadt, 2005; Schadt *et al.*, 2005a,b) (Fig. 21.2A). On the contrary, in the absence of genetic or other directed perturbation information, the flow of information is not unambiguous, so that many indistinguishable relationships are possible and therefore additional data are required to establish the correct relationships between genes and traits of interest. Given two traits linked to a common DNA locus and the constraint that these variables are linearly related, there are three basic relationships one can infer among the traits: (1) the two traits are independently driven by the DNA locus, (2) the first trait is causal for

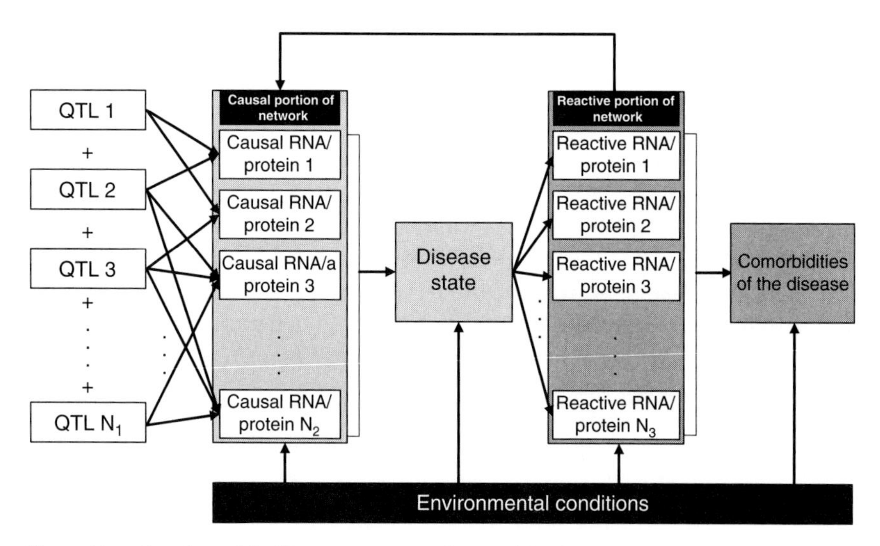

Figure 21.1. Simple model of how genetic perturbations (QTL), gene expression or protein traits, and disease traits interact. The QTL represents variations in DNA that in turn lead to changes in RNA or protein activity. As a result of these changes, broader perturbations to the transcriptional network are induced that ultimately lead to common human diseases like obesity. The disease state can in turn lead to additional changes in the transcriptional network that may give rise to comorbidities of the disease. In addition to the genetic causes of disease, environmental factors also play a critical role, as common human diseases are the result of complex interactions among multiple genetic loci and between the genetic loci and environmental factors.

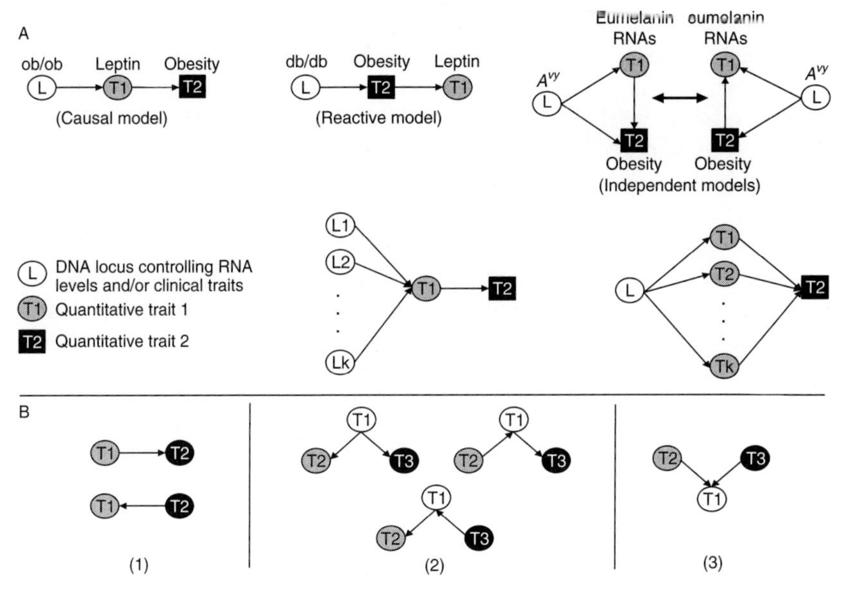

Figure 21.2. Possible relationships between phenotypes driven by common genetic loci. (A) The first three graphs represent the set of possible relationships between two traits and a controlling genetic locus, when feedback mechanisms are ignored. For the independent models, the two graphs depicted are Markov equivalent given simple linear relationships between the nodes (Chickering 1995). However, these two graphs may not be the same for more sophisticated models (Kulp and Jagalur 2006). The final two graphs represent more complicated scenarios in which multiple genetic loci control a given trait that in turn drives a second trait, or a single genetic locus drives multiple traits that collectively drive another trait. (B) Markov equivalent structures. Panels (1) and (2) represent graphs that are Markov equivalent. Panel (3) represents a graph that is not Markov equivalent to any other graph.

the second trait with respect to the DNA locus, and (3) the first trait is reactive to the second trait with respect to the DNA locus. For example, in Fig. 21.2A, the ob/ob mouse [leptin gene (*Lep*) knockout strain] has no expression of the leptin gene (first trait), and the absence of leptin in this mouse leads to an obesity phenotype (second trait). Thus, leptin is causal for obesity with respect to the *Lep* locus. On the contrary, the db/db mouse [leptin receptor (*Lepr*) knockout strain] develops an obesity phenotype because it is not sensitive to leptin. Leptin levels increase because of the obese state of the animal, and so in this instance leptin expression would be reacting to the obesity phenotype instead of causing the phenotype, with respect to *Lpr* locus. Finally, the agouti A^{vy} mouse provides an example of correlations between eumelanin RNA levels and obesity phenotypes induced by an allele that acts independently on these different traits, simultaneously, but independently causing both decreased levels of eumelanin RNA and an obesity phenotype. More generally, a clinical and expression trait for a particular gene may depend on the activity of other genes, such that conditional on these other genes the clinical and expression traits are independent.

It is important to note here that when we use the term causality throughout the rest of this chapter, it is perhaps meant in a more nonstandard sense than most researchers in the life sciences may be accustomed to. In the molecular biology or biochemistry setting, claiming a causal relationship between, say, two proteins usually means that one protein has been determined experimentally to physically interact with or to induce processes that directly affect another protein, and that in turn leads to a phenotypic change of interest. In such instances, an understanding of the causal factors relevant to this activity is known, and careful experimental manipulation of these factors subsequently allows for the identification of genuine causal relationships. As shown above, whether leptin is causal or reactive to obesity is context dependent, and the term causal in this case denotes a known, biological cause–effect relationship. However, for causal relationships we seek to detect by fitting data to mathematical models, the term "causal" is used from the standpoint of statistical inference, where statistical associations between changes in DNA, changes in expression (or other molecular phenotypes), and changes in complex phenotypes like disease are examined for patterns of statistical dependency among these variables that allows directionality to be inferred among them, where the directionality then provides the source of causal information (highlighting putative regulatory control as opposed to physical interaction). The graphical models (networks) described here, therefore, are necessarily probabilistic structures that use the available data to infer the correct structure of relationships among genes and between genes and clinical phenotypes (Schadt and Lum, 2006). In a single experiment, with one time point measurement, these methods cannot easily model more complex regulatory structures that are known to exist, like negative feedback control. However, the methods can be useful in providing a broad picture of correlation and causative relationships, and while the more complex structures may not be explicitly represented in this setting, they are captured nevertheless given they represent observed states that are reached as a result of more complicated processes like feedback control.

The integration of the genetic information enables the dissection of the covariance structure for two traits of interest into genetic and nongenetic components, where the genetic component can then be leveraged to support whether two traits (e.g., an expression and disease trait) are related in a causal, reactive, or independent manner. There are a number of ways in which these data can be considered in assessing which of these models best fits the data. One of the first methods published was a likelihood-based causality model selection procedure originally proposed by Schadt *et al.* (2005a). As discussed above, there are three basic ways to order two quantitative traits T_1 and T_2 that are both found to link to the same locus L in an F2 intercross population. Assuming standard Markov properties for these basic relationships, the joint probability distributions corresponding to these three models, respectively, are:

$$P(L, T_1, T_2) = P(L)P(T_1|L)P(T_2|T_1) \tag{M1}$$

$$P(L, T_1, T_2) = P(L)P(T_2|L)P(T_1|T_2) \tag{M2}$$

$$P(L, T_1, T_2) = P(L)P(T_2|L)P(T_1|T_2, L) \tag{M3}$$

where the final term on the right-hand side of Eq. (M3) reflects that the correlation between T_1 and T_2 may be explained by other shared loci or common environmental influences, in addition to locus L. $P(L)$ is the genotype probability distribution for locus L and is based on a previously described recombination model (Jiang and Zeng, 1995). Under constraint of linear relationship, $T_1 \rightarrow T_2$ and $T_2 \rightarrow T_1$ for model M3 are Markov equivalent (Chickering, 1995) so that $P(T_2|L)P(T_1|T_2, L) = P(T_1|L)P(T_2|T_1, L)$. Thus, M3 actually represents more than one pleiotropic model as depicted in Fig. 21.2A. In any case, the primary utility of this type of test is to establish whether two traits are causally related, so that distinguishing between the independent state and the causal–reactive states is one of the main aims. The likelihoods for each model are formed by multiplying the densities for each of the component pieces across all of the individuals in the population (Schadt *et al.*, 2005a). Once the likelihoods associated with each of the three possible models are constructed, they are maximized with respect to the model parameters, and the model with the smallest Akaike Information Criterion (AIC) (Sakamoto *et al.*, 1986) is identified as the model best supported by the data.

The causality test outlined above takes into account only linear interactions. Kulp and Jagalur (2006) extended this method to infer causal relationships by including locus–gene expression interactions so that independent models represented in Fig. 21.2A can be further divided into a model where T_1 and T_2 are truly independent or where T_1 and T_2 are causally related. Under the model where T_1 is causal for T_2, the trait T_2 in Fig. 21.2A can be modeled as $p(T_2|L, T_1) = N(\beta_0 + \beta_1 T_1 + \beta_2 L + \beta_3 T_1 L, \sigma)$, where N indicates a standard normal distribution. Then, the full model is compared with null model, $\log\left[p(T_2|L, T_1)\big/p(T_2|L, T_1 : \beta_1 = \beta_2 = \beta_3 = 0)\right]$, as well as the reduced models $\log\left[p(T_2|L, T_1)\big/p(T_2|L, T_1 : \beta_2 = \beta_3 = 0)\right]$ and $\log\left[p(T_2|L, T_1)\big/p(T_2|L, T_1 : \beta_1 = \beta_3 = 0)\right]$. If all three likelihood ratios are high, then the causal–reactive relationship in Fig. 21.2A is inferred as the most likely relationship. Composite likelihood ratio tests and model selection will be discussed in more detail in Moreno *et al.*, this volume.

A number of other methods to infer causal relationships among traits by integrating expression and genetic data have been subsequently developed. For example, Li *et al.* (2006) used structural equation models (SEMs) to infer causal relationships among clinical traits by considering multiple genetic loci simultaneously. However, because of the complexity of this type of modeling, only a small number of traits (expression or clinical) can be considered at a time. In practice, if less complicated models can explain a similar percentage of the variance in traits of interest compared with more complicated models, then less complicated models will be preferred and selected based on AIC (Sakamoto *et al.*, 1986) or Bayesian Information Criterion (BIC) (Schwarz, 1978). Thus, it is reasonable to start with simple linear models (Schadt *et al.*, 2005a), build up to models that consider nonlinear effects between a single locus and gene if necessary, and then move into more complicated models if these simple models cannot capture a significant amount of the variance.

III. COEXPRESSION NETWORKS

The classic reductionist approach to elucidating biological systems as applied to genetics has motivated the identification of single genes associated with disease as one means of getting a foot into disease pathways. However, even in cases where genes are involved in pathways that are well known, it is unclear whether the gene causes disease via the known pathway or whether the gene is involved in other pathways or more complex networks that lead to disease. This was the case with *Tgfbr2*, a recently identified and validated obesity susceptibility gene (Schadt *et al.*, 2005a). While *Tgfbr2* plays a central role in the well-studied transforming growth factor-beta (TGF-β) signaling pathway (Ramji *et al.*, 2006), *Tgfbr2* and other genes in this signaling pathway are correlated with hundreds of other genes (Schadt *et al.*, 2005a,b), so that it is possible that perturbations in these other genes or in *Tgfbr2* itself may drive diseases like obesity by influencing other parts of the network beyond the TGF-β signaling pathway. Therefore, considering single genes in the context of a whole gene network may provide the necessary context within which to interpret the disease role a given gene may play.

Networks generally provide a convenient framework for exploring the context within which single genes operate. Networks are simply graphical models composed of nodes and edges. For gene networks associated with biological systems, the nodes in the network typically represent genes, gene products, or other important molecular entities like metabolites, and edges (links) between any two nodes indicate a relationship between the two corresponding genes.

For example, an edge between two genes may indicate that the corresponding expression traits are correlated in a given population of interest (Zhu *et al.*, 2004), that the corresponding proteins interact (Rual *et al.*, 2005), or that changes in the activity of one gene lead to changes in the activity of the other gene (Schadt *et al.*, 2005a). Interaction or association networks have recently gained more widespread use in the biological community, where networks are formed by considering only pairwise relationships between genes, including protein interaction relationships (Han *et al.*, 2004), coexpression relationships (Gargalovic *et al.*, 2006; Ghazalpour *et al.*, 2006), as well as other straightforward measures that may indicate association between two genes. In all cases, these networks have been demonstrated to exhibit a scale-free and hierarchical connectivity structures (Barabasi and Oltvai, 2004; Ghazalpour *et al.*, 2006; Lum *et al.*, 2006), providing higher-level insights into how biological networks may be ordered. The scale-free property exhibited by most biological networks implies that, like the Internet, most genes in a biological system are strongly connected to a small number of genes, while a smaller set of genes (often referred to as hub nodes) are connected to many other genes. The hierarchical property implies that biological networks are highly modular, with genes clustering into groups that are highly interconnected with each other, but not as highly connected with genes in other groups.

A. Constructing weighted and unweighted coexpression networks

Depending on whether the interaction strength of two genes is considered, there are two different approaches for analyzing gene coexpression networks: (1) an unweighted network analysis that involves setting hard thresholds on the significance of the interactions and (2) a weighted approach that avoids hard thresholding. For unweighted gene coexpression networks, gene–gene relationships are encoded in a binary form. Two genes in the network are connected by an edge if the correlation coefficient or the significance level of the correlation measure meets some predetermined threshold (Bergmann *et al.*, 2004; Butte and Kohane, 2000; Carter *et al.*, 2004; Davidson *et al.*, 2003; Lum *et al.*, 2006). The drawbacks of an unweighted network analysis are that the determination of the hard threshold is somewhat arbitrary and the resulting networks may be sensitive to the threshold selected. More importantly, the binary encoding actually destroys information regarding the interaction strength between two genes, resulting in a loss of power to establish higher-order relationships among the genes in the network. In contrast, the weighted gene coexpression network analysis assigns a connection weight to all pairs of genes by employing soft-thresholding functions whose parameters are estimated based on a biologically motivated scale-free topology criterion (Zhang and Horvath, 2005). Weighted gene coexpression networks preserve the continuous nature of gene–gene interaction at the transcriptional level and are robust to parameter selection. However, constructing these networks is more computationally intensive as all pairs of nodes are simultaneously considered, so that as the number of nodes grows, the number of pairs to consider grows quadratically.

Gene coexpression network analyses on yeast microarray data have revealed that gene network connectivity is strongly correlated with gene essentiality (gene deletion lethality) and sequence conservation. That is, the more central a gene is in the network, the more likely it is to be essential to yeast survival and the more likely its DNA sequence is to be conserved (Carlson *et al.*, 2006). For drug development, these findings may suggest that targeting hub genes would be more effective in disrupting disease-related networks for the purpose of therapy. On the contrary, if a gene is very central to a biological system, then affecting the activity of that gene may impact many higher-order phenotypes in addition to the desired disease phenotype, resulting in adverse events that ultimately doom a drug to failure. Therefore, before any recommendations can be made as to whether targeting of highly connected nodes is desirable or not in a pharmaceutical setting, a more comprehensive characterization of network connectivity and its association to disease in human systems will be required.

B. Using genetics in constructing coexpression networks

Multiple traits driven by common quantitative trait locus (QTL) is a central idea that can be leveraged to construct networks. The construction of coexpression networks can be aided by the introduction of genetic data, which at the very least can serve as a filter to help reduce artifactual correlations between expression traits. Significant artifactual correlations can arise in larger-scale gene expression studies because of correlated noise structures between the array-based experiments in such studies. Therefore, one of the more straightforward ways to leverage the eQTL data in this setting is to simply filter out gene–gene correlations in which the expression traits are not at least partially explained by common genetic effects (Lum *et al.*, 2006). For example, we can connect two genes with an edge in an unweighted coexpression network if (1) the p value for the Pearson correlation coefficient between the two genes is less than some prespecified threshold and (2) the two genes have at least one common eQTL. This can be taken a step further by formally assessing whether two expression traits driven by a common QTL are related in a causal or reactive fashion, filtering out correlations driven by expression traits that are independently driven by common or closely linked QTL (Doss *et al.*, 2005; Schadt *et al.*, 2005a).

One intuitive way to establish whether two genes share at least one eQTL is to carry out single-trait eQTL mapping for each expression trait and then consider eQTL for each trait overlapping if the corresponding logarithm of odd(LOD) score for the eQTLs are above some threshold and if the eQTL are in close proximity to one another. The significance of the statistic corresponding to the strength of association between two genes in the coexpression networks is then chosen such that the resulting network exhibits the scale-free property (Barabasi and Albert, 1999; Ghazalpour *et al.*, 2006; Lum *et al.*, 2006) and the false discovery rate for the gene–gene pairs represented in the network is constrained. Beyond this simple, albeit intuitively appealing eQTL overlap method, we can formally test whether two overlapping eQTL represent a single eQTL or closely linked eQTL by employing a pleiotropy effects test (PET), such as that originally described by Jiang and Zeng (1995) and Zeng (1993). The formation of gene clusters by simultaneously considering gene–gene and marker–gene correlations also promises to provide a more comprehensive characterization of shared genetic effects (Lee *et al.*, 2006).

C. Identifying modules of highly interconnected genes in coexpression networks

Given the scale-free and hierarchical nature of coexpression networks (Barabasi and Oltvai, 2004; Ghazalpour *et al.*, 2006; Lum *et al.*, 2006), one of the key problems is to identify the network modules, or functional units, in the network that represents those hub nodes (nodes that are significantly correlated with many other nodes) that are highly interconnected with one another, but that are not as highly connected with other hub nodes. Figure 21.3 illustrates a topological connectivity map for the most highly connected genes in the adipose tissue of the F2 cross C57BL/6J×C3H/HeJ(BXH) (Ghazalpour *et al.*, 2006; Chen *et al.*, 2008). After hierarchically clustering both dimensions of this plot, the network is seen to break out into clearly identifiable modules. Gene–gene coexpression networks are highly connected, and the clustering results shown in Fig. 21.3 illustrate that there are gene modules arranged hierarchically within these networks.

Ravasz *et al.* (2002) used manually selected height cutoff to separate tree branches after hierarchical clustering, in contrast to Lee *et al.* (2004) who formed maximally coherent gene modules with respect to gene ontology (GO) functional categories. Another strategy is to employ a measure similar to that used by Lee *et al.* (2004), but without the dependence on the GO functional annotations, given it is of interest to determine independently whether coexpression modules are enriched for GO functional annotations (Lum *et al.*, 2006). An emerging trend for module identification is to uncover alternative network structures such as cores and cliques and high-level organization forms like overlapping modules (communities) (Palla *et al.*, 2005; Wuchty and Almaas, 2005). The modules identified in this way are informative for identifying the functional components of the network that are associated with disease (Lum *et al.*, 2006).

Figure 21.3. A weighted, gene coexpression network constructed using the gene expression data profiled from the adipose tissue of 148 F2 male mice constructed from the B6 and C3H strains discussed in the text. The symmetric heat map with rows and columns as genes represents the network connection strength (indicated by the blackness, where black means two nodes are tightly connected) between any pair of nodes (genes) in the network. The network connection strength is measured as the topological overlap between genes. The network modules highlighted as grey color block along the rows and columns (each grey block represents a module) were identified via an average linkage hierarchical clustering algorithm using topological overlap as the dissimilarity metric.

It has been demonstrated that the types of modules depicted in Fig. 21.3 are enriched for known biological pathways, for genes that associate with disease traits, and for genes that are linked to common genetic loci (Ghazalpour *et al.*, 2006; Lum *et al.*, 2006). In this way, one can identify those key groups of genes that are perturbed by genetic loci that lead to disease, and that therefore define the intermediate steps that actually define disease states. For additional examples using coexpression networks to elucidate complex traits like disease, see the chapter by Farber and Lusis (2008) in this volume.

IV. PROBABILISTIC CAUSAL NETWORKS: BAYESIAN NETWORKS AS A FRAMEWORK FOR DATA INTEGRATION

With the completion of the sequencing of whole genomes of *Saccharomyces cerevisiae* (Foury *et al.*, 1998), *Caenorhabditis elegans* (C. *elegans Sequencing Consortium*, 1998), *Drosophila* (Adams, 2000), mouse (Mouse Genome Sequencing Consortium, 2002), human (Venter, 2001, International Human Genome Sequencing Consortium, 2004), and many other species (Aparicio *et al.*, 2002), the next challenge is to study the biological functions driven by the different regions of the genome, determining whether such regions encode for a protein or noncoding RNA, the functional role played by a given protein or RNA, the biological processes related to this function, and so on. There are continually growing numbers of systematically generated data, including gene expression (transcriptomics); protein–protein interaction assessed by yeast two hybrid; protein identification, quantification, and posttranslation modification identification by mass spectrometry (proteomics); and more recently metabolite levels measured by nuclear magnetic resonance (NMR) or mass spectrometry (metabolomics). To assess the function of individual genes, compendiums of yeast gene knockout (Hughes *et al.*, 2000) and mouse gene knockout (e.g., DeltaBase) have been constructed, in addition to global synthetic fitness or lethality (epistatic interaction) screens (Pan *et al.*, 2006). Further, there are efforts to systematically collect and manually curate knowledge and to represent such data into easily accessible databases, such as Kyoto Encyclopedia of Genes and Genomes (KEGG; Kanehisa *et al.*, 2006), BioCarta (http://www.biocarta.com), MetaCore (http://www.genego.com), and Ingenuity (http://www.ingenuity.com). With all of these efforts, integrating these types of high-throughput data is critical if we are to construct models that are predictive of complex biological systems. Ideker *et al.* (2001b) integrated genomics, gene expression, and proteomics data to study small networks, refining such networks in a trial-and-error manner using experimental approaches. Systematically integrating different types of data into probabilistic networks using Bayesian networks has been proposed and applied for the purpose of predicting protein–protein interactions (Jansen *et al.*, 2003) and protein function (Lee *et al.*, 2004). However, these Bayesian networks are still based on association between nodes in the network as opposed to causal relationships. From these types of networks we cannot infer whether a specific perturbation will affect a complex disease trait. To make such predictions we need networks capable of representing causal relationships. Probabilistic causal networks are one way to model such relationships, where causality in this context reflects a probabilistic belief that one node in the network affects the behavior of another. Bayesian networks (Pearl, 1988) are one type of probabilistic causal network that provide a natural framework for integrating highly dissimilar types of data.

A. Bayesian networks

Bayesian networks (Jensen, 1996) consist of a set of nodes, X, and a set of directed edges between pairs of nodes. The nodes represent random variables (either continuous or discrete), and the directed edges represent dependencies between nodes. The nodes and edges together form a graph, G, that is directed, where the connections that are allowed between nodes are constrained such that no cycles form in the graph. Given this constraint, Bayesian networks are often referred to as directed acyclic graphs (DAG). The graphical structure for a Bayesian network can be represented as a table of conditional probabilities when the nodes of the network are discrete random variables, or as a conditional probability density function when the nodes of the network are continuous random variables. The conditional probabilities represent the relationship any given node has with its parent nodes (a parent node is a node in the network with edges that are directed to other nodes in the network). In general, a probabilistic causal network (a graph G over a set of variables X) can be described by a joint probability distribution $p(X_1, \ldots, X_n)$. If the graph G is a DAG, then the distribution can be decomposed into the product:

$$p(X_1, \ldots, X_n) = \prod_{i=1}^{n} p(X_i | Pa_i)$$

where Pa_i^G is the set of parents of X_i in G. Given this form of the probability distribution we can learn network structures and estimate parameters locally, rather than fitting structures and estimating parameters globally (a less tractable problem).

Bayesian networks can be reconstructed based on a diversity of data, including expression data. These types of models have been fit before on expression data and have led to networks that were not at all predictive. To understand why certain types of data may fail to give rise to predictive networks in this setting, while other types of data may lead to predictive networks, it is important to understand the concept of Markov equivalence. If two network structures have the same backbone (undirected graph) and represent the same set of conditional independencies between variables of interest, they are referred to as being Markov equivalent (Glymour, 2001). For example, the two structures in Fig. 21.2B.1 as well as the structures in Fig. 21.2B.2 are Markov equivalent, whereas the V structure (two parents pointing to the same child node) in Fig. 21.2B.3 has no Markov equivalent structure.

Given enough data the V structures in a given graph can be uniquely defined (Verma and Pearl, 1990). However, because of the limited expression data typically available for any particular system in a given state, network reconstruction processes typically result in the identification of multiple networks that explain the data equally well. In fact, in most cases causal relationships cannot be reliably inferred from gene expression data alone, since for any particular network changing the direction of an edge between any two genes has little effect on the model fit. To break this Markov equivalence and reliably infer causal relationships, additional information is required. Bayes formula allows us to determine the probability of a network model M given observed data D as a function of our prior belief that the model is correct and the probability of the observed data given the model: $p(M|D) \sim p(D|M) * p(M)$, where:

$$p(M) = \prod_{(X_i->Y_j)^G} p(X_i \rightarrow Y_j)$$

is the structure prior. One way to differentiate Markov equivalent classes is to assign different structure priors $p(X_i \rightarrow Y_j)$ and $p(X_j \rightarrow Y_i)$. As is now discussed, the genetic data is a good source from which to construct structure priors (Zhu *et al.*, 2004, 2007).

B. Deriving structure priors from genetic data

In segregating populations like an F2 intercross, variations in DNA are the ultimate cause of traits under genetic control. As defined above, an expression trait that gives rise to a *cis*-acting eQTL corresponds to a structural gene that harbors a DNA variant in the gene region that affects transcript levels. In constructing a network, genes with *cis*-acting eQTLs can be allowed to serve as parent nodes of genes with *trans*-acting eQTLs, $p(X_{cis} \rightarrow X_{trans}) = 1$, whereas genes with *trans*-acting eQTLs cannot be parents of genes with *cis*-acting eQTLs, $p(X_{trans} \rightarrow X_{cis}) = 0$. Thus, genes with *cis*-acting eQTLs represent the top layers of the network.

We can further extend this concept by leveraging the genetic architecture more generally. Expression traits that are found to be under genetic control (giving rise to either *cis* or *trans* eQTL) can be tested individually for pleiotropic effects (PE) at each of their eQTL to determine whether any other genes in the set are driven by the same genetic loci (Jiang and Zeng, 1995). If two genes, X_a and X_b, are found to be driven by common genetic loci, the gene pair and the corresponding locus can be used to infer a causal–reactive or independent relationship based on a formal causality test (Schadt *et al.*, 2005a). The reliability of each possible relationship between gene X_a and gene X_b at locus l_i, $p(X_a \rightarrow X_b|l_i)$, $p(X_b \rightarrow X_a|l_i)$, and $p(X_a \perp X_b|l_i)$, are estimated by a standard bootstrapping procedure. If an independent relationship is inferred, $p(X_a \perp X_b|l_i) > 0.5$, then the prior probability that gene A is a parent of gene B is scaled as:

$$p(X_a \rightarrow X_b) = 1 - \frac{\sum_{i \in PE} p(X_a \perp X_b|l_i)}{\sum_{i \in PE} 1}$$

by considering all loci where the two genes are detected as having PE. If a causal or reactive relationship is inferred, $p(X_a \rightarrow X_b|l_i)$ or $p(X_b \rightarrow X_a|l_i)$ is >0.5, then the prior probability is scaled as:

$$p(X_a \rightarrow X_b) = \frac{2 * \sum_i p(X_a \rightarrow X_b|l_i)}{\sum_i p(X_a \rightarrow X_b|l_i) + p(X_b \rightarrow X_a|l_i)}$$

If the causal–reactive relationship between genes X_a and X_b cannot be reasonably inferred, then the complexity of the eQTL signature for each gene can be taken into consideration. Genes with a simpler, albeit stronger eQTL signature (i.e., a small number of eQTL explain the genetic variance component for the gene, with a significant proportion of the overall variance explained by the genetic effects) can be taken to be more likely to be causal compared with genes with a more complex, possibly weaker eQTL signatures (i.e., a larger number of eQTL explaining the genetic variance component for the gene, with less of the overall variance explained by the genetic effects). In this case, the structure prior that gene X_a is a parent of gene X_b can be taken as:

$$p(X_a \rightarrow X_b) = 2 * \frac{1 + n(X_b)}{2 + n(X_a) + n(X_b)}$$

where $n(X_a)$ and $n(X_b)$ are the number of eQTLs with LOD scores greater than a threshold for X_a and X_b, respectively.

Given these structure priors and the Bayesian network reconstructions, it is of interest to assess how much improvement is realized in the network reconstruction accuracy by incorporation of the genetic information. More generally it is of interest to provide a framework to assess network reconstruction algorithms more generally, so that different algorithms can be compared and different types of data can be directly compared to determine what has the most significant impact on network reconstruction accuracy. Beginning steps were recently taken by Zhu *et al.* (2007) to quantify the improvement in network reconstruction accuracy realized by incorporating the genetic data as prior information. To start this process, an F2 intercross data set, including genotypes and gene expression data, was simulated following the scheme shown in Fig. 21.4. Bayesian networks were reconstructed based on the simulated data with and without including genetic data. The resulting networks were then directly compared with the networks used in the simulation (so in simulating the data the true network is known). Results showed that structure priors derived from genetic data help recover not only true causal relationship but also true relationships regardless of edge direction (Zhu *et al.*, 2007).

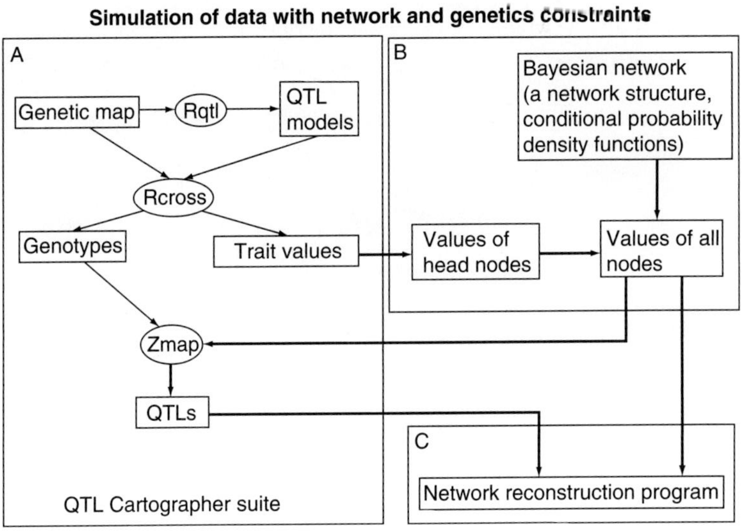

Simulation of data with network and genetics constraints

A
Genetic map
Rqtl
QTL models
Rcross
Genotypes
Trait values
Zmap
QTLs
QTL Cartographer suite
B
Bayesian network (a network structure, conditional probability density functions)
Values of head nodes
Values of all nodes
C
Network reconstruction program

Figure 21.4. Data simulation scheme with genetic and network constraints as previously published (Zhu et al., 2007) and as discussed in the text. (A) A segregating population (an F2 intercross in this case) is simulated using the QTL Cartographer software suite (Rqtl, Rcross, and Zmapqtl). The QTL model for a trait is defined using the Rqtl program, and the heritability of the QTL is defined using the Rcross program. (B) The traits simulated by Rcross are used as the head nodes in the simulated network. The remaining traits are simulated based on the values of the head nodes according to the DAG structure and the set of conditional probability density functions associated with this structure. (C) After traits for all nodes in the network are simulated, they are scanned for QTLs using the Zmapqtl program. The traits and the associated QTL are then input into the network reconstruction program.

The improvement of reconstruction accuracy due to structure priors derived from genetics is also assessed using real data (Zhu *et al.*, 2004). C57BL/6J (B6) and BDA were crossed to generate an F2 population; 111 female mice were genotyped and their livers were harvested and profiled on a 23,000 gene microarray (Schadt *et al.*, 2003). Bayesian networks reconstructed with and without structure priors derived from genetic data were compared with the corresponding correlation-based association network. A gene in the network, *Hsd11b1*, was perturbed *in silico* so that the perturbation signatures were predicted based on the three networks, and these predicted signatures were then compared with a signature derived experimentally by knocking down the activity of *Hsd11b1*. Only 10% of the predicted signature based on the correlation network overlapped the experiment signature, while 20 and 50% of the predicted signatures based on Bayesian networks without and with genetic data, respectively,

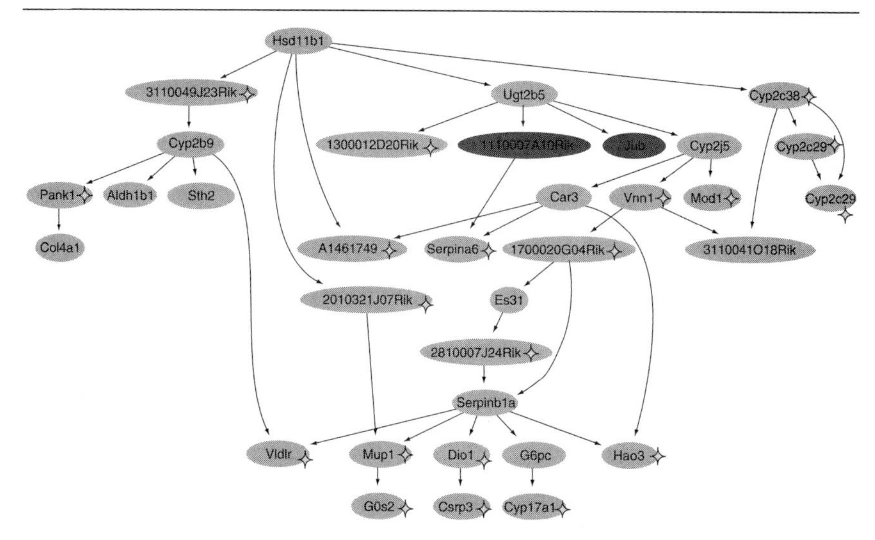

Figure 21.5. Bayesian network constructed from liver expression data collected from a previously described F2 intercross population (Schadt et al., 2003) using genetic data as a prior. The network depicted here is a subnetwork of the full liver gene network, defined by those nodes in the complete network whose gene expression is predicted to change given the downregulated state of *Hsd11b1*, based on the Bayesian network. From this network, 33 genes were predicted to change, given *Hsd11b1* in the downregulated state. The color indicates the predicted expression states (dark grey: upregulated, light grey: downregulated) as defined by the network. The stars indicate the 20 genes in the *Hsd11b1* experimentally determined perturbation signature.

leveraged as priors overlapped the experimental signature (Zhu *et al.*, 2004). The Bayesian network reconstructed with genetic data and its overlap with the experimental *Hsd11b1* signature is shown in Fig. 21.5. Therefore, in this instance, the genetic data significantly improved the prediction accuracy of the Bayesian network.

C. Structure priors derived from other data sources

In addition to genetics data, there are many types of data complementary to gene expression data, including protein–protein interaction data, protein state data, protein–DNA binding data, RNA–DNA binding data, and so on. Gene expression is regulated by transcription factors (TFs), and many gene–gene expression correlations can be explained by coregulation by the same TFs. When high-confidence protein–protein interactions are identified, the corresponding pair of genes tend to correlate at the expression level. To complement the use of genetic data, TF binding site data and protein–protein interaction data are also easily integrated as priors in the Bayesian network reconstruction process. TF binding site predictions for mammals are often complicated by the uncertainty of the transcription start sites and promoter regions. TF–TF interactions (*cis*-regulated modules) and TF–environment interactions are also important. ChIP-on-Chip experiments can provide experimental support as to whether TF binding site predictions are good or not, but even these experimental data do not represent absolute truth. For example, it has been clearly demonstrated that TF binding sites vary under different stimulations to experimental systems (Schreiber *et al.*, 2006). However, progress has been made by applying comparative genomics filter to the TF binding predictions (Defrance and Touzet, 2006) as well as enhancing experimental designs so that the detection of TF–DNA interactions are not condition specific (Berger *et al.*, 2006). Still, protein–protein interactions for human and mouse are incomplete, with only 10% of the true interactions detected from protein–protein interaction experiments (Bader and Chant, 2006), where true interactions are embedded in a much larger number of false-positive interactions. Higher throughput protein–protein interaction experiments are in progress (Rual *et al.*, 2005) and are being combined with cross-species information to achieve better accuracy (Gandhi *et al.*, 2006).

The poor reliability of protein–protein and protein–DNA binding data in humans and mouse make it difficult to leverage these data as priors for network reconstruction. On the contrary, TF binding site predictions and protein–protein interactions are better characterized in yeast. Yeast make for a nice demo system to assess the improvement in network accuracy in networks reconstructed using different types of data. In one study, 112 segregants from a yeast cross were expression profiled and genotyped (Brem *et al.*, 2002; Yvert *et al.*, 2003). Bayesian networks were constructed with expression data alone and with genetic data as priors. In addition, high-quality TF binding site data were derived from high-quality ChIP-on-Chip experiments and phylogenetic conservation filter (MacIsaac *et al.*, 2006), and protein–protein interaction data were derived from manually curated protein complexes (Guldener *et al.*, 2006) as well as from complexes identified by clique community analysis. These data were then combined together to form a structure prior used to reconstruct Bayesian networks. As discussed below, these combined data were demonstrated to significantly enhance network reconstruction accuracy.

Another interesting prior that can be leveraged to reconstruct networks relates to the scale-free property, a general property of biological networks (Albert *et al.*, 2000). Scale-free priors have been used to integrate genetics and expression data (Lee *et al.*, 2006). This same type of scale-free prior can also be used to integrate TF and protein–protein complex data. For example, given a transcription factor T, and a set of genes G that contain the binding site of T, the transcription factor prior p_{tf} can be defined so that it is proportional to the number of responders that are correlated with the TF's expression level:

$$\log\left(p_{\mathrm{tf}}(T \rightarrow g)\right) \propto \log\left(\sum_{g_i \in G} p_{\mathrm{qtl}}(T \rightarrow g_i)^{*}\delta\right)$$

where $p_{\mathrm{qtl}}(T \rightarrow g)$ is the prior for the QTL and $\delta = \begin{cases} 1, \text{if } \operatorname{corr}(T, g_i) \geq r_{\mathrm{cutoff}} \\ 0, \text{if } \operatorname{corr}(T, g_i) < r_{\mathrm{cutoff}} \end{cases}$. The correlation cutoff r_{cutoff} can be determined using permuted data to minimize the false-discovery rate. For protein–protein interactions that cannot be integrated using a scale-free prior, a constant pairwise prior can be used: $p_{\mathrm{pc}}(X_i \rightarrow X_j) = p_{\mathrm{pc}}(X_j \rightarrow X_i) = c$. When the Bayesian networks based on these yeast data were reconstructed using these priors and compared to the yeast knockout compendium data (Hughes *et al.*, 2000), 125, 139, and 152 knockout signatures were enriched in the networks reconstructed using only expression data; using only expression and genetic data; and using expression, genetic, TF binding site, and protein–protein interaction data, respectively. These results indicate that the integration of orthogonal experimental data improves the quality of reconstructed networks, and these more predictive networks will have greater utility in refining the definition of disease, identifying disease subtypes, identifying targets for disease, and identifying biomarkers for disease and drug response.

In disease models constructed from mouse liver data, it is not uncommon to observe at least half of genes in the mouse genome as expressed. Further, the expression levels of nearly half of the expressed genes are observed to change significantly across different genetic backgrounds. That is, significant intersample variation is observed for at least half of the genes detected as expressed. Therefore, to capture this extent of variation in practice, a Bayesian network of 6000 nodes is needed to model the system. Bayesian networks of this size are routinely built for different F2 intercross populations using supercomputers. Heuristic search methods for constructing Bayesian networks have been proposed to utilize limited computational resources (Chen *et al.*, 2006; Pena *et al.*, 2005).

Bayesian networks are just one of many methods that can be used to construct graphical networks. If we constraint to linear interactions only, multiple linear regression (Gardner *et al.*, 2003) or SEMs (Li *et al.*, 2006) can be used to construct networks with a small number of nodes. Bayesian networks on the contrary are capable of capturing linear and nonlinear interactions, and they are also suitable for reconstructing larger-scale networks (Pe'er *et al.*, 2001; Zhu *et al.*, 2004). However, there are a number of clear drawbacks for representing general biological networks using Bayesian networks. In any dynamic, stable biological system there will be multiple feedback loops in the molecular networks that underlie these systems. These feedback loops cannot be captured by standard Bayesian networks. A number of alternative models have been proposed to address this deficiency, including using dynamic Bayesian networks to represent feedback loops (Dojer *et al.*, 2006; Husmeier, 2003; Zou and Conzen, 2005), probabilistic Boolean networks (Hashimoto *et al.*, 2004; Shmulevich *et al.*, 2002), and probabilistic factor graphs (Gat-Viks *et al.*, 2006; Yeang *et al.*, 2004). The methods described for integrating diverse biological data into Bayesian networks are also readily applied to general probabilistic graphical models, which include probabilistic Boolean models and factor graph model, among several others.

V. AN INSIGHTFUL EXAMPLE: INTEGRATING DATA LEADS TO THE IDENTIFICATION OF GENE AFFECTING PLASMA CHOLESTEROL LEVELS

Applying the types of methods discussed above in the appropriate setting can lead directly to the identification of genes that drive disease traits of interest. Looking at disease traits in a population setting is key, since variations in the disease traits of interest observed among individuals in the population can be examined to determine if they covary with genotypic, gene expression, protein expression, and other related molecular phenotypes. Given changes in DNA and environment are ultimately responsible for the state of molecular networks that drive disease-associated traits, integrating these data to elucidate such networks is a natural evolution from the classic genetic approaches pursued in years past. Starting with a segregating population, classic genetic approaches can be employed to map the genetic loci controlling for molecular and clinical traits of interest, and these data can then be integrated as discussed above to infer causality among the traits of interest. Further, these data can be more generally integrated to get at networks that underlie disease, and then leveraged to understand the context in which genes that drive disease operate. This general path was taken by Cervino *et al.* (2005) to identify *Insig2* as a gene controlling for plasma cholesterol levels in a segregating mouse population. *Insig2* was identified as a causal gene for plasma cholesterol levels (Cervino *et al.*, 2005) by integrating genetic and systems biology approaches in mouse. *Insig2* was subsequently inferred as a key causal gene for obesity based on human association studies (Herbert *et al.*, 2006), and mutations in *Insig2* were found to significantly increase the risk of obesity in an already overweight population (Rosskopf *et al.*, 2007). Here we review the steps that led to the identification and characterization of *Insig2* as a key node in networks associated with metabolic traits, where these steps are illustrative of the integrative approaches discussed in previous sections.

The first steps leading to the identification of *Insig2* as a key cholesterol gene were the identification of genetic loci controlling for liver gene expression traits and cholesterol traits, and then application of a causality test to identify genes supported as causal for the cholesterol traits. Common approaches for the genetic study of complex human diseases are association and linkage analyses. Linkage studies typically seek to associate disease traits with large chromosome areas using family data. Once linkage to a region has been identified, genetic association studies can be used to fine map the region under the linkage peak to identify the true causal gene for the trait of interest. In the study by Cervino *et al.* (2005), linkage analysis for metabolic and liver gene expression traits was carried out in an F2 intercross population constructed from C57BL/6J (B6) and C3H/HeJ (C3H) mice on an ApoE null background (Wang *et al.*, 2006). Metabolic traits were scored in each animal, and liver samples from each animal were collected and profiled on a gene expression microarray representing roughly 23,000 transcripts. The strongest LOD scores observed in the analysis of the linkage results were located on chromosome 1 for free fatty acids, plasma glucose levels, plasma high-density lipoprotein (HDL) cholesterol levels, plasma low-density lipoprotein (LDL) + very low-density lipoprotein (VLDL) cholesterol levels, total plasma cholesterol levels, and triglyceride levels. Total cholesterol was the trait most strongly linked to this chromosome 1 locus, which covered an interval spanning nearly 40 cM (typical for an F2 cross). However, the region defined by a one LOD drop (99% confidence interval for the true QTL location) covered only 5.5 cM (between 58.9 cM and 64.4 cM, corresponding to the physical interval 120–128 Mb). Given the linkage results for total cholesterol were restricted to this 8-Mb region, the liver expression data were used to identify genes that (1) resided in the 8-Mb region; (2) had liver gene expression levels that correlated with the cholesterol trait; (3) gave rise to a *cis*-acting eQTL, suggesting that DNA variations within the gene itself lead to variations in liver gene expression levels; and (4) were supported as causal for the cholesterol levels with respect to the genetic locus using a causality test. The only gene in the 8-Mb region to meet all of these criteria was *Insig2*.

Given the strong support for *Insig2* as a causal candidate for cholesterol levels in the cross, the next step involved carrying out a genetic association test to validate whether DNA variations in the *Insig2* region were in fact associated with cholesterol levels. Because a different mouse population was used in the association study compared to that used in the linkage study, the association study not only allows direct testing of candidate genes in the linkage region but also serves to validate specific hypotheses generated from the integrative genomics approach employed to identify *Insig2* as a candidate. To test for association, 62 inbred strains of mice were genotyped using a high-density single nucleotide polymorphism (SNP) array, and cholesterol traits associated with the strains were obtained from the Jax phenome database (Paigen and Eppig 2000). From the association analysis, an SNP in the *Insig2* gene region was identified as significantly associated with total cholesterol in three different data sets (Cervino *et al.*, 2005), validating the linkage result obtained in the cross.

Up to this point, *Insig2* was identified and validated as causal for total plasma cholesterol levels without considering the network context within which this gene is predicted to operate in liver. However, given the liver expression data in the cross we can directly observe genes interacting with *Insig2* as a way to enhance confidence that *Insig2* could be a key player in cholesterol-related traits. Toward this end, a liver gene network was reconstructed from the liver gene expression data in the cross, using the genetic data as priors, as discussed above. This network was used to support the hypothesis that *Insig2* is involved in regulating cholesterol levels, as well as to generate new hypotheses related to the mechanism by which this gene operates. A Bayesian network for the female liver data in the above cross was constructed (Zhu *et al.*, 2004). From this network (Fig. 21.6), *Insig2* is seen to causally regulate well-known genes involved in cholesterol synthesis, including *Sqle*, *Hmgcs1*, *Cyp51*, and several others. *Insig2* prevents proteolytic activation of TFs sterol-regulatory element binding proteins

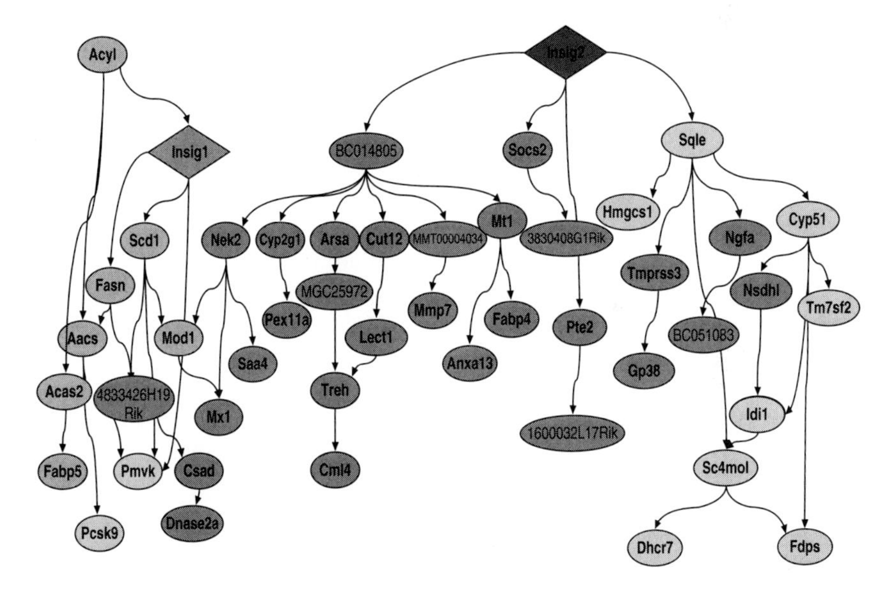

Figure 21.6. Subnetwork around *Insig2* from the complete Bayesian network constructed from the liver gene expression data of the F2 intercross population described in the text (Cervino et al., 2005). This *Insig2* network indicates that *Insig2* regulates genes involved in cholesterol synthesis (light grey on the right) and coregulates genes involved in fatty acid synthesis (light grey on the left), a possible mechanism for *Insig2* affecting plasma cholesterol levels via the regulation of SREBP targets, such as *Sqle* and *Hmgcs1*.

(SREBPs) and enhances degradation of *Hmgcr* (Engelking *et al.*, 2006). *Sqle* and *Hmgcs1* are well-known SREBP targets (Maxwell *et al.*, 2003). Thus, SREBPs are the intermediate through which *Insig2* is causal for *Sqle* and total plasma cholesterol levels. The network reconstructed in this instance not only supports *Insig2* as a key driver of cholesterol levels in this particular population but also provides hints of how *Insig2* may cause variations in cholesterol levels.

The *Insig2* network is not limited only to generating this type of hypothesis. It can also be used to select alternative drug targets around the candidate genes in cases where the candidate genes themselves are not easily targeted. In this present case, it is difficult to find *Insig2* activators. Other genes in the subnetwork, such as *Tm7sf2*, *Fdps*, and *Acyl*, in addition to *Sqle* and *Hmgcs1*, are also well-known SREBP targets (Maxwell *et al.*, 2003). Instead of targeting these genes individually, it is possible to target the regulation mechanism of this subnetwork by targeting SREBPs. It has been shown that knocking down SREBPs has desirable effects on metabolic pathways (Guillet-Deniau *et al.*, 2004). For additional examples on employing these types of integrative approaches to identify novel genes for disease traits of interest, see the chapter by Farber and Lusis (2008) in this volume.

VI. CONCLUSIONS

Diseases like obesity are diseases of the system, potentially involving many different pathways operating in many different tissues, and ultimately giving rise not only to different disease subtypes but to different comorbidities of the disease as well. The integration of large-scale molecular profiling data (e.g., gene expression data), genotypic data, clinical data, and other biologically relevant information will be critical if we ever hope to understand more fully how genetic and environmental perturbations to a given system lead to complex traits like disease. If common forms of diseases like obesity, diabetes, and heart disease represent states of a network, then focusing on single gene perturbations will likely never reveal the most effective ways to treat or prevent these diseases. Integration of a diversity of data in this setting is the key, since no single dimension will provide the complete answer. For example, the identification of DNA polymorphisms that associate with diseases like obesity and diabetes represents only the beginning in a long series of steps needed to elucidate disease pathways and to establish the specific role individual genes may play in the process. While these types of genetic discoveries provide a peek into pathways that underlie disease, they are usually devoid of context, so that elucidating the functional role such genes play in disease can linger for years, or even decades, as has been the case for ApoE, an Alzheimer's susceptibility gene identified nearly 15 years ago (Peacock *et al.*, 1993).

At present, the reconstruction of complete biological networks is an intractable problem, partly because today's methods and data are inadequate for representing biological systems in a comprehensive fashion. Complex systems involve diverse interacting parts, and these parts involve a complex hierarchy of networks that underlie the complex phenotypes manifested by the system (Schadt and Lum, 2006). Genomic or genetic networks interacting with transcriptional networks that in turn interact with protein, metabolic, and signaling networks present challenges to elucidating common human disease. At this point, we must restrict attention to local network behavior in specific contexts (e.g., a given tissue under a set of environmental constraints) related to disease traits of interest, where the aim is not complete reconstruction of the system, but rather a partitioning of the network space into component pieces that explain variations in the traits of interest. Several groups have shown that it is feasible to reconstruct a reasonably accurate view of the molecular processes underlying disease, to identify key drivers of disease, and to identify key therapeutic intervention points as a result. Until highly predictive networks can be constructed, inferences drawn from biological networks should be considered as hypotheses that can be tested, where the results can then illuminate the representation of the system. Given the various types of data needed to construct an accurate network, any framework that aims to model the complexity of biological systems needs to be flexible and to allow incorporation of greatly diverse data, as well as the ability to update the nature of the relationships as rules are learned or as the extent and diversity of data increases. While this approach may not perfectly represent the complexity of a biological system, the beginning steps may lead to an understanding that is sufficient to establish a set of "rules" that elucidate the simplifying assumptions that in turn lead to a more tractable problem and more predictive networks, and as a result, more effective treatments for disease. Until recently most drug discovery had been based on the concept that drugs could be designed against targets identified along simple linear pathways in treating broad disease categories such as depression, obesity, and all forms of breast cancer. However, most often, after these drugs have been tested in general human populations, the frequency of beneficial responses is observed to be less than expected from the experimental models for the disease. Taking a more systems biology approach to this problem can lead to causal relationships that are context specific, and as a result can reveal the true complexity of disease, and therefore should lead to therapeutics that are more effective and better able to target those most likely to benefit (Schadt *et al.*, 2005b).

The integration of the diverse sets of molecular data now being generated in population settings is only in its infancy. Many of the methods employed to date toward this end are more heuristical in nature and so will benefit from a more formal treatment. In addition, little to date has been done to integrate expression data from multiple tissues to dissect how modules in one tissue may communicate with modules in another tissue. The type of interactions considered along with eQTL data so far have been restricted to RNA–RNA association data, despite the availability of large-scale DNA–protein and protein–protein interaction data. The predictive power of the types of networks discussed in this chapter could be enhanced by more systematically integrating protein–protein interactions, protein–DNA interactions, protein–RNA interactions, RNA–RNA interactions, protein state information, methylation state, and interactions with metabolites, as these types of data become available. These developments promise to take us beyond the single gene view of disease and move us closer to the type of systems-level view depicted in Fig. 21.1 that may be needed to fully understand the complexity of common human diseases like obesity and diabetes. Of course, further study and experimentation are needed to demonstrate more convincingly that systems biology approaches to understanding the state of a given molecular network, interactions among molecular networks, and how the states of such networks change in response to different genetic and environmental contexts are tractable enough to take us beyond a reductionist approach that has to date enabled great success in elucidating the complexity of living systems more generally.

References

Adams, M. D. (2000). The genome sequence of Drosophila melanogaster. *Science* **287,** 2185–2195.

Adams, M. D., Celniker, S. E., Holt, R. A., Evans, C. A., Gocayne, J. D., Amanatides, P. G., Scherer, S. E., Li, P. W., Hoskins, R. A., Galle, R. F., George, R. A., Lewis, S. E., *et al.* (2000). The genome sequence of Drosophila melanogaster. *Science* **287**(5461), 2185–2195.

Albert, R., Jeong, H., and Barabasi, A. L. (2000). Error and attack tolerance of complex networks. *Nature* **406**(6794), 378–382.

Aparicio, S., Chapman, J., Stupka, E., Putman, N., Chia, J. M., Dehal, P., Christoffels, A., Rash, S., Hoon, S., Smit, A., *et al.* (2002). Whole-genome shotgun assembly and analysis of the genome of Fugu rubripes. *Science* **297**(5585), 1301–1310.

Bader, J. S., and Chant, J. (2006). Systems biology. When proteomes collide. *Science* **311**(5758), 187–188.

Barabasi, A. L., and Albert, R. (1999). Emergence of scaling in random networks. *Science* **286**(5439), 509–512.

Barabasi, A. L., and Oltvai, Z. N. (2004). Network biology: Understanding the cell's functional organization. *Nat. Rev. Genet.* **5**(2), 101–113.

Berger, M. F., Philippakis, A. A., Qureshi, A. M., He, F. S., Estep, P. W., and Bulyk, M. L. (2006). Compact, universal DNA microarrays to comprehensively determine transcription-factor binding site specificities. *Nat. Biotechnol.* **24**(11), 1429–1435.

Bergmann, S., Ihmels, J., and Barkai, N. (2004). Similarities and differences in genome-wide expression data of six organisms. *PLoS Biol.* **2**(1), e9.

Brem, R. B., Yvert, G., Clinton, R., and Kruglyak, L. (2002). Genetic dissection of transcriptional regulation in budding yeast. *Science* **296**(5568), 752–755.

Butte, A. J., and Kohane, I. S. (2000). Mutual information relevance networks: Functional genomic clustering using pairwise entropy measurements. *Pac. Symp. Biocomput.* 418–429.

Bystrykh, L., Weersing, E., Dontje, B., Sutton, S., Pletcher, M. T., Wiltshire, T., Su, A. I., Vellenga, E., Wang, J., Manly, K. F., *et al.* (2005). Uncovering regulatory pathways that affect hematopoietic stem cell function using 'genetical genomics'. *Nat. Genet.* **37**(3), 225–232.

C. *elegans* Sequencing Consortium (1998). Genome sequence of the nematode C. elegans: A platform for investigating biology. *Science* **282**(5396), 2012–2018.

Carlson, M. R., Zhang, B., Fang, Z., Mischel, P. S., Horvath, S., and Nelson, S. F. (2006). Gene connectivity, function, and sequence conservation: Predictions from modular yeast co-expression networks. *BMC Genomics*. **7**, 40.

Carter, S. L., Brechbuhler, C. M., Griffin, M., and Bond, A. T. (2004). Gene co-expression network topology provides a framework for molecular characterization of cellular state. *Bioinformatics* **20**(14), 2242–2250.

Cervino, A. C., Li, G., Edwards, S., Zhu, J., Laurie, C., Tokiwa, G., Lum, P. Y., Wang, S., Castellini, L. W., Lusis, A. J., *et al.* (2005). Integrating QTL and high-density SNP analyses in mice to identify *Insig2* as a susceptibility gene for plasma cholesterol levels. *Genomics* **86**(5), 505–517.

Chen, X. W., Anantha, G., and Wang, X. (2006). An effective structure learning method for constructing gene networks. *Bioinformatics* **22**(11), 1367–1374.

Chen, Y., Zhu, J., Lum, P. Y., Yang, X., Pinto, S., MacNeil, D. J., Zhang, C., Lamb, J., Edwards, S., Sieberts, S. K., Leonardson, A., Castellini, L. W., *et al.* (2008). Variations in DNA Elucidate Molecular Networks that Cause Disease. *Nature* (in press).

Chesler, E. J., Lu, L., Shou, S., Qu, Y., Gu, J., Wang, J., Hsu, H. C., Mountz, J. D., Baldwin, N. E., Langston, M. A., *et al.* (2005). Complex trait analysis of gene expression uncovers polygenic and pleiotropic networks that modulate nervous system function. *Nat. Genet.* **37**(3), 233–242.

Chickering, D. M. (1995). A transformational characterization of equivalent Bayesian network structures. *Eleventh Conference on Uncertainty in Artificial Intelligence, Morgan Kaufmann.*

Davidson, E. H., McClay, D. R., and Hood, L. (2003). Regulatory gene networks and the properties of the developmental process. *Proc. Natl. Acad. Sci. USA* **100**(4), 1475–1480.

Defrance, M., and Touzet, H. (2006). Predicting transcription factor binding sites using local over-representation and comparative genomics. *BMC Bioinformatics*. **7**, 396.

Dojer, N., Gambin, A., Mizera, A., Wilczynski, B., and Tiuryn, J. (2006). Applying dynamic Bayesian networks to perturbed gene expression data. *BMC Bioinformatics*. **7**, 249.

Doss, S., Schadt, E. E., Drake, T. A., and Lusis, A. J. (2005). Cis-acting expression quantitative trait loci in mice. *Genome Res.* **15**(5), 681–691.

Engelking, L. J., Evers, B. M., Richardson, J. A., Goldstein, J. L., Brown, M. S., and Liang, G. (2006). Severe facial clefting in Insig-deficient mouse embryos caused by sterol accumulation and reversed by lovastatin. *J. Clin. Invest.* **116**(9), 2356–2365.

Farber, C. R., and Lusis, A. J. (2008). Integrating global gene expression analysis and genetics. *In* "Genetic Dissection of Complex Traits" (D. C. Rao and C. C. Gu, eds.), 2nd Edition, pp. 573–598. Academic Press, San Diego.

Foury, F., Roganti, T., Lecrenier, N., and Purnelle, B. (1998). The complete sequence of the mitochondrial genome of Saccharomyces cerevisiae. *FEBS Lett.* **440**(3), 325–331.

Gandhi, T. K., Zhong, J., Mathivanan, S., Karthick, L., Chandrika, K. N., Mohan, S. S., Sharma, S., Pinkert, S., Nagaraju, S., Periaswamy, B., *et al.* (2006). Analysis of the human protein interactome and comparison with yeast, worm and fly interaction datasets. *Nat. Genet.* **38**(3), 285–293.

Gardner, T. S., di Bernardo, D., Lorenz, D., and Collins, J. J. (2003). Inferring genetic networks and identifying compound mode of action via expression profiling. *Science* **301**(5629), 102–105.

Gargalovic, P. S., Imura, M., Zhang, B., Gharavi, N. M., Clark, M. J., Pagnon, J., Yang, W. P., He, A., Truong, A., Patel, S., *et al.* (2006). Identification of inflammatory gene modules based on variations of human endothelial cell responses to oxidized lipids. *Proc. Natl. Acad. Sci. USA* **103**(34), 12741–12746.

Gat-Viks, I., Tanay, A., Raijman, D., and Shamir, R. (2006). A probabilistic methodology for integrating knowledge and experiments on biological networks. *J. Comput. Biol.* **13**(2), 165–181.

Ghazalpour, A., Doss, S., Zhang, B., Wang, S., Plaisier, C., Castellanos, R., Brozell, A., Schadt, E. E., Drake, T. A., Lusis, A. J., *et al.* (2006). Integrating genetic and network analysis to characterize genes related to mouse weight. *PLoS Genet.* **2**(8), e130.

Glymour, C. N. (2001). "The Mind's Arrows: Bayes Nets and Graphical Causal Models in Psychology." Bradford Book, Cambridge, Mass.

Gourevitch, B., Bouquin-Jeannes, R. L., and Faucon, G. (2006). Linear and nonlinear causality between signals: Methods, examples and neurophysiological applications. *Biol. Cybern.* **95**(4), 349–369.

Granger, C. (1969). Investigating causal relations by econometric models and cross-spectral methods. *Econometrica* **37**, 424–438.

Guillet-Deniau, I., Pichard, A. L., Kone, A., Esnous, C., Nieruchalski, M., Girard, J., and Prip-Buus, C. (2004). Glucose induces de novo lipogenesis in rat muscle satellite cells through a sterol-regulatory-element-binding-protein-1c-dependent pathway. *J. Cell Sci.* **117**(Pt. 10), 1937–1944.

Guldener, U., Munsterkotter, M., Oesterheld, M., Pagel, P., Ruepp, A., Mewes, H. W., and Stumpflen, V. (2006). MPact: The MIPS protein interaction resource on yeast. *Nucleic Acids Res.* **34**(Database issue), D436–D441.

Han, J. D., Bertin, N., Hao, T., Goldberg, D. S., Berriz, G. F., Zhang, L. V., Dupuy, D., Walhout, A. J., Cusick, M. E., Roth, F. P., *et al.* (2004). Evidence for dynamically organized modularity in the yeast protein-protein interaction network. *Nature* **430**(6995), 88–93.

Hartwell, L. H., Hopfield, J. J., Leibler, S., and Murray, A. W. (1999). From molecular to modular cell biology. *Nature* **402**(Suppl. 6761), C47–C52.

Hashimoto, R. F., Kim, S., Shmulevich, I., Zhang, W., Bittner, M. L., and Dougherty, E. R. (2004). Growing genetic regulatory networks from seed genes. *Bioinformatics* **20**(8), 1241–1247.

Herbert, A., Gerry, N. P., McQueen, M. B., Heid, I. M., Pfeufer, A., Illig, T., Wichmann, H. E., Meitinger, T., Hunter, D., Hu, F. B., *et al.* (2006). A common genetic variant is associated with adult and childhood obesity. *Science* **312**(5771), 279–283.

Hughes, T. R., Marton, M. J., Jones, A. R., Roberts, C. J., Stoughton, R., Armour, C. D., Bennett, H. A., Coffey, E., Dai, H., He, Y. D., *et al.* (2000). Functional discovery via a compendium of expression profiles. *Cell* **102**(1), 109–126.

Husmeier, D. (2003). Sensitivity and specificity of inferring genetic regulatory interactions from microarray experiments with dynamic Bayesian networks. *Bioinformatics* **19**(17), 2271–2282.

Ideker, T., Galitski, T., and Hood, L. (2001a). A new approach to decoding life: Systems biology. *Annu. Rev. Genomics Hum. Genet.* **2**, 343–372.

Ideker, T., Thorsson, V., Ranish, J. A., Christmas, R., Buhler, J., Eng, J. K., Bumgarner, R., Goodlett, D. R., Aebersold, R., and Hood, L. (2001b). Integrated genomic and proteomic analyses of a systematically perturbed metabolic network. *Science* **292**(5518), 929–934.

International Human Genome Sequencing Consortium (2004). Finishing the euchromatic sequence of the human genome. *Nature* **431**(7011), 931–945.

Jansen, R. C., and Nap, J. P. (2001). Genetical genomics: The added value from segregation. *Trends Genet.* **17**(7), 388–391.

Jansen, R., Yu, H., Greenbaum, D., Kluger, Y., Krogan, N. J., Chung, S., Emili, A., Snyder, M., Greenblatt, J. F., and Gerstein, M. (2003). A Bayesian networks approach for predicting protein-protein interactions from genomic data. *Science* **302**(5644), 449–453.

Jensen, F. V. (1996). "An introduction to Bayesian Networks." UCL Press Limited, London.

Jiang, C., and Zeng, Z. B. (1995). Multiple trait analysis of genetic mapping for quantitative trait loci. *Genetics* **140**(3), 1111–1127.

Kanehisa, M., Goto, S., Hattori, M., Aoki-Kinoshita, K. F., Itoh, M., Kawashima, S., Katayama, T., Araki, M., and Hirakawa, M. (2006). From genomics to chemical genomics: New developments in KEGG. *Nucleic Acids Res.* **34**(Database issue), D354–D357.

Kulp, D. C., and Jagalur, M. (2006). Causal inference of regulator-target pairs by gene mapping of expression phenotypes. *BMC Genomics* **7**, 125.

Lee, I., Date, S. V., Adai, A. T., and Marcotte, E. M. (2004). A probabilistic functional network of yeast genes. *Science* **306**(5701), 1555–1558.

Lee, S. I., Pe'er, D., Dudley, A. M., Church, G. M., and Koller, D. (2006). Identifying regulatory mechanisms using individual variation reveals key role for chromatin modification. *Proc. Natl. Acad. Sci. USA* **103**(38), 14062–14067.

Li, R., Tsaih, S. W., Shockley, K., Stylianou, I. M., Wergedal, J., Paigen, B., and Churchill, G. A. (2006). Structural model analysis of multiple quantitative traits. *PLoS Genet.* **2**(7), e114.

Lum, P. Y., Chen, Y., Zhu, J., Lamb, J., Melmed, S., Wang, S., Drake, T. A., Lusis, A. J., and Schadt, E. E. (2006). Elucidating the murine brain transcriptional network in a segregating mouse population to identify core functional modules for obesity and diabetes. *J. Neurochem.* **97**(Suppl. 1), 50–62.

Luscombe, N. M., Babu, M. M., Yu, H., Snyder, M., Teichmann, S. A., and Gerstein, M. (2004). Genomic analysis of regulatory network dynamics reveals large topological changes. *Nature* **431**(7006), 308–312.

MacIsaac, K. D., Wang, T., Gordon, D. B., Gifford, D. K., Stormo, G. D., and Fraenkel, E. (2006). An improved map of conserved regulatory sites for Saccharomyces cerevisiae. *BMC Bioinformatics.* **7**, 113.

Mani, S., and Cooper, G. F. (2000). Causal discovery from medical textual data. *Proc. AMIA. Symp.* 542–546.

Maxwell, K. N., Soccio, R. E., Duncan, E. M., Sehayek, E., and Breslow, J. L. (2003). Novel putative SREBP and LXR target genes identified by microarray analysis in liver of cholesterol-fed mice. *J. Lipid. Res.* **44**(11), 2109–2119.

Monks, S. A., Leonardson, A., Zhu, H., Cundiff, P., Pietrusiak, P., Edwards, S., Phillips, J. W., Sachs, A., and Schadt, E. E. (2004). Genetic inheritance of gene expression in human cell lines. *Am. J. Hum. Genet.* **75**(6), 1094–1105.

Mootha, V. K., Lindgren, C. M., Eriksson, K. F., Subramanian, A., Sihag, S., Lehar, J., Puigserver, P., Carlsson, E., Ridderstrale, M., Laurila, E., *et al.* (2003). PGC-1alpha-responsive genes involved in oxidative phosphorylation are coordinately downregulated in human diabetes. *Nat. Genet.* **34**(3), 267–273.

Morley, M., Molony, C. M., Weber, T. M., Devlin, J. L., Ewens, K. G., Spielman, R. S., and Cheung, V. G. (2004). Genetic analysis of genome-wide variation in human gene expression. *Nature* **430**(7001), 743–747.

Mouse Genome Sequencing Consortium (2002). Initial sequencing and comparative analysis of the mouse genome. *Nature* **420**(6915), 520–562.

Mukhopadhyay, N. D., and Chatterjee, S. (2007). Causality and pathway search in microarray time series experiment. *Bioinformatics* **23**(4), 442–449.

Paigen, K., and Eppig, J. T. (2000). A mouse phenome project. *Mamm. Genome.* **11**(9), 715–717.

Palla, G., Derenyi, I., Farkas, I., and Vicsek, T. (2005). Uncovering the overlapping community structure of complex networks in nature and society. *Nature* **435**(7043), 814–818.

Pan, X., Ye, P., Yuan, D. S., Wang, X., Bader, J. S., and Boeke, J. D. (2006). A DNA integrity network in the yeast Saccharomyces cerevisiae. *Cell* **124**(5), 1069–1081.

Pe'er, D., Regev, A., Elidan, G., and Friedman, N. (2001). Inferring subnetworks from perturbed expression profiles. *Bioinformatics* **17**(Suppl. 1), S215–S224.

Peacock, M. L., Warren, J. T., Jr., Roses, A. D., and Fink, J. K. (1993). Novel polymorphism in the A4 region of the amyloid precursor protein gene in a patient without Alzheimer's disease. *Neurology* **43**(6), 1254–1256.

Pearl, J. (1988). "Probabilistic Reasoning in Intelligent Systems: Networks of Plausible Inference." Morgan Kaufmann Publishers, San Mateo, California.

Pena, J. M., Bjorkegren, J., and Tegner, J. (2005). Growing Bayesian network models of gene networks from seed genes. *Bioinformatics* **21**(Suppl. 2), ii224–ii229.

Petretto, E., Mangion, J., Dickens, N. J., Cook, S. A., Kumaran, M. K., Lu, H., Fischer, J., Maatz, H., Kren, V., Pravenec, M., *et al.* (2006a). Heritability and tissue specificity of expression quantitative trait loci. *PLoS Genet.* **2**(10), e172.

Petretto, E., Mangion, J., Pravanec, M., Hubner, N., and Aitman, T. J. (2006b). Integrated gene expression profiling and linkage analysis in the rat. *Mamm. Genome.* **17**(6), 480–489.

Pinto, S., Roseberry, A. G., Liu, H., Diano, S., Shanabrough, M., Cai, X., Friedman, J. M., and Horvath, T. L. (2004). Rapid rewiring of arcuate nucleus feeding circuits by leptin. *Science* **304** (5667), 110–115.

Ramji, D. P., Singh, N. N., Foka, P., Irvine, S. A., and Arnaoutakis, K. (2006). Transforming growth factor-beta-regulated expression of genes in macrophages implicated in the control of cholesterol homoeostasis. *Biochem. Soc. Trans.* **34**(Pt 6), 1141–1144.

Ravasz, E., Somera, A. L., Mongru, D. A., Oltvai, Z. N., and Barabasi, A. L. (2002). Hierarchical organization of modularity in metabolic networks. *Science* **297**(5586), 1551–1555.

Rosen, D. R., Siddique, T., Patterson, D., Figlewicz, D. A., Sapp, P., Gentati, A., Donaldson, D., Goto, J., O'Regan, J. P., Deng, H. X., *et al.* (1993). Mutations in Cu/Zn superoxide dismutase gene are associated with familial amyotrophic lateral sclerosis. *Nature* **362**(6415), 59–62.

Rosskopf, D., Bornhorst, A., Rimmbach, C., Schwahn, C., Kayser, A., Kruger, A., Tessman, G., Geissler, I., Kroemer, H. K., and Volzke, H. (2007). Comment on a common genetic variant is associated with adult and childhood obesity. *Science* **315**(5809), 187; author reply 187.

Rual, J. F., Venkatesan, K., Hao, T., Hirozane-Kishikawa, T., Dricot, A., Li, N., Berriz, G. F., Gibbons, F. D., Dreze, M., Ayivi-Guedehoussou, N., *et al.* (2005). Towards a proteome-scale map of the human protein-protein interaction network. *Nature* **437**(7062), 1173–1178.

Sakamoto, Y., Ishiguro, M., and Kitagawa, G. (1986). "Akaike Information Criterion Statistics." D. Reidel Publishing Company, Dordrecht, Holland.

Schadt, E. E. (2005). Exploiting naturally occurring DNA variation and molecular profiling data to dissect disease and drug response traits. *Curr. Opin. Biotechnol.* **16**(6), 647–654.

Schadt, E. E., and Lum, P. Y. (2006). Reverse engineering gene networks to identify key drivers of complex disease phenotypes. *J. Lipid. Res.* **47**(12), 2601–2613.

Schadt, E. E., Monks, S. A., Drake, T. A., Lusis, A. J., Che, N., Colinayo, V., Ruff, T. G., Milligan, S. B., Lamb, J. R., Cavet, G., *et al.* (2003). Genetics of gene expression surveyed in maize, mouse and man. *Nature* **422**(6929), 297–302.

Schadt, E. E., Lamb, J., Yang, X., Zhu, J., Edwards, S., Guhathakurta, D., Sieberts, S. K., Monks, S., Reitman, M., Zhang, C., *et al.* (2005a). An integrative genomics approach to infer causal associations between gene expression and disease. *Nat. Genet.* **37**(7), 710–717.

Schadt, E. E., Sachs, A., and Friend, S. (2005b). Embracing complexity, inching closer to reality. *Sci. STKE* **205**(295), pe40.

Schreiber, J., Jenner, R. G., Murray, H. L., Gerber, G. K., Gifford, D. K., and Young, R. A. (2006). Coordinated binding of NF-kappaB family members in the response of human cells to lipopolysaccharide. *Proc. Natl. Acad. Sci. USA* **103**(15), 5899–5904.

Schwarz, G. (1978). Estimating the dimension of a model. *Ann. Stat.* **6**(2), 461–464.

Shmulevich, I., Dougherty, E. R., Kim, S., and Zhang, W. (2002). Probabilistic Boolean Networks: A rule-based uncertainty model for gene regulatory networks. *Bioinformatics* **18**(2), 261–274.

van 't Veer, L. J., Dai, H., van de Vijver, M. J., He, Y. D., Hart, A. A., Mao, M., Peterse, H. L., van der Kooy, K., Marton, M. J., Witteveen, A. T., *et al.* (2002). Gene expression profiling predicts clinical outcome of breast cancer. *Nature* **415**(6871), 530–536.

Van Eerdewegh, P., Little, R. D., Dupuis, J., Del Mastro, R. G., Falls, K., Simon, J., Torrey, D., Pandit, S., McKenny, J., Braunschweiger, K., *et al.* (2002). Association of the ADAM33 gene with asthma and bronchial hyperresponsiveness. *Nature* **418**(6896), 426–430.

Venter, J. C. (2001). The sequence of the human genome. *Science* **291,** 1304–1351.

Verma, T., and Pearl, J. (1990). Equivalence and synthesis of causal models. UAI '90. *"Proceedings of the Sixth Annual Conference on Uncertainty in Artificial Intelligence."*. pp. 255–270.

Wang, S., Yehya, N., Schadt, E. E., Wang, H., Drake, T. A., and Lusis, A. J. (2006). Genetic and genomic analysis of a fat mass trait with complex inheritance reveals marked sex specificity. *PLoS Genet.* **2**(2), e15.

Wuchty, S., and Almaas, E. (2005). Peeling the yeast protein network. *Proteomics* **5**(2), 444–449.

Yeang, C. H., Ideker, T., and Jaakkola, T. (2004). Physical network models. *J. Comput. Biol.* **11**(2–3), 243–262.

Yvert, G., Brem, R. B., Whittle, J., Akey, J. M., Foss, E., Smith, E. N., Mackelprang, R., and Kruglyak, L. (2003). Trans-acting regulatory variation in Saccharomyces cerevisiae and the role of transcription factors. *Nat. Genet.* **35**(1), 57–64.

Zeng, Z. B. (1993). Precision mapping of quantitative trait loci. *Genetics* **121,** 185–199.

Zerhouni, E. (2003). Medicine. The NIH roadmap. *Science* **302**(5642), 63–72.

Zhang, B., and Horvath, S. (2005). A general framework for weighted gene co-expression network analysis. *Stat. Appl. Genet. Mol. Biol.* **4,** Article17.

Zhu, J., Lum, P. Y., Lamb, J., GuhaThakurta, D., Edwards, S. W., Thieringer, R., Berger, J. P., Wu, M. S., Thompson, J., Sachs, A. B., *et al.* (2004). An integrative genomics approach to the reconstruction of gene networks in segregating populations. *Cytogenet. Genome. Res.* **105**(2–4), 363–374.

Zhu, J., Wiener, M. C., Zhang, C., Fridman, A., Minch, E., Lum, P. Y., Sachs, J. R., and Schadt, E. E. (2007). Increasing the power to detect causal associations by combining genotypic and expression data in segregating populations. *PLoS Comput. Biol.* **3**(4), e69.

Zou, M., and Conzen, S. D. (2005). A new dynamic Bayesian network (DBN) approach for identifying gene regulatory networks from time course microarray data. *Bioinformatics* **21**(1), 71–79.

22 The Promise of Composite Likelihood Methods for Addressing Computationally Intensive Challenges

Na Li

Division of Biostatistics, School of Public Health, University of Minnesota, Minneapolis, Minnesota 55455

ABSTRACT

High-dimensional genetic data, due to its complex correlation structure, poses an enormous challenge to standard likelihood–based methods for making statistical inference. As an approximation, composite likelihood has proved to be a successful strategy for some genetic applications. It has the potential to see even wider application and much research is needed. We first give a brief description of composite likelihood. The advantage of this method and potential challenges in inference are noted. Next, its applications in genetic studies are

Advances in Genetics, Vol. 60 0065-2660/08 $35.00

DOI: 10.1016/S0065-2660(07)00422-1

reviewed, specifically in estimating population genetics parameters such as recombination rate, and in multi-locus linkage disequilibrium mapping of disease genes with some discussion about future research directions.

I. INTRODUCTION

For high-dimensional correlated data, the joint distribution may be difficult to specify or calculate. One such example is population-based molecular genetic data. Genetic markers on the same chromosome from apparently "unrelated" individuals are correlated in a very complex way. The individual chromosomes are connected through an unobserved genealogy that may extend many generations and is essentially of infinite dimension even for very small samples. Many stochastic populations or evolutionary processes can influence the genealogy such as subdivision, migration, and selection. On a chromosome, genetic polymorphisms first arose due to mutation. While marker alleles on the same chromosome tend to be passed from generation to generation together, recombination and gene conversions allow the exchange of genetic material among chromosomes. As an approximation to the standard complete likelihood, composite likelihood methods (Lindsay, 1988) provide a good alternative strategy for modeling the data and making statistical inference. The composite likelihood is formed by the products of component likelihoods which may be correlated. The approach is sometimes referred to as pseudo-likelihood (Besag, 1974; Cox and Reid, 2004). However, the term pseudo-likelihood is also used in different contexts. To avoid confusion, we will adhere to the nomenclature of Lindsay (1988).

In addition to genetic applications reviewed here, composite likelihood methods are often used in image analysis, spatial analysis (ecology), time series, and social network modeling.

II. COMPOSITE LIKELIHOOD METHODS

Suppose $\mathbf{Y}$ is a vector random variable with density $f(\mathbf{y};\ \theta)$, where θ is an unknown vector parameter. Under certain regularity conditions, the maximum likelihood estimator (MLE) of θ is obtained by setting the score function (the derivative of the log likelihood) to zero and solving the resulting equation:

$$\mathbf{U}(\theta) \equiv \frac{\partial \log f(\mathbf{y};\theta)}{\partial \theta} \equiv \frac{\partial \ell(\theta)}{\partial \theta} = 0$$

The score function $\mathbf{U}(\theta)$ is unbiased with $\mathrm{E}\mathbf{U}(\theta) = \mathbf{0}$, and satisfies:

$$\mathrm{E}(\mathbf{U}\mathbf{U}^T) = -\mathrm{E}\left(\frac{\partial \mathbf{U}}{\partial \theta}\right) \tag{1}$$

The above quantity is also known as the Fisher information $I(\theta)$. If we observe n independent realizations from the distribution $f(\mathbf{y};\ \theta)$, the MLE $\hat{\theta}$ from the score equation $\sum_{i=1}^{n} \mathbf{U}_i(\theta) = 0$ is consistent and has an asymptotic normal distribution:

$$\sqrt{n}(\hat{\theta} - \theta_0) \rightarrow N(0, I^{-1}(\theta_0))$$

as n goes to infinity, where θ_0 corresponds to the true value of the parameter.

In some situations, the joint distribution f may be difficult to express analytically and difficult to calculate. It could be easier to find a component log likelihood $\ell_i(\theta) = \log f(y_i; \theta)$, for a subset of the data Y_i (say, $i = 1, \ldots, m$). The components Y_i need not be mutually exclusive and the density may be expressed marginally or conditionally upon other parts of data. The choice depends on the application. For example, in a Markov random field, the model is expressed in terms of conditional densities $f(x_i|x_{n(i)})$ where $n(i)$ denotes the neighborhood of location i. It is then natural to use the conditional likelihoods as components. In other cases, the marginal likelihood function $f(\mathbf{y}_i;\ \theta)$ for mutually exclusive Y_i is more convenient. When there is little information about θ in the marginal likelihoods, we can also use the marginal pairwise likelihood $f(\mathbf{y}_i, \mathbf{y}_j;\ \theta)$. Examples will be shown later on.

In any case, the composite log likelihood is simply the sum of the log component likelihoods:

$$\ell_C(\theta) = \sum_{i=1}^{m} \ell_i(\theta)$$

The composite likelihood is not the true likelihood function unless the components are independent. Nevertheless, we can estimate parameter θ by maximizing this composite log likelihood. Under suitable regularity conditions, we can exchange the differentiation and summation. It can be shown that the maximum composite likelihood estimator is also consistent since the composite score function:

$$\mathbf{U}_C(\theta) = \frac{\partial \ell_C(\theta)}{\partial \theta} = \sum_{i=1}^{m} \frac{\partial \ell_i(\theta)}{\partial \theta} = \sum_{i=1}^{m} \mathbf{U}_i(\theta)$$

is an unbiased estimating function if ℓ_i is a genuine log likelihood for component i such that $\mathrm{E}(\mathbf{U}_i) = 0$. Note however in population genetics applications, due to the complex dependence structure, the regularity conditions may not hold or it might be very difficult to prove that they do, as we will see later.

The consistency of a composite likelihood estimator only requires the correct specification of component likelihoods (e.g., the marginal distribution of a univariate Y_i) so it is more robust to model misspecification. The composite likelihood estimator can be consistent even when the complete likelihood estimator is not, since the complete likelihood is more prone to be misspecified.

Assuming the necessary conditions hold, the composite likelihood estimator $\tilde{\theta}$ is asymptotically normal with mean θ_0. The asymptotic variance is the inverse of an analog of Fisher's information defined as (Lindsay, 1988):

$$\begin{aligned} I_C(\theta) &= E\left(\frac{\partial \mathbf{U}_C}{\partial \theta}\right)(\mathrm{Var}(\mathbf{U}_C))^{-1}E\left(\frac{\partial \mathbf{U}_C}{\partial \theta}\right)^T \\ &= \left(\sum_{i=1}^{m}\mathrm{Var}(\mathbf{U}_i)\right)(E(\mathbf{U}_C\mathbf{U}_C^T))^{-1}\left(\sum_{i=1}^{m}\mathrm{Var}(\mathbf{U}_i)\right)^T \end{aligned} \tag{2}$$

The middle term involves $\mathrm{E}(\mathbf{U}_i\mathbf{U}_j)$ which can be difficult to calculate, especially since the joint distribution of $\mathbf{Y}_i$ and $\mathbf{Y}_j$ may not even be specified. If independent realizations from $f(\mathbf{y};\theta)$ are available, it is possible to estimate it using the empirical variance in the line of generalized estimating equations (Liang and Zeger, 1986; Zeger *et al.*, 1988) which would be consistent as the number of independent units goes to infinity. Otherwise, it might be possible to use weighted empirical variance estimates (Lumley and Heagerty, 1999) or resampling-based methods (Heagerty and Lele, 1998). In addition, the composite likelihood ratio statistic (CLRT),

$$\mathrm{CLRT} = 2(\ell_C(\tilde{\theta}) - \ell_C(\theta_0))$$

no longer has the usual asymptotic χ^2 distribution. Making valid inference based on composite likelihood is a major difficulty. Generally because the components of the composite likelihood are likely to be positively correlated, ignoring the correlation (assuming $\mathrm{E}(\mathbf{U}_i\mathbf{U}_j) = 0$ or using standard χ^2 distribution) is expected to be anti-conservative (type I error rate inflated). In practice, as least in the genetic examples, inference based on composite likelihood is done informally and often simulation studies are conducted to calibrate the results.

It can be shown that the information for composite likelihood (2) is smaller than that of the complete likelihood (1). Except for special cases (e.g., at specific parameter values), composite likelihood is less efficient than the complete likelihood. Lindsay (1988) noted that it might be possible to improve efficiency by using proper weights for individual components. The idea is similar to the use of a suitable working correlation matrix (instead of the identity matrix) in generalized estimating equations (Liang and Zeger, 1995). In practice, the optimal weights may be difficult to find. However, when the components are equal in marginal information, using equal weights is the optimal strategy.

III. APPLICATIONS IN POPULATION GENETICS

One of the problems in population genetics that has attracted tremendous interest recently is to make inference about population parameters based on a sample of DNA sequence or haplotype data. Consider a random sample of n haplotypes at L loci h_{ij}, $i = 1, \ldots, n$, $j = 1, \ldots, L$. For simplicity, we will assume that loci are diallelic and ignore the fact that humans are diploids and haplotypes often have to be inferred from unphased genotype data.

The haplotypes are related because they are descended from common ancestors. Mutations arise to create new polymorphic loci. Adjacent alleles are likely to be passed together from generation to generation (i.e., in linkage). The haplotypes can also exchange alleles through recombination which breaks down the linkage. In the current generation, however, there might not be enough historic recombination events, especially between close loci, so the loci may be still correlated, that is, in linkage disequilibrium.

One quantity that is of particular interest is the scaled recombination rate $\rho = 4N_e r$ where r is the per-site recombination rate per meiosis and N_e is the effective population size. It is a natural measure of multi-locus linkage disequilibrium (Pritchard and Przeworski, 2001). Coalescent-based models (Hudson, 1983; Kingman, 1982) are commonly used to describe this stochastic population process and form the basis for statistical inference. For a given genealogical history of a sample, it is trivial to determine the probability of the data under the coalescent model for fixed parameter values. However, calculation of the likelihood requires integration over the huge set of possible genealogical history compatible with the sample. Despite the development of sophisticated Monte Carlo algorithms (e.g., Fearnhead and Donnelly, 2001; Griffiths and Marjoram, 1996; Kuhner *et al.*, 2000; Stephens and Donnelly, 2000), it remains an impossible task for all but the smallest data sets.

To get around the computational challenge, several methods have been proposed based on approximation of the likelihood. The first such approach to estimate ρ was proposed in Hey and Wakeley (1997). They first derived the likelihood function for a pair of loci of four haplotypes. Then for three informative (e.g., polymorphic) loci, a composite likelihood is formed by the product of the two probabilities from the loci 1 and 2, and loci 2 and 3. A maximum likelihood estimate of ρ is then obtained. For a larger sample with more than four haplotypes and three loci, an overall estimate of ρ is taken to be the average estimate for every possible set of two intervals in all possible subsets of four haplotypes. In practice, this can only be done using a (relative) small number of random subsets. They also discussed the possibility of using composite likelihood from all possible subsets instead of simply averaging the estimates. However, it was too computationally intensive to be feasible.

In Fearnhead and Donnelly (2001), an importance sampling–based approach was developed to estimate ρ based on haplotype data. However, it was very computationally intensive, especially for large number of segregating loci. The authors later proposed a composite likelihood approach (Fearnhead and Donnelly, 2002) for longer haplotypes. In this approximation, long sequences are divided into small disjoint subregions, and the likelihoods for each subregion are combined to form the composite likelihood. Through simulations they showed that the composite likelihood estimator performed fairly well. Despite the dependence of the components, standard asymptotic distributional results are useful approximates if the number of subregions is small with relatively large subregions.

Hudson (2001) proposed another composite likelihood method based on the likelihoods of all possible pairs of loci. For two diallelic loci, the data can be summarized by a 2×2 table. Under certain model assumptions, the probability of the data can then be calculated either analytically (for small number of haplotypes) or by a Monte Carlo method. These probabilities can be tabulated and saved to serve as a look-up table for future computation. As a result, this method is extremely fast. This pairwise likelihood method was later extended to allow for recurrent mutations (McVean *et al.*, 2002) where importance sampling is used to calculate the pairwise likelihoods, and applied to virus genetic data. With emphasis on modeling fast-evolving virus population (e.g., HIV-1), Carvajal-Rodriguez *et al.* (2006) further extended this composite likelihood estimator with a more general mutation model that allows more than two states and variable mutation rates among loci. Wall (2004) considered estimation of gene conversion rate using composite likelihood formed by three-locus likelihoods.

Most of these authors emphasized point estimation only and used simulations to empirically evaluate the performance (Smith and Fearnhead, 2005). Fearnhead (2003) presented some theoretical results for composite likelihood–based estimators of ρ. Note that because of shared ancestry and linkage, independent realizations from the population process are not available. In particular, since the individuals are highly correlated, the information content increases very slowly with larger number of haplotypes (individuals), roughly on the order of $\log(n)$ (Felsenstein, 2006; Pluzhnikov and Donnelly, 1996). Therefore, Fearnhead (2003) considered only the asymptotics when the number of loci goes to infinity. It was shown that the composite likelihood estimator of Fearnhead and Donnelly (2002) is consistent, whereas the consistency of Hudson's pairwise likelihood estimator cannot be proved. Intuitively, this is because the correlation between two subregions decreases and eventually goes to zero as the distance between them increases. Furthermore, assuming equal subregion length and constant r, data from each of the subregion is identically distributed. On the other hand, when all possible pairwise likelihoods are included, the components are not identically distributed and the correlation structure is complicated. Recently, Wiuf (2006) proved the consistency of the subregion-based composite likelihood methods in estimating population parameters in general (not just recombination rates) and also under more general population models.

The estimator based on a truncated version of the pairwise likelihood method, however, was shown to be consistent (Fearnhead, 2003). In this version, only pairs of loci that are within a certain fixed distance apart are included. Again, this ensures that correlation decreases to zero when the components (pairs of loci) are sufficiently far away. It was also noted that this truncated composite likelihood corresponds to using weighted composite likelihoods with weight 0 if the pair of single nucleotide polymorphisms (SNPs) are too far away from each other. The weighting scheme ensures the consistency of the estimator with little if any efficiency loss.

Another topic related to estimating recombination rate is detection of recombination hotspots (Crawford *et al.*, 2004; Fearnhead *et al.*, 2004; Li and Stephens, 2003). Fearnhead and Smith (2005) proposed a penalized likelihood model:

$$\sum_{i=1}^{S-5} \ell_i(\rho_i) - \lambda h$$

where S is the total number of loci, h is the (unknown) number of hotspots. Each component $\ell_i(\rho_i)$ is the log-likelihood function [based on an approximated version of Fearnhead and Donnelly (2002)] for a subregion of 5 loci with ρ_i the average recombination rate in the subregion. Note that here the subregions are overlapping and λ is a penalty parameter designed to prevent overfitting. This is also an example of composite likelihood where equal (but not fixed) weights are used.

Another application of composite likelihood methods in population genetics is to detect the effects of selection (e.g., Jensen *et al.*, 2005; Kim and Stephan, 2002; Meiklejohn *et al.*, 2004; Zhu and Bustamante, 2005). In particular, Kim and Nielsen (2004) extended the pairwise likelihood of Hudson (2001) to a selective sweep model. No theoretical results have been proved regarding the properties of these estimators. In particular, Wiuf (2006) noted that the theory developed in the paper does not apply to models with selection.

IV. APPLICATIONS IN FINE MAPPING OF DISEASE MUTATIONS

A simple example of the use of composite likelihood dated from the earliest days of linkage analysis. Exactly multi-point linkage analysis software is very limited in the maximum size of the pedigree. Even Markov chain Monte Carlo–based algorithms have their limits. When a very large pedigree is encountered, it is broken down to small pieces to satisfy the constraints of the software.

The likelihoods (LOD scores) from the sub-pedigrees are then summed as if they are independent. It was shown that the lost of efficiency can be reduced when larger components are used (Chapman *et al.*, 2001).

In traditional genetic linkage and association studies, founders or population-based cases and controls are assumed to be independent. Although this is generally a good approximation, it is not always appropriate (Leutenegger *et al.*, 2002). Care must be taken to properly control the type 1 error rate.

One of the challenges in linkage disequilibrium–based association analysis is to combine information from multiple marker loci. Several composite likelihood methods have been developed that were based on the association between a marker and the putative disease locus. The primary difference among them is the parametrization of the association parameter and the model used to relate the association parameter to genetic distance or recombination rate, which provides a basis for combining information across markers that are of varying distance away from the disease locus.

A. Terwilliger's method

In Terwilliger (1995), the association between the disease and a (potentially multi-allelic) marker is characterized by the λ statistic, the relative increase in frequency of the marker allele among chromosomes carrying the disease mutation compared to the general population. Consider a multi-allelic marker and a sample of chromosomes containing either the disease allele D or normal allele N. Let p_i be the population frequency of marker allele i, and $q_i = \Pr(i|D)$, the allele frequency of i on the D-bearing chromosomes. Assuming that allele i is associated with the disease, that is, it occurred on the ancestral haplotype where disease mutation D first appeared, we have $q_i = p_i + \lambda(1 - p_i)$, which defines association parameter λ. Using Bayes Theorem, it is easy to see that $r_i = \Pr(i|N) = p_i - \lambda(1 - p_i)[p_D/(1 - p_D)]$ where p_D is the frequency of the disease mutation. Similarly for the allele j not associated with the disease, we can derive $q_j = \Pr(j|D) = p_j - \lambda p_i$ and $r_j = \Pr(j|N) = 1 - \Pr(i|N)$.

Thus, the data is reduced to be a $m \times 2$ table with m being the number of alleles. A likelihood function for λ can be written, assuming allele i is associated with the disease. Let X_j denote the number of haplotypes containing alleles j and allele D, Y_j denote the number of haplotypes containing alleles j and allele N (the wild-type allele), the likelihood function is computed as:

$$L_i = \sum_{j=1}^{m} q_j^{X_j} r_j^{Y_j}$$

where q_j and r_j were given above, and the sum is taken over all m alleles of the marker, including the allele i assumed to be associated with D. If we assume in addition that the first disease mutation occurred independently of any marker alleles, the probability that it was on the same chromosome as i is then p_i, the frequency of allele i. The overall likelihood is written as $L = \sum_{i=1}^{m} p_i L_i$ which can be maximized to provide a maximum likelihood estimate of λ. Note that here it is assumed that the marker allele is older than the disease mutation and the marker allele frequencies do not change over time. One consequence of this assumption is that common alleles are more likely to be associated with disease mutation.

Another consequence of the model is that $\lambda \geq 0$ (only positive correlation is allowed). To test the null hypothesis $\lambda = 0$, the likelihood ratio statistic asymptotically has a 1:1 mixture distribution of a point mass 0 and χ_1^2.

Likelihoods for multiple markers are then combined to form a composite likelihood. The relationship $\lambda_k = \alpha(1 - \theta_k)^n$ is used to pool information across markers. Here, α is the proportion of the disease chromosomes bearing the same ancestral mutation D. It is a heterogeneity parameter to allow for uncertainty in disease locus genotype (i.e., not all chromosomes from affected individuals carry the same mutation at the locus of interest), where n is the number of generations since the disease mutation D occurred, θ is the known recombination fraction between the marker k and the test position.

The author proposed to use fixed values of α and n (which determines λ for each marker since θ_k is known) and use the sum of likelihood test statistics for all the markers as the overall test statistic which is again assumed to have a 50:50 mixture of χ_0^2 and χ_1^2 distribution. The test becomes increasingly conservative with larger number of markers.

Alternatively, a composite likelihood approach can be used to estimate the unknown disease location $L(\theta, \alpha, n) = \sum_k L(\lambda_k)$ where λ_k is expressed in terms of the distance between marker k and location of the disease locus θ_k using the formula as above. For a given disease locus position θ, the likelihood is maximized with respect to α and n. The profile likelihood can then be plotted as a function of the putative disease locus position, similar to the location LOD score in linkage analysis. It was noted that estimate of the disease locus position was not very reliable because the model assumptions might not be valid.

B. Devlin *et al.*'s method

For very rare alleles only, Devlin *et al.* (1996) proposed to use a different measure of association, the robust attributable risk (δ) which for a 2 × 2 table of marker-disease combination equals:

$$\delta = \frac{\pi_{11}\pi_{22} - \pi_{12}\pi_{21}}{\pi_{11}\pi_{22} + \pi_{21}\pi_{22}} \quad (3)$$

Then, the sample statistic $Y = -\log \hat{\delta}$ is modeled approximately by a gamma distribution with mean μ and variance μ/λ. This model assumes a Wright–Fisher population model with nonoverlapping generations. In addition, the marker polymorphism is assumed to be older than the disease mutation and its frequency does not change over time. The disease mutation arose only once, and had only one copy thus was initially in complete disequilibrium with nearby marker alleles. The mean μ is approximately $n\theta$ where n is again the number of generations since the disease mutations occurred and θ is the recombination fraction between the marker and disease locus. The variance of Y is harder to formulate directly. Rather it is expressed as $Var(Y_j) = r_j\nu$ for the j-th marker where $\nu = Var(Y_1)$ and r_j, the relative variance, is assumed to depend only on θ and the allele frequencies. By summing over the likelihood function of all the markers, a composite likelihood is formed with parameters (n,θ,ν). This model was also extended in the same paper by adding an additional heterogeneity parameter to allow for two disease mutations instead of just one.

C. Malécot model for linkage disequilibrium

Morton and his colleagues have developed a rich literature for modeling linkage disequilibrium and LD mapping based on the association measure:

$$\hat{\rho} = \frac{ad - bc}{(a+b)(c+d)}$$

where a, b, c, d are the counts of four haplotypes between two diallelic loci and are arranged such that $ad \geq bc$ and $b \leq c$ (Maniatis *et al.*, 2002; Morton *et al.*, 2001; Morton *et al.*, 2007; Morton, 2005). Under Malécot's model for isolation by distance (Malécot, 1973), its expectation can be written as:

$$E(\hat{\rho}) = \rho = (1 - L)M \exp\left(-\sum_i \varepsilon_i d_i\right) + L$$

where L is a parameter to account for the bias and can be interpreted as the association between unlinked loci; parameter M allows for multiple origins of the disease mutation and equals 1 if there was only a single mutation (i.e., complete initial LD when the mutation occurred). $\sum \varepsilon_i d_i$ is the distance between the two loci where the sum is taken over intervals between pairs of adjacent loci. ε_i is assumed to be a constant within each interval and d_i is the physical length of the interval (in kilobytes). $\varepsilon_i d_i$ provides an additive measure of LD, as genetic distance, measured in LD units (LDU) (Maniatis *et al.*, 2002). ε_i is specific to each adjacent marker pair and can be estimated using population data in similar way as construction of linkage maps. Population-specific LD maps have been constructed using genome-wide SNP data from the HapMap project (Tapper *et al.*, 2003, 2005) and their properties have been studied extensively (Maniatis *et al.*, 2007; Zhang *et al.*, 2002, 2004).

In the context of association mapping, the association measure $\hat{\psi}$ between the marker and disease phenotypes can be represented either through the regression coefficient for continuous phenotypes or correlation coefficient for binary affectation status (Collins and Morton, 1998; Maniatis *et al.*, 2004). The expected value ψ_i of the association statistic $\hat{\psi}_i$ between marker i and the phenotype is again modeled by the Malécot equation,

$$\psi_i = (1 - L)Me^{-\varepsilon|S_i - S|} + L \tag{4}$$

where S_i is the location of the marker and S is the unknown position of the causal mutation. To combine information across multiple markers, a composite log likelihood is formed by effectively assuming $\hat{\psi}_i$ to have a Gaussian distribution with mean ψ_i and variance $1/K_{\psi_i}$.

$$\ell = \frac{-\sum_i K_{\psi_i}(\hat{\psi}_i - \psi_i)^2}{2}$$

Under the null hypothesis that none of the SNPs in the region is associated with the disease, the parameter M is constrained to be equal to 0, S does not exist, and L can be estimated from external information. Under the alternative hypothesis, both M and S are parameters to be estimated from the data. The parameter L, on the other hand, can either be estimated from the data or using external information. A likelihood ratio test based on the composite likelihood can then be applied which has either 2 or 3 degrees of freedom respectively depending on whether L is estimated from current data or not.

In applications of composite likelihoods, statistical inference is often done informally. The primary difficulty relates to composite likelihood–based inference in general as discussed in Section III. It is more difficult in a genetic setting since it is not possible to define the correlation of two components, and there is essentially only one realization from the population process. In addition, because approximations and heuristic arguments are used in multiple parts of the model, it is difficult to evaluate the performance of the methods and provide practical guidelines about under what situations these methods might be more useful or can be improved upon.

Maniatis *et al.* (2004) used simulations to evaluate the type I error rate of likelihood ratio tests based on the composite likelihood. They concluded that the type I error rates are acceptable and the test is slightly anti-conservative for the most complex model (all parameters M, S, and L estimated from the data). More recently, Morton *et al.* (2007) proposed a permutation-based method to better control the type I error rate and also considered estimation of the false discovery rate (FDR) in the setting of genome-wide association scan. The advantage of using composite likelihood–based methods is that they cannot only provide an estimate of the location, but also give confidence intervals (Maniatis *et al.*, 2005; Zhang *et al.*, 2006).

D. Other methods

A composite likelihood approach was also adopted by Xiong and Guo (1997). For a single marker, the log likelihood is defined as:

$$\ell(\theta) = E\left(\prod_{i=1}^{m} p_{id}^{k_{id}} \right) \tag{5}$$

where θ is the recombination fraction between the marker and disease mutation, p_{id} is the frequency of the marker allele i on the haplotypes bearing disease mutation, k_{id} is the observed number of haplotypes harboring the disease mutation. The population frequency of the marker allele is assumed to be constant but its frequency in the disease population changes over time and is a function of θ. The expectation in (5) is taken over all possible genealogy of the sample since the disease mutation first arose. This expectation is difficult to evaluate. Under certain assumptions, Xiong and Guo (1997) derived first- and second-order approximations of the likelihood. For multiple markers, the composite likelihood is simply $\ell = \sum \ell_i$. The recombination fractions between markers and the disease locus at an unknown position x are given by Haldane's map function.

The composite likelihood methods reviewed thus far are all based on modeling the linkage disequilibrium between a single marker and the disease locus. A common drawback of these methods is that they do not account for the dependence of adjacent markers. This is not satisfactory because when densely spaced markers are available, they are likely to be highly correlated. Ignoring the correlation will result in loss of efficiency and incorrect inference. For a rare, fully penetrant disease mutation, Garner and Slatkin (2002) developed a theoretical model of the sampling distribution of two-marker haplotypes that are linked to the disease mutation. They studied the use of maximum likelihood methods compared to the composite likelihood methods based on two single-marker likelihoods and concluded that the full two-marker likelihood method gave much better localization of the disease mutation. Their composite likelihood methods sometimes placed the disease mutation in the wrong interval. This two-marker model could not be extended to include more markers and it leaves out the question of which two markers to use when multiple markers are available. It is conceivable that a composite likelihood can be constructed based on the two-marker and single-marker likelihoods, for example (Cox and Reid, 2004):

$$\ell(\theta) = \sum_{s<t} \ell(Y_s, Y_t; \theta) - aK \sum_{s} \ell(Y_s; \theta)$$

where K is the number of markers, θ is the parameter of interest (i.e., the position of the disease mutation), and a is a constant chosen to achieve optimality (e.g., $a = 0$ corresponds to a pairwise likelihood). It is likely that some truncation, that is, ignoring pairs of markers that are too far away, would improve efficiency.

A different approach to multi-locus linkage disequilibrium mapping is based on haplotypes (McPeek and Strahs, 1999). A likelihood model is defined for the probability of an observed haplotype, given an ancestral haplotype and the disease status. The ancestral haplotype is a parameter to be estimated. Given the ancestral haplotype, the observed haplotypes are assumed to be independent, which is to assume a star-shape genealogy. The likelihood inference is then based on combining the individual component likelihoods, ignoring possible correlations among the haplotypes. This method can also be considered as an application of composite likelihood methods. This approach was later extended to a Bayesian framework (Liu *et al.*, 2001).

V. OTHER APPLICATIONS

The application of composite likelihood is not limited to correlated genotype data but also to pedigree data and high-dimensional phenotype data in general.

For a case-control study, a haplotype-based likelihood ratio test of association can be conducted using multinomial likelihoods. However, when the subjects are related, as is often the case where the samples were originally collected for linkage studies, standard likelihood method does not apply. Browning *et al.* (2005) proposed a weighted composite likelihood method for haplotype-based case-control association testing using pedigrees. In the simple case, suppose haplotypes can be observed, the likelihood for one individual is $L_i = p_{h_{i1}} p_{h_{i2}}$ where h_{i1} and h_{i2} are the two haplotypes and $p_{h_{i1}}$, $p_{h_{i2}}$ are the corresponding population frequencies. A composite likelihood is formed by ignoring the fact the individuals may be related: $L = \prod_{i=1}^{n} (L_i)^{\alpha_i}$. Here α_i is a weight based on kinship coefficients matrix K, $\alpha = 1/2\ (1^T K^{-1})$, which was shown to minimize the variance of haplotype frequency estimate. The composite likelihood for unphased data can be developed similarly, and different weighting schemes were also discussed.

In addition to increasingly high-dimensional genotype data, there is a trend toward collecting more and more phenotype data as the technology becomes increasingly available, such as multiple structural measures from medical imaging data, measurement of lipid fractions via NMR. These measures can also be collected over time and thus pose a major challenge in the analysis, especially considering the fact that many genetic analyses are complicated enough and often need special software. In a nongenetic setting, the approach adopted by Fieuws and Verbeke (2006) provides an instructive example. The problem was to model hearing thresholds at 22 different frequencies measured over 20 years. The joint model $L(Y_1, \ldots, Y_{22}; \Theta^*)$ for all 22 outcomes was intractable. Instead, the log likelihoods for all pairs of outcomes $L(Y_r, Y_s; \Theta_{rs})$ were maximized separately to give estimates of $\Theta = \{\Theta_{rs}, r < s\}$. Some of the parameters in the joint model Θ^* will appear multiple times in the set of parameters Θ, whereas others that are specific to outcomes r and s will appear only once. Under suitable conditions, the estimator $\hat{\Theta}$ has an asymptotic normal distribution, and the parameter of interest Θ^* is estimated by transformation $\hat{\Theta}^* = \mathbf{A}\hat{\Theta}$ which effectively takes average estimates of those common parameters over all pairs. Although not exactly a composite likelihood method, it shares similar characteristics. Simulation studies showed that the pairwise approach yielded unbiased estimates with robust standard errors. Although there was some loss of efficiency, especially for the parameters common to multiple pairs of outcome variables, it is not important in practice when the fully efficient likelihood estimator is not available.

VI. PROSPECTIVES AND DISCUSSION

With the fast-pacing revolution of technology, more and more data become readily available for genetic researchers. Currently studies are underway to genotype up to a million SNPs using high-throughput platforms. By pooling resources across studies, a "mega-study" may include tens of thousands of subjects. Developments in genomics and proteomics make it feasible to measure the mRNA or protein levels for thousands of genes at the same time (Cheung *et al.*, 2005). To make good use of the data, both statistical efficiency and computational feasibility have to be considered in modeling and inference. Composite likelihood may prove to be at least a good starting point in untangling the large volume of data.

Composite likelihood–based approaches have been very successfully applied in the estimation of population genetics parameters. However, its usage in linkage disequilibrium mapping is not as widespread. There are a few reasons why association mapping is more difficult. First, the disease loci (mostly like more than one for complex disease) are unknown thus cannot be modeled directly. The observed phenotypes only provide limited information regarding the underlying genotypes. To model the unobserved disease mutation genotypes, it is often necessary to specify a penetrance model which is typically difficult. It is often implicitly assumed that there is one disease mutation in the region of interest but in reality, several different mutations may contribute to the disease for different subjects in the sample. Or perhaps two mutations in the region together increase the risk of disease. These complications make the idea of "localizing" the disease mutation not a very well-defined problem. Some models of linkage disequilibrium require parameters such as the age of the disease mutation that cannot be reliably estimated without knowing the frequency of the disease allele, the genetic model (recessive or multiplicative, etc.) and the penetrance. Second, the theoretical results derived for population genetics parameters all assumed that on the length of the region (chromosome) goes to infinity. Aside from the practical limitation of finite chromosome length, for gene mapping only markers in close neighborhood of the disease mutation (whose positions are unfortunately unknown) are likely to provide useful information. Third, in genetic mapping studies, various ascertainment schemes are used in attempts to increase the statistical power. It is not clear how to model ascertainment in the population genetics framework.

Nevertheless, there is a lot of potential for applications of composite likelihood methods in genetic studies of complex diseases. Several areas would be of particular interest. First, single-marker–based models (e.g., Graham and Thompson, 1998) can be further developed to allow for combining information across markers (e.g., by the Malécot model) and thus form the basis of composite likelihoods. Second, we need to develop methods for multi-locus association mapping that incorporate coalescent models with more than just two loci or one haplotype. One example is the three-locus likelihood model based on Wright–Fisher model of population genetics (Boitard *et al.*, 2006). Based on existing examples, it is expected that using larger, higher-order, components will lead to increased efficiency of the composite likelihood estimator and likely better asymptotic behavior as well. Third, the validity of inference based on composite likelihoods needs to be carefully studied, through theoretical development or extensive simulations. Resampling-based methods (Morton *et al.*, 2007) are very promising and can be further explored. Recently theoretical development in this area includes Varin and Vidoni (2005).

Finally, composite likelihood methods can also be used to model high-dimensional phenotype data. For example, bivariate linkage analysis is available in existing software (Almasy *et al.*, 1997) but higher dimensional phenotypes are more difficult to model directly due to the large number of variance parameters. A similar strategy as in Fieuws and Verbeke may provide an attractive alternative.

References

Almasy, L., Dyer, T. D., and Blangero, J. (1997). Bivariate quantitative trait linkage analysis: Pleiotropy versus co-incident linkages. *Genet. Epidemiol.* **14,** 953–958.

Besag, J. E. (1974). Spatial interaction and the statistical analysis of lattice systems (with discussion). *J. R. Stat. Soc. B* **36,** 192–236.

Boitard, S., Abdallah, J., de Rochambeau, H., Cierco-Ayrolles, C., and Mangin, B. (2006). Linkage disequilibrium interval mapping of quantitative trait loci. *BMC Genomics* **7,** 54.

Browning, S. R., Briley, J. D., Briley, L. P., Chandra, G., Charnecki, J. H., Ehm, M. G., Johansson, K. A., Jones, B. J., Karter, A. J., Yarnall, D. P., and Wagner, M. J. (2005). Case-control single-marker and haplotypic association analysis of pedigree data. *Genet. Epidemiol.* **28,** 110–122.

Carvajal-Rodriguez, A., Crandall, K. A., and Posada, D. (2006). Recombination estimation under complex evolutionary models with the coalescent composite-likelihood method. *Mol. Biol. Evol.* **23,** 817–827.

Chapman, N. H., Leutenegger, A. L., Badzioch, M. D., Bogdan, M., Conlon, E. M., Daw, E. W., Gagnon, F., Li, N., Maia, J. M., Wijsman, E. M., and Thompson, E. A. (2001). The importance of connections: Joining components of the Hutterite pedigree. *Genet. Epidemiol.* **21**(Suppl. 1), S230–S235.

Cheung, V. G., Spielman, R. S., Ewens, K. G., Weber, T. M., Morley, M., and Burdick, J. T. (2005). Mapping determinants of human gene expression by regional and genome-wide association. *Nature* **437,** 1365–1369.

Collins, A., and Morton, N. E. (1998). Mapping a disease locus by allelic association. *Proc. Natl. Acad. Sci. USA* **95,** 1741–1745.

Cox, D. R., and Reid, N. (2004). A note on pseudolikelihood constructed from marginal densities. *Biometrika* **91,** 729–737.

Crawford, D. C., Bhangale, T., Li, N., Hellenthal, G., Rieder, M. J., Nickerson, D. A., and Stephens, M. (2004). Evidence for substantial fine-scale variation in recombination rates across the human genome. *Nat. Genet.* **36,** 700–706.

Devlin, B., Risch, N., and Roeder, K. (1996). Disequilibrium mapping: Composite likelihood for pairwise disequilibrium. *Genomics* **36,** 1–16.

Fearnhead, P. (2003). Consistency of estimators of the population-scaled recombination rate. *Theor. Popul. Biol.* **64,** 67–79.

Fearnhead, P., and Donnelly, P. (2001). Estimating recombination rates from population genetic data. *Genetics* **159,** 1299–1318.

Fearnhead, P., and Donnelly, P. (2002). Approximate likelihood methods for estimating local recombination rates. *J. R. Stat. Soc. B* **64,** 657–680.

Fearnhead, P., and Smith, N. G. C. (2005). A novel method with improved power to detect recombination hotspots from polymorphism data reveals multiple hotspots in human genes. *Am. J. Hum. Genet.* **77,** 781–794.

Fearnhead, P., Harding, R. M., Schneider, J. A., Myers, S., and Donnelly, P. (2004). Application of coalescent methods to reveal fine-scale rate variation and recombination hotspots. *Genetics* **167,** 2067–2081.

Felsenstein, J. (2006). Accuracy of coalescent likelihood estimates: Do we need more sites, more sequences, or more loci? *Mol. Biol. Evol.* **23,** 691–700.

Fieuws, S., and Verbeke, G. (2006). Pairwise fitting of mixed models for the joint modeling of multivariate longitudinal profiles. *Biometrics* **62,** 424–431.

Garner, C., and Slatkin, M. (2002). Likelihood-based disequilibrium mapping for two-marker haplotype data. *Theor. Popul. Biol.* **61,** 153–161.

Graham, J., and Thompson, E. A. (1998). Disequilibrium likelihoods for fine-scale mapping of a rare allele. *Am. J. Hum. Genet.* **63,** 1517–1530.

Griffiths, R. C., and Marjoram, P. (1996). Ancestral inference from samples of DNA sequences with recombination. *J. Comput. Biol.* **3,** 479–502.

Heagerty, P. J., and Lele, S. R. (1998). A composite likelihood approach to binary spatial data. *J. Am. Stat. Assoc.* **93,** 1099–1111.

Hey, J., and Wakeley, J. (1997). A coalescent estimator of the population recombination rate. *Genetics* **145,** 833–846.

Hudson, R. R. (1983). Properties of a neutral allele model with intragenic recombination. *Theor. Popul. Biol.* **23,** 183–201.

Hudson, R. R. (2001). Two-locus sampling distribution and their application. *Genetics* **159,** 1805–1817.

Jensen, J. D., Kim, Y., DuMont, V. B., Aquadro, C. F., and Bustamante, C. D. (2005). Distinguishing between selective sweeps and demography using DNA polymorphism data. *Genetics* **170,** 1401–1410.

Kim, Y., and Nielsen, R. (2004). Linkage disequilibrium as a signature of selective sweeps. *Genetics* **167,** 1513–1524.

Kim, Y., and Stephan, W. (2002). Detecting a local signature of genetic hitchhiking along a recombining chromosome. *Genetics* **160,** 765–777.

Kingman, J. F. C. (1982). The coalescent. *Stochastic Process. Appl.* **13,** 235–248.

Kuhner, M. K., Yamato, J., and Felsenstein, J. (2000). Maximum likelihood estimation of recombination rates from population data. *Genetics* **156,** 1392–1401.

Leutenegger, A. L., Génin, E., Thompson, E. A., and Clerget-Darpoux, F. (2002). Impact of parental relationships in maximum LOD score affected sib-pair method. *Genet. Epidemiol.* **23,** 413–425.

Li, N., and Stephens, M. (2003). Modeling linkage disequilibrium and identifying recombination hotspots using single-nucleotide polymorphism data. *Genetics* **165,** 2213–2233.

Liang, K. Y., and Zeger, S. L. (1986). Longitudinal data analysis using generalized linear models. *Biometrika* **73,** 13–22.

Liang, K. Y., and Zeger, S. L. (1995). Inference based on estimating functions in the presence of nuisance parameters (with discussion). *Stat. Sci.* **10,** 158–173.

Lindsay, B. (1988). Composite likelihood methods. *Contemp. Math.* **80,** 221–239.

Liu, J. S., Sabatti, C., Teng, J., Keats, B. J., and Risch, N. (2001). Bayesian analysis of haplotypes for linkage disequilibrium mapping. *Genome Res.* **11,** 1716–1724.

Lumley, T., and Heagerty, P. (1999). Weighted empirical adaptive variance estimators for correlated data regression. *J. R. Stat. Soc. B* **61,** 459–477.

Malécot, G. (1973). Isolation by distance. *In* "Genetic Structure of Populations" (N. E. Morton, ed.), pp. 72–75. University of Hawaii Press, Honolulu.

Maniatis, N., Collins, A., Xu, C. F., McCarthy, L. C., Hewett, D. R., Tapper, W., Ennis, S., Ke, X., and Morton, N. E. (2002). The first linkage disequilibrium (LD) maps: Delineation of hot and cold blocks by diplotype analysis. *Proc. Natl. Acad. Sci. USA* **99,** 2228–2233.

Maniatis, N., Collins, A., Gibson, J., Zhang, W., Tapper, W., and Morton, N. E. (2004). Positional cloning by linkage disequilibrium. *Am. J. Hum. Genet.* **74,** 846–855.

Maniatis, N., Morton, N., Gibson, J., Xu, C. F., Hosking, L., and Collins, A. (2005). The optimal measure of linkage disequilibrium reduces error in association mapping of affection status. *Hum. Mol. Genet.* **14,** 143–153.

Maniatis, N., Collins, A., and Morton, N. E. (2007). Effects of single SNPs, haplotypes, and whole-genome LD maps on accuracy of association mapping. *Genet. Epidemiol.* **31,** 179–188.

McPeek, M. S., and Strahs, A. (1999). Assessment of linkage disequilibrium by the decay of haplotype sharing, with application to fine-scale genetic mapping. *Am. J. Hum. Genet.* **65,** 858–875.

McVean, G., Awadalla, P., and Fearnhead, P. (2002). A coalescent-based method for detecting and estimating recombination from gene sequences. *Genetics* **160,** 1231–1241.

Meiklejohn, C. D., Kim, Y., Hartl, D. L., and Parsch, J. (2004). Identification of a locus under complex positive selection in *Drosophila simulans* by haplotype mapping and composite-likelihood estimation. *Genetics* **168,** 265–279.

Morton, N. (2005). Linkage disequilibrium maps and association mapping. *J. Clin. Invest.* **115,** 1425–1430.

Morton, N. E., Zhang, W., Taillon-Miller, P., Ennis, S., Kwok, P. Y., and Collins, A. (2001). The optimal measure of allelic association. *Proc. Natl. Acad. Sci. USA* **98,** 5217–5221.

Morton, N., Maniatis, N., Zhang, W., Ennis, S., and Collins, A. (2007). Genome scanning by composite likelihood. *Am. J. Hum. Genet.* **80,** 19–28.

Pluzhnikov, A., and Donnelly, P. (1996). Optimal sequencing strategies for surveying molecular genetic diversity. *Genetics* **144,** 1247–1262.

Pritchard, J. K., and Przeworski, M. (2001). Linkage disequilibrium in humans: Models and data. *Am. J. Hum. Genet.* **69,** 1–14.

Smith, N. G. C., and Fearnhead, P. (2005). A comparison of three estimators of the population-scaled recombination rate: Accuracy and robustness. *Genetics* **171,** 2051–2062.

Stephens, M., and Donnelly, P. (2000). Inference in molecular population genetics (with discussion). *J. R. Stat. Soc. B* **62,** 605–655.

Tapper, W. J., Maniatis, N., Morton, N. E., and Collins, A. (2003). A metric linkage disequilibrium map of a human chromosome. *Ann. Hum. Genet.* **67,** 487–494.

Tapper, W., Collins, A., Gibson, J., Maniatis, N., Ennis, S., and Morton, N. E. (2005). A map of the human genome in linkage disequilibrium units. *Proc. Natl. Acad. Sci. USA* **102,** 11835–11839.

Terwilliger, J. D. (1995). A powerful likelihood method for the analysis of linkage disequilibrium between trait loci and the one or more polymorphic marker loci. *Am. J. Hum. Genet.* **56,** 777–787.

Varin, C., and Vidoni, P. (2005). A note on composite likelihood inference and model selection. *Biometrika* **92,** 519–528.

Wall, J. D. (2004). Estimating recombination rates using three-site likelihoods. *Genetics* **167,** 1461–1473.

Wiuf, C. (2006). Consistency of estimators of population scaled parameters using composite likelihood. *J. Math. Biol.* **53,** 821–841.

Xiong, M., and Guo, S. W. (1997). Fine-scale genetic mapping based on linkage disequilibrium: Theory and applications. *Am. J. Hum. Genet.* **60,** 1513–1531.

Zeger, S. L., Liang, K. Y., and Albert, P. S. (1988). Models for longitudinal data: A generalized estimating equation approach. *Biometrics* **44,** 1049–1060.

Zhang, W., Collins, A., Maniatis, N., Tapper, W., and Morton, N. E. (2002). Properties of linkage disequilibrium (LD) maps. *Proc. Natl. Acad. Sci. USA* **99,** 17004–17007.

Zhang, W., Collins, A., Gibson, J., Tapper, W. J., Hunt, S., Deloukas, P., Bentley, D. R., and Morton, N. E. (2004). Impact of population structure, effective bottleneck time, and allele frequency on linkage disequilibrium maps. *Proc. Natl. Acad. Sci. USA* **101,** 18075–18080.

Zhang, W., Maniatis, N., Rodriguez, S., Miller, G. J., Day, I. N. M., Gaunt, T. R., Collins, A., and Morton, N. E. (2006). Refined association mapping for a quantitative trait: Weight in the H19-IGF2-INS-TH region. *Ann. Hum. Genet.* **70,** 848–856.

Zhu, L., and Bustamante, C. D. (2005). A composite-likelihood approach for detecting directional selection from DNA sequence data. *Genetics* **170,** 1411–1421.

23

Comparative Genomics for Detecting Human Disease Genes

Carol Moreno, Jozef Lazar, Howard J. Jacob, and Anne E. Kwitek

Human and Molecular Genetics Center,
Medical College of Wisconsin, Milwaukee, Wisconsin 53226

ABSTRACT

Originally, comparative genomics was geared toward defining the synteny of genes between species. As the human genome project accelerated, there was an increase in the number of tools and means to make comparisons culminating in having the genomic sequence for a large number of organisms spanning the evolutionary tree. With this level of resolution and a long history of comparative biology and

Advances in Genetics, Vol. 60

0065-2660/08 $35.00
DOI: 10.1016/S0065-2660(07)00423-3

comparative genetics, it is now possible to use comparative genomics to build or select better animal models and to facilitate gene discovery. Comparative genomics takes advantage of the functional genetic information from other organisms, (vertebrates and invertebrates), to apply it to the study of human physiology and disease. It allows for the identification of genes and regulatory regions, and for acquiring knowledge about gene function. In this chapter, the current state of comparative genomics and the available tools are discussed in the context of developing animal model systems that reflect the clinical picture.

I. INTRODUCTION

Comparative genomics is a relatively new field of biological research in which the genome sequences of different species are compared. However, it should be noted that comparative genomics, to some degree, began nearly 200 years ago when animal models were first sought to mimic human disease and to help determine physiological mechanisms related to humans (Desnick *et al.*, 1982). The initial studies could be defined more as comparative biology, which remains an important field of biomedical research. A premise behind this research is that if a model shares a phenotype, some part of the mechanism underlying it may be common to humans. The idea that not only phenotypes but also genomic organization tends to be evolutionarily conserved between species was postulated in the early 1900s (Castle and Wachter, 1924; Haldane, 1927). The history and development of comparative genomics are relevant to its potential future role. Over the past 20 years, there has been an evolution from comparative biology and comparative genetics to comparative genomics; an additional branch—comparative medicine—has quickly emerged over the last decade. To assess the number of publications associated with these topics, we used PubMed to search the literature for the following keywords: comparative biology, comparative genetics, comparative genomics, and comparative medicine; the results by decade are shown in Fig. 23.1.

There has been a dramatic increase in publications for all of the areas since the 1960s. In the 1960s and 1970s, there was relative parity between comparative biology and genetics. Indeed, the first report of using a computer program to compare genes across species, albeit using protein sequence, was by Dr. Margaret Dayhoff in 1964. The 1970s saw the development of sequencing (Maxam and Gilbert, 1977; Sanger *et al.*, 1977) and the expansion of recombinant DNA technology, which resulted in a concordant upsurge in comparative genetics and genomics in the 1980s that continues today. The term genomics is attributed to T. H. Roderick of the Jackson Laboratory, Bar Harbor, Maine, for suggesting the term for a new discipline that involves mapping/sequencing (McKusick and Ruddle, 1987). However, the first comparison of whole-genome

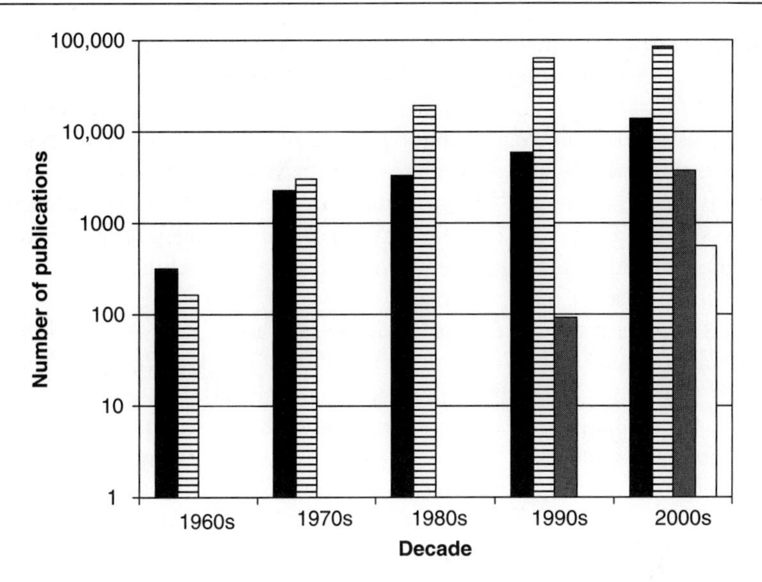

Figure 23.1. Chronological distribution of publications in Pubmed with the keywords: Comparative biology (black), comparative genetics (horizontal stripes), comparative genomics (grey), and comparative medicine (white).

sequence was with two different tobacco viruses and was published in 1984 by Ohno *et al.* (1984). This paper was the first we could identify that captures the use of whole-genome sequence alignments between two organisms to study gene structure and to infer function. To some degree, the use of comparative genomics for identifying functional elements in DNA, based on evolutionary conservation between species, has become standard practice in the field of molecular genetics, and has been one of the major drivers for the Human Genome Project to focus large efforts in genomics, genetics, and physiological infrastructure in multiple model organisms. Nadeau and Taylor (1984) and Cheng *et al.* (1988) helped bring the concept of using conserved synteny between mouse and human to help define gene function. Studies involving chromosome-banding conservation and chromosome painting (ZOO-FISH) have since shown that large stretches of DNA are conserved in mammalian species as divergent as humans and fin whales (Nash and O'Brien, 1982; Sawyer and Hozier, 1987; Scherthan *et al.*, 1994; Weinberg and Stanyon, 1995). While these studies showed conservation of chromosomal segments, they could not demonstrate the explicit conserved gene order at high resolution; such detail can be accomplished only at the genetic/physical mapping or at the sequence level. As genetic and physical maps of human and model organisms developed with the advent of the Human Genome Project in the 1990s, and as the number of identified genes increased, the number of possible integration points dramatically enhanced the quality and density of comparative maps (O'Brien *et al.*, 1999). The increased number of mapped genes, cDNAs, and ESTs has led to sequence comparisons to identify orthologous genes (homologous genes in different species evolving from a common ancestral gene) (Clark, 1999; Fitch, 2000). When mapped in both species, these orthologues serve as "anchors" useful in identifying conserved segments between species. They also potentiate the use of animal models for direct study of gene function and regulation, often translatable to human biological pathways. EST projects exist for multiple organisms; these sequences can be clustered by sequence alignment to create a predicted "gene" transcript, even prior sequencing of the genomic DNA or full-length cDNAs. The UniGene set from the National Center for Biotechnology Information (NCBI) contains clusters from multiple mammalian species, ranging from human, mouse, and rat to cow, pig, sheep, rabbit, and macaque. The sequencing of 78 genomes, including 38 mammals, provides many additional anchors between species as genomic sequence can be directly aligned to determine gene models and conserved elements. Thus far, mammalian comparative maps based on sequence have been generated between human, mouse, rat, dog, and cow (Gibbs *et al.*, 2004; Lindblad-Toh *et al.*, 2005; Waterston *et al.*, 2002); although information on conserved synteny for many others are available through the UCSC, Ensembl, and/or NCBI browsers, the alignments are at the scaffold level rather than alignments of the assembled genome.

While comparative genomics research has built a powerful tool set for identifying the genes and regulatory elements contained within the genomes, it is the nascent development of "comparative medicine" that is the likely future of comparative genomics. While the concept of comparative medicine is clearly a recent descendent of comparative biology (the first PubMed result for a "comparative medicine" query is in 2000), we think it will have a broader basis and application due to its ability to use genetic and genomic tools to generate animal models that more accurately emulate the human disease. The ability to manipulate the genomes of many different species via transgenesis has enabled investigators to modify the genomic and genetic basis of an organism's biology. In 2000, we reported that quantitative trait loci (QTLs) in the rat could be used to predict the locations where human QTLs were likely to exist (Stoll *et al.*, 2000), an approach that has since been confirmed by others (Jacob and Kwitek, 2002; Korstanje and DiPetrillo, 2004; Sugiyama *et al.*, 2001). The importance of these findings is that animal models can be used to positionally clone a gene causing a "disease" trait and then translate that information to the human disease. Conversely, a disease gene might be cloned in the human, but an animal model is required to study the gene function. This concept was extended toward "humanized" animal models, whereby a gene(s) in a pathway from a human is integrated into another genome, typically that of mouse. In short, gene manipulations and positional cloning studies in animals, translated to human, has led to the gene identification or validation of 350 single-locus disorders and over 1700 diseases or traits in cat, cattle, dog, horse, pig, or sheep, many of which have human counterparts according to the Online Mendelian Inheritance in Animals database (http://www.angis.org.au/Databases/BIRX/omia/) (Nicholas, 2003). While its power for single-gene traits is well-established, comparative medicine is perhaps more essential for understanding common multifactorial diseases (multiple genes with environmental interactions) and likely for drug studies over the next decade. In this chapter, we discuss the tools and strategies pointing to this ensuing change.

II. THE POWER AND PROMISE OF COMPARATIVE GENOMICS

Analysis of the sequence of the human genome itself has yielded a vast amount of information. But there are many regions to which no features have been assigned (Waterston *et al.*, 2002). Other regions clearly represent genes, but their functional roles remain unknown. Comparing the human genome sequence to that of a second genome can answer many of these questions. Sequence alignments enable the identification of conserved sequence segments and assess their functional significance. In choosing a species for comparison to human, one might be tempted to pick a close relative, such as chimpanzee. But in some cases, we can learn more from a distant ancestor such as the mouse or pufferfish. The genes responsible for common structures or functions critical for life should be highly conserved and we might expect them to stand out against a background of sequences that have gone through random drift. This assertion is based on the hypothesis that important biological sequences are conserved between species through evolutionary pressure requiring a particular function to be preserved. The availability of sequence from a large number of species in recent years has radically changed the types of comparisons that can be made across multiple species, which facilitates the detection of putative functionally conserved sequences. This is most readily apparent for protein-encoding sequences, but also holds true for the sequences involved in the regulation of gene expression and other elements with currently unknown functions. It also helps determining the most appropriate evolutionary distance for identification of different coding and regulatory motifs (Dubchak *et al.*, 2000; Gottgens *et al.*, 2002; Hood *et al.*, 1995; Koop and Hood, 1994; Pennacchio and Rubin, 2001).

Putative orthologous relationships between human genes and those from different model species, ranging from primates to *Drosophila*, can be determined by aligning gene and protein sequences using complex computer algorithms. Databases algorithms such as Basic Local Alignment Search Tool (BLAST) (www.ncbi.nlm.nih.gov/Education/blasttutorial.html) are able to perform sequence comparison between many organisms, including comparing proteins to nucleic acids and vice versa. These algorithms allow the prediction of orthologous sequences using measures of sequence identity across a given fragment length as well as the evolutionary distance between the organisms. For instance, Makalowski and Boguski (1998) found human and rodent genes share 85% sequence identity in gene-coding sequence and ~68% identity in the untranslated regions.

The concept of comparative genomics can be extended beyond aligning and identifying conserved regions of putative function between genomes. The identification of these functional elements, be it genes or regulatory regions, promise a more comprehensive integration of biology and pathobiology between human and model organisms. The ultimate model might be considered one that shares not only a disease phenotype but also an altered component of the genetic pathway or mechanism that is common to a clinical subpopulation. Thus, comparative genomics and genetics offer great momentum toward comparative medicine.

A. Characterization of genes and their regulation

The first application of comparative sequence analysis relates to the discovery of new genes which have previously been invisible to extensive computational and experimental investigation within given species. While a significant fraction of the genes in the human genome have likely already been identified, genome-wide searches for conserved sequences across multiple organisms should aid in the identification of genes missed in the initial annotation of human sequence.

Indeed, there have been several recent examples in which comparative sequence data have led to the discovery and functional understanding of previously undefined genes. The complete human/mouse orthologous-sequence data set proved particularly valuable in the characterization of gene families in humans and mice (Dehal *et al.*, 2001). It also allowed the discovery of apolipoprotein A5 (*APOA5*), which participates in the control of plasma triglycerides and has been associated with increased triglycerides in different human populations [Caucasian, African-American, Hispanic, and Asian populations (Endo *et al.*, 2002; Nabika *et al.*, 2002; Pennacchio *et al.*, 2001, 2002; Talmud *et al.*, 2002)]. This strategy can also be used to identify predicted genes in the human and mouse genome.

Comparative sequence analysis is also used for the identification of conserved gene-regulatory sequences. Although coding regions comprise not more than 1.5% of the human genome, the minimal amount of the human genome estimated to be under evolutionary constraint is ~5% (Abrahams *et al.*, 2002; Schwartz *et al.*, 2000), which suggests that a respectable fraction of the human genome may consist of noncoding functional elements. Although some of these may be noncoding RNAs (Waterston *et al.*, 2002), a large number are likely to be gene-regulatory elements. Many regulatory regions have been identified using comparative genomics between humans and different species (Cooper and Sidow, 2003), such as the *CD4* gene silencer, first identified in the chicken (Koskinen *et al.*, 2002), as well as enhancers for the *Dlx* gene cluster, important in the development of forebrain and jaws (Depew *et al.*, 2002; Ghanem *et al.*, 2003).

When using comparative analyses for identifying biologically relevant sequence elements, one must remember to take into account that too little conservation could lead to missed functional regions in the genome. A lack of conservation between distant species does not necessarily mean that the region is not functionally important. One must carefully balance between evolutionary distance and biological similarities in order to identify conserved sequences that could be relevant to a particular biological process, as well as staying focused on the question being asked using comparative genomics.

1. Comparing nonmammalian models

It is generally thought that humans and rodents (mouse and rat) occupy a privileged position for cross-species sequence comparisons, but they alone cannot capture all functionally relevant sequences. In practice, human–rodent comparisons are not always feasible for deriving biological insights for a given genomic region. One might choose to compare mouse to human for mapping the regulation of genes involved in a common pathway or disease in both organisms. However, some functional regions are masked within regions of low divergence (Nobrega *et al.*, 2003). To study regions of the human genome where human–mouse/rat sequence comparisons are not ideal, examination of species occupying different evolutionary distances may be useful. Genomic sequence comparisons between distant species can be used to identify not only genes and their intron–exon structure but also regulatory elements present in the large noncoding fraction of the genome (Loots *et al.*, 2000; Pennacchio *et al.*, 2001; Pennacchio and Rubin, 2001). The release of the genome sequence for many organisms has enabled the use of multiple species for genome alignment and identification of functional regions, both genes and enhancers/repressors of gene transcription. Sequence information for some of these species may have little use as model organisms for experimental biology, but play a pivotal role for annotating the human genome through comparative genomics. The evolutionary distance between human and chicken, for instance, is ~3–4 times more than that between humans and mice. There is, therefore, a higher likelihood that the sequence conservation between humans and birds is due to the selection of a functional element. Furthermore, in addition to the identification of enhancer elements, comparative genomics is also useful for the detailed characterization of their composition. For instance, *phylogenetic footprinting* uses multispecies sequence alignments to identify highly conserved motifs at a fine scale (6–12 bp), comparable to the size of transcription factor binding sites (TFBSs). By using this strategy, Gumucio *et al.* (Hardison *et al.*, 1997) identified such "footprints" in the nestin enhancer. Orthologous chicken sequence was used in the identification of the homeobox gene *Nkx2–5* (Lien *et al.*, 2002). Further dissection of this enhancer through phylogenetic footprinting revealed the precise TFBSs responsible for the enhancer activity, aided by having the chicken genomic sequence.

Human–fish sequence comparisons also provide a useful evolutionary position for comparative sequence–based discovery. The fish is the most distant vertebrate (over 400 million years of evolution) with available genomic sequence for comparison with humans (Brenner *et al.*, 1993). Several species of fish have been fully sequenced, including zebrafish, and the two pufferfish, *Fugu rubripes* and *Tetraodon nigroviridis* (Brenner *et al.*, 1993). Their genome sequence has been used to identify functional elements in the human genome, for example, gene-regulatory sequences for the stem cell leukemia (*SCL*) gene, which was conserved in all the exons and eight regulatory regions between humans and mice, but only some were conserved between humans and fish. On the one hand, these data question the utility of sequence comparisons beyond mammals in thoroughly identifying gene-regulatory elements. On the other hand, the simultaneous use of deep sequence comparison across species of the highly conserved *SCL* intervals enabled the discovery of key regulatory regions (Gottgens *et al.*, 2002; Hardison *et al.*, 1997). This phylogenetic footprinting (Hardison *et al.*, 1997) identified two highly conserved promoter sequences that were shown to be necessary for full SCL expression in erythroid cells. This study showed that deep sequence alignments could reveal highly conserved functional motifs. Importantly, the annotation of conserved sequences between the human and *Fugu rubripes* genomes led to the rapid identification of over 1000 previously unidentified human genes (Abrahams *et al.*, 2002; Brenner *et al.*, 1993). While the majority of conserved sequences between human and pufferfish represent orthologous genes, thousands of conserved sequences lie in intergenic regions (between genes), suggesting the discovery of functionally important noncoding sequences in the human genome. Since the pufferfish has a compacted genome (365 million bp), the gene-regulatory regions are predicted to be much closer together in the pufferfish than in humans (Gilligan *et al.*, 2002). For example, eight nonexonic conserved sequences were identified 750 kb from the human *SOX9* transcription factor gene, while in the pufferfish genome they were found within 80 kb of *SOX9*, and were shown to represent distant sequences that regulate SOX9 expression (Bagheri-Fam *et al.*, 2001). Thus, the use of *Fugu rubripes* sequences in genomic comparisons may be a useful tool for the identification of both local and distant regulatory elements, particularly important in early development. These examples highlight the power of comparative sequence analysis in discovering various functional regions of the human genome. Based on the evolutionary relationship among vertebrates, conservation provides a blueprint to our shared genomic machinery. However, given the extreme evolutionary distance between fish and mammals, and the extensive biological differences that separate these species, it is likely that these conserved sequences represent only a subset of the functional elements in the human genome.

Finally, it is possible to extend the comparative studies well beyond vertebrates. Interestingly, a variety of *Drosophila* studies have found homologies and models in the fruit fly genome for various human diseases. Of the currently 2309 human disease gene, 75% have homologues in *Drosophila* and nearly a third of these genes are as highly conserved as genes known to be functionally equivalent between flies and human (Bier, 2005). The yeast has proved a valuable knockout organism for a wide variety of eukaryotic studies, especially on basic cellular growth and metabolism, as well as providing critical insights into mechanisms of common disease. Knockout studies in "lower" eukaryotes such as yeast have elucidated fundamental aspects of protein aggregation, amyloid formation, and toxicity, which can contribute to understanding diseases such as cystic fibrosis, prion-related disease, Alzheimer's, Parkinson's, and Huntington's (Coughlan and Brodsky, 2005; Lashuel and Hirling, 2006; Outeiro and Muchowski, 2004).

2. Interprimate comparisons

It must be recognized that not all sequences are conserved across great evolutionary distances. Some are highly species specific. For these, there was a need to develop strategies to characterize the number of genes and regulatory elements (up to 20%) that do not have a corresponding orthologue between humans and rodents, and that reflect more recent changes in DNA sequence that account for biology in humans. This has been achieved by comparing human sequence to closer evolutionary species, such as other primates. Since nonhuman primates share most genes present in the human genome, due to their phylogenetic proximity, sequence comparison results in high levels of identity, which may limit the ability to detect key functional elements. In order to overcome this excessive sequence identity, a strategy named "phylogenetic shadowing" was introduced (Boffelli *et al.*, 2003). With this approach, orthologous sequence from a dozen or more primate species is compared, therefore increasing the evolutionary distance of the sequence comparisons. This allows the identification of regions of increased variation as well as "shadows," representing conserved segments. Phylogenetic shadowing has been used for the identification of coding regions and putative gene-regulatory elements. Furthermore, as few as four to six primate species, in addition to humans, are sufficient for the identification of a large fraction of putative functional elements in the human genome (Boffelli *et al.*, 2003).

3. Comparative genomic visualization tools and genome browsers

Many comparative genomic tools have been developed to compare the sequence of multiple species, and plot the comparison visually. The two most commonly used are Visualization Tool for Alignment (VISTA) and Percent Identity Plot Maker (PipMaker) (Mayor *et al.*, 2000; Schwartz *et al.*, 2000). Both tools compare multiple megabases of sequence simultaneously from two or more species and have web accessibility. The output is a plot with identification of conserved coding and noncoding sequences.

VISTA combines a global-alignment program (AVID) (Couronne *et al.*, 2003) with a running-plot graphical tool to display the alignment (Mayor *et al.*, 2000) (http://www-gsd.lbl.gov/vista/). Global alignments are produced when two DNA sequences are compared and an optimal similarity score is determined over the entire length of the two sequences. A useful aspect of VISTA is the easily interpretable peak-like features depicting conserved DNA sequences. In contrast, PipMaker uses BLASTZ, a modified local-alignment program, and displays plots with solid horizontal lines to indicate ungapped regions of conserved sequence (i.e., blocks of alignments that lack insertions or deletions) (http://bio.cse.psu.edu/pipmaker/) (Schwartz *et al.*, 2000). Local alignments are generated when two DNA sequences are compared and optimal similarity scores are determined over numerous subregions along the length of the two sequences. Both programs provide the tools to visualize comparative sequence data for conserved regions that, therefore, have a putative function in their interval of interest. VISTA is a suite of tools which includes, among others, mVISTA, for multiple species comparison (Mayor *et al.*, 2000); rVISTA, which searches for transcription factor binding sites (TFBSs) within conserved sequence (Ovcharenko *et al.*, 2004); and GenomeVISTA, which enables comparison of a particular sequence with whole-genome assemblies to find the orthologues automatically (Couronne *et al.*, 2003). There are other tools commonly used for comparative genomics alignments (Jareborg and Durbin, 2000; Szafranski *et al.*, 2006; Wu and Watanabe, 2005).

Recently, a database of evolutionary conserved regions (ECRs) in vertebrate genomes was created (Loots and Ovcharenko, 2007), entitled ECRbase, which is constructed from a collection of whole-genome alignments produced by the ECR Browser. ECRbase features a database of syntenic blocks that recapitulate the evolution of rearrangements in vertebrates and a comprehensive collection of promoters in all vertebrate genomes generated using multiple sources of gene annotation. The database also contains a collection of annotated genes, TFBSs in evolutionary conserved and promoter elements (http://ecrbase.dcode.org), as well as the location of single nucleotide polymorphisms (SNPs) identified and reported in dbSNP (http://www.ncbi.nlm.nih.gov/SNP/index.html).

The completion of the genomic sequence of human and draft sequence assemblies of many additional organisms has enabled large-scale analysis of individual genomes as well as genome-to-genome comparisons. These whole-genome analyses, accessible through web-based genome browsers, provide preprocessed databases for the scientific community (Kent *et al.*, 2002; Waterston *et al.*, 2002). While these sequence data were initially useful for researchers seeking additional genomic sequence for individual genes of interest based on homology searches, large computational projects have focused on the detailed annotation of the human genome, followed by the annotation of the rat and mouse genomes.

In addition to gene annotation for the entire human genome, online resources have also recently become available for whole-human/whole-mouse comparative sequence data. Several important advances have made whole-genome comparisons possible. Whole-genome assemblies, in addition to satisfying the obvious need for sequence data for a given genome, have provided the substrates for genome-to-genome comparisons. Furthermore, the successful whole-genome annotation of genes, including their chromosomal location, serves as a reference for the position of a given alignment in the genome; previous gene-by-gene comparisons required the user to painstakingly input these annotation features. For mammals, this gene annotation is most detailed for the human genome, although progress is being made in annotating the pufferfish, mouse, and rat genomes. As a consequence, current whole-genome comparisons primarily use the human genome as the base reference sequence. Three major resources are currently available for preprocessed human/mouse whole-genome comparisons: UCSC Genome Browser, VISTA Genome Browser, and PipMaker (Couronne *et al.*, 2003; Kent *et al.*, 2002; Schwartz *et al.*, 2000).

III. ANIMAL MODELS FOR HUMAN DISEASE

The use of laboratory animals is an integral part of the medical and biological research. The efficient transfer of knowledge from animal to human and vice versa depends on identifying and studying animal models that mimic human disease processes. It is clear that biomedical science in the "postgenomic age" can be advanced only through the use of appropriate model systems. Importantly, one must keep in mind that a model organism is not a human and cannot represent the entire clinical spectrum of disease. Just as common disease is highly heterogeneous in human populations, the translation of experiments performed solely in one animal species, one genetic background, one time point during disease, or one gender may not accurately predict outcomes in all clinical scenarios (Roep, 2007).

Although different mammals share diseases with humans, rat and mouse remain the premier models for the genetic basis of human disease. Many rat and mouse strains have been selectively bred for both Mendelian and multifactorial diseases (polygenic with environmental influence) such as diabetes, atherosclerosis, heart disease, cancer, glaucoma, anemia, hypertension, obesity, osteoporosis, bleeding disorders, asthma, and neurological disorders. In some cases, there are multiple inbred strains for a single multifactorial disease. For example, five different rat strains (BUF, DA, F344, LEW, and PVG) have an increased risk of multiple sclerosis (MS). Crosses between two of these disease strains (DA and LEW) and resistant control strains have resulted in the identification of 18 QTLs involved in experimental allergic encephalomyelitis, an animal model of MS (Bergsteinsdottir *et al.*, 2000; Dahlman *et al.*, 1999a,b; Roth *et al.*, 1999). Some of the mapped QTL share a common genome location (overlapping confidence intervals) for the same trait in multiple strains (e.g., Eae2 and Eae11).

The zebrafish is another animal model with excellent future for modeling human disease. Some of the obvious advantages are external development and optical clarity during embryogenesis, high reproductive capacity, short generation time, amenable to transgenesis, and genetic screens. Moreover, there is high homology between zebrafish and human genome not only in coding regions but also in conserved noncoding elements (Shin *et al.*, 2005). Already several examples of clinically relevant zebrafish mutant phenotypes are available (for review, see Dooley and Zon, 2000; Igarashi, 2005; Tanoue *et al.*, 2002; Ward and Lieschke, 2002). The zebrafish is particularly suitable to the study of hematopoetic disorders (Ross *et al.*, 2005), cardiovascular diseases (North and Zon, 2003; Sehnert and Stainier, 2002), aging process (Keller and Murtha, 2004), cancer (Berghmans *et al.*, 2005), muscle disease (Guyon *et al.*, 2007; Kunkel *et al.*, 2006), kidney diseases (Drummond, 2005; Hentschel *et al.*, 2005), glaucoma (McMahon *et al.*, 2004), and diseases of other organs (Sprague *et al.*, 2006).

The genomic resources that have been generated over the past years, such as high-density maps of multiple rodent strains, may lead to the identification of shared haplotypes between the disease strains, which can facilitate positional cloning of the disease allele. Results of these studies can then be translated to human disease to identify genes and/or pathways in common between the species.

A. QTL to gene

1. QTL mapping

QTL mapping is a proven useful resource to assign the biology onto the genomic sequence by identifying chromosomal regions that contain genes affecting complex phenotypes. While a QTL is a rather large genetic locus, the gene(s) within this interval is responsible for a component of the trait variation, enabling the genome to be annotated with physiology in advance of the gene being identified. The ultimate goal of QTL mapping is to identify the genes, by positional cloning, that underlie complex phenotypes and diseases and to gain a better understanding of their physiology and pathophysiology. To date, there have been more than 2900 published QTL studies in the mouse [according to the Mouse Genome Informatics (MGI) database] (Flint *et al.*, 2005) and 536 QTL papers published with over 1000 QTLs reported for different physiological and pathophysiological traits in the rat. These papers include investigations of the genetic basis of blood pressure (Rapp, 2000), diabetes (Galli *et al.*, 1996; Jacob *et al.*, 1992; Pravenec *et al.*, 1996), cardiovascular disease (Moreno *et al.*, 2003; Stoll *et al.*, 2001), stroke (Rubattu *et al.*, 1996), ethanol preference (Murphy *et al.*, 2002), behavioral conditioning and anxiety (Fernandez-Teruel *et al.*, 2002; Flint, 2003), fat accumulation (Tanomura *et al.*, 2002), arthritis (Olofsson *et al.*, 2003a), copper metabolism (de Wolf *et al.*, 2002), pituitary tumor growth (Wendell and Gorski, 1997), aerobic capacity (Ways *et al.*, 2002), and chemical carcinogenesis (De Miglio *et al.*, 2002). QTL mapping has been applied to many other organisms as well: zebrafish for QTL mapping of phenotypes with clinical implications such as behavior (Wright *et al.*, 2006); cows for mapping of agriculturally important traits such as milk quality and quantity (Grisart *et al.*, 2002; Malek *et al.*, 2006), which may shed light on fat content and nutrition in human milk; hip dysplasia in dogs (Todhunter *et al.*, 2003) with relevance to human arthritis. Most of the QTLs have been mapped in the past 3 years, mainly due to advances in technologies that allow for high-throughput genotyping and an accelerated development of genetically modified strains of mice, rats, and zebrafish.

Following this trend, it is expected that there will be an acceleration in the rate of gene discovery in rat and mouse, due to dramatic increases in high-throughput sequencing for detecting sequence variants, as well as microarray technologies for gene identification, pathway analysis, and mapping of *cis* and *trans* regulatory elements. These resources will greatly facilitate the identification of genes underlying the hundreds of QTLs mapped for complex diseases and phenotypes.

a. Congenic and consomic strains

QTL mapping is often followed by confirmation of the loci by the development of congenic lines (a region from one strain transferred to the genome background of another strain), in order to evaluate the QTL in the absence of other mapped QTL, and as a step in positional cloning (Flint *et al.*, 2005). Congenic animals are generated by initially intercrossing two strains, and then systematically backcrossing for 10 generations to the strain carrying the desired genomic background. Marker-assisted selection is used for ensuring the introgressed (Lande and Thompson, 1990) region is passed to each subsequent generation. After 10 generations, the strains are intercrossed to generate a congenic that is homozygous for the introgressed interval (Snell, 1948). Since congenic strains differ only in a short genome segment from their background strain, it is possible to investigate the phenotypic effect of the locus, isolated from other effects caused by other loci on the original genetic background. In complex diseases, with multiple QTLs determining a trait, generation of double or triple congenic strains may be necessary in order to confirm a susceptibility locus introgressed onto a resistant genome background. To date, congenic strains for the rat number 381, many of which have narrowed the critical genomic interval to a handful of candidate genes. More than 50% of these congenics were developed for studying blood pressure control, followed by congenics for non–insulin-dependent diabetes mellitus. Most of these congenics have been developed since 2002 (Twigger *et al.*, 2007).

Chromosome substitution strains (CSSs or consomics) are animal strains in which a whole chromosome from one strain is transferred to the genomic background of another strain by a methodology similar to that of congenic generation. This method was first described in plants (Law, 1966) and *Drosophila* (Caligari and Mather, 1975), and later described in mice (Nadeau *et al.*, 2000) and rats (Cowley *et al.*, 2004a). The first complete CSS set in mice was created from A/J and C57BL/6 strains in 2004 and was used to detect QTLs across the mouse genome (Singer *et al.*, 2004). A second set with PWD/Ph strain chromosomes on the B6 background was recently constructed. The Medical College of Wisconsin has assembled two complete rat panels of consomic strains, using the BN strain that was sequenced as the donor strain. In these consomic strains, a chromosome from the BN/NHsdMcwi rat was substituted, one at a time, into the genetic background of the SS/JrHsdMcwi (Dahl salt-sensitive; SS) or FHH/EurMcwi (Fawn Hooded Hypertensive; FHH) rats. The SS rat is a model for salt-sensitive hypertension (Rapp, 1982), insulin resistance (Kotchen *et al.*, 1991), hyperlipidemia (Reaven *et al.*, 1991), endothelial dysfunction (Luscher *et al.*, 1987), cardiac hypertrophy (Ganguli *et al.*, 1979), and glomerulosclerosis (Roman and Kaldunski, 1991). The FHH rat is a model for systolic hypertension, renal disease, pulmonary hypertension, a bleeding disorder, alcoholism, and depression (Provoost, 1994). One major advantage of using consomic strains is to rapidly generate congenic lines. This is possible because the genome backgrounds have already been fixed, enabling a consomic to be converted into a congenic in two generations (Cowley *et al.*, 2004b). The contribution of genes on each chromosome to the observed traits can not only be assessed by phenotyping but also by expression profiling. Comparisons between the consomic and parental strains provide valuable insights into the genomic pathways (clustered gene expression patterns) that differ between strains and how these differences might be connected to a particular pathogenic phenotype of the animal. From a genetic mapping perspective, the consomics offer another advantage: QTL detection using consomic strains is more powerful because the variation due to a mixed genome background is eliminated, in contrast to F2 intercross offspring; therefore, each QTL explains a greater proportion of the total phenotypic variation (Nadeau *et al.*, 2000). The disadvantage of this strategy is that it prevents the study of how gene–gene interactions affect a phenotype.

b. Recombinant inbred strains

Recombinant inbred (RI) strains provide an additional tool for mapping phenotypes to the genome. This strategy is based on the generation of a panel of inbred strains derived from an F2 population (Bailey, 1971; Pravenec *et al.*, 1996; Swank and Bailey, 1973). The RI strategy generates lines that contain more than one QTL per line, permitting the analysis of gene interaction and detection of weak loci. One of the largest rodent RI panels is the reciprocal HXB/BXH RI rat strains, derived from the spontaneously hypertensive (SHR) and the BN rat strains (Pravenec *et al.*, 1989, 1999). These strains are a great resource for genetic analysis of cardiovascular and metabolic phenotypes. The use of this RI strain panel has facilitated the mapping of several traits, including blood pressure (Pravenec *et al.*, 1995), reproductive traits (Zidek *et al.*, 1999), metabolic traits (Pravenec *et al.*, 2002), behavior (Conti *et al.*, 2004), and cancer susceptibility (Bila and Kren, 1996). Other rat RIs are available, such as the LExF (Shisa *et al.*, 1997), although phenotypic characterization of this panel has not been as extensive as that for the HXB/BXH. RI mouse strains have also been developed to map complex diseases. An example is the reciprocal AXB, BXA strains, derived from the A/J and C57BL/6 (B6) inbred strains. Recently, expanded RI panels (Abiola *et al.*, 2003; Williams *et al.*, 2004) and advanced intercross lines (AIL), in which generations beyond F2 are generated to increase recombination frequency (Darvasi and Soller, 1995), have increased the power of QTL mapping. A variation of the RI strain concept is the Collaborative Cross, proposed by the Complex Trait Consortium (Abiola *et al.*, 2003).

c. Heterogeneous stocks

Heterogeneous stocks (HS) are derived from crossing eight inbred strains followed by continuous outbreeding for several generations (Hansen and Spuhler, 1984). This collection of strains is extremely powerful and another alternative for investigators using rodents in their research programs. The chromosomes of the HS progeny represent a random mosaic of the founding animals with an average distance between recombination events close to a single centiMorgan (cM). This high degree of recombination enables the fine-mapping of QTLs to sub-cM intervals and the identification of multiple QTLs within what was previously identified as a single QTL (Ariyarajah *et al.*, 2004; Stylianou *et al.*, 2004). The first use of HS mice for QTL mapping was described by Talbot *et al.* (1999). Fine-mapping of QTL to sub-cM intervals in HS mice has been successful for traits of anxiety (Mott *et al.*, 2000), ethanol-induced locomotor activity (Demarest *et al.*, 2001), and conditioned fear (Mott *et al.*, 2000), and to identify serine–threonine kinase 6 (Stk6) as a candidate skin-tumor susceptibility gene (Ewart-Toland *et al.*, 2003). An HS rat colony was derived by the National Institute of Health (NIH) in 1984 for alcohol studies (Pandey *et al.*, 2002) and is available for genetic studies in both Europe and the United States.

d. Gene expression profiling

Despite more than two decades of intensive research, investigators have had, so far, limited success in elucidating the genetic basis of complex disease. One reason is that many investigators have used only one of the above-mentioned approaches, rather than integrating information gathered by multiple means. A powerful alternative may be to simultaneously integrate genomics and gene expression (transcriptomics) (Yagil *et al.*, 2005). This approach integrates gene expression data with genetic linkage data from segregating populations by treating gene expression as a quantitative trait and mapping expression QTL (eQTL) for those traits (Hubner *et al.*, 2005). This strategy is based on the assumption that genes which are both differentially expressed and map to a disease-related QTL are more likely to be involved in the pathophysiology of that disease than genes in the interval that are not differentially expressed. This approach allows for the dissection of primary and secondary genetic determinants of gene expression. The *cis*-acting QTLs (those that lie nearby the gene that is expressed) are more straightforward to investigate because they are under simpler genetic control and are likely to be due to sequential variants within the gene itself or in neighboring regulatory elements. A *trans*-acting QTL maps to a genomic region that is distant (usually on different chromosome) from the physical location of the gene being transcribed (Petretto *et al.*, 2006). Several important findings have resulted from gene expression profiling: the *CD36* gene was identified as a QTL for insulin fatty-acid metabolism (Aitman *et al.*, 1999); complement factor 5 has been suggested as a gene at a susceptibility QTL in a model of asthma (Karp *et al.*, 2000); lipoxygenase (Alox15) has been proposed as a candidate gene for bone mass (Gerhard *et al.*, 2004); an interferon-inducible gene for susceptibility to an autoimmune condition (systemic lupus) (Rozzo *et al.*, 2001); and glutathione S-transferase M2 has been identified as a gene that might be involved in hypertension (McBride *et al.*, 2003).

Recently, a goal to accelerate and complete the characterization of the transcriptome of an organism led to the development of "Genome tiling arrays" (array-based hybridization to scan the entire genome of an organism at once) (Bertone *et al.*, 2005; Johnson *et al.*, 2005; Mockler *et al.*, 2005). These arrays provide an opportunity for (a) detection of rare transcripts, (b) simultaneous analysis of numerous samples and genomic sequences, (c) independence from current genome annotations, and (d) the discovery and characterization of differentially/alternatively spiced transcripts. Genomic tiling arrays in combination with computational methods can provide insight toward a more comprehensive understanding of basic molecular and cellular processes.

Taken together, none of the strategies described provide a comprehensive solution to gene identification after QTL mapping, but together they provide a powerful approach to find the genes responsible for particular phenotype.

B. Sequence to function

Genetic studies for complex disease are currently undergoing a significant shift, from linkage studies to that of genome-wide association studies using high-density SNPs maps. This is due in part to the development of the HapMap, a detailed genome-wide haplotype map of the human genome (International HapMap Project, 2003). SNP data obtained in any haplotype map project will undoubtedly be informative for a variety of additional purposes, including genetic mapping and cloning approaches that do not depend on haplotype-block definition (Liao *et al.*, 2004; Pletcher *et al.*, 2004). Rapid development of high-throughput SNP-typing strategies, combinations of mapping gene expression QTL (eQTL) and physiological QTLs (pQTL) (Hubner *et al.*, 2005; Malek *et al.*, 2006), and the increased efficiency of resequencing candidate regions might overcome many of the bottlenecks that are associated with the traditional mapping and cloning approaches. The combination of the genomic sequence and a larger SNP database will further enhance the tool kit for investigators using animal models to study human disease.

The success of the HapMap project has extended to genome-wide SNP genotyping projects in additional organisms. The most comprehensive data set is in the mouse. As the NIH succeeded in the goal to sequence the genomes of 15 additional mouse strains (http://www.niehs.nih.gov/crg/cprc.htm), SNP identification becomes a straightforward process of comparing sequence between the various strains. This information led to large SNP genotyping efforts in the mouse. In 2004, Pletcher *et al.* (2004) published the typing of over 10,000 SNPs in 48 common mouse strains, generating nearly 500,000 genotypes. This data set has grown to 140,000 SNPs in a panel of 16 inbred mouse strains, released in 2005 by Wade and Daly (2005), which will soon be greatly expanded to over 8 million SNPs in collaboration with Perlegen. Similar efforts by members of the Complex Trait Consortium (Churchill *et al.*, 2004) (www.complextrait.org) generated six million SNP genotypes for several lines of mice with data accessible through Web site (www.well.ox.ac.uk/mouse/INBREDS; www.genenetwork.org).

As the rat is likely the most common model for complex disease developed via a selective phenotyping strategies, a haplotype map of this organism is also a strong resource for linking genotype to phenotype. The rat genome sequencing consortium has an ongoing SNP discovery effort based upon shotgun sequencing of eight commonly used inbred rat strains, aiming at the discovery of more about 280,000 SNPs for each of the eight strains. Furthermore, Celera produced genomic shotgun sequences for the Sprague-Dawley rat strain at about 1.5× coverage, providing sufficient numbers of SNPs for dense genome coverage (Kaiser, 2005). To fully encompass the power of the recent rat genome sequence, there are initiatives underway to initiate the complete genetic dissection of the ancestral segments making up the most commonly used inbred lines. STAR, funded by the European Union, generated a haplotype map for the rat (http://www.snp-star.eu and http://euratools.csc.mrc.ac.uk) and is genotyping about 100,000 SNPs in at least 200 different inbred strains.

A powerful application of these high-density genotype data sets across inbred mouse and rat strains is to link biological information via *in silico* mapping to identify disease-associated haplotypes in different inbred mouse strains, equating a strain to an individual in a population (Grupe *et al.*, 2001). Although relatively few genes have been identified by this approach, the exponentially growing amount of genotype data in multiple rodent strains offers future success to mapping complex traits (Cervino *et al.*, 2005, 2006; Hillebrandt *et al.*, 2005; Pletcher *et al.*, 2004). These genotyping data will be useful to correlate phenotypes with the ancestral origin of a strain, allowing the identification and fine-mapping of the critical regions responsible for specific traits. Through the inclusion of essentially all important strains that are used around the world for QTL-mapping experiments, researchers will be able to reduce the critical interval by linkage analysis using shared segments between their strains of interest, or select ideal strain combinations for intercross/backcross experiments.

A potentially enlightening use of the haplotype maps across species is to investigate the possibility that haplotype structures might have functional importance and may actually be evolutionarily conserved. Guryev *et al.* developed a high-resolution haplotype map for 5 Mb of the rat genome on chromosome 1, and compared the haplotype structure of this region with the orthologous human and mouse segments (Fig. 23.2). The 5-Mb conserved interval, which contains common QTL across the species, shows a significant similarity in haplotype structure between the species, with a tendency for blocks to encompass complete genes (Guryev *et al.*, 2006). This study suggests that QTL may not be caused by a single causal sequence variant, rather a complex of variants potentially involving multiple genes. Although sequence variants themselves may not be conserved, it is quite possible that the genome context in regions with conserved haplotypes has functional importance. These types of comparative data may assist in prioritizing SNP associations in genome-wide association studies for analyses of locus interactions in humans.

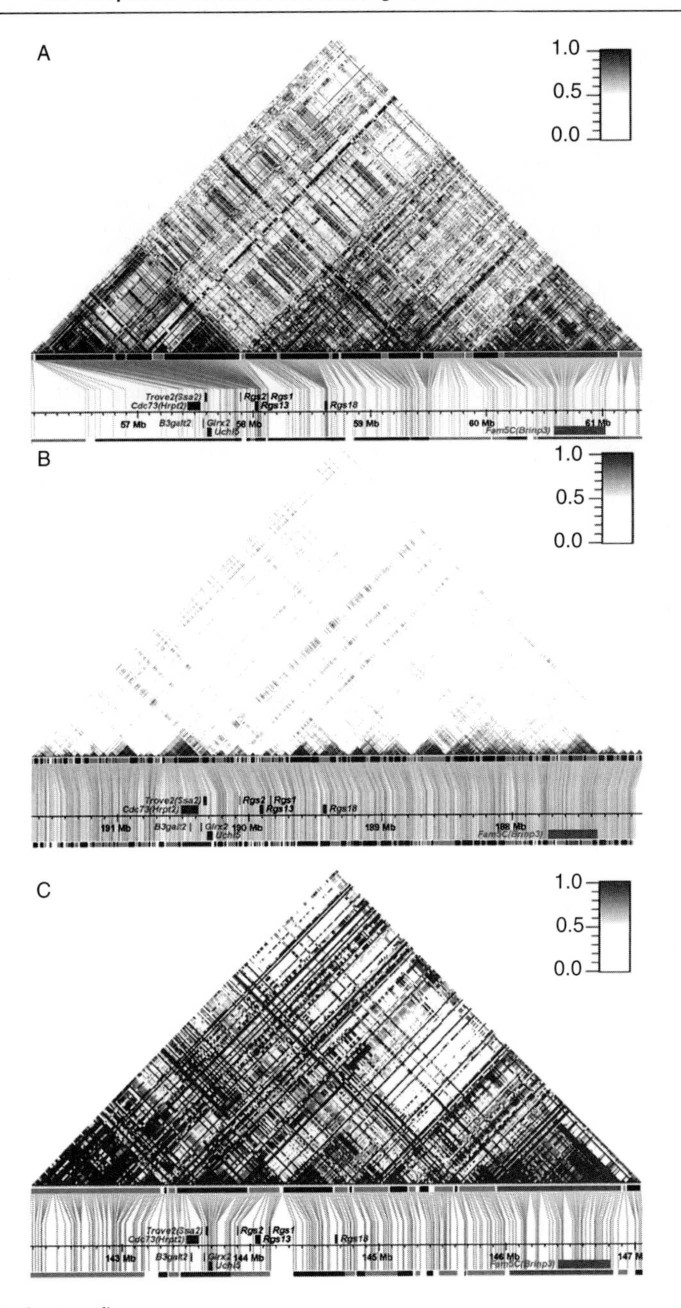

Figure 23.2. (*continued*)

C. Comparing phenomes

As outlined in previous chapters, it is now possible to perform genome-wide association studies generating hundreds of thousands of genotypes in thousands of patient DNA samples. The identification of genes associated with many complex diseases, such as macular degeneration, diabetes, and cancer, is being published at an ever-growing rate (Hanson *et al.*, 2007; Schramm *et al.*, 2005; Sladek *et al.*, 2007; Spinola *et al.*, 2007) and the numbers will continue to increase dramatically over the next 3 years. The bottleneck in identifying genes for common diseases in humans is no longer the capacity and cost of the technology. One remaining bottleneck is the availability of large populations with consistent and well-characterized descriptions of clinical traits, or multiple relevant phenotypes, required to provide power to detect genes with modest clinical effects. Large efforts are underway to develop BioBanks that merge hundreds of thousands of DNA samples available for genotyping with electronic medical records systems that have the ability for cross-talk (Anderlik, 2003; Kaiser, 2002). If sufficient clinical information can be collected on a global scale, one could envision a "human phenome" effort (Freimer and Sabatti, 2003), providing an incredibly powerful resource for linking genotype to phenotype. However, even with the ability to link genotype to phenotype via BioBanks and genome-wide association studies, the need to validate candidate-associated genes and determine their role in disease is critical. The genome sequencing and biological characterization of model organisms provide a very powerful means for generating a multispecies phenome. The data for each species are common—genotypes and phenotypes. While the genotyping is standardized across species, the description of traits or clinical measurements is not, even between different physicians or between different investigators in common research fields. The necessity for standardized nomenclature has been recognized and efforts are underway to generate methodologies to describe traits and measurements in an organized fashion, for example, PATO—phenotype and trait ontology (Smith *et al.*, 2005). Sequenced model organisms such as yeast, mouse, and rat have initiated their own phenome projects (Bogue, 2003; Fernandez-Ricaud *et al.*, 2007; Mashimo *et al.*, 2005), providing an expansive, centralized resource of well-annotated phenotypic measures as well as the genetic variation across different strains. Together these resources will allow for gene validation in the model that

Figure 23.2. Patterns of Linkage Disequilibrium (LD) for orthologous genomic segments of ~5 Mb in rat, human, and mouse. LD plots for orthologous genomic segments in rat (A), human (B), and mouse (C) are shown. For each panel, the following information is shown: Linkage Disequilibrium (LD) plot (top), haplotype blocks in SNP coordinates (middle), and physical map and haplotype blocks in physical coordinates (bottom). Figures and legend reproduced with permission from Guryev *et al.* (2006).

best fits the clinical phenotype at the genetic level and for leveraging the strengths of each organism. For example, yeast may be the simplest and most practical model to understand the role of a gene in basic cell function or metabolism, while the rat may be the best model to understand the mechanisms regulating cerebral vascular flow. In the not too distant future, one could essentially envision an approach of entering standard queries to both the electronic medical records and the various model organism databases to generate a cross-species platform to identify the genetics of complex disease and to have integrated resources to study the pathobiological mechanisms as well as develop clinical treatments.

Comparative genomics studies at the whole-genome scale that would have been headline news a couple of years ago are now available via public Web sites, enabling the investigator to utilize an enormous data set and powerful analysis tools. Unfortunately, we have an imbalance in the other direction, and face a challenge in acquiring large data sets of phenotypes to match millions of genotypes (Freimer and Sabatti, 2003).

With the ever increasing genotype resources across multiple strains in several species comes a parallel effort to generate large amounts of detailed physiological measurements. Public phenome resources have been generated for species ranging from yeast (Fernandez-Ricaud *et al.*, 2007) to mouse (Bogue, 2003; Bogue *et al.*, 2007) and rat (Mashimo *et al.*, 2005).

PROPHECY is a centralized database for collection and visualization of phenotypes related to growth and environmental adaptation from 984 heterozygous diploid strains as well as from 574 yeast strains overexpressing membrane proteins (Fernandez-Ricaud *et al.*, 2007). The diploid strains represent gene-deletions for nearly every gene in the yeast genome. It provides a query tool for obtaining phenotype and genotype information as well as informatic tools to link the two. The Mouse Phenome Database (MPD) is a well-developed phenotype/genotype resource for the mouse. Investigators from several fields of research are submitting data on commonly used mouse strains, ranging from anatomy to behavior, cancer to immunology, and pharmacology to reproduction (Bogue *et al.*, 2007). In the rat, there are two major phenome projects; Mashimo *et al.* (2005) from the National Bio Resource Project (NBRP) for the rat have characterized 109 physiological traits in over 100 inbred rat strains. To link the phenoytpes to the rat genome, they have genotyped 122 strains with a standardized set of 357 microsatellite markers (Mashimo *et al.*, 2006). These resources can be accessed through the NBRP Web site. PhysGen, one of the National Heart, Lung, and Blood Institute's (NHLBI's) Programs for Genomic Applications, is another major project which has phenotyped over 16,000 rats to link over 280 different traits to the rat genome (http://pga.mcw.edu) (Kwitek *et al.*, 2006; Malek *et al.*, 2006) using both common inbred and outbred strains as well as two panels of consomic strains.

Three important areas of technology development are also changing the situation in the phenotyping field: transcriptome RNA analysis, proteomic profiling, and high-throughput cell-based assays. These are not the usual phenotypes with which geneticists are accustomed to work, but they are arguably just as critical to our understanding of health, disease processes, and treatment. Large microarray and proteome studies have been regarded as expensive science; however, when applied to mapping panels they are actually the least expensive and most efficient genetic studies in terms of cost per trait. A recent trio of papers (Abiola *et al.*, 2003; Bystrykh *et al.*, 2005; Hubner *et al.*, 2005) each report on ~10,000 mRNA phenotypes for brain, hematopoetic stem cells, kidney, and fat for 30 strains of mice and rats. The multiplicative power of studying large constellations of phenotypes (Abiola *et al.*, 2003; Threadgill *et al.*, 2002) makes this a new kind of genetics. The same trend is true in proteomics for mammalian tissues (McRedmond *et al.*, 2004; Mijalski *et al.*, 2005; Ruse *et al.*, 2004) and will soon be applied to large mapping panels for the first transcriptome-proteome-genetics studies. The third member of the new-wave phenotyping is a collection of rapidly maturing, high-throughput, cell-based assays (Williams, 2006). The main challenge we are now facing is how to construct appropriate analytical methods and shared resources for exploring, exploiting, and validating this expanded world of phenotypes (Churchill *et al.*, 2004; Freimer and Sabatti, 2003).

IV. COMPARATIVE GENOMICS AND BUILDING BETTER ANIMAL MODELS

A. Transgenesis and mutagenesis

The development of "knockout" technology [for review, see Detrich *et al.* (Goldstein, 2001)] to selectively disable target genes in mice in the 1980s has proven to be a powerful tool for revealing gene function. Many companies are banking on the potential benefits of knockout mice for drug discovery. For example, Lexicon Genetics, Inc., uses patented gene-targeting technologies to alter specific DNAs in mouse embryonic stem (ES) cells, which are then cloned to create their knockout mice. Generating more than 1500 new knockouts per week, the company has a living database of over 170,000 knockout mice in which randomly chosen gene sequences have been altered. These activities will deliver valuable reagents, phenotypic data, and better knowledge about common laboratory strains. A similar NIH-funded endeavor, the Knock Out Mouse Project (KOMP), has also been launched, with the goal to systematically knock out each mouse gene in the C57BL/6 strain and to make the resulting ES lines publicly available (Austin *et al.*, 2004). While these resources are extremely useful, it is clear that these knockout animal models will not be sufficient to explain all the complexities of multifactorial disease. For example, it is becoming increasingly evident that genome background effects on phenotypic outcome of a gene knockout can be dramatic, as evidenced by Magnuson *et al.* using an epidermal-growth factor (EGF) knockout (Threadgill *et al.*, 1995). Moreover, knockout experiments frequently have embryonic lethal effect, adaptive gene expression compensating for the absence or overexpression of a specific gene, and an inappropriate gene expression.

To address the complexities of temporal, spatial, and genome background effects, additional model organisms and technical advances have been made. Both transgenesis and gene targeting can be more tissue or temporally specific by introducing conditional models that allow for more precise control of transgene induction, preferably by pharmacological means (for review, see Bockamp *et al.*, 2002; Szulc *et al.*, 2006). As recently developed in zebrafish, using recombinase (Dong and Stuart, 2004; Thummel *et al.*, 2006) opens new opportunities to study temporal and spatial control of transgene expression in this animal model. Although there are many models for complex disease in rat developed by phenotypic selection strategies, confirmation using homologous recombination is not possible in the rat, as no one has been able to develop ES cells for them. However, transgenesis using classical DNA microinjection, DNA transfer by lentivirus, sperm-mediated transgenesis, embryo cloning by nuclear transfer, and germline mutagenesis (Tesson *et al.*, 2005) have shown success in the rat and can be used to validate a candidate gene or to study a gene's function. A number of inbred strains of rats (582) as well as congenic (381), consomic (45), and more recently transgenic (31) strains (numbers registered in the Rat Genome Database; http://rgd.mcw.edu/) provide a wide variety of genetic backgrounds for transgenic experiments.

An alternative approach to using ES cells in the rat is to use chemical mutagens such as *N*-ethyl-*N*-nitrosourea (ENU) to knock out a gene. While ENU mutagenesis has been utilized for many years, the past 10 years have seen an increase in its use to generate gene mutations (Guenet, 2004). There has been an increasing emphasis on large-scale ENU screens for systematic and comprehensive gene function analyses of the mouse genome (Brown and Balling, 2001; Hrabe de Angelis, 1998; Nadeau, 2000; Nolan *et al.*, 2000). The mouse mutagenesis programs are vastly phenotype-driven, employing appropriate phenotypic screens on mutagenized animals to identify novel mutant phenotypes, particularly those that model human disease, followed by mapping and isolation of the underlying causal genes. An alternative strategy is to screen mutagenized animals for a gene(s) of interest, keeping only the animals having a particular mutation of interest. This strategy has been used to knock out several genes in the rat (Homberg *et al.*, 2005; Zan *et al.*, 2003). We expect this methodology to gain in popularity until such time as there are ES cells for rats, and remain an alternative approach when genome backgrounds affect a phenotype.

Silencing of gene expression by RNA interference (RNAi) has become a powerful tool for functional genomics and a potential therapeutic approach to human disease (Kim and Rossi, 2007). One can expect in coming years an explosion of studies focusing on the design, effective delivery, and mechanism of RNAi as effector molecules of sequence-specific gene silencing in the whole animal.

B. Humanizing disease models

The ability to manipulate the genome via genetic engineering enables the alteration of specific gene expression as well as to use animals as surrogate hosts for expression of genes from other species. This is particularly relevant for the study of specific alleles of a gene or if a gene is not part of the model system's genome.

Humanized animal models have been developed as an important research tool for the *in vivo* studies as well as ancillary studies of cells and tissues from these animals (Shultz *et al.*, 2007). Importantly, modeling in transgenic animals has been shown to provide a bridge between genetic linkage studies in humans and functional association of a (mutant) gene with particular pathological feature. This approach was used to study diseases such as narcolepsy (Beuckmann *et al.*, 2004), Alzheimer's disease (Carvajal *et al.*, 2004; Czech *et al.*, 1998; Echeverria *et al.*, 2004; Gibbs *et al.*, 2004), Hirschsprung's disease (Gariepy *et al.*, 1998), Huntington's disease (Petrasch-Parwez *et al.*, 2007), amyotrophic lateral sclerosis (Howland *et al.*, 2002; Matsumoto *et al.*, 2006), and infectious diseases (Lassnig *et al.*, 2005). These examples provide a proof-of-principle that human disease modeling in animals is valuable and one can expect that the etiology of other diseases will be similarly elucidated using comparative genomics and transgenesis. Moreover, transgenic models expressing a human gene enable additional studies such as longitudinal *in vivo* studies of disease progression through the natural life span of the model system, as well as help to monitor the effects of long-term treatments, cell implantation, microsurgery, or antisense approaches along the course of the induced disease. With almost a complete sequence of >350 organisms (most are single cell organisms) (Yutzey and Robbins, 2007), comparative genomics may enable the use of tissue-specific promoters/enhancer sequences to recapitulate normal or altered expression in a gene of interest. Mice are currently the preferred model organism (Peters *et al.*, 2007) for the development of humanized models due to the ability to use ES cells.

Over the past 10 years, many groups have tried to engineer minichromosomes or human artificial chromosome (Basu and Willard, 2005; Otsuki *et al.*, 2005; Suzuki *et al.*, 2006) as an alternative vector for nonintegrating gene-delivery platform. While many obstacles remain for their use in gene delivery, one syndrome model mouse was created harboring human chromosome 21 in addition to the normal mouse karyotype (O'Doherty *et al.*, 2005) demonstrating how the proposed therapeutic strategy of gene delivery can be used to engineer new models.

The role of animal models will still be crucial for understanding the physiology and pathophysiology of disease. It is important to keep in mind that each model offers only partial information with undefined clinical value in advance of translational studies. Therefore, a more critical attitude and repetition of crucial observations will be necessary in the future, for example, in multiple model systems or in different environments/genome backgrounds.

C. Comparative QTL

Most rat and mouse models of common, multifactorial disease reflect a clinical phenotype, and several comparative mapping studies have determined that common phenotypes often mapped to conserved genomic regions between rodents and human (Korstanje and DiPetrillo, 2004; Stoll *et al.*, 2000; Sugiyama *et al.*, 2001; Wang and Paigen, 2005), increasing the potential to identify animal models that better match human disease both at the phenotypic level and at the genotypic level. The availability of the genomic sequence of rat and many other species introduces the possibility of comparative genome analysis at a nucleotide level rather than the previous approach of identifying conserved synteny based upon low resolution ordering of orthologous genes. In 2000, Stoll, *et al.* (2000) reported that QTLs in the rat could be used to predict the locations that human QTLs for similar traits were likely to exist, by identifying traits that map to evolutionarily conserved regions in both organisms. Since then, numerous other studies have demonstrated that there are evolutionarily conserved regions in human, mouse, and rat that are linked to the same phenotype in all three species (Jacob and Kwitek, 2002; Korstanje and DiPetrillo, 2004; Stoll *et al.*, 2000; Sugiyama *et al.*, 2001). These include disorders such as obesity, hypertension, autoimmunity, and coronary heart disease, to name a few (Jacob and Kwitek, 2002; Rollins *et al.*, 2006). Over the next 5 years, we can expect a large increase in the number of studies using cross-species comparisons to find causes of common complex diseases (Glazier *et al.*, 2002; Korstanje and Paigen, 2002). Approximately 100 papers report a particular disease trait maps to the same conserved region across rodents and human, illustrating the near-term benefits of the rat genome project. However, one must keep in mind that QTLs are large and numerous for multifactorial traits, such as behavior and metabolic syndrome. Therefore, overlapping QTL may sometimes be merely a chance event. To address this issue, animal models can be more extensively evaluated for disease subphenotypes to better match QTLs by intermediate phenotype. Furthermore, with the emergence of high-density SNP maps in rodents, fine resolution mapping may reduce the size of a QTL. Finally, comparative mapping data from additional species such as dog, cow, or other models might be used to confirm conserved QTL and assist in refining the critical genomic region. For example, hip dysplasia can be a risk factor for arthritis in dog. Two QTL have been mapped on canine chromosome 1 for degenerative joint disease (DJD) (Chase *et al.*, 2004). By selecting the flanking markers from the linkage results, we can see that DJD QTL FH2598 is conserved with a fragment of human chromosome 19. Interestingly, the dog QTL aligns with two QTL for osteoarthritis in human (Demissie *et al.*, 2002). Furthermore, two QTL for phenotypes related to human arthritis map to the conserved regions of rat on chromosome 1: Cia2—collagen-induced arthritis 2 (Remmers *et al.*, 1996) and Pia11—pristine-induced arthritis 11 (Lu *et al.*, 2002).

Integration of the genome sequence with existing mapping data and the biological data attached to those maps, plus the creation and annotation of a comprehensive catalog of gene products, will increase the use of such comparative studies in translational research.

D. Genes positionally cloned

Traditional positional cloning efforts in the rat have been coming to fruition in identifying disease genes. Numerous genes have now been identified in the rat by positional cloning, concurrent with the great increase in rat genomic resources. These include genes for cancer (BHD, Tsc2) (Okimoto *et al.*, 2004; Yeung *et al.*, 1994); type 1 diabetes (Gimap5, Cblb) (MacMurray *et al.*, 2002; Yokoi *et al.*, 2002); type 2 diabetes (Cd36) (Aitman *et al.*, 1999); neurological disorders (Cct4, Reln, Unc5h3) (Kuramoto *et al.*, 2004; Lee *et al.*, 2003; Yokoi *et al.*, 2003); arthritis (Ncf1) (Olofsson *et al.*, 2003b); renal disease (Pkhd1, Rab38) (Ward *et al.*, 2002), (Fcgr3) (Petretto *et al.*, 2006); bleeding disorders (Rab38, VKOR) (Oiso *et al.*, 2004; Rost *et al.*, 2004); retinal degeneration (Mertk) (Gal *et al.*, 2000); and hypotrichosis (Dsg4, Whn) (Jahoda *et al.*, 2004; Segre *et al.*, 1995). Many of these genes were cloned from spontaneous mutants with Mendelian inheritance of disease, for example, the Pkdh1 mutation in the PCK rat causes autosomal recessive polycystic kidney disease (ARPKD). Several genes underlying QTL have been identified in the mouse as well, for traits relating to growth, plasma cholesterol levels, inflammation related to asthma, neprotic syndrome, and irritable bowel disease (Cho *et al.*, 2006; de Buhr *et al.*, 2006; Oliver *et al.*, 2005; Petretto *et al.*, 2006; Suto, 2005). Utilizing an integrated approach, using genetic mapping, gene expression, and comparative genetics/genomics, promises a bright future for identifying genes predisposing risk for common human disease.

V. DISCUSSION

Comparative genomics and comparative medicine are relatively new fields that complement the deep and rich history of comparison-based disciplines in biology and genetics. The recent determination of the genomic sequence for multiple organisms has aided in global gene predictions and determining gene regulation, as well as in the identification of elements of the sequences that are highly conserved but whose functions are currently undefined. The availability of dense genetic maps in multiple species, the genomic sequence for 38 mammals, and the strong evidence of QTLs being conserved across multiple species suggest that comparative medicine/genomics will begin to unravel the biological basis of common complex disease. When the genes are identified, it is reasonable to assume that some of these genes will play a role in human disease.

The need for a multiple species platform, integrated at both the genotypic level and the phenotypic level, will become increasingly apparent as whole-genome-wide association studies begin to implicate large numbers of loci, each contributing a small portion to the overall phenotype. Given this complexity, it will not be economical to generate "humanized" models with current technologies; furthermore, multifactorial disease is likely to be a more complex scenario than the additive action of single SNPs. We believe that the study of conserved QTL, which has the ability to integrate biology, physiology, pathobiology, and pharmacology across related genomes, has the potential to dissect complex interactions, including the environment. In this process, investigators will finally integrate genetics, genomics, and biology into a format that increases our understanding of human disease—the ultimate comparative medicine platform.

One example of how genomic and phenotypic information in the rat is integrated and related to the human is provided in Fig. 23.3. The availability of many genetically characterized strains of rat, each with unique phenotypic characteristics, allows for the selection of the model closest to the disease of interest to study the genetic basis of complex disease, and to further identify the genomic regions and ultimately the identification of the causative disease gene (s). Through the use of comparative genomics at the level of QTL or sequence, and data integration, results can be related to human systems and to the human genome, allowing a direct biological connection between species.

We can expect numerous discoveries coming from many different organisms over the next several decades. Looking at the landscape of comparative genomics and the growing number of tools and data sets, it would be tempting to predict the need for additional genome sequencing efforts will fade, that comparative genomics will be equivalent to additional computational analyses. However, current genome sequences are really references, missing the vast sequence variation inherent between humans as well as between model

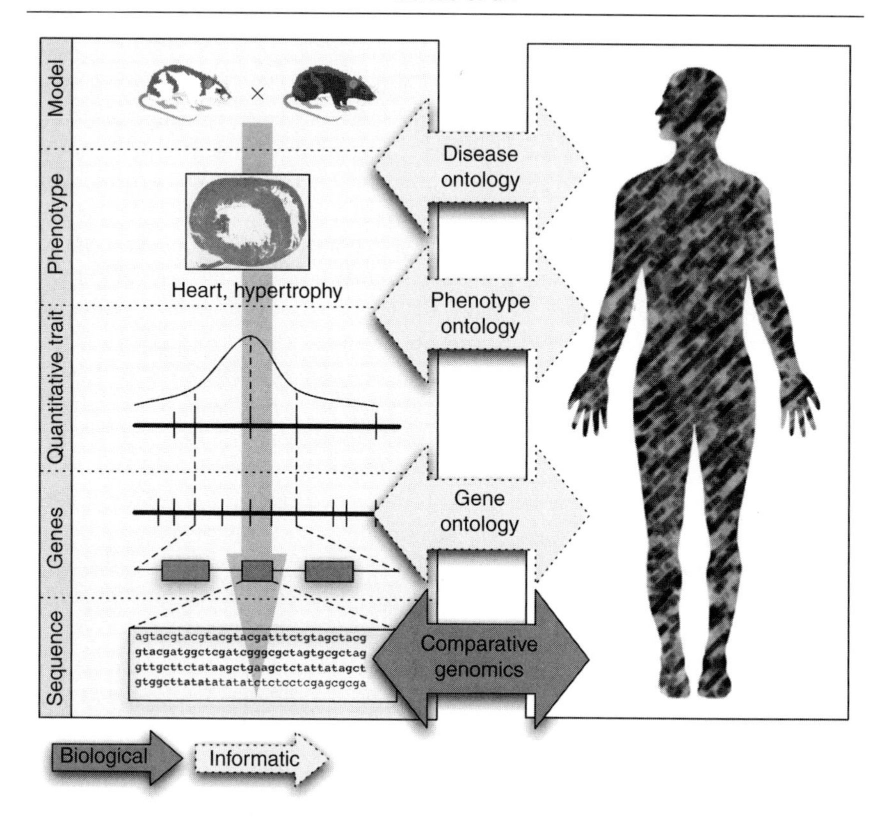

Figure 23.3. Overview of physiological genomics in the rat and the various translational technologies available to relate rat research to human systems. *Left*: Traditional positional cloning techniques whereby two rats, one possessing the phenotype of interest and the other being a nonaffected control strain, are crossed, and the progeny are genotyped and phenotyped leading to the definition of quantitative trait loci (QTLs) linking the phenotype to specific regions of the genome. The regions are then examined for potential candidate genes, and comparative genomics is used to integrate evidence from other organisms or to translate results to the human genome. Translation of information to the human system is also becoming possible using informatic tools such as ontologies. Various ontologies are shown connecting elements of the experimental paradigm to related human data. Figure and legend reproduced with permission of Twigger *et al.* (2005).

strains. The likelihood that sequencing costs drop to a point in which the entire genomic sequence can be determined for every person or individual animals makes it feasible to further understand the function of the genome, much of which we still consider "junk DNA."

The possibility of having individual sequenced genomes truly potentiates comparative medicine within and between species. With this, the Rosetta stone in genomics will finally be complete. However, the interpretation of the stone with respect to a gene's function and role in the pathogenesis of disease will still be required. Now is the time to reconsider how we use animal models to understand human disease.

In 2007, animal models are selected primarily on one of two criteria: (1) a gene or its expression is modified or can be modified and its implication in a human disease can be studied in a particular model system; (2) the physiological characteristics of the model systems reflect some aspect of the clinical picture. There has been some movement toward other considerations such as genome background and effects of the environment (e.g., diet). It appears that the added benefit of comparative genomics is seldom considered when a model is selected. With a large number of QTL genetically mapped in many different species, and many mapping to the same evolutionarily conserved regions, it seems reasonable that many of the same genes will play a role in the same disease process. Comparative mapping indicates that QTL for hypertension map to evolutionarily conserved regions in the rat and two human populations to the same conserved region of the genome (Cowley *et al.*, 2001; Hamet *et al.*, 2005). While it is unlikely that this single model system reflects all of human hypertension, we would suggest that it could be considered model for a subset of hypertension. We predict that these types of considerations based on comparative genomics are particularly relevant to the development of new therapeutics.

Within the pharmaceutical industry, preclinical drug studies for efficacy and safety are carried out in one strain of a given species, typically a single outbred strain for safety testing and perhaps two different models of the disease. These data are then used to extrapolate not only across the species as a whole (e.g., all rats or all mice) but also across to human. It is time to reconsider how animal models are selected for drug studies. As outlined in this chapter, comparative tools are available to markedly advance the selection of animal models to understand human disease. The dawn of comparative medicine is here, how long will it take to be adopted is the only remaining question.

References

The International HapMap Project (2003). *Nature* **426**(6968), 789–796.

Abiola, O., Angel, J. M., Avner, P., Bachmanov, A. A., Belknap, J. K., Bennett, B., Blankenhorn, E. P., Blizard, D. A., Bolivar, V., Brockmann, G. A., Buck, K. J., Bureau, J. F., *et al.* (2003). The nature and identification of quantitative trait loci: A community's view. *Nat. Rev. Genet.* **4**(11), 911–916.

Abrahams, B. S., Mak, G. M., Berry, M. L., Palmquist, D. L., Saionz, J. R., Tay, A., Tan, Y. H., Brenner, S., Simpson, E. M., and Venkatesh, B. (2002). Novel vertebrate genes and putative regulatory elements identified at kidney disease and NR2E1/fierce loci. *Genomics* **80**(1), 45–53.

Aitman, T. J., Glazier, A. M., Wallace, C. A., Cooper, L. D., Norsworthy, P. J., Wahid, F. N., Al-Majali, K. M., Trembling, P. M., Mann, C. J., Shoulders, C. C., *et al.* (1999). Identification of Cd36 (Fat) as an insulin-resistance gene causing defective fatty acid and glucose metabolism in hypertensive rats. *Nat. Genet.* **21**(1), 76–83.

Anderlik, M. (2003). Commercial biobanks and genetic research: Ethical and legal issues. *Am. J. Pharmacogenomics* 3(3), 203–215.

Ariyarajah, A., Palijan, A., Dutil, J., Prithiviraj, K., Deng, Y., and Deng, A. Y. (2004). Dissecting quantitative trait loci into opposite blood pressure effects on Dahl rat chromosome 8 by congenic strains. *J. Hypertens.* **22**(8), 1495–1502.

Austin, C. P., Battey, J. F., Bradley, A., Bucan, M., Capecchi, M., Collins, F. S., Dove, W. F., Duyk, G., Dymecki, S., Eppig, J. T., *et al.* (2004). The knockout mouse project. *Nat. Genet.* **36**(9), 921–924.

Bagheri-Fam, S., Ferraz, C., Demaille, J., Scherer, G., and Pfeifer, D. (2001). Comparative genomics of the SOX9 region in human and Fugu rubripes: Conservation of short regulatory sequence elements within large intergenic regions. *Genomics* **78**(1–2), 73–82.

Bailey, D. W. (1971). Recombinant-inbred strains. An aid to identify, linkage, and function of histocompatibility and other genes. *Transplantation* **11,** 325–327.

Basu, J., and Willard, H. F. (2005). Artificial and engineered chromosomes: Non-integrating vectors for gene therapy. *Trends Mol. Med.* **11**(5), 251–258.

Berghmans, S., Jette, C., Langenau, D., Hsu, K., Stewart, R., Look, T., and Kanki, J. P. (2005). Making waves in cancer research: New models in the zebrafish. *Biotechniques* **39**(2), 227–237.

Bergsteinsdottir, K., Yang, H. T., Pettersson, U., and Holmdahl, R. (2000). Evidence for common autoimmune disease genes controlling onset, severity, and chronicity based on experimental models for multiple sclerosis and rheumatoid arthritis. *J. Immunol.* **164**(3), 1564–1568.

Bertone, P., Gerstein, M., and Snyder, M. (2005). Applications of DNA tiling arrays to experimental genome annotation and regulatory pathway discovery. *Chromosome Res.* **13**(3), 259–274.

Beuckmann, C. T., Sinton, C. M., Williams, S. C., Richardson, J. A., Hammer, R. E., Sakurai, T., and Yanagisawa, M. (2004). Expression of a poly-glutamine-ataxin-3 transgene in orexin neurons induces narcolepsy-cataplexy in the rat. *J. Neurosci.* **24**(18), 4469–4477.

Bier, E. (2005). *Drosophila*, the golden bug, emerges as a tool for human genetics. *Nat. Rev. Genet.* **6**(1), 9–23.

Bila, V., and Kren, V. (1996). The teratogenic action of retinoic acid in rat congenic and recombinant inbred strains. *Folia Biol. (Praha)* **42**(4), 167–173.

Bockamp, E., Maringer, M., Spangenberg, C., Fees, S., Fraser, S., Eshkind, L., Oesch, F., and Zabel, B. (2002). Of mice and models: Improved animal models for biomedical research. *Physiol. Genomics* **11**(3), 115–132.

Boffelli, D., McAuliffe, J., Ovcharenko, D., Lewis, K. D., Ovcharenko, I., Pachter, L., and Rubin, E. M. (2003). Phylogenetic shadowing of primate sequences to find functional regions of the human genome. *Science* **299**(5611), 1391–1394.

Bogue, M. (2003). Mouse Phenome Project: Understanding human biology through mouse genetics and genomics. *J. Appl. Physiol.* **95**(4), 1335–1337.

Bogue, M. A., Grubb, S. C., Maddatu, T. P., and Bult, C. J. (2007). Mouse Phenome Database (MPD). *Nucleic Acids Res.* **35**(Database issue), D643–D649.

Brenner, S., Elgar, G., Sandford, R., Macrae, A., Venkatesh, B., and Aparicio, S. (1993). Characterization of the pufferfish (Fugu) genome as a compact model vertebrate genome. *Nature* **366**(6452), 265–268.

Brown, S. D., and Balling, R. (2001). Systematic approaches to mouse mutagenesis. *Curr. Opin. Genet. Dev.* **11**(3), 268–273.

Bystrykh, L., Weersing, E., Dontje, B., Sutton, S., Pletcher, M. T., Wiltshire, T., Su, A. I., Vellenga, E., Wang, J., Manly, K. F., *et al.* (2005). Uncovering regulatory pathways that affect hematopoietic stem cell function using 'genetical genomics'. *Nat. Genet.* **37**(3), 225–232.

Caligari, P. D., and Mather, K. (1975). Genotype–environment interaction. III. Interactions in *Drosophila* melanogaster. *Proc. R. Soc. Lond. B Biol. Sci.* **191**(1104), 387–411.

Carvajal, C. C., Vercauteren, F., Dumont, Y., Michalkiewicz, M., and Quirion, R. (2004). Aged neuropeptide Y transgenic rats are resistant to acute stress but maintain spatial and non-spatial learning. *Behav. Brain Res.* **153**(2), 471–480.

Castle, W., and Wachter, W. (1924). Variations of linkage in rats and mice. *Genetics* **9**, 1–12.

Cervino, A. C., Li, G., Edwards, S., Zhu, J., Laurie, C., Tokiwa, G., Lum, P. Y., Wang, S., Castellini, L. W., Lusis, A. J., *et al.* (2005). Integrating QTL and high-density SNP analyses in mice to identify Insig2 as a susceptibility gene for plasma cholesterol levels. *Genomics* **86**(5), 505–517.

Cervino, A. C., Gosink, M., Fallahi, M., Pascal, B., Mader, C., and Tsinoremas, N. F. (2006). A comprehensive mouse IBD database for the efficient localization of quantitative trait loci. *Mamm. Genome* **17**(6), 565–574.

Chase, K., Lawler, D. F., Adler, F. R., Ostrander, E. A., and Lark, K. G. (2004). Bilaterally asymmetric effects of quantitative trait loci (QTLs): QTLs that affect laxity in the right versus left coxofemoral (hip) joints of the dog (Canis familiaris). *Am. J. Med. Genet.* A **124**(3), 239–247.

Cheng, S. V., Nadeau, J. H., Tanzi, R. E., Watkins, P. C., Jagadesh, J., Taylor, B. A., Haines, J. L., Sacchi, N., and Gusella, J. F. (1988). Comparative mapping of DNA markers from the familial Alzheimer disease and Down syndrome regions of human chromosome 21 to mouse chromosomes 16 and 17. *Proc. Natl. Acad. Sci. USA* **85**(16), 6032–6036.

Cho, A. R., Uchio-Yamada, K., Torigai, T., Miyamoto, T., Miyoshi, I., Matsuda, J., Kurosawa, T., Kon, Y., Asano, A., Sasaki, N., *et al.* (2006). Deficiency of the tensin2 gene in the ICGN mouse: An animal model for congenital nephrotic syndrome. *Mamm. Genome* **17**(5), 407–416.

Churchill, G. A., Airey, D. C., Allayee, H., Angel, J. M., Attie, A. D., Beatty, J., Beavis, W. D., Belknap, J. K., Bennett, B., Berrettini, W., *et al.* (2004). The Collaborative Cross, a community resource for the genetic analysis of complex traits. *Nat. Genet.* **36**(11), 1133–1137.

Clark, M. (1999). Comparative genomics: The key to understanding the Human Genome Project. *BioEssays* **21**(2), 121–130.

Conti, L. H., Jirout, M., Breen, L., Vanella, J. J., Schork, N. J., and Printz, M. P. (2004). Identification of quantitative trait Loci for anxiety and locomotion phenotypes in rat recombinant inbred strains. *Behav. Genet.* **34**(1), 93–103.

Cooper, G. M., and Sidow, A. (2003). Genomic regulatory regions: Insights from comparative sequence analysis. *Curr. Opin. Genet. Dev.* **13**(6), 604–610.

Coughlan, C. M., and Brodsky, J. L. (2005). Use of yeast as a model system to investigate protein conformational diseases. *Mol. Biotechnol.* **30**(2), 171–180.

Couronne, O., Poliakov, A., Bray, N., Ishkhanov, T., Ryaboy, D., Rubin, E., Pachter, L., and Dubchak, I. (2003). Strategies and tools for whole-genome alignments. *Genome Res.* **13**(1), 73–80.

Cowley, A. W., Jr., Roman, R. J., Kaldunski, M. L., Dumas, P., Dickhout, J. G., Greene, A. S., and Jacob, H. J. (2001). Brown Norway chromosome 13 confers protection from high salt to consomic Dahl S rat. *Hypertension* **37**(2 Part 2), 456–461.

Cowley, A. W., Jr., Liang, M., Roman, R. J., Greene, A. S., and Jacob, H. J. (2004a). Consomic rat model systems for physiological genomics. *Acta Physiol. Scand.* **181**(4), 585–592.

Cowley, A. W., Jr., Roman, R. J., and Jacob, H. J. (2004b). Application of chromosomal substitution techniques in gene-function discovery. *J. Physiol.* **554**(Pt 1), 46–55.

Czech, C., Lesort, M., Tremp, G., Terro, F., Blanchard, V., Schombert, B., Carpentier, N., Dreisler, S., Bonici, B., Takashima, A., *et al.* (1998). Characterization of human presenilin 1 transgenic rats: Increased sensitivity to apoptosis in primary neuronal cultures. *Neuroscience* **87**(2), 325–336.

Dahlman, I., Jacobsson, L., Glaser, A., Lorentzen, J. C., Andersson, M., Luthman, H., and Olsson, T. (1999a). Genome-wide linkage analysis of chronic relapsing experimental autoimmune encephalomyelitis in the rat identifies a major susceptibility locus on chromosome 9. *J. Immunol.* **162**(5), 2581–2588.

Dahlman, I., Wallstrom, E., Weissert, R., Storch, M., Kornek, B., Jacobsson, L., Linington, C., Luthman, H., Lassmann, H., and Olsson, T. (1999b). Linkage analysis of myelin oligodendrocyte glycoprotein-induced experimental autoimmune encephalomyelitis in the rat identifies a locus controlling demyelination on chromosome 18. *Hum. Mol. Genet.* **8**(12), 2183–2190.

Darvasi, A., and Soller, M. (1995). Advanced intercross lines, an experimental population for fine genetic mapping. *Genetics* **141**(3), 1199–1207.

de Buhr, M. F., Mahler, M., Geffers, R., Hansen, W., Westendorf, A. M., Lauber, J., Buer, J., Schlegelberger, B., Hedrich, H. J., and Bleich, A. (2006). Cd14, Gbp1, and Pla2g2a: Three major candidate genes for experimental IBD identified by combining QTL and microarray analyses. *Physiol. Genomics* **25**(3), 426–434.

De Miglio, M. R., Pascale, R. M., Simile, M. M., Muroni, M. R., Calvisi, D. F., Virdis, P., Bosinco, G. M., Frau, M., Seddaiu, M. A., Ladu, S., *et al.* (2002). Chromosome mapping of multiple loci affecting the genetic predisposition to rat liver carcinogenesis. *Cancer Res.* **62**(15), 4459–4463.

de Wolf, I. D., Bonne, A. C., Fielmich-Bouman, X. M., van Oost, B. A., Beynen, A. C., van Zutphen, L. F., and van Lith, H. A. (2002). Quantitative trait loci influencing hepatic copper in rats. *Exp. Biol. Med. (Maywood)* **227**(7), 529–534.

Dehal, P., Predki, P., Olsen, A. S., Kobayashi, A., Folta, P., Lucas, S., Land, M., Terry, A., Ecale Zhou, C. L., Rash, S., *et al.* (2001). Human chromosome 19 and related regions in mouse: Conservative and lineage-specific evolution. *Science* **293**(5527), 104–111.

Demarest, K., Koyner, J., McCaughran, J., Jr., Cipp, L., and Hitzemann, R. (2001). Further characterization and high-resolution mapping of quantitative trait loci for ethanol-induced locomotor activity. *Behav. Genet.* **31**(1), 79–91.

Demissie, S., Cupples, L. A., Myers, R., Aliabadi, P., Levy, D., and Felson, D. T. (2002). Genome scan for quantity of hand osteoarthritis: The Framingham Study. *Arthritis Rheum.* **46** (4), 946–952.

Depew, M. J., Lufkin, T., and Rubenstein, J. L. (2002). Specification of jaw subdivisions by Dlx genes. *Science* **298**(5592), 381–385.

Desnick, R., Patterson, D., and Scarpelli, D. (eds.) (1982). "Animal Models of Inborn Errors of Metabolism." Alan R. Liss, New York.

Dong, J., and Stuart, G. W. (2004). Transgene manipulation in zebrafish by using recombinases. *Methods Cell Biol.* **77**, 363–379.

Dooley, K., and Zon, L. I. (2000). Zebrafish: A model system for the study of human disease. *Curr. Opin. Genet. Dev.* **10**(3), 252–256.

Drummond, I. A. (2005). Kidney development and disease in the zebrafish. *J. Am. Soc. Nephrol.* **16**(2), 299–304.

Dubchak, I., Brudno, M., Loots, G. G., Pachter, L., Mayor, C., Rubin, E. M., and Frazer, K. A. (2000). Active conservation of noncoding sequences revealed by three-way species comparisons. *Genome Res.* **10**(9), 1304–1306.

Echeverria, V., Ducatenzeiler, A., Alhonen, L., Janne, J., Grant, S. M., Wandosell, F., Muro, A., Baralle, F., Li, H., Duff, K., *et al.* (2004). Rat transgenic models with a phenotype of intracellular A beta accumulation in hippocampus and cortex. *J. Alzheimers Dis.* **6**(3), 209–219.

Endo, K., Yanagi, H., Araki, J., Hirano, C., Yamakawa-Kobayashi, K., and Tomura, S. (2002). Association found between the promoter region polymorphism in the apolipoprotein A-V gene and the serum triglyceride level in Japanese schoolchildren. *Hum. Genet.* **111**(6), 570–572.

Ewart-Toland, A., Briassouli, P., de Koning, J. P., Mao, J. H., Yuan, J., Chan, F., MacCarthy-Morrogh, L., Ponder, B. A., Nagase, H., Burn, J., *et al.* (2003). Identification of Stk6/STK15 as a candidate low-penetrance tumor-susceptibility gene in mouse and human. *Nat. Genet.* **34**(4), 403–412.

Fernandez-Ricaud, L., Warringer, J., Ericson, E., Glaab, K., Davidsson, P., Nilsson, F., Kemp, G. J., Nerman, O., and Blomberg, A. (2007). PROPHECY—a yeast phenome database, update (2006). *Nucleic Acids Res.* **35**(Database issue), D463–D467.

Fernandez-Teruel, A., Escorihuela, R. M., Gray, J. A., Aguilar, R., Gil, L., Gimenez-Llort, L., Tobena, A., Bhomra, A., Nicod, A., Mott, R., *et al.* (2002). A quantitative trait locus influencing anxiety in the laboratory rat. *Genome Res.* **12**(4), 618–626.

Fitch, W. (2000). Homology: A personal view on some of the problems. *Trends Genet.* **16**(5), 227–231.

Flint, J. (2003). Analysis of quantitative trait loci that influence animal behavior. *J. Neurobiol.* **54**(1), 46–77.

Flint, J., Valdar, W., Shifman, S., and Mott, R. (2005). Strategies for mapping and cloning quantitative trait genes in rodents. *Nat. Rev. Genet.* **6**(4), 271–286.

Freimer, N., and Sabatti, C. (2003). The human phenome project. *Nat. Genet.* **34**(1), 15–21.

Gal, A., Li, Y., Thompson, D. A., Weir, J., Orth, U., Jacobson, S. G., Apfelstedt-Sylla, E., and Vollrath, D. (2000). Mutations in MERTK, the human orthologue of the RCS rat retinal dystrophy gene, cause retinitis pigmentosa. *Nat. Genet.* **26**(3), 270–271.

Galli, J., Li, L. S., Glaser, A., Ostenson, C. G., Jiao, H., Fakhrai-Rad, H., Jacob, H. J., Lander, E. S., and Luthman, H. (1996). Genetic analysis of non-insulin dependent diabetes mellitus in the GK rat. *Nat. Genet.* **12**(1), 31–37.

Ganguli, M., Tobian, L., and Iwai, J. (1979). Cardiac output and peripheral resistance in strains of rats sensitive and resistant to NaCl hypertension. *Hypertension* **1**(1), 3–7.

Gariepy, C. E., Williams, S. C., Richardson, J. A., Hammer, R. E., and Yanagisawa, M. (1998). Transgenic expression of the endothelin-B receptor prevents congenital intestinal aganglionosis in a rat model of Hirschsprung disease. *J. Clin. Invest.* **102**(6), 1092–1101.

Gerhard, D. S., Wagner, L., Feingold, E. A., Shenmen, C. M., Grouse, L. H., Schuler, G., Klein, S. L., Old, S., Rasooly, R., Good, P., *et al.* (2004). The status, quality, and expansion of the NIH full-length cDNA project: The Mammalian Gene Collection (MGC). *Genome Res.* **14**(10B), 2121–2127.

Ghanem, N., Jarinova, O., Amores, A., Long, Q., Hatch, G., Park, B. K., Rubenstein, J. L., and Ekker, M. (2003). Regulatory roles of conserved intergenic domains in vertebrate Dlx bigene clusters. *Genome Res.* **13**(4), 533–543.

Gibbs, R. A., Weinstock, G. M., Metzker, M. L., Muzny, D. M., Sodergren, E. J., Scherer, S., Scott, G., Steffen, D., Worley, K. C., Burch, P. E., *et al.* (2004). Genome sequence of the Brown Norway rat yields insights into mammalian evolution. *Nature* **428**(6982), 493–521.

Gilligan, P., Brenner, S., and Venkatesh, B. (2002). Fugu and human sequence comparison identifies novel human genes and conserved non-coding sequences. *Gene* **294**(1–2), 35–44.

Glazier, A. M., Nadeau, J. H., and Aitman, T. J. (2002). Finding genes that underlie complex traits. *Science* **298**(5602), 2345–2349.

Goldstein, J. L. (2001). Laskers for 2001: Knockout mice and test-tube babies. *Nat. Med.* **7**(10), 1079–1080.

Gottgens, B., Barton, L. M., Chapman, M. A., Sinclair, A. M., Knudsen, B., Grafham, D., Gilbert, J. G., Rogers, J., Bentley, D. R., and Green, A. R. (2002). Transcriptional regulation of the stem cell leukemia gene (SCL)—comparative analysis of five vertebrate SCL loci. *Genome Res.* **12**(5), 749–759.

Grisart, B., Coppieters, W., Farnir, F., Karim, L., Ford, C., Berzi, P., Cambisano, N., Mni, M., Reid, S., Simon, P., *et al.* (2002). Positional candidate cloning of a QTL in dairy cattle: Identification of a missense mutation in the bovine DGAT1 gene with major effect on milk yield and composition. *Genome Res.* **12**(2), 222–231.

Grupe, A., Germer, S., Usuka, J., Aud, D., Belknap, J., Klein, R., Ahluwalia, M., Higuchi, R., and Peltz, G. (2001). In silico mapping of complex disease-related traits in mice. *Science* **292**(5523), 1915–1918.

Guenet, J. L. (2004). Chemical mutagenesis of the mouse genome: An overview. *Genetica* **122**(1), 9–24.

Guryev, V., Smits, B. M., van de Belt, J., Verheul, M., Hubner, N., and Cuppen, E. (2006). Haplotype block structure is conserved across mammals. *PLoS Genet.* **2**(7), e121.

Guyon, J. R., Steffen, L. S., Howell, M. H., Pusack, T. J., Lawrence, C., and Kunkel, L. M. (2007). Modeling human muscle disease in zebrafish. *Biochim. Biophys. Acta* **1772**(2), 205–215.

Haldane, J. (1927). The comparative genetics of color in rodents and carnivora. *Biol. Rev. Camb. Philos. Soc.* **2,** 199–212.

Hamet, P., Merlo, E., Seda, O., Broeckel, U., Tremblay, J., Kaldunski, M., Gaudet, D., Bouchard, G., Deslauriers, B., Gagnon, F., *et al.* (2005). Quantitative founder-effect analysis of French Canadian families identifies specific loci contributing to metabolic phenotypes of hypertension. *Am. J. Hum. Genet.* **76**(5), 815–832.

Hansen, C., and Spuhler, K. (1984). Development of the National Institutes of Health genetically heterogeneous rat stock. *Alcohol. Clin. Exp. Res.* **8**(5), 477–479.

Hanson, R. L., Craig, D. W., Millis, M. P., Yeatts, K. A., Kobes, S., Pearson, J. V., Lee, A. M., Knowler, W. C., Nelson, R. G., and Wolford, J. K. (2007). Identification of PVT1 as a candidate gene for end-stage renal disease in type 2 diabetes using a pooling-based genome-wide single nucleotide polymorphism association study. *Diabetes* **56**(4), 975–983.

Hardison, R., Slightom, J. L., Gumucio, D. L., Goodman, M., Stojanovic, N., and Miller, W. (1997). Locus control regions of mammalian beta-globin gene clusters: Combining phylogenetic analyses and experimental results to gain functional insights. *Gene* **205**(1–2), 73–94.

Hentschel, D. M., Park, K. M., Cilenti, L., Zervos, A. S., Drummond, I., and Bonventre, J. V. (2005). Acute renal failure in zebrafish: A novel system to study a complex disease. *Am. J. Physiol. Renal Physiol.* **288**(5), F923–F929.

Hillebrandt, S., Wasmuth, H. E., Weiskirchen, R., Hellerbrand, C., Keppeler, H., Werth, A., Schirin-Sokhan, R., Wilkens, G., Geier, A., Lorenzen, J., *et al.* (2005). Complement factor 5 is a quantitative trait gene that modifies liver fibrogenesis in mice and humans. *Nat. Genet.* **37**(8), 835–843.

Homberg, J. R., Olivier, J. D., Smits, B., Mudde, J., Cools, A. R., Ellenbroek, B. A., and Cuppen, E. (2005). O9 phenotyping of the serotonin transporter knockout rat. *Behav. Pharmacol.* **16**(Suppl 1), S21.

Hood, L., Rowen, L., and Koop, B. F. (1995). Human and mouse T-cell receptor loci: Genomics, evolution, diversity, and serendipity. *Ann. N. Y. Acad. Sci.* **758,** 390–412.

Howland, D. S., Liu, J., She, Y., Goad, B., Maragakis, N. J., Kim, B., Erickson, J., Kulik, J., DeVito, L., Psaltis, G., *et al.* (2002). Focal loss of the glutamate transporter EAAT2 in a transgenic rat model of SOD1 mutant-mediated amyotrophic lateral sclerosis (ALS). *Proc. Natl. Acad. Sci. USA* **99**(3), 1604–1609.

Hrabe de Angelis, M. B. R. (1998). Large scale ENU screens in the mouse: Genetics meets genomics. *Mutat. Res.* **400**(1–2), 25–32.

Hubner, N., Wallace, C. A., Zimdahl, H., Petretto, E., Schulz, H., Maciver, F., Mueller, M., Hummel, O., Monti, J., Zidek, V., *et al.* (2005). Integrated transcriptional profiling and linkage analysis for identification of genes underlying disease. *Nat. Genet.* **37**(3), 243–253.

Igarashi, P. (2005). Overview: Nonmammalian organisms for studies of kidney development and disease. *J. Am. Soc. Nephrol.* **16**(2), 296–298.

Jacob, H. J., and Kwitek, A. E. (2002). Rat genetics: Attaching physiology and pharmacology to the genome. *Nat. Rev. Genet.* **3**(1), 33–42.

Jacob, H. J., Pettersson, A., Wilson, D., Mao, Y., Lernmark, A., and Lander, E. S. (1992). Genetic dissection of autoimmune type I diabetes in the BB rat. *Nat. Genet.* **2**(1), 56–60.

Jahoda, C. A., Kljuic, A., O'Shaughnessy, R., Crossley, N., Whitehouse, C. J., Robinson, M., Reynolds, A. J., Demarchez, M., Porter, R. M., Shapiro, L., *et al.* (2004). The lanceolate hair rat phenotype results from a missense mutation in a calcium coordinating site of the desmoglein 4 gene. *Genomics* **83**(5), 747–756.

Jareborg, N., and Durbin, R. (2000). Alfresco—a workbench for comparative genomic sequence analysis. *Genome Res.* **10**(8), 1148–1157.

Johnson, J. M., Edwards, S., Shoemaker, D., and Schadt, E. E. (2005). Dark matter in the genome: Evidence of widespread transcription detected by microarray tiling experiments. *Trends Genet.* **21** (2), 93–102.

Kaiser, J. (2002). Biobanks. Private biobanks spark ethical concerns. *Science* **298**(5596), 1160.

Kaiser, J. (2005). GENOMICS: Celera to end subscriptions and give data to public GenBank. *Science* **308**(5723), 775.

Karp, C. L., Grupe, A., Schadt, E., Ewart, S. L., Keane-Moore, M., Cuomo, P. J., Kohl, J., Wahl, L., Kuperman, D., Germer, S., *et al.* (2000). Identification of complement factor 5 as a susceptibility locus for experimental allergic asthma. *Nat. Immunol.* **1**(3), 221–226.

Keller, E. T., and Murtha, J. M. (2004). The use of mature zebrafish (Danio rerio) as a model for human aging and disease. *Comp. Biochem. Physiol. C Toxicol. Pharmacol.* **138**(3), 335–341.

Kent, W. J., Sugnet, C. W., Furey, T. S., Roskin, K. M., Pringle, T. H., Zahler, A. M., and Haussler, D. (2002). The human genome browser at UCSC. *Genome Res.* **12**(6), 996–1006.

Kim, D. H., and Rossi, J. J. (2007). Strategies for silencing human disease using RNA interference. *Nat. Rev. Genet.* **8**(3), 173–184.

Koop, B. F., and Hood, L. (1994). Striking sequence similarity over almost 100 kilobases of human and mouse T-cell receptor DNA. *Nat. Genet.* **7**(1), 48–53.

Korstanje, R., and DiPetrillo, K. (2004). Unraveling the genetics of chronic kidney disease using animal models. *Am. J. Physiol. Renal Physiol.* **287**(3), F347–F352.

Korstanje, R., and Paigen, B. (2002). From QTL to gene: The harvest begins. *Nat. Genet.* **31**(3), 235–236.

Koskinen, R., Salomonsen, J., Tregaskes, C. A., Young, J. R., Goodchild, M., Bumstead, N., and Vainio, O. (2002). The chicken CD4 gene has remained conserved in evolution. *Immunogenetics* **54**(7), 520–525.

Kotchen, T. A., Zhang, H. Y., Covelli, M., and Blehschmidt, N. (1991). Insulin resistance and blood pressure in Dahl rats and in one-kidney, one-clip hypertensive rats. *Am. J. Physiol.* **261**(6 Pt 1), E692–E697.

Kunkel, L. M., Bachrach, E., Bennett, R. R., Guyon, J., and Steffen, L. (2006). Diagnosis and cell-based therapy for Duchenne muscular dystrophy in humans, mice, and zebrafish. *J. Hum. Genet.* **51**(5), 397–406.

Kuramoto, T., Kuwamura, M., and Serikawa, T. (2004). Rat neurological mutations cerebellar vermis defect and hobble are caused by mutations in the netrin-1 receptor gene Unc5h3. *Brain Res. Mol. Brain Res.* **122**(2), 103–108.

Kwitek, A. E., Jacob, H. J., Baker, J. E., Dwinell, M. R., Forster, H. V., Greene, A. S., Kunert, M. P., Lombard, J. H., Mattson, D. L., Pritchard, K. A., Jr., *et al.* (2006). BN phenome: Detailed characterization of the cardiovascular, renal, and pulmonary systems of the sequenced rat. *Physiol. Genomics* **25**(2), 303–313.

Lande, R., and Thompson, R. (1990). Efficiency of marker-assisted selection in the improvement of quantitative traits. *Genetics* **124,** 743–756.

Lashuel, H. A., and Hirling, H. (2006). Rescuing defective vesicular trafficking protects against alpha-synuclein toxicity in cellular and animal models of Parkinson's disease. *ACS Chem. Biol.* **1**(7), 420–424.

Lassnig, C., Kolb, A., Strobl, B., Enjuanes, L., and Muller, M. (2005). Studying human pathogens in animal models: Fine tuning the humanized mouse. *Transgenic Res.* **14**(6), 803–806.

Law, C. N. (1966). The location of genetic factors affecting a quantitative character in wheat. *Genetics* **53**(3), 487–498.

Lee, M. J., Stephenson, D. A., Groves, M. J., Sweeney, M. G., Davis, M. B., An, S. F., Houlden, H., Salih, M. A., Timmerman, V., de Jonghe, P., *et al.* (2003). Hereditary sensory neuropathy is caused by a mutation in the delta subunit of the cytosolic chaperonin-containing t-complex peptide-1 (Cct4) gene. *Hum. Mol. Genet.* **12**(15), 1917–1925.

Liao, G., Wang, J., Guo, J., Allard, J., Cheng, J., Ng, A., Shafer, S., Puech, A., McPherson, J. D., Foernzler, D., *et al.* (2004). In silico genetics: Identification of a functional element regulating H2-Ealpha gene expression. *Science* **306**(5696), 690–695.

Lien, C. L., McAnally, J., Richardson, J. A., and Olson, E. N. (2002). Cardiac-specific activity of an Nkx2–5 enhancer requires an evolutionarily conserved Smad binding site. *Dev. Biol.* **244**(2), 257–266.

Lindblad-Toh, K., Wade, C. M., Mikkelsen, T. S., Karlsson, E. K., Jaffe, D. B., Kamal, M., Clamp, M., Chang, J. L., Kulbokas, E. J., Zody, M. C., *et al.* (2005). Genome sequence, comparative analysis and haplotype structure of the domestic dog. *Nature* **438**(7069), 803–819.

Loots, G., and Ovcharenko, I. (2007). ECRbase: Database of evolutionary conserved regions, promoters, and transcription factor binding sites in vertebrate genomes. *Bioinformatics* **23**(1), 122–124.

Loots, G., Locksley, R., Blankespoor, C., Wang, Z., Miller, W., and Rubin, E. (2000). Identification of a coordinate regulator of interleukins 4, 13, and 5 by cross-species sequence comparisons. *Science* **288**(7 April), 2319.

Lu, S., Nordquist, N., Holmberg, J., Olofsson, P., Pettersson, U., and Holmdahl, R. (2002). Both common and unique susceptibility genes in different rat strains with pristane-induced arthritis. *Eur. J. Hum. Genet.* **10**(8), 475–483.

Luscher, T. F., Raij, L., and Vanhoutte, P. M. (1987). Endothelium-dependent vascular responses in normotensive and hypertensive Dahl rats. *Hypertension* **9**(2), 157–163.

MacMurray, A. J., Moralejo, D. H., Kwitek, A. E., Rutledge, E. A., Van Yserloo, B., Gohlke, P., Speros, S. J., Snyder, B., Schaefer, J., Bieg, S., *et al.* (2002). Lymphopenia in the BB rat model of type 1 diabetes is due to a mutation in a novel immune-associated nucleotide (Ian)-related gene. *Genome Res.* **12**(7), 1029–1039.

Makalowski, W., and Boguski, M. S. (1998). Evolutionary parameters of the transcribed mammalian genome: An analysis of 2,820 orthologous rodent and human sequences. *Proc. Natl. Acad. Sci. USA* **95**(16), 9407–9412.

Malek, R. L., Wang, H. Y., Kwitek, A. E., Greene, A. S., Bhagabati, N., Borchardt, G., Cahill, L., Currier, T., Frank, B., Fu, X., *et al.* (2006). Physiogenomic resources for rat models of heart, lung and blood disorders. *Nat. Genet.* **38**(2), 234–239.

Mashimo, T., Birger, V., Kuramoto, T., and Serikawa, T. (2005). Rat Phenome Project: The untapped potential of existing rat strains. *J. Appl. Physiol.* **98,** 371–379.

Mashimo, T., Voigt, B., Tsurumi, T., Naoi, K., Nakanishi, S., Yamasaki, K., Kuramoto, T., and Serikawa, T. (2006). A set of highly informative rat simple sequence length polymorphism (SSLP) markers and genetically defined rat strains. *BMC Genet.* **7,** 19.

Matsumoto, A., Okada, Y., Nakamichi, M., Nakamura, M., Toyama, Y., Sobue, G., Nagai, M., Aoki, M., Itoyama, Y., and Okano, H. (2006). Disease progression of human SOD1 (G93A) transgenic ALS model rats. *J. Neurosci. Res.* **83**(1), 119–133.

Maxam, A. M., and Gilbert, W. (1977). A new method for sequencing DNA. *Proc. Natl. Acad. Sci. USA* **74**(2), 560–564.

Mayor, C., Brudno, M., Schwartz, J. R., Poliakov, A., Rubin, E. M., Frazer, K. A., Pachter, L. S., and Dubchak, I. (2000). VISTA: Visualizing global DNA sequence alignments of arbitrary length. *Bioinformatics* **16**(11), 1046–1047.

McBride, M. W., Carr, F. J., Graham, D., Anderson, N. H., Clark, J. S., Lee, W. K., Charchar, F. J., Brosnan, M. J., and Dominiczak, A. F. (2003). Microarray analysis of rat chromosome 2 congenic strains. *Hypertension* **41**(3 Pt 2), 847–853.

McKusick, V. A., and Ruddle, F. H. (1987). A new discipline, a new name, a new journal (editorial). *Genomics* **1**(Sep), 1–2.

McMahon, C., Semina, E. V., and Link, B. A. (2004). Using zebrafish to study the complex genetics of glaucoma. *Comp. Biochem. Physiol. C Toxicol. Pharmacol.* **138**(3), 343–350.

McRedmond, J. P., Park, S. D., Reilly, D. F., Coppinger, J. A., Maguire, P. B., Shields, D. C., and Fitzgerald, D. J. (2004). Integration of proteomics and genomics in platelets: A profile of platelet proteins and platelet-specific genes. *Mol. Cell. Proteomics* **3**(2), 133–144.

Mijalski, T., Harder, A., Halder, T., Kersten, M., Horsch, M., Strom, T. M., Liebscher, H. V., Lottspeich, F., de Angelis, M. H., and Beckers, J. (2005). Identification of coexpressed gene clusters in a comparative analysis of transcriptome and proteome in mouse tissues. *Proc. Natl. Acad. Sci. USA* **102**(24), 8621–8626.

Mockler, T. C., Chan, S., Sundaresan, A., Chen, H., Jacobsen, S. E., and Ecker, J. R. (2005). Applications of DNA tiling arrays for whole-genome analysis. *Genomics* **85**(1), 1–15.

Moreno, C., Dumas, P., Kaldunski, M. L., Tonellato, P. J., Greene, A. S., Roman, R. J., Cheng, Q., Wang, Z., Jacob, H. J., and Cowley, A. W., Jr. (2003). Genomic map of cardiovascular phenotypes of hypertension in female Dahl S rats. *Physiol. Genomics* **15**(3), 243–257.

Mott, R., Talbot, C. J., Turri, M. G., Collins, A. C., and Flint, J. (2000). A method for fine mapping quantitative trait loci in outbred animal stocks. *Proc. Natl. Acad. Sci. USA* **97**(23), 12649–12654.

Murphy, J. M., Stewart, R. B., Bell, R. L., Badia-Elder, N. E., Carr, L. G., McBride, W. J., Lumeng, L., and Li, T. K. (2002). Phenotypic and genotypic characterization of the Indiana University rat lines selectively bred for high and low alcohol preference. *Behav. Genet.* **32**(5), 363–388.

Nabika, T., Nasreen, S., Kobayashi, S., and Masuda, J. (2002). The genetic effect of the apoprotein AV gene on the serum triglyceride level in Japanese. *Atherosclerosis* **165**(2), 201–204.

Nadeau, J., and Taylor, B. (1984). Lengths of chromosomeal segments conserved since divergence of man and mouse. *Proc. Natl. Acad. Sci. USA* **81**, 814–818.

Nadeau, J., Singer, J., Matin, A., and Lander, E. (2000). Analysing complex genetic traits with chromosome substitution strains. *Nat. Genet.* **24**(3), 221–225.

Nadeau, J. H. (2000). Muta-genetics or muta-genomics: The feasibility of large-scale mutagenesis and phenotyping programs. *Mamm. Genome* **11**(7), 603–607.

Nash, W., and O'Brien, S. (1982). Conserved regions of homologous G-banded chromosomes between orders in mammlaian evolution: Carnivores and primates. *Proc. Natl. Acad. Sci. USA* **79**, 6631–6635.

Nicholas, F. W. (2003). Online Mendelian Inheritance in Animals (OMIA): A comparative knowledgebase of genetic disorders and other familial traits in non-laboratory animals. *Nucleic Acids Res.* **31**(1), 275–277.

Nobrega, M. A., Ovcharenko, I., Afzal, V., and Rubin, E. M. (2003). Scanning human gene deserts for long-range enhancers. *Science* **302**(5644), 413.

Nolan, P. M., Peters, J., Strivens, M., Rogers, D., Hagan, J., Spurr, N., Gray, I. C., Vizor, L., Brooker, D., Whitehill, E., *et al.* (2000). A systematic, genome-wide, phenotype-driven mutagenesis programme for gene function studies in the mouse. *Nat. Genet.* **25**(4), 440–443.

North, T. E., and Zon, L. I. (2003). Modeling human hematopoietic and cardiovascular diseases in zebrafish. *Dev. Dyn.* **228**(3), 568–583.

O'Brien, S., Menotti-Raymond, M., Murphy, W., Nash, W., Wienberg, J., Stanyon, R., Copeland, N., Jenkins, N., Womack, J., and Graves, J. (1999). The promise of comparative genomics in mammals. *Science* **286,** 458–481.

O'Doherty, A., Ruf, S., Mulligan, C., Hildreth, V., Errington, M. L., Cooke, S., Sesay, A., Modino, S., Vanes, L., Hernandez, D., *et al.* (2005). An aneuploid mouse strain carrying human chromosome 21 with down syndrome phenotypes. *Science* **309**(5743), 2033–2037.

Ohno, T., Aoyagi, M., Yamanashi, Y., Saito, H., Ikawa, S., Meshi, T., and Okada, Y. (1984). Nucleotide sequence of the tobacco mosaic virus (tomato strain) genome and comparison with the common strain genome. *J. Biochem. (Tokyo)* **96**(6), 1915–1923.

Oiso, N., Riddle, S. R., Serikawa, T., Kuramoto, T., and Spritz, R. A. (2004). The rat Ruby (R) locus is Rab38: Identical mutations in Fawn-hooded and Tester-Moriyama rats derived from an ancestral Long Evans rat sub-strain. *Mamm. Genome* **15**(4), 307–314.

Okimoto, K., Sakurai, J., Kobayashi, T., Mitani, H., Hirayama, Y., Nickerson, M. L., Warren, M. B., Zbar, B., Schmidt, L. S., and Hino, O. (2004). A germ-line insertion in the Birt-Hogg-Dube (BHD) gene gives rise to the Nihon rat model of inherited renal cancer. *Proc. Natl. Acad. Sci. USA* **101**(7), 2023–2027.

Oliver, F., Christians, J. K., Liu, X., Rhind, S., Verma, V., Davison, C., Brown, S. D., Denny, P., and Keightley, P. D. (2005). Regulatory variation at glypican-3 underlies a major growth QTL in mice. *PLoS Biol.* **3**(5), e135.

Olofsson, P., Holmberg, J., Pettersson, U., and Holmdahl, R. (2003a). Identification and isolation of dominant susceptibility loci for pristane-induced arthritis. *J. Immunol.* **171**(1), 407–416.

Olofsson, P., Holmberg, J., Tordsson, J., Lu, S., Akerstrom, B., and Holmdahl, R. (2003b). Positional identification of Ncf1 as a gene that regulates arthritis severity in rats. *Nat. Genet.* **33**(1), 25–32.

Otsuki, A., Tahimic, C. G., Tomimatsu, N., Katoh, M., Chen, D. J., Kurimasa, A., and Oshimura, M. (2005). Construction of a novel expression system on a human artificial chromosome. *Biochem. Biophys. Res. Commun.* **329**(3), 1018–1025.

Outeiro, T. F., and Muchowski, P. J. (2004). Molecular genetics approaches in yeast to study amyloid diseases. *J. Mol. Neurosci.* **23**(1–2), 49–60.

Ovcharenko, I., Nobrega, M. A., Loots, G. G., and Stubbs, L. (2004). ECR Browser: A tool for visualizing and accessing data from comparisons of multiple vertebrate genomes. *Nucleic Acids Res.* **32**(Web Server issue), W280–W286.

Pandey, J., Cracchiolo, D., Hansen, F. M., and Wendell, D. L. (2002). Strain differences and inheritance of angiogenic versus angiostatic activity in oestrogen-induced rat pituitary tumours. *Angiogenesis* **5**(1–2), 53–66.

Pennacchio, L. A., and Rubin, E. M. (2001). Genomic strategies to identify mammalian regulatory sequences. *Nat. Rev. Genet.* **2**(2), 100–109.

Pennacchio, L. A., Olivier, M., Hubacek, J. A., Cohen, J. C., Cox, D. R., Fruchart, J. C., Krauss, R. M., and Rubin, E. M. (2001). An apolipoprotein influencing triglycerides in humans and mice revealed by comparative sequencing. *Science* **294**(5540), 169–173.

Pennacchio, L. A., Olivier, M., Hubacek, J. A., Krauss, R. M., Rubin, E. M., and Cohen, J. C. (2002). Two independent apolipoprotein A5 haplotypes influence human plasma triglyceride levels. *Hum. Mol. Genet.* **11**(24), 3031–3038.

Peters, L. L., Robledo, R. F., Bult, C. J., Churchill, G. A., Paigen, B. J., and Svenson, K. L. (2007). The mouse as a model for human biology: A resource guide for complex trait analysis. *Nat. Rev. Genet.* **8**(1), 58–69.

Petrasch-Parwez, E., Nguyen, H. P., Lobbecke-Schumacher, M., Habbes, H. W., Wieczorek, S., Riess, O., Andres, K. H., Dermietzel, R., and Von Horsten, S. (2007). Cellular and subcellular localization of Huntington aggregates in the brain of a rat transgenic for Huntington disease. *J. Comp. Neurol.* **501**(5), 716–730.

Petretto, E., Mangion, J., Dickens, N. J., Cook, S. A., Kumaran, M. K., Lu, H., Fischer, J., Maatz, H., Kren, V., Pravenec, M., *et al.* (2006). Heritability and tissue specificity of expression quantitative trait loci. *PLoS Genet.* **2**(10), e172.

Pletcher, M. T., McClurg, P., Batalov, S., Su, A. I., Barnes, S. W., Lagler, E., Korstanje, R., Wang, X., Nusskern, D., Bogue, M. A., *et al.* (2004). Use of a dense single nucleotide polymorphism map for in silico mapping in the mouse. *PLoS Biol.* **2**(12), e393.

Pravenec, M., Klir, P., Kren, V., Zicha, J., and Kunes, J. (1989). An analysis of spontaneous hypertension in spontaneously hypertensive rats by means of new recombinant inbred strains. *J. Hypertens.* **7**(3), 217–221.

Pravenec, M., Gauguier, D., Schott, J. J., Buard, J., Kren, V., Bila, V., Szpirer, C., Szpirer, J., Wang, J. M., Huang, H., *et al.* (1995). Mapping of quantitative trait loci for blood pressure and cardiac mass in the rat by genome scanning of recombinant inbred strains. *J. Clin. Invest.* **96**(4), 1973–1978.

Pravenec, M., Gauguier, D., Schott, J. J., Buard, J., Kren, V., Bila, V., Szpirer, C., Szpirer, J., Wang, J. M., Huang, H., *et al.* (1996). A genetic linkage map of the rat derived from recombinant inbred strains. *Mamm. Genome* **7**(2), 117–127.

Pravenec, M., Kren, V., Krenova, D., Bila, V., Zidek, V., Simakova, M., Musilova, A., van Lith, H. A., and van Zutphen, L. F. (1999). HXB/Ipcv and BXH/Cub recombinant inbred strains of the rat: Strain distribution patterns of 632 alleles. *Folia Biol. (Praha)* **45**(5), 203–215.

Pravenec, M., Zidek, V., Musilova, A., Simakova, M., Kostka, V., Mlejnek, P., Kren, V., Krenova, D., Bila, V., Mikova, B., *et al.* (2002). Genetic analysis of metabolic defects in the spontaneously hypertensive rat. *Mamm. Genome* **13**(5), 253–258.

Provoost, A. P. (1994). Spontaneous glomerulosclerosis: Insights from the fawn-hooded rat. *Kidney Int. Suppl.* **45,** S2–S5.

Rapp, J. P. (1982). Dahl salt-susceptible and salt-resistant rats. A review. *Hypertension* **4**(6), 753–763.

Rapp, J. P. (2000). Genetic analysis of inherited hypertension in the rat. *Physiol. Rev.* **80**(1), 135–172.

Reaven, G. M., Twersky, J., and Chang, H. (1991). Abnormalities of carbohydrate and lipid metabolism in Dahl rats. *Hypertension* **18**(5), 630–635.

Remmers, E. F., Longman, R. E., Du, Y., O'Hare, A., Cannon, G. W., Griffiths, M. M., and Wilder, R. L. (1996). A genome scan localizes five non-MHC loci controlling collagen-induced arthritis in rats. *Nat. Genet.* **14**(1), 82–85.

Roep, B. O. (2007). Are insights gained from nod mice sufficient to guide clinical translation? Another inconvenient truth. *Ann. N. Y. Acad. Sci.* **1103,** 1–10.

Rollins, J., Chen, Y., Paigen, B., and Wang, X. (2006). In search of new targets for plasma high-density lipoprotein cholesterol levels: Promise of human-mouse comparative genomics. *Trends Cardiovasc. Med.* **16**(7), 220–234.

Roman, R. J., and Kaldunski, M. (1991). Pressure natriuresis and cortical and papillary blood flow in inbred Dahl rats. *Am. J. Physiol.* **261**(3 Pt 2), R595–R602.

Ross, M. T., Grafham, D. V., Coffey, A. J., Scherer, S., McLay, K., Muzny, D., Platzer, M., Howell, G. R., Burrows, C., Bird, C. P., *et al.* (2005). The DNA sequence of the human X chromosome. *Nature* **434**(7031), 325–337.

Rost, S., Fregin, A., Ivaskevicius, V., Conzelmann, E., Hortnagel, K., Pelz, H. J., Lappegard, K., Seifried, E., Scharrer, I., Tuddenham, E. G., *et al.* (2004). Mutations in VKORC1 cause warfarin resistance and multiple coagulation factor deficiency type 2. *Nature* **427**(6974), 537–541.

Roth, M. P., Viratelle, C., Dolbois, L., Delverdier, M., Borot, N., Pelletier, L., Druet, P., Clanet, M., and Coppin, H. (1999). A genome-wide search identifies two susceptibility loci for experimental autoimmune encephalomyelitis on rat chromosomes 4 and 10. *J. Immunol.* **162** (4), 1917–1922.

Rozzo, S. J., Allard, J. D., Choubey, D., Vyse, T. J., Izui, S., Peltz, G., and Kotzin, B. L. (2001). Evidence for an interferon-inducible gene, Ifi202, in the susceptibility to systemic lupus. *Immunity* **15**(3), 435–443.

Rubattu, S., Volpe, M., Kreutz, R., Ganten, U., Ganten, D., and Lindpaintner, K. (1996). Chromosomal mapping of quantitative trait loci contributing to stroke in a rat model of complex human disease. *Nat. Genet.* **13**(4), 429–434.

Ruse, C. I., Tan, F. L., Kinter, M., and Bond, M. (2004). Intregrated analysis of the human cardiac transcriptome, proteome and phosphoproteome. *Proteomics* **4**(5), 1505–1516.

Sanger, F., Nicklen, S., and Coulson, A. R. (1977). DNA sequencing with chain-terminating inhibitors. *Proc. Natl. Acad. Sci. USA* **74**(12), 5463–5467.

Sawyer, J., and Hozier, H. (1987). High resolution of mouse chromosomes: Banding conservation between man and mouse. *Science* **232,** 1632–1635.

Scherthan, H., Cremer, T., Arnason, U., Weier, H.-U., Lima-de-Feria, A., and Fronicke, L. (1994). Comparative chromosome painting discloses homologous segments in distantly related organisms. *Nat. Genet.* **6,** 342–347.

Schramm, A., Schulte, J. H., Klein-Hitpass, L., Havers, W., Sieverts, H., Berwanger, B., Christiansen, H., Warnat, P., Brors, B., Eils, J., *et al.* (2005). Prediction of clinical outcome and biological characterization of neuroblastoma by expression profiling. *Oncogene* **24**(53), 7902–7912.

Schwartz, S., Zhang, Z., Frazer, K. A., Smit, A., Riemer, C., Bouck, J., Gibbs, R., Hardison, R., and Miller, W. (2000). PipMaker—a web server for aligning two genomic DNA sequences. *Genome Res.* **10**(4), 577–586.

Segre, J. A., Nemhauser, J. L., Taylor, B. A., Nadeau, J. H., and Lander, E. S. (1995). Positional cloning of the nude locus: Genetic, physical, and transcription maps of the region and mutations in the mouse and rat. *Genomics* **28**(3), 549–559.

Sehnert, A. J., and Stainier, D. Y. (2002). A window to the heart: Can zebrafish mutants help us understand heart disease in humans? *Trends Genet.* **18**(10), 491–494.

Shin, J. T., Priest, J. R., Ovcharenko, I., Ronco, A., Moore, R. K., Burns, C. G., and MacRae, C. A. (2005). Human-zebrafish non-coding conserved elements act *in vivo* to regulate transcription. *Nucleic Acids Res.* **33**(17), 5437–5445.

Shisa, H., Lu, L., Katoh, H., Kawarai, A., Tanuma, J., Matsushima, Y., and Hiai, H. (1997). The LEXF: A new set of rat recombinant inbred strains between LE/Stm and F344. *Mamm. Genome* **8**(5), 324–327.

Shultz, L. D., Ishikawa, F., and Greiner, D. L. (2007). Humanized mice in translational biomedical research. *Nat. Rev. Immunol.* **7**(2), 118–130.

Singer, J. B., Hill, A. E., Burrage, L. C., Olszens, K. R., Song, J., Justice, M., O'Brien, W. E., Conti, D. V., Witte, J. S., Lander, E. S., *et al.* (2004). Genetic dissection of complex traits with chromosome substitution strains of mice. *Science* **304**(5669), 445–448.

Sladek, R., Rocheleau, G., Rung, J., Dina, C., Shen, L., Serre, D., Boutin, P., Vincent, D., Belisle, A., Hadjadj, S., *et al.* (2007). A genome-wide association study identifies novel risk loci for type 2 diabetes. *Nature* **445**(7130), 881–885.

Smith, C. L., Goldsmith, C. A., and Eppig, J. T. (2005). The Mammalian Phenotype Ontology as a tool for annotating, analyzing and comparing phenotypic information. *Genome Biol.* **6**(1), R7.

Snell, G. D. (1948). Methods for study of histocompatibility genes. *J. Genet.* **49,** 87–108.

Spinola, M., Leoni, V. P., Galvan, A., Korsching, E., Conti, B., Pastorino, U., Ravagnani, F., Columbano, A., Skaug, V., Haugen, A., *et al.* (2007). Genome-wide single nucleotide polymorphism analysis of lung cancer risk detects the KLF6 gene. *Cancer Lett.* **251**(2), 311–316.

Sprague, J., Bayraktaroglu, L., Clements, D., Conlin, T., Fashena, D., Frazer, K., Haendel, M., Howe, D. G., Mani, P., Ramachandran, S., *et al.* (2006). The Zebrafish Information Network: The zebrafish model organism database. *Nucleic Acids Res.* **34**(Database issue), D581–D585.

Stoll, M., Kwitek-Black, A. E., Cowley, A. W., Jr., Harris, E. L., Harrap, S. B., Krieger, J. E., Printz, M. P., Provoost, A. P., Sassard, J., and Jacob, H. J. (2000). New target regions for human hypertension via comparative genomics. *Genome Res.* **10**(4), 473–482.

Stoll, M., Cowley, A. W., Jr., Tonellato, P. J., Greene, A. S., Kaldunski, M. L., Roman, R. J., Dumas, P., Schork, N. J., Wang, Z., and Jacob, H. J. (2001). A genomic-systems biology map for cardiovascular function. *Science* **294**(5547), 1723–1726.

Stylianou, I. M., Christians, J. K., Keightley, P. D., Bunger, L., Clinton, M., Bulfield, G., and Horvat, S. (2004). Genetic complexity of an obesity QTL (Fob3) revealed by detailed genetic mapping. *Mamm. Genome* **15**(6), 472–481.

Sugiyama, F., Churchill, G. A., Higgins, D. C., Johns, C., Makaritsis, K. P., Gavras, H., and Paigen, B. (2001). Concordance of murine quantitative trait loci for salt-induced hypertension with rat and human loci. *Genomics* **71**(1), 70–77.

Suto, J. (2005). Apolipoprotein gene polymorphisms as cause of cholesterol QTLs in mice. *J. Vet. Med. Sci.* **67**(6), 583–589.

Suzuki, N., Nishii, K., Okazaki, T., and Ikeno, M. (2006). Human artificial chromosomes constructed using the bottom-up strategy are stably maintained in mitosis and efficiently transmissible to progeny mice. *J. Biol. Chem.* **281**(36), 26615–26623.

Swank, R. T., and Bailey, D. W. (1973). Recombinant inbred lines: Value in the genetic analysis of biochemical variants. *Science* **181**(106), 1249–1252.

Szafranski, K., Jahn, N., and Platzer, M. (2006). tuple_plot: Fast pairwise nucleotide sequence comparison with noise suppression. *Bioinformatics* **22**(15), 1917–1918.

Szulc, J., Wiznerowicz, M., Sauvain, M. O., Trono, D., and Aebischer, P. (2006). A versatile tool for conditional gene expression and knockdown. *Nat. Methods* **3**(2), 109–116.

Talbot, C. J., Nicod, A., Cherny, S. S., Fulker, D. W., Collins, A. C., and Flint, J. (1999). High-resolution mapping of quantitative trait loci in outbred mice. *Nat. Genet.* **21**(3), 305–308.

Talmud, P. J., Hawe, E., Martin, S., Olivier, M., Miller, G. J., Rubin, E. M., Pennacchio, L. A., and Humphries, S. E. (2002). Relative contribution of variation within the APOC3/A4/A5 gene cluster in determining plasma triglycerides. *Hum. Mol. Genet.* **11**(24), 3039–3046.

Tanomura, H., Miyake, T., Taniguchi, Y., Manabe, N., Kose, H., Matsumoto, K., Yamada, T., and Sasaki, Y. (2002). Detection of a quantitative trait locus for intramuscular fat accumulation using the OLETF rat. *J. Vet. Med. Sci.* **64**(1), 45–50.

Tanoue, A., Nasa, Y., Koshimizu, T., Shinoura, H., Oshikawa, S., Kawai, T., Sunada, S., Takeo, S., and Tsujimoto, G. (2002). The alpha(1D)-adrenergic receptor directly regulates arterial blood pressure via vasoconstriction. *J. Clin. Invest.* **109**(6), 765–775.

Tesson, L., Cozzi, J., Menoret, S., Remy, S., Usal, C., Fraichard, A., and Anegon, I. (2005). Transgenic modifications of the rat genome. *Transgenic Res.* **14**(5), 531–546.

Threadgill, D. W., Dlugosz, A. A., Hansen, L. A., Tennenbaum, T., Lichti, U., Yee, D., LaMantia, C., Mourton, T., Herrup, K., Harris, R. C., *et al.* (1995). Targeted disruption of mouse EGF receptor: Effect of genetic background on mutant phenotype. *Science* **269**(5221), 230–234.

Threadgill, D. W., Hunter, K. W., and Williams, R. W. (2002). Genetic dissection of complex and quantitative traits: From fantasy to reality via a community effort. *Mamm. Genome* **13**(4), 175–178.

Thummel, R., Bai, S., Sarras, M. P., Jr., Song, P., McDermott, J., Brewer, J., Perry, M., Zhang, X., Hyde, D. R., and Godwin, A. R. (2006). Inhibition of zebrafish fin regeneration using *in vivo* electroporation of morpholinos against fgfr1 and msxb. *Dev. Dyn.* **235**(2), 336–346.

Todhunter, R. J., Casella, G., Bliss, S. P., Lust, G., Williams, A. J., Hamilton, S., Dykes, N. L., Yeager, A. E., Gilbert, R. O., Burton-Wurster, N. I., *et al.* (2003). Power of a Labrador Retriever-Greyhound pedigree for linkage analysis of hip dysplasia and osteoarthritis. *Am. J. Vet. Res.* **64**(4), 418–424.

Twigger, S. N., Pasko, D., Nie, J., Shimoyama, M., Bromberg, S., Campbell, D., Chen, J., Dela Cruz, N., Fan, C., Foote, C., *et al.* (2005). Tools and strategies for physiological genomics—the rat genome database. *Physiol. Genomics* **23,** 246–256.

Twigger, S. N., Shimoyama, M., Bromberg, S., Kwitek, A. E., and Jacob, H. J. (2007). The Rat Genome Database, update 2007—easing the path from disease to data and back again. *Nucleic Acids Res.* **35**(Database issue), D658–D662.

Wade, C. M., and Daly, M. J. (2005). Genetic variation in laboratory mice. *Nat. Genet.* **37**(11), 1175–1180.

Wang, X., and Paigen, B. (2005). Genome-wide search for new genes controlling plasma lipid concentrations in mice and humans. *Curr. Opin. Lipidol.* **16**(2), 127–137.

Ward, A. C., and Lieschke, G. J. (2002). The zebrafish as a model system for human disease. *Front. Biosci.* **7,** d827—d833.

Ward, C. J., Hogan, M. C., Rossetti, S., Walker, D., Sneddon, T., Wang, X., Kubly, V., Cunningham, J. M., Bacallao, R., Ishibashi, M., *et al.* (2002). The gene mutated in autosomal recessive polycystic kidney disease encodes a large, receptor-like protein. *Nat. Genet.* **30**(3), 259–269.

Waterston, R. H., Lindblad-Toh, K., Birney, E., Rogers, J., Abril, J. F., Agarwal, P., Agarwala, R., Ainscough, R., Alexandersson, M., An, P., *et al.* (2002). Initial sequencing and comparative analysis of the mouse genome. *Nature* **420**(6915), 520–562.

Ways, J. A., Cicila, G. T., Garrett, M. R., and Koch, L. G. (2002). A genome scan for Loci associated with aerobic running capacity in rats. *Genomics* **80**(1), 13–20.

Weinberg, J., and Stanyon, R. (1995). Chromosome painting in mammals as an approach to comparative genomics. *Curr. Opin. Genet. Dev.* **5,** 792–797.

Wendell, D. L., and Gorski, J. (1997). Quantitative trait loci for estrogen-dependent pituitary tumor growth in the rat. *Mamm. Genome* **8**(11), 823–829.

Williams, R. W. (2006). Expression genetics and the phenotype revolution. *Mamm. Genome* **17**(6), 496–502.

Williams, R. W., Bennett, B., Lu, L., Gu, J., DeFries, J. C., Carosone-Link, P. J., Rikke, B. A., Belknap, J. K., and Johnson, T. E. (2004). Genetic structure of the LXS panel of recombinant inbred mouse strains: A powerful resource for complex trait analysis. *Mamm. Genome* **15**(8), 637–647.

Wright, D., Nakamichi, R., Krause, J., and Butlin, R. K. (2006). QTL analysis of behavioral and morphological differentiation between wild and laboratory zebrafish (Danio rerio). *Behav. Genet.* **36**(2), 271–284.

Wu, T. D., and Watanabe, C. K. (2005). GMAP: A genomic mapping and alignment program for mRNA and EST sequences. *Bioinformatics* **21**(9), 1859–1875.

Yagil, C., Hubner, N., Monti, J., Schulz, H., Sapojnikov, M., Luft, F. C., Ganten, D., and Yagil, Y. (2005). Identification of hypertension-related genes through an integrated genomic-transcriptomic approach. *Circ. Res.* **96**(6), 617–625.

Yeung, R. S., Xiao, G. H., Jin, F., Lee, W. C., Testa, J. R., and Knudson, A. G. (1994). Predisposition to renal carcinoma in the Eker rat is determined by germ-line mutation of the tuberous sclerosis 2 (TSC2) gene. *Proc. Natl. Acad. Sci. USA* **91**(24), 11413–11416.

Yokoi, N., Komeda, K., Wang, H. Y., Yano, H., Kitada, K., Saitoh, Y., Seino, Y., Yasuda, K., Serikawa, T., and Seino, S. (2002). Cblb is a major susceptibility gene for rat type 1 diabetes mellitus. *Nat. Genet.* **31**(4), 391–394.

Yokoi, N., Namae, M., Wang, H. Y., Kojima, K., Fuse, M., Yasuda, K., Serikawa, T., Seino, S., and Komeda, K. (2003). Rat neurological disease creeping is caused by a mutation in the reelin gene. *Brain Res. Mol. Brain Res.* **112**(1–2), 1–7.

Yutzey, K. E., and Robbins, J. (2007). Principles of genetic murine models for cardiac disease. *Circulation* **115**(6), 792–799.

Zan, Y., Haag, J. D., Chen, K. S., Shepel, L. A., Wigington, D., Wang, Y. R., Hu, R., Lopez-Guajardo, C. C., Brose, H. L., Porter, K. I., *et al.* (2003). Production of knockout rats using ENU mutagenesis and a yeast-based screening assay. *Nat. Biotechnol.* **21**(6), 645–651.
Zidek, V., Pintir, J., Musilova, A., Bila, V., Kren, V., and Pravenec, M. (1999). Mapping of quantitative trait loci for seminal vesicle mass and litter size to rat chromosome 8. *J. Reprod. Fertil.* **116**(2), 329–333.

Part V

OUTSTANDING CHALLENGES

24

From Genetics to Mechanism of Disease Liability

Andreas Rohrwasser, Paul Lott, Robert B. Weiss, and Jean-Marc Lalouel

Department of Human Genetics, The University of Utah School of Medicine, Salt Lake City, Utah

Advances in Genetics, Vol. 60
0065-2660/08 $35.00
DOI: 10.1016/S0065-2660(07)00424-5

ABSTRACT

With each advance in genomic technology, new statistical methods have regularly emerged to test genetic hypotheses in complex inheritance, as evidenced throughout this book. Notwithstanding the approach used, the greatest challenge in the genetics of complex traits remains the identification of the gene(s) and the molecular variant(s) accounting for a genetic inference based on statistical testing. We take the example of quantitative trait locus (QTL) mapping for blood pressure (BP) and related phenotypes in rodents to review the current landscape. Traditional approaches to refined mapping are typically hampered by the small effect and the small proportion of the variance attached to individual QTLs. The alternative of functional screens in intact animals, whether by chemical mutagenesis or gene targeting, remains a daunting undertaking. Such limitations account for the slow progress to date of inferences from QTL to gene(s). We select a QTL for differential sodium sensitivity between two mouse inbred lines to propose an approach that can be used in relatively large genomic regions (1) by optimizing the selection of candidate genes and (2) by subjecting such genes to high-throughput functional screens. While this is still work in progress, we think it abundantly illustrates what is ahead of us in delineating genetic variation that underlie complex disease.

I. INTRODUCTION

In the initial edition of this textbook on the Genetic Dissection of Complex Traits, we were pleased to be invited to write a final chapter that would take the reader from the focus of the book, namely the statistical approaches and tools to incriminate genes in complex disease, to a discussion of the treacherous and elusive aspects of the "end game" whereby a statistical inference about a candidate genomic region or a gene could be translated into actual understanding of pathophysiology at all levels of observation, including molecular, cellular, organ, and whole body level (Lalouel, 2001). We also felt that solving the genetic puzzle should include an analysis of the evolutionary genetics that may have been at play to bring about genetic diversity in disease liability in current populations.

While our focus in this new chapter will remain similar, marked developments in both molecular and statistical methods have considerably altered the landscape, so that a new perspective needs to be developed about the challenges that arise in pursuing this end game. As before, we will focus on essential hypertension as our model to drive such discussion. The condition is an excellent example of common disease with complex inheritance, and it is an area where we have more direct experience.

Whereas a variety of approaches can be applied to obtain support for a genetic hypothesis, the most challenging issue still remains the indictment of genes and molecular variants from a functional standpoint. We will focus more specifically on proceeding from inference of a quantitative trait locus (QTL) to one or more underlying gene functionally accounting for the QTL. Although LD mapping based on case-control studies is conceptually distinct from linkage-inferred QTL, the conceptual issue of functional identification remains the same.

II. IDENTIFICATION OF QTL BY STATISTICAL ANALYSIS

A. QTL mapping in human genetics

The analysis of quantitative traits and QTL mapping has been an integral part of the development of statistical methods in animal and plant breeding. Although human genetics was for many years primarily concerned with the inheritance of discrete phenotypes, the extension of methods of segregation and linkage analysis from single Mendelian disorders to complex inheritance and the inference of QTL has a long history that conceptually goes back at least to John Edwards (Edwards, 1960). Complex models of inheritance applied to segregation analysis (Elston and Stewart, 1971; Lalouel and Morton, 1981; Lalouel *et al.*, 1983; MacLean *et al.*, 1975; Morton and MacLean, 1974) were simplified in linkage analysis, but the earliest models and programs for genome-wide, multilocus (or interval) linkage mapping, such as LINKAGE for pedigrees of arbitrary complexity (Lathrop *et al.*, 1984) or MAPMAKER for more restricted family structures (Lander *et al.*, 1987) explicitly addressed mapping of QTLs.

B. Mapping determinants of essential hypertension in humans

Whether treated as a discrete or as a continuous phenotype, common essential hypertension (EH) is typical of the field of complex disease. Gain- or loss-of-function mutations have been identified that elucidate the molecular basis of rare Mendelian hypertension syndromes (Lifton *et al.*, 2001). Most point to a primary mechanism that rests in the kidney, particularly emphasizing the dominance of the aldosterone-sensitive distal nephron (ASDN) and the dominant role of the epithelial sodium channel (ENaC) in controlling sodium excretion. EH, by contrast, has witnessed only a few notable advances, with significant progress toward a functional validation for angiotensinogen (Jeunemaitre *et al.*, 1992; Lalouel *et al.*, 2001), α-adducin (Bianchi and Tripodi, 2003), and the $\beta3$ subunit of the G-coupled receptor protein (Siffert, 2005). In all three instances, the proximal tubule appears implicated.

Genome-wide scans, even with considerable sample size, have been remarkable by the marginal significance and the lack of consistency of the linkage support across studies (Liu *et al.*, 2004). The largest such enterprise, the Family Blood Pressure Program (2002), including over 11,000 study subjects, has so far yielded two encouraging leads. A LOD score of 2.84 on chromosome 2, followed by dense single nucleotide polymorphism (SNP) typing of multiple candidate genes in the region, has implicated the sodium-bicarbonate transporter SLC4A5 as a candidate in this region (Barkley *et al.*, 2004). An initial LOD score of 2.6 with the phenotype of postural change in systolic BP (Pankow *et al.*, 2000) was followed by an estimate of 4.2 when additional data were included (Pankow *et al.*, 2005). Extensive resequencing of the positional candidate gene NEDD4L, a major regulator of ENaC, allowed us to identify a common variant that affects splicing and the ability to generate in ASDN an abundant transcriptional isoform that includes a unique functional domain (Dunn *et al.*, 2002). Although association studies support a role for this variant in EH (Pankow *et al.*, 2005; Russo *et al.*, 2005), the statistical support for this variant remains modest at best. Association with the SLC4A5 polymorphism was replicated in one study (Hunt *et al.*, 2006). Efforts are now under way to identify genes underlying promising LOD scores for the phenotypes of EH or intermediate phenotypes that are usually quantitative traits (Hunt *et al.*, 2006).

C. The predominant role of the kidney in EH

Experimental physiology supports two possible mechanisms for essential hypertension. Considerable experimental evidence contributed in large part by Guyton and his colleagues (Cowley and Roman, 1996; Guyton, 1980; Hall and Guyton, 1990) have stressed that EH could develop through one of two fundamental mechanisms affecting either volume or compliance of the circulatory system. Volume-dependent hypertension results from primary sodium and water retention through the kidney. General vasoconstriction including the renal arteries represents the other mechanism. Factors that predispose to one or the other mechanisms need not exhibit cooperative interaction. Indeed, we have advanced this hypothesis as a possible explanation for the lack of success of genetic mapping attempts applied to a phenotype that confounds these two pathophysiological mechanisms (Lalouel, 2003).

Much experimental evidence documents the predominant role of the kidney in volume-dependent hypertension. It is remarkable that molecular variants and hypotheses advanced thus far to account for either rare Mendelian or common forms of human hypertension (Lifton *et al.*, 2001) implicate volume-dependent mechanisms in the development of the disease. This confirms the dominant role of this organ in effecting plasma volume regulation. That it may also be the site of the primary alteration leading to hypertension is supported by the fact that, in all experimental models where renal transplantation has been performed, it confers hypertension to the recipient (Grisk and Rettig, 2001). These observations critically support an initial focus on the kidney, acknowledging the potential dominant role of a vascular mechanism in forms of hypertension associated with other manifestations such as atherosclerosis.

D. From QTL to genes in rodents

QTL mapping has developed over a longer period of time and at a much faster pace in animal and plant genetics, if only because markers that tag QTLs can in themselves be used to design artificial selection and breeding programs in the absence of knowledge of the underlying, causal genetic diversity.

Despite the experimental control afforded in these models, progress from QTL to genes in rodents has proven challenging (Abiola *et al.*, 2003; Darvasi and Pisante-Shalom, 2002; Flint *et al.*, 2005; Glazier *et al.*, 2002; Korstanje and Paigen, 2002; Nadeau and Frankel, 2000; Nadeau *et al.*, 2000). Indeed, the complex trait of BP has yielded a very large number of QTLs that, if likely to be individually of lower complexity, nonetheless remain statistical entities that may be the manifestation of yet unidentified molecular variant(s). It is widely recognized that the greatest challenge rests in this endgame. Contrasted views have been expressed by experts in the field on the manner in which this could be achieved (Glazier *et al.*, 2002; Korstanje and Paigen, 2002; Nadeau and Frankel, 2000; Nadeau *et al.*, 2000), leading a consortium of 80 scientists to issue a loosely formulated consensus view (Abiola *et al.*, 2003). Progress from QTL to genes is promising (Korstanje and Paigen, 2002) but remains limited to a small fraction of QTLs that typically exhibit unique phenotypes and/or large phenotypic effects (Glazier *et al.*, 2002).

A review of the literature in April 2005 (Flint *et al.*, 2005) makes the clearest case. It lists 700 QTLs in rat and 2,050 QTLs in rodents. These are statistical entities based only on significant linkage in segregating crosses. Among these, the authors note that "if we uncritically accept all the claims as correct, only about 20 genes have been identified." They add that "at the present rate of progress (about 20 genes identified in 15 years) it will take 1,500 years to find all the genes that underlie known QTLs." Furthermore, "it is likely that only a small fraction of the total number of QTLs that segregate in inbred strain crosses are currently known to us" and therefore that "we urgently need advances in gene, not QTL, identification."

These authors list 21 QTLs that would qualify as success. They do note that defining success, namely proving the causality of the molecular-to-phenotype relationship, remains challenging, citing two recent reviews (Glazier *et al.*, 2002; Korstanje and Paigen, 2002) where catalogs of purported successes reveal marked discrepancies. It is also evident, and generally acknowledged, that success is limited to traits with unique phenotypic characteristics and/or major genetic effects that approach Mendelian inheritance.

E. Few genes, if any, explaining a blood pressure QTL have been identified by the traditional approach

Many QTLs underlying BP have been identified in rodents, with one or more on most rat chromosomes (Rapp, 2000). These vary considerably with background, crosses, and phenotyping methods. Of the 21 genes nominated as successes by Flint *et al.* (2005), three purportedly involve BP QTLs.

The first one listed, CYP11B1, was first identified as a biochemical Mendelian trait segregating in the Dahl rats in the 1970s (Rapp and Dahl, 1972, 1976) accounting for a BP difference of as much as 23 mmHg. Mutations in the gene differentiating Dahl salt-resistant from Dahl salt-sensitive rats were identified in 1993 (Cicila *et al.*, 1993; Matsukawa *et al.*, 1993). Decades of refined mapping through congenics eventually led to the "demonstration" in 2003 that CYP11B1 and a BP difference could be retained in a 177-kb introgressed fragment, leading the authors to conclude "We believe that we have reached the limit of resolution for congenic analysis of a QTL in a rodent animal model, and we conclude that CYP11B1 causes the observed QTL on rat chromosome 7 in Dahl rats" (Garrett and Rapp, 2003). This was indeed a remarkable feat, but the flow of inference was basically reversed: from gene to QTL.

A second entry (Flint *et al.*, 2005) is ACE2. It is true that, in an initial report, the gene was located in a QTL region in several rats models of hypertension and that both protein and mRNA were lower in hypertensive compared to normotensive rats (Crackower *et al.*, 2002). The same publication also reported that targeted disruption of the gene induced a defect in cardiac contractility but no effects on BP or renal morphology and function. Consequently, the causal implication of ACE2 in models of hypertension remains a hypothesis (Yagil and Yagil, 2003).

The third gene listed, CD36, offers an even more insightful example of the challenge of mapping genes underlying complex phenotypes. A QTL for both high BP and dyslipidemia on the SHR rat chromosome 4 was further isolated in a congenic animal that expressed both phenotypic components and retained a 40-cM segment of rat chromosome 4 (Aitman *et al.*, 1999). In subsequent microarray screening, a single gene in the region exhibited profound differential expression and a molecular lesion accounting for loss of function in SHR. In subsequent work, it became evident that while CD36 accounts for the dyslipidemia phenotype, it did not explain the increased BP preserved in the congenic animal (Pravenec *et al.*, 2003). In more recent work, reviewed by Pravenec and Kurtz (2007), a relationship between CD36 deficiency and the hypertension of the metabolic syndrome has become more apparent. This was the bleak reality of the situation at the time of this writing (late 2005).

III. A STRONG QTL FOR SODIUM SENSITIVITY IN THE C57BL/6J MOUSE INBRED LINE

Given the acknowledged lack of specificity of the hypertension phenotype, intermediate phenotypes that would more closely reflect an underlying genetic mechanism should serve as insightful substitutes for genetic studies. Sodium sensitivity (NAS) is one such commonly proposed phenotype. By convention, it is defined as an increase in BP—usually above some arbitrary threshold—in response to sodium loading, the thresholds and the maneuvers applied admitting substantial variation. It is commonly understood as a phenotype reflecting volume-dependent hypertension. As the kidney is the most powerful servo-controller of BP (Guyton, 1980, 1991), it is agreed that an underlying mechanism for this phenotype should first be sought after in the kidney.

The extensive phenotypic variability displayed among inbred lines of mouse, as documented in the Phenome Database (http://phenome.jax.org/phenome) offers an excellent opportunity to proceed with gene identification in simple crosses between select lines.

A. QTL mapping of NaS between C57BL/6J and A/J

The NaS of the inbred line C57BL/6J (denoted thereafter as B6) has long been recognized in the field, to the point that it is commonly mentioned without explicit reference (Meneton *et al.*, 2000). The genetic basis of this B6 phenotype is now supported by two publications. Sugiyama *et al.* (2001) have reported six QTL affecting the salt-sensitivity of BP in a backcross between B6 × AJ and B6 animals. The most significant QTL by far was located on chromosome 4. While initial analyses were compatible with the existence of two putative QTL in this region, the final model (Table 2 in Sugiyama *et al.*, 2001) provided support for only one QTL at this site, with $F_{1,242} = 56.0$, $p = 1.3 \times 10^{-12}$.

Independent confirmation of the chromosome 4 QTL, thereafter referred to as QTL_4, was obtained in a large (A/J × B6) F2 intercross, with a LOD score of 9.8 (Woo and Kurtz, 2003). When we pool and analyze jointly the individual data from the two studies, we obtain a maximum LOD score in excess of 22 (see Section II.B). It follows that support for this QTL is very strong and has been replicated. It is remarkable that, of a total of 10 QTLs identified in these two studies contrasting B6 and AJ, concordance was likely for only a second QTL of lesser significance on chromosome 1. QTL_4 behaves as a recessive B6 trait: (1) it was detected in the backcross to B6 (Sugiyama *et al.*, 2001); (2) in the intercross (Woo and Kurtz, 2003), the estimated effects on BP of QTL_4 yielded similar, lower values for heterozygous and AJ homozygous animals compared to homozygous carriers of B6 alleles.

Further evidence of the NaS of B6 is substantiated by technical differences between the two experiments. Woo *et al.* (Woo and Kurtz, 2003), measuring tail-cuff BP in animals warmed at 31 °C (so as to ensure blood flow through the ventral tail artery), identified a strain difference under standard diet. By prewarming the animals to 37 °C, Sugiyama *et al.* (2001) induced global body vasodilatation that masked this baseline strain difference; and consequently they had to use 2 weeks of sodium loading to re-express this BP difference. This is a clear sign of the volume dependence of BP regulation in B6, and that the physiological difference is manifest as a strain-specific effect even in the absence of sodium loading. If follows that differential expression as a function of strain alone should prove meaningful in our screen for genes underlying QTL_4 (Section III.A and Table 24.1).

B. Combining data for the chromosome 4 QTL

We have combined the individual data from these two studies (Sugiyama *et al.*, 2001; Woo and Kurtz, 2003) to estimate the overall linkage support for QTL_4 using R/QTL (Broman *et al.*, 2003). A LOD score in excess of 22 is obtained, as shown in Fig. 24.1, where the LOD score curve is plotted with reference to the genomic sequence. Despite the high LOD score achieved overall due to large sample size, QTL_4 accounts for a 5–7 mmHg increase in SBP, less than 10% of the total variance of the trait.

C. A QTL that accounts for a small proportion of the total variance of a quantitative phenotype of low specificity, such as BP, may not be amenable to fine mapping and positional cloning

Once a QTL has been identified, the traditional approach of "positional cloning" relies on refined mapping of the QTL through further crosses until a "reasonable" critical region has been identified. Rodents afford various alternatives such as

Table 24.1. Summary of Genes with Significant Differential Expression (2) as a Function of (a) Diet, (b) Strain, and (c) Strain × Diet Interaction With and Without Bonferroni Correction for Multiple Comparisons

GCRMA-ANOVA analysis of MU430									
Probe set ID	Gene title	Chr	Start position	Diet	Bonferroni *p*-value	Min	Max	Fold AJ	Fold B6
(a) Diet (1)									
1431085_at	RIKEN cDNA D630039A603 gene	4	57184366	0.0014	0.0233	6.3471	8.0345	1.0030	0.1622
1425150_at	RIKEN cDNA C730036D15 gene	4	48624436	0.0014	0.0236	5.8173	9.0745	1.9605	1.8974
1453193_s_at	Kinesin family member 12	4	61493128	0.0035	0.0596	7.6221	8.6767	−0.5253	−0.5393
1448288_at	Nuclear factor I/B	4	80569065	0.0079	0.1342	5.2219	7.6041	0.8556	0.8902
1450613_x_at	–	4	86966352	0.0110	0.1872	2.6068	3.2606	−0.3440	−0.2373
1427633_a_at	Pregnancy-associated plasma protein A	4	63452281	0.0177	0.3017	3.5382	6.4062	1.2256	1.0395
1424083_at	ROD1 regulator of differentiation 1 (*S. pombe*)	4	58742766	0.0184	0.3130	3.9843	5.5570	−0.3300	1.1527
1438244_at	–	4	80770764	0.0185	0.3137	4.8026	6.7376	1.0129	1.0082
1419182_at	Polydomain protein	4	57310550	0.0186	0.3155	3.6919	6.1918	−0.7350	−0.8344
1423679_at	RIKEN cDNA 2810432L 12 gene	4	48830454	0.0240	0.4078	6.3061	7.4307	−0.7012	−0.3842
1453191_at	"Procollagen, type XXVII, alpha 1"	4	61543329	0.0279	0.4744	8.1037	9.0075	0.4608	0.1921
1448755_at	"Procollagen, type XV"	4	46453285	0.0391	0.6648	5.5918	7.0975	−0.8478	−0.3338
1424877_a_at	"Aminolevulinate, delta-, dehydratase"	4	60836670	0.0395	0.6709	8.5311	10.8574	0.4792	0.0688

(*Continues*)

Table 24.1. (*Continued*)

GCRMA-ANOVA analysis of MU430									
Probe set ID	**Gene title**	**Chr**	**Start position**	**Diet**	**Bonferroni *p*-value**	**Min**	**Max**	**Fold AJ**	**Fold B6**
1449168_a_at	A kinase (PRKA) anchor protein 2	4	57122257	0.0417	0.7089	6.0406	7.5660	1.0536	0.1356
1429882_at	RIKEN cDNA 2610005L07 gene	4	60274112	0.0423	0.7190	3.5205	6.3387	1.7506	0.6952
1423636_at	WD repeat domain 31	4	60781189	0.0490	0.8322	3.2924	3.6031	−0.2072	−0.0141
1425343_at	RIKEN cDNA 2810435D12 gene	4	60826509	0.0494	0.8404	4.6650	7.0504	−0.1272	0.6629
Probe set ID	**Gene title**	**Chr**	**Start position**	**Strain**	**Bonferroni *p*-value**	**Min**	**Max**	**Fold norm**	**Fold high**
(*b*) *Strain* (3)									
1448318_at	Adipose differentiation-related protein	4	84931013	0.0000	0.0000	6.9490	9.4411	2.1877	2.1286
1424877_a_at	"Aminolevulinate, delta-, dehydratase"	4	60836670	0.0000	0.0003	8.5311	10.8574	−1.4302	−1.8406
1425343_at	RIKEN cDNA 2810435D12 gene	4	60826509	0.0002	0.0031	4.6650	7.0504	−1.7137	−0.9236
1437085_at	RIKEN cDNA D630039A03 gene	4	57184366	0.0004	0.0056	6.3477	8.0345	−0.2806	−1.1214
1450903_at	RAD23b homolog (*S. cerevisiae*)	4	54595095	0.00009	0.0111	7.7804	8.9448	−0.5709	0.8191
1425124_at	RIKEN cDNA 5830442J gene	4	60755042	0.0027	0.0357	6.9827	8.4453	−0.2969	−0.9990
1417777_at	Leukotriene B4 12-hydroxydehydrogenase	4	58234587	0.0169	0.2202	7.0145	8.1080	−0.3997	−0.6289
1416142_at	–	4	85128547	0.0175	0.2276	5.0360	7.6610	1.3481	1.1879
1423636_at	WD repeat domain 31	4	60781189	0.0221	0.2876	3.2924	3.6031	−0.2172	−0.0242

1437610_x_at	Ribosomal protein S8	4	80405602	0.0317	0.4125	10.9499	12.2295	−0.0927	−0.5753
1453191_a_at	"Procollagen, type XXVII, alpha 1"	4	61543329	0.0382	0.4964	8.1037	9.0075	−0.1688	−0.4375
1426315_a_at	RIKEN cDNA 6330416G13 gene	4	61887729	0.0446	0.5793	5.3321	6.0797	−0.3314	−0.2529
1428384_at	"DNA segment, Chr 4, Brigham and Women's Genetics 0951 expressed"	4	79185996	0.0452	0.5880	5.4937	7.3125	−0.2950	−0.6906

Probe set ID	Gene title	Chr	Start position	Diet:strain	Bonferroni p-value	Min	Max
(c) Strain × diet							
1424083_at	ROD1 regulator of differentiation 1 (*S. pombe*)	4	58742766	0.0019	0.0134	3.9843	5.5570
1437085_at	RIKEN cDNA D630039A03 gene	4	57184366	0.0102	0.0715	6.3477	8.0345
1455591_at	"Neural precursor cell expressed, developmentally downregulated gene 10"	4	61466022	0.0201	0.1407	5.6600	6.9260
1424426_at	Methylthioadenosine phosphorylase	4	87446105	0.0355	0.2484	5.2122	5.7733
1451469_at	RIKEN cDNA B430108F07 gene	4	83372585	0.0371	0.2596	2.8080	4.1451
1423636_at	WD repeat domain 31	4	60781189	0.0380	0.2657	3.2924	3.6031
1453077_a_at	"Small nuclear RNA activating complex, polypeptide 3"	4	81692212	0.0389	0.2726	2.8340	3.7050

(1) Fold change calculated high-norm.
(2) Log base 2.
(3) Fold change calculated B6-AJ.

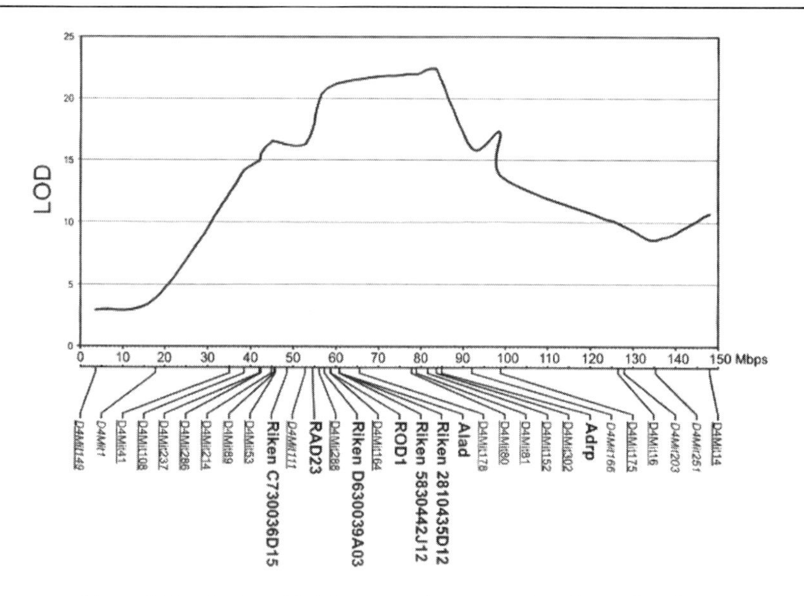

Figure 24.1. Combined LOD score on chromosome 4 with typed markers.

introgression of a region that maintains the phenotype from one strain to the other, of which chromosome segregating strains and the congenic approach are typical examples. Whereas this is rather straightforward for single Mendelian trait with complete penetrance, the approach has proven most challenging for quantitative trait with complex inheritance, as typically the phenotype under investigation disappears as the critical region is refined through iterative crosses. A variety of reasons can be advanced to account for this phenomenon.

First, multiple genes usually underlie a given QTL for a complex trait. "Perhaps the most important finding to emerge from attempts to clone genes that underlie rodent QTLs is that a single genetic effect, as detected in an inbred strain cross, turns out in most cases to be due to several physically linked small effects" (Flint *et al.*, 2005). Further, they add "Classical approaches to gene identification generally assume that a single QTL is just that, a single genetic effect. By not allowing for the fractionation of the QTL, gene identification becomes much more frustrating". They cite a large number of reports in support of this conclusion. Among these, Ron Davis' work in Yeast (Steinmetz *et al.*, 2002) and Wayne Frankel's experience with a seizure QTL in mouse (Legare *et al.*, 2000) stand out for the clarity of their demonstration. No statistical argument can resolve this issue *a priori*.

Second, chromosome substitution strains (CSS) are no solution to this problem. "The main drawback of the method is that it makes no allowances for the fractionation of a large QTL effect into many loci with smaller effects"

(Flint *et al.*, 2005). It is remarkable that a leading modern proponent of what has been a long-standing method in plants and Drosophila (Nadeau *et al.*, 2000), frustrated by the "long and bumpy road" leading from QTL discovery to gene identification, has advocated mutagenesis as a global alternative to gene identification (Nadeau and Frankel, 2000). This statement speaks for itself.

Third, positional cloning may not apply to complex traits such as BP. Whether CSS or other crosses are used, loci exerting small effects on a continuous trait such as BP cannot be mapped by observing overlap regions or separation by single crossing-over, as can be done for single Mendelian traits. True QTL isolation and positional cloning do not apply. Indeed, introgression into a small interval would require a double recombinant and an obvious segregating phenotype. This is just not feasible here.

Fourth, background remains challenging in all genetic crosses, just as ethnic heterogeneity is in human studies. QTL isolation critically depends on the genetic background selected, and this choice may be problematic. The considerable experience developed in the Dahl rat, where it is recognized that "the genetic background of the R [sodium-resistant] rats is not very permissive for expressing BP differences" (Rapp, 2000), is most relevant in our situation. Bringing sodium-sensitive (NaS) QTLs into a sodium-resistant (NaR) background usually lead to their suppression, even when accounting for their recessivity. The alternative used in rat, namely transferring regions from the NaR to NaS, and therefore attempting to follow and isolate genes leading to a decrease in BP has so far not afforded gene identification.

Lastly, guessing candidate underlying QTLs has proven most elusive. It has been a long established tradition to guess the implication of an apparently obvious candidate gene underlying a QTL on the basis of expression or known function. Examples abound where these educated guesses have not been sustained by experimental proof. Most notorious examples in rats include the Sa gene on chromosome 1 (Iwai and Inagami, 1991), the ACE gene on chromosome 10 (Hilbert *et al.*, 1991; Jacob *et al.*, 1991), discoveries reported in *Nature and Cell* with the triumphant editorial accolades of "ACE is it" and "ACE in the hole", renin on chromosome 13 reported in *Science* (Rapp *et al.*, 1989), and more recently ACE2 on the X chromosome in *Nature* (Crackower *et al.*, 2002). These predictions have not borne fruit.

IV. FROM QTL TO GENES: GENE PRIORITIZATION

A panel of 80 experts in the field agree: "It is now feasible to identify the gene (or other functional element) that is responsible for a quantitative trait, even when the critical region is relatively large. The availability of known polymorphisms among strains also facilitates the identification of candidate genes in larger critical regions" (Abiola *et al.*, 2003). The continuous likelihood of linkage can be used to select candidates without resorting to arbitrary or elusive "critical" boundaries (see below).

The end-game of gene identification involves two conceptual elements: (1) the selection of genes for further investigation, or gene prioritization, and (2) the application of high-throughput functional screens.

A. Gene prioritization based on differential expression in select tissues or cells

Expression profiling affords tests of the hypothesis that genes underlying a QTL may be differentially expressed in the effector tissue. To test whether we could identify genes that are differentially expressed in the kidneys of B6 and AJ as a function of dietary sodium, we performed three successive screens. Because microarray technology, genome coverage and analytical methods have considerably improved over the past three years, we will summarize only our latest experiments using Affymetrix Mu430 2.0 mouse microarrays. All screens involved 6-month-old B6 and AJ males fed diet with either standard (0.5%) or high sodium (3.6%) for two weeks; six animals were used for each combination of strain and diet, for a total of 24 animals. After removal of renal capsule and surrounding fat tissue, kidneys were pooled for the six animals in each of the four groups. RNA was isolated using Trizol reagent, and further purified with RNeasy columns (Qiagen). RNA quality was verified by gel electrophoresis and OD measurements.

1. Probe generation and array hybridization

cDNAs were prepared from each RNA pool using reverse transcriptase (GIBCO-BRL SuperScript II). Each cDNA was subsequently used as a template in triplicate reactions to generate biotin-labeled cRNA using an *in vitro* transcription reaction (Enzo). Each cRNA (three per pool) was hybridized with an individual Affymetrix Mu430 oligonucleotide array, which was subsequently processed and scanned according to the manufacturer's instructions (Affymetrix). All arrays were hybridized in the same batch to avoid variability in hybridization conditions. Data were saved as raw image files and converted into probe set data files using Microarray Suite (Affymetrix, MAS 5.0).

The Mu430 gene chip represents a collection of transcripts compiled from dbEST (NCBI, June 2002), GenBank (NCBI, Release 129, April 2002), and RefSeq (NCBI, June 2002) and the Whitehead Institute Center for Genome Research (April 2002). The array encompasses around 39,000 transcripts of 34,000 genes. In Affymetrix GeneChip arrays, each gene is probed by eleven

25-mer oligonucleotides that perfectly match to mRNA sequence of the gene, referred to as perfect match (PM) probes. In order to account for optical noise and cross-hybridization, the standard Affymetrix algorithm (MAS 5.0) pairs each PM probe with a mismatch probe (MM) that is identical to the PM except for the middle base. The intensities observed on the MM probes are subtracted from the PM intensities to give background-corrected intensities. Alternative algorithms have been developed and shown to outperform MAS 5.0. Among those, the robust multiarray analyses RMA and GCRMA have become popular substitutes for MAS 5.0 (Bolstad *et al.*, 2003, 2004; Cope *et al.*, 2004; Irizarry *et al.*, 2003a,b).

2. Quality control

Quality control was performed at all steps. RNA quality prior to *in vitro* transcription was assessed by gel electrophoresis and OD measurements. Following chip hybridization, overall assessment of 5′ to 3′ representation of all probe sets indicates fragmentation and integrity of the starting material.

3. Microarray analysis methods

We applied both RMA and GCRMA to analyze Affymetrix probe set data. Software for RMA and GCRMA is freely available (*bioconductor.org*) for use in the R-package for statistical computing (*r-project.org*) (Gentleman *et al.*, 2004). In RMA, the probe set data from all arrays were simultaneously normalized using quantile normalization, which eliminates systematic differences between chips, without significantly altering the relative intensity of probes within a chip. Thereafter, a mean optical background level for each array is estimated, and the intensity for each probe is adjusted to remove this background or noise. The normalized, background-corrected data is then transformed to a log-based 2 scale. A median-polish procedure is used to combine multiple probes into a single measure of expression for each gene on each array. This global background adjustment gains substantially in precision with only minor sacrifices in accuracy. GCRMA (Bolstad *et al.*, 2003; Cope *et al.*, 2004; Irizarry *et al.*, 2003a,b; Wu and Irizarry, 2004) integrates the base composition of probes and same summarization and normalization algorithms as in RMA. Assessments on the Latin-square data set provided by Affymetrix suggested that with comparable precision to RMA, GCRMA has much better accuracy than MAS 5.0 and some of the widely used alternatives including RMA, dChip, and PerfectMatch. The improvement in accuracy is most obvious for moderate expression levels. The expression measures obtained using GCRMA can be used for further analysis, such as differential expression estimation. GCRMA provides estimates of differential expression and standard errors for those estimates directly from the probe level data. In this application, we only provide the summary tables obtained with the GCRMA algorithm.

4. Microarray analysis: results

Expression data obtained with Mu430 were analyzed with GCRMA. Expression levels of genes mapping between markers D4Mit214 and D4Mit166 (Fig. 24.1) were subjected to two analysis of variance (ANOVA) testing for strain, diet, and strain × diet interaction. Results were sorted based on statistical significance. Bonferroni correction was applied to account for multiple comparisons. Table 24.1 summarizes genes with significant differential expression as a function of strain (a), diet (b), and strain × diet interaction (c). Gray background highlights genes that retain significance after Bonferroni correction. Validation through RT-PCR and subsequent functional screens would be prioritized by taking into account the likelihood ratio in support of linkage at each candidate gene relative to the maximum likelihood.

B. Gene prioritization based on *in silico* search of functional annotations and genomic differences among strains

1. Search of functional annotations

Genes can be investigated in terms of function and included or excluded based on tissue- or cell-specific expression thereby prioritizing candidate status for functional testing. As genetic variants accounting for a QTL may not lead to differential expression, an alternative to select genes for functional screens is to use either (1) *in silico* searches of functional annotations, (2) *in silico* examination of gene expression in target tissue(s), and (3) tests for genomic differences among the strains contrasted, analogous to SNP differences in case-control studies.

With advances in gene annotations, clustering differentially expressed genes into functional classes has become a routine task in microarray analyses or QTL mapping. Based on Gene Ontology, a large number of tools are available (http://www.geneontology.org/GO.tools.shtml). Most approaches "interpret" genes of interest separately using categories from the three ontologies: "biological processes", "molecular function", and "cellular component". Current efforts focus on integrating individual modules in multidimensional functional modules of integrated cellular functions (Zhu *et al.*, 2007). Symatlas (http://symatlas.gnf.org/SymAtlas/) allows data mining of candidate genes for tissue- and cell-specific expression patterns and provides functional annotations. The MfunGD tool (http://mips.gsf.de/genre/proj/mfungd) provides a resource for annotated mouse proteins and their occurrence in protein networks integrating manual protein annotations and protein–protein interaction. Multiple commercial tools complement this compendium and are far too numerous to cover in this review. However, the breath of these emerging, increasingly sophisticated tools emphasizes their usefulness as unique resources to probe potential function of candidate genes defined by a QTL. Based on such searches, genes can be further prioritized by taking into account the likelihood ratio of linkage at a locus relative to the maximum LOD score.

2. Genomic differences among strains

The resequencing of 15 genetically diverse strains of inbred laboratory mice by Perlegen funded by the National Institute of Environmental Health Sciences (NIEHS) has provided a uniform and high density SNP data set of nucleotide polymorphisms (SNPs, $>8.2 \times 10^7$) relative to the reference strain already sequenced (C57BL/6J). Curation of this data set through the Jackson laboratory (http://phenome.jax.org) allows strain-specific comparisons between two or more strains and identification of differentiating SNPs. SNPs can be categorized based on their positions relative to gene structures. Priorities can be assigned based on whether the SNP results in (1) synonymous or (2) nonsynonymous codon changes, (3) functional changes in elements necessary for splicing, (4) changes in 5′ or 3′ UTRs, or (5) changes in regulatory elements controlling gene transcription. Differentiating SNPs, with or without potential functional effects, can be further investigated based on whether they are "private" and specific for only one strain. The following graphic illustrates how we are using this information to select candidate genes. Figure 24.2A depicts a genome wide comparison of SNPs differentiating C57BL6J and A/J with the QTL on chromosome 4. Pair wise, strain-specific comparisons of SNPs from C57BL/6J versus all 15 other strains, using SNP data from Perlegen2, Celera2, and Broad1 mapped to NCBI mouse build 36.1, allows ascertainment based on whether the differentiating polymorphic SNP is strain specific or also occurs in other comparisons (Fig. 24.2B). Light gray bars represent polymorphic sites where allelic data is available. Dark gray bars represent the number of SNP locations where the reference strain (C57BL/6J) and the query strain are polymorphic. Differentiating SNPs (Fig. 24.2C) can be extracted and mapped as custom tracks on more versatile genome browsers (http://genome.ucsc.edu/cgi-bin/hgGateway) allowing more in-depth analysis and exploration (Fig. 24.2D).

A notable feature emerging from this examination is that the expectation of large blocks of strain-specific differences generated by recombination between ancestral differences among founders of laboratory strains (Wade *et al.*, 2002), as predicted by initial analysis at low density, is not confirmed once this density is markedly increased (Yalcin *et al.*, 2004). What actual situation holds in humans remains to be determined.

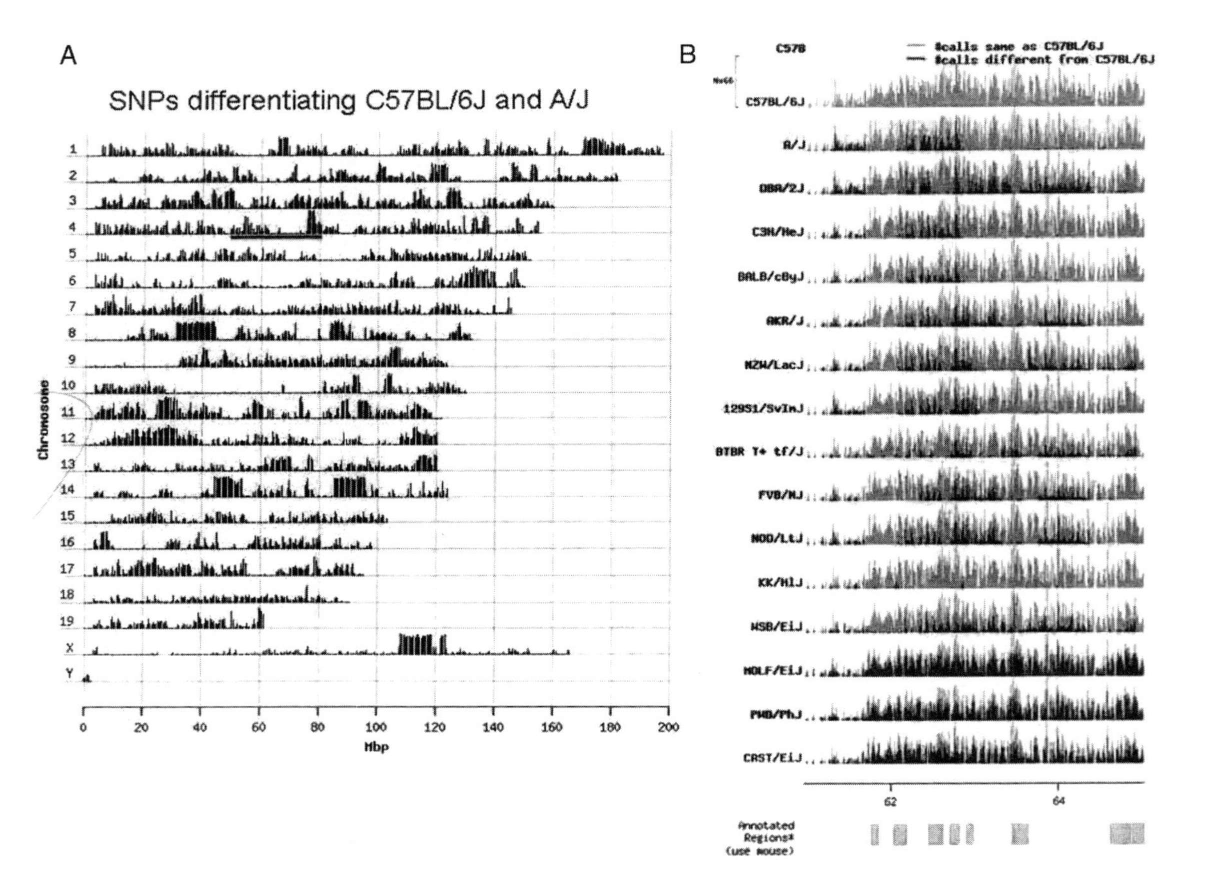

Figure 24.2. Continued

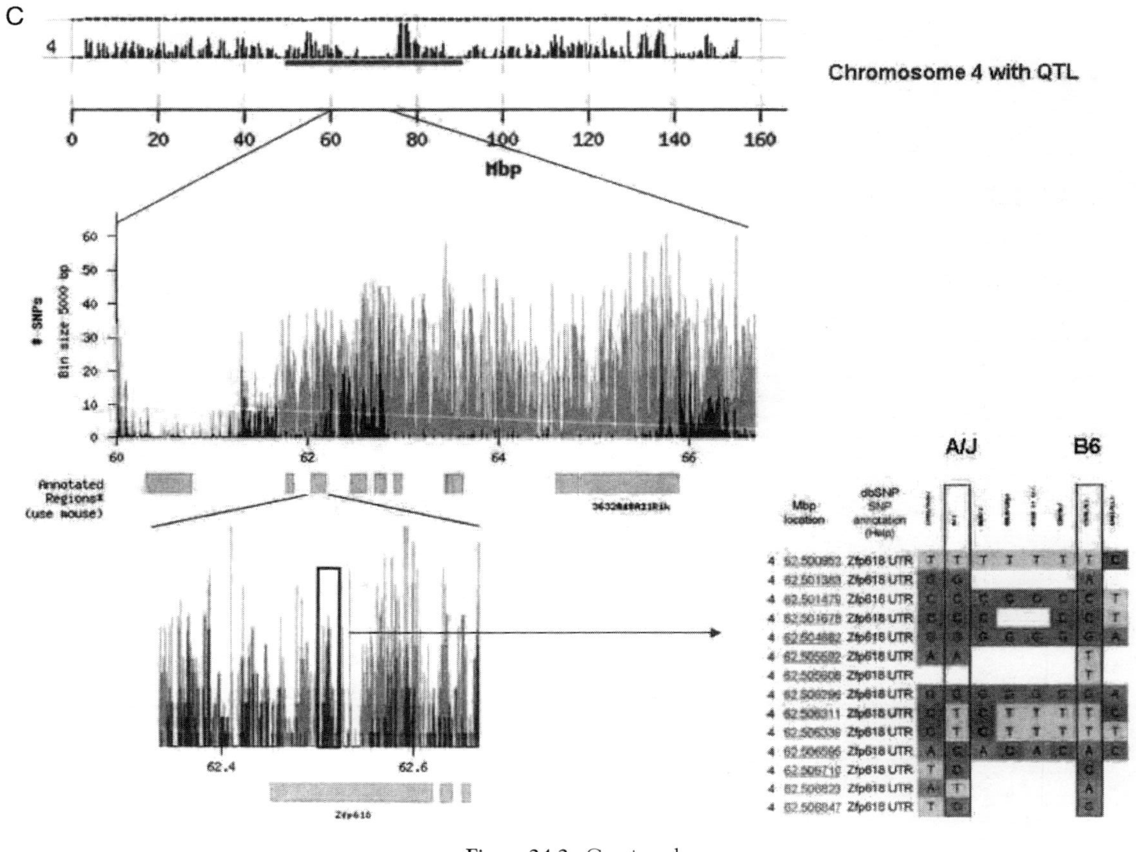

Figure 24.2. Continued

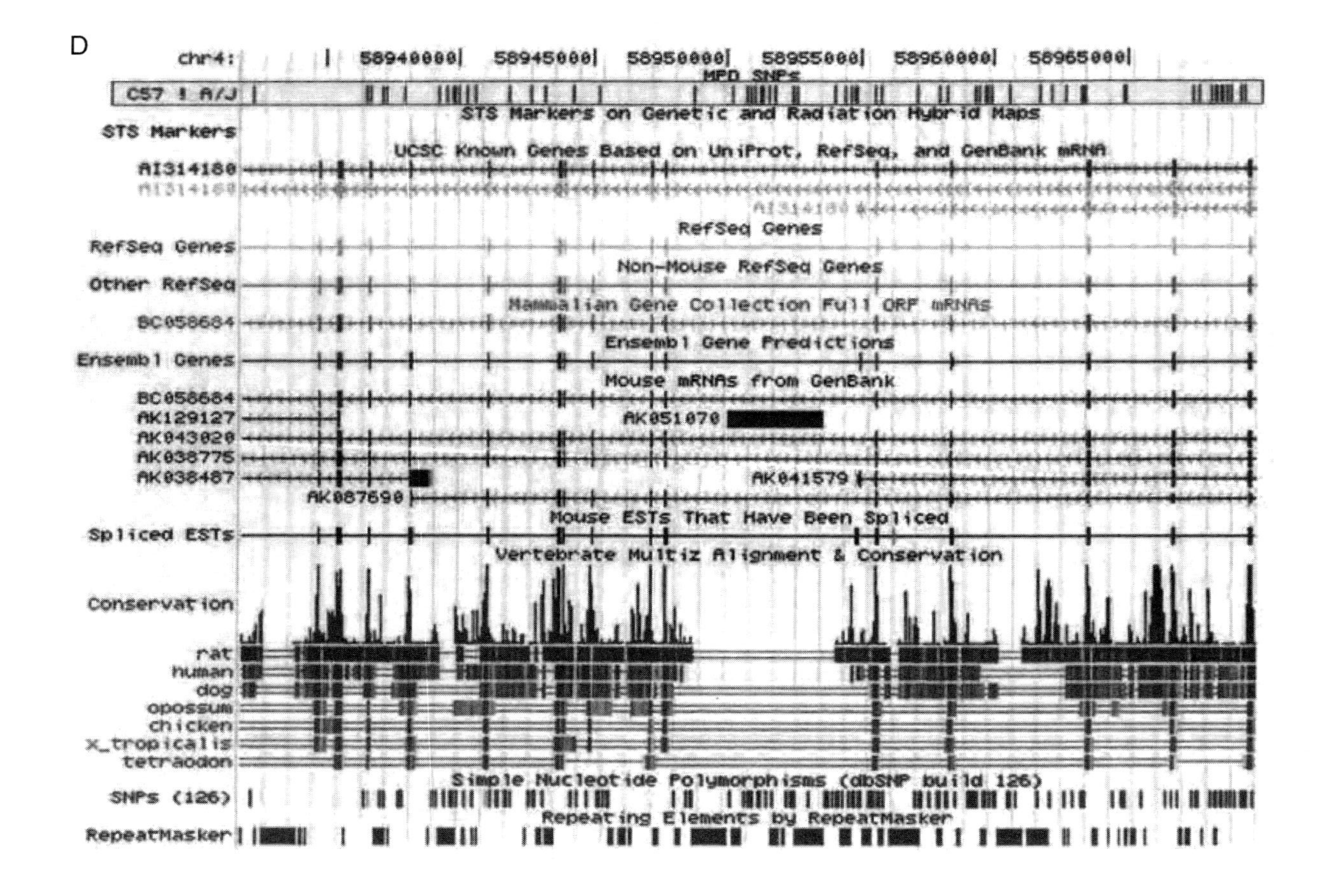

Figure 24.2. Continued

V. FROM QTL TO GENES: HIGH-THROUGHPUT FUNCTIONAL SCREENS

We will argue that recent developments in functional genomics should enable high-throughput functional screens that would do away both with the guessing game about candidate genes and the high cost in animal life and low yield of refined mapping approaches.

A. The ultimate need for a functional proof

It is generally acknowledged that the ultimate proof of the involvement of a molecular variant in a complex phenotype will consist of a demonstration that replacement of one variant by the other reproduces the complex phenotype (Abiola *et al.*, 2003; Nadeau and Frankel, 2000). Ideally this would require gene targeting, a powerful but laborious enterprise not amenable to large-scale functional screening. Even this approach is not without potential pitfalls for the analysis of complex traits. Even a clear-cut trait such as the lung phenotype of cystic fibrosis has exhibited marked attenuation by genetic background in knockout mice (Haston *et al.*, 2002).

B. The power of high-throughput functional screens

Experts agree that *in vitro* functional tests, particularly of cellular phenotypes, would constitute an efficient screening tool for putative candidate genes and variants (Abiola *et al.*, 2003; Glazier *et al.*, 2002). Evidently, this short loop affords considerable economy and high throughput by contrast with phenotypic assessment in genetically engineered animals. Rather than using BP, which requires large number of intact animals to analyze a small effect, we propose an *in vitro* screen of net sodium transport by epithelial cells of two major nephron segments that have been incriminated in human hypertension. It affords high throughput and tests a specific hypothesis, namely that the sodium sensitivity of BP of B6 compared to AJ reflects chronic activation of sodium-retention mechanisms in the kidney.

Figure 24.2. (A) Genome wide polymorphic SNPs differentiating C57BL/6J and A/J strain with underlined (gray) QTL on chromosome 4. (B) Comparison of a short segment of chromosome 4 (61–61 Mbp) for all 15 strains relative C57BL6J, using SNP data from Perlegen2, Celera2, and Broad1 mapped to NCBI mouse build 36.1. Light gray bars represent polymorphic sites where allelic data is available. Dark gray bars represent the number of SNP locations where the reference strain (C57BL/6J) and the query strain are polymorphic. (C) From an entire chromosome view to SNP or bp resolution around 62.5 Mbp. (D) SNP data extracted and mapped as custom tracks (boxed window) on UCSC genome browser (http://genome.ucsc.edu/).

C. Functional analysis of sodium transport in epithelial cells of the kidney

Genes prioritized by the methods delineated above can be subjected to high-throughput screens of their putative effect on sodium transport in proximal or distal segments of the nephron. The screening assay should fulfill the following conditions: (1) it must represent a valid biological phenotype, (2) readout conditions must allow throughput of all candidates and comparison and standardization to internal references or controls (e.g., relative 100% and/or 0% activity references), and (3) biological inference must be possible from loss-of-function or gain-of-function experiments.

We will outline here functional tests of candidate genes that are likely to affect sodium reabsorption in the collecting duct. As the amiloride-sensitive sodium channel ENaC is the final effector of sodium reabsorption in this segment of the nephron and is implicated directly or indirectly in multiple Mendelian forms of hypertension and hypotension, candidate genes will be investigated based on their effects on this channel. Functional read-outs rely on ENaC-specific short-circuit current measured directly or indirectly using fluorescent dyes.

Novel isolation techniques of intact tubules from transgenic mice expressing fluorescent reporter genes in connecting tubule and collecting duct allow large-scale isolation of tubules and examinations in intact tubules or derived primary cultures (Miller *et al.*, 2006). We recently described this novel cell isolation procedure, combining a transgenic mouse model overexpressing EGPF in connecting and collecting tubule and large particle-based flow cytometry. We showed that isolated tubules express segment-specific markers and could be used for studies ex vivo. While this approach relies on continuous animal use (albeit modest compared to congenic approaches), it provides an experimental assay system very close to *in vivo* function. Alternatively, several permanent cell lines derived from this segment of interest are available. A possible drawback of continuous cell lines is that they may not display the differentiated status of the primary cells from which they were derived.

Loss-of-function assays using siRNA afford parallel and high-throughput functional analyses of sodium transport in this segment. Recent advances in siRNA technologies, the availability of validated siRNA sequences against a large number of biological targets, and reduced "off-target" effects, make this technology amenable to functional screens. Conceivably, such approach would even allow testing of all candidates underlying the QTL. Candidate genes can also be tested for potential gain-of-function effects. Using cell culture systems again derived from isolated tubules or employing a cell line, the candidate gene can be transfected and the effect on sodium reabsorption activity monitored.

While these proposed functional screens rely on artificial experimental systems, they provide the functional tools to further prioritize tests in intact animals, ultimately providing proof that the candidate gene or its molecular variants indeed mediate the genetic effect underlying the QTL. They can accommodate the involvement of multiple genes underlying a QTL, and they can be applied without prior identification of the functional mutations accounting for the phenotypic effect. Based on our experience with the angiotensinogen gene (Lalouel and Rohrwasser, 2007), we believe formal inferences can be drawn about haplotypes and their potential relationship with phenotypes before mutational analysis has been completed.

VI. PERSPECTIVE

Although QTL mapping among inbred lines takes advantage of a situation that does not directly translate in humans, analogies can be made to situations that will arise from human experiments when linkage is followed by saturation mapping with SNPs and haplotypes begin to be delineated. Because of the extensive nature of linkage disequilibrium (LD), the complexity of LD patterns in most regions, genetic heterogeneity, and the likelihood that multiple genes and multiple alleles be involved, we believe functional screens should increasingly be considered an integral part of this end-game toward disease gene discovery. A unique advantage of such tests is that, rather than trying to identify the modest effect of an actual functional variant at a locus, they can test the more manifest role of gain- or loss-of-function at this locus. We hope the perspective developed throughout this chapter will naturally complement the presentation and discussion of statistical methods that attempt to address the formidable challenge of elucidating genetic determinants of common disease.

References

Abiola, O., Angel, J. M., Avner, P., Bachmanov, A. A., Belknap, J. K., Bennett, B., Blankenhorn, E. P., Blizard, D. A., Bolivar, V., Brockmann, G. A., Buck, K. J., Bureau, J. F., *et al.* (2003). The nature and identification of quantitative trait loci: A community's view. *Nat. Rev. Genet.* **4,** 911–916.

Aitman, T. J., Glazier, A. M., Wallace, C. A., Cooper, L. D., Norsworthy, P. J., Wahid, F. N., Al-Majali, K. M., Trembling, P. M., Mann, C. J., Shoulders, C. C., Graf, D., St Lezin, E., *et al.* (1999). Identification of Cd36 (Fat) as an insulin-resistance gene causing defective fatty acid and glucose metabolism in hypertensive rats. *Nat. Genet.* **21,** 76–83.

Barkley, R. A., Chakravarti, A., Cooper, R. S., Ellison, R. C., Hunt, S. C., Province, M. A., Turner, S. T., Weder, A. B., and Boerwinkle, E. (2004). Positional identification of hypertension susceptibility genes on chromosome 2. *Hypertension* **43,** 477–482.

Bianchi, G., and Tripodi, G. (2003). Genetics of hypertension: The adducin paradigm. *Ann. N. Y. Acad. Sci.* **986,** 660–668.

Bolstad, B. M., Irizarry, R. A., Astrand, M., and Speed, T. P. (2003). A comparison of normalization methods for high density oligonucleotide array data based on variance and bias. *Bioinformatics* **19,** 185–193.

Bolstad, B. M., Collin, F., Simpson, K. M., Irizarry, R. A., and Speed, T. P. (2004). Experimental design and low-level analysis of microarray data. *Int. Rev. Neurobiol.* **60,** 25–58.

Broman, K. W., Wu, H., Sen, S., and Churchill, G. A. (2003). R/qtl: QTL mapping in experimental crosses. *Bioinformatics* **19,** 889–890.

Cicila, G. T., Rapp, J. P., Wang, J. M., St Lezin, E., Ng, S. C., and Kurtz, T. W. (1993). Linkage of 11 beta-hydroxylase mutations with altered steroid biosynthesis and blood pressure in the Dahl rat. *Nat. Genet.* **3,** 346–353.

Cope, L. M., Irizarry, R. A., Jaffee, H. A., Wu, Z., and Speed, T. P. (2004). A benchmark for affymetrix genechip expression measures. *Bioinformatics* **20,** 323–331.

Cowley, A. W., Jr., and Roman, R. J. (1996). The role of the kidney in hypertension. *JAMA* **275,** 1581–1589.

Crackower, M. A., Sarao, R., Oudit, G. Y., Yagil, C., Kozieradzki, I., Scanga, S. E., Oliveira-dos-Santos, A. J., da Costa, J., Zhang, L., Pei, Y., Scholey, J., Ferrario, C. M., *et al.* (2002). Angiotensin-converting enzyme 2 is an essential regulator of heart function. *Nature* **417,** 822–828.

Darvasi, A., and Pisante-Shalom, A. (2002). Complexities in the genetic dissection of quantitative trait loci. *Trends Genet.* **18,** 489–491.

Dunn, D. M., Ishigami, T., Pankow, J., von Niederhausern, A., Alder, J., Hunt, S. C., Leppert, M. F., Lalouel, J. M., and Weiss, R. B. (2002). Common variant of human NEDD4L activates a cryptic splice site to form a frameshifted transcript. *J. Hum. Genet.* **47,** 665–676.

Edwards, J. H. (1960). The simulation of mendelism. *Acta Genet. Stat. Med.* **10,** 63–70.

Elston, R. C., and Stewart, J. (1971). A general model for the genetic analysis of pedigree data. *Hum. Hered.* **21,** 523–542.

Flint, J., Valdar, W., Shifman, S., and Mott, R. (2005). Strategies for mapping and cloning quantitative trait genes in rodents. *Nat. Rev. Genet.* **6,** 271–286.

Garrett, M. R., and Rapp, J. P. (2003). Defining the blood pressure QTL on chromosome 7 in Dahl rats by a 177-kb congenic segment containing Cyp11b1. *Mamm. Genome* **14,** 268–273.

Gentleman, R. C., Carey, V. J., Bates, D. M., Bolstad, B., Dettling, M., Dudoit, S., Ellis, B., Gautier, L., Ge, Y., Gentry, J., Hornik, K., Hothorn, T., *et al.* (2004). Bioconductor: Open software development for computational biology and bioinformatics. *Genome Biol.* **5,** R80–R80.15.

Glazier, A. M., Nadeau, J. H., and Aitman, T. J. (2002). Finding genes that underlie complex traits. *Science* **298,** 2345–2349.

Grisk, O., and Rettig, R. (2001). Renal transplantation studies in genetic hypertension. *News Physiol. Sci.* **16,** 262–265.

Guyton, A. C. (1980). "Arterial Pressure and Hypertension." W.B. Saunders, Philadelphia.

Guyton, A. C. (1991). Blood pressure control—Special role of the kidneys and body fluids. *Science* **252,** 1813–1816.

Hall, J. E., and Guyton, A. C. (1990). Control of sodium secretion and arterial pressure by intrarenal mechanisms and the renin-angiotensin system. *In* "Hypertension: Pathophysiology, Diagnosis and Management" (J. H. Laragh and B. M. Brenner, eds.), pp. 1105–1129. Raven Press, New York.

Haston, C. K., McKerlie, C., Newbigging, S., Corey, M., Rozmahel, R., and Tsui, L. C. (2002). Detection of modifier loci influencing the lung phenotype of cystic fibrosis knockout mice. *Mamm. Genome* **13,** 605–613.

Hilbert, P., Lindpaintner, K., Beckmann, J. S., Serikawa, T., Soubrier, F., Dubay, C., Cartwright, P., De Gouyon, B., Julier, C., Takahasi, S., Vincent, M., Ganten, D., *et al.* (1991). Chromosomal mapping of two genetic loci associated with blood-pressure regulation in hereditary hypertensive rats. *Nature* **353,** 521–529.

Hunt, S. C., Xin, Y., Wu, L. L., Cawthon, R. M., Coon, H., Hasstedt, S. J., and Hopkins, P. N. (2006). Sodium bicarbonate cotransporter polymorphisms are associated with baseline and 10-year follow-up blood pressures. *Hypertension* **47,** 532–536.

Irizarry, R. A., Bolstad, B. M., Collin, F., Cope, L. M., Hobbs, B., and Speed, T. P. (2003a). Summaries of affymetrix genechip probe level data. *Nucleic Acids Res.* **31,** e15, 1–8.

Irizarry, R. A., Hobbs, B., Collin, F., Beazer-Barclay, Y. D., Antonellis, K. J., Scherf, U., and Speed, T. P. (2003b). Exploration, normalization, and summaries of high density oligonucleotide array probe level data. *Biostatistics* **4,** 249–264.

Iwai, N., and Inagami, T. (1991). Isolation of preferentially expressed genes in the kidneys of hypertensive rats. *Hypertension* **17,** 161–169.

Jacob, H. J., Lindpaintner, K., Lincoln, S. E., Kusumi, K., Bunker, R. K., Mao, Y. P., Ganten, D., Dzau, V. J., and Lander, E. S. (1991). Genetic mapping of a gene causing hypertension in the stroke-prone spontaneously hypertensive rat. *Cell* **67,** 213–224.

Jeunemaitre, X., Soubrier, F., Kotelevtsev, Y. V., Lifton, R. P., Williams, C. S., Charru, A., Hunt, S. C., Hopkins, P. N., Williams, R. R., Lalouel, J. M., *et al.* (1992). Molecular basis of human hypertension: Role of angiotensinogen. *Cell* **71,** 169–180.

Korstanje, R., and Paigen, B. (2002). From QTL to gene: The harvest begins. *Nat. Genet.* **31,** 235–236.

Lalouel, J. M. (2001). From genetics to mechanism of disease liability. *In* "Genetic Dissection of Complex Traits" (D. C. Rao and M. Province, eds.), pp. 517–533. Academic Press, San Diego.

Lalouel, J. M. (2003). Large-scale search for genes predisposing to essential hypertension. *Am. J. Hypertens.* **16,** 163–166.

Lalouel, J. M., and Morton, N. E. (1981). Complex segregation analysis with pointers. *Hum. Hered.* **31,** 312–321.

Lalouel, J. M., and Rohrwasser, A. (2007). Genetic susceptibility to essential hypertension: Insight from angiotensinogen. *Hypertension* **49,** 597–603.

Lalouel, J. M., Rao, D. C., Morton, N. E., and Elston, R. C. (1983). A unified model for complex segregation analysis. *Am. J. Hum. Genet.* **35,** 816–826.

Lalouel, J. M., Rohrwasser, A., Terreros, D., Morgan, T., and Ward, K. (2001). Angiotensinogen in essential hypertension: From genetics to nephrology. *J. Am. Soc. Nephrol.* **12,** 606–615.

Lander, E. S., Green, P., Abrahamson, J., Barlow, A., Daly, M. J., Lincoln, S. E., and Newburg, L. (1987). MAPMAKER: An interactive computer package for constructing primary genetic linkage maps of experimental and natural populations. *Genomics* **1,** 174–181.

Lathrop, G. M., Lalouel, J. M., Julier, C., and Ott, J. (1984). Strategies for multilocus linkage analysis in humans. *Proc. Natl. Acad. Sci. USA* **81,** 3443–3446.

Legare, M. E., Bartlett, F. S., II, and Frankel, W. N. (2000). A major effect QTL determined by multiple genes in epileptic EL mice. *Genome Res.* **10,** 42–48.

Lifton, R. P., Gharavi, A. G., and Geller, D. S. (2001). Molecular mechanisms of human hypertension. *Cell* **104,** 545–556.

Liu, W., Zhao, W., and Chase, G. A. (2004). Genome scan meta-analysis for hypertension. *Am. J. Hypertens.* **17,** 1100–1106.

MacLean, C. J., Morton, N. E., and Lew, R. (1975). Analysis of family resemblance. IV. Operational characteristics of segregation analysis. *Am. J. Hum. Genet.* **27,** 365–384.

Matsukawa, N., Nonaka, Y., Higaki, J., Nagano, M., Mikami, H., Ogihara, T., and Okamoto, M. (1993). Dahl's salt-resistant normotensive rat has mutations in cytochrome P450(11 beta), but the salt-sensitive hypertensive rat does not. *J. Biol. Chem.* **268,** 9117–9121.

Meneton, P., Ichikawa, I., Inagami, T., and Schnermann, J. (2000). Renal physiology of the mouse. *Am. J. Physiol. Renal Physiol.* **278,** F339–F351.

Miller, R. L., Zhang, P., Chen, T., Rohrwasser, A., and Nelson, R. D. (2006). Automated method for the isolation of collecting ducts. *Am. J. Physiol. Renal. Physiol.* **291,** F236–F245.

Morton, N. E., and MacLean, C. J. (1974). Analysis of family resemblance. 3. Complex segregation of quantitative traits. *Am. J. Hum. Genet.* **26,** 489–503.

Nadeau, J. H., and Frankel, W. N. (2000). The roads from phenotypic variation to gene discovery: Mutagenesis versus QTLs. *Nat. Genet.* **25,** 381–384.

Nadeau, J. H., Singer, J. B., Matin, A., and Lander, E. S. (2000). Analysing complex genetic traits with chromosome substitution strains. *Nat. Genet.* **24,** 221–225.
Pankow, J. S., Rose, K. M., Oberman, A., Hunt, S. C., Atwood, L. D., Djousse, L., Province, M. A., and Rao, D. C. (2000). Possible locus on chromosome 18q influencing postural systolic blood pressure changes. *Hypertension* **36,** 471–476.
Pankow, J. S., Dunn, D. M., Hunt, S. C., Leppert, M. F., Miller, M. B., Rao, D. C., Heiss, G., Oberman, A., Lalouel, J. M., and Weiss, R. B. (2005). Further evidence of a quantitative trait locus on chromosome 18 influencing postural change in systolic blood pressure: The hypertension genetic epidemiology network (HyperGEN) study. *Am. J. Hypertens.* **18,** 672–678.
Pravenec, M., and Kurtz, T. W. (2007). Molecular genetics of experimental hypertension and the metabolic syndrome. From gene pathways to new therapies. *Hypertension.* **49,** 941–952.
Pravenec, M., Landa, V., Zidek, V., Musilova, A., Kazdova, L., Qi, N., Wang, J., St Lezin, E., and Kurtz, T. W. (2003). Transgenic expression of CD36 in the spontaneously hypertensive rat is associated with amelioration of metabolic disturbances but has no effect on hypertension. *Physiol. Res.* **52,** 681–688.
Rapp, J. P. (2000). Genetic analysis of inherited hypertension in the rat. *Physiol. Rev.* **80,** 135–172.
Rapp, J. P., and Dahl, L. K. (1972). Mendelian inheritance of 18- and 11 beta-steroid hydroxylase activities in the adrenals of rats genetically susceptible or resistant to hypertension. *Endocrinology* **90,** 1435–1446.
Rapp, J. P., and Dahl, L. K. (1976). Mutant forms of cytochrome P-450 controlling both 18- and 11beta-steroid hydroxylation in the rat. *Biochemistry* **15,** 1235–1242.
Rapp, J. P., Wang, S. M., and Dene, H. (1989). A genetic polymorphism in the renin gene of Dahl rats cosegregates with blood pressure. *Science* **243,** 542–544.
Russo, C. J., Melista, E., Cui, J., DeStefano, A. L., Bakris, G. L., Manolis, A. J., Gavras, H., and Baldwin, C. T. (2005). Association of NEDD4L ubiquitin ligase with essential hypertension. *Hypertension* **46,** 488–491.
Siffert, W. (2005). G protein polymorphisms in hypertension, atherosclerosis, and diabetes. *Annu. Rev. Med.* **56,** 17–28.
Steinmetz, L. M., Sinha, H., Richards, D. R., Spiegelman, J. I., Oefner, P. J., McCusker, J. H., and Davis, R. W. (2002). Dissecting the architecture of a quantitative trait locus in yeast. *Nature* **416,** 326–330.
Sugiyama, F., Churchill, G. A., Higgins, D. C., Johns, C., Makaritsis, K. P., Gavras, H., and Paigen, B. (2001). Concordance of murine quantitative trait loci for salt-induced hypertension with rat and human loci. *Genomics* **71,** 70–77.
Wade, C. M., Kulbokas, E. J., III, Kirby, A. W., Zody, M. C., Mullikin, J. C., Lander, E. S., Lindblad-Toh, K., and Daly, M. J. (2002). The mosaic structure of variation in the laboratory mouse genome. *Nature* **420,** 574–578.
Woo, D. D., and Kurtz, I. (2003). Mapping blood pressure loci in (A/J x B6)F2 mice. *Physiol. Genomics* **15,** 236–242.
Wu, Z., and Irizarry, R. A. (2004). Preprocessing of oligonucleotide array data. *Nat. Biotechnol* **22,** 656–658; author reply 658.
Yagil, Y., and Yagil, C. (2003). Hypothesis: ACE2 modulates blood pressure in the mammalian organism. *Hypertension* **41,** 871–873.
Yalcin, B., Fullerton, J., Miller, S., Keays, D. A., Brady, S., Bhomra, A., Jefferson, A., Volpi, E., Copley, R. R., Flint, J., *et al.* (2004). Unexpected complexity in the haplotypes of commonly used inbred strains of laboratory mice. *Proc. Natl. Acad. Sci. USA* **101,** 9734–9739.
Zhu, J., Wang, J., Guo, Z., Zhang, M., Yang, D., Li, Y., Wang, D., and Xiao, G. (2007). GO-2D: Identifying 2-dimensional cellular-localized functional modules in gene ontology. *BMC Genomics* **8**(30), 1–27.

25

Into the Post-HapMap Era

Newton E. Morton
Human Genetics Division, University of Southampton,
Southampton SO16 6YD, United Kingdom

ABSTRACT

The HapMap Project has shifted genetic epidemiology of complex inheritance away from linkage into association mapping of genes affecting disease and response to therapy. Starting with a physical map produced by the Human Genome Project and recent investigation of structural polymorphisms in HapMap samples, population-specific linkage disequilibrium (LD) maps that accurately reflect the fine structure of blocks and steps have been created for use in association mapping, and by interpolation to increase the resolution of linkage maps. All this evidence can be integrated by meta-analysis if expressed as an estimated location and its standard error, a property apparently unique to

Advances in Genetics, Vol. 60 0065-2660/08 $35.00

DOI: 10.1016/S0065-2660(07)00425-7

composite likelihood, recently freed from autocorrelation by permutation of affection status. Methods that do not estimate a standard error are easier to apply, but may be misleading if a causal marker has not been typed. The month of June 2007 saw advances in genome-wide association scans (GWAS) for several diseases. Many questions remain to be answered if genetic epidemiology is to continue the significant contribution to medicine that its definition promises and its history illustrates.

I. INTRODUCTION

A succession of advances occurred in every branch of genetics during the last century. William Bateson recognized mutation and coined the vocabulary of the classical period (including *genetics*, *allele*, *epistasis*, *homozygous*, *heterozygous*, *coupling*, and *repulsion*), but refused to accept the chromosome theory that led to cytogenetics, recognition of the double helix, development of human gene mapping, and ultimately to the Human Genome Project. Archibald Garrod, a contemporary of Bateson, created biochemical genetics that reached the human genome by a different path. Wilhelm Weinberg introduced statistical genetics, which like population genetics did not realize its potential until the double helix was interrogated by computers. Along the way Aravinda Chakravarti discovered the block and step structure of linkage disequilibrium (LD) in the β hemoglobin region, which could not be generalized to the whole genome until the HapMap Project in this century.

However, my task is not to review the post-classical history of genetics, which has yet to be written, but to discuss recent developments in genetic epidemiology, the science that deals with etiology, distribution, and control of disease in groups of relatives and with inherited causes of disease in populations (Table 25.1). First, the Human Genome Project ended with virtual completion of a high-resolution map in base pairs (bp), the exemplar for other genomes and the template on which genetic determinants in our species are defined. Second, the HapMap Project established diversity in single nucleotide polymorphisms (SNPs) within and among four samples, effectively ruling out the possibility that any two individuals other than identical twins have the same genotype, or that any individual has the consensus sequence immortalized in the Human Genome Project. Finally, a diversity of variable insertions, deletions, duplications, and repeats has been demonstrated not only with sequences that produce proteins, but also in sequences that have other or unknown functions. This flood of information, still in its early stages, has pushed genomes to the forefront of genetics whether applied to evolution, geographic dispersal, disease, or other attributes.

Table 25.1. Unfolding of Complex Inheritance (After Morton, 2006)

Period	Years	Linkage	Association	kb map	LD map	SNPs	Theory
Before Human Genome	<1998	+++	±	0	0	0	0
Before HapMap	1998–2001	++	+	±	0	±	±
After HapMap	>2001	+	++	+	+	+	+

Today, the adjective *genomic* is often used carelessly as in *genomic epidemiology* to include gene interactions and other phenotypic effects that to a large degree are extranuclear and involve post-genomic products. It would be illogical to call them *genomic*, and pointless to partition *genetic* into genomic and nongenomic compartments.

II. LINKAGE MAPPING AND CYTOGENETIC ASSIGNMENT

Early success in creating a linkage map came from two approaches. One aimed at rare dominant or recessive genes with high penetrance whose genetic parameters could be determined by segregation analysis of pedigrees with specified ascertainment, leaving only recombination with a mapped marker to be estimated by linkage. These markers were fully penetrant with two or more common alleles, preferably codominant. Their cytogenetic location was determined within a chromosome band by immunology or starch-gel electrophoresis, subsequently replaced by restriction fragment length polymorphisms and later by microsatellites and other DNA markers. Deletions, inversions, and chromosome rearrangements were used in some cases for cytogenetic assignment.

Before the end of the century, these methods had proven to be highly successful, and common oligogenes (see Fig. 25.1) that were not fully penetrant became a target for linkage mapping. Affected relatives often had different causes, and methods were developed to use multiple markers (usually microsatellites) in a model expressed as variance components without allele frequency, penetrance, or ascertainment probability. Cost has been high relative to success, and gene assignments are rarely accurate within a chromosome band, at much lower resolution than the physical map. Complexity of the linkage data and model poses multiple dilemmas. Autocorrelation of multiple markers can be compensated by permutation tests, but point locations are sacrificed and

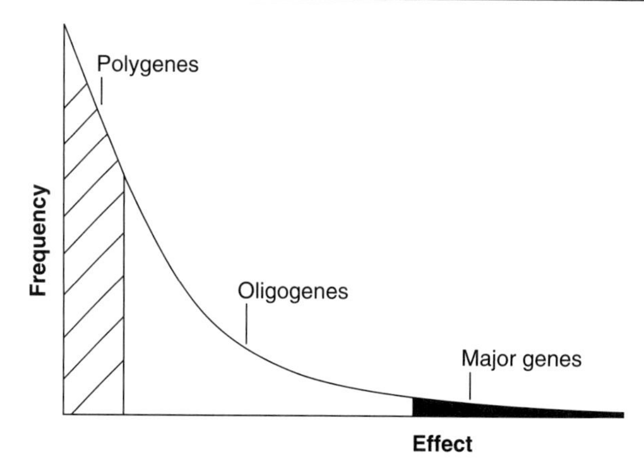

Figure 25.1. Summed causal gene frequency by mean effect (Morton, 1998).

meta-analysis loses power. Assuming no interference makes multiple markers feasible, but this Haldane model violates overwhelming evidence of interference. Conversion of estimates to approximate the Kosambi model improves location, but still underestimates interference. Diseases of late onset give many ambiguities for linkage analysis, and the sample size is usually small. Exaggerated concern that familial structure could in some way violate patient anonymity creates prejudice against linkage mapping, which continues to be useful for rare genes of high penetrance either in homozygotes with inbreeding or in heterozygotes otherwise. When neither condition is met, association mapping stimulated by the HapMap Project is replacing linkage for complex inheritance except to salvage some information from bin assignment in earlier studies. Meanwhile, linkage maps at high resolution are not created by recombination, which to be reliable would be prohibitively costly, but by interpolation from LD maps into linkage maps at much lower resolution (Tapper *et al.*, 2005).

III. ASSOCIATION MAPPING AND FUNCTIONAL ASSIGNMENT

HapMap stands like Janus, looking in both directions at once. It abandons the microsatellites of linkage maps for SNPs that are individually less informative, but cheaper and easier to type at high density. It accommodates haplotype blocks, but is indecisive about linkage maps and does not extend to maps in LD units (LDU) or to polymorphisms based on chromosomal deletions, insertions, or rearrangements. This is analogous to a linkage map that refuses to accept linkage units (cM) and certain marker classes. The great virtues of HapMap are its close association with the DNA sequence, utility for association mapping (especially for complex inheritance), and application to population structure, gene selection, and evolution.

Maps in LDU accurately reflect the block and step pattern that gives association mapping greater efficiency than physical maps provide. This is strikingly evident when selecting or analyzing multiple markers in a region and estimating confidence intervals. Approximate proportionality to the sex-averaged linkage maps in cM is distorted by selection, migration, and drift during and after population bottlenecks. Consequently, populations differ to an extent that can be determined only by construction of a regional LD map, whereas a single linkage map is satisfactory for all populations, albeit at low density. In the absence of a population-specific LD map based on a large sample, synthesis of two or more regional maps into a cosmopolitan consensus, each scaled to the same LDU length, gives shorter blocks and more accurate association mapping than a single small sample provides. In principle, population stratification poses a greater problem for mapping by association in random samples than by linkage in pedigrees, but if necessary stratification can be overcome by assigning cases and controls to more homogeneous groups defined by markers that best differentiate subpopulations.

Methods for association mapping fall into two broad classes. The first (discussed in this book by Li) uses composite likelihood with permutation to control autocorrelation of neighboring markers. The Malecot model (Collins and Morton, 1998) is based on descent from 1 or 2 markers with two to four haplotypes, whereas coalescence theory tries to infer a single ancestral haplotype for multiple markers (Nordborg, 2001). Error in that inference increases with the number of markers and inversely with population size (Watterson and Guess, 1977). Proponents have expressed doubt that extension to whole genomes is feasible (Morris *et al.*, 2003) and no example has been published. Composite likelihood methods estimate the location S of a causal marker and its information K that other methods do not provide. These numerous alternatives include the most significant SNP in a region (msSNP), haplotypes variously defined, products of probabilities, and identification by linkage of significant regions without a point location. There has been little comparison of alternative methods for genome-wide scans, a large number of which are being made. Computing time can be greatly reduced, especially for complex methods, by parallel processing (Lau *et al.*, 2007).

The goal for association mapping is to locate a causal locus more accurately and reliably than linkage mapping. Success measured by meta-analysis may be confirmed by functional assignment, which at present demands a small target and is not feasible for genome scans. The two approaches are complementary but have not been integrated in a statistical test.

IV. FULLY PARAMETRIC ANALYSIS

As noted by Devlin *et al.* (1996) for complex inheritance, "the full likelihood model would be very difficult to specify without unrealistically stringent assumptions about population history." Composite likelihood methods respond to this limitation by a parameter S for location and one or two ancillary parameters to describe association, which for the Malecot model are M at the causal site and L for large distance, an artefact of constraining pairwise association to nonnegative values. Nonparametric analysis often sacrifices the S parameter, greatly reducing value for meta-analysis. Other parameters may be added for correlation of relatives and ascertainment probability. Permutation tests may be invoked to allow for autocorrelation. In constructing LD maps, a parameter ε_k is estimated for the k-th interval to convert kb to LDU. Whatever the model, fully parametric analysis is incompatible with complex inheritance until causal loci and their principal modifiers are characterized.

V. META-ANALYSIS

The process of using statistical methods to combine results of independent studies, called *meta-analysis*, is mandatory when many studies are inconclusive because of small size, a high number of tests, population differences, or other factors: perhaps none is powerful enough to establish statistically significant conclusions, but the aggregate may be convincing or at least point the way to a definitive test. Meta-analysis should summarize the relevant data in a useful form for reexamination and at the same time identify the data and methods that contribute little or nothing to meta-analysis. Preliminary databases do not identify metrics that provide reliable combination of evidence (Ioannidis *et al.*, 2006; Kaiser, 2006).

Association mapping and genome scanning have led to new perspectives in meta-analysis. Fisher (1950) considered the general problem of combining probabilities from independent tests of significance. He introduced his revolutionary concept in this way: "When a number of quite independent tests of significance have been made, it sometimes happens that although few or none can be claimed individually as significant, yet the aggregate gives an impression that the probabilities are on the whole lower than would have been obtained by chance. It is sometimes desired, taking account only of these probabilities, and not of the detailed composition of the data from which they are derived, which may be of very different kinds, to obtain a single test of the significance of the aggregate, based on the product of the probabilities individually observed."

Classical meta-analysis is based on r independent samples, the i-th of which contributes a P_i value that on the null hypothesis H_0 is uniformly distributed, and so $-2\sum_{i=1}^{r}\ln P_i = \chi^2_{2r}$. This simple test has one advantage: it allows pooling of evidence based on different statistics, and under H_1 their significance increases with r (Littell and Folks, 1973). However, there are several disadvantages:

1. Equal weight is given to samples with different standard errors.
2. There is no test of homogeneity.
3. There is no point estimate to become more precise as n increases.

Currently linkage analysis often uses this approach, contributing to its poor localization of a causal gene (Levinson *et al.*, 2003; Wise *et al.*, 1999).

A much superior meta-analysis is based on an estimate S_i of location in the i-th independent sample, together with an information weight K_i for $i = 1, \ldots, r$ (Morton *et al.*, 2007). If these are accurately calculated, the values of S need not be based on the same statistic, but may include parametric and nonparametric tests, dichotomous and continuous phenotypes, random individuals and pedigrees, random or hypernormal controls, and other differences that affect K_i but not S_i. Of course, it is essential that the S_i refer to an identical map for all i, whether the units are physical (bp), linkage (cM), or linkage disequilibrium (LD). The physical map is most stable and so provides by interpolation a convenient although inefficient meta-analysis. However expressed, the point estimate is $\Sigma S_i K_i / \Sigma K_i$ with nominal variance $1/\Sigma K_i$ and residual variance $\chi^2_{r-1} = \Sigma K_i S_i^2 - (\Sigma K_i S_i)^2 / \Sigma K_i$. In this way, the mean location and its consistency are examined independently and efficiently, without the subjectivity of hypothesis weighting (Roeder *et al.*, 2006). The nominal variance is increased and residual variance correspondingly decreased if there is significant heterogeneity in S_i.

A limitation of meta-analysis is that genomic regions defined for composite likelihood are in a particular study nonoverlapping and jointly exhaustive for a larger target, which ideally is a whole genome. The algorithm used to define regions depends on marker density and other factors, making variation among studies inevitable. The problems this poses for meta-analysis have not yet been addressed. They are likely to lead to different regions for each disease, with apparent wastelands separated by established loci that alone repay examination for gene–gene and gene–environment interactions.

VI. ANALYSIS OF msSNPs

The distribution of the most significant SNP in a region provides a different application of χ^2_2 than Fisher foresaw. CHROMSCAN assumes additivity, since a causal SNP may well not have been tested, and nonadditivity degrades with recombination (Morton *et al.*, 2007). Additivity should be tested in a candidate region but is an appropriate assumption for a phase 1 genome scan. Reduction of a 2 × 3 table (case/control, three genotypes) to a 2 × 2 table on the additivity assumption gives a nominal χ^2_1 for the unique msSNP in each region. However, selection as the msSNP from at least 30 SNPs in a defined region greatly biases this nominal χ^2_1 and the conventional P value computed on the null hypothesis. Using the Fisherian principle that this P should correspond to $\chi^2_2 = -2 \ln P$, let the variance of this nominal χ^2_2 among regions in a genome scan be V_n with mean μ_n. Under H_0, the correct values are $V = 4$ and $\mu = 2$. If selection of msSNPs were unbiased, adjustment of V_n would give an estimate of μ near 2, whereas adjustment of μ_n is less sensitive to small values of P, and therefore would not provide a good estimate of V. We must reduce the bias in μ_n before adjusting V_n.

Step 1. The i-th region provides one msSNP with nominal χ^2_{1ni} and nominal P_{ni}. Let R be the effective number of independent SNPs in a subset of m regions with limited diversity in total number of SNPs. The Bonferroni model assumes a corrected value $P_{ci} = 1 - (1 - P_{ni})^R$. To obtain a mean of $\chi^2_{2c} = 2$ when $\chi^2_{ci} = -2 \ln P_{ci}$, we take $\left[\frac{-2\sum \ln P_{ci}}{m}\right] = \frac{-2\sum \ln [1-(1-P_{ni})^R]}{m} = 2$ and solve the equation $m = \Sigma \ln [1 - (1 - P_{ni})^R]$ by regula falsi to give the Bonferroni P_{ci} with the desired mean of χ^2_{2c}.

Step 2. To set the variance of χ^2_{2c} to 4 requires dividing both χ^2_{2c} and μ by $\beta = \sqrt{\frac{\sum (\chi^2_{2ci}-2)^2}{4(m-1)}}$ to give the desired variance with $\mu = 2/\beta$, which is acceptable only if β is $\sim$1.

This greatly reduces the significance of msSNPs, but conserves the order of the nominal P values without consideration of smaller effects of number of SNPs tested in the region and length of the region in kb or LDU. Therefore msSNPs, although requiring much less calculation than composite likelihood, may be less powerful when their bias is allowed for, as expected from neglect of all other SNPs in the region. Since msSNPs provide no standard error or location, their meta-analysis must be by P logic, which cumulates regional significance but does not refine point location within the region. Some instances must remain where an erroneous LD map, a SNP in an LD step, or a low regional density gives greater power to a tested msSNP. Conversely, in other instances, the ability of composite likelihood to combine location evidence on all markers in a region and over all samples must be an advantage. Optimal combination of evidence remains an illusive goal.

VII. THE FALSE DISCOVERY RATE

Among rejections of a null hypothesis H_0, the false discovery rate (FDR) is the proportion where H_0 is true. Benjamini and Hochberg (1995) rediscovered FDR, which has been developed further by Storey and Tibshirani (2003). Following

Table 25.2. Stages in Association Mapping (Morton, 2006)

Stage	SNP density	Rare SNPs	Functional tests
1. Genome scan	+	−	−
2. Candidate region	++	−	−
3. Candidate locus	+++	+	+
4. Causal	All SNPs	++	++

the precedent of linkage (Morton, 1955), it is conventional to set FDR $\leq .05$, and so the critical significance level α must satisfy $\alpha < .05/N$, where N is the number of regions scanned. Assuming about 6,500 regions of 10 LDU with at least 30 markers, as might be observed at high marker density, the required α is 7.7×10^{-6}. A power of $\Omega > .15$ is sufficient to keep the FDR $< .05$ when the assumptions of the model are valid (Morton *et al.*, 2007). Support from one or more independent samples is usually required to exclude a false discovery (conventionally called a type I error), especially when the significance level is suggestive but falls far short of α. Then the FDR is inseparable from meta-analysis. However, as regional sequencing becomes cheaper and functional assignment easier, multiple samples will be less critical. Inevitably the lifespan of meta-analysis as we know it will be short, and its successors cannot yet be foreseen. In the interim, the reliability of a genome-wide association scan to identify candidate regions depends on a powerful method of analysis, applied to markers of high density in a large sample of cases and controls, which together constitute stage 1, the object of which is to make optimal selection of a much smaller number of candidate regions in stage 2 (Table 25.2). For validity of the FDR, these regions must be studied in an independent sample preferably larger and with more markers. If all these assumptions are met and the FDR is satisfied, the confidence interval will be small and progression to functional tests in stage 3 is likely to be rewarding. In current practice, these conditions are not met, and a substantial fraction of stage 2 studies lead to rejection of the region.

VIII. EARLY-HAPMAP PROJECTS

Explosive development of genome-wide scanning in the last few years is comparable to the impact of linkage early in the last century. Both advances rapidly changed genetics, but with significant differences. Early linkage studies were the contribution of a few people working at Columbia University on highly penetrant genes in Drosophila, with no anticipation that their research could be applied to complex inheritance in humans. In contrast, the pioneers before the International HapMap Project have been joined by many researchers in several countries who initially directed their studies to theory, SNPs, and association without regard to genome-wide mapping. They established that the HapMap samples, although small, give surprisingly reliable maps of haplotypes and LD that help to interpret population differences and should enhance association mapping. However, this preliminary research was just the first step toward its major objective: "to create a resource that would accelerate the identification of genetic factors that influence medical traits . . . . An important application of the HapMap data is to help make possible comprehensive, genome-wide association studies" (International HapMap Consortium, 2005).

Since virtual completion of the HapMap, its goal is driving two kinds of research. One, typified by BioBank (Wright *et al.*, 2002), collects epidemiological data on a cohort of several hundred thousand ostensibly random individuals and then genotypes case and control samples for selected diseases. By definition, such a design is costly of time and money, and the likelihood of achieving random samples of affected and controls is at best uncertain. The appeal of this design is primarily to epidemiologists interested in gene–environment interaction and not deterred by concern that many of the genes are unknown and most environmental factors ascertained through questionnaires are poorly measured.

A better design is used in Japan with the goal of creating *personalized medicine* in which treatment is determined by typing hospital inpatients and outpatients for genes that affect relevant drug response. A preliminary success obtained evidence for several genes that indicate the appropriate warfarin dose for individual patients receiving anti-coagulant therapy (Mushiroda *et al.*, 2006; Nakamura, 2005). This design minimizes cost, maximizes immediate utility, and obtains more reliable data than questionnaires.

A different strategy uses a cohort studied for a particular disease or group of related diseases. This increases the quality of data, reduces its volume, and assures that controls come from the same population. Genome scanning comes after data collection, as with BioBank, but costs and time are less. Currently many examples of this design are being typed for public release in the near future, after a grace period of half a year or more to allow contributors to precede later investigators.

The impact of the HapMap Project is best appreciated by returning to the first edition of this book. Nearly everything has changed in the few subsequent years. Before HapMap, almost every paper assumed linkage data in families, with association in related cases and controls no more than a marginal addition. Population stratification was generally assumed to be a serious problem, without citing any unanticipated example. Much attention has subsequently been devoted to detecting and controlling stratification, most simply by partitioning cases and controls into unstratified samples for meta-analysis. The same approach (and others) can be used for interactions with environments and other genes if evidence for their existence is compelling. For both linkage and association, attention was often directed to most significant single markers rather than composite likelihood of multiple markers as exemplified by the Malecot model (Collins and Morton, 1998) and coalescence theory (Nordborg, 2001). Because of their novelty, these composite likelihoods were omitted from the index of the first edition, together with other concepts that the flowering of association mapping under stimulus from HapMap has made familiar. As a consequence of this rapid transition, a remarkably high proportion of the methods advocated in the first edition have seen little use that might guide the genome-wide analyses now turning to genetic epidemiology for solution.

IX. GENOME-WIDE ASSOCIATION SCANS

The month of June 2007 saw striking successes of the GWAS approach to regionally most significant SNPs (msSNPs). Steinthorsdottir *et al.* (2007) reported on more than 300,000 SNPs in 5,275 controls and 1,399 cases of type 2 diabetes (T2D) from Iceland. Four other samples of European origin were also studied, as well as single Chinese and West African samples. Three previously reported loci were confirmed, two others were not replicated, and a novel locus was validated in the European and Chinese samples. All reported analyses after the genome scan were of msSNPs in a "fast-tracking strategy [that] was not expected to be comprehensive," leaving many more variants yet to be identified. In its design and successes, this study provides a benchmark for genome scans.

Three papers in *Science* cast light on T2D. The first (Diabetes Genetics Institute of Broad Institute of Harvard and MIT, Lund University, and Novartis Institutes for Biomedical Research, 2007) preferred whole-genome association studies (WGAS) as a synonym of genome-wide association scans (GWAS). Genome-wide association analysis has yet to be abbreviated as GWAA. They used quantitative traits to provide evidence for associations at two loci previously unknown to be at risk for T2D, and one simultaneously reported in the paper summarized above, as well as the first replication of two additional T2D loci. Two papers from the Wellcome Trust Case Control Consortium (WTCCC) in the UK (Zeggini *et al.*, 2007) and the investigation of non-insulin-dependent diabetes by the FUSION Consortium in Finland and the United States (Scott *et al.*, 2007) confirm that the number of T2D loci now confidently identified is at least 10. There is a general agreement that these loci identified from msSNPs explain only a small proportion of individual and familial risk for T2D. However, they represent a major advance by association mapping over the results obtained from linkage analysis of families over a longer period.

The most diverse and impressive contribution to GWAS so far was published in *Nature*. It is best understood by beginning with a brief but thoughtful examination of the hopes and risks when "genome miners rush to stake claims" (Wadman, 2007), followed by Bowcock (2007) describing the "50 research teams using 500,000 markers for each of 17,000 individuals to identify 24 genetic risk factors for 7 common human diseases." Next, there is a thoughtful analysis of replicating genotype–phenotype associations (Chanock *et al.*, 2007), and finally three papers describing GWAS of 14,000 cases in seven common diseases (The Wellcome Trust Case Control Consortium, 2007).

Books can and certainly will be written on this subject and its ramifications. There is heavy reliance on msSNPs, assuming that all causal SNPs are tested or represented by a surrogate in tight LD and with the same frequency. Composite likelihood does not depend on this assumption, but loses power when a causal SNP or a good surrogate is tested in a relevant region but neighboring markers useful for composite likelihood are omitted. Recent experience has shown that the two approaches are complementary, converging only when all informative SNPs are tested. That ideal has not yet been realized, since one million markers in a microarray is a small fraction of SNPs and an even smaller fraction of other relevant markers. Evidence from composite likelihood allows meta-analysis whether or not the same markers are used in different studies, but meta-analysis of msSNPs requires that they be invariant.

The history of association mapping is very short compared with linkage, which itself has major unsolved problems. Remarkable progress has been made with GWAS. In the feasible future, its progress will be explosive unless forced into an arbitrary mould that ignores the limitations of first-generation methods.

X. WHAT NEXT?

The rapid pace of genomic development from the Human Genome Project to HapMap and now to more complex polymorphisms has created a hothouse in which meta-analysis of whole-genome scans is perceived as the most reliable path to identification of genes in complex inheritance, since independent studies of smaller regions are noncomparable and give incomplete coverage. By the end of 2007, there will be many dozens of whole-genome scans, but the statistical sophistication for efficient and reliable meta-analysis will still be in its infancy, neglected by most genetic epidemiologists. Since its inception 30 years ago, the Genetic Analysis Workshop (GAW) that meets in alternate years has depended on real or simulated data, and therefore lags at least two years behind the need for their methods. At GAW15, not more than 1% of effort was devoted to meta-analysis, none of it genome-wide. If the pace of current data collection is any guide, genome-wide meta-analysis will occupy 99% of genetic epidemiology in the next decade. This should (and perhaps will) shift current emphasis on exotica like gene-gene and gene-environment interactions to the immediate problem of precisely specifying genomic locations important for disease in some if not all populations. Nothing is less rewarding than looking for seldom replicated interactions between nonsignificant candidates. Only when localization is confirmed can functional and interaction analysis profitably begin.

During the localization phase, several advances should be attempted.

1. Linkage evidence would be more useful if the twin goals of reliable likelihood and point estimation were both satisfied. Current procedures aim for the former without the latter, thereby losing power in meta-analysis.
2. Efficient selection of SNPs and short sequences for genome scanning remains controversial, and so the evolution of microarrays cannot be foreseen. Inclusion of nonsignificant regions in a later stage is ideal but costly. Their exclusion is parsimonious, but could well be erroneous unless the preceding stage was extravagantly large. Pooling of DNA within cases and controls is frugal but of uncertain reliability, and extension from msSNPs to composite likelihood is risky (Pearson *et al.*, 2007).
3. Sound meta-analysis would be fostered if a reliable estimate of $\chi^2{}_1$ associated with the deviation of an estimated location in sample i from the weighted mean can be obtained without access to the raw data, giving a LOD_1 that can be summed over independent samples without increasing degrees of freedom, just as in linkage analysis (Morton, 1955, 1956). Complex inheritance makes that more difficult, but not impossible.
4. Many of the metrics currently advocated have not been shown to estimate location S and information K for each sample, and are therefore not competitive with composite likelihood for meta-analysis. The next few years will see rejection of metrics and databases with this limitation. In particular, msSNPs and haplotypes are vulnerable unless incorporated into composite likelihood.

Against this background, any attempt to predict the post-HapMap future will be flawed. Consensus must be reached on the merits of two or more alternative formulations of composite likelihood and on the supporting roles (if any) to be played by tag SNPs, most significant single SNPs, other polymorphisms, and haplotypes. Utility of alternative sampling designs and cosmopolitan maps in kb, cM, and LD units will be determined. A balance must be struck for responsible access to publicly funded data without violating or exaggerating the rights of an inner circle that may not represent the cutting edge of genetic epidemiology. By the time this book is published, there will be a flood of data from the NIH and elsewhere on many diseases requiring whole-genome association mapping and meta-analysis. Only a handful of genetic epidemiologists have any experience in this area, and they will be arguing over largely untested opinions in the absence of more than one program capable of analyzing steps as well as blocks by composite likelihood with allowance for autocorrelation and estimation both of a point location and its standard error. Moreover, the conditions under which analysis will be attempted are uncertain. How long will data be reserved for investigators who contributed the evidence, before becoming accessible to investigators with experience in relevant analysis? How much will release of covariates be crippled by unjustified fear of anonymized data? How can a rational decision be made by granting agencies to expand or terminate a project that has not been completely analyzed? How useful will an untested method or database be? What metrics contribute most to meta-analysis? So long as these and many related questions remain unanswered, the value of the Genetic Association Information Network (GAIN) and similar consortia is jeopardized. These ambiguities must be resolved and their solution applied to multiple data sets. Genetic epidemiology is at a crossroads never before faced, where it must either make significant contributions to medicine as its definition promises and its history illustrates, or be dismissed as a dilettante playing with simulated data. Scientists since Benjamin Franklin have replied to dismissal of innovation by asking "of what use is a newborn child?", but they must avoid the self-satisfaction satirized in Glasbergen's cartoon as "scientists have isolated the gene that makes scientists want to isolate genes." In this conflict workshops, scientific societies, journals, books like this, and comprehension by granting agencies will play a critical role.

References

Benjamini, Y., and Hochberg, Y. (1995). Controlling the false discovery rate: A practical and powerful approach to multiple testing. *J. R. Stat. Soc. B* **87,** 289–300.

Bowcock, A. M. (2007). Guilt by association. *Nature* **447**(7145), 645–646.

Chanock, S. J., Manolio, T., Boehnke, M., Boerwinkle, E., Hunter, D. J., Thomas, G., Hirschhorn, J. N., Abecasis, G., Altshuler, D., Bailey-Wilson, J. E., Brooks, L. D., Cardon, L. R., *et al.* (2007). Replicating genotype-phenotype associations. *Nature* **447**(7145), 655–660.

Collins, A., and Morton, N. E. (1998). Mapping a disease gene by allelic association. *Proc. Natl. Acad. Sci. USA* **95,** 1741–1745.

Devlin, B., Risch, N., and Roeder, K. (1996). Disequilibrium mapping: composite likelihood for pairwise disequilibrium. *Genomics* **36,** 1–16.

Diabetes Genetics Institute of Broad Institute of Harvard and MIT, Lund University and Novartis Institutes for Biomedical Research (2007). Genome-wide association analysis identifies loci for type 2 diabetes and triglyceride levels. *Science* **316**(5829), 1331–1336.

Fisher, R. A. (1950). "Statistical Methods for Research Workers." 11th Edition, pp. 99–101. Oliver and Boyd, Edinburgh.

International HapMap Consortium (2005). A haplotype map of the human genome. *Nature* **437,** 1299–1320.

Ioannidis, J. P. A., Gwinn, M., Little, J., Higgins, J. P. T., Bernstein, J. I., Boffetta, P., Bondy, M., Bray, M. S., Brenchley, P. E., Buffler, P. A., Casas, J. P., Chokkalingam, A., *et al.* (2006). A road map for efficient and reliable human genome epidemiology. *Nat. Genet.* **38,** 3–5.

Kaiser, J. (2006). NIH goes after whole genome in search of disease genes. *Science* **311,** 933.

Lau, W., Kuo, T. Y., Tapper, W., Cox, S., and Collins, A. (2007). Exploiting large scale computing to construct high resolution linkage maps of the human genome. *Bioinformatics* **23,** 517–519.

Levinson, D. F., Levinson, M. D., Segurado, R., and Lewis, C. M. (2003). Genome scan meta-analysis of schizophrenia and bipolar disorder, part 1: Methods and power analysis. *Am. J. Hum. Genet.* **73,** 17–33.

Littell, R. C., and Folks, J. L. (1973). Asymptotic optimality of Fisher's method of combining independent tests. *J. Am. Stat. Assoc.* **68,** 193–194.

Morris, A. P., Whittaker, J. C., Xu, G. F., Hosking, L. K., and Balding, D. J. (2003). Multipoint linkage-disequilibrium mapping narrows location and identifies mutation heterogeneity. *Proc. Natl. Acad. Sci. USA* **100,** 13442–13446.

Morton, N. E. (1955). Sequential tests for the detection of linkage. *Am. J. Hum. Genet.* **7,** 277–318.

Morton, N. E. (1956). The detection and estimation of linkage between the genes for elliptocytosis and the Rh blood type. *Am. J. Hum. Genet.* **8,** 80–96.

Morton, N. E. (1998). Significance tests in complex inheritance. *Am. J. Hum. Genet.* **62,** 690–697.

Morton, N. E. (2006). Fifty years of genetic epidemiology, with special reference to Japan. *J. Hum. Genet.* **51,** 269–277.

Morton, N. E., Maniatis, N., Zhang, W., Ennis, S., and Collins, A. (2007). Genome scanning by composite likelihood. *Am. J. Hum. Genet.* **80,** 19–28.

Mushiroda, T., Ohnishi, Y., Saito, S., Takahashi, A., Kikuchi, Y., Saito, S., Shimomura, H., Wanibuchi, Y., Susuki, I., Kamatani, N., and Nakamura, Y. (2006). Association of VKORC1 and CYP209 polymorphisms with warfarin dose requirements in Japanese patients. *J. Hum. Genet.* **51,** 249–253.

Nakamura, Y. (2005). Scientists create Japan's largest BioBank for genetic studies of 47 common diseases (Interview). *Affymetrix Microarray Bull.* **1,** 16–19.

Nordborg, M. (2001). Coalescent theory. *In* "Handbook of Statistical Genetics" (D. J. Balding, M. Bishop, and C. Cannings, eds.), pp. 179–212. John Wiley and Sons, Chichester, UK.

Pearson, J. V., Huentelman, M. J., Halperin, R. F., Tembe, W. D., Melquist, S., Homer, N., Brun, M., Szelinger, S., Coon, K. D., Zismann, V. L., Webster, J. A., Beach, T., *et al.* (2007). Identification of the genetic basis for complex disorders by use of pooling-based genomewide single-nucleotide-polymorphism association studies. *Am. J. Hum. Genet.* **80,** 126–139.

Roeder, K., Bacanu, S. A., Wasserman, L., and Devlin, B. (2006). Using linkage genome scans to improve power of association in genome scans. *Am. J. Hum. Genet.* **78,** 243–252.

Scott, L. J., Mohlke, K. L., Bonnycastle, L. L., Willer, C. J., Li, Y., Duren, W. L., Erdos, M. R., Stringham, H. M., Chines, P. S., Jackson, A. U., Prokunina-Olsson, L., Ding, C.-J., *et al.* (2007). A genome-wide association study of type 2 diabetes in Finns detects multiple susceptibility variants. *Science* **316**(5829), 1341–1345.

Steinthorsdottir, V., Thorleifsson, G., Reynisdottir, I., Benediktsson, R., Jonsdottir, T., Walters, G. B., Styrkarsdottir, U., Gretarsdottir, S., Emilsson, V., Ghosh, S., Baker, A., Snorradottir, S., *et al.* (2007). A variant in CDKAL1 influences insulin response and risk of type 2 diabetes. *Nat. Genet.* **39**(3241), 770–775.

Storey, J. D., and Tibshirani, R. (2003). Statistical significance for genomewide studies. *Proc. Natl. Acad. Sci. USA* **100,** 9440–9445.

Tapper, W., Collins, A., Gibson, J., Maniatis, N., Ennis, S., and Morton, N. E. (2005). A map of the human genome in linkage disequilibrium units. *Proc. Natl. Acad. Sci. USA* **102,** 11835–11839.

The Wellcome Trust Case Control Consortium (2007). Genome-wide association study of 14,000 cases of seven common diseases and 3,000 shared controls. *Nature* **447**(7145), 661–678.

Wadman, N. (2007). Genome miners rush to stake claims. *Nature* **447**(7145), 623.

Watterson, G. A., and Guess, H. A. (1977). Is the most frequent allele the oldest? *Theor. Pop. Biol.* **11,** 141–160.

Wise, L. H., Lanchbury, J. S., and Lewis, C. M. (1999). Meta-analysis of genome searches. *Ann. Hum. Genet.* **63,** 263–272.

Wright, A. F., Carothers, A. D., and Campbell, H. (2002). Gene-environment interactions—The BioBank UK study. *Pharmacogenom. J.* **2,** 75–82.

Zeggini, E., Weedon, M. N., Lindgren, C. M., Frayling, T. M., Elliott, K. S., Lango, H., Timpson, N. J., Perry, J. R., Rayner, N. W., Freathy, R. M., Barrett, J. C., Shields, B., *et al.* (2007). Replication of genome-wide association signals in UK samples reveals risk loci for type 2 diabetes. *Science* **316**(5829), 1336–1341.

Index

E

F

K

L

M

Q